Legionella

STATE OF THE ART 30 YEARS AFTER
ITS RECOGNITION

Legionella

STATE OF THE ART 30 YEARS AFTER ITS RECOGNITION

Edited by

Nicholas P. Cianciotto
Department of Microbiology-Immunology
Northwestern University Medical School
Chicago, Illinois

Yousef Abu Kwaik
Department of Microbiology and
 Immunology
University of Louisville College of
 Medicine
Louisville, Kentucky

Paul H. Edelstein
Department of Pathology and
 Laboratory Medicine
University of Pennsylvania School of
 Medicine
Philadelphia, Pennsylvania

Barry S. Fields
Respiratory Diseases Branch
Centers for Disease Control and
 Prevention
Atlanta, Georgia

David F. Geary
D. F. Geary Consultants, LLC
Annapolis, Maryland

Timothy G. Harrison
Respiratory and Systemic Infection
 Laboratory
Health Protection Agency
Centre for Infections
London, United Kingdom

Carol A. Joseph
Respiratory Diseases Department
Health Protection Agency
Centre for Infections
London, United Kingdom

Rodney M. Ratcliff
Infectious Diseases Laboratories
Institute of Medical and Veterinary
 Science
Adelaide, Australia

Janet E. Stout
VA Pittsburgh Healthcare System
Pittsburgh, Pennsylvania

and

Michele S. Swanson
Department of Microbiology and
 Immunology
University of Michigan Medical School
Ann Arbor, Michigan

ASM PRESS

Washington, DC

Address editorial correspondence to ASM Press, 1752 N St. NW, Washington, DC 20036-2904, USA

Send orders to ASM Press, P.O. Box 605, Herndon, VA 20172, USA
Phone: (800) 546-2416 or (703) 661-1593
Fax: (703) 661-1501
E-mail: books@asmusa.org
Online: estore.asm.org

Library of Congress Cataloging-in-Publication Data

International Conference on Legionella (6th : 2005 : Chicago, Ill.)
    Legionella : state of the art 30 years after its recognition / [edited
by] Nicholas P. Cianciotto . . . [et al.].
        p. ; cm.
    "Researchers from around the world gathered in Chicago, Illinois,
USA on October 16-20, 2005 for the 6th International Conference
on Legionella"—Pref.
    Includes bibliographical references and index.
    ISBN-13: 978-1-55581-390-1 (alk. paper)
    ISBN-10: 1-55581-390-9 (alk. paper)
    1. Legionnaire's disease—Congresses.   I. Cianciotto, Nicholas P.
II.   Title.
    [DNLM: 1. Legionellosis—Congresses.   2. Legionella—Congresses.
3. Legionnaires' Disease—Congresses.   WC 200 I61L 2006]
RC152.7.L4422 2006
362.196'241—dc22

2006023616

10 9 8 7 6 5 4 3 2 1

Cover photo: Legionella pneumophila obtained from a log-phase broth culture. After air drying on a glass slide, a phase contrast image was collected, treated with a bas relief filter, and then colorized using Adobe Photoshop software. Photo courtesy of Ari B. Molofsky, an M.D./Ph.D. student in Michele Swanson's laboratory.

CONTENTS

PREFACE

In the summer of 1976, a mysterious outbreak of pneumonic disease occurred at the Legionnaires' convention in Philadelphia, Pennsylvania, USA. Soon thereafter, the medical and scientific communities realized that a new disease, Legionnaires' disease, and a new bacterial genus, *Legionella*, had emerged. Since that time, there has been much progress toward understanding *Legionella* and diagnosing, treating, monitoring, and preventing legionellosis. Nonetheless, there remains much to be learned about the organism and the ways of addressing its impact on public health. Thus, researchers from around the world gathered in Chicago, Illinois, USA on October 16–20, 2005 for the 6th International Conference on *Legionella* (www.legionellaconf.org). The Conference had approximately 400 attendees who participated in nearly 40 invited lectures, 200 poster presentations, 3 panel discussions, and numerous other opportunities for scientific exchange and collaboration. This book summarizes the contributions presented at the meeting and as such represents the current state-of-the-art of *Legionella* research and investigation. We thank the many individuals who contributed to the success of the 6th International Conference, including our fellow members of the Scientific Organizing Committee, our numerous, generous sponsors, the managed meetings group of the ASM, and the Conference attendees. Finally, we thank all who have further contributed to the completion of this publication, including the many authors who submitted manuscripts, Jenny Dao who assisted us with processing the manuscripts, and ASM Press who once again served as our expert publisher.

NICHOLAS CIANCIOTTO
YOUSEF ABU KWAIK
PAUL EDELSTEIN
BARRY FIELDS
DAVID GEARY
TIMOTHY HARRISON
CAROL JOSEPH
RODNEY RATCLIFF
JANET STOUT
MICHELE SWANSON

CLINICAL AND DIAGNOSTIC ASPECTS OF LEGIONNAIRES' DISEASE

I

CLINICAL FEATURES OF LEGIONNAIRES' DISEASE: A SELECTIVE REVIEW

Paul H. Edelstein

I

Legionnaires' disease was thought to be an atypical and easily recognized form of pneumonia for several years after the 1976 Philadelphia epidemic. News accounts of the time characterized the disease as a novel mystery illness that had not previously been seen by physicians. The first book written on Legionnaires' disease highlighted what were thought to be these novel and atypical features (Table 1), and as late as the early 1980s physicians wrote that "the clinical presentation of Legionnaires' disease is distinctive enough that the diagnosis should be considered in most cases" (7, 8).

Astute clinicians began to question the characterization of the illness as clinically unique, and in a seminal prospective study of community-acquired and nosocomial pneumonia Yu and colleagues showed that Legionnaires' disease could not be distinguished on presentation from non-Legionnaires' disease cases of pneumonia (19). Only hyponatremia was more common in Legionnaires' disease, occurring in 44% of these patients versus 14% of pneumonia patients without the disease. This study was followed by several others showing that neither community-acquired nor

nosocomial Legionnaires' disease could be accurately distinguished on presentation from other common pneumonias on clinical, laboratory, and roentgenographic grounds. Each study found one or two differences between Legionnaires' disease and other pneumonias, often differing from one study to another (15, 17). Although one finding, most commonly diarrhea or hyponatremia, might have been more common in Legionnaires' disease, there was enough overlap in the frequency of these findings to prevent accurate differentiation of the pneumonia etiology in a single patient, as opposed to distinguishing large populations of patients with and without Legionnaires' disease.

Convinced that Legionnaires' disease was indeed unique, two groups formulated scoring systems designed to aid in the specific diagnosis of Legionnaires' disease at presentation, based on a number of epidemiologic, clinical, and nonspecific laboratory findings. The Winthrop-University scoring system used 22 weighted clinical and laboratory parameters to derive a score, with patients divided into "unlikely," "probable," and "highly probable" categories of the likelihood of having Legionnaires' disease based on their score (1). Ohio State and Centers for Disease Control investigators proposed a second scoring system, termed the CBPIS system, but never published

Paul H. Edelstein, Department of Pathology and Laboratory Medicine, University of Pennsylvania School of Medicine, Philadelphia, PA 19104-4283.

TABLE 1 Legionnaires' disease features thought to be typical in 1980

Clinical	Laboratory
Long incubation period with prodrome	Hypophosphatemia
Low cough frequency	Hyponatremia
Small amounts of nonpurulent sputum	Elevation of liver-associated enzymes
Pulse-temperature dissociation	Urine myoglobinuria
Diarrhea and abdominal pain	Increased serum creatinine kinase
Headache and myalgia; confusion	Paucity of neutrophils in sputum
Multiple rigors	
High hospitalization rate	

the details. Fernández-Sabé and colleagues constructed and tested the CBPIS scoring system based on the original description in abstract form and communications with one of the original authors (3).

The Winthrop-University criteria were tested by Gupta and colleagues in a retrospective study of 68 patients with and without Legionnaires' disease (5). Included in those scored unlikely to have Legionnaires' disease were 8% of patients with Legionnaires' disease and 32% of patients with pneumococcal pneumonia. In the probable and highly probable categories were 14 and 78% of Legionnaires' disease patients, respectively, but also 32 and 35% of pneumococcal pneumonia patients, respectively. Using the combined probable and highly probable categories to determine whether therapy should be given for Legionnaires' disease has a sensitivity of 92% but a specificity of only 32%. Based on an average Legionnaires' disease prevalence of 5%, the positive and negative predictive values for the unlikely category would be 52 and 99%, respectively. Using these criteria would result in treating about 10 patients without Legionnaires' disease for every one with the disease and not treating about 1% of patients with the disease, making it unclear if application of these criteria would be useful.

The CBPIS scoring system uses fewer and different criteria than the Winthrop-University system but performs about the same (3). Fernández-Sabé and colleagues prospectively studied 81 patients with Legionnaires' disease and 136 patients with pneumococcal pneu-

monia. In the low likelihood for Legionnaires' disease category were 4% of patients with the disease and 17% of pneumococcal pneumonia patients. The high likelihood category constituted 51% of Legionnaires' disease patients and 14% of pneumococcal pneumonia patients. Combining the moderate and high categories resulted in a sensitivity of 96% and specificity of less than 17%. Again, treating everyone for Legionnaires' disease who had other than a low score would result in treating many more patients with non-Legionnaires' disease than those with the disease.

An important finding of the CBPIS study was that emergency department physicians were quite good at diagnosing Legionnaires' disease in the absence of knowledge of specific test results. They considered 64% of Legionnaires' disease patients and only 6% of those with pneumococcal pneumonia to most likely have Legionnaires' disease. Specific therapy for Legionnaires' disease was given to 89% of those who ended up having the disease but to only 18% of those with pneumococcal pneumonia. Finally, *L. pneumophila* urine antigen testing was ordered for 73% of those with Legionnaires' disease but only for 9% of those with pneumococcal pneumonia. These findings mean that experienced clinicians can distinguish between Legionnaires' disease and pneumococcal pneumonia at the time of initial clinical presentation and in the absence of specific laboratory test results. Modeling this diagnostic ability using a scoring system has been a challenge, but the findings suggest that this is possible. Moreover, these findings indicate that the majority of

Legionnaires' disease patients have a characteristic clinical presentation. It is important to note that 11% of those with Legionnaires' disease were not treated for the disease by the emergency department clinicians and stress the importance of empiric therapy for this disease despite atypical clinical findings.

Another recent approach that has been taken to detect Legionnaires' disease is the analysis of blood acute phase reactant levels. Prat and colleagues measured blood procalcitonin and neopterin levels in pneumonia patients with and without Legionnaires' disease (16). Procalcitonin is inflammatory cytokine driven and generally increased in pyogenic infections, whereas neopterin is interferon-gamma driven and generally increased in intracellular infections. This study showed that patients with pneumococcal pneumonia had higher levels of procalcitonin and lower levels of neopterin than did Legionnaires' disease patients and that procalcitonin levels increased with greater disease severity regardless of diagnosis. However, there was too much overlap to distinguish between Legionnaires' disease and pneumococcal pneumonia using these acute phase reactant assays. Calculation of neopterin:procalcitonin ratios showed that Legionnaires' disease patients had significantly greater test ratios than did patients with pneumococcal pneumonia but that overlap of these ratios was too great to create a specific test (C. Prat, personal communication).

The clinical spectrum of Legionnaires' disease has been expanded since the 1976 epidemic. More is now known about coinfection with other microorganisms, cavitary lung disease, extrapulmonary disease, nonpulmonary disease, and late sequelae of Legionnaires' disease.

About 2 to 10% of patients with Legionnaires' disease are coinfected with other microorganisms (2). These coinfecting organisms include typical pulmonary pathogens such as *Streptococcus pneumoniae* and *Haemophilus influenzae*. In one study of nosocomial Legionnaires' disease 11% of patients having blood cultures had positive cultures for *Staphylococcus aureus*, *Escherichia coli*, *S. pneumoniae*, or *H. in-*

fluenzae (10). These coinfections may result in the misdiagnosis of Legionnaires' disease and failure to treat it, as well as the failure to treat for the coinfecting organism. Multiple case reports of superinfections or coinfections with opportunistic organisms also exist, including infections caused by cytomegalovirus, a variety of fungi, and several types of bacteria. Failure to respond to treatment for Legionnaires' disease, or relapse, should prompt consideration of coinfection or superinfection.

Cavitary lung disease is now acknowledged to be relatively common in immunocompromised patients with Legionnaires' disease, especially in some groups of solid organ transplant recipients (2, 4). Cavitation develops in lung areas affected by the pneumonia, often within days of onset of symptoms. Cure requires prolonged therapy, sometimes for months, and relapses may occur if nonbactericidal therapy, such as erythromycin, is administered. Response to what is ordinarily highly effective therapy may be very slow, with persistently positive sputum cultures for *Legionella* bacteria.

Extrapulmonary infection caused by *Legionella* bacteria, a very rare occurrence, is now recognized to occur in the presence and absence of Legionnaires' disease (12). Sites of infection disseminated from the lung have included brain, kidneys, thyroid, pancreas, liver, spleen, prostate, heart, pericardium, bowel, peritoneum, testes, bone and bone marrow, muscles and other soft tissues, and other sites. Local or systemic immunodeficiency is a common theme in these infections. Bacteremia without pneumonia has been reported, as well as sinusitis, wound infections, prosthetic heart valve infections, and even a perirectal abscess. Response to specific therapy may be slow, with surgical drainage of abscesses required in some situations.

Several types of probably noninfectious extrapulmonary disease have also been recognized since the 1976 Legionnaires' disease epidemic. Pancreatitis has been prominent, with more than 10 such cases reported, and probably many more occurring based on personal experience (13). Muscle involvement, first recognized in 1976, continues to be a common extrapulmonary disease, ranging in severity from simple myalgia to rhabdomyolysis with accompanying renal failure. Up to 30% of Legionnaires' disease patients have elevated levels of serum creatine kinase, although a very small fraction of such patients have clinically important muscle disease (17). While cytokine-driven mechanisms seems likely for both pancreatitis and muscle disease in Legionnaires' disease patients, rare cases of direct tissue infection exist (18).

Long-term follow-up of large numbers of Legionnaires' disease patients infected during a large epidemic in the Netherlands has afforded some answers on possible sequelae of the disease (6, 9). Of 86 non-randomly selected former Legionnaires' disease patients, 33 had some evidence of pulmonary disease by chest roentgenography, carbon monoxide diffusion measurements (DLCO), or both tests; 64% of these 33 patients had some abnormality on high resolution computerized tomographic lung scanning. None of the 33 patients had abnormal total lung capacity or vital capacity measurements, but 24 of them had some minor DLCO abnormality. Receipt of mechanical ventilation was the largest risk factor for radiographic or DLCO abnormalities. About 60% of the 86 patients complained of shortness of breath, regardless of roentgenographic or DLCO abnormalities. Whether these complaints and minor abnormalities were due to the Legionnaires' disease is unanswered by this or any other study of postpneumonia sequelae in the absence of an adequately controlled study. In addition, no physiological explanation for the shortness of breath was found, making it unlikely that the pneumonia was the direct cause of the symptoms. An adequately controlled study is required, preferably with known premorbid lung function test results, before firm conclusions can be made as to whether Legionnaires' disease causes chronic lung disease, especially at the frequency noted in this study. At most, it appears that pulmonary sequelae of Legionnaires' disease, if they occur at all, are minor in most cases.

Unresolved persistent symptoms after an episode of acute pneumonia are well recognized (11). This has also been shown to be the case for Legionnaires' disease in a follow-up study of 122 Legionnaires' disease patients infected during the Bovenkarspel Flower Exhibition epidemic (9). Two months after the outbreak 81% of patients complained of fatigue; 75% were dizzy, had headache, poor memory, or difficulty in concentrating; and 79% had paresthesias, myalgia, or weakness. Even after 17 months most of these symptoms persisted for 63 to 75% of patients, significantly greater than those of a matched Dutch population and more frequent than reported in studies of mild community acquired pneumonia (11, 14). Of note, the methods used in the Dutch study could have allowed a patient selection bias toward those with more severe symptoms or a greater frequency of susceptibility to psychological stress. The authors considered that this symptom complex could be attributed to a post-traumatic stress disorder but could not determine whether the symptoms were due to Legionnaires' disease, severe pneumonia in general, or the psychological impact of the outbreak itself. Controlled studies of this phenomenon are required to determine true frequencies of long term symptoms due to Legionnaires' disease for both outbreak-related and sporadic Legionnaires' disease.

The importance of diagnosing Legionnaires' disease is just as great now as it was in 1976. In contrast to the situation in 1976, we now know that prompt and proper therapy leads to cure in the majority of patients, and we have specific diagnostic tests. The 1976 clinicians were about 80% correct in concluding that Legionnaires' disease is an atypical and recognizable illness, but it is that missed 20% that gives us problems. The challenges for the next decade are to develop better clinical tools for diagnosing Legionnaires' disease, most likely aided by computer modeling of a large number of clinical and laboratory parameters, and to establish specific diagnostic tests that are cheaper and more reliable than our present ones. Once these two goals are achieved, it is likely that Legionnaires' disease will be recognized to cause a broader spectrum of disease than is currently known, apart from a probable beneficial effect on cure rates and recognition of small epidemics.

REFERENCES

1. **Cunha, B. A.** 1998. Clinical features of Legionnaires' disease. *Semin. Resp. Infect.* **13:**116–127.
2. **Edelstein, P. H. and N. P. Cianciotto.** 2005. Legionella, p. 2711–2724. *In* G. L. Mandell, J. E. Bennett, and R. Dolin (ed.), *Principles and practice of infectious diseases.* Elsevir Churchill Livingstone, Philadelphia.
3. **Fernández-Sabé, N., B. Rosón, J. Carratalà, J. Dorca, F. Manresa, and F. Gudiol.** 2003. Clinical diagnosis of *Legionella* pneumonia revisited: evaluation of the Community-Based Pneumonia Incidence Study Group scoring system. *Clin. Infect. Dis.* **37:**483–489.
4. **Fraser, T. G., T. R. Zembower, P. Lynch, J. Fryer, P. R. Salvalaggio, A. V. Yeldandi, and V. Stosor.** 2004. Cavitary *Legionella* pneumonia in a liver transplant recipient. Transplant *Infec. Dis.* **6:**77–80.
5. **Gupta, S. K., T. F. Imperiale, and G. A. Sarosi.** 2001. Evaluation of the Winthrop-University Hospital criteria to identify *Legionella* pneumonia. *Chest* **120:**1064–1071.
6. **Jonkers, R. E., K. D. Lettinga, T. H. Pels Rijcken, J. M. Prins, C. M. Roos, O. M. van Delden, A. Verbon, P. Bresser, and H. M. Jansen.** 2004. Abnormal radiological findings and a decreased carbon monoxide transfer factor can persist long after the acute phase of *Legionella pneumophila* pneumonia. *Clin. Infect. Dis.* **38:**605–611.
7. **Kirby, B. D., K. M. Snyder, R. D. Meyer, and S. M. Finegold.** 1980. Legionnaires' disease: report of sixty-five nosocomially acquired cases and review of the literature. *Medicine* (Baltimore) **59:**188–205.
8. **Lattimer, G. L., and R. A. Ormsbee.** 1981. *Legionnaires' disease.* Dekker, New York.
9. **Lettinga, K. D., A. Verbon, P. T. Nieuwkerk, R. E. Jonkers, B. P. Gersons, J. M. Prins, and P. Speelman.** 2002. Health-related quality of life and posttraumatic stress disorder among survivors of an outbreak of Legionnaires disease. *Clin. Infect. Dis.* **35:**11–17.
10. **Marrie, T. J., D. Haldane, and G. Bezanson.** 1992. Nosocomial Legionnaires' disease: clinical and radiographic patterns. *Can. J. I. D.* **3:**253–260.
11. **Marrie, T. J., C. Y. Lau, S. L. Wheeler, C. J. Wong, and B. G. Feagan.** 2000. Predictors of symptom resolution in patients with community-acquired pneumonia. *Clin. Infect. Dis.* **31:**1362–1367.
12. **McClelland, M. R., L. T. Vaszar, and F. T. Kagawa.** 2004. Pneumonia and osteomyelitis due to *Legionella longbeachae* in a woman with systemic lupus erythematosus. *Clin. Infect. Dis.* **38:**e102–e106.
13. **Megarbane, B., S. Montambault, I. Chary, M. Guibert, O. Axler, and F. G. Brivet.** 2000. Acute pancreatitis caused by severe *Legionella pneumophila* infection. *Infection* **28:**329–331.
14. **Metlay, J. P., M. J. Fine, R. Schulz, T. J. Marrie, C. M. Coley, W. N. Kapoor, and D. E. Singer.** 1997. Measuring symptomatic and functional recovery in patients with community-acquired pneumonia. *J. Gen. Intern. Med.* **12:**423–430.
15. **Mulazimoglu, L. and V. L. Yu.** 2001. Can Legionnaires disease be diagnosed by clinical criteria? A critical review. *Chest* **120:**1049–1053.
16. **Prat, C., J. Dominguez, F. Andreo, S. Blanco, A. Pallarés, F. Cuchillo, C. Ramil, J. Ruiz-Manzano, and V. Ausina.** Procalcitonin and neopterin correlation with aetiology and severity of pneumonia. *J. Infect.*, in press.
17. **Sopena, N., M. Sabria-Leal, M. L. Pedro-Botet, E. Padilla, J. A. Domínguez, J. Morera, and P. Tudela.** 1998. Comparative study of the clinical presentation of *Legionella* pneumonia and other community-acquired pneumonias. *Chest* **113:**1195–1200.
18. **Warner, C. L., P. B. Fayad, and R. R. Heffner, Jr.** 1991. Legionella myositis. *Neurology* **41:**750–752.
19. **Yu, V. L., F. J. Kroboth, J. Shonnard, A. Brown, S. McDearman, and M. Magnussen.** 1982. Legionnaires' disease: new clinical perspective from a prospective pneumonia study. *Am. J. Med.* **73:**357–361.

TREATMENT OF LEGIONNAIRES' DISEASE

Jorge Roig and Jordi Rello

2

EMPIRIC COVERAGE OF LEGIONNAIRES' DISEASE (LD) IN COMMUNITY-ACQUIRED PNEUMONIA (CAP)

The coverage of atypical agents in the empiric therapeutic approach of CAP has been a matter of controversy for the past decade. Recent data suggest that the potential benefit of covering atypical pathogens in the initial, empiric therapy of CAP comes basically from the subset of patients with LD (10, 19). A recent meta-analysis shows that in nonsevere CAP the relative risk for treatment failure is significantly lower (0.40, 95% confidence interval 0.19 to 0.85) when patients with *Legionella pneumophila* do receive active antibiotics against atypical pathogens (10). Shefet et al. have also shown that the empiric atypical coverage for patients with CAP who require hospitalization is mostly beneficial on the basis of those patients with LD (19). LD has also been identified as one of the factors that are significantly associated with early treatment failure in hospitalized CAP (17). Finally, *Legionella* spp. usually ranks second to pneumococcus on the list of causative agents of severe CAP (SCAP) (1).

ANTIBIOTIC THERAPY

LD treatment recommendations are supported by data obtained from in vitro and cellular studies, experimental studies with animal models, and observational studies, some of which come from prospective clinical studies of CAP (1, 5, 6, 15). Optimal therapy against *Legionella* spp. is then based on antibiotics with high intrinsic activity, an appropriate pharmacokinetic and pharmacodynamic profile, including the ability to enter and concentrate in phagocytic cells and alveolar exudates, a low incidence of adverse reactions, and an advantageous cost-efficacy relationship (1, 5, 6, 15, 16). In general lines, the treatment of choice for atypical agents has changed from erythromycin to the newer macrolides and fluoroquinolones (1, 5, 6, 15, 16, 20). However, erythromycin continues to be an effective antibiotic against *Legionella* spp. and it has been shown to still be effective in a recent series of LD (8).

Other more recent macrolides share with erithromycin the ability to penetrate phagocytic cells, with the advantage of showing an overall better intrinsic activity against *Legionella*. Besides this superior in vitro activity against *Legionella*, they offer pharmacokinetic and pharmacodynamic advantages. Relatively minor differences in the comparative in vitro

Jorge Roig, Pulmonary Division, Nostra Senyora de Meritxell Hospital, Escaldes-Engordany, Andorra, AD700. *Jordi Rello,* Intensive Care Department, Joan XXIII University Hospital, Tarragona, Spain.

Legionella: State of the Art 30 Years after Its Recognition
Edited by Nicholas P. Cianciotto et al.
©2006 ASM Press, Washington, D.C.

activity among the new macrolides are also found in different studies (5, 6, 15, 16). Intravenous forms of azithromycin and clarithromycin are commercially available in most countries. The intravenous form of spiramycin has been widely used in France (1). Because of a better pharmacokinetic profile (once-a-day administration, shorter course of therapy than usual in mild to moderate cases) and the ability to cause an irreversible inhibition of *Legionella* growth, azithromycin is probably the most advantageous of the macrolides for treating hospitalized LD (6, 13, 15).

Fluoroquinolones show a good ability to achieve good intracellular levels and high in vitro activity against *Legionella* (5, 6, 20). A few experts have recommended a loading dose consisting of a double dose of levofloxacin (as has also been suggested for azithromycin), but the rationale for this is unclear. No consistent evidence supports this approach. New fluoroquinolones, such as gemifloxacin, gatifloxacin and pazufloxacin, also show an excellent intrinsic activity against *Legionella* spp. (16). Clinically relevant experience with gemifloxacin in LD has already been reported (16).

Clinical experience with cotrimoxazole is quite limited although it has shown good effectiveness in the guinea pig model but with less impressive results than those of erythromycin or rifampicin (16). Prystynamycin has been found to be useful in a small number of patients with CAP caused by *Legionella pneumophila* (16). Table 1 summarizes a general recommendation for treating LD (15).

MACROLIDES VERSUS FLUOROQUINOLONES

There is some controversy about the use of new fluoroquinolones versus macrolides as the treatment of choice for LD. Among macrolides, azithromycin has usually been superior to clarithromycin or erythromycin in many cell models and most animal models (5, 6). In the animal model, the killing is most rapid with some of the new potent quinolones, but there are striking differences in the degree of lung inflammation with different antimicro-

bials. The least inflammation is found in the animal model with azithromycin, while the most is observed with erythromycin, with the quinolones being intermediate. Moreover, azithromycin and quinolones are both able to cause an irreversible inhibition of *Legionella* intracellular growth even after drug removal in the macrophage model (6).

Our experience on the topic of macrolides versus fluorquinolones as the treatment of choice for LD has been recently published. In an observational study, we compared two homogeneous groups of patients with legionellosis (18). Fifty-four received fluoroquinolones and seventy-six who were treated with macrolides. Although the time to apyrexia was significantly longer in the macrolide group (77.1 hours versus 48 hours), statistically significant differences were not found in length of hospital stay, mortality, and clinical or radiological complications. Two other similar studies have been recently published, but their conclusions are limited since their methodological drawbacks, as also happened in our study, are relevant (2, 11). However, it seems that for the great majority of patients with LD who require hospitalization there are no significant differences in fatality rate or the number of serious complications with whatever form of therapy is chosen.

OTHER THERAPEUTIC MEASURES

Respiratory failure, particularly when adult respiratory distress syndrome (ARDS) is present, is a major cause of fatality (1, 7, 15). In patients who require mechanical ventilation, the goal is to improve gas interchange and avoid causing ventilatory-induced lung injury, maintaining plateau pressures under 25. A lung-protective strategy of ventilation with low tidal volumes (<7 ml/kg) protects the lung in acute lung injury. Patients with LD and ARDS would most likely benefit from this approach. FiO_2 should be minimized to target an acceptable SaO_2 up to 90%. Recruitment maneuvers may prevent alveolar collapse and improve oxygenation. Positioning patients in a prone position may be used as rescue therapy for the

TABLE 1 Recommended therapy in Legionnaires' disease[a]

Antimicrobial agents	Dosage	Route
Macro–azalides[b]		
Azithromycin[c]	500 mg every 24 hours	IV, p.o.
Clarithromycin	500 mg every 12 hour	IV, p.o.
Spiramycin	1,5 M IU every 8 hours	IV
	6 to 9 M IU (total daily dose)	p.o.
Erythromycin[d]	1 g every 6–8 hours	IV, p.o.
Tetracyclines		
Doxycycline[e]	200 mg every 24 hours	IV, p.o.
Fluoroquinolones		
Levofloxacin[c]	500–750 mg every 24 hours	IV, p.o.
Moxifloxacin[c]	400 mg every 24 hours	IV, p.o.
Gemifloxacin[f]	320 mg every 24 hours	p.o.
Gatifloxacin[f]	200–400 mg every 24 hours	IV, p.o.
Ciprofloxacin	400 mg every 8 to 12 hours	IV
	500–750 mg every 12 hours	p.o.
Ofloxacin	400 mg every 12 hours	IV, p.o
Ketolides		
Telithromycin[f]	800 mg every 24 hours	p.o.

[a]Oral therapy is recommended only in those mild cases that do not require hospitalization. Some antibiotics are commercially available in selected countries.

[b]In mild cases other oral macrolides are also effective: josamycin (1 g every 12 hours), roxithromycin (150 mg every 12 hours), dirithromycin (500 mg every 24 hours).

[c]Recommended in more severe cases, particularly in the immunocompromised.

[d]Less active than other macrolides; risk of fluid overload, phlebitis, and transitory deafness with IV administration. Oral dosage of 1 g every 6 to 8 hours is often poorly tolerated.

[e]Given in one or two divided doses.

[f]Because of short accumulated clinical experience, their use is recommended only in mild to moderate cases.

most severe episodes. Preliminary studies in animal models raise some concern about the risk of hyperoxia in severe legionellosis. Extracorporeal membrane oxygenation has been anecdotally reported as a successful therapeutic option in treating severe *Legionella*-associated ARDS. As many patients may recover, even completely, after many days of needing mechanical ventilation, an aggressive approach is mandatory in the face of respiratory failure.

Shock and acute renal failure are both associated with a high risk of death (7, 14, 15). Hemodynamic control is the cornerstone of therapy in patients with hemodynamic instability. In cases of deterioration of renal function, corrective measures and diligent administration of substitutive treatment are mandatory until complete recovery of renal function is achieved.

Steroid Therapy

Some nonimmunocompromised patients with severe LD may benefit from a short course of steroid therapy, as has been suggested for other types of SCAP. However, there is no good evidence to recommend this approach on a routine basis. Steroids may also be useful in the proliferative phase of diffuse alveolar damage (in patients with ARDS), in some reactive extrapulmonary manifestations (arthritis; myocarditis; renal, neurological, or hemathological features), and when an inflammatory pattern is identified in representative samples of lung tissue in patients with a protracted course (15, 16).

Duration of Therapy

The length of therapy has to be decided on an individualized basis (15, 16). Clinical judgment must be used to establish how long treatment should be administered, but usually a 7 to 14 day course of therapy is sufficient to cure most patients. With azithromycin, even 5 days may be sufficient in mild cases (13). It has also been reported that some patients with mild LD may be cured with just a 5-day course of therapy with levofloxacin at a dosage of 500 to 750 mg daily (4, 20). A short course of therapy should be avoided in immunocompromized patients and in cases of endocarditis (15).

Extrapulmonary Features

Extrapulmonary manifestations of legionellosis, which are rare but sometimes observed in the immunocompromised host, may indicate significant therapeutic connotations (14). Misdiagnosis of herpetic encephalitis or acute crisis of ulcerative colitis has been reported when *Legionella* involvement of brain or colon occurs. Some of the extrapulmonary localizations of *Legionella* may present as an abscess or a purulent collection that will need to be drained by means of catheter insertion or even surgical removal (15). Some extrapulmonary manifestations of legionellosis may be only reactive, most likely immunologically mediated, and they can respond to a short course of steroid therapy.

Polymicrobial Infection

Mixed infections in legionellosis should be kept in mind concerning the inmunocompromized population since there are many reports of death when clinicians fail to identify and treat the dual component of infection (14, 15). A list of these mixed infections is enumerated in Table 2.

Nonresolving, Severe Legionellosis

In Fig. 1, a proposal of an algorhytmic approach to severe legionellosis with poor clinical resolution is suggested. In patients with delayed resolution, superinfection by *Pseudomonas aeruginosa* should be suspected early on. The development of antibiotic resistance in patients with persisting or relapsing *Legionella* infections is not an issue (14, 15).

COMBINED THERAPY

Some international guidelines recommend combined therapy for severe episodes but no consistent evidence supports this suggestion. For most patients monotherapy with a macrolide or a selected fluoroquinolone usually leads to a more cost-effective outcome (5, 6). Possible toxicities of adding more than one antibiotic may be a concern, particularly in the intensive care unit setting. During a prospective, observational, multicenter Spanish study on 529 cases of SCAP that required admission to the ICU, 23 cases of confirmed, extremely severe legionellosis were found (3). Among all SCAP cases with a microbiological documentation of etiology ($n = 276$), *Streptococcus pneumoniae* was the causative agent of 143 cases, and a variety of other microorganisms were responsible for the others ($n = 110$). Immunocompetent patients accounted for 20 out of 23 cases of legionellosis. The overall mortality rate of the *Legionella* group was 26% (6/23). Two cases of death, one in an immunocompetent individual, occurred in patients with inadequate initial treatment. It is worth pointing out that, unlike previously published studies (7), our patients presented with extremely severe

TABLE 2 Polymicrobial infection[a] in legionellosis

Other *Legionella* spp.	Dual infections by different species of *Legionella* and different serotypes of *L. pneumophila*
Other bacteria	*Streptococcus pneumoniae, Proteus mirabilis, Staphylococcus aureus, Escherichia coli, Prevotella intermedia, Neisseria meningitidis, Enterococcus facium, Enterobacter cloacae, Klebsiella pneumoniae, Haemophilus influenzae, Streptococcus mitis, Listeria monocytogenes, Nocardia asteroides*
Mycobacteria	*Mycobacterium tuberculosis*
Virus	*Herpesvirus, Influenza, Cytomegalovirus*
Fungus	*Aspergillus, Cryptococcus*
Parasites	*Pneumocystis jiroveci, Leishmania*

[a]Alleged mixed infections with *Mycoplasma pneumoniae, Chlamydia pneumoniae,* and *Coxiella burnettii* have been reported on the basis of serology, which raises much concern about specificity.

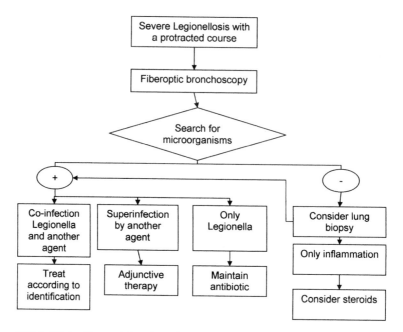

FIGURE 1 Proposal for algorithmic approach to management of intubated patients with nonresolving legionellosis. From reference 15 with permission of the British Society for Antimicrobial Therapy.

legionellosis, since only 8% of them did not require to be ventilated. In our series of patients with extremely severe LD, the comparison of Kaplan-Meier survival curves between patients who received monotherapy and those treated with combination therapy showed a clear trend to a better survival rate in the combined therapy group. However, these differences in survival (fatality rate of 43.3% with monotherapy versus 14.2% with combined therapy) were not statistically significant, probably because of a lack of statistical potency owing to the small number of cases. Mean survival (log rank = 0.203) with combined therapy was 67.7 days (95% CI 50.9 to 84.4), and with monotherapy was 29.7 days (95% CI 17.6 to 41.8). Therefore, these data suggest that combined therapy is most likely associated with a better outcome when compared to monotherapy in the subset of patients with very severe LD who underwent mechanical ventilation.

The most frequently used combined therapy in our study was clarithromycin with rifampin. It is not clear which combined antibi-

otic approach is preferable, although rifampin is the most commonly used agent in combination therapy. In vitro and in vivo evidence shows a highest potency of rifampin against *Legionella* (5), although the theoretical possibility of rapid induction of rifampin-resistant strains when it is administered alone precludes its use as sole therapy. Because the risk of transient hyperbilirrubinemia related to rifampin is high and develops early, we recommend limiting rifampin to the first 5 days (9).

PROGNOSTIC FACTORS

An early, appropriate treatment usually predicts a better outcome and the achievement of a lower mortality rate, particularly in cases with severe clinical presentation that require admission to the intensive care unit (7). Severe disease itself, acute renal failure, a smoking habit, and immunocompromise are the most consistently identified prognostic factors of death in LD (14,15).

Based on in our experience (data from the CAPUCI study presented at the 6th International Conference on Legionella, Chicago,

2005), we identified (T- Fisher and χ^2 tests) the following variables to be significantly associated with death: immunocompromise (3 deaths out of 6 occurred in immunocompromised patients, while no immunocompromise was observed among the 17 survivors; $P = 0.01$), shock (observed in the six patients who died, $P = 0.04$), acute renal failure (odds ratio [OR] = 23, p = 0.008), and APACHE II score >15 (p = 0.05). Diabetes mellitus was another variable associated with a trend to lower survival (OR = 7.5, p = 0.08). On univariate logistic regression analysis the following variables were also found to be associated with death: diabetes mellitus (score = 5.2, p = 0.022), APACHE (score = 9.4, p = 0.002) and acute physiologic score (score = 3.9, p = 0.047). The only variable that remained statistically significant on multivariate logistic regression analysis was the APACHE score (OR 1.86) at intensive care unit admission.

ACKNOWLEDGMENT

We acknowledge investigators of the CAP Intensive Care Units (CAPUCI) Study.

REFERENCES

1. Benhamou, D., J. P. Bru, C. Chidiac, J. Étienne, P. Léophonte, N. Marty, R. Poirier, and R. M. Rouquet. 2005. Legionnaires' disease: definition, diagnosis and treatment. *Med. Mal. Infect.* **35:**1–5.

2. Blázquez, R. M., F. J. Espinosa, L. Alemany, R. L. Ramos, J. M. Sánchez-Nieto, M. Segovia, J. A. Serrano, and F. Herrero. 2005. Antimicrobial chemotherapy for Legionnaires disease: levofloxacin versus macrolides. *Clin. Infect. Dis.* **40:**800–806.

3. Bodí, M., A. Rodríguez, J. Solé-Violan, M. C. Gilavert, J. Garnacho, J. Blanquer, J. Jiménez, M. V. de la Torre, J. M. Sirvent, J. Almirall, A. Doblas, J. R. Badía, F. Garcia, A. Mendía, R. Jordà, F. Bobillo, J. Vallés, M. J. Broch, N. Carrasco, M. A. Herranz, and J. Rello, for the Community-Acquired Pneumonia Intensive Care Units (CAPUCI) Study Investigators. 2005. Antibiotic prescription for community-acquired pneumonia in the intensive care unit. Impact of adherence to IDSA guidelines and outcome. *Clin. Infect. Dis.* **41:**1709–1716.

4. Dunbar, L. M., M. M. Khashab, J. B. Kahn, N. Zadeikis, J. X. Xiang, A. M. Tennenberg. 2004. Efficacy of 750-mg, 5-day levofloxacin in the treatment of community-acquired pneumonia caused by atypical pathogens. *Curr. Med. Res. Opin.* **20:**555–563.

5. Edelstein, P. H. 1998. Antimicrobial chemotherapy for Legionnaires disease: time for a change. *Ann. Intern. Med.* **129:**328–330.

6. Edelstein, P. H. 2002. Chemotherapy of Legionnaires' disease with macrolide or quinolone antimicrobial agents, p. 183–188. *In* R. Marre et al. (ed.) *Legionella.* ASM Press, Washington, D.C.

7. Gacouin, A., Y. Le Tulzo, S. Lavoue, C. Camus, J. Hoff, R. Bassen, C. Arvieux, C. Heurtin, and R. Thomas. 2002. Severe pneumonia due to *Legionella pneumophila* : prognostic factors, impact of delayed appropriate antimicrobial therapy. *Inten. Care Med.* **28:**686–691.

8. Howden, B. P., R. L. Stuart, G. Tallis, M. Bailey, and P. D. Johnson. 2003. Treatment and outcome of 104 hospitalized patients with Legionnaires' disease. *Intern. Med. J.* **33:**484–488.

9. Hubbard, R. B., R. M. Mathur, and J. T. MacFarlane. 1993. Severe community acquired *Legionella* pneumonia: treatment, complications and outcome. *Q. J. Med.* **86:**327–332.

10. Mills, G. D., M. R. Oehley, and B. Arrol. 2005. Effectiveness of beta lactam antibiotics compared with antibiotics active against atypical pathogens in non-severe community-acquired pneumonia: meta-analysis. *Br. Med. J.* **330:**460.

11. Mykietiuk, A., J. Carratalà, N. Fernández-Sabe, J. Dorca, R. Verdaguer, F. Manresa, and F. Gudiol. 2005. Clinical outcomes for hospitalized patients with *Legionella* pneumonia in the antigenuria era: the influence of levofloxacin therapy. *Clin. Infect. Dis.* **40:**794–799.

12. Nara, C., K. Tateda, T. Matsumoto, A. Ohara, S. Miyazaki, T. J. Standiford, and K. Yamaguchi. 2004. *Legionella*-induced acute lung injury in the setting of hyperoxia: protective role of tumour necrosis factor-α. *J. Med. Microb.* **53:**727–733.

13. Plouffe, J. F., R. F. Breiman, B. S. Fields, M. Herbert, J. Inverso, C. Knirsch, A. Kolokathis, T. J. Marrie, L. Nicolle, and D. B. Schwartz. 2003. Azithromycin in the treatment of *Legionella* pneumonia requiring hospitalization. *Clin. Infect. Dis.* **37:**1475–1480.

14. Roig, J., M. Sabria, and M. L. Pedro-Botet. 2003. *Legionella* spp.: community-acquired and nosocomial infections. *Curr. Opin. Infect. Dis.* **16:**145–151.

15. Roig, J., and J. Rello. 2003. Legionnaires' disease: a rational approach to therapy. *J. Antimicrob. Chemother.* **51:**1119–1129.

16. Roig, J., J. Casal, P. Gispert, and E. Gea. 2006. Antibiotic therapy of community-acquired

pneumonia caused by atypical agents. *Med. Mal. Infect.* Conférence de Consensus en Thérapeutique Anti-infectieuse. Institute Pasteur, Paris, France.

17. **Roson, B., J. Carratala, N. Fernandez-Sabe, F. Tubau, F. Manresa, and F. Gudiol.** 2004. Causes and factors associated with early failure in hospitalized patients with community-acquired pneumonia. *Arch. Intern. Med.* **164:**502–508.

18. **Sabrià, M., M. L. Pedro-Botet, J. Gómez, J. Roig, B. Vilaseca, N. Sopena, V. Baños, and Legionnaires' Disease Therapy Group.** 2005. Fluoroquinolones versus macrolides in the treatment of Legionnaires Disease. *Chest* **128:**1401–1405.

19. **Shefet, D., E. Robenshtok, M. Paul, and L. Leibovici.** 2005. Empirical atypical coverage for inpatients with community-acquired pneumonia. *Arch. Intern. Med.* **165:**1992–2000.

20. **Yu, V. L., R. N. Greenberg, N. Zadeikis, J. E. Stout, M. M. Khashab, W. H. Olson, and A. M. Tennenberg.** 2004. Levofloxacin efficacy in the treatment of community-acquired legionellosis. *Chest* **125:**2135–2139.

DIAGNOSTICS AND CLINICAL DISEASE TREATMENT: USEFULNESS OF MICROBIOLOGICAL DIAGNOSTIC METHODS FOR DETECTION OF *LEGIONELLA* INFECTIONS

Paul Christian Lück, Jürgen H. Helbig, Heike von Baum, and Reinhard Marre

3

After transmission from the environment, *Legionella* spp. may cause pneumonias or influenza-like respiratory infections. Since the clinical presentation is not specific for *Legionella* infections, microbiological diagnostic methods play the key role in establishing the etiological diagnosis. Currently available are culture, urinary antigen detection, direct fluorescent antigen testing, detection of nucleic acid, and serology. Based on data published in the literature as well as data from an ongoing study of community-acquired pneumonia in adults (www.capnetz.de), the usefulness of each method will be discussed briefly.

The distribution of *Legionella* spp. and serogroups isolated from patients with pneumonia showed that approximately 60 to 80% of cases are caused by *L. pneumophila* serogroup 1, 15 to 35% are caused by other serogroups of this species, and about 10% are caused by other *Legionella* spp. (19). MAb 3-1 positive strains are associated with outbreaks and are seldom found in unselected environmental samples. They are responsible, however, for the majority of community- and travel-associated cases. In contrast, the distribution of serogroups of nosocomial legionellosis is similar to the distribution in water samples (Fig. 1). This information is relevant since the sensitivity of some diagnostic assays depends on the species and/or serogroup of the causative strain.

CULTURE OF LEGIONELLAE FROM CLINICAL SAMPLES

Culture is still the gold standard among the diagnostic methods for *Legionella* infections. All cultivable *Legionella* spp. can be detected by using culture, although non-*pneumophila* spp. grow more slowly. The medium necessary for the cultivation of legionellae is buffered charcoal yeast extract agar supplemented with antibiotics. Some *Legionella* strains might be susceptible to the antibiotics used in selective media. Therefore, antibiotic-free agar should be used as an additional culture medium (4). Because the quality control of media designed for the recovery of *Legionella* spp. is difficult to achieve within the laboratory, the use of ready-to-use plates is recommended (13).

Since legionellae are environmental, aquatic organisms which do not colonize humans and were never isolated from healthy persons, the specificity of culture is near 100% (5). False-positive culture results may occur if clinical

Paul Christian Lück Institut für Medizinische Mikrobiologie und Hygiene, Technische Universität Dresden, Fiedlerstrasse 42, D-01307 Dresden, Germany. *Heike von Baum* Department of Hospital Hygiene, University of Ulm, Germany. *Reinhard Marre* Department of Medical Microbiology and Hygiene, University of Ulm, Germany.

Legionella: State of the Art 30 Years after Its Recognition
Edited by Nicholas P. Cianciotto et al.
©2006 ASM Press, Washington, D.C.

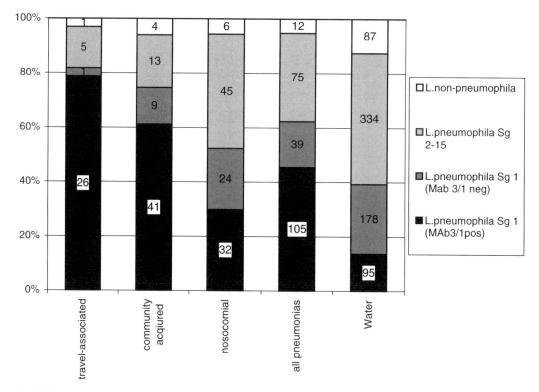

FIGURE 1 Prevalence of *Legionella* spp. and serogroups among clinical and environmental *Legionella* isolates isolated in Germany from 1986 to 2005 and characterized at the Institute of Medical Microbiology and Hygiene, University of Dresden.

samples are contaminated with water containing legionellae, although reports of such cases are very rare.

The sensitivity of culture for the diagnosis of Legionnaires' disease has been estimated to be in the range of 11 to 65% in retrospective studies usually performed in reference laboratories (2, 5, 7, 12; P. C. Lück, unpublished data). In contrast, the percentage of positive cultures tends to be lower if studies are performed prospectively in routine laboratories (1) (CAPNETZ, Table 1). One of the limiting factors for cultivation of *Legionella* spp. seems to be the experience of the laboratory staff (5). Sometimes *Legionella* colonies have unusual colony morphology and might be overlooked (Fig. 2). Bronchoalveolar lavage fluid, bronchial aspirates, lung biopsies, postmortem tissue specimens, and sputum are particularly suitable for culture, whereas pleural fluid is less suitable (2, 4, 5). *Legionella* has seldom been cultured from

extrapulmonary sites, but the frequency of extrapulmonary legionellosis has not been studied extensively.

The most important technique for the identification of legionellae in the clinical laboratory is the serological characterization of isolated strains. A fluorescein-conjugated monoclonal antibody (MAb) which recognizes an outer membrane protein of *L. pneumophila* is commercially available. This species-specific MAb detects all serogroups and can be used for rapid identification of *L. pneumophila* from clinical samples.

Serogroup-specificity is based on the lipopolysaccharide, and the division into serogroups (sg) is based on the reactivity with polyclonal antisera (8). A few serogroups, mainly sg 1 of *L. pneumophila*, can be divided into MAb subgroups which are useful for epidemiological investigations.

Polyclonal antisera, either fluorescein-conjugated or coupled to Latex-beads, for some

TABLE 1 Sensitivity of diagnostic methods in 112 cases of community-acquired *Legionella* pneumonia in adults detected in the German pneumonia study CAPNETZ (http://www.CAPNETZ.de)

Clinical features, number of patients	Culture[a] positive/tested	PCR in respiratory specimens[b] positive/tested	PCR in serum samples[c] positive/tested	PCR in urine samples[c] positive/tested	Antigen in urine samples[d] positive/tested	Serum antibodies in the acute phase serum[e] positive/tested
Patients hospitalized, lethal outcome, n = 13	2/2	7/8	8/8	1/8	8/12	1/12
Patients hospitalized, survived, n = 66	1/9	37/42	29/33	9/30	32/61	0/63
Patients treated in ambulant settings, n = 33	0/7	24/27	8/15	9/33	4/21	1/29
All n = 112	3/16	68/77	45/56	14/59	49/106	2/104

[a]Cultures were performed in routine laboratories in the vicinity of the local clinical centers.
[b]Genus-specific, seminested PCR with sequencing confirmation (14).
[c]Genus-specific TaqMan PCR.
[d]Biotest ELISA.
[e]Indirect immunofluorescence test with formalin inactivated antigens (*L. pneumophila* sg 1 to 15, *L. bozemanii, L. dumoffii, L. jordanis, L. micdadei, L. longbeachae* Sg 1 and 2).

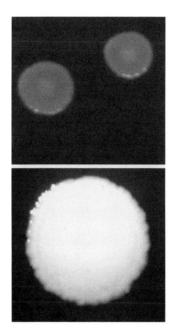

FIGURE 2 *L. pneumophila* colonies with atypical colony appearance grown on GVPC agar.

clinically relevant, but not all, *Legionella* spp. are commercially available. These polyclonal antisera may cross-react with a variety of other bacteria, but these cross-reactions present few problems for the identification of pure cultures isolated on artificial media (5). By using antisera, it is often difficult or even impossible to identify isolated strains to the species level. Therefore, the most accurate identification of *Legionella* isolates can be obtained by sequence analysis of the macrophage infectivity potentiator (*mip*) gene (16). A web-based tool for the identification is available from the European Working Group on *Legionella* Infections (http://www.ewgli.org).

So far, a positive culture is the only method that allows the comparison of patient and environmental *Legionella* strains, thus confirming or excluding a given environmental reservoir as the source of infection.

DETECTION OF *LEGIONELLA* ANTIGEN IN URINE

The antigen excreted with urine has been characterized as heat-stable and resistant to enzymatic cleavage with a molecular weight of about 10 kDa. These characteristics are typical for lipopolysaccharide components. A problem in general is that *L. pneumophila* is divided into 15 serogroups and several monoclonal subgroups. Indeed, all assays for the detection of *L. pneumophila* urinary antigens show sufficient recognition of the antigens which are homologous to the serogroup/monoclonal subgroup used as immunogen for preparation of the antisera, i.e., *L. pneumophila* sg 1 (3, 5, 6, 9, 20).

Several enzyme-linked immunosorbent assays are commercially available from Binax, Biotest, Bartels, Trinity Biotech, and other companies. Apart from these, a rapid immunochromatographic assay (BinaxNow) has been developed and validated in clinical studies which makes it possible to detect antigenuria within 15 min and requires no laboratory equipment. The specificities of all assays, which were mostly evaluated by testing urine specimens from patients with urinary tract infections or pneumonias caused by other pathogens, have been reported to be >99.9% (2, 3, 5, 9).

Concerning the sensitivity for *L. pneumophila* sg 1 infections, all assays are able to detect more than 71 to 92% of cases proven by culture (2, 5, 6, 9, 10). If culture-proven cases caused by other serogroups and legionellosis diagnosed by serology were included, the reported sensitivities would amount to 75 to 90% because all assays available have sensitivities for non-serogroup 1 infections of less than 35% (8). The Biotest enzyme-linked immunosorbent assay seems to be slightly more sensitive in diagnosing legionellosis caused by serogroups 2 to 15. Consequently, the sensitivity is higher for community-acquired and travel-associated *Legionella* infections, which are often caused by serogroup 1 strains (Mab3-1 positive) than for nosocomial cases, which are frequently caused by strains belonging to MAb 3-1 negative sg 1, serogroups 2 to 14 and *Legionella* species (10).

Several investigators recently reported that the sensitivity of the urinary antigen correlates with the severity of illness (1, 20, CAPNETZ, Table 1). This seems to be plausible since the antigen load is higher in severe cases. The use of concentrated urine samples increased the sensitivity without a significant decrease in specificity (3, 6). Since this concentration step is time- and labor-intensive, we use this approach if the urine assay gives borderline results.

The advantages of urinary antigen detection are obvious: specimens are easy to obtain and can be investigated repeatedly; antigenuria is detectable very early and, therefore, often gives the first evidence for *Legionella* infection. Furthermore, testing is very rapid and has a very high specificity. In addition, positive urinary antigen results can also be obtained in cases of nonpneumonic, Pontiac fever–like illness.

In most cases, the antigenuria ends after 10 to 14 days. Despite the initiation of appropriate treatment, antigenuria may persist for some weeks or months. The persistence of antigenuria does not reflect a failure of treatment, but is significantly associated with immunosuppressive therapy (11, 18). Despite the advantages of the urinary antigen detection, one should keep in mind that a negative urinary antigen result does not necessarily exclude a *Legionella* infection.

DETECTION OF LEGIONELLAE BY DIRECT FLUORESCENT ANTIBODY TESTING

Direct fluorescent antibody (DFA) testing of respiratory specimens is a rapid method for the detection of *Legionella* antigen. The previously mentioned fluorescein-conjugated MAb against an outer membrane protein of *L. pneumophila* reacts with all serogroups of *L. pneumophila*. Since 90 to 95% of all cases of legionellosis are caused by this species, this seems to be acceptable. The sensitivity of DFA testing has been reported to range from 22 to 70% (7, 12; P. C. Lück, unpublished data). In contrast to urinary antigen detection, there is no difference regarding the sensitivity in cases of community acquired, travel-associated, and nosocomial legionellosis (P. C. Lück, unpublished data). However, the sensitivity depends mainly on the type of specimen used, the technical equipment (e.g., a cytospin centrifuge for concentration of diluted specimens), and the experience of the laboratory staff. Since the antigen detected

by this assay is not degraded after fixation with formalin, the test allows an etiologic diagnosis in fixed lung tissue, which cannot be done by using the other available methods. Although the reagent is supposed to be highly specific, we observed cases of *Staphylococcus aureus* pneumonia with false-positive fluorescending bacteria, probably owing to binding to protein A. Another important issue is the prevention of false-positive results of DFA testing owing to clinical specimens coming in contact with water containing legionellae as a result of *Legionella* contamination in buffers or legionellae washed off from positive control slides.

DETECTION OF *LEGIONELLA* NUCLEIC ACIDS IN CLINICAL SAMPLES

Over the past few years, PCR has been increasingly used to detect fastidious bacteria in clinical samples. Depending on the primers used, the PCR assays detect *L. pneumophila*, several or all published species of the genus *Legionella*. Accordingly, the specificities of the tests are near 100% (2, 5).

The hitherto available data allow a preliminary estimation of sensitivity and specificity values for PCR in comparison to those of other methods. Since *Legionella* is not considered to be part of the normal human flora, the detection of the presence or absence of *Legionella* DNA in clinical specimens is the main criterion rather than the quantification of the pathogen. The high sensitivity for the detection of *Legionella* DNA in respiratory samples demonstrated by several studies suggests that PCR may exceed culture in its ability to detect *Legionella* in the above-mentioned specimens (2, 5, 7, 11, 17). For respiratory specimens an external quality assurance system has been running in Germany since 2004 (http://www.instand.de). The results of this external quality assurance scheme revealed certain problems in terms of sensitivity and specificity that occur in routine laboratories, but this should be overcome in the future.

The excretion of DNA fragments in the urine is described for several bacterial pathogens, suggesting the suitability of urine PCR for the detection of *Legionella* DNA. Whereas the sensitivity of PCR seems not to be superior to the detection of *Legionella* antigens in urine, the possibility of using primer systems with a broader specificity than the serogroup level appears to be advantageous for diagnostic purposes (2, 11, 15).

The detection of *Legionella* DNA in serum has also been also described (12, 15, CAPNETZ, Table 1). *Legionella* DNA was detectable in acute phase and convalescent phase sera from patients with confirmed legionellosis, but not in the sera from patients without evidence of legionellosis. The sensitivity reported is high and suggests that this method might be used more broadly and evaluated in clinical studies (12, 15).

ANTIBODY DETECTION IN HUMAN SERA

The indirect fluorescent antibody (IFA) test is the only method for antibody detection that has been evaluated and standardized. It is the test format offered by most commercial suppliers.

The sensitivity of serology is generally limited by the time required to develop a detectable antibody response during the course of the infection. Thus, only two patients had detectable antibody levels at the time of examination during the CAPNETZ study (Table 1). Approximately 25 to 40% of patients seroconvert within the first week after the onset of symptoms, and 10 to 15% of patients seroconvert as late as 6 to 9 weeks after onset of the illness (2). Approximately 20 to 30% of patients do not develop significantly elevated antibody titers, even after prolonged observation. This limits the overall sensitivity of serology to 70 to 80% (5). The specificity of seroconversion (fourfold titer rise) using *L. pneumophila* sg 1 antigen in the IFA test has been reported to be approximately 99%. In general, it is recommended to perform *Legionella* serology with polyvalent conjugates detecting immunoglobulin (Ig)G, IgM, and IgA antibodies since the immune response in the course of *Legionella* infections varies with respect to different immunoglobulin classes. IgM-specific tests have a limited specificity and are of no help in differentiating between recent and past infection (2).

It must be emphasized that the exact sensitivity and specificity of serology has only been thoroughly evaluated for IFA tests using *L. pneumophila* sg 1 antigen. Sensitivity and specificity of other serogroups of *L. pneumophila* and of non-*pneumophila* legionellae are not known exactly. Probably, they are lower. Infections with other serogroups of *L. pneumophila* might be detected with sg 1 antigen in some, but not all, cases. Therefore, many laboratories perform *Legionella* serology with a number of antigens. Pools containing different antigens are commonly used for screening purposes, but positive reactions must be confirmed by monovalent antigens, as nonspecific reactions occur frequently with pool antigens.

The background frequency of elevated anti-*Legionella* titers in the normal population has been found to vary from 1 to 36% (2, 5). This should be considered if single elevated titers are used for diagnosis.

None of the diagnostic tests presently available offers the desired quality with respect to sensitivity and specificity. Therefore, the standard technique is to use several diagnostic tests in parallel. Culture should be obligatory, especially if hospitalized patients with underlying diseases are investigated. Urinary antigen detection is a valuable tool in the majority of community-acquired cases when *L. pneumophila* sg 1 is the causative agent. The detection of antibodies in a patient's sera is still a good laboratory test but is of little use in the acute phase of the illness. The detection of nucleic acid specific for *Legionella* is very promising but needs further validation.

ACKNOWLEDGMENT

Our work was supported by the Federal Ministry of Education and Research of Germany, Network of Competence in Medicine CAPNETZ.

We thank Carolin Dix, Sigrid Gäbler, Jutta Paasche, Anna Rein, and Kerstin Seeliger and Ines Wolf for excellent technical assistance.

REFERENCES

1. **Blazquez, R. M., F. J. Espinosa, C. M. Martnez-Toldos, L. Alemany, M. C. Garca-Orenes, and M. Segovia.** 2005. Sensitivity of urinary antigen test in relation to clinical severity in a large

outbreak of *Legionella* pneumonia in Spain. *Eur. J. Clin. Microbiol. Infect. Dis.* **24:**800–806.

2. **DenBoer, J. W. and E. P. Yzerman.** 2004. Diagnosis of *Legionella* infection in Legionnaires' disease. *Eur. J. Clin. Micobiol. Infect. Dis.* **23:**871–878.

3. **Dominguez, J. A., N. Gali, P. Pedroso, A. Fargas, E. Padilla, J. M. Manterola, and L. Matas.** 1998. Comparison of the Binax *Legionella* Urinary Antigen Enzyme Immunoassay (EIA) with the Biotest *Legionella* Urin Antigen EIA for detection of *Legionella* antigen in both concentrated and nonconcentrated urine samples. *J. Clin. Microbiol.* **36:**2718–2722.

4. **Edelstein, P. H.** 2000. Detection of selected fastidious bacteria. *Clin. Infect. Dis.* **31:**846.

5. **Fields, B. S., R. F. Benson, and R. E. Besser.** 2002. *Legionella* and Legionnaires' disease: 25 years of investigation. *Clin. Microbiol. Rev.* **15:**506–526.

6. **Guerrero, C., C. M. Toldos, G. Yague, C. Ramirez, T. Rodriguez, and M. Segovia.** 2004. Comparison of diagnostic sensitivities of three assays (Bartels Enzyme Immunoassay [EIA], Biotest EIA, and Binax NOW Immunochromatographic Test) for detection of *Legionella pneumophila* serogroup 1 antigen in urine. *J. Clin. Microbiol.* **42:**467–468.

7. **Hayden, R. T., J. R. Uhl, X. Qian, M. K. Hopkins, M. C. Aubry, A. H. Limper, R. V. Lloyd, and F. R. Cockerill.** 2001. Direct detection of *Legionella* species from bronchoalveolar lavage and open lung biopsy specimens: comparison of lightcycler pcr, in situ hybridization, direct fluorescence antigen detection, and culture. *J. Clin. Microbiol.* **39:**2618–2626.

8. **Helbig, J. H., J. B. Kurtz, M. C. Pastoris, C. Pelaz, and P. C. Lück.** 1997. Antigenic lipopolysaccharide components of *Legionella pneumophila* recognized by monoclonal antibodies: possibilities and limitations for division of the species into serogroups. *J. Clin. Microbiol.* **35:**2841–2845.

9. **Helbig, J. H., S. A. Uldum, P. C. Lück, and T. G. Harrison.** 2001. Detection of *Legionella pneumophila* antigen in urine samples by the BinaxNOW immunochromatographic assay and comparison with both Binax *Legionella* Urinary Enzyme Immunoassay (EIA) and Biotest *Legionella* Urin Antigen EIA. *J. Med. Microbiol.* **50:**509–516.

10. **Helbig, J. H., S. A. Uldum, S. Bernander, P. C. Lück, G. Wewalka, B. Abraham, V. Gaia, and T. G. Harrison.** 2003. Clinical utility of urinary antigen detection for diagnosis of community-acquired, travel-associated, and nosocomial Legionnaires' disease. *J. Clin. Microbiol.* **41:**838–840.

11. **Koide, M., F. Higa, M. Tateyama, H. Sakugawa, and A. Saito.** 2004. Comparison of polymerase chain reaction and two urinary antigen detection kits for detecting *Legionella* in clinical samples. *Eur. J. Clin. Microbiol. Infect. Dis.* **23:**221–223.

12. **Lindsay, D. S. J., W. H. Abraham, W. Findlay, P. Christie, F. Johnston, and G. F. S. Edwards.** 2004. Laboratory diagnosis of Legionnaires' disease due to *Legionella pneumophila* serogroup 1: comparison of phenotypic and genotypic methods. *J. Med. Microbiol.* **53:**183–187.

13. **Lück, P. C., L. Igel, J. H. Helbig, E. Kuhlisch, and L. Jatzwauk.** 2004. Comparison of commercially available media for the recovery of *Legionella* species. *Int. J. Hyg. Environ. Health* **207:**589–593.

14. **Miyamoto, H., H. Yamamoto, K. Arima, J. Fujii, K. Maruta, K. Izu, T. Shiomori, and S. Yoshida.** 1997. Development of a new seminested PCR method for detection of *Legionella* species and its application to surveillance of legionellae in hospital cooling tower water. *Appl. Environ. Microbiol.* **63:**2489–2494.

15. **Murdoch, D. R., E. J. Walford, L. C. Jennings, G. J. Light, M. I. Schousboe, A. Y. Chereshsky, S. T. Chambers, and G. I. Town.** 1996. Use of the polymerase chain reaction to detect *Legionella* DNA in urine and serum samples from patients with pneumonia. *Clin. Infect. Dis.* **23:**475–480.

16. **Ratcliff, R. M., J. A. Lanser, P. A. Manning, and M. W. Heuzenroeder.** 1998. Sequence-based classification scheme for the genus *Legionella* targeting the *mip* gene. *J. Clin. Microbiol.* **36:**1560–1567.

17. **Reischl, U., H. J. Linde, N. Lehn, O. Landt, K. Barratt, and N. Wellinghausen.** 2002. Direct detection and differentiation of *Legionella* spp. and *Legionella pneumophila* in clinical specimens by dual-color real-time PCR and melting curve analysis. *J. Clin. Microbiol.* **40:**3814–3817.

18. **Sopena, N., M. Sabri, M. L. Pedro-Botet, E. Reynaga, M. Garca-Nez, J. Domnguez, and L. Matas.** 2002. Factors related to persistence of *Legionella* urinary antigen excretion in patients with Legionnaires disease. *Eur. J. Clin. Microbiol. Infect. Dis.* **21:**845–848.

19. **Yu, V. L., J. F. Plouffe, M. C. Pastoris, J. E. Stout, M. Schousboe, A. Widmer, J. Summersgill, T. File, C. M. Heath, D. L. Paterson, and A. Chereshsky.** 2002. Distribution of *Legionella* species and serogroups isolated by culture in patients with sporadic community-acquired legionellosis: an international collaborative survey. *J. Infect. Dis.* **186:**127–128.

20. **Yzerman, E. P., B. Den, K. D. Lettinga, J. Schellekens, J. Dankert, and M. Peeters.** 2002. Sensitivity of three urinary antigen tests associated with clinical severity in a large outbreak of Legionnaires' disease in The Netherlands. *J. Clin. Microbiol.* **40:**3232–3236.

HOSPITAL- AND COMMUNITY-ACQUIRED *LEGIONELLA* PNEUMONIA: TWO FACES OF THE SAME DISEASE?

M. L. Pedro-Botet, L. Mateu, N. Sopena, S. Roure, I. Casas, M. García-Núñez, C. Rey-Joly, and M. Sabrià

4

Legionella pneumophila has been recognized as an important cause of community- and hospital-acquired pneumonia. Information concerning clinical differences between these two settings of *Legionella* infection is scarce and is limited to the preantigenuria period, during which time the clinical aspects of nosocomial and community cases were reported to be similar (3, 4).

However, recent studies suggest that, in fact, *Legionella* infection may differ in these two settings, since the sources of infection, the modes of transmission, the population characteristics, and the delay in the initiation of an appropriate treatment are probably different. (5, 6).

The aim of this study was to compare the individual risk factors, clinical features, and mortality of *Legionella* pneumonia in hospital and community scenarios.

The Hospital Germans Trias i Pujol is a 630-bed research hospital for 700,000 inhabitants located in Badalona (Catalonia, Spain). Our hospital has medicine and surgery departments; obstetric and pediatric departments; dialysis, kidney, and bone marrow transplant units; and an

intensive care unit. Since 1983, we have had an endemic situation of hospital-acquired *Legionella* infection which has been controlled with the implementation of copper and silver ionization since 1999. We have systematically used urinary antigen testing as a specific diagnostic test for *Legionella* infection since 1994. It should also be pointed out that the reporting of Legionnaires' disease has been mandatory in Catalonia since 1987.

A comparative study was performed from 1983 to 2005, with 425 cases of pneumonia due to *L. pneumophila* being diagnosed during this time. Only 320 of these cases, however, fulfilled the criteria for inclusion in this study.

The definition of Legionella infection was as established by Den Boer et al. in 2004 (2), and criteria for a diagnosis of nosocomial infection were those established in the definition of the Centers for Disease Control and Prevention in 1997 (1).

The patients were divided into two groups based on the site of acquisition. Group 1 included 197 patients with community-acquired *Legionella* pneumonia (CALP), and group 2 included 123 patients with hospital-acquired *Legionella* pneumonia (HALP).

According to microbiological data, *L. pneumophila* serogroup 1 urinary antigen test was positive in 87.5% of the patients, and 49.7% of

M. L. Pedro-Botet, L. Mateu, N. Sopena, S. Roure, I. Casas, M. García-Núñez, C. Rey-Joly, and M. Sabrià   Legionella Study Group (GELeg), Autonomous University of Barcelona, Hospital Universitari Germans Trias i Pujol de Badalona, Catalonia, Spain.

the patients seroconverted. *L. pneumophila* serogroup 1 was isolated in 52 out of 217 patients, and direct fluorescent antibody was positive in 6 patients.

The demographic characteristics, individual and aspiration risk factors, clinical manifestations, analytical and therapeutic data, radiological manifestations, and outcome of all the patients included were collected.

The variables were analyzed using the Student's t test for quantitative variables and the chi-square test for qualitative variables with an SPSS program for Windows®. Multivariate logistic regression analysis was performed for variables found to be significant on univariate analysis.

Univariate analysis showed that patients with HALP were older than those with community-acquired disease (49.6% of patients were over 65 years versus 33.5% in the CALP group; $P = 0.005$). Male gender was significantly more frequent in patients with CALP (81.2%) than in those with HALP (70.7%; $P = 0.04$). Individual risk factors including smoking, alcoholism, underlying diseases, and immunosuppressive therapy, especially corticotherapy, were more prevalent in the hospital than in the community setting (86.6 versus 55.3%; $P = 0.000$).

Regarding clinical manifestations at presentation, univariate analysis showed that respiratory symptoms such as cough (73.5 versus 49.1%; $P = 0.000$), gastrointestinal (26 versus 14.3%; $P = 0.02$), and neurological manifestations (36.2 versus 19%; $P = 0.001$) were more common in patients with CALP.

Among the analytical data, neither peripheral white blood cell count, natremia, aspartate aminotransferase, nor respiratory failure discriminated between CALP and HALP. Univariate analysis showed that blood urea nitrogen was more commonly altered in the hospital setting (14% in CALP versus 25.3% in HALP; $P = 0.04$), while creatine kinase tended to be more often altered in the community setting (26.5% in CALP versus 10.7% in HALP; $P = 0.09$). It is of note that 14 patients with CALP developed rabdomyolysis, while

none of the patients with HALP did ($P = 0.00$).

At presentation, the radiological features were similar in both groups.

With regard to treatment, patients with HALP were more rapidly and adequately treated.

There were no differences in time to apyrexia or admission to the intensive care unit. Respiratory failure was the most common complication in both groups, followed by renal failure and septic shock. Considering radiological complications, patients with CALP developed radiological extension more frequently (8.2% in CALP versus 1.8% in HALP; $P = 0.02$). Mortality was significantly higher in the hospital group (14.8%) compared to the community-acquired group (5.1%; $P = 0.006$).

Logistic regression analysis indicated that age >65 years, underlying diseases such as chronic lung diseases, chronic heart failure, morbid obesity, cancer, corticotherapy, aspiration risk factors, elevation in blood urea nitrogen, and mortality were significant independent variables in the HALP group. To the contrary, cough and radiological extension were variables independently related to community-acquired cases.

Legionnaires' disease affects different population groups according to the site of acquisition, reflecting the hospital population characteristics. Preexisting chronic diseases in CALP do not seem to be essential to acquire *Legionella* pneumonia, since 45% of patients with CALP had no underlying disease and most of the patients were smokers and/or alcoholics. Moreover, 15% of the patients with CALP did not have any individual risk factor. Respiratory symptoms, gastrointestinal, neurological manifestations, and laboratory abnormalities prevailed in the community setting probably for two reasons: first, the CALP patients entered the hospital late because of the lack of specificity in initial symptomatology, allowing extrapulmonary symptoms to develop. Second, in our hospital and in hospitals with endemic legionellosis, the diagnosis and the onset of treatment for nosocomial *Legionella* pneumonia is immediate because of the high level of suspicion. This makes the extrapulmonary manifestations of *Legionella* infection less likely to be observed. Mortality was significantly higher in HALP due to underlying diseases rather than virulence of the *Legionella* infection. Nevertheless, this value was lower than in other series due to the rapid initiation of an appropriate antibiotic treatment in our center.

In summary, the high level of suspicion of Legionnaires' disease in our hospital allows a rapid and appropriate treatment for nosocomial *Legionella* pneumonia, and consequently extrarespiratroy symptoms do not develop and mortality is lower compared to other studies. The best way to prevent HALP is by culturing the hospital water supply, and, if colonized, Legionella tests should be used in any case of hospital-acquired pneumonia. Hospital practices such as the use of tap water for oral hygiene, nasogastric tubes, enteric nutrition, pureed diet, medication, and respiratory devices should be prohibited in a health care center colonized by *Legionella* because of the high risk of aspiration for inpatients.

ACKNOWLEDGMENTS

Manuscript supported by Red Respira (RTIC C03/11), Fondo de Investigación sanitaria, Instituto de Salud Carlos III), and Associació IBEMI.

REFERENCES

1. **Anonymous.** 1997. CDC guidelines focus on prevention of nosocomial pneumonia. *Am. J. Health Syst. Pharm.* **54:**1022–1025.
2. **Den Boer, J. W. and E. P. F. Yzerman.** 2004. Diagnosis of Legionella infection in Legionnaires' disease. *Eur. J. Clin. Microbiol. Infect. Dis.* **23:**871–878.
3. **Helms, C. M., J. P. Viner, D. D. Weisenburger, L. C. Chiu, E. D. Renner and W. Jonson.** 1984. Sporadic Legionnaires' disease: clinical observations on 87 nosocomial and community-acquired cases. *Am. Med. J. Sci.* **288:**2–12.
4. **Pedro-Botet, M. L., M. Sabriá-Leal, M. Haro, C. Rubio, G. Gimenez, N. Sopena, and J. Tor.** 1995. Nosocomial and community-acquired Legionella pneumonia: clinical comparative analysis. *Eur. Respir. J.* **8:**1929–1933.
5. **Pedro-Botet, M. L., M. Sabriá Leal, N. Sopena, J. M. Manterola, J. Morera, R. Blavia, E. Padilla, L. Matas, and J. M. Gimeno.** 1998. Role of immunosuppression in the evolution of legionnaires' disease. *Clin. Infect. Dis.* **26:**14–19.
6. **Sabria, M., and V. L. Yu.** 2002. Hospital-acquired legionellosis: solutions for a preventable infection. *Lancet* ID **2:**368–373.

RISK FACTORS FOR MORTALITY BY LEGIONNAIRES' DISEASE (1983–2005)

N. Sopena, M. L. Pedro-Botet, L. Mateu, S. Roure, I. Casas,
M. Esteve, S. Ragull, C. Rey-Joly, and M. Sabrià

5

Legionella species is a frequent cause of community-acquired pneumonia and hospital-acquired pneumonia in centers colonized by this microorganism (3, 7, 9, 10). The diagnosis of Legionnares' disease had increased in the past decade since the introduction of urinary antigen detection for Legionella pneumophila serogroup 1 (1, 10). Delay in the initiation of antibiotic treatment, hospital-acquisition, and immunosuppression have been associated with a worse prognosis in Legionnaires' disease (LD) (2, 4–6, 8). Nevertheless, changes in the epidemiology, diagnostic procedures, and therapy have been made in the past decade that may have influenced prognostic factors of LD. The objective of our study was to determine variables related to mortality of LD.

The study was performed in "Germans Trias i Pujol" Hospital, a 630-bed tertiary care center in Badalona (Barcelona) that serves an urban area of 700,000 inhabitants and has 22,000 admissions annually. The hospital has medicine and surgery departments; obstetric and pediatric departments; dialysis, kidney, and bone marrow transplant units; and an intensive care unit. Reporting of LD has been mandatory in Catalonia (Spain) since 1987. The urinary antigen assay Legionella pneumophila serogroup 1 has been systematically used as a specific diagnosis test for Legionella infections in our hospital since 1994. Since 1983, we have had an endemic situation of hospital-acquired Legionella infection. Hyper-chlorination and superheating have not been effective in eradicating this microorganism from the water supply or decreasing the number of cases of hospital-acquired LD. In September 1999 a copper/silver ionization system was implemented.

An observational, comparative study was performed from 1983 to 2005, with 408 cases of Legionella pneumonia being diagnosed during this time. Patients were classified into two groups: group 1 included 50 patients who died of Legionella pneumonia, and group 2 included 358 patients who survived Legionella pneumonia or died of other causes. The variables studied included demographic characteristics, individual and aspiration risk factors, clinical manifestations, radiological manifestations, diagnostic data, treatment, and outcome. Data were collected from our Legionella database.

The variables were analyzed using univariate and multivariate logistic regression analysis by SPSS version 12.0.

N. Sopena, M. L. Pedro-Botet, L. Mateu, S. Roure, I. Casas, M. Esteve, S. Ragull, C. Rey-Joly, and M. Sabrià Legionella Study Group (GELeg), Hospital Universitario Germans Trias i Pujol, Autonomous University of Barcelona, Catalonia, Spain.

Regarding demographic data and risk factors, patients who died had a higher frequency
of underlying diseases than the control group
(89.8 versus 66.9%): mainly chronic respiratory disease (42.9 versus 29.6%), chronic heart
failure (22.4 versus 10.2%), neoplasm (38.8
versus 15.8%), neurological sequelae (20 versus
7.6%), and inmunosuppressive therapy—
mainly corticoids (45.8 versus 14.6%). Hospital acquisition was more frequent among those
who died (48 versus 26.2%). However, there
was no difference in mean age or male gender
between the two groups.

According to clinical presentation, fever was
significantly less frequent in patients who died
(85.4 versus 96.33%), and, to the contrary, dyspnea prevailed in those who survived or died
of other causes (71.9 versus 27.6%). Regarding
analytical data, blood urea nitrogen elevation
higher than 13 mmol/liter (35 versus 13.2%)
and respiratory failure (72.2 versus 42.5%)
were significantly more frequent in patients
who died that in the other group. Positive sputum culture for Legionella was significantly
more frequent in patients who died (32 versus
11.1%). No differences between the two groups
were found in the delay of initiation of an appropriate antibiotic treatment or in the rate of
treatment with quinolones.

The rate of complications (85.7 versus 44%)
including respiratory failure (72.3 versus
42.5%) and mechanical ventilation (31.8 versus 6.6%), renal failure (35.4 versus 6.6%), and
septic shock (41.7 versus 2.8%) was higher in
patients who died. Moreover, the patients who
died were admitted more frequently to the intensive care unit (35 versus 10.1%). Mortality
of the series was 12.2%.

On multivariate analysis, a markedly elevated
risk of death was identified for patients with
chronic heart failure (odds ratio [OR], 4.9; 95%
confidence interval [CI], 1.3 to 18.1), hematological cancer (OR 19.7; 95% CI, 4.5 to 86.6),
or corticotherapy (OR 3.9; 95% CI, 1.2 to
12.4), and those who developed shock (OR
9.4; 95% CI, 2.4, 36.8) or needed mechanical
ventilation (OR 7.5; 95% CI, 1.7, 31.4).

In conclusion, despite the improvement in
the diagnostic procedures of LD and the use of
more efficacious antibiotics against Legionella,
chronic heart failure, hematological cancer,
and corticotherapy are still bad prognostic factors of LD. Recommendations for prevention
of LD should focus on settings in which there
are persons at greatest risk for illness or serious
outcome.

ACKNOWLEDGMENTS
Manuscript supported by Red Respira (RTIC
C03/11), Fondo de Investigación sanitaria, Instituto de
Salud Carlos III) and Associació IBEMI.

REFERENCES
1. Benin, A. L., R. F. Benson, and R. E. Bessser.
2002. Trends in Legionnaires' disease, 1980–1998.
Declining mortality and new patterns of diagnosis. Clin. Infect. Dis. 35:1039–1046.
2. Gacouin, A., Y. Le Tulzo, S. Lavoue, C. Camus,
H. Hoff, R. Bassen, C. Arvieux, C. Heurtin,
and R. Thomas. 2002. Severe pneumonia due
to Legionella pneumophila: prognostic factors, impact of delayed appropriate antimicrobial therapy.
Intensive Care Med. 28:686–691.
3. Gupta, S. K., and G. A. Sarosi. 2001. The role
of atypical pathogens in community-acquired
pneumonia. Med. Clin. North Am. 85:1349–1365.
4. Health, C. H., D. I. Grove, and D. F. M. Looke.
1996. Delaying appropriate therapy of Legionella
pneumonia associated with increased mortality.
Eur. J. Clin. Microbiol. Infect. Dis. 15:286–290.
5. Marston, B. J., H. B. Lipman, and R. F.
Breiman. 1994. Surveillance for Legionnaires'
disease: risk factors for morbidity and mortality.
Arch. Intern. med. 154:2417–2422.
6. Mykietiuk, A., J. Carratala, N. Fernandez-
Sabe, J. Dorca, R. Verdaguer, F. Manresa, and
F. Gudiol. 2005. Clinical outcomes for hospitalized patients with Legionella pneumonia in the
antigenuria era: the influenza of levofloxacin therapy. Clin. Infect. Dis. 40:794–799.
7. Pedro-Botet, M. L., J. E. Stout, and V. L. Yu.
2002. Legionnaires' disease contracted from patient homes: the coming of the third plague? Eur.
J. Clin. Microbiol. Infect. Dis. 21:699–705.
8. Pedro-Botet, M. L., M. Sabrià, N. Sopena,
J. M. Manterola, J. Morera, R. Blavia, E.
Padilla, L. Matas, and J. M. Gimeno. 1998.
Role of immunosuppression in the evolution of
Legionnaries' disease. Clin. Infect. Dis. 26:14–19.

9. **Sabrià, M., J. M. Modol, M. Garcia-Núñez, E. Reynaga, M. L. Pedro-Bote, N. Sopena, and C. Rey-Joly.** 2004. Environmental cultures and hospital-acquired Legionnaires' disease. A five-year prospective study in twenty hospitals in Catalonia. Spain. *Infect. Control Hosp. Epidemiol.* **25:**1072–1076.

10. **Sopena, N., M. Sabrià, M. L. Pedro-Botet, J. M. Manterola, L. Matas, J. Domínguez, J. M. Mòdol, P. Tudela, V. Ausina, and M. Foz.** 1999. Prospective study of community-acquired pneumonia of bacterial etiology in adults. *Eur. J. Clin. Microbiol. Infect. Dis.* **18:**852–858.

TRENDS OBSERVED IN LEGIONNAIRES' DISEASE IN A HOSPITAL IN CATALONIA, SPAIN, 1983–2005

*I. Casas, M. L. Pedro-Botet, N. Sopena, M. Esteve, L. Mateu,
S. Roure, M. García-Núñez, C. Rey-Joly, and M. Sabrià*

6

Since its first description in 1976, *Legionella pneumophila* has been recognized as an important cause of community- and hospital-acquired pneumonia in both healthy and immunosuppressed patients (2–5). At the Hospital Germans Trias i Pujol, since 1983, we have had an endemic situation of hospital-acquired *Legionella* infection which has been controlled with the implementation of copper and silver ionization since 1999. We have systematically used urinary antigen testing as a specific diagnostic test for *Legionella* infection since 1994. It should also be pointed out that the reporting of Legionnaires' disease (LD) has been mandatory in Catalonia since 1987.

Earlier diagnosis and changes in treatment of LD may have allowed the recognition of this infection in patients with different demographic characteristics and risk factors and may have influenced clinical appearance and outcome (1). The increase in surveillance and preventive measures may have modified the frequency of community- and hospital-acquired LD.

The aim of this study was to describe demographic data; clinical, analytical, and radio-logical characteristics; diagnostic methods; and therapeutic data of a series of community- and hospital-acquired cases of LD collected from 1983 to 2005 and to determine trends in demographic and clinical data, individual risk factors, diagnostic methods, and outcome over time.

The study was performed in Germans Trias i Pujol Hospital, a 630-bed tertiary care center sited in Badalona (Barcelona) that serves an urban area of 700,000 inhabitants and has 22,000 admissions annually. The hospital has medicine and surgery departments; obstetric and pediatric departments; dialysis, kidney, and bone marrow transplant units; and an intensive care unit.

We performed a descriptive study of the *Legionella* database with 425 patients collected from 1983 to May 2005.

We collected demographic data (age, gender), place of acquisition (community, hospital, long-term care center), risk factors (smoking, alcoholism, and underlying diseases), clinical presentation (respiratory and extrarespiratory symptoms), analytical data and radiological presentation, diagnostic techniques, and outcome.

Trends in cases per year, demographic and clinical data, individual risk factors, diagnostic methods, and outcome per year were calculated by linear regression using the SPSS software to

I. Casas, M. L. Pedro-Botet, N. Sopena, M. Esteve, L. Mateu, S. Roure, M. García-Núñez, C. Rey-Joly, and M. Sabrià Legionelosis Study Group (GELeg), Hospital Universitario Germans Trias i Pujol, Autonomous University of Barcelona, Catalonia, Spain.

determine whether the slope of the trend line differed from zero. The slope of the trend line was estimated, turning the dependent variable into the neperian logarithm. $P \leq 0.05$ was considered to be significant. In the calculations we only included the years and the variables with a significant number of registered cases of LD: mean age, rate of male gender, rate of underlying diseases, rate of complications, rate of culture-based diagnosis, rate of mortality, rate of extrarespiratory symptoms, and rate of urinary antigen test-based diagnosis.

We obtained data from 425 cases from 1983 to May 2005. According to diagnostic methods, 246 (57.6%) patients had a positive *L. pneumophila* serogroup 1 urinary antigen test, 52 (12.1%) had a positive culture of respiratory sample for *L. pneumophila* serogroup 1, and 135 (31.6%) seroconverted to *L. pneumophila* serogroup 1 to 6.

The mean age was 59.8 years (range, 0 to 98), with males making up 73.5%. Two-hundred and eighty-six (69.8%) patients had underlying disease, with chronic lung disease being of note in 128 (31.4%); 190 (46.6%) were smokers, and 46 (14.6%) had aspiration risk factors. Extrarespiratory symptoms were found in 171 (40.2%) patients, with complications on evolution in 198 (49.1%); 38 (9.6%) patients needed mechanical ventilation, and the mortality rate was 12.2% (53/425). Mean time to apyrexia was 53.7 hours. According to the analytical data, 73 (19.9%) had hyponatremia, 147 (40.2%) had a creatine kinase increase, and 54 (23.1%) had an aspartate aminotransferase increase. Concerning radiological manifestations, 395 (99.7%) patients had an alveolar infiltrate in chest X ray. Infiltrates were multilobar or bilateral in 88 (22.1%) patients.

In the global series 48.2% of the cases were community acquired. We grouped the cases in 6-year periods, and the community cases rose from 22.3% (1983 to 1988) to 77.9% (2001 to 2005).

The mean delay in the initiation of an appropriate antibiotic treatment was 4.33 days (range, 0 to 30 days).

From 1983 to 2005 the proportion of males and the rate of extrarespiratory symptoms increased over time, but only the urinary antigen test–based diagnosis significantly increased ($P < 0.001$).

The median age, rate of underlying diseases, proportion of smokers, proportion of complications, and rate of mortality decreased over time, but only the proportions of culture-based diagnosis and serologic testing significantly decreased ($P = 0.018$ and $P < 0.0001$).

Since 1983, the incidence of community-acquired LD has increased remarkably. Mortality by nosocomial and community-acquired LD combined has almost significantly decreased over time, which may be due to the earlier diagnosis and changes in treatment. There have also been important changes in diagnostic methods during the study period, and the significant decrease in culture-based diagnosis limits the recognition of non-*L. pneumophila* 1 disease and impairs epidemiological studies.

ACKNOWLEDGMENTS

Manuscript supported by Red Respira (RTIC C03/11), Fondo de Investigación sanitaria, Instituto de Salud Carlos III), and Associació IBEMI.

REFERENCES

1. **Benin, A. L., R. F. Benson, and R. E. Besser.** 2002. Trends in Legionnaires's disease, 1980-1998. Declining mortality and new patterns of diagnosis. *Clin. Infect. Dis.* **35:**1039–1046.
2. **Gupta, S. K., and G. A. Sarosi.** 2001. The role of atypical pathogens in community-acquired pneumonia. *Med. Clin. North Am.* **85:**1349–1365.
3. **Pedro-Botet, M. L., J. E. Stout, and V. L. Yu.** 2002. Legionnaires' disease contracted from patient homes: the coming of the third plague? *Eur. J. Clin. Microbiol. Infect. Dis.* **21:**699–705.
4. **Sabrià, M., J. M. Modol, M. Garcia-Núñez, E. Reynaga, M. L. Pedro-Botet, N. Sopena, and C. Rey-Joly.** 2004. Environmental cultures and hospital-acquired legionnaires' disease. A five-year prospective study in twenty hospitals in Catalonia. Spain. *Infect. Control Hosp. Epidemiol.* **25:**1072–1076.
5. **Sopena, N., M. Sabrià, M. L. Pedro-Botet, J. M. Manterola, L. Matas, J. Domínguez, J. M. Mòdol, P. Tudela, V. Ausina, and M. Foz.** 1999. Prospective study of community-acquired pneumonia of bacterial etiology in adults. *Eur. J. Clin. Microbiol. Infect. Dis.* **18:**852–858.

COMMUNITY-ACQUIRED PNEUMONIA IN HUMAN IMMUNODEFICIENCY VIRUS-INFECTED PATIENTS: COMPARATIVE STUDY OF *STREPTOCOCCUS PNEUMONIAE* AND *LEGIONELLA PNEUMOPHILA* SEROGROUP 1

M. L. Pedro-Botet, N. Sopena, A. García-Cruz, L. Mateu, S. Roure, M. J. Dominguez, I. Sanchez, C. Rey-Joly, and M. Sabrià

7

Bacterial pneumonia continues to be the most frequent pulmonary complication in human immunodeficiency virus (HIV)-infected patients, although its prevalence has decreased with the use of highly active antiretroviral therapy (3, 4). The most common etiologic agent of community-acquired pneumonia (CAP) in infected and uninfected patients is *Streptococcus pneumoniae* (5). *Legionella* is at present one of the leading causes of CAP in the general population (8). However, the incidence of Legionnaires' disease varies from 0 to 8% of microbiologically confirmed pneumonias in HIV-positive patients, probably since most of the series of CAP in HIV-infected patients did not routinely include techniques to diagnose Legionnaires' disease (1). Although scarce, several reports have indicated that the course of this infection in HIV-infected patients is different from that in noninfected patients (7).

To our knowledge there are no clinical studies comparing *S. pneumoniae* pneumonia (SPP) and *Legionella* pneumonia (LP) in HIV-infected patients. The aim of this study was to compare the epidemiological data, clinical features, outcome, and mortality of pneumonia by *S. pneumoniae* and by *Legionella* spp. in HIV-infected patients.

An observational, comparative study was performed in 15 HIV patients with CAP by *Legionella* (group 1) and 46 by *S. pneumoniae* (group 2). Patients with only a serologic diagnosis of LP were excluded. Patients with only a positive sputum culture for *S. pneumoniae* were excluded. Sixteen (34.7%) patients had a positive *S. pneumoniae* urinary antigen test. *S. pneumoniae* was isolated in sputum in 30 patients (65.2%), in blood culture in 27 patients (58.6%), and in pleural fluid in 5 patients (10.8%). In one patient (2.1%) protected specimen brush culture was positive for *S. pneumoniae*. Twelve (80%) patients had a positive *L. pneumophila* serogroup (sg) 1 urinary antigen test. Six (40%) seroconverted to *L. pneumophila* sg 1 to 6, and in four (26.6%) cases culture of sputum ($n = 3$) or bronchoalveolar lavage ($n = 1$) were positive for *L. pneumophila* sg 1.

No statistically significant differences were observed between the two groups in either the use of an appropriate antibiotic treatment in the emergency department or in the delay from the presentation of the symptoms until the initiation of an appropriate antibiotic therapy.

M. L. Pedro-Botet, N. Sopena, A. García-Cruz, L. Mateu, S. Roure, M. J. Dominguez, I. Sanchez, C. Rey-Joly, and M. Sabrià Legionella Study Group (GELeg), Autonomous University of Barcelona, Hospital Universitari Germans Trias i Pujol de Badalona, Catalonia, Spain.

Legionella: State of the Art 30 Years after Its Recognition
Edited by Nicholas P. Cianciotto et al.
©2006 ASM Press, Washington, D.C.

No statistically significant differences were observed in age or sex. Among individual risk factors, smoking was significantly more frequent among the patients with Legionnaires' disease ($P = 0.03$). Other chronic diseases were significantly more frequent in patients with LP (46.6 versus 8.7%) ($P = 0.002$), with cancer being of note ($P = 0.00009$). Patients with *Legionella* infection received chemotherapy (20%) more frequently than patients with SPP (0%) ($P = 0.01$).

Regarding HIV infection–related data, a history of intravenous drug use was significantly more frequent in patients with SPP than in those with LP ($P = 0.03$). The use of highly active antiretroviral therapy in patients with Legionnaires' disease was significantly more frequent than in those with SPP (71.4 versus 27.3%) ($P = 0.004$). The mean value of CD4-positive lymphocytes was significantly higher in patients with Legionnaires' disease (408.6/mm^3 [range, 7 to 889]) than in those with SPP (162/mm^3 [range, 0 to 620]) ($P = 0.04$). Twenty-eight patients (68.3%) with SPP had a CD4 count less than 200/mm^3 versus four (28.6%) with LP ($P = 0.01$). The viral load was undetectable significantly more frequently in patients with LP (71.4%) than in those with SPP (22.2%; $P = 0.01$). Finally, AIDS was significantly less frequent in patients with LP than in those with SPP ($P = 0.03$).

A Fine score higher than 3 was significantly more common in patients with LP ($P = 0.007$), as were dyspnea ($P = 0.04$) and extrarespiratory symptoms (gastrointestinal and neurological) ($P = 0.02$). There were no differences in radiological manifestations between the two groups. Regarding analytical parameters, hyponatremia ($P = 0.002$) and an increase in creatine phosphokinase ($P = 0.006$) and aspartate aminotransferase ($P = 0.09$) were significantly more frequent in the *Legionella* group.

Mean time to apyrexia was 103.5 h (range, 6 to 240 h) for patients with LP and 95.6 h (range, 12 to 408 h) for those with SPP. Five (33.3%) patients with LP developed respiratory failure on evolution, with a prevalence of 2.2% in the *S. pneumoniae* group ($P = 0.003$).

Invasive or noninvasive ventilation was required in four patients (26.6%) with LP versus none of those with SPP ($P = 0.002$). Concerning radiological manifestations, four (26.7%) patients with LP developed bilateral pulmonary infiltrates, while none of the patients with SPP did so ($P = 0.002$). Hospital stay was longer for patients with SPP ($P = 0.06$). Among patients with SPP, four patients (8.7%) developed pleural empyema requiring drainage, and three (6.5%) developed other complications such as endocarditis (n = 1), hospital-acquired *Pseudomonas aeruginosa* pneumonia (n = 1) and visceral Leishmaniasis (n = 1). Finally, mortality related to pneumonia was greater in patients with LP, although without statistical significance.

LP had a more severe presentation, worse evolution, and higher mortality than SPP, although no differences were found in the delay in the initiation of an appropriate antibiotic treatment in either group, and a better immunological status was observed in the former group. The intracellular replication of *Legionella* and the need for integrity of cellular immunity to control this infection probably justify the data observed in HIV-infected patients. According to the results of this study, differential clinical (extrarespiratory symptoms) and analytical (increases in aspartate amino transferase, creatine phosphokinase, and hyponatremia) data of legionellosis are especially prevalent in HIV-infected patients and seem to be more consistent than in the general population (6). Thus, LP should be clinically easier to differentiate from SPP in HIV-infected than in uninfected patients with the use of the Winthrop-University Hospital scale (2).

LP had a more severe presentation than SPP in HIV-infected patients according to the Fine score and the need for invasive or noninvasive ventilation. Furthermore, on evolution, radiological extension and respiratory failure were significantly more frequent in LP than in SPP, with a higher, albeit not significant, mortality. The greater number of patients with cancer and chemotherapy in the *Legionella* group could justify a worse evolution of these patients, but the exclusion of these patients did not significantly influence the outcome of this infection in a previous study (7).

Mean hospital stay was longer for patients with SPP and was mainly related to the development of pleural empyema.

Based on the results of our study, classic risk factors of LP described in early series such as cigarette smoking, cancer, and chemotherapy prevailed in HIV-infected patients with LP. Moreover, an atypical presentation of CAP in HIV-infected patients should strongly signal the possibility of LP and, finally, *Legionella* should be included among the main etiologies of severe CAP in HIV-infected patients. Nevertheless, patients with SPP had a longer hospital stay because of the appearance of pleural empyema and other complications related to a severe immunosuppression. LP should be considered as an opportunistic infection in HIV-infected patients, and the *Legionella* urinary antigen test should be mandatory whenever we are faced with CAP in an HIV-positive patient.

ACKNOWLEDGMENTS

Manuscript supported by Red Respira (RTIC C03/11), Fondo de Investigación sanitaria, Instituto de Salud Carlos III), and Associació IBEMI.

REFERENCES

1. **Casau, N. C.** 2004. Low prevalence of Legionnaires' disease in HIV-infected patients. *AIDS Read.* **14:**269.
2. **Cunha, B. A.** 1998. Clinical features of Legionnaires' disease. *Semin. Respir. Infect.* **13:**116–127.
3. **De Gaetano Donati, K., M. Tumbarello, E. Tacconelli, S. Bertagnolio, R. Rabagliati, G. Scoppetuolo, R. Citton, M. Cataldo, E. Rastrelli, G. Fadda, and R. Cauda.** 2003. Impact of highly active antiretroviral therapy (HAART) on the incidence of bacterial infections in HIV-infected subjects. *J. Chemother.* **16:**60–65.
4. **Feikin, D., C. Feldman, A. Schuchat, and E. N. Janoff.** 2004. Global strategies to prevent bacterial pneumonia in adults with HIV disease. *Lancet Infect. Dis.* **4:**445–455.
5. **Feldman, C.** 2005. Pneumonia associated with HIV infection. *Curr. Opin. Infect.* **18:**165–170.
6. **Mulazimoglu, L., and V. L. Yu.** 2001. Can Legionniares' disease be diagnosed by clinical criteria? A critical review. *Chest* **120:**1049–1053.
7. **Pedro-Botet, M. L., M. Sabriá, N. Sopena, M. García-Núñez, M. J. Domínguez, E. Reynaga, and C. Rey-Joly.** 2003. Legionnaires' disease and HIV infection. *Chest* **124:**543–547.
8. **Sopena, N., M. Sabriá, M. L. Pedro-Botet, J. M. Manterola, L. Matas, J. Dominguez, J. M. Mòdol, P. Tudela, V. Ausina, and M. Foz.** 1999. Prospective study of community-acquired pneumonia of bacterial etiology in adults. *Eur. J. Clin. Microbiol. Infect. Dis.* **18:**852–858.

NOSOCOMIAL *LEGIONELLA* INFECTION IN THE COUNTY OF COPENHAGEN, 2000–2004

Jette M. Bangsborg, Jens Otto Jarløv, and Søren A. Uldum

8

In Denmark (5,400,000 inhabitants), *Legionella* infection is a notifiable disease. Two departments at Statens Serum Institut (SSI) in Copenhagen are involved in collecting data about these cases, since (i) notifications based on clinical and local laboratory data are sent to the Department of Epidemiology; (ii) all cultured *Legionella* patient isolates are referred to the Department of Bacteriology, Mycology, and Parasitology (ABMP) from local departments of clinical microbiology for confirmatory identification and serotyping; and (iii) water samples from environmental investigations are sent to AMBP for culture and DNA typing (Fig. 1).

Every local clinical microbiology department in Denmark has its own infectious disease control unit, which is responsible for the surveillance and investigation of nosocomial infections. In the County of Copenhagen (619,000 inhabitants), one department (the Department of Clinical Microbiology at Herlev University Hospital [DCM Herlev]) serves three hospitals in the major Copenhagen area (Gentofte, Herlev, and Glostrup Hospitals, with a total of 2,400

beds) with microbiological services, including diagnostic methods for *Legionella* infection, and infection control. We sought to evaluate our diagnostic and environmental results for a 5-year period (2000 to 2004), as well as the available clinical data for the patients.

The diagnosis of *Legionella* infection was made by the detection of *Legionella* species in respiratory specimens by PCR, a positive urinary antigen test, or a significant rise in antibody titre. If onset of pneumonia symptoms occurred more than 10 days postadmission or less than 2 days postdischarge, the patient was considered a nosocomial (N) case of infection; with an onset 2 to 10 days postadmission or postdischarge, the patient was considered a possible nosocomial (PN) case.

PCR was performed as an in-house multiplex method detecting all *Legionella* species by 5S rDNA, and *L. pneumophila* (*Lp*) by the Mip gene, from 1998. As of 2004, the method was converted to real-time PCR (LightCycler, Roche) (J. Bangsborg, B. Erlang, and H. Vinner, Abstr. 15th Eur. Congr. Clin. Microbiol. Infect. Dis., abstract 639, 2005). All PCR positive patient samples were subjected to culture. Urinary antigen detection was performed with a commercially available ELISA method (BioTest, Dreieich, Germany); from 2003 on, the BinaxNow Legionella urinary antigen test

Jette M. Bangsborg, and Jens Otto Jarløv Department of Clinical Microbiology, Herlev University Hospital, DK-2730 Herlev, Denmark. *Søren A. Uldum* Department of Bacteriology, Mycology, and Parasitology, Statens Serum Institut, DK-2300 Copenhagen S, Denmark.

Legionella: State of the Art 30 Years after Its Recognition
Edited by Nicholas P. Cianciotto et al.
©2006 ASM Press, Washington, D.C.

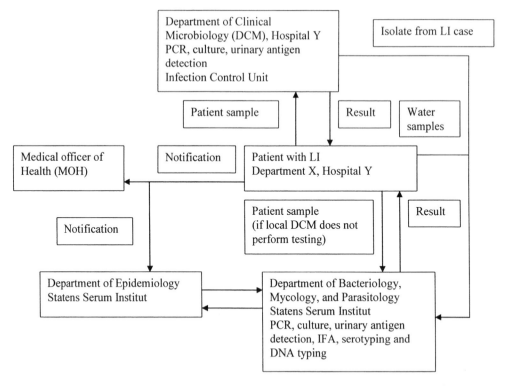

FIGURE 1 Reporting procedure for cases of *Legionella* infection (LI) in Denmark.

(Binax, Scarborough, ME) was used. Antibody detection was performed at ABMP using indirect immunofluorescence.

Diagnostic results and clinical data, including onset of disease and location of the patients, were retrieved from the laboratory information system at DCM Herlev (ADBakt, Autonik, Sweden).

In most cases of culture verified nosocomial infection, water samples (1 liter) were collected from cold and hot tap water in the patient's room and from the shower in the adjoining bathroom. Water samples were sent and received on the day of sampling at SSI. Filtration and culture was performed in compliance with ISO 17025 (http://www.iso.org). In DNA typing of patient isolates and related environmental isolates by restriction enzyme analysis, the enzymes HaeIII and HhaI were used (1).

During 2000 to 2004, 81 patients were diagnosed with *Legionella* infection at DCM Herlev. Twenty-seven patients (33%) were classified as either N (21 patients) or PN (6 patients) cases. Six patients (22%) died. Underlying medical conditions were present in 22 patients (hematologic malignancy, solid cancer, COLD or pulmonary fibrosis, stroke, kidney transplant recipient). Three patients were recovering from recent abdominal surgery. The mean age of the patients was 64.1 ± 13.8 years. During the same period, 74 of 479 cases (15%) reported to SSI were judged to be nosocomial (N or PN) with a mortality of 31% (23/74) (Table 1). Of the 27 nosocomial cases diagnosed at DCM Herlev, 18 were verified by culture, a positive urinary antigen test, or a significant increase in antibody titre, and 9 were patients with a positive PCR result only.

Culture-verified cases yielded six *Lp* serogroup (sg) 1, one *Lp* sg 3, two *Lp* sg 6, two *Lp* sg 12, one *Lp* sg 15, and one *L. micdadei*. From one patient, both *L. bozemanii* and *Lp* sg 3 were cultured.

TABLE 1 National (Statens Serum Institut, SSI) and local (Department of Clinical Microbiology at Herlev University Hospital [DCM Herlev]) data showing nosocomial and possible nosocomial cases of *Legionella* infection as compared to the total number of cases.

	2000	2001	2002	2003	2004	Total
Reported cases of *Legionella* infection in Denmark (SSI)	92	103	96	90	98	479
Reported cases of nosocomial and possible nosocomial cases of *Legionella* infection (SSI)	15	16	18	15	10	74
Laboratory confirmed cases of *Legionella* infection (PCR and/or urinary antigen positive) (DCM Herlev)	8	22	21	15	15	81
Nosocomial cases and possible nosocomial cases of *Legionella* infection (DCM Herlev)	1	4	8	7	7	27

Two clusters occurred in two hospitals (six and three patients, respectively). The cluster with six patients was related to the Department of Hematology and included patients with *Lp* sg 1, 6, and 12. The other cluster was related to The Department of Oto-Rhino-Laryngology, and all three patients were infected with *Lp* sg 1. The remaining 18 (sporadic) patients were in 9 other departments. Environmental investigation was performed in 11 cases at 2 hospitals, and several *Legionella* serogroups and species were cultured. In both hospitals, *Legionella* was found in both cold and hot tap water with colony counts of 10 to 10^4 CFU/liter. In all but one case, the same DNA type was identified in patient and water strains.

Compared to the country as a whole, nosocomial cases seem to be overrepresented in patients diagnosed with *Legionella* infection (33 versus 15%) in our area but have a lower mortality (22 versus 31%). The national figures, however, do not take into account the cases that are not reported. An estimate of the number of unreported cases based upon the number of laboratory-confirmed cases at SSI each year is about 11% (3.2 to 15.6% from 2000 to 2004). Even with this reservation, the discrepancy remains great. One explanation for this could be a higher risk of nosocomial infection in our county hospitals (higher colony counts, more virulent strains, more susceptible patients, etc.) compared to the country as a whole. Another explanation could be a higher awareness and diagnostic efficiency with more prompt treatment, reflected in the lower mortality.

Since the national (SSI) data also include the DCM Herlev results, the proper comparison is between the mortality of nosocomial cases of 6 of 27 (22%) in the County of Copenhagen versus the mortality found for nosocomial cases in Denmark outside the County of Copenhagen, i.e., (23 minus 6)/(74 minus 27) (36%). By the chi-square test, the difference between the two mortality rates, however, is not statistically significant ($P = 0.30$). Comparing the proportion of nosocomial cases in the County of Copenhagen, i.e., 27/81 (33%), to the corresponding figure for Denmark outside the County of Copenhagen, i.e., (74 minus 27)/(479 minus 81) (12%), this difference is highly significant ($P < 0.001$).

As more than one half of the culture-verified cases were *Lp* other than sg 1 and other species, PCR followed by culture was, and still is, the most useful diagnostic method. A culture positivity rate of 13/27 cases (48%) is in the high end of many reports on the sensitivity of culture (11 to 59%) [summarized in (3)]. Both estimates, however, together with the time required for a positive result, make culture of little practical use in guiding immediate patient management. The high proportion of non–sg 1 *Lp* infections in Denmark (Scandinavia) has been reported elsewhere (4) and precludes the use of urinary antigen detection as a reliable diagnostic test in other than travel-associated *Legionella* infections. Within a cluster of epidemiologically related patients, as illustrated by the outbreak in the Department of Hematology, it is remarkable that different serogroups and DNA types can occur. This finding is in contrast to many other reports of nosocomial *Legionella* infection, where one predominant subtype of *Lp* has been endemic [e.g., as in (2) and (5)].

During the period of 2000 to 2004, several efforts were made to combat nosocomial *Legionella* infection at our hospitals, such as increasing the hot water temperature and replacing hot water tanks with heat exchangers. Restrictions on the use of potable water, probably the most efficient precaution (6), have been extended from immunocompromised patients to other risk groups, e.g., patients who are prone to aspiration and patients in the intensive care unit.

REFERENCES

1. **Bangsborg, J. M., S. Uldum, J. S. Jensen, and B. G. Bruun.** 1995. Nosocomial legionellosis in three heart-lung transplant patients: case reports and environmental observations. *Eur. J. Clin. Microbiol. Infect. Dis.* **14:**99–104.

2. **Darelid, J., S. Bernander, K. Jacobson, and S. Löfgren.** 2004. The presence of a specific genotype of Legionella pneumophila serogroup 1 in a hospital and municipal water distribution system over a 12 year period. *Scand. J. Infect. Dis.* **36:**417–423.

3. **Den Boer, J. W. and E. P. F. Yzerman.** 2004. Diagnosis of Legionella infection in Legionnaires' disease. *Eur. J. Clin. Microbiol. Infect. Dis.* **23:**871–878.

4. **Helbig, J. H., S. Bernander, M. Castellani Pastoris, J. Etienne, V. Gaia, S. Lauwers, D. Lindsay, P. C. Lück, T. Marques, S. Mentula, M. F. Peeters, C. Pelaz, M. Struelens, S. A. Uldum, G. Wewalka, and T. G. Harrison.** 2002. Pan-European study on culture-proven Legionnaires' disease: distribution of Legionella pneumophila serogroups and monoclonal subgroups. *Eur. J. Clin. Microbiol. Infect. Dis.* **21:**710–716.

5. **Kool, J. L., A. E. Fiore, C. M. Kioski, E. W. Brown, R. F. Benson, J. M. Pruckler, C. Glasby, J. C. Butler, G. D. Cage, J. C. Carpenter, R. M. Mandel, B. England, and R. F. Breiman.** 1998. More than 10 years of unrecognized nosocomial transmission of Legionnaires' disease among transplant patients. *Infect. Contr. Hosp. Epidemiol.* **19:**898–904.

6. **Marrie, T. J., D. Haldane, S. MacDonald, K. Clarke, C. Fanning, S. Le Fort-Jost, G. Bezanson, and J. Joly.** 1991. Control of endemic nosocomial legionnaires' disease by using sterile potable water for high risk patients. *Epidemiol. Infect.* **107:**591–605.

A QUESTION OF TIME: A SHORT REVIEW OF DATA ON THE INCUBATION PERIOD BETWEEN EXPOSURE AND SYMPTOM ONSET FOR LEGIONNAIRES' DISEASE

Thomas W. Armstrong

9

The literature on Legionnaires' disease (LD) widely cites 2 to 10 days (less frequently to 14 days) as the incubation period for the disease. This is the lag period between exposure to the infectious *Legionella* organism and appearance of illness symptoms. During an attempt to identify the data substantiating that widely cited range, some inconsistencies appeared. Although reports support most cases manifesting in the 2- to 10-day window, there is potential for a variable proportion of related cases to develop later. This short review summarizes the findings of reports with data adequate to examine the potential for incubation periods longer than 10 to 14 days. The findings of this review suggest that while the majority of cases will develop by 14 days, 19 days may be the upper end of the incubation period. Presently, data are not available to determine if the period extends a few days beyond that apparent limit.

The premise for the widely cited 2- to 10-day incubation period for LD may have started with the first report of the disease (5) and carried forward from there. Many subsequent investigations used a case cutoff date of 10 days

following presumptive exposure. By definition, such reports cannot capture cases with longer incubation periods. For resource reasons and to minimize false-positive cases, it is understood that epidemiologic surveillance in outbreaks must use a reasonable but not over-long incubation period definition. However, the widely cited 2 to 10 days may be insufficient for a reasonably complete capture of outbreak related cases. Based on a report that provides data on the distribution of time to symptoms (3), 2 to 14 days appears to capture 90% of the range, and 2 to 19 days captures 100% of the cases in that outbreak. However, the question of the true upper end of the range of the incubation period may require additional research and data to fully resolve.

The scientific literature was searched, primarily via PubMed and Scirus, for reports on outbreaks of LD. The Legionella publication database of Paul Edelstein (4) and the LD web pages maintained by Denis Green (6) were supplemental sources. Although the latter is not restricted to the scientific literature, it is an extensive compilation of reports on LD outbreaks worldwide and thus served as a cross-check on the capture of the other search efforts. The recovered reports were then reviewed for information on the incubation period considered for case inclusion and for data on the

Thomas W. Armstrong   Occupational and Public Health Division, ExxonMobil Biomedical Sciences, Inc., Annandale, NJ 08801.

actual periods between exposure and symptom onset for the cases. For studies to provide data to establish the range of the incubation period between exposure and symptoms, the following conditions needed to be met:

- Known source of exposure
- Known timing of exposure
- Adequately long follow-up systems for case enrollment
- Matching of the case *Legionella* strain to the source strain by serotype or other method

Studies seldom met all of these conditions rigorously for all cases, so dependable data to resolve the lag time question are currently sparse.

The upper range on the time to develop symptoms after exposure to *Legionella* may extend to 19 days. Table 1 summarizes the six reports that provided information to support this conclusion.

The widely quoted 2- to 10-days period is thought to be the typical range, with 10 days not necessarily meant as an absolute upper limit. However, many investigators have used 10 days as a cutoff for case inclusion in LD

outbreak investigations, and this limits availability of data relevant to the longer incubation period question. Analysis of the data from den Boer et al. (3) (who used a 20 day cutoff) suggests the following distribution for days to symptom onset (percentages are approximate):

Onset within 3 to 10 days following exposure   50% of cases

Onset within 3 to 14 days following exposure   90% of cases

Onset within 3 to 17 days following exposure   99% of cases

Onset within 3 to 19 days following exposure   100% of cases

See Fig. 1 for a graphic presentation of the data on which the summary above is based. Note that the 19 day limit is derived from actual case reported exposure dates given by the investigators (3). The epidemic curve on which Fig. 1 is based does not provide the actual date of presumptive exposure, which ranged from 23 February to as late as 28 February.

These data reflect the outcome for the population exposed and the bacterial strain involved.

TABLE 1 Summary of published data relevant to the incubation period for legionnaires' disease[a]

Reported Time Period	Reference	Comment
To 19 days	3	The investigation used a cutoff period of 20 days after the cases visited the exposure source at the Westfriese Floral Show for case inclusion, so the study is inadequate to determine longer onset periods. Not all cases provided specimens for genetic matching of strains. However, visit to the floral show was a requirement for enrollment as a presumptive case.
Possibly to 17 days	7	The investigation of a cruise ship outbreak indicated a period of 2 to 17 days postembarkation for the onset of symptoms. However, the exposures occurred over an interval of an uncertain number of days following embarkation.
To 14 days	2	The investigative team used a 2 week limit on case inclusion, so this report is inadequate for determining longer incubation times. Genetic matching was not completed for all cases, but attendance at the fair was a requirement for enrollment.
To 16 days	1	The report includes discussion of a case manifesting 16 days after the source cooling tower was disinfected. No information was given in this report with respect to strain matching.
Up to 18 days	6	The segment "A Review of Notified Cases of Legionellosis in Western Australia, 1994" summarizes a case in which symptoms appeared 18 days after the likely exposure event.

[a]Note that this list does not include reports (many available) with defined case cutoffs of 10 days or 14 days.

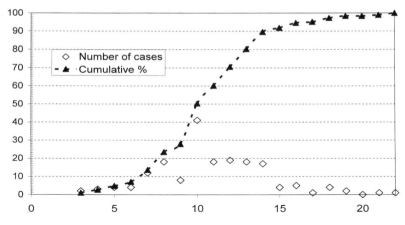

Days to Symptoms Following 23 February Start of Potential Exposure

FIGURE 1 The cumulative percent of cases accrued reaches 50% at day 10, 90% at day 14, and 95% at day 16. A 10-day cutoff could miss 50% of cases. Note that exposure may have occurred as late as 28 February.

The possibility remains that different population demographics or a different *Legionella* species or strains could produce a somewhat different time distribution.

The five studies summarized in Table 1 are based on clinically defined cases, with reasonably certain source identification, exposure time, and, for the most part, matching of strains from the source to varying proportions of the identified cases. There are many more case reports and outbreak investigations than the five in Table 1, but this short review by design focused specifically on the reports that speak to the question of infective periods longer than 10 days.

Investigators of outbreaks with well-defined sources and exposure windows should consider a longer case inclusion period from 14 to perhaps as long as 21 days to ensure good data capture and to help better define the latency period between exposure and symptoms for LD.

REFERENCES

1. **Centers for Disease Control.** 1994. Legionnaires' disease associated with cooling towers—Massachusetts, Michigan, and Rhode Island, 1993. *Morb. Mortal. Wkly. Rep.* **43:**491–493.
2. **De Schrijver, K., K. Dirven, K. Van Bouwel, L. Mortelmans, P. Van Rossom, T. De Beuke-** laar, C. Vael, M. Fajo, O. Ronveaux, M. F. Peeters, Z. A. van der, A. Bergmans, M. Leven, and H. Goossens. 2003. An outbreak of Legionnaire's disease among visitors to a fair in Belgium in 1999. *Public Health* **117:**117–124.
3. **den Boer, J. W., E. P. Yzerman, J. Schellekens, K. D. Lettinga, H. C. Boshuizen, J. E. Van Steenbergen, A. Bosman, H. S. Van den, H. A. Van Vliet, M. F. Peeters, R. J. van Ketel, P. Speelman, J. L. Kool, and M. A. Conyn-van Spaendonck.** 2002. A large outbreak of Legionnaires' disease at a flower show, the Netherlands, 1999. *Emerg. Infect. Dis.* **8:**37–43.
4. **Edelstein, P. H.** 11 August 2005 accession date. Legionnaires' Disease Database. [Online.] Available at: http://www.uphs.upenn.edu/bugdrug/legionella/ldrefman.html.
5. **Fraser, D. W., T. R. Tsai, W. Orenstein, W. E. Parkin, H. J. Beecham, R. G. Sharrar, J. Harris, G. F. Mallison, S. M. Martin, J. E. McDade, C. C. Shepard, and P. S. Brachman.** 1977. Legionnaires' disease: description of an epidemic of pneumonia. *N. Engl. J. Med.* **297:**1189–1197.
6. **Green, D.** 11 August 2005 accession date. Legionnaires' Disease. [Online.] Available at: http://www.q-net.net.au/~legion/.
7. **Jernigan, D. B., J. Hofmann, M. S. Cetron, C. A. Genese, J. P. Nuorti, B. S. Fields, R. F. Benson, R. J. Carter, P. H. Edelstein, I. C. Guerrero, S. M. Paul, H. B. Lipman, and R. Breiman.** 1996. Outbreak of Legionnaires' disease among cruise ship passengers exposed to a contaminated whirlpool spa. *Lancet* **347:**494–499.

SEVERE LEGIONNAIRES' DISEASE SUCCESSFULLY TREATED WITH LEVOFLOXACIN AND AZITHROMYCIN

M. L. Pedro-Botet, N. Sopena, C. Tural, A. García-Cruz,
L. Mateu, S. Ragull, M. García-Núñez, S. Roure,
C. Rey-Joly, and M. Sabrià

10

Fluoroquinolones are the most active drugs against *Legionella* in intracellular and animal models, achieving high intracellular levels and low MICs (9). In patients with human immunodeficiency virus infection, Legionnaires' disease (LD) has a more severe clinical presentation and worse evolution, with a mortality rate of up to 20% (7). Combination therapies, especially erythromycin plus rifampin, have been used in severe cases of LD, although no clear benefit has been demonstrated in clinical studies (9). The combination of macrolides and fluoroquinolones has shown promising effects against *Legionella* in in vitro studies. Nonetheless, there are few data on the clinical efficacy of this combination in the treatment of severe LD.

We present a case of a severely inmunosuppressed woman with LD in whom treatment failed with levofloxacin and who recovered after the addition of azithromycin. The patient was a 47-year-old human immunodeficiency virus–infected female with a history of stage IV rectal cancer with hepatic and pulmonary metastasis for which she underwent local irradiation and chemotherapy up until admission (last cycle May 12, 2004). Prior to admission, the patient was on highly active antiretroviral therapy, had a CD4 count of $130/\mu l$ and an undetectable viral load. On May 19, 2004, she was admitted to the hospital with community-acquired pneumonia and amoxicillin-clavulanate was begun. On day 4 after admission, amoxicillin-clavulanate was switched to levofloxacin (500 mg every 12 h intravenously) because of positivity for *L. pneumophila* serogroup 1 urinary antigen. Fever did not decrease and the patient developed severe respiratory failure and bilateral lung extension. Fibrobronchoscopy was performed and other pathogens were excluded. Noninvasive mechanical ventilation (NIMV) was required on day 7. As the patient did not improve, azithromycin was added on day 10 (500 mg four times a day intravenously). The patient improved dramatically within 24 hours. NIMV was discontinued on day 12. At discharge (day 15), the bilateral infiltrates had disappeared, and only an alveolar infiltrate in the right lower lobe persisted.

This clinical description suggests the clinical benefit of combination therapy with levofloxacin and azithromycin in the treatment of severe *Legionella* pneumonia in a severely immunosuppressed patient.

The impact of new antibiotic options on the prognosis of LD is not well known. The

M. L. Pedro-Botet, N. Sopena, C. Tural, A. García-Cruz, L. Mateu, S. Ragull, M. García-Núñez, S. Roure, C. Rey-Joly, and M. Sabrià   Legionella Study Group (GELeg), Autonomous University of Barcelona, Hospital Universitari Germans Trias i Pujol de Badalona, Catalonia, Spain.

intracellular location of the pathogen is relevant to the efficacy of the antibiotic. Antibiotics capable of achieving intracellular concentrations higher than the MIC are more clinically effective than antibiotics with poor intracellular penetration. At present, new macrolides such as clarithromycin and azithromycin and fluoroquinolones are the most active drugs in the treatment of *Legionella* pneumonia. Despite the lack of randomized studies comparing the efficacy of both treatments, three recently published observational studies demonstrated that patients with LD treated with quinolones improved more rapidly than those receiving macrolides. Nevertheless, no differences in complications or mortality were observed (1, 4, 8).

Therapeutic combinations have been used in patients with severe LD, with the most frequent being the use of erythromycin–rifampicin, which, despite demonstrating synergic effects in experimental models, have never shown a clear benefit in clinical studies. Moreover, the use of this combination is limited by its potential toxicity.

As far as we know, five anecdotal cases in the literature have reported the efficacy of combination therapy in patients with severe LD (3, 5, 6, 11, 12). Conclusions based on only one clinical observation obviously have some limitations. First, azithromycin could be more efficacious than levofloxacin, but neither in vitro *Legionella* susceptibility studies (10) nor animal models (2) of LD have demonstrated a greater activity of azithromycin compared with levofloxacin or other quinolones. Second, improvement could be casual and due to levofloxacin and not to the addition of azithromycin. The severe respiratory failure and the need for NIMV despite three days of double doses of intravenous levofloxacin and the dramatic improvement of the patient 24 h after the addition of azithromycin suggest the benefit of the combination therapy.

Even though the advances in rapid specific diagnosis and the greater efficacy of the new macrolides and quinolones have led to a better prognosis of *Legionella* pneumonia, mortality continues to be very high in immunosuppressed patients. Thus, randomized trials comparing the efficacy of combination therapies with new macrolides and quinolones versus monotherapy should be performed to evaluate their real impact in reducing morbidity and mortality of *Legionella* pneumonia with bad prognostic factors.

REFERENCES

1. **Blazquez-Garrido, R. M., F. J. Espinosa Parra, L. Alemany Francés, R. M. Ramos Guevara, J. M. Sánchez-Nieto, M. Segovia Hernández, J. A. Serrano Martínez, and F. Herrero Huerta.** 2005. Antimicrobial chemotherapy for Legionnaires' disease: levofloxacin versus macrolides. *Clin. Infect. Dis.* **40:**800–806.
2. **Edelstein, P. H. and M. A. C. Edelstein.** 1991. In vitro activity of azithromycin against clinical isolates of Legionella species. *Antimicrob. Agents Chemother.* **35:**180–181.
3. **Ishii, Y., M. Bando, S. Ohno, and Y. Sugiyama.** 2005. A travel abroad-associated case of Legionella pneumonia diagnosed by urinary antigen detection test. *Kansenshogaku Zasshi* **79:**290–293.
4. **Mykietiuk, A., J. Carratalà, N. Fernández-Sabé, J. Dorca, R. Verdaguer, F. Manresa, and F. Gudiol.** 2005. Clinical outcomes for hospitalized patients with Legionella pneumonia in the antigenuria era: the influence of levofloxacin therapy. *Clin. Infect. Dis.* **40:**794–799.
5. **Oguma, A., T. Kojima, D. Himeji, Y. Arinobu, I. Kikuchi, and A. Ueda.** 2004. A case of Legionella pneumonia complicated with acute respiratory distress syndrome treated with methylprednisolone and silvelestat sodium in combination with intravenous erythromycin and ciprofloxacin. *Nihon Kokyuki Gakkai Zasshi* **42:**956–960.
6. **Okano, Y., T. Motoki, M. Miki, N. Hatakeyama, Y. Iwahara, Y. Nakamura, and F. Ogushi.** 2001. A case of Legionella pneumonia successfully treated intravenously with both erythromycin and ciprofloxacin. *Nihon Kokyuki Gakkai Zasshi* **39:**949–954.
7. **Pedro-Botet, M. L., M. Sabrià, N. Sopena, M. García-Núñez, M. J. Domínguez, E. Reynaga, and C. Rey-Joly.** 2003. Legionnaires' disease and HIV infection. *Chest* **124:**543–547.
8. **Sabrià, M., M. L. Pedro-Botet, J. Gómez, J. Roig, B. Vilaseca, N. Sopena, and V. Baños, for the Legionnaires' Disease Therapy Group.** 2005. Fluoroquinolones versus macrolides in the treatment of Legionnaires' disease. *Chest* **128:**1401–1405.
9. **Sabrià, M., and V. L. Yu.** 2002. *Legionella* species (Legionnaires' disease), p. 395–417. *In* V. L. Yu, R. Weber, and D. Raoult, (ed.), *Antimicrobial Therapy*

and Vaccines, 2nd ed. Apple Trees Productions, LLC, New York, N.Y.

10. **Stout, J. E., K. Sens, S. Mietzner, A. Obman, and V. L. Yu.** 2005. Comparative activity of quinolones, macrolides and ketolides against Legionella species using in vitro broth dilution and intracellular susceptibility testing. *Int. J. Antimicrob. Agents* **25:**302–307

11. **Stroup, J. S., S. E. Hendrickson, and M. Neil.** 2004. *Legionella* pneumonia and HIV infection: a case report. *AIDSRead* **14:**267–271.

12. **Trubel, H. K., H. G. Meyer, B. Jahn, M. Knuf, W. Kamin, and R. G. Huth.** 2002. Complicated nosocomial pneumonia due to *Legionella pneumophila* in an immunocompromised child. *Scand. J. Infect. Dis.* **34:**219–221.

IN VITRO ACTIVITIES OF VARIOUS ANTIBIOTICS AGAINST *LEGIONELLA PNEUMOPHILA*

A. Seher Birteksöz, Z. Zeybek, and A. Çotuk

▌▐

It is known that species of *Legionella* live in natural waters and can colonized in man-made water systems. Bacterial transmission to humans occurs through droplets generated from cooling towers, shower heads, and other man-made devices that generate aerosols. *Legionella pneumophila*, the cause of Legionnaires' disease, is a frequently isolated species in immuno-compromised patients.

L. pneumophila infections may progress among these patients rapidly. Optimal results in the treatment of Legionnaires' disease may only be achieved with prompt effective antimicrobial therapy (1, 2, 4). We studied the in vitro activities of erythromycin, azithromycin, ciprofloxacin, ofloxacin, levofloxacin, doxycyclin and rifampin against *L. pneumophila* strains isolated from several water systems of different buildings in Istanbul. The legionella analysis was conducted according to ISO 1998 (5). All samples were concentrated by the filtration method and buffered charcoal-yeast extract (BCYE) agar supplemented with glycine, vancomycine, and polymyxine B, and cyclohex-imide was used for the cultivation. Serological

identification was done by latex aglutination test kits (Oxoid) (3).

In this study, in vitro activities of some antibiotics were assessed against *L. pneumophila* isolated from water samples. *L. pneumophila* ATCC 33152 and *Staphylococcus aureus* ATCC 29213 were used as control strains. All antimicrobial agents were obtained from their manufacturers. Except for rifampin, stock solutions from dry powders were prepared in concentration of 2,000 mg/liter and stored frozen at $-80°C$. Frozen solutions of antibiotics were used within 6 months. Rifampin solutions were prepared on the day of use. Buffered yeast extract broth was used for MICs determinations, and BCYE agar was used for minumum bactericidal concentration (MBC) determinations.

MICs were determined by microbroth dilution technique as described in Ref. 6. Serial twofold dilutions ranging from 16 to 0.0156 mg/liter for erythromycin and azithromycin; 2 to 0.00097 mg/liter for ciprofloxacin, ofloxacin, levofloxacin; 64 to 0.0625 mg/liter for doxycyclin; and 0.0625 to 0.00006 mg/liter for rifampin were prepared in buffered yeast extract broth. The inoculum was prepared using a 24-h broth culture of each isolate adjusted to a turbidity equivalent to a 0.5 Mc-Farland standard and diluted to give a final concentration of 5×10^5 CFU/ml in the test

A. Seher Birteksöz Department of Pharmaceutical Microbiolog, Faculty of Pharmacy, University of Istanbul, Turkey 34116. *Z. Zeybek and A. Çotuk* Department of Biology, Faculty of Science, University of Istanbul, Turkey 34120.

TABLE 1 MICs and MBCs of antibiotics against *L. pneumophila* strains

Antibiotics	MIC (mg/liter[a])		MBC (mg/liter[b])	
	50	90	50	90
Erythromycin	0.125	2	0.5	8
Azithromycin	0.0625	0.25	0.25	4
Ciprofloxacin	0.0156	0.125	0.0312	0.5
Ofloxacin	0.0312	0.125	0.0312	0.5
Levofloxacin	0.0156	0.125	0.0312	0.25
Doxycycline	1	4	8	32
Rifampin	0.00048	0.00097	0.00097	0.0156

[a]50 and 90, MIC_{50} and MIC_{90}.
[b]50 and 90, MBC_{50} and MBC_{90}.

tray. The trays were covered and placed in plastic bags to prevent evaporation and incubated at 35°C for 48 h. The MIC was defined as the lowest concentration of antibiotic giving complete inhibition of visible growth.

MBCs were determined at the conclusion of the incubation period by removing two 10-μl samples from each well demonstrating no visible growth and plated onto BCYE agar (7). Resultant colonies were counted after a 48-h

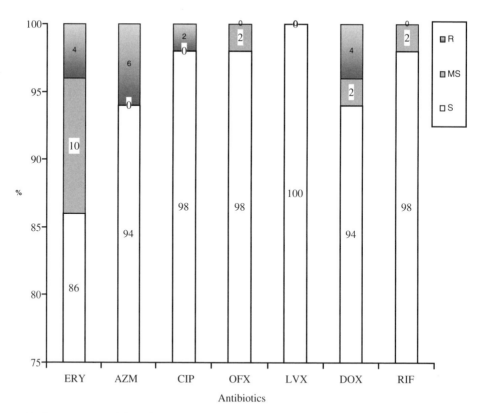

FIGURE 1 Susceptibility of *L. pneumophila* strains to the antibiotics. R, resistant; MS, moderately susceptible; S, susceptible; ERY, erythromycin; AZM, azithromycin; CIP, ciprofloxacin; OFX, ofloxacin; LVX, levofloxacin; DOX, doxycycline; RIF, rifampin.

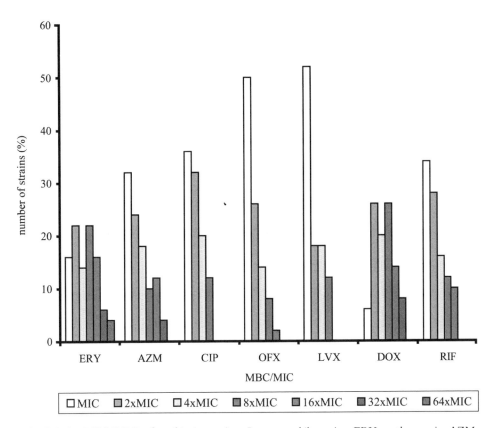

FIGURE 2 MBC/MIC of antibiotics against *L. pneumophila* strains. ERY, erythromycin; AZM, azithromycin; CIP, ciprofloxacin; OFX, ofloxacin; LVX, levofloxacin; DOX, doxycycline; RIF, rifampin.

incubation at 35°C. The MBC was defined at the lowest concentration of antibiotic giving at least a 99.9% killing of the initial inoculum.

In our study, 50 *L. pneumophila* strains have been isolated from hot and cold water systems of different buildings. Thirty-two percent of these strains have been identified as *L. pneumophila* serogrup 1 and 68% identified as *L. pneumophila* serogroup 2 to 14. Twelve percent of isolated strains have been obtained from the cooling tower, 4% from cold water tanks, 24% from hot water tanks, 34% from hot water faucets, and 26% from hot water shower heads. High levels of *L. pneumophila* identified in hot water samples have revealed that the conditions in these systems have an increasing effect on the growth of *L. pneumophila*. Also, these findings have shown that there is a relationship

between the presence of bacteria and water temperature.

The MICs of tested antibiotics against all *L. pneumophila* strains are in Table 1. A total of 86% of the strains have been found to be susceptible to erythromycin; 94% are susceptible to azithromycin and doxycyclin; 98% are susceptible to ciprofloxacin, ofloxacin, and rifampin; and 100% are susceptible to levofloxacin (Fig. 1). None of the *L. pneumophila* strains is resistant against levofloxacin and ofloxacin, so these antibiotics can be used safely only in the treatment of *L. pneumophila* infections.

It has been determined that *L. pneumophila* strains obtained from the cooling tower are sensitive to all antibiotics tested. On the other hand, it has been found that 8% of tested *L. pneumophila* strains isolated from the cold water tank are resistant to azithromycin. Other findings have shown that 8% of strains isolated from the hot water tank are resistant to erythromycin, azithromycin, and doxycycline; 6% of strains isolated from hot water faucets are resistant to azithromycin; and 8% of strains isolated from hot water shower heads are resistant to erythromycin, azithromycin, ciprofloxacin, and doxycycline. According to these findings, it is clear that tested strains which have been found resistant to antibiotics have been isolated mostly from hot water shower heads.

Sensitivity tests used for the selection of the antibiotics have measured general inhibitor effect of antibiotics. In practice, these data can be sufficient in the treatment of many infections. However, it is also important to determine bactericidal activity in the treatment of severe infections that result with mortality. Therefore, the determination of MBCs beside MICs and the rate of MBC/MIC aids in selection of the treatment. Our study has shown that the MBC_{50} and the MBC_{90} values of ciprofloxacin, ofloxacin, and levofloxacin that have been obtained against $L.\ pneumophila$ strains have similar concentrations with the MIC_{50} and MIC_{90} values determined (Table 1). When the rate of the MBC/MIC of tested antibiotics were compared, it was determined that doxycycline and erythromycin have killed an equal number of $L.\ pneumophila$ strains, and ofloxacin and levofloxacin have killed half of the strains at concentrations equal to the MIC.

Consequently, this study has shown that ofloxacin and levofloxacin can be preferable antibiotics in the treatment of $L.\ pneumophila$ infections where bactericidal activity is needed (Fig. 2). We think our findings can be helpful concerning the selection of antibiotics in the treatment of the severe infections of $L.\ pneumophila$.

ACKNOWLEDGMENTS

This work was supported by the Research Fund of the University of Istanbul. Project number: T-1089/18062001.

REFERENCES

1. **Addiss, D. G., J. P. Davis, M. Laventure, P. J. Wand, M. A. Hutchinson, and R. M. Mckinney.** 1992. Community-acquired Legionnaires' disease associated with a cooling tower: evidence for longer-distance transport of *Legionella pneumophila. Am. J. Epidemiol.* **130:**557–568.
2. **Bollin, G. E., J. F. Plouffe, M. F. Para, and B. Hackman.** 1985. Aerosols containing *Legionella pneumophila* generated by shower heads and hot-water faucets. *Appl. Environ. Microbiol.* **50:**1128–1131.
3. **Fields, B. S.** 1997. Legionella and protozoa: interaction of a pathogen and its natural host. *In* J. M. Barbaree, R. F. Breiman, and A. P. Dufour (ed.). *Manual of Environmental Microbiology*, 2nd ed. American Society for Microbiology, Washington, D.C.
4. **Health & Safety Series Booklet.** 1991. *The Control of Legionellosis Including Legionellaires' Disease*, HMSO HS(G) 70, London.
5. **International Organization for Standardization (ISO).** 1998. *Water Quality Detection and Enumeration of Legionella.* ISO 11731: (E).
6. **National Committee For Clinical Laboratory Standards.** 2000. *Methods for Dilution Antimicrobial Susceptibility Tests for Bacteria that Grow Aerobically*, 5th ed. Approved Standard M7-A5 NCCLS, Wayne.
7. **National Committee For Clinical Laboratory Standards.** 1999. *Methods for Determining Bactericidal Activity of Antimicrobial Agents.* Approved Guideline. M26, NCCLS, Wayne.

DETECTION OF *LEGIONELLA PNEUMOPHILA* DNA IN SERUM SAMPLES FROM PATIENTS WITH LEGIONNAIRES' DISEASE

Bram M. W. Diederen, Caroline M. A. de Jong, Faïcal Marmouk, Jan A. J. W. Kluytmans, Marcel F. Peeters, and Anneke van der Zee

12

Legionnaires' disease (LD) is an acute pneumonia caused by *Legionella* spp., gram-negative bacilli ubiquitous in both man-made and natural aquatic reservoirs. Although 48 *Legionella* species have been described, more than 90% of culture-confirmed clinical cases are caused by *Legionella pneumophila* (10). Diagnosis of LD in patients with pneumonia is based on phenotypic (culture, serologic testing, antigen detection in urine) and genotypic PCR methods. Diagnostic PCR assays have principally targeted specific regions within 16S rRNA genes (8, 9), 5S rDNA (4, 6), or the macrophage inhibitor potentiator (*mip*) gene (3) and have the potential to provide a rapid diagnosis of LD with use of readily obtainable specimens such as serum or urine (4, 5, 6). The aim of this study was to assess the sensitivity and specificity of *Legionella* DNA detection using PCR on serum samples from patients with proven LD and patients with infections other than *Legionella*.

Samples were collected between June 1999 and May 2005. A proven case of LD was de-fined as a patient who suffered from symptoms compatible with pneumonia, who showed radiological signs of infiltration, and who showed laboratory evidence of infection with *L. pneumophila*. Laboratory evidence included (i) a single high value of IgM and/or IgG antibodies or seroconversion for *L. pneumophila* serogroup 1 to 7 in paired acute-phase and convalescent-phase sera (Serion ELISA; Institut Virion/Serion GmbH, Würzburg, Germany), (ii) plus one or more of the following criteria: isolation of *L. pneumophila* from a lower respiratory tract sample, a positive urinary antigen test (Binax now; Binax, USA), and a positive PCR result for *L. pneumophila* on a lower respiratory tract sample using a 16S rRNA assay (9). Cases included were 67 adults (48 males and 19 females, male/female ratio 2.5) between 32 and 79 years old (mean age, 54.8 years). The laboratory results of these patients were as follows: (positive/number tested [%]) serology 67 of 67 (100%), isolation 15 of 29 (52%), PCR 26 of 26 (100%), urinary antigen test 53 of 59 (90%).

In addition, serum samples from 36 patients with respiratory tract infections other than *Legionella* were tested in a similar manner to serve as controls. All patients tested negative for *Legionella* antigen in urine (Binax now; Binax) and *L. pneumophila* serology (Serion ELISA). These samples were obtained from 5 patients

Bram M. W. Diederen, Caroline M. A. de Jong, Faïcal Marmouk, Marcel F. Peeters, and Anneke van der Zee Laboratory of Medical Microbiology and Immunology, St. Elisabeth Hospital, 5000 AS Tilburg, The Netherlands. *Jan A. J. W. Kluytmans* Laboratory of Microbiology and Infection Control, Amphia Hospital, 4800 RK Breda, The Netherlands.

with community-acquired pneumonia and positive blood cultures with *Streptococcus pneumoniae* (5 samples) and 31 patients with respiratory tract infections and a fourfold increase in (complement-fixating) antibody titer against influenza A virus, adenovirus, *Chlamydia psittaci*, or *Mycoplasma pneumoniae* (55 samples).

For DNA extraction, 200 μl of serum was processed with MagNA Pure, using the Total Nucleic Acid Kit (Roche Diagnostics) with an elution volume of 50 μl. Five microliters of the eluate was used as a template in PCR.

For the detection of *Legionella* in serum samples, three separate assays were used, targeted at specific regions within the 16S rRNA gene, 5S rRNA, and the *mip* gene (Table 1). The primers of the first PCR-probe assay were based on the 16S rRNA gene as was described previously (9). The second PCR was based on the primers described by Lindsay et al. (4) and detected in real-time using a Taq-Man probe Leg5S. The third PCR was an *L. pneumophila*-specific PCR based on the sequences of the *mip* gene. Real-time PCR was performed on an Abiprism 7900HT Sequence Detection System (Applied Biosystems, Foster City, Calif.).

Serum samples from two healthy volunteers were included as negative controls after every four samples. In each run a no-template control (mix control) was added. Sensitivity controls consisted of 10-fold dilutions of *L. pneumophila* DNA ranging from 1,000 fg to 10 fg. A thousand fg of *L. bozemanii* DNA served as a control in the 16S rRNA-based PCR for discrimination between *L. pneumophila* and other *Legionella* species. As an internal control, phocine herpes virus (PhHV) was added to the samples to monitor processing as well as PCR inhibition, and detection of phocine herpes virus was included in the *mip* PCR.

The chi-square test was used to compare categorical data. A result of $P < 0.05$ was considered to be significant.

We included 151 serum samples from 67 patients with proven LD and 60 samples from 36 patients with infections caused by other pathogens. The samples of patients with infections

TABLE 1   16S rRNA, 5S rRNA, and *mip* primer and probe sequences used in PCR (bp; basepairs)

Primer or probe	Sequence	Product size (bp)
Primers		
16 S rRNA PCR	Leg1 (forward 5'-TACCTACCCTTGACATACAGTG-3')	
	Leg2 (reverse 5'-CTTCCTCCGGTTTGTCAC-3')	200
Mip PCR	Mip-F1 (forward 5'-GCCAAGTGGTTTGCAATACAAA-'3)	
	Mip-R1 (reverse 5'-CTCGACAGTGACTGTATCCGATTT-'3)	80
5S PCR	5S1 (forward 5'-ACTATAGCGATTTGGAACCA-'3)	
	5S2 (reverse 5'-GCGATGACCTACTTTCGCAT-'3)	76
PhHv PCR	PhHV-F1 (forward 5'-GGCGAATCACAGATTGAATC-'3)	
	PhHV-R1 (reverse 5'-GCGGTTCCAAACGTACCAA-'3)	80
Probes		
16S rRNA PCR	LPN (5'-GAGTCCCCACCATCACATG-3')	
	LSPP (5'-GGTTGCGTCGTTACG-3')	
Mip PCR	LPN-mip (FAM-5'-TAATCAATGCTGGAAATGGTGTTAA-ACCCG-3'-TAMRA)	
Real-time 5S PCR	Leg5S (FAM-5'-CCGCGCCAATGATAGTGTGAGGC-3'-TAMRA)	
PhHv PCR	PhHV-1 (VIC-5'-TTTTTTATGTGTCCGCCACCATCTGG-ATC-'3-TAMRA)	

other than *Legionella* all tested negative in 16S PCR, real-time 5S PCR, and real-time *mip* PCR. Of the 67 patients with proven LD, 68.6% (46 of 67) tested positive in one or more PCR assays and in at least one serum sample. In two samples (obtained from two patients) inhibition of PCR occured.

The relative sensitivity of the separate assays to diagnose a patient with LD was 53.7% (36 of 67), 59.7% (40 of 67) and 58.2% (39 of 67) for 16S PCR, real-time 5S PCR, and real-time *mip* PCR, respectively. The differences in sensitivity and detection rate between the different PCR assays were not statistically significant ($P > 0.5$).

From 37 of 46 PCR-positive patients, both acute and convalescent serum was obtained, and the average interval between sample collection was 18.6 days (range: 0 to 80 days). Of those patients, 81% (30 of 37) were PCR positive only in the acute serum sample. We found that in 90.5% (38 of 42) of PCR-positive patients of whom a urinary antigen test had been performed, the urinary antigen test was positive.

One patient was admitted to the emergency department of our hospital with proven LD caused by *L. pneumophila* serogroup 1 (urinary antigen test positive, culture positive, and sputum PCR positive). From this patient, consecutive serum samples were collected for

Legionella-specific, real-time PCR. The results of PCR showed an increase in Ct values (corresponding with a logarithmic decrease of bacterial DNA) in the course of time (Fig. 1) and were found to mirror the clinical condition of the patient and c-reactive protein values (not shown) during the acute stage of infection.

Existing laboratory tests lack sensitivity in detecting all cases of LD or provide only a retrospective diagnosis. Estimated sensitivities of sputum culture range from less than 10% to 80% and vary according to different comparison standards and by individual laboratories. Moreover, more than 50% of LD patients produce no suitable specimens for culture (1, 7). The *Legionella* urinary antigen test does not reliably detect infection due to *Legionella* species other than that due to *L. pneumophila* serogroup 1, and relying solely on urinary antigen can lead to missed diagnoses (2, 7). Serologic methods to diagnose *L. pneumophila* infections are highly sensitive (7), but their utility is generally limited to epidemiologic studies due to the time lag needed to detect seroconversion.

The advantages of PCR for the detection of *L. pneumophila* DNA in serum are evident; serum samples are readily obtainable and can be processed within a working day. In PCR-positive patients from whom both acute and convalescent serum was obtained, 81% (30 of

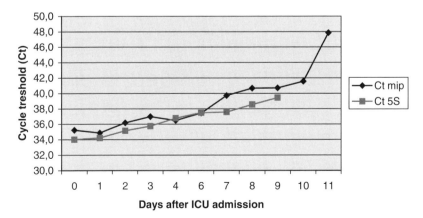

FIGURE 1 Results of real-time *Legionella* PCR on serum from a patient with proven LD requiring intensive care unit admission. An increase in Ct value corresponds with a decrease in bacterial DNA.

37) were positive only in the acute serum sample. We found that in 90.5% (38 of 42) of PCR-positive patients of whom a urinary antigen test had been performed, the urinary antigen test was positive. In other words, an attributable yield in sensitivity of 10% seems feasible with PCR on a serum sample collected in the acute phase of disease. PCR analysis of daily collected serum samples of a seriously ill patient were found to correspond with the course of illness. Detection of *Legionella* DNA in serum might reflect changes in bacterial load over time and may allow the assessment of the response of the patient to treatment. Our findings suggest that detection of *Legionella* DNA in serum is a valuable tool in addition to existing diagnostic tests for the rapid diagnosis of LD. Further work is needed to establish a standardized (commercial) PCR method that will be robust eenoog to be used outsider the setting of a research laboratory.

REFERENCES

1. **Fields, B. S., R. F. Benson, and R. E. Besser.** 2002. *Legionella* and Legionnaires' disease: 25 years of investigation. *Clin. Microbiol. Rev.* **15**:506–526.
2. **Helbig, J. H., S. A. Uldum, S. Bernander, P. C. Lück, G. Wewalka, B. Abraham, V. Gaia, and T. G. Harrison.** 2003. Clinical utility of urinary antigen detection for diagnosis of community-acquired, travel-associated, and nosocomial Legionnaires' disease. *J. Clin. Microbiol.* **41**:838–840.
3. **Lindsay, D. S., W. H. Abraham, and R. J. Fallon.** 1994. Detection of mip gene by PCR for diagnosis of Legionnaires' disease. *J. Clin. Microbiol.* **32**:3068–3069.
4. **Lindsay, D. S., W. H. Abraham, W. Findlay, P. Christie, F. Johnston, and G. F. Edwards.** 2004. Laboratory diagnosis of Legionnaires' disease due to *Legionella pneumophila* serogroup 1: comparison of phenotypic and genotypic methods. *J. Med. Microbiol.* **53**:183–187.
5. **Matsiota-Bernard, P., S. Waser, and G. Vrioni.** 2000. Detection of *Legionella pneumophila* DNA in urine and serum samples from patients with pneumonia. *Clin. Microbiol. Infect.* **6**:223–225.
6. **Muder, R. R. and V. L. Yu.** 2002. Infection due to *Legionella* species other than *L. pneumophila*. *Clin. Infect. Dis.* **35**:990–998.
7. **Murdoch, D. R.** 2003. Diagnosis of *Legionella* infection. *Clin. Infect. Dis.* **36**:64–69.
8. **Reischl, U., H. J. Linde, N. Lehn, O. Landt, K. Barratt, and N. Wellinghausen.** 2002. Direct detection and differentiation of *Legionella* spp. and *Legionella pneumophila* in clinical specimens by dual-color real-time PCR and melting curve analysis. *J. Clin. Microbiol.* **40**:3814–3817.
9. **van der Zee, A., H. Verbakel, C. de Jong, R. Pot, A. Bergmans, and M. F. Peeters.** 2002. Novel PCR-probe assay for detection of and discrimination between *Legionella pneumophila* and other *Legionella* species in clinical samples. *J. Clin. Microbiol.* **40**:1124–1125.
10. **Yu, V. L., J. F. Plouffe, M. C. Pastoris, J. E. Stout, M. Schousboe, A. Widmer, J. Summersgill, T. File, C. M. Heath, D. L. Paterson, and A. Chereshsky.** 2002. Distribution of *Legionella* species and serogroups isolated by culture in patients with sporadic community-acquired legionellosis: an international collaborative study. *J. Infect. Dis.* **186**:127–128.

SPECIFIC DETECTION OF *LEGIONELLA* IN SAMPLES FROM PATIENTS WITH COMMUNITY-ACQUIRED PNEUMONIA BY PCR AND A COLORIMETRIC DETECTION SYSTEM (REVERSE DOT BLOT)

Jörn Kircher, Alexander Kirchhoff, and Arndt Rolfs

13

Despite extensive efforts, the etiology of about 40 to 60% of community-acquired pneumonia remains unclear; a number of these cases are attributed to *Legionella* (1). The incidence in this group may be as high as 38% (2). Cultivation is still supposed to be the gold standard, but there is doubt about the sensitivity in clinical practice (4).

We established a new method for the detection of *Legionella* by using PCR to amplify genomic ribosomal DNA and a reverse dot blot for the detection of *Legionella pneumophila* and non-*pneumophila* species.

We used a multiplex-PCR and the coamplification of human DNA (pyruvate dehydrogenase gene) as an internal control to avoid false-positive results. DNA was extracted using the Qiaamp Tissue Kit or Blood Kit (Qiagen GmbH, Germany). For amplification of *Legionella*-specific DNA, primers (ggctgattgt-cttgacca and aggaagcctcacacta) for the highly conserved region of the 23 S and 5 S ribosomes were used, creating amplificates between 312 and 366 bp in length (3). Digoxigenin-dUTP was used for labeling of amplificates for reverse dot blotting. To avoid false-positive results, incubation with uracil-N-glycosylase was used, and for avoidance of false-negative results the coamplification of the pyruvate dehydrogenase gene was used, creating an amplificate of 185 bp in length in each PCR.

Amplificates were incubated with dotted oligo-dT-oligonucleotides UV-cross-linked on nylon membranes. Three oligonucleotides specific for the genus *Legionella* (aaccacctgataccat-ctcgaactcagaa), the species *Legionella pneumophila* (acgtatcgtgtaaactctgactc), and the pyruvate dehydrogenase gene (agtatgatggggcatacaag) were used in each reaction. Every fifth reaction was a negative control using pure water instead of sample DNA, and only results with clear positive pyruvate dehydrogenase gene and negative water signal were involved.

We were able to demonstrate a sensitivity of 50 fg DNA, which compares to about 10 copies of the Legionella genome.

We used the established method to investigate clinical specimens of 93 patients taken out of a prospective randomized study about the epidemiology of community-acquired pneumonia in Berlin, Germany, from 1991 to 1992. Inclusion criteria followed suggestions of the German Society of Pneumology (5).

Jörn Kircher University of Rostock, Department of Orthopaedic Surgery, 18057 Rostock, Germany. *Alexander Kirchhoff* University of Rostock, Department of Ophthalmology, 18057 Rostock, Germany. *Arndt Rolfs* University of Rostock, Department of Neurology, Neurobiological Lab, 18147 Rostock, Germany.

Legionella: State of the Art 30 Years after Its Recognition
Edited by Nicholas P. Cianciotto et al.
©2006 ASM Press, Washington, D.C.

Group 1 consisted of 47 patients with 3 types of clinical specimens: blood or buffy coat, sputum or bronchoalveolar lavage material, and urine. Group 2 consisted of 46 patients with urine specimens only.

We found 9 of 47 patients in group 1 positive for *Legionella pneumophila*, corresponding to an incidence of 19.1%, and 15 patients positive for non-*pneumophila Legionella*, corresponding to an incidence of 31.9%. Three of 46 patients in group 2 were positive for *Legionella pneumophila*, corresponding to an incidence of 6.5%, and 15 patients were positive for non-*pneumophila Legionella*, corresponding to an incidence of 32.6%. The statistical analysis of all obtained data (anamnestic data such as COLD, human immunodeficiency virus status, etc.; paraclinic data such as hemoglobin, CRP, etc.; chest X ray; cultivation results of specimens) revealed no significant difference of any investigated factor for *Legionella*-positive-tested patients (neither *Legionella pneumophila* nor non-*pneumophila Legionella*).

We were able to demonstrate the ability to detect *Legionella* using PCR in conjunction with reverse dot blotting. This method allows the enhanced detection of non-*pneumophila* species in addition to established methods. Results in both groups are higher than those published in the literature. This is due to enhanced sensitivity of the method compared to conventional investigation methods. It needs to be pointed out that genomic DNA was detected, whereas the source of it (either vital infective bacteria or other) cannot be differentiated. Interpretation of the results for the incidence of non-*pneumophila* infections remains controversial because of a lack of facts about pathogenity of this group and comparable results in other studies.

REFERENCES

1. **Bartlett, J. G.** 2004. Diagnostic test for etiologic agents of community-acquired pneumonia. *Infect. Dis. Clin. North Am.* **18:**809–827.
2. **Marston, B. J., J. F. Plouffe, T. M. File, Jr., B. A. Hackman, S. J. Salstrom, H. B. Lipman, M. S. Kolczak, and R. F. Breiman, and The Community-Based Pneumonia Incidence Study Group.** 1997. Incidence of community-acquired pneumonia requiring hospitalization. Results of a population-based active surveillance Study in Ohio. *Arch. Intern. Med.* **157:**1709–1718.
3. **Robinson, P. N., B. Heidrich, F. Tiecke, F. J. Fehrenbach, and A. Rolfs.** 1996. Species-specific detection of Legionella using polymerase chain reaction and reverse dot-blotting. *FEMS Microbiol. Lett.* **140:**111–119.
4. **Roig, J., M. Sabria, and M. L. Pedro-Botet.** 2003. Legionella spp.: community acquired and nosocomial infections. *Curr. Opin. Infect. Dis.* **16:**145–151.
5. **Schaberg, T. and H. Lode.** 1991. Clinical aspects and diagnosis of community-acquired pneumonia. *Dtsch. Med. Wochenschr.* **116:**1877–1880. [In German.]

TRANSCRIPTION-MEDIATED AMPLIFICATION ASSAY FOR DETECTION OF *LEGIONELLA PNEUMOPHILA* IN SAMPLES FROM PATIENTS WITH COMMUNITY-ACQUIRED PNEUMONIA

*Marie K. Hudspeth, Kathleen Clark-Dickey,
Elizabeth M. Marlowe, Laura G. Schindler,
Karen Campbell, and James T. Summersgill*

14

Diagnosis of acute *Legionella pneumophila* in patients with community-acquired pneumonia can be time-consuming and difficult to confirm using current diagnostic tests. A research prototype assay using Gen-Probe's proprietary transcription-mediated amplification (TMA) technology was evaluated for detection of *L. pneumophila* in throat swabs from patients with documented community-acquired pneumonia. A patient was considered positive for *Legionella* pneumonia if any of the following occurred: positive culture from a respiratory specimen, positive urinary antigen (*L. pneumophila* serogroup 1), positive PCR from oropharyngeal swab or bronchoalveolar lavage, positive direct fluorescent antibody (DFA) assay on respiratory specimen, or a fourfold increase in anti–*L. pneumophila* IgG/M/A >1:128. A probable case of *Legionella* pneumonia included a single IgG/M/A antibody titer of IgG/M/A ≥1:256.

The format of the TMA *L. pneumophila* assay was similar to that of Gen-Probe's AP-

TIMA family of assays. The TMA *L. pneumophila* assay procedure includes target capture, TMA, and the hybridization protection assay. Target rRNA is separated from the other specimen components and the transport media by the addition of target capture reagent, magnetic bead separation and washing using a target capture system. Amplification reagent, oil, and enzyme reagent are added to the rRNA target on the magnetic beads. Isothermal TMA amplification occurs at 42°C. Detection occurs by hybridization protection assay, and the reaction is read as relative light units in a LEADER HC+ luminometer. For these data, a result over 50,000 relative light units for the TMA *L. pneumophila* assay was considered positive. For comparison, a *Legionella* PCR was performed as described by Ramirez et al. (1).

Amplified methods are generally more sensitive than other laboratory methods, particularly for fastidious organisms such as *Legionella*. Therefore, in addition to the initial TMA *L. pneumophila* assay that targets a specific region of *L. pneumophila* rRNA, another target was utilized in an alternate region of *L. pneumophila* rRNA as a confirmatory TMA-based assay.

Serial dilutions of *L. pneumophila* serogroup 1 rRNA and *L. pneumophila* serogroup 1 bacteria

Marie K. Hudspeth, Kathleen Clark-Dickey, and Elizabeth M. Marlowe Gen-Probe Inc., San Diego, CA 92121. *Laura G. Schindler, Karen Campbell, and James T. Summersgill* Division of Infectious Diseases, Department of Medicine, University of Louisville School of Medicine, Louisville, KY 40292.

Legionella: State of the Art 30 Years after Its Recognition
Edited by Nicholas P. Cianciotto et al.
©2006 ASM Press, Washington, D.C.

(ATCC VR-2282) were prepared in swab transport medium. Five replicates at each concentration were tested with both TMA *L. pneumophila* assays. Both assays detected 1 CFU and as little as 20 copies of *L. pneumophila* rRNA. Fifteen serotypes of *L. pneumophila* and 46 strains of other bacteria, representing closely related organisms and a cross-section of phylogeny, were tested at ~10^6 organisms per assay. All *L. pneumophila* serotypes tested were detected using the TMA *L. pneumophila* assay, and no cross-reactivity was observed with the 46 other non-*L. pneumophila* bacteria tested, with the exception of *L. wadsworthii*.

One hundred samples from patients with documented community-acquired pneumonia were subjected to multiple laboratory tests including TMA, PCR, DFA, a urinary antigen assay, serology, and culture. Of the 100 specimens, 75 were negative by all methods. One specimen was positive by PCR, TMA, and the urinary antigen test. Three samples were positive by PCR and TMA. One specimen was PCR positive and TMA negative, but was positive with the confirmatory TMA assay. Twenty samples were positive by TMA only. Sixty-two of the 75 TMA *L. pneumophila* assay negatives were confirmed by the alternate TMA *L. pneumophila* assay. Twenty-one of the 24 TMA *L. pneumophila* positives were confirmed by the alternate TMA *L. pneumophila* assay. These results are summarized in Table 1.

In this study, nucleic acid amplification tests appear to be more sensitive than other methods for the detection of *L. pneumophila*, since results were negative for all assays except PCR

TABLE 1   Results of PCR and TMA *L. pneumophila* assays on community-acquired pneumonia patient samples

Assay and result	No. of samples
PCR	
Positive	5[a]
Negative	95
TMA *L. pneumophila*	
Positive	24[a]
Negative	76
Alternate TMA *L. pneumophila*	
Positive	35[a]
Negative	65

[a]One sample was also positive by the urinary antigen test.

and TMA, with the exception of one sample that was positive by urinary antigen. The TMA *L. pneumophila* assay detected more samples as positive than did PCR. Future studies need to be performed to determine the correlation of amplified assay results, such as the TMA *L. pneumophila* assay, with other diagnostic methods for the detection of *L. pneumophila* from patients with community-acquired pneumonia. Further development and testing must be done to confirm the clinical utility of the TMA *L. pneumophila* assay.

REFERENCE

1. **Ramirez, J. A., S. Ahkee, A. Tolentino, R. D. Miller and J. T. Summersgill.** 1996. Diagnosis of *Legionella pneumophila, Mycoplasma pneumoniae*, or *Chlamydia pneumoniae* lower respiratory infection using the polymerase chain reaction on a single throat swab. *Diagn. Microbiol. Infect. Dis.* **24:**7–14.

DETECTION OF *LEGIONELLA* SPP. AND *LEGIONELLA PNEUMOPHILA*–SPECIFIC DNA IN RESPIRATORY SECRETIONS BY PCR-ENZYME-LINKED IMMUNOSORBENT ASSAY AND COMPARISON WITH CONVENTIONAL METHODS

Diane S. J. Lindsay, William H. Abraham, Alistair W. Brown, and Giles F. S. Edwards

15

Since the discovery of *Legionella* in 1977 (4), a number of diagnostic tests have been used to help in the detection and isolation of this micro-organism. Historically, these have included animal passage, culture, fluorescent-antibody assays and more recently urinary antigen (UA) and molecular-based techniques (1,5). Culture is still seen as definitive in case diagnosis, but the sensitivity of culture can vary dramatically between <10 and 80% (6). Unfortunately, culture sampling has dramatically decreased with the onset of UA testing. UA is detected in the acute phase of disease by an enzyme immunoassay (EIA) and is highly specific for *Legionella pneumophila* serogroup (Sg) 1 but has lower sensitivity in nosocomial cases where the causative organism is likely to be a non–*L. pneumophila* sg 1 (2). Molecular methods have utilized PCR to detect *Legionella*-specific DNA in a number of specimen types (5) and has shown varying degrees of sensitivity and specificity. In this study, a PCR-enzyme-linked immunosorbent assay (ELISA) was devised for the detection of *Legionella* and *Legionella pneumophila*–specific DNA in human respiratory secretions. The

PCR-ELISA incorporates a solid-phase probe hybridization step to verify the PCR reaction. The sensitivity of the technique was compared with 5sRNA PCR with Southern blotting (SB) confirmation and conventional direct fluorescent antibody (DFA) testing and culture which were performed as previously described (3).

16SRNA PCR

Briefly, DNA was extracted from a total of 343 routine respiratory secretions using Nucleospin tissue columns. Each PCR reaction contained 1x PCR buffer, 4 mM MgCl$_2$, 10 pmol of 16S RNA primer 1 (5′ TAC CTA CCC TTG ACA TAC AGT G 3′) and 16S RNA primer 2 (5′ CTT CCT CCG GTT TGT CAC 3′) (7), 200 µM digoxigenin dNTP, 1.25 U of *Taq* polymerase, and 0.25 U of uracil DNA glycosylase and nuclease-free water (Promega). Ten microliters of DNA was added to the PCR reaction, and positive (heat-killed suspension of *L. pneumophila* Sg 1) and negative (nuclease-free water) controls were incorporated into each assay. PCR amplification was performed on a Hybaid PCR Express at 25°C for 10 min then 95°C for 5 min; followed by 40 cycles of 94°C for 30 s, 60°C for 1 min, and 70°C for 30 s; then 1 cycle of 72°C for 5 min. PCR products were visualized on a 1.5% agarose gel at 100 V for 30 min.

Diane S. J. Lindsay, William H. Abraham, Alistair W. Brown, and Giles F. S. Edwards Scottish *Legionella* Reference Laboratory, Stobhill Hospital, Glasgow, UK G21 3UW.

Legionella: State of the Art 30 Years after Its Recognition
Edited by Nicholas P. Cianciotto et al.
©2006 ASM Press, Washington, D.C.
55

PCR-ELISA

The PCR-ELISA was performed using reagents from the Roche Diagnostic PCR-ELISA (Dig Detection) kit Cat No. 1636111 following the manufacturers' instructions. A *Legionella*-specific biotin–labeled 16S RNA probe (5′ CGT AAC GAG CGC AAC CC 3′) (7) and *Legionella pneumophila*–specific biotin–labeled 16S RNA probe (5′ ATG TGA TGG TGG GGA CTC T 3′) (7) were titrated, and the optimal concentration was calculated empirically. Hybridization was performed at 60°C with shaking for 3 h, and then the plate was washed and anti-digoxigenin AP conjugate was added at 37°C for 30 min. The plate was then washed and substrate was added and left in the dark at 37°C for 30 min. with shaking. The absorbance at 405 nm was read on a Lab-system multiskan spectrophotometer. A positive result was classified as having an optical density (OD) reading of ≥0.200 + negative OD, and a negative result as ≤0.200 + negative OD, assuming that the positive and negative controls were within the QC ranges.

A total of 343 respiratory samples were tested, of which 21 patients were identified as cases of legionellosis by fulfilling one or more of the criteria for a definitive case. Of the 21 cases: 42.9% were DFA positive; 57.1% were culture positive and 85.7% were 5S RNA/SB and 16S RNA PCR-ELISA positive, respectively, for *Legionella* (Table 1). Of those positive in the 16S RNA PCR-ELISA, the majority (94.4%) were identified as *L. pneumophila*. Only one case that was serological positive (showing a greater than fourfold rise in titer to *L. micdadei*) was positive

in the *Legionella* PCR-ELISA alone. A further five presumptive cases were identified by the 5S RNA and 16S RNA PCR alone as *L. pneumophila*. However, these cases could not be classed as definitive using the current case definition due to lack of corroborating specimens. With PCR-based assays becoming more sensitive and specific, perhaps the distinction of a definitive case should incorporate more molecular techniques. At the moment a PCR positive is only seen as presumptive. Given time, the theory that we are under-detecting cases because of inadequacies in current methods could result in more sensitive and specific methods such as PCR, along with strong clinical evidence, being acceptable to define a case. Also, with the advent of gene sequencing techniques, there may be no requirement in the future for culture, as microorganisms will be identified directly from the original patient sample.

The 16S RNA PCR-ELISA has the same sensitivity as the 5S RNA PCR/SB technique but can also identify *L. pneumophila* and is more sensitive than either DFA or culture. The lack of suitable samples, i.e., respiratory secretions, should not inhibit investigation, as *Legionella*-specific DNA has been found in serum and urine (5). Moreover, the simplicity and ease of automation of this technique could be easily modified to identify a range of respiratory organisms in a routine laboratory setting.

This study has highlighted the benefits of a PCR-ELISA technique if applied in the acute phase of disease when culture is slow and UA only able to detect *L. pneumophila* with any level of sensitivity or specificity.

TABLE 1 DFA, culture and PCR results on the 21 definitive and 5 presumptive cases

Technique	No. of positive samples/total samples tested (%)	
	Definitive	Presumptive
DFA	9/21 (42.9)	0/5 (0)
Culture	12/21 (57.1)	0/5 (0)
5S RNA PCR	18/21 (85.7)	5/5 (100)
16S RNA PCR	18/21 (85.7)	5/5 (100)
16S RNA *L. pneumophila* PCR	17/18 (94.4)[a]	5/5 (100)

[a]Percent positive of those positive in the 16S RNA PCR.

REFERENCES

1. **Fallon, R. J.** 1981. Laboratory diagnosis of Legionnaires' disease. *Assoc. Clin. Pathol.* **99:**1–15.
2. **Helbig, J. H., S. A. Uldum, S. Bernander, C. L. Luck, G. Wewalka, B. Abraham, V. Gaia, and T. G. Harrison.** 2003. Clinical utility of urinary antigen detection for diagnosis of community-acquired, travel-associated and nosocomial Legionnaires' disease. *J. Clin. Microbiol.* **41:** 838–840.
3. **Lindsay, D. S. J., W. H. Abraham, W. Findlay, P. Christie, F. Johnston, and G. F. S. Edwards.** 2004. Laboratory diagnosis of Legionnaires' disease due to *Legionella pneumophila* serogroup 1: comparison of phenotypic and genotypic methods. *J. Med. Microbiol.* **53:**183–187.
4. **McDade, J. E., C. C. Shepard, and D. W. Fraser.** 1977. Legionnaires' disease: isolation of a bacterium and demonstration of its role in respiratory disease. *N. Engl. J. Med.* **297:**1197–1203.
5. **Murdoch, D. R.** 2003. Nucleic acid amplification tests for the diagnosis of pneumonia. *Clin. Infect. Dis.* **36:**1162–1170.
6. **Sopena, N., M. Sabria-Leal, M. L. Pedro-Botet, E. Padilla, J. Dominguez, J. Morera, and P. Tudela.** 1998. Comparative study of the clinical presentation of *Legionella pneumophila* and other community-acquired pneumonias. *Chest* **113:** 1195–1200.
7. **van der Zee, A., H. Verbakel, C. de Jong, R. Pot, M. Peeters, J. Schellekens, and A. Bergmans.** 2002. Clinical validation of diagnosis of *Legionella* infections, p. 189–192. *In* R. Marre, Y. Abu Kwaik, C. Bartlett, N. P. Cianciotto, B. S. Fields, M. Frosch, J. Hacker, and P. C. Luck (ed.), *Legionella*, ASM Press, Washington, DC.

SEROLOGICAL VERSUS SEQUENCE-BASED METHODS FOR *LEGIONELLA* IDENTIFICATION

B. Baladrón, V. Gil, and C. Pelaz

16

The genus *Legionella* comprises 48 species and 70 different serogroups, and approximately half of them have been associated with human disease. Serological methods have been widely used for species and serogroup identification, but the progressive characterization of new species has established that antigen cross-reactivity limits specificity and restricts their use to a few frequently isolated species. Recently, genotypic methods have been proposed to characterize strains of *Legionella* at the species level, including ribotyping (1) and sequencing of different genes, such as 16S rRNA, *dot*A, *mip*, *rpo*B, and 23S-5S ISR (2, 3, 4, 7). Sequence analysis of the 16S rRNA gene has been used for phylogenetic relatedness of species (2), and a *mip* sequence-based classification scheme has been developed for *Legionella* (7).

Species and serogroup distribution of *Legionella* Spanish isolates (500 clinical and 4,000 environmental), using home-made serological reagents and immunofluorescence assays, was as follows: 83.8% of clinical isolates and 50% of environmental ones were identified as *L. pneumophila* serogroup 1, 12.7% of clinical isolates and 42.7% of environmental ones were *L.*

pneumophila serogroups 2 to 14, and 3.4% of clinical isolates and 7.2% of environmental ones were other species of *Legionella* (6).

In the present study serological and sequence-based methods used for *Legionella* identification were compared, with 93 *Legionella* strains including clinical and environmental ones, *L. pneumophila,* and other species. The aims of the study were to determine the correlation between serological and sequence-based methods and to establish a better strategy to increase the number of species of *Legionella* that can be identified.

A total of 93 isolates of *Legionella* were analyzed—28 *L. pneumophila* isolates and 65 *Legionella* isolates of other species, including 30 clinical and 63 environmental isolates. They were recovered between 1991 and 2004 from different regions of Spain and previously identified in the Reference Laboratory of *Legionella* (Majadahonda, Madrid). Strains were grown on buffered charcoal yeast extract agar, and they were identified by three serological methods and by sequence-based methods using two genes, *mip* and 16S rRNA. The immunofluorescence test (IF) was performed using home-made rabbit antisera against *L. pneumophila* serogroups (sg) 1 to 14 and eight other *Legionella* species (*L. longbeachae* sg 1 and 2, *L. micdadei, L. gormanii, L. wadsworthii, L. jordanis, L. bozemanii* sg 1 and 2,

B. Baladrón, V. Gil, and C. Pelaz Legionella Laboratory, Centro Nacional Microbiología, Instituto de Salud Carlos III, Majadahonda, 28220 Madrid, Spain.

Legionella: State of the Art 30 Years after Its Recognition
Edited by Nicholas P. Cianciotto et al.
©2006 ASM Press, Washington, D.C.

L. dumoffii, L. oakridgensis) (5). Strains presenting a negative IF reaction were called *Legionella* spp., and some of them were identified in other European laboratories (Legionella Reference Laboratory in Lyon and PHLS in London). Direct fluorescent assay was performed using a commercial kit (Monofluo *L. pneumophila* IFA test Bio-Rad IFD) and latex agglutination using the OXOID kit. *mip* sequencing was performed following European Working Group on Legionella Infections (EWGLI) protocol (7), and the EWGLI database was used for species allocation. 16S rRNA sequencing was performed as previously described (2), and GeneBank was used for species allocation.

Serological and sequence identification results are summarized in Table 1 for clinical strains and Table 2 for environmental ones. Comparing serological methods, 90 of 93 (96.7%) strains presented agreeing results with the three methods used. The oxoid agglutination test was negative with two strains (*L. pneumophila* sg 1 and *L. longbeachae*), and one was negative with *L. pneumophila* sg 1 home-made antiserum (Tables 1 and 2). Ten strains were *Legionella* spp. with our current home-made antisera; some of them were identified by other laboratories.

Comparing gene sequencing methods, 85 of 93 (96.7%) strains presented agreeing results with both genes. Discrepancies were only detected among environmental strains other than *L. pneumophila*. The percentage of similarity for species allocation was 99 to 100% with the *mip* gene and 97 to 100% with the 16S rRNA gene. In 26 of 30 (86.6%) clinical strains, species allocation was done with 100% similarity for each gene, but among environmental ones this percentage decreased to 42 of 53 (79.2%) with *mip* and 39 of 53 (73.5%) with 16S rRNA.

Of the 93 strains, 63 (67.7%) were allocated to the same species by all methods used (serological and sequence-based). Among clinical strains this percentage was 96.6% (26 of 30), and the only discrepancy was a strain identified as *L. jordanis*/*L. bozemanii* by serology and *L. parisiensis* by sequencing with both genes. However, the agreement was lower among en-

vironmental strains; 27 of 63 (42.8%) of these strains presented identical species identification by serology and *mip*, and 34 of 63 (53.9%) by serology and 16S rRNA.

Serological methods have been the most frequently used for *Legionella* identification, but the number of species that could be identified is limited. Commercial antisera have been produced for a limited number of *Legionella* species by using either direct fluorescent antibody or latex agglutination. IF reactions with home-made antisera have been used in reference laboratories, but a limited number of laboratories have antisera for all *Legionella* species. Our current IF panel of antisera against *Legionella* species was enough to identify clinical strains as well as *L. pneumophila* strains. However, it would be necessary to increase the panel in a large number of infrequent species antibodies to identify environmental strains other than *L. pneumophila*, in which there is higher diversity, more frequent cross-reactions, and frequently not clear species identification.

Sequencing methods allow the identification of any strain, detecting all species described and new ones. In this study all strains were identified with both genes, but disagreements were detected when both gene results were compared. These disagreements (8% of strains) were detected among environmental strains other than *L. pneumophila*, but not among clinical strains. *L. londiniensis* was frequently involved (identified as *L. nautarum, L. erytra*, or *L. parisiensis* by the second gene). Similar *Legionella* identifications have been shown by other authors (3) when different gene sequencing results were comapared, but *mip* was shown to be more variable and discriminative than 16S rRNA (7).

Discrepant results were detected in 32.2% of strains when serological and sequence-based methods were compared, and *L. anisa* was frequently involved. These discrepancies were less frequent among clinical strains (1 of 30, 3.3%) than among environmental ones (29 of 63, 46%), and disagreements were not detected among *L. pneumophila* strains. 16S rRNA showed more agreement with serological results than did *mip*. A similar correlation has been detected by

TABLE 1 Serological and sequence identification

Number of strains	Serological methods			Sequencing methods	
	Oxoid agglutination	BioRad IFD	Home-made IF	*mip* gene (100% similarity)	16s rRNA gene (100% similarity)
4	*L. pneumophila* sg 2–14	*L. pneumophila* sg 1–14	*L. pneumophila* sg 4, 8, 12	*L. pneumophila* (99.6–100)	*L. pneumophila* (99–100)
5	*L. pneumophila* sg 1	*L. pneumophila* sg 1–14	*L. pneumophila* sg 1	*L. pneumophila* (99.8–100)	*L. pneumophila*
1	*L. pneumophila* sg 2–14	*L. pneumophila* sg 1–14	*L. pneumophila* sg 3, 6	*L. pneumophila* (99.81)	*L. pneumophila*
5	*Legionella* spp.	Negative	*L. micdadei*	*L. micdadei*	*L. micdadei*
11	*Legionella* spp.	Negative	*L. longbeachae*	*L. longbeachae* (99.7–100)	*L. longbeachae* (98–100)
1	*Legionella* spp.	Negative	*L. dumoffii*	*L. dumoffii*	*L. dumoffii*
1	*Legionella* spp.	Negative	*L. jordanis/bozemanii*[b]	*L. parisiensis*[b]	*L. parisiensis*
1	Negative[a]	Negative	*L. longbeachae*	*L. longbeachae*	*L. longbeachae*
1	*Legionella* spp.	Negative	*L. jordanis*	*L. jordanis*	*L. jordanis* (99)

[a]Disagreement results between serological methods.
[b]Disagreement results between serological and molecular methods.

TABLE 2 Serological and sequence identification

Number of strains	Serological methods			Sequencing methods	
	Oxoid agglutination	BioRad IFD	Home-made IF	mip gene (100% similarity)	16s rRNA gene (100% similarity)
13	L. pneumophila sg 1	L. pneumophila	L. pneumophila sg 1	L. pneumophila (99.2–100)	L. pneumophila
1	L. pneumophila sg 2–14	L. pneumophila	L. pneumophila sg 4, 8, 12	L. pneumophila	L. pneumophila
1	L. pneumophila sg 2–14	L. pneumophila	L. pneumophila sg 3	L. pneumophila (99)	L. pneumophila
1	L. pneumophila sg 2–14	L. pneumophila	L. pneumophila sg 5, 8, 9	L. pneumophila (99.7)	L. pneumophila
1	L. pneumophila sg 1	L. pneumophila	Legionella spp.[a]	L. pneumophila (99.8)	L. pneumophila
1	Negative[a]	L. pneumophila	L. pneumophila sg 1	L. pneumophila	L. pneumophila
1	Negative	Negative	L. londiniensis[d]	L. erythra[b]	L. londiniensis[b]
2	Negative	Negative	L. rubrilucens[d]	L. taurinensis (99)	L. rubrilucens/taurinensis (99)
1	Legionella spp.	Negative	L. bozemanii/jordanis[c]	L. londiniensis[b,c]	L. parisiensis[b]
1	Legionella spp.	Negative	L. longbeachae	L. sandcrusis[b]	L. longbeachae[b] (99)
1	Negative	Negative	L. rubrilucens[d]	LD-4748[b] (99.7)	L. rubrilucens[b] (99)
1	Legionella spp.	Negative	L. bozemanii/longbeachae[c]	L. parisiensis[b,c]	L. anisa[b] (98)
1	Negative	Negative	L. rubrilucens[d,c]	L. erythra[c]	L. erythra
1	Negative	Negative	L. nautarum[d]	L. londiniensis[b] (99.8)	L. nautarum[b] (97)
11	Legionella spp.	Negative	L. micdadei/longbeachae[c]	L. anisa[c]	L. anisa
1	Negative	Negative	L. oakridgensis[c]	L. anisa[b,c]	L. nautarum[b]
1	Negative	Negative	L. londiniensis[d]	L. londiniensis	L. londiniensis
11	Legionella spp.	Negative	L. bozemanii/longbeachae[c]	L. anisa[c]	L. anisa (99–100)
1	Legionella spp.	Negative	L. longbeachae[c]	L. anisa[c]	L. anisa
3	Legionella spp.	Negative	L. dumoffii	L. dumoffii	L. dumoffii (99)
1	Legionella spp.	Negative	L. erythra[d,c]	L. taurinensis[c]	L. taurinensis/rubrilucens (99)
1	Negative	Negative	L. taurinensis[d]	L. taurinensis	L. taurinensis
1	Negative	Negative	Legionella spp.	L. taurinensis	L. taurinensis
1	Legionella spp.	Negative	L. erythra	L. erythra	L. erythra
1	Legionella spp.	Negative	L. micdadei	L. micdadei	L. micdadei
1	Negative	Negative	L. wadsworthii[c]	L. oakridgensis[c]	L. oakridgensis
1	Negative	Negative	Legionella spp.	L. londiniensis[b]	L. nautarum[b]
1	Negative	Negative	L. oakridgensis	L. oakridgensis	L. oakridgensis

[a] Disagreement results between serological methods.
[b] Disagreement results between molecular methods.
[c] Disagreement results between serological and molecular methods.
[d] Identification carried out in another laboratory (N. Borstein, Lyon and T. Harrison, PHLS, London)

other authors when phenotypic and sequence-based *Legionella* classification methods were compared (3, 7). However, the three red auto-fluorescent species *L. rubrilucens, L. erythra,* and *L. taurinensis,* were shown to be indistinguishable on 16S rDNA sequences and, in these cases, it would be necessary to use a panel of antibodies or another sequence method to be able to identify them (3).

Using the IFD *L. pneumophila* kit (Bio-RAd), followed by IF with home-made antisera in case of an *L. pneumophila*–positive reaction and a sequence-based method in case of a negative IFD reaction, could be a good strategy to identify any *Legionella* isolate.

ACKNOWLEDGMENTS

This work was partially financiated by Desinfecciones Alcora S.A.

REFERENCES

1. **Cordevant, C., J. S. Tang, D. Cleland, and M. Lange.** 2003. Characterization of members of the Legionellaceae family by automated ribotyping. *J. Clin. Microbiol.* **41**:34–43.
2. **Fry, N. K., T. J. Rowbotham, N. A. Saunder, and T. M. Embley.** 1991. Direct amplification and sequencing of 16S ribosomal DNA of an intracellular *Legionella* species recovered by amoeba enrichment from the sputum of a pacient with pneumonia. *FEMS Microbiol. Lett.* **83**:165–168.
3. **Grattard, F., C. Ginevra, S. Riffard, A. Ros, S. Jarraud, J. Etienne, and B. Pozzetto.** 2006. Analysis of the genetic diversity of *Legionella* by sequencing the 23S-5S ribosomal intergenic spacer region: from phylogeny to direct identification of isolates at the species level from clinical specimens. *Microbes Infect.* **8**:73–83.
4. **Ko, K. S., H. K. Lee, M. Y. Park, M. S. Park, K. H. Lee, S. Y. Woo, Y. J. Yun, and Y. H. Kook.** 2002. Population genetic structure of *Legionella pneumophila* inferred from RNA polymerase gene (*rpo*B) and DotA gene (*dot*A) sequences. *J. Bacteriol.* **184**:2123–2130.
5. **Pelaz, C., L. García, and C. Martín-Bourgon.** 1992. Legionellae isolated from clinical and environmental samples in Spain (1983-1990): monoclonal typing of *Legionella pneumophila* serogroup 1 isolates. *Epidemiol. Infect.* **108**:397–402.
6. **Pelaz, C. and C. Martín-Bourgon.** 2000. Infección por Legionella en España: Análisis de las cepas humanas y ambientales aisladas entre 1980 y 1999. *Enf. Emerg..* **2**:214–219.
7. **Ratcliff, R. M., J. A. Lanser, P. A. Manning, and M. W. Heuzenroeder.** 1998. Sequence-based classification scheme for the genus *Legionella* targering the *mip* gene. *J. Clin. Microbiol.* **36**:1560–1567.

SEROLOGIC STUDY OF AN OUTBREAK OF LEGIONNAIRES' DISEASE: VARIATION OF SENSITIVITY ASSOCIATED WITH THE SUBGROUP OF *LEGIONELLA PNEUMOPHILA* SG 1 ANTIGEN USED AND EVIDENCE OF CONCURRENT REACTIVITY TO OTHER ATYPICAL PNEUMONIA AGENTS

Sverker Bernander, Berndt E. B. Claesson, Eva Hjelm,
Nils Svensson, and Martin Hjorth

17

An outbreak of Legionnaires' disease (LD) occurred in August 2004 in Sweden in the town of Lidköping, by Lake Vänern. An industrial cooling tower was found to be the most likely source of infection. Two genotypes of *Legionella pneumophila* serogroup (sg) 1 were found by culture in patient samples, and one belonging to subgroup Benidorm (genotype A) and the other to subgroup Bellingham (genotype B) (Table 1). Both genotypes were also found in the suspected cooling tower. At the time of the outbreak, diagnosis could be confirmed in 15 patients by using urinary antigen tests and culture. However, during the same month there was an overall threefold increase in the number of hospitalized cases of pneumonia at the local hospital (Fig. 1).

The indirect immunofluorescent antibody technique (IFAT) is still a standard method for antibody assay in patients with legionellosis, al-

lowing for the screening of antibodies against several species and subtypes of *Legionella*. However, it is time-consuming, and experience is needed to run tests. Cross-reactions with other bacterial genera have been described (8, 10). Confirmed cases of legionellosis might be serologically negative (11). Thus, even for *L. pneumophila* sg 1, the sensitivity and specificity of serologic methods is not optimal. They are, however, important as an adjunct to other diagnostic procedures, which are even less sensitive, and as a tool in epidemiological investigations (6, 7, 10). A serological study was therefore undertaken to assess the impact of the legionella outbreak on all pneumonia cases. Patient samples were also tested for antibodies against other atypical pneumonia agents, i.e. *Mycoplasma pneumoniae* and *Chlamydophila pneumoniae*.

Fourteen hospitalized LD cases, confirmed in an early stage of the outbreak, and 56 other pneumonia cases from the outbreak period were included in the study. In addition, 32 outpatients with suspected pneumonia were tested. Acute and convalescent sera were obtained if possible. IFAT was performed using a modification of earlier methods (6, 10). All serum samples were screened with antigens

Sverker Bernander and Eva Hjelm Department of Clinical Microbiology, Uppsala University Hospital, SE-751 85 Uppsala, Sweden. *Berndt E. B. Claesson* Department of Clinical Microbiology, Capio diagnostik, Kärnsjukhuset, SE-541 85 Skövde, Sweden. *Nils Svensson* Regional Unit for Communicable Disease Control, SE-541 85 Skövde, Sweden. *Martin Hjorth* Department of Medicine, Lidköping Hospital, SE-531 85 Lidköping, Sweden.

TABLE 1 Results of MAb subgrouping and sequence-based typing of six genes

Gene	Patient 1 Benidorm	Patient 2 Benidorm	Patient 3 Bellingham	C-tower 1[a] Benidorm	C-tower 2[a] Benidorm	Humidifier[b] Benidorm
flaA	6	6	7	6	6	7
pilE	10	10	6	10	10	6
asd	2	2	17	2	2	17
mip	28	28	3	28	28	3
mompS	9	9	13	9	9	13
proA	4	4	11	4	4	11
Genotype	A	A	B	A	A	B

[a]Cooling tower at an ethanol-producing plant.
[b]Humidifier moistening vegetables (mist machine) in a shopping center. An isolate identical in AFLP with this strain was also found in cooling tower 2 and in a cooling tower belonging to a different industrial facility, but they were not genotyped using sequence-based typing. Genotype B is either a subgroup Bellingham or a Benidorm.

prepared from four subgroups of heat-killed *L. pneumophila* sg 1, i.e. Bellingham (ATCC 43111), Knoxville (ATCC 42793), OLDA (ATCC 43109), and Philadelphia (ATCC 33152). Local experience in the laboratory using more than one subgroup of antigen had been shown earlier to increase sensitivity. All patients with positive or borderline titers were also tested by using a Benidorm subgroup antigen prepared from one of the patient isolates in the outbreak. A fourfold titer rise to 1/128 or more was considered significant, as was a single titer of the same magnitude if the patient was part of the outbreak. A titer of 1/64 was considered to be borderline, indicating a presumptive diagnosis. Samples from 69

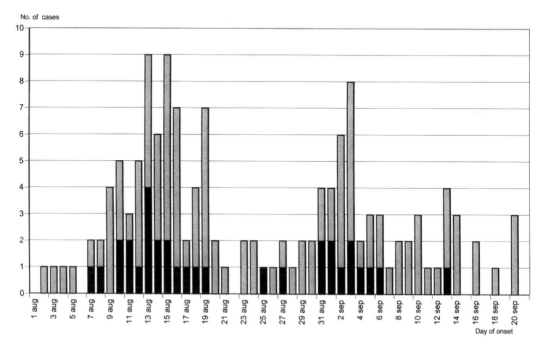

FIGURE 1 Diagram showing the course of the Lidköping outbreak. Black: 32 confirmed and Presumptive cases of LD. Shaded: 106 other pneumonia cases form the same period. The *y* axis shows number of cases, and the *x* axis shows day of onset.

hospitalized cases of pneumonia were also tested for IgG, IgM, and IgA antibodies against *M. pneumoniae* and *C. pneumoniae* using commercial enzyme-linked immunosorbent assay methods (ThermoLabsystems Oy, Vantaa, Finland). Significant titer changes were defined according to package inserts. The genotype of the patient strains was confirmed by sequence-based typing, and the serological subgroup was confirmed by using the Dresden MAb panel (3, 4).

Of the 15 cases initially confirmed by urinary antigen test and/or culture, 8 were positive by serology, 2 were borderline, and 4 were negative (Table 2). Two of the negative-tested patients had been fully sampled in the acute and convalescent phase. One of the 15 patients was never sampled for serological testing. Eight of the other 56 patients were positive for *L. pneumophila* sg 1, and 7 were considered to have a borderline titer level (1/64). Of the outpatients, one was found to have had a confirmed LD and one a presumptive legionella infection. Seven of these 17 new cases of both categories would have been missed if a Bellingham subgroup antigen had not been used. Of the 17 patients with significant antibody levels according to the above-mentioned criteria (≥ 1/128), 10 achieved the highest titer only with the Bellingham subgroup, 3 with OLDA/Bellingham, and 4 with Philadelphia/Knoxville.

Thus, in all there were 24 patients with confirmed legionellosis and 8 with a presumptive diagnosis. The sensitivity of the test was increased appreciably by using a Bellingham antigen. The low reactivity with the Benidorm antigen is surprising considering that isolates from two patients were Benidorm. However, the Bellingham and Benidorm subgroups resemble each other antigenically, the main difference being the presence of an acetylated epitope encoded by the *lag-1* gene. This hydrophobic epitope, which is reactive with the Dresden panel Mab 3/1 (MAb 2 in the international panel), covers the other polysaccharide epitopes beneath, which might dominate in the immunologic patient response (5). Sim-

ilar conditions are conceivable in a setting where Knoxville and Philadelphia strains are used as antigen instead of antigenically related Oxford and OLDA strains. In this study the cut-off titer levels used were lower than those recommended in other studies (Centers for Disease Control and Prevention, Atlanta, Ga.). However, they are in accordance with experience in the laboratory where the tests were conducted and with results obtained in another study in a nearby geographical area (2). Our results show that at least in the context of the Lidköping outbreak the use of both MAb 3/1 positive and negative strains as antigen is necessary for adequate sensitivity. Diagnostic tests performed in the acute phase of disease, i.e., culture, urinary antigen test, and early serology, seem to have detected <50% of tested cases. PCR methods were not utilized in the study.

A significant titer increase for *M. pneumoniae* was seen in 4 cases, *C. pneumoniae* in 2 cases, and both agents in 7 of 29 patients with confirmed or presumptive legionellosis. In the control group of 40 other hospitalized pneumonia patients from the same period the corresponding numbers were 9, 1, and 3, respectively. Positive tests for both *M. pneumoniae* and *C. pneumoniae* in the same patient had not been observed earlier at the laboratory conducting the tests. Results of 10 patients who had a confirmed LD by urinary antigen test and/or culture are given in Table 2.

Since 13 cases with confirmed or presumptive legionella infection showed significant titer rises for *M.* and/or *C. pneumoniae*, a polyclonal immunologic response must be suspected. Furthermore, among the patients who were positive in urinary antigen testing and/or culture for *L. pneumophila* six showed a significant increase in antibody level against either one or both of these infectious agents (Table 2). There are a few early reports of concomitant increases in antibody levels for other atypical agents in connection with legionellosis, even though genuine cross-reactions could not be proven (9, 10). To our knowledge, though, no formal studies have been conducted to assess

TABLE 2 Results of testing 12 patients serologically with IFAT using 5 subgroups of *L. pneumophila* sg 1 (Bellingham, Benidorm, Knoxville, OLDA, and Philadelphia)[a]

Patient	A/C	Belling.	Benid.	Knoxv.	OLDA	Philad.	U.Antig.	Culture	M. pneum	C. pneum	Age	Sex
1	C	1,024	256	64	256	256	+	—	IgG	—	59	M
	C	**2,048**	256	128	256	256						
2	A	16	16	16	16	16	+		—	—	84	F
	C	32	64	16	16	16						
	C	**128**	64	16	16	16						
3	A	**64**	32	16	16	16	+	—	—	—	75	M
	A	32	16	16	16	16						
	C	16	16	16	16	16						
4	A	16	16	16	16	16	+	—	—	—	73	F
	C	16	16	16	16	16						
	C	32	32	16	16	16						
5	C	**128**	32	16	16	16	+	+	IgG IgA	—	59	F
6	C	32	32	16	16	16	+	—	IgG IgA	IgG IgA	48	F
	C	**512**	64	32	128	32						
9	C	**2,048**	256	128	256	128	+	+	—	—	45	M
	C	1,024	16	256	32	128						
10	A	16	16	16	16	16	—	+	IgG IgA	IgA	60	M
	A	16	16	16	16	16						
	C	16	16	16	16	16						
12	A	16	16	16	16	16	+	+	—	—	51	F
	C	**256**	32	64	128	128						
13	A	16	16	16	16	16	+		—	IgA	36	M
	C	256	128	256	128	**512**						
14	A	16	16	16	16	16	+	—	IgG	—	71	M
	C	16	16	64	16	**128**						
	C	16	16	64	16	128						
15	A	16	16	16	16	16	+	—	IgG IgA	IgG IgA	73	M
	C	16	16	16	16	16						
	C	**64**	16	16	32	**64**						

[a]LD confirmed in all patients by legionella urinary antigen test and/or culture. Numerals in bold italics show the highest titer obtained in a patient. A = acute serum sample, C = convalescent serum sample. Age in years. M = male, F = female. Significant titer rises of antibodies against *M.* and *C. pneumoniae* are noted as "IgG" and "IgA."

the existence of a polyclonal immunologic response. Cross-reactivity with legionellae has been suggested in infections caused by *Campylobacter* spp. and *M. pneumoniae* (8, 10). However, an ongoing outbreak of pneumonia caused by *M.* or *C. pneumoniae* in adjacent geographical areas could not be established during the outbreak and is unlikely to occur in a summer month.

The specificity of the commercial enyzme-linke immunosorbent assay test used for assaying antibody levels against *M.* and *C. pneumoniae* is crucial, and further ongoing studies to check this are therefore in progress. However, there are no indications of inadequate specificity in earlier validations of these tests (1). If legionellosis is capable of triggering a polyclonal immunologic response, it is conceivable that several pneumonia cases during the outbreak period in Lidköping, i.e., the ones showing rising antibody levels against *M.* and *C. pneumoniae,* suffered from LD. Furthermore, the increase in the number of hospitalized pneumonia cases coinciding with the legionella cases seen in Fig. 1 also confirms such a conclusion.

Optimal sensitivity of immunofluorescent antibody tests in serological diagnosis of *L. pneumphila* sg 1 infections requires the use of an antigen subgroup that is in agreement with the antigenic setup of the epidemic strain. The use in such a test of an antigen from a subgroup of *L. pneumophila* containing MAb 3/1-reactive epitopes might decrease the sensitivity of IFAT. Legionella infection can probably induce a polyclonal immunologic response that may give rise to false-positive results in testing for *Mycoplasma pneumoniae* and *Chlamydophila pneumoniae.* Diagnostic tests performed in the acute phase of LD might only detect a minority of the true cases occurring in an outbreak.

REFERENCES

1. **Claesson, B. E. B., H. Enroth, S. Elowson, and M. Hellgren-Leonardsson.** 2004. Evaluation of diagnostic methods for Mycoplasma pneumoniae and Chlamydophila pneumoniae using serology, PCR and the BD Probe Tec ET System. ECCMID, Prague, May 2004. *Clin. Microbiol. Infect.* **10:**6–7.

2. **Darelid, J., H. Hallander, S. Löfgren, B.-E. Malmvall, and A.-M. Olinder-Nielsen.** 2001. Community spread of *Legionella pneumophila* serogroup 1 in temporal relation to a nosocomial outbreak. *Scand. J. Infect. Dis.* **33:**194–199.

3. **Gaia, V., N. K. Fry, B. Afshar, P. C. Lück, H. Meugnier, J. Etienne, R. Peduzzi, and T. G. Harrison.** 2005. Consensus sequence-based scheme for epidemiological typing of clinical and environmental isolates of *Legionella pneumophila.* *J. Clin. Microbiol.* **43:**2047–2052.

4. **Helbig, J. H., J. B. Kurtz, M. Castellani Pastoris, C. Pelaz, and P. C. Lück.** 1997. Antigenic polysaccharide components of *Legionella pneumophila* recognized by monoclonal antibodies: possibilities and limitations for division of the species into serogroups. *J. Clin. Microbiol.* **35:**2841–2845.

5. **Helbig, J. H., P. C. Lück, Y. A. Kniel, W. Witzleb, and U. Zähringer.** 1995. Molecular characterization of a virulence-associated epitope on the lipopolysaccharide of Legionella pneumophila serogroup 1. *Epidemiol. Infect.* **115:**71–78.

6. **Kallings, I. and K. Nordström.** 1983. The pattern of immunoglobulins with special reference to IgM in Legionnaires' disease patients during a 2 year follow-up period. *Zbl. Bakt. Hyg., I. Abt. Orig. A* **255:**27–32.

7. **Lindsay, D. S. J., W. H. Abraham, W. Findlay, P. Christie, F. Johnston, and G. F. S. Edwards.** 2004. Laboratory diagnosis of Legionnaires' disease due to *Legionella pneumophila* serogroup 1: comparison of phenotypic and genotypic methods. *J. Med. Microbiol.* **53:**183–187.

8. **Marshall, L. E., T. C. Boswell, and G. Kudesia.** 1994. False positive legionella serology in campylobacter infection: campylobacter serotypes, duration of antibody response and elimination of cross-reactions in the indirect fluorescent antibody test. *Epidemiol. Infect.* **112:**347–357.

9. **Wentworth, B. B. and H. E. Stiefel.** 1982. Studies on the specificity of *Legionella* serology. *J. Clin. Microbiol.* **15:**961–963.

10. **Wilkinson, H. W., A. L. Reingold, B. J. Brake, D. L. McGiboney, G. W. Gorman, and C. V. Broome.** 1983. Reactivity of serum from patients with suspected legionellosis against 29 antigens of Legionellaceae and *Legionella*-like organisms by indirect immunofluorescense assay. *J. Infect. Dis.* **147:**23–31.

11. **Zuravleff, J. J., V. L. Yu, J. W. Shonnard, B. K. Davis, and J. D. Rihs.** 1983. Diagnosis of Legionnaires' disease: an update of laboratory methods with new emphasis on isolation by culture. *JAMA* **250:**1981–1985.

SEROTYPING OF *LEGIONELLA PNEUMOPHILA* IN EPIDEMIOLOGICAL INVESTIGATIONS: LIMITATIONS IN THE ERA OF GENOTYPING

Jürgen H. Helbig and Paul Christian Lück

18

The end of the 1980s was the "golden age" of *Legionella pneumophila* serotyping. In recent years molecular genotyping methods have been increasingly applied to the classification of legionellae. Nevertheless, for differentiating between *Legionella* isolates at the species level by DNA sequencing of rRNA or *mip* genes (3, 10) and the fingerprint level for characterization of clonal specific genotypes (2, 4), *L. pneumophila* can only be differentiated on the basis of serological markers. So far, the species *L. pneumophila* contains 15 serogroups (1) and 10 monoclonal subgroups within serogroup 1 (7). For recognition of these 24 serovariants, 21 sera and monoclonal antibodies (MAbs) are necessary. In the laboratory praxis, however, the recognition and typing on this high level of diversity is seldom carried out and can only be done in a few specialized laboratories worldwide, because the whole panel of antibodies is not commercially available or cross-reactions obtained by rabbit sera prevent the correct typing. Recently, members of the European Working Group on *Legionella* Infections demonstrated the dissimilar distribution of serotypes among patient isolates (Fig. 1).

Sixty-seven percent of all patient isolates belong to four monoclonal subgroups of serogroup 1 (Philadelphia, Benidorm, France/Allentown, Knoxville), which are characterized by carrying the virulence-associated epitope recognized by MAb 3/1 of the Dresden Panel (5). In contrast, other studies (8, 9) have demonstrated that the distribution of serotypes in water systems has another profile, with less than 20% of MAb 3/1–posive strains, which demonstrates the correlation between serotypes and virulence.

Many water systems are colonized with more than one strain. Therefore, in the context of timely outbreak investigation serotyping is an ideal screening tool for examination of large numbers of colonies from agar plates to select which isolates should be further investigated by molecular typing methods (2). Here we describe a short typing system for *L. pneumophila* that might be useful in epidemiological investigations.

SELECTION OF REAGENTS FOR SEROTYPING BY ENZYME-LINKED IMMUNOSORBENT ASSAY (ELISA)

Out of the Dresden Panel for serotyping of *L. pneumophila* (5, 6), 12 MAbs were selected to create 10 *L. pneumophila* reagents for serotyping (Table 1). In addition, the MAb 3 that is commercially available from the American Type

Jürgen H. Helbig and Paul Christian Lück   Medical Faculty TU Dresden, Institute of Medical Microbiology and Hygiene, D-01307 Dresden, Germany.

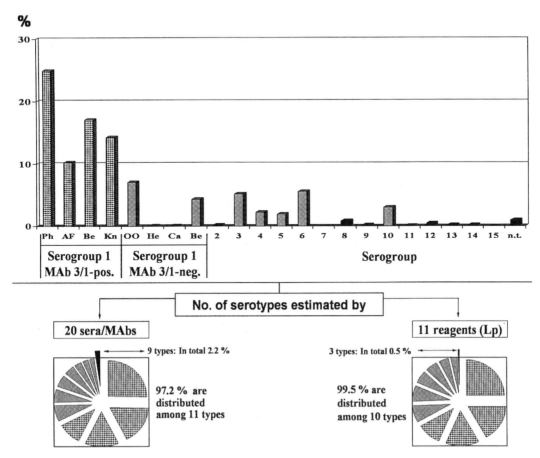

FIGURE 1   Distribution of *L. pneumophila* serogroups and MAb subgroups among 1,335 human isolates (Reprinted from Ref. 5 with kind permission of Springer Science and Business Media). Abbreviations for monoclonal subgroups of serogroup 1: Ph, Philadelphia; Be, Benidorm; FA, France/Allentown; Kn, Knoxville; OO, OLDA/Oxford; He, Heysham; Ca, Camoerdown; Bh, Bellingham. n.t., not typeable into the known serogroups.

Culture Collection (1767-CRL) was included and designated in this study as Lp1c. Lp1a contains two MAbs for recognition of MAb 3/1–positive and MAb 3/1–negative serogroup 1 strains. Using all of the 6 reagents for subtyping of serogroup 1 (Lp1a to Lp1f), 8 of the 10 monoclonal subgroups proposed by Joly et al. (7) can be distinguised (Table 1). Serogroups 3 and 6, being the most frequent non–serogroup 1 groups (Fig. 1), were recognized on the serogroup level by MAbs Lp3 and Lp6, respectively. MAbs LpH and LpK recognize two serogroup-cross-reactive epitopes. At least one of them is located on all strains not belonging to serogroups 1, 7, and 11. Strains of sero-

groups 7 and 11 carry no common epitope with other serogroups. Therefore, the reagent LpG was generated out of two MAbs recognizing these two serogroups.

Serotyping of *L. pneumophila* is mostly done by using indirect immunofluorescence tests. However, in a European multicenter study (for participants, see Acknowledgment) for typing of *L. pneumophila*, it was demonstrated that this technique produces variable results, especially for the recognition of serogroup-cross-reactive epitopes by reagents LpH and LpK. These problems can be avoided by using an ELISA that is a more reproducible technique because of reading of positive results with respect to the

TABLE 1 Reaction pattern of *L. pneumophila* reagents ($n = 11$) with respect to serogroup and monoclonal subgroup destination

Serogroup(s)	MAb subgroup	Reactivity[a] of reagent										
		Lp1a	Lp1b	Lp1c[b]	Lp1d	Lp1e	Lp1f	Lp3	Lp6	LpG	LpH	LpK
1	Philadelphia	+	+	o	o	+	o	o	o	o	o	o
1	Benidorm	+	+	o	+	o	o	o	o	o	o	o
1	France/Allentown	+	+	o	o	o	o	o	o	o	o	o
1	Knoxville	+	+	+	±	o	o	o	o	o	o	o
1	OLDA/Oxford	+	o	o	o	+	o	o	o	o	o	o
1	Heysham	+	o	+	o	o	o	o	o	o	o	o
1	Camperdown	+	o	o	o	o	o	o	o	o	o	o
1	Bellingham	+	o	o	+	o	+	o	o	o	o	o
3		o	o	o	o	o	o	+	o	o	o	o
6		o	o	o	o	o	o	o	+	+	+	o
7, 11		o	o	o	o	o	o	o	o	o	o	o
non 1, 7, or 11		o	o	o	o	o	o	o	o	o	+	±

[a]+, positive; o, negative; ±, positive or negative
[b]MAb 3 of the typing scheme by Joly et al. (1986). The hybridoma is commercially available by the American Type Culture Collection.

blank value, whereas microscopically judging is sometimes equivocal and influenced by the laboratory staff. The detailed ELISA procedure is described elsewhere (6). Briefly, colonies on agar plates were suspended in phosphate-buffered saline, giving an optical density of approximately McFarland 3. This suspension has to be heat-treated for 10 min at 95°C and coated on microplate wells following an incubation time of 2 h at 37°C. Afterwards, the *L. pneumophila* reagents, anti-mouse-IgG, and substrate are added. This technique allows the screening of many colonies in less than six h. In addition, the selection of only 11 reagents combined with a blank without MAbs as the 12th value was done to adapt the typing scheme to the microplate array.

DISCRIMINATION POWER BY USING *L. PNEUMOPHILA* REAGENTS

Fig. 1 shows the distribution of *L. pneumophila* serogroups and MAb subgroups among 1,335 human isolates (5). After summarizing the distribution of serotypes estimated by using either the whole panel of sera (15 rabbit sera and serogroup-specific MAbs [6], plus 6 MAbs for subtyping of serogroup 1) or the short panel of *L. pneumophila* reagents, the difference in discrimination power between human isolates is not much lower and can be neglected. The major subtypes were recognized at nearly the same level; only minor types were separated more strongly (nine serotypes for 2.2% of isolates versus three types for 0.5%). In connection to this, it has to be mentioned that these data obtained with the whole panel are only interesting in special cases, but not in epidemiological investigations which are caused mainly by MAb 3/1–positive strains, whereas in water systems MAb 3/1–negative strains are dominant (8, 9).

In our opinion, the classification of all LPS variants on the basis of serogroups is not useful for proving an epidemiologic link. Our selection of 11 reagents allows the differentiation of the main serotypes causing legionellosis on or below the serogroup level, whereas all other serotypes dominant in man-made and natural water systems were detected above the serogroup level. The development of this short typing scheme based on a low number of reagents but still having a high level of discrimination for patient versus environmental isolates is an excellent preliminary tool for epidemiological investigations.

ACKNOWLEDGMENTS

The authors thank B. Abraham (Glasgow, UK), S. Bernander (Uppsala, Sweden), J. Dominguez (Barcelona, Spain), F. Fendukly (Stockholm, Sweden), L. Franzin (Turin, Italy), V. Gaia (Lugano, Switzerland), T. G. Harrison (London, UK), N. Maes (Brussels, Belgium), T. Marques (Carnaxide, Portugal), S. Mentula (Helsinki, Finland), C. Pelaz Antolin (Madrid, Spain), M. L. Ricci (Roma, Italy), M. Reyrolle (Lyon, France), S. A. Uldum (Copenhagen, Denmark), G. Wewalka (Vienna, Austria), and I. Wybo (Brussels, Belgium) for testing the *L. pneumophila* reagents and for giving information about their use.

REFERENCES

1. **Brenner, T. U., A. G. Steigerwalt, P. Epple, W. F. Bibb, R. M. McKinney, R. W. Starnes, J. M. Colville, R. K. Selander, P. A. Edelstein, and C. W. Moss**. 1988. *Legionella pneumophila* serogroup Lansing 3 isolated from a patient with fatal pneumonia, and descriptions of *L. pneumophila* subsp. *pneumophila* subsp. nov., *L. pneumophila* subsp. *fraseri* subsp. nov., and *L. pneumophila* subsp. *pascullei* subsp. nov. *J. Clin. Microbiol.* **26:**1695–1703.

2. **Fry, N. K., S. Alexiou-Daniel, J. M. Bangsborg, S. Bernander, M. Castellani Pastoris, J. Etienne, M. Forsblom, V. Gaia, J. H. Helbig, D. Lindsay, P. C. Lück, C. Pelaz, S. A. Uldum, and T. G. Harrison**. 1999. A multicenter evaluation of genotypic methods for the epidemiologic typing of *Legionella pneumophila* serogroup 1: results of a pan-European study. *Clin. Microbiol. Infect.* **5:**462–477.

3. **Fry, N. K. and T. G. Harrion**. 1998. An evaluation of intergenic rRNA gene sequence length polymorphism analysis for the identification of *Legionella* species. *J. Med. Microbiol.* **47:**667–678.

4. **Gaia, V., N. K. Fry, B. Afshar, P. C. Lück, H. Meugnier, J. Etienne, R. Peduzzi, and T. G. Harrison.** 2005. Consensus sequence-based scheme for epidemiological typing of clinical and environmental isolates of *Legionella pneumophila*. *J. Clin. Microbiol.* **43:**2047–2052.

5. **Helbig, J. H., S. Bernander, M. Castellani Pastoris, J. Etienne, V. Gaia, S. Lauwers, D. Lindsay, P. C. Lück, T. Marques, S. Mentula, M. F. Peeters, C. Pelaz, M. Struelens, S. A.**

Uldum, G. Wewalka, and T. G. Harrison (2002). Pan-European study on culture-proven Legionnaires' disease: distribution of *Legionella pneumophila* serogroups and monoclonal subgroups. *Eur. J. Clin. Microbiol Infect. Dis.* **21:**710–716.

6. **Helbig, J. H., J. B. Kurtz, M. Castellani Pastoris, C. Pelaz, and P. C. Lück**. 1997. The antigenic lipopolysaccharide components of *Legionella pneumophila* recognized by monoclonal antibodies: possibilities and limitations for division of the species into serogroups. *J. Clin. Microbiol.* **35:**2841–2845.

7. **Joly J. R., R. M. McKinney, O. J. Tobin, W. F. Bibb, I. D. Watkins, and D. Ramsay.** 1986. Development of a standardized subgrouping scheme for *Legionella pneumophila* serogroup 1 using monoclonal antibodies. *J. Clin. Microbiol.* **23:**768–771.

8. **Lück, P. C. and J. H. Helbig**. 2002. Typing of *Legionella* strains isolated from patients and environmental sources in Germany, 1990-2000, p. 267–270. *In* R. Marre et al. (ed.), Legionella. ASM Press, Washington, D.C.

9. **Pringler, N., P. Brydov, and S. A. Uldum.** 2002. Occurrence of *Legionella* in Danish hot water systems, p. 298–301. *In* Legionella. R. Marre et al. (ed.), ASM Press, Washington, D.C.

10. **Ratcliff, R. M., J. A. Lanser, P. A. Manning, and M. W. Heuzenroeder.** 1998 Sequence-based classification scheme for the genus *Legionella* targeting the *mip* gene. *J. Clin. Microbiol.* **36:**1560–1567.

DUOPATH *LEGIONELLA*: A NEW IMMUNOCHROMATOGRAPHIC TEST FOR SIMULTANEOUS IDENTIFICATION OF *LEGIONELLA PNEUMOPHILA* AND *LEGIONELLA* SPECIES

Jürgen H. Helbig, Paul Christian Lück,
Britta Kunz, and Andreas Bubert

19

Several molecular biological methods (3) are increasingly being used for the identification of legionellae in man-made aquatic environments. Nevertheless, the gold standard method in a routine laboratory is still the enrichment on GVPC agar plates with a subsequent confirmation that is mostly carried out by serological methods. Currently, the genus *Legionella* is known to include 51 species (Enzéby, http://www.cict.fr). Some of the species have been isolated only from environmental sources up to now, but it is generally accepted that all species may cause pneumonia, especially in immunocompromised persons. These huge numbers of *Legionella* spp. represent a very wide serological heterogeneity that can lead to unsatisfying sensitivity and specificity of serological identification tools. Already more than 20 years ago a slide agglutination test for identification of 22 *Legionella* spp. was described (4). Nevertheless, commercially available serological assays are rare and the frequently used *Legionella* Latex Test assay (Oxoid, Basingstoke, United Kingdom) recognizes only seven of the most important pneumonia-causing *Legionella* non-*pneumophila* species. DNA sequencing could circumvent this limitation, but these methods are generally neither user-friendly nor convenient. Moreover, rather than the identification of the species, it is the confirmation of the presence of *Legionella* spp. that is most important for the monitoring of water systems, and that is often required according to different official regulations. In addition, the differentiation between *L. pneumophila* and all other *Legionella* spp. is recommended and often stipulated. A new immunochromatographic identification (lateral flow) assay named Duopath *Legionella* recently developed by Merck KGaA is intended for the simultaneous recognition of *L. pneumophila* or *Legionella* spp. on the same test device. Here we present the first set of data discussing the usefulness of this assay. The procedure was as follows:

All *Legionella*-type strains or water samples were grown on GVPC agar (Merck, Darmstadt, Germany). For testing, one *Legionella* colony (approx. colony diameter 1 to 2 mm) or suspect *Legionella* colony was picked up and resuspended in a 0.9% NaCl solution containing 1% Tween 20. After the addition of polymyxin B (*Bacillus cereus* Selective Supplement, Merck, Darmstadt, Germany), the suspension

Jürgen H. Helbig and Paul Christian Lück Medical Faculty TU Dresden, Institute of Medical Microbiology and Hygiene, D-01307 Dresden, Germany. *Britta Kunz and Andreas Bubert* Merck KGaA, R&D Life Science & Analytics, D-64293 Darmstadt, Germany.

Legionella: State of the Art 30 Years after Its Recognition
Edited by Nicholas P. Cianciotto et al.
©2006 ASM Press, Washington, D.C.

was incubated 2 to 5 min at room temperature followed by 5 min at 95°C. The suspension was then quickly cooled to room temperature, and a 150-µl sample was pipetted into the sample port of Duopath *Legionella*. Results at test L.pn (for *L. pneumophila*) and Spec (for *Legionella* spp.) and control zones were read after 30 min without the use of a magnifying glass. Fig. 1 demonstrates how Duopath *Legionella* is read.

The specificity of Duopath *Legionella* was calculated by testing 50 bacterial strains isolated from water samples and grown on GVPC agar plates (Merck, Darmstadt, Germany). Non-*Legionella* isolates were characterized by their ability to grow on blood agar. None of those bacteria that were able to grow on blood agar were positive by testing with Duopath *Legionella*, thus giving a specificity of 100%.

To determine the sensitivity of Duopath *Legionella* for *L. pneumophila,* 69 *L. pneumophila* strains were tested altogether. Details are listed in Table 1. *L. pneumophila* water isolates were confirmed as *L. pneumophila* by using MONOFLUO anti-*Legionella* staining reagent (BIO-RAD, Munich, Germany) and serotyped (1) as serogroups 1, 4, 5, 6, 8, or 10. All of them were correctly identified by providing bands at the detection zones for both *L. pneumophila* and *Legionella* spp. (Fig. 1B), thus leading to a sensitivity of 100%.

To determine the sensitivity of Duopath *Legionella* for *Legionella* spp., 31 of the named *Legionella* non-*pneumophila* species able to grow on GVPC agar were tested with Duopath *Legionella*. Environmental isolates (*n* = 45) were involved from 15 *Legionella* spp., with one to four strains per species. As the gold standard, these isolates were classified to species level by *mip* gene sequencing (see chapter 87) according to the guidelines and database of the European Working Group for *Legionella* Infections (http://www.ewgli.org). Duopath *Legionella* recognized 29 of the 31 *Legionella* non-*pneumophila*-type strains, resulting in a sensitivity of 94% (Table 1). Out of the environmental isolates, Duopath *Legionella* was positive for 43 of the 45 (96%) strains tested. All of the 76 *Legionella* non-*pneumophila* strains were negative for Duopath *Legionella* at the *L. pneumophila*–specific L.pn test zone (Figure 1A).

The latex agglutination assay *Legionella* spp. recognizes seven of the most frequent *Legionella* non-*pneumophila* species causing Legionnaires' disease but not the wide range of legionellae found in water systems, which are also suspected to be pneumonia pathogens. In many countries, water or environmental samples have to be analyzed for the presence of *Legionella* spp. instead of only *L. pneumophila*. Here, Duopath *Legionella* revealed a significant advantage over the latex assay and makes the

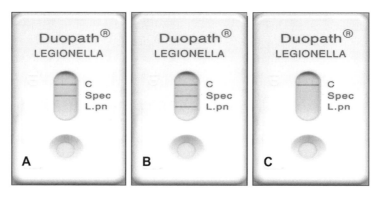

FIGURE 1 Reading of Duopath *Legionella*. (A) Sample is positive for *Legionella* spp. but not for *L. pneumophila*; (B) sample is positive for *L. pneumophila*; (C) sample is negative.

TABLE 1 Sensitivity of Duopath *Legionella*

Species/serogroups	Sources	No.	Duopath *Legionella* spp. positive (%)	Duopath *L. pneumophila* positive (%)
L. non-*pneumophila*	ATCC reference strains	31	29 (94%)	0
L. non-*pneumophila*	Water isolates	45	43 (96%)	0
L. pneumophila serogroup 1 to 15[a]	ATCC reference strains	24	24 (100%)	24 (100%)
L. pneumophila Serogroups 1, 4, 5, 6, 8, 10	Water isolates	45	45 (100%)	45 (100%)

[a]Including all reference strains of monoclonal subtyping scheme for serogroup 1 (2) ATCC, American Type Culture Collection

phenotypic diagnostic gap significantly smaller. Therefore, Duopath *Legionella* can be considered a user-friendly, simple and reliable test for the simultaneous identification of *L. pneumophila* and non-*pneumophila*.

ACKNOWLEDGMENTS

The authors thank Birgit Bubert for excellent consultancy and Sigrid Gäbler, Jutta Paasche, Kerstin Seeliger, and Ines Wolf for excellent technical assistance.

REFERENCES

1. **Helbig, J. H., J. B. Kurtz, M. Castellani Pastroris, C. Pelaz, and P. C. Lück.** 1997. The antigenic lipopolysaccharide components of *Legionella pneumophila* recognized by monoclonal antibodies: possibilities and limitations for division of the species into serogroups. *J. Clin. Microbiol.* **35:**2841–2845.

2. **Joly, J. R., R. M. McKinney, O. J. Tobin, W. F. Bibb, I. D. Watkins, and D. Ramsay.** 1986. Development of a standardized subgrouping scheme for *Legionella pneumophila* serogroup 1 using monoclonal antibodies. *J. Clin. Microbiol.* **23:**768–771.

3. **Ratcliff, R. M., J. A. Lanser, P. A. Manning, and M. W. Heuzenroeder.** 1998. Sequence-based classification scheme for the genus *Legionella* targeting the *mip* gene. *J. Clin. Microbiol.* **36:**1560–1567.

4. **Thacker, W. L., B. B. Plikaytis, and H. W. Wilkinson.** 1985. Identification of 22 *Legionella* species and 33 serogroups with the slide agglutination test. *J. Clin. Microbiol.* **21:**779–782.

ANTIGENIC DIVERSITY OF A 19-KILODALTON PEPTIDOGLYCAN-ASSOCIATED LIPOPROTEIN AMONG *LEGIONELLA* SPECIES DETERMINED BY REACTIVITY PATTERNS TO MONOCLONAL ANTIBODIES

Jin-Hee Moon, Jin-Ah Yang, Hee-Sun Sim, Dae Won Park, Jang Wook Sohn, Hae Kyung Lee, Mi Yeon Park, and Min Ja Kim

20

Legionella is an important cause of both community-acquired and nosocomial pneumonias. *Legionella* pneumonia may be severe and is potentially fatal in elderly and immunocompromised patients, and rapid diagnosis and early antibiotic treatment are required (3, 4). A 19-kDa peptidoglycan-associated lipoprotein (PAL) of *Legionella* species is a species-common, surface exposed, soluble antigen. We have recently shown that *Legionella* PAL is a highly immunogenic and potential broad-spectrum urinary diagnostic antigen detected by PAL-specific polyclonal IgG-based enzyme-linked immunosorbent assay (ELISA) (2). Since the genus *Legionella* includes high number of species and serotypes, it is desirable to investigate the antigenic diversity of the PAL protein to improve the reliability of the diagnostic test.

In this study, we produced 17 monoclonal antibodies (mAbs) against the recombinant PAL

(rPAL) cloned from *Legionella pneumophila* serogroup (sg) 1 and investigated antigenic diversity of the PAL proteins from soluble fractions of 16 *Legionella* species including 22 serogroups by mAb reactivity patterns in ELISA. Soluble antigens from *Legionella* strains (*L. pneumophila* sg 1, 3, 4, 5, and 6; *L. annisa*; *L. dumoffii*; *L. gormanii*; *L. jordanis*; *L. micdadei*; *L. oakridgensis*; *L. sainthelensi*; *L. bozemanii* sg 1 and 2; *L. longbeachae* sg 1 and 2; *L. Hackeliae* sg 1) were prepared by the method of Berdal et al. (1).

To demonstrate the *Legionella* PAL antigen as a component of soluble antigens of *Legionella* species, soluble antigens extracted form several *Legionella* species (*L. pneumophila* sg 1, 3 to 6; *L. anisa*; *L. dumoffii*; *L. gormanii*; *L. jordanis*; *L. micdadei*; *L. oakridgensis*; *L. sainthelens*; *L. bozemanii* sg 1 and 2; *L. longbeachae* sg 1 and 2; *L. Hackeliae* sg 1) were analyzed by sodium dodecyl sulfate-polyacrylamide gel and immunobloting with the anti-PAL IgG antibody of *L. pneumophila* sg 1. All species tested showed an 18 to 19kDa prominent protein that reacted strongly to anti-PAL IgG..

Seven *Legionella* strains including *L. pneumophila* sg 1 and 3 to 6, *L. anisa*, and *L. sainthelensi* were reacted to all 17 mAbs, and the remaining representative strains of 8 *Legionella* species reacted in the range of 5.8 to 70.6%.

Jin-Hee Moon, Jin-Ah Yang, and Hee-Sun Sim Research Institute of Emerging Infectious Disease, Korea University, Seoul, 136-705, Republic of Korea. *Dae Won Park and Jang Wook Sohn* Div. of Infectious Disease, Dept. of Internal Medicine, Korea University College of Medicine, Seoul, 136-705, Republic of Korea. *Hae Kyung Lee and Mi Yeon Park* Div. of Bacterial Respiratory Infections, Infectious Disease Research Center, National Institutes of Health of Korea, Seoul, 122-701, Republic of Korea. *Min Ja Kim* Research Institute of Emerging Infectious Disease, and Div. of Infectious Disease, Dept. of Internal Medicine, Seoul 136-705, Republic of Korea.

Legionella: State of the Art 30 Years after Its Recognition
Edited by Nicholas P. Cianciotto et al.
©2006 ASM Press, Washington, D.C.

TABLE 1 Reactivity patterns of the PAL antigens from soluble fractions of 11 *Legionella* species to 17 monoclonal antibodies by ELISA

	mAb	3-7	4-27	4-23	4-22	4-19	4-31	3-17	3-22	4-1	3-15	4-38	4-16	4-2	4-15	4-30	3-19	3-16
Legionella Pneumophila	sg 1	+3	+3	+3	+2	+2	+3	+3	+2	+3	+2	+3	+3	+3	+2	+3	+3	+3
	sg 3	+3	+3	+2	+1	+1	+3	+3	+1	+3	+2	+2	+2	+1	+2	+3	+2	+3
	sg 4	+3	+3	+3	+3	+3	+3	+3	+2	+3	+2	+3	+2	+3	+2	+3	+2	+3
	sg5	+3	+3	+3	+3	+3	+3	+3	+1	+3	+2	+3	+2	+3	+2	+3	+2	+3
	sg 6	+3	+3	+3	+3	+3	+3	+3	+2	+3	+2	+3	+3	+3	+3	+3	+3	+3
Non-*Legionella Pneumophila*	*L. sainthelensi*	+3	+2	+2	+3	+2	+2	+2	+1	+3	+1	+2	+2	+3	+1	+3	+1	+2
	L. anisa	+3	+3	+2	+2	+1	+3	+3	+1	+3	+2	+1	+2	+1	+2	+3	+2	+3
	L. micdadei	+2	+2	+2	+1	−	+2	+2	+1	+1	−	+3	+1	+1	+1	−	−	−
	L. oakridgensis	+3	+3	+2	−	−	+1	−	−	+3	+1	+3	+2	+3	+2	+1	−	−
	L. dumoffii	+1	+2	−	−	+2	+2	−	−	+1	+1	−	−	−	−	−	−	−
	L. bozemania sg 1	+3	+3	+3	+3	+1	+3	+3	+3	+1	+1	−	−	−	−	−	−	−
	L. bozemania sg 2	+3	+3	+3	+2	+1	+3	+3	+3	+1	+1	−	−	−	−	−	−	−
	L. longbeachae sg 1	+3	+2	+3	+1	+2	+2	+2	+3	−	−	−	−	−	−	−	−	−
	L. longbeachae sg 2	+2	+2	+3	+3	+2	−	−	−	−	−	−	−	−	−	−	−	−
	L. hackelia sg 1	+1	+1	−	−	−	−	−	−	−	−	−	−	−	−	−	−	−
	L. jordanis	+2	−	+3	−	−	−	−	−	−	−	−	−	−	−	−	−	−
	L. gormanii	+1	−	−	−	−	−	−	−	−	−	−	−	−	−	−	−	−

+3 ≥1.00; +2 0.99 to 0.50; +1 0.24 to 0.49.

Cut-off value: *P* > 0.29; mAb, monoclonal antibody; sg, serogroup.

In particular, *L. hackelia* sg 1, *L. jordanis*, and *L. gormanii* were found to be reactive only to one or two mAbs. For the reactivity patterns with individual mAbs, 12 different reactivity patterns were found among 11 *Legionella* species tested (Table 1), indicating the potential multiple-PAL epitopes and the presence of the structural differences.

We concluded that antigenic diversity of a *Legionella* species–common PAL protein is present among *Legionella* species and that developing diagnostic agents using a *Legionella* antigen, even if it is conserved, might be better for separately detecting *L. pneumophila* species and non-pneumophila *Legionella* species.

REFERENCES

1. **Berdal, B. P., C. E. Farshy, J. C. Farshy, and J. C. Feeley.** 1979. Detection of *Legionella* pneumophila antigen in urine by enzyme-linked immunospecific assay. *J. Clin. Microbiol.* **9:**575–578.

2. **Kim, M. J., J. W. Sohn, D. W. Park, S. C. Park, and B. C. Chun.** 2003. Characterization of a lipoprotein common to Legionella species as a urinary broad-spectrum antigen for diagnosis of Legionnaires' disease. *J. Clin. Microbiol.* **41:**2974–2979.

3. **Kirby, B. D., K. M. Snyder, R. D. Meyer, and S. M. Finegold**. 1980. Legionnaires' disease: report of sixty-five nosocomially acquired cases of review of the literature. *Medicine* **59:**188–205.

4. **Stout, J. E. and V. L. Yu.** 1997. Legionellosis. *N. Engl. J. Med.* **337:**682–687.

EVALUATION OF A NEW RAPID IMMUNOCHROMATOGRAPHIC TEST USING PEPTIDOGLYCAN-ASSOCIATED LIPOPROTEIN FOR DETECTION OF *LEGIONELLA* ANTIGEN IN URINE SAMPLES FROM ADULTS WITH PNEUMONIA

Jang Wook Sohn, Hee-Sun Sim, Hye Won Jeong,
Dae Won Park, Hee Jin Cheong, Woo Ju Kim,
Sun Ae Kim, Young Sik Cho, and Min Ja Kim

21

Since antigen detection in urine has proved to be a sensitive and rapid method for detecting *Legionella pneumophila*, this technique has become one of the most used tools for the diagnosis of Legionnaires' disease. However, the currently available urinary antigen tests are limited in their ability to detect serogroups other than *L. pneumophila* serogroup 1 (sg 1). Different assay kits have been evaluated, with the sensitivity for detection of other serogroups ranging from 14 to 69% (1–3).

We have previously reported the *Legionella* species–common peptidoglycan-associated lipoprotein as a broad-spectrum antigen for urinary diagnostic testing (4). In this study, we developed a new immunochromatographic test (ICT), the SD Bioline *Legionella* antigen test (Standard Diagnostics, Inc., Korea), which uses antibodies specific to the peptidoglycan-associated lipoprotein antigen, and evaluated

the performance of the test kit by comparing it with the Binax NOW *Legionella* urinary antigen test (Binax, Portland, Maine) and Biotest *Legionella* urine antigen EIA (Biotest AG, Dreieich, Germany).

We tested urine samples from 99 adult patients with pneumonia. *Legionella* infection was defined as isolation of the organism, serology with a fourfold or greater rising IFA antibody titer for *L. pneumophila* in the acute and convalescent phase or stationary high IFA antibody titers ($\geq$1:256), or by positive *Legionella* PCR on respiratory sample plus single high IFA antibody titer ($\geq$1:256) in patients with clinically compatible illness.

Controls consisted of 88 patients in whom legionellosis was excluded by negative results of the PCR assay and low IFA antibody titers ($\leq$1:64) in acute- and convalescent-phase sera for *L. pneumophila*. We also prepared soluble antigens from 15 *Legionella* species (*L. pneumophila* sg 1, 3, 4, 5, and 6; *L. gormanii*; *L. longbeachae* sg 1 and 2; *L. bozemanae* sg 1 and 2; *L. micdadei*; *L. sainthelensi*; *L. hackelia*; *L. oakridgensis*; and *L. jordanis*), as described previously (5). Extracted soluble antigen samples that were spiked artificially into urine from a healthy

Hee-Sun Sim Institute of Emerging Infectious Diseases, Korea University, Seoul, 136-705, Republic of Korea. *Min Ja Kim, Hye Won Jeong, Dae Won Park, Hee Jin Cheong, and Woo Ju Kim* Div. of Infectious Disease, Dept. of Internal Medicine, Korea University College of Medicine, Seoul, 136-705, Republic of Korea. *Sun Ae Kim and Young Sik Cho* Standard Diagnostics, Inc., Kyung-gi do, Republic of Korea.

TABLE 1 Results of 3 *Legionella* urinary antigen tests in 11 patients with Legionnaires' disease

No.	*Legionella* culture	*Legionella* sputum PCR	*Legionella* IFA		Urinary Antigen test		
			Acute phase	Convalescent phase	Biotest EIA	SD Bioline ICT	Binax NOW ICT
1	*L. pneumophila* sg 2–14	P	1:512	NT	P	P	P
2	*L. pneumophila* sg 1	P	1:512	NT	P	P	P
3	No growth	P	1:256	1:256	P	P	P
4	No growth	P	1:512	1:512	P	P	P
5	No growth	P	1:256	NT	N	N	P
6	No growth	N	1:512	NT	N	P	P
7	No growth	N	1:512	1:512	N	N	N
8	No growth	P	<1:16	1:128	N	P	P
9	No growth	P	1:512	NT	P	N	P
10	*L. pneumophila* sg 1	NT	NT	NT	N	P	P
11	*L. pneumophila* sg 2–14	P	1:1024	NT	N	P	N

P, positive; N, negative; NT, not tested

person at a concentration of 100 ng/120 μl were tested.

The results of the three urinary diagnostic kits in 11 patients with laboratory-diagnosed Legionnaires' disease were shown in Tables 1 and 2. Sensitivities for the SD Bioline ICT, the Binax NOW ICT, and the Biotest EIA were 81.8% (9 of 11), 81.8% (9 of 11), and 45.5% (5 of 11), respectively. None of the 88 patients with nonlegionella pneumonia was positive for all three diagnostic kits tested. Specificities for the three test kits were all 100%. The agreement between the two ICTs was 96.3%. Urine samples from two patients with isolation of *L. pneumophila* non-sg 1 were positive only with the SD Bioline. Test results of the SD Bioline and the Binax Now in urine samples artificially inoculated with extracted antigens of 5 *L. pneumophila* species and 10 nonpneumophila *Legionella* species are shown in Table 3. The new SD Bioline detected 13 *Legionella* species except for *L. jordanis,* and *L. hackelia,* whereas the Binax NOW only detected *L. pneumophila* sg 1.

In this study we have shown that the new SD Bioline ICT is a rapid, broad-spectrum *Legionella* urinary antigen assay to detect both pneumophila and nonpneumophila species and serogroups. The sensitivity and specificity of the new ICT test were similar to those of Binax Now ICT in the clinical samples tested. Further evaluation is required, especially for pneumonia due to species or serogroups other than *L. pneumophila* sg 1.

TABLE 2 Comparison of test performance of 3 *Legionella* urinary antigen assays

	% Sensitivities	% Specificities	% PPV[a]	% NPV[b]
SD bioline ICT	72.7 (8 of 11)	100 (0 of 70)	100	95.9
Binax NOW ICT	81.8 (9 of 11)	100 (0 of 70)	100	97.2
Biotest EIA	45.5 (5 of 11)	100 (0 of 70)	100	92.1

[a]PPV, positive prediction value
[b]NPV, negative prediction value

TABLE 3 Detection of *Legionella* soluble antigens among those artificially inoculated in normal urine samples

Legionella species	SD Bioline ICT	Binax Now ICT
L. pneumophila sg 1	Positive	Positive
L. pneumophila sg 3	Positive	Negative
L. pneumophila sg 4	Positive	Negative
L. pneumophila sg 5	Positive	Negative
L. pneumophila sg 6	Positive	Negative
L. gormanii	Positive	Negative
L. longbeachae sg 1	Positive	Negative
L. longbeachae sg 2	Positive	Negative
L. bozmania sg 1	Positive	Negative
L. bozmania sg 2	Positive	Negative
L. micdade	Positive	Negative
L. sainthelensi	Positive	Negative
L. hackelia	Negative	Negative
L. oakridgensis	Positive	Negative
L. jordanis	Negative	Negative

REFERENCES

1. **Benson, R. F., P. W. Tang, and B. S. Fields.** 2000. Evaluation of the Binax and Biotest urinary antigen kits for detection of Legionnaires' disease due to multiple serogroups and species of Legionella. *J. Clin. Microbiol.* **38:**2763–2765.
2. **Dominguez, J. A., N. Gali, P. Pedroso, A. Fargas, E. Padilla, J. M. Manterola, and L. Matas.** 1998. Comparison of the Binax Legionella urinary antigen enzyme immunoassay (EIA) with the Biotest Legionella Urine antigen EIA for detection of Legionella antigen in both concentrated and nonconcentrated urine samples. *J. Clin. Microbiol.* **36:**2718–2722.
3. **Helbig, J. H., S. A. Uldum, P. C. Luck, and T. G. Harrison.** 2001. Detection of Legionella pneumophila antigen in urine samples by the BinaxNOW immunochromatographic assay and comparison with both Binax Legionella Urinary Enzyme Immunoassay (EIA) and Biotest Legionella Urin Antigen EIA. *J. Med. Microbiol.* **50:**509–516.
4. **Kim, M. J., J. W. Sohn, D. W. Park, S. C. Park, and B. C. Chun.** 2003. Characterization of a lipoprotein common to Legionella species as a urinary broad-spectrum antigen for diagnosis of Legionnaires' disease. *J. Clin. Microbiol.* **41:**2974–2979.
5. **Tang, P. W., and S. Toma.** 1986. Broad-spectrum enzyme-linked immunosorbent assay for detection of Legionella soluble antigens. *J. Clin. Microbiol.* **24:**556–558.

RAPID IDENTIFICATION OF LEGIONELLA PNEUMOPHILA, LEGIONELLA ANISA, AND LEGIONELLA TAURINENSIS WITH LATEX AGGLUTINATION REAGENTS

M. Reyrolle, C. Ratat, J. Freney, M. Leportier,
S. Jarraud, and J. Etienne

22

The clinical distribution of *Legionella* species and serogroups does not correspond to the environmental distribution. *L. pneumophila* serogroup 1 is more prevalent among clinical isolates (95.8%) than environmental isolates (28.2%), whereas *L. anisa* and *L. taurinensis* are more frequent in water samples. So, concerning the lack of commercial reagents for rapid identification, we developed latex reagents to identify the majority of the clinical and environmental strains.

The research latex product was done to identify *L. pneumophila*, *L. anisa*, and *L. taurinensis*. The comparison with two commercialized latex reagents was made using 190 *Legionella* wild-type strains (96 *L. pneumophila*, 63 *L. anisa*, 31 *L. taurinensis*); they corresponded to the French clinical and environmental distribution of the isolates.

In conclusion, the three latex kits tested efficiently identified all the *L. pneumophila* isolates within 15 min. Only the research latex

product is able to identify *L. anisa* and *L. taurinensis*. The research latex product is a rapid and easy tool to identify the majority of the clinical and environmental *Legionella* isolates.

The test panel consisted of 190 Legionella wild strains: 96 *L. pneumophila* serogroups (sg) (30 sg 1, 5, sg 2, 10 sg 3, 5 sg 4, 4 sg 5, 10 sg 6, 1 sg 7, 10 sg 8, 4 sg 9, 5 sg 10, 5 sg 12, 1 sg 13, 5 sg 14, 1 sg 15), 63 *L. anisa* isolates, and 31 *L. taurinensis* isolates. The strains were either isolated by or referred to the French *Legionella* Reference Center. They were previously identified with a classical identification scheme. The majority of them were environmental strains; only 15 of 30 sg 1 were clinical strains.

Three latex Kits were tested. The latex Kit Oxoid allowed the identification of *L. pneumophila* sg 1 and 2 to 14 and the identification of seven *Legionella* species. The latex Kit AES allowed the identification of *L. pneumophila* sg 1 and 2 to 15. The research latex product allowed the identification of *L. pneumophila* sg 1 and 2 to 15 and the separate identification of *L. anisa* and *L. taurinensis*.

The first step of classical strain identification was the cultural discrimination (exigence of *Legionella* growth factors), followed by direct fluorescent-antibody assay (DFA) (1). The *Lp* serogroup identification was then performed

M. Reyrolle, S. Jarraud, and J. Etienne Centre National de Référence des *Legionella* (CNRL), INSERM E-0230, Laboratoire de Bactériologie, Faculté de Médecine Laennec IFR 62, 7 rue Guillaume Paradin, 69372 Lyon Cedex 08, France. *C. Ratat and M. Leportier* Service Immunologie Research Department, bioMérieux, 69280 Marcy l'Etoile, France. *J. Freney* Laboratoire de Microbiologie, EA3090, ISPB, Université Claude Bernard Lyon1, 8 avenue Rockefeller, 69373 Lyon, France.

TABLE 1 Results of comparative identifications with the three latex kits

Strain identified	No. of strains tested	No. of isolates identified		
		AES kit	Oxoid kit	Research latex
L. pneumophila sg 1–15	96	96	95	95
L. anisa	63	0	63[a]	62
L. taurinensis	31	0	0	31
Identification rate				99.4%

[a]Identified as *Legionella* spp.

using home-made rabbit polyclonal sera. The DFA technique took 3 h.

To identify the non-pneumophila *Legionella* strains, the RAPD-PCR technique was used as described by Lopresti et al. (2). This technique took about 2 days.

For the three latex kits the same tube method was used. The tube method was the emulsion of 4 or 10 colonies (fresh subculture issued from one colony) in a tube containing 400 µl of phosphate-buffered saline pH 7.2. The reading was done after rocking the card for 1 min. A result was positive if the agglutination of latex particules occurred within 1 min. A negative result was obtained if no agglutination occurred. A granular or stringy agglutination was an uninterpretable result. The result with latex agglutination was obtained within 15 min.

The results of the three latex kits were similar for the identification of *L. pneumophila* as shown in Table 1. The *L. pneumophila* sg 15 isolate was identified by two of three kits. The research latex product allowed the identification of *L. anisa* and *L. taurinensis* within 15 min., with a total rate of identification of 99.4% for the 190 tested isolates.

Latex agglutination is a rapid technique to identify *Legionella* isolates. Therefore, we have developed latex reagents which allow the identification of the majority of the clinical and environmental *Legionella* in 15 min. The specificity of the research latex reagents was previously described (3). The technique is easy to carry out and is up to 95% reliable. It also works better and is quicker than the usual DFA and RAPD-PCR techniques.

The research latex product is a reliable and rapid tool to identify the most frequent *Legionella* in clinical and water samples.

REFERENCES

1. **Bornstein, N., D. Marmet, and J. Fleurette.** 1981. Isolation and characterization of the three first strains of *Legionella pneumophila* found in France. *Ann. Microbiol.* **132:**405–417.
2. **Lopresti, F., S. Riffard, F. Vandenesch, and J. Etienne.** 1998. Identification of *Legionella* species by random amplified polymorphic DNA profiles. *J. Clin. Microbiol.* **36:**3193–4000.
3. **Reyrolle, M., C. Ratat, M. Leportier, S. Jarraud, J. Freney, and J. Etienne.** 2004. Rapid identification of *Legionella pneumophila* serogroups by latex agglutination. *Eur. J. Clin. Microbiol. Infect. Dis.* **23:**864–866.

CLINICAL PRESENTATION, LABORATORY DIAGNOSIS, AND TREATMENT OF LEGIONNAIRES' DISEASE

David R. Murdoch, Thomas J. Marrie, and Paul H. Edelstein

23

The panel discussion on clinical aspects of Legionnaires' disease (LD) posed several controversial questions to the audience after a brief discussion of the current knowledge concerning the question by one of the panelists.

WHO SHOULD BE TESTED FOR LEGIONNAIRES' DISEASE?

Specialized tests are required to diagnose LD, and these need to be specifically requested by clinicians (7). The main diagnostic methods are culture, urinary antigen detection, serology, and PCR, all of which are costly to perform and variably subject to false-positive results in low-prevalence populations. Although improvements have been made in recent years, no single test is reliable on its own to detect all cases of LD. The decision to test for legionella infection is complicated by the fact that LD cannot be clinically or radiographically distinguished from other causes of pneumonia. Scoring systems have been devised to predict patients with LD, but they have not proven to be reli-

able (6). Interestingly, a recent study showed that experienced clinicians were able to suspect LD with a sensitivity of about 64% (1), but it is unclear whether this translates to other regions with a lower incidence of LD. Despite the problems of clinically diagnosing LD, it is still possible to select a subset of patients for whom the yield from legionella diagnostic tests is likely to be relatively high. This group includes the following who are admitted to hospital with pneumonia: elderly persons, smokers, immunosuppressed individuals, and those with chronic lung disease. Possible testing strategies include testing all hospitalized patients with community-acquired pneumonia (CAP); testing only select populations, such as those with severe pneumonia or those at higher risk for LD; or no testing at all. Regardless of the testing strategy, testing for *Legionella* spp. should be done as part of the workup of outbreaks of pneumonia.

There was no uniformity of opinion in the audience regarding testing strategies. Some test all patients with pneumonia, while some test only certain patients. Many would test all adults hospitalized with CAP. Some objections to testing all patients were the cost of testing, the low frequency of LD as the cause of CAP, and the less than optimal sensitivity of the urine antigen test. Points raised that favor broader

David R. Murdoch Department of Pathology, Christchurch School of Medicine & Health Sciences, University of Otago, Christchurch, New Zealand. *Thomas J. Marrie* Faculty of Medicine and Dentistry, Walter C. MacKenzie Health Sciences Center, Edmonton, Alberta, Canada. *Paul H. Edelstein* Department of Pathology and Laboratory Medicine, University of Pennsylvania School of Medicine, Philadelphia, PA 19104.

testing included the clinical inability to distinguish LD from other causes of pneumonia and the clinical and epidemiologic value of knowing the diagnosis. Some concluded that a cost-effectiveness study is needed to determine if such an approach would reduce health-care costs or save lives. There was little support for no testing, even when empiric therapy for LD is given, as testing was regarded as important for epidemiologic and clinical reasons.

Similarly there was no agreement in the audience regarding which tests to use. Many use the urinary antigen test as the primary method. People from areas where *L. longbeachae* (and others) are important would include culture in addition to the urinary antigen test, as most commercially available urinary antigen tests detect only *L. pneumophila* serogroup 1. The necessity of culture for epidemiologic reasons was stressed by some.

THE ROLE OF DIAGNOSTIC NUCLEIC ACID AMPLIFICATION TESTS

There is now a considerable amount of data on nucleic acid amplification tests (NAATs), particularly PCR, for diagnosing LD. NAATs enable specific amplification of minute amounts of *Legionella* DNA, provide results within a short time frame, and have the potential to detect infections caused by any *Legionella* species and serogroup. Many studies have shown that PCR has a similar (if not better) sensitivity to culture when testing lower respiratory tract samples (8). The utility of testing nonrespiratory samples is unresolved. There are no standardized protocols (like for most NAATs for infectious diseases), although commercial assays (often in multiplex format with other respiratory tract pathogens) are available (3, 4). Discussion points included whether PCR can now replace culture and the role of PCR for testing nonrespiratory samples.

Members of the audience had variable access to *Legionella* PCR, although it is clear that many places (especially in Europe) have ready access. Mixed views were presented about the importance of PCR, although the general view was that PCR is a useful method for testing lower respiratory samples. Many thought that PCR could replace culture. Some dissenters brought up the cost of PCR testing, the variable performance of the test, the lack of standardization, and the need to offer testing out of hours. The need for rigorous assessment of nonrespiratory sample types was emphasized. Some European groups favored the use of PCR for testing serum samples. A clear need for standardization of PCR protocols and quality control schemes was a major point of discussion.

IS EMPIRIC THERAPY FOR LEGIONNAIRES' DISEASE ALWAYS INDICATED?

The choice of empiric antibiotic therapy for CAP depends on several important considerations. These include the site of care (home, hospital ward, or intensive care unit), the endemic rate of LD in the local population, local practice guidelines, and clinical judgment. The prevalence of LD among adults with CAP varies with study design and geographic location, although it is commonly about 2 to 8% (occasionally >15%). The 2003 IDSA guidelines for the management of CAP recommend antibiotic therapy active against *Legionella* spp., *Mycoplasma pneumoniae*, and *Chlamydia pneumoniae* in most situations, apart from suspected aspiration pneumonia or *Pseudomonas* infection (5). Randomized controlled trials of treatment for CAP show that clinical success is significantly higher and mortality lower when antibiotics are used that have activity against *Legionella* spp. (2, 9). Clinical judgment is still a critical part of patient care, and empiric therapy for CAP should include coverage for LD if risk factors for LD are elicited from the history.

Most audience members supported empiric therapy for LD in most cases of adult CAP, especially for those with moderate to severe disease. However, some clearly felt that low local rates of LD, combined with the risk of development of antibiotic resistance among more common respiratory pathogens, mandated against uniform empiric treatment for LD in mild cases of CAP.

WHAT ANCILLARY TREATMENTS FOR LEGIONNAIRES' DISEASE ARE HELPFUL OR HARMFUL?

Various ancillary treatments have been proposed for LD, including corticosteroids, oxygen, cytokine inhibitors, activated protein C, and glucose control. For some of these treatments there is evidence indicating benefits for treating severe infection/sepsis (10), although data are lacking for the specific treatment of LD. Corticosteroid therapy may be indicated for adult respiratory distress syndrome caused by LD or bronchiolitis obliterans organizing pneumonia, but this has not been systematically studied.

There was no clear consensus on the use of ancillary therapies for LD in the absence of studies specific to the disease. Many agreed that ancillary therapies shown to be beneficial for other types of pneumonia would likely be applicable to the treatment of LD.

REFERENCES

1. **Fernández-Sabé, N., B. Rosón, J. Carratalà, J. Dorca, F. Manresa, and F. Gudiol.** 2003. Clinical diagnosis of *Legionella* pneumonia revisited: evaluation of the Community-Based Pneumonia Incidence Study Group scoring system. *Clin. Infect. Dis.* **37**:483–489.
2. **García Vázquez, E., J. Mensa, J. A. Martínez, M. A. Marcos, J. Puig, M. Ortega, and A. Torres.** 2005. Lower mortality among patients with community-acquired pneumonia treated with a macrolide plus a beta-lactam agent versus a beta-lactam agent alone. *Eur. J. Clin. Microbiol. Infect. Dis.* **24**:190–195.
3. **Ginevra, C., C. Barranger, A. Ros, O. Mory, J.-L. Stephan, F. Freymuth, M. Johannès, B. Pozzetto, and F. Grattard.** 2005. Development and evaluation of Chlamylege, a new commercial test allowing simultaneous detection and identification of *Legionella, Chlamydophila pneumoniae,* and *Mycoplasma pneumoniae* in clinical respiratory specimens by multiplex PCR. *J. Clin. Microbiol.* **43**:3247–3254.
4. **Khanna, M., J. Fan, K. Pehler-Harrington, C. Waters, P. Douglass, J. Stallock, S. Kehl, and K. J. Henrickson.** 2005. The Pneumoplex assays, a multiplex PCR-enzyme hybridization assay that allows simultaneous detection of five organisms, *Mycoplasma pneumoniae, Chlamydia (Chlamydophila) pneumoniae, Legionella pneumophila, Legionella micdadei,* and *Bordetella pertussis,* and its real-time counterpart. *J. Clin. Microbiol.* **43**: 565–571.
5. **Mandell, L. A., J. G. Bartlett, S. F. Dowell, T. M. File, D. M. Musher, and C. Whitney.** 2003. Update of practice guidelines for the management of community-acquired pneumonia in immunocompetent adults. *Clin. Infect. Dis.* **37**: 1405–1433.
6. **Mulazimoglu, L. and V. L. Yu.** 2001. Can Legionnaires disease be diagnosed by clinical criteria? A critical review. *Chest* **120**:1049–1053.
7. **Murdoch, D. R.** 2003. Diagnosis of *Legionella* infection. *Clin. Infect. Dis.* **36**:64–69.
8. **Murdoch, D. R.** 2004. Molecular genetic methods in the diagnosis of lower respiratory tract infections. *APMIS* **112**:713–727.
9. **Shefet, D., E. Robenshtok, M. Paul, and L. Leibovici.** 2005. Empirical atypical coverage for inpatients with community-acquired pneumonia. *Arch. Intern. Med.* **165**:1992–2000.
10. **Vincent, J. L., E. Abraham, D. Annane, G. Bernard, E. Rivers, and G. Van den Berghe.** 2002. Reducing mortality in sepsis: new directions. *Crit. Care* **6**:S1–S18.

EPIDEMIOLOGY AND STRAIN TYPING METHODS

LEGIONNAIRES' DISEASE IN EUROPE 1995–2004: A TEN-YEAR REVIEW

Carol A. Joseph and Katherine D. Ricketts[*]

24

The identification of *Legionella pneumophila* in the late 1970s and the awareness of its potential to negatively impact on countries with a large tourist industry led European countries to join forces in the 1980s to share scientific knowledge and to coordinate action against this environmentally based disease. The European Working Group for Legionella Infections (EWGLI) was formed in 1986 by 14 countries, initially supported by the World Health Organization (WHO). In 1987, the group launched its European surveillance scheme for travel-associated Legionnaires' disease (EWGLINET), with Sweden acting as the coordinating center. England took over this role in 1993 when funding for the scheme transferred to the European Commission as part of its funding for public health in Europe.

An international surveillance scheme can only function well if it obtains data from participating countries that adhere to common epidemiological and microbiological case definitions, reporting procedures, and standardized responses to cases, particularly when the source of infection is frequently in one country and diagnosis of the disease in another. In this regard EWGLINET has been incredibly successful. It now has 37 collaborating countries that between them report almost 700 cases each year and carry out over 100 environmental investigations annually into cases and clusters associated with international accommodation sites. The scheme also supports ongoing microbiological research and development for improved laboratory detection of clinical and environmental sources of infection using standardized methods within European countries (5, 7). Adherence to the EWGLI guidelines, introduced in 2002, for control and prevention of travel-associated cases of Legionnaires' disease is leading to greater protection of travelers in all European countries (8).

How has this work impacted on national surveillance of Legionnaires' disease within the participant countries? This review covers EWGLINET's role in the collection of national legionella data that enables trends over time at the aggregate level to be studied as well as historical trends within and between European countries. It covers the 10-year period 1995 to 2004 and mainly focuses on trends in aggregate data. More detailed reports showing trends within and between countries from 1995 to 2004 have been published on a regular basis (1–4, 8, 11).

Carol A. Joseph and Katherine D. Ricketts Health Protection Agency, Centre for Infections, 61 Colindale Avenue, London NW9 5EQ.
[*]On behalf of the European Working Group for Legionella Infections.

Legionella: State of the Art 30 Years after Its Recognition
Edited by Nicholas P. Cianciotto et al.
©2006 ASM Press, Washington, D.C.

DATA COLLECTION WITHIN EUROPE

Each year EWGLINET countries are requested to complete a set of standard tables that provide epidemiological and microbiological information for all cases in their annual national dataset. The tables present information on age and sex distribution; deaths; whether the reported cases are hospital-, community-, or travel-associated; and details of national outbreaks by source and type. Microbiological data include method of diagnosis for all cases, and the species, serogroup, and subgroup of culture-confirmed cases. The data are analyzed at the EWGLINET coordinating center in London and presented annually at the EWGLI scientific conference.

EUROPEAN TRENDS

The European data set from contributing countries between 1995 and 2004 is composed of 27,244 cases of Legionnaires' disease. The data were reported by an average of 29 countries (range 24 to 35) and represent a mean incidence rate of 6.29 cases per million population (range 3.7 to 10.1) based on a mean average population of 412 million (range 339 to 557 million). The total number of reported legionella deaths in this period was 2,241—a mean case fatality rate of 8.22% (range 4.9 to 13.1%).

Altogether, 68% of the reported cases were male, 26% were female, and 6% were sex un-known. Cases rose by 73% over the 10-year period, from 1,255 in 1995 to 4,588 in 2004, with some of the increase due to more countries contributing data in recent years (Table 1) and the remainder to improved ascertainment and the occurrence of large community outbreaks.

In this 10-year dataset, nosocomially acquired legionella infection was reported for 8.8% of the cases. This proportion has been falling steadily since 2000, from 12.8% in that year to 6.7% in 2004. Community-acquired cases comprise the largest proportion of cases in Europe, at 38% overall, but vary from year to year depending on the number and size of reported outbreaks. Travel-associated cases account for 20.2% of cases, 13.8% as a result of travel abroad and 6.4% as a result of travel in the country of residence. Cases associated with travel abroad range between 19.3% in 1997 and 10.3% in 2001, with the proportion in the most recent years hovering around 12.4%. Internal travel-associated cases, in contrast, have risen from 1.9% in 1995 to 1996 to 8.6% in 2004. The remaining 33% of cases in this dataset were reported as being of unknown category of exposure. Although high, this proportion has been falling during this 10-year period from a peak of 50% in 1995 (Table 2).

Outbreak detection is an important objective of surveillance, and in the period under review 549 outbreaks or clusters were recog-

TABLE 1[a]   Total reported cases and rate per million population, 1995 to 2004

Year	No. of cases	No. of countries contributing data	Population (millions)	Rate per million population
1995	1,255	24	339	3.70
1996	1,563	24	350	4.46
1997	1,360	24	351	3.87
1998	1,442	28	333	4.33
1999	2,136	28	398	5.38
2000	2,156	28	400	5.38
2001	3,470	29	455	7.60
2002	4,696	32	467	10.1
2003	4,578	34	468	9.8
2004	4,588	35	557	8.2

[a]Reprinted from the journals *Epidemiology and Infection* (9) and *Eurosurveillance* (11) with permission of the publishers.

TABLE 2 Cases and proportion by category of infection, 1995 to 2004

Year	Nosocomial	Community	Travel abroad/home	Travel unknown/other	Not known	Total
1995	163	264	176/12	–	640	1,255
1996	187	594	186	–	596	1,563
1997	219	387	263/34	–	457	1,360
1998	209	478	245/52	–	458	1,442
1999	195	679	288/151	–	823	2,136
2000	275	659	357/143		722	2,156
2001	333	1,475	482/185	5/–	988	3,470
2002	280	1,782	585/359	2/–	1,688	4,696
2003	348	2,110	561/369	–	1,190	4,578
2004	308	1,884	589/395	–/42	1,370	4,588

nized by national systems. Of these, 90 outbreaks (16.4%) were linked to hospitals, 320 (58.3%) were linked to travel abroad or travel at home, 135 (24.6%) were associated with community settings, and 4 (0.7%) were related to other settings such as private homes. The number of cases attributed to individual outbreaks is low for most nosocomial and travel-associated outbreaks compared with community settings, where the number affected has frequently exceeded 100, with one outbreak in Spain in 2001 affecting more than 400 people (6) and another in England affecting almost 200 (10).

Between 1995 and 2001, the number of reported outbreaks doubled from 21 to 41 per year. However, in 2002, a change to the EWG-LINET case definition for travel-associated clusters was made which resulted in the overall number of annual outbreaks and clusters rising to 100 or more each year. Cooling towers were the confirmed source of 47 community outbreaks, hot and cold water systems for 178 outbreaks, and spa pools for 27 outbreaks. Recent investigations of the travel-associated clusters show that about 59% of the implicated hotels and other accommodation sites were positive for *Legionella*.

Microbiological diagnoses between 1995 and 2004 show an average of 352 clinical isolates of *Legionella* were obtained each year (range 236 to 491), representing 13% of the total cases. Of these isolates, 95% were reported to be *Legionella pneumophila,* and the remainder were reported as *L. longbeachae, L. bozemanae, L. micdadei, L. dumoffii, L. gormanii, L. anisa, L. cincinatensis, L. feeleii, L. parisiensis, and L. sainthelensi.* For 100 isolates the *Legionella* species was not given.

The proportion of cases diagnosed by urinary antigen detection has seen a dramatic increase since 1998. In 1995, they represented 14.6% of diagnosed cases; in 1998 this had risen to 33%, and in 2004, to 74% of all cases. Serological diagnoses, either by single high titer or by a fourfold increase in antibody levels, represent 24% of the 10-year dataset but only about 11% in recent years because of the relative change in diagnostic methods toward greater use of urinary antigen detection (Table 3).

This short review of surveillance data on cases of Legionnaires' disease in Europe has mainly focused on descriptive aggregated data for the period 1995 to 2004. Major differences between countries in number of cases, category of exposure, number and range of outbreaks, and methods of diagnosis can only be highlighted rather than examined in depth. However, although an increase in case detections is occurring generally in Europe as a consequence of countries adopting the urinary antigen detection test as the main method of diagnosis, incidence rates continue to vary between countries. Based on community studies, EWGLI estimates that incidence rates of 15 to 20 cases per million population should be expected in most industrialized European countries (9). In the past few years

TABLE 3 Cases and proportion by main method of diagnosis, 1995 to 2004

Main method of diagnosis	L. pneumophila sg 1	L. pneumophila other serogroup or serogroup not determined	Other legionella species or species not known	All legionella cases
Isolation	2,564	796	168	3,528
Antigen detection: urinary	14,000	1,474	393	15,867
Serology: sero-conversion	1,523	1,289	455	3,267
Serology: single high titer	1,480	1,279	425	3,184
Antigen detection: respiratory	128	104	21	253
PCR	24	169	54	247
Other	59	90	27	176
Not known	294	147	281	722
Total (each case counted only once)	20,072	5,348	1,824	27,244

countries such as Belgium, Croatia, Denmark, France, Luxembourg, Spain, and Switzerland have reached or exceeded this level; in a few others case ascertainment is increasing rapidly (Italy, The Netherlands), while for many others (most of the 2004 new European Union [EU] member states, England, Germany, and Greece), incidence rates are well below what should be expected. For most of the new EU member states, zero or very low incidence rates are related to lack of laboratory resources and expertise for legionella diagnosis and weak surveillance of this disease at the national level. In England, which has a strong surveillance system, the problem is more one of under-diagnosis, since many clinicians do not request the appropriate tests when managing patients with pneumonia.

Differences between countries in category of exposure are also apparent in the dataset. For instance, in northern European countries such as Denmark, the United Kingdom, and The Netherlands, between 33 and 50% of all their reported annual cases are associated with travel abroad, usually in warmer countries of southern Europe, and between 3 and 7% with travel at home. In contrast southern European countries such as France, Italy, and Spain report only 2 to 12% of their cases as contracted abroad, whereas 12 to 27% of their cases are due to travel at home by their own residents. The number of outbreaks detected by individ-

ual countries also varies and ranges from none to over 30 per year and is not necessarily related to incidence rates in the associated countries. Switzerland, for example, has reported no outbreaks between 2000 and 2004 yet has one of the highest incidence rates in the dataset. In France and Spain the outbreaks are mainly due to community- or travel-associated infection, whereas in Italy they are mainly due to travel or nosocomial infection. Outbreaks rarely occur in The Netherlands, but a diagnosis of Legionnaires' disease in Dutch residents often leads to the recognition of many outbreaks associated with travel abroad. In England the outbreaks are evenly attributed to community- or travel-associated infections, the latter mostly due to travel abroad by English residents.

All countries with data in this review acknowledge the importance of international collaborations since these also benefit microbiological and epidemiological developments at the national level. Reporting of cases to a national center in each country facilitates outbreak detection as well as monitoring of local and regional trends and is important for developing and auditing national control and prevention programs. The sharing of these data each year at the international level enables countries to compare surveillance activities and outbreak investigation results. It also helps raise awareness of the need for improved diagnosis and reporting of legionella cases where relevant.

REFERENCES

1. **Anonymous.** 1997. Legionnaires' disease in Europe, 1996. *Wkly. Epidemiol. Rec.* **72:**253–260.
2. **Anonymous.** 1998. Legionnaires' disease in Europe, 1997. *Wkly. Epidemiol. Rec.* **73:**257–264.
3. **Anonymous.** 1999. Legionnaires' disease in Europe, 1998. *Wkly. Epidemiol. Rec.* **74:**273–280.
4. **Anonymous.** 2000. Legionnaires' disease in Europe, 1999. *Wkly. Epidemiol. Rec.* **75:**347–352.
5. **Fry, N. K, J. M. Bangsborg, A. Bergmans, S. Bernander, J. Etienne, L. Franzin, V. Gaia, P. Hasenberger, B. Baladrón Jiminéz, D. Jonas, D. Lindsay, S. Mentula, A. Papoutsi, M. Struelens, S. A. Uldum, P. Visca, W. Wannet, and T. G. Harrison.** 2002. Designation of European Working Group on Legionella Infections amplified length polymorphism types of *Legionella pneumophila* serogroup 1 and results of intercenter proficiency testing using a standard protocol. *Eur. J. Clin. Microbiol. Infect. Dis.* **21:**722–728.
6. **García-Fulgueiras, A., C. Navarro, D. Fenoll, J. García, P. González-Diego, T. Jimenéz-Buñuales, et al.** 2003. Legionnaires' disease outbreak in Murcia, Spain. *Emer. Infect. Dis.* **9:**915–921.
7. **Helbig, J. H., S. A. Uldum, S. Bernander, P. C. Luck, G. Wewalka, W. Abraham, V. Gaia, and T. G. Harrison.** 2003. Clinical utility of urinary antigen detection for diagnosis of community-acquired, travel-associated and nosocomial Legionnaires' disease. *J. Clin. Microbiol.* **41:**838–840.
8. **Joseph, C. A., J. V. Lee, J. Van Wijngaarten, M. Castellani Pastoris, and V. Drasar on behalf of the European Working Group for Legionella Infections.** 2002. European Guidelines for Control and Prevention of Travel Associated Legionnaires' Disease. Public Health Laboratory Service, London and http://www.ewgli .org.
9. **Joseph, C. A., on behalf of the European Working Group for Legionella Infections.** 2004. Legionnaires' Disease in Europe 2000-2002. *Epidemiol. Infect.* **132:**417–424.
10. **Joseph, C. A.** 2002. New outbreak of Legionnaires' disease in the United Kingdom. Editorial. *Br. Med. J.* **325:**347–348.
11. **Ricketts, K. D. and C. A. Joseph on behalf of the European Working Group for Legionella Infections.** 2005. Legionnaires' Disease in Europe 2003-2004. http://www.eurosurveillance.org/em/v10n12/1012-226.asp.

TYPING OF *LEGIONELLA PNEUMOPHILA* AND ITS ROLE IN ELUCIDATING THE EPIDEMIOLOGY OF LEGIONNAIRES' DISEASE

Timothy G. Harrison, Norman K. Fry, Baharak Afshar, William Bellamy, Nita Doshi, and Anthony P. Underwood

25

The main reason for typing *Legionella pneumophila* is to help identify environmental sources giving rise to cases of legionellosis, thus allowing control measures to be implemented and further cases to be prevented in a timely manner. As *L. pneumophila* is very common in the environment, a wide range of methods, both phenotypic and genotypic, has been developed in an attempt to differentiate between strains. The suitability of these various methods depends, in large part, on the context within which they are to be applied. Three distinct contexts can be identified:

INVESTIGATION OF EPIDEMIC DISEASE

In large outbreaks the source of infection is often strongly indicated by epidemiological data obtained early in the investigation, and the causative strain is typically present in high numbers (5). Here the speed with which results can be obtained is the key factor to be considered in selecting an appropriate method. The aim is simply to find a match between clinical and environmental isolates to confirm the epidemiological data implicating a source. Finding a match reassures that the source has been found, while failure to find a match strongly suggests the source has not been located and prompts further urgent action.

Almost any highly discriminatory, appropriately validated method that offers reasonable intraexperimental reproducibility (same laboratory, same time) is likely to yield "fingerprints" of sufficient quality to provide supporting microbiological evidence implicating a particular source. Monoclonal antibody (mAB) subgrouping (5, 12) is probably the most rapid and reliable phenotypic method available and is excellent for initial screening of isolates. However, its utility is limited by poor reagent availability (outside Europe) and the fact that outbreaks are caused by only a few mAb subgroups, so in this context, discrimination is poor. Thus, genotypic methods are almost always required. Over the years many such methods have been developed, ranging from the simple restriction-endonuclease analysis, with frequent (18), or rare (3) DNA-cutters to PCR-based typing methods such as arbitrarily primed PCR (5) or amplified fragment-length polymorphism typing (7).

Timothy G. Harrison, Norman K. Fry, Baharak Afshar, William Bellamy, and Nita Doshi Health Protection Agency, Respiratory and Systemic Infection Laboratory, Center for Infections, 61 Colindale Ave., London NW9 5EQ. *Anthony P. Underwood* Health Protection Agency, Statistics, Modelling and Bioinformatics Department, Center for Infections, 61 Colindale Ave., London NW9 5EQ.

Legionella: State of the Art 30 Years after Its Recognition
Edited by Nicholas P. Cianciotto et al.
©2006 ASM Press, Washington, D.C.

INVESTIGATION OF
ENDEMIC DISEASE

In contrast to epidemic legionellosis, in institutional situations such as hospital- or hotel-associated legionellosis, cases may occur sporadically and periodically over many months or years (4, 13, 16). Although the ecosystems of these large buildings are relatively stable, they are often very complex, containing multiple *Legionella* species, serogroups, and/or subgroups. In these situations typing methods must still be highly discriminatory but also must offer excellent intralaboratory reproducibility (same laboratory, different time) if they are to be capable of designating locally defined types sufficient to allow comparison of isolates over a number of years. Only a few of the many genotypic methods described have been shown to meet these exacting requirements. Of these, pulsed-field gel electrophoresis is the most widely applied (14, 15), but others have also been used. For example, a restriction-fragment length polymorphism (RFLP) typing method was used as the standard method in the United Kingdom (UK) for almost 20 years (11).

INVESTIGATION OF TRAVEL-
ASSOCIATED LEGIONELLOSIS

Investigation of travel-associated legionellosis requires even more rigorously standardized systems, as not only must the types be clearly defined but they also must be recognizable and reproducible in laboratories in different countries (different laboratories, different times). In addition to good intralaboratory reproducibility, this requires excellent interlaboratory reproducibility. To meet this need the European Working Group for Legionella Infections (EWGLI) embarked on a program of method assessment and standardization in 1996. Although much progress has been made, problems of reproducibility and standardization have limited the utility of most genotypic typing methods so far described (6, 8).

SEQUENCE-BASED TYPING

Recently a DNA sequence-based typing (SBT) method was developed by members of EWGLI (8, 9). This, together with the availability of a dedicated on-line database and analysis tools (http://www.ewgli.org), finally offers a solution to the problems of reproducibility and standardization. Being a PCR-based method, SBT data may be obtained very rapidly from isolates and, in the absence of isolates, directly from clinical samples (see chapter 41). The method has now been used in national and international outbreak investigations, and proficiency testing shows it to be very robust.

OVERVIEW OF SBT DATA
OBTAINED TO DATE

The standardized SBT method targets part of six genes (*flaA*, *pilE*, *asd*, *mip*, *mompS*, *proA*), each of which shows considerable allelic variation, thus providing an extremely discriminatory typing system. Although it has been in use for only a short time, to date, 102 distinct allelic profiles (comprising a string of the allele numbers for each of the six targets e.g., 1,4, 3,1,1,1) have been identified among the 275 *L. pneumophila* isolates in the EWGLI database. Our preliminary analyses of the cumulative data for each of the six alleles reveal a high degree of linkage disequilibrium, suggesting that there is little recombination between, or within, the alleles. This finding is consistent with multilocus enzyme electrophoresis data from earlier studies by Selander and colleagues (17) and indicates that *L. pneumophila* is a highly clonal organism (Fig. 1).

Furthermore, even though there are only a few data available for isolates of non-European origin, it is already clear that some SBTs are globally distributed (Table 1.). Again this observation is consistent with earlier multilocus enzyme electrophoresis data (17) and suggests that *L. pneumophila* genotypes are relatively stable with regard to both time and place of isolation. Furthermore, our data confirm and extend previous studies showing that genotype is essentially independent of phenotype (10). Thus, isolates of a single allelic profile can express different serogroups (sg) (e.g., SBT 3,10,1,28,14,9

FIGURE 1 Minimum spanning tree of 124 unrelated L. pneumophila isolates from the EWGLI culture collection.

isolates can be sg 6, sg 8, or sg 12) and different mAb subgroups (e.g., SBT 1,4,3,1,1,1 isolates can be mAb Philadelphia, Allentown, Benidorm, Oxford, OLDA, or Camperdown).

ARE SOME STRAINS MORE LIKELY TO CAUSE INFECTION THAN OTHERS?

The common international language provided by the SBT methodology opens a way to address some long-standing epidemiological questions such as, Are some strains more likely than others to cause infection? and Are the reported differences in incidence of legionellosis in various countries, in part, a reflection of the different distribution of strains in these countries?

Studies in France identified a strain of *L. pneumophila*, characterized by a distinct pulsed-field gel electrophoresis profile, which was endemic in the Paris area (14). Subsequent work showed this Paris strain was responsible for 12.2% of all cases of legionellosis in France (1). However, the strain was also found to be common in the environment, and recent work from the same group failed to find any significant overall difference between clinical and environmental strains (2). In the UK, RFLP typing studies showed that 40% of all unrelated clinical isolates were either RFLP type 1, type 5, or type 14. While RFLP type 1 was also common among the UK environmental isolates examined, RFLP types 5 and 14 were uncommon or absent. Thus, RFLP types 5 and 14 appear to be significantly overrepresented among clinical isolates relative to their environmental frequency, suggesting they have an increased propensity to cause human infection (11).

Using SBT typing, we now know that the Paris strain and RFLP type 1 are SBT 1,4,3,1,1,1 (or, in a few instances, single locus variants of it) and RFLP type 5 isolates are SBT 3,4,1,1,14,9. Given the global distribution of SBT 1,4,3,1,1,1 (Table 1), it seems probable that it is a commonly occurring clone of worldwide distribution which not infrequently gives rise to human infection. In contrast, at least from these preliminary studies, while SBT 3,4,1,1,14,9 is also globally dis-

TABLE 1 Examples of global distribution of some *L. pneumophila* clones

SBT profile	Detected in
1,4,3,1,1,1	Australia
	Austria
	Canada
	Denmark
	France
	Germany
	Greece
	Italy
	Japan
	Spain
	Sweden
	Switzerland
	UK
	United States
3,4,1,1,14,9	Germany
	Spain
	UK
	United States
4,7,11,3,11,12	France
	Germany
	Italy
	Japan
	Spain
	Sweden
	UK

tributed, it appears to be underrepresented in the environment. Furthermore, of the seven major-outbreak-associated strains for which SBT data are available, five are SBT 3,4,1,1,14,9 (Table 2). Taken together, these data indicate that SBT 3,4,1,1,14,9 strains may have an enhanced ability to cause human disease compared to other *L. pneumophila* strains. Clearly these are very preliminary conclusions, and a much larger body of data, particularly from environmental isolates, is needed to confirm or refute these early findings. However, if substantiated, these observations have important and widespread implications for our understanding of the epidemiology of legionellosis and for the monitoring and control of *L. pneumophila* in the built environment.

TABLE 2 Available SBT data for *L. pneumophila* strains causing major outbreaks

Year	Outbreak	SBT profile
1976	Philadelphia	3,4,1,1,14,9
1988	London	3,4,1,1,14,9
1989	London	3,4,1,1,14,9
2001	Murcia, Spain	3,4,1,1,14,9
2002	Barrow, UK	2,3,6,25,2,1
2003	Lens, France	12,9,26,5,26,17
2003	Hereford, UK	3,4,1,1,14,9

ACKNOWLEDGMENTS

We thank all members of EWGLI who have submitted isolates and/or data to the EWGLI database. In addition, specific thanks to Carmen Pelaz-Antolin for SBT data for the Murcia outbreak strain, and to Junko Amemura-Maekawa, John Savill, Sallene Wong and Kanti Pabbaraju for SBT data relating to isolates in Japan, Australia, and Canada.

REFERENCES

1. Aurell, H., J. Etienne, F. Forey, M. Reyrolle, P. Girardo, P. Farge, B. Decludt, C. Campese, F. Vandenesch, and S. Jarraud. 2003. *Legionella pneumophila* serogroup 1 strain Paris: endemic distribution throughout France. *J. Clin. Microbiol.* **41:**3320–3322.
2. Aurell, H., P. Farge, H. Meugnier, M. Gouy, F. Forey, G. Lina, F. Vandenesch, J. Etienne, and S. Jarraud. 2005. Clinical and environmental isolates of *Legionella pneumophila* serogroup 1 cannot be distinguished by sequence analysis of two surface protein genes and three housekeeping genes. *Appl. Environ. Microbiol.* **71:** 282–289.
3. Bangsborg, J. M., P. Gerner-Smidt, H. Colding, N.-E. Fiehn, B. Bruun, and N. Høiby. 1995. Restriction fragment length polymorphism of rRNA genes for molecular typing of members of the family Legionellaceae. *J. Clin. Microbiol.* **33:**402–406.
4. Bernander, S., K. Jacobson, J. H. Helbig, P. C. Lück, and M. Lundholm. 2003. A hospital-associated outbreak of Legionnaires' disease caused by *Legionella pneumophila* serogroup 1 is characterized by stable genetic fingerprinting but variable monoclonal antibody patterns. *J. Clin. Microbiol.* **41:**2503–2508.
5. Brown, C. M., P. J. Nuorti, R. F. Breiman, A. L. Hathcock, B. S. Fields, H. B. Lipman, G. C. Llewellyn, J. Hofmann, and M. Cetron. 1999. A community outbreak of Legionnaires' disease linked to hospital cooling towers: an epidemiological method to calculate dose of exposure. *Int. J. Epidemiol.* **28:**353–359.
6. Fry, N. K., S. Alexiou-Daniel, J. M. Bangsborg, S. Bernander, M. Castellani-Pastoris, J. Etienne, B. Forsblom, V. Gaia, J. H. Helbig, D. Lindsay, L. P. Christian, P. C. Lück, C. Pelaz, S. A. Uldum, and T. G. Harrison. 1999. A multicenter evaluation of genotypic methods for the epidemiologic typing of *Legionella pneumophila* serogroup 1: results of a pan-European study. *Clin. Microbiol. Infect.* **5:**462–477.
7. Fry, N. K., J. M. Bangsborg, A. Bergmans, S. Bernander, J. Etienne, L. Franzin, V. Gaia, P. Hasenberger, B. Baladrón Jiménez, D. Jonas, D. Lindsay, S. Mentula, A. Papoutsi, M. Struelens, S. A. Uldum, P. Visca, W. Wannet, and T. G. Harrison. 2002. Designation of the European Working Group on Legionella Infection (EWGLI) amplified fragment length polymorphism types of *Legionella pneumophila* serogroup 1 and results of intercentre proficiency testing using a standard protocol. *Eur. J. Clin. Microbiol. Infect. Dis.* **21:**722–728.
8. Gaia, V., N. K. Fry, T. G. Harrison, and R. Peduzzi. 2003. Sequence-based typing of *Legionella pneumophila* serogroup 1 offers the potential for true portability in legionellosis outbreak investigation. *J. Clin. Microbiol.* **41:**2932–2939.
9. Gaia, V., N. K. Fry, B. Afshar, P. C. Lück, H. Meugnier, J. Etienne, R. Peduzzi, and T. G. Harrison. 2005. Consensus sequence-based scheme for epidemiological typing of clinical and environmental isolates of *Legionella pneumophila*. *J. Clin. Microbiol.* **43:**2047–2052.
10. Harrison, T. G., N. A. Saunders, A. Haththotuwa, N. Doshi, and A. G. Taylor. 1990. Typing of *Legionella pneumophila* serogroups 2-14 strains by analysis of restriction fragment length polymorphisms. *Lett. Appl. Microbiol.* **11:** 189–192.
11. Harrison, T. G., N. Doshi, N. K. Fry, and C. A. Joseph. 2002. Subtyping of *Legionella pneumophila*: a review of 19 years of data from England and Wales. Proceedings of the 22nd Annual Meeting of the European Working Group on Legionella Infections. 26–28 May 2002, St Julian's, Malta, p. 28.
12. Helbig, J. H., S. Bernander, M-Pastoris Castellani, J. Etienne, V. Gaia, S. Lauwers, D. Lindsay, P. C. Lück, T. Marques, S. Mentula, M. F. Peeters, C. Pelaz, M. Struelens, S. A. Uldum, G. Wewalka, and T. G. Harrison. 2002. Pan-European study on culture-proven Legionnaires' disease: distribution of *Legionella pneumophila* serogroups and monoclonal subgroups. *Eur. J. Clin. Microbiol. Infect. Dis.* **21:**710–716.
13. Joseph, C. A. 2004. Legionnaires' disease in Europe 2000-2002. *Epidemiol. Infect.* **132:**417–424.

14. **Lawrence, C., M. Reyrolle, S. Dubrou, F. Forey, B. Decludt, C. Goulvestre, P. Matsiota-Bernard, J. Etienne, and C. Nauciel.** 1999. Single clonal origin of a high proportion of *Legionella pneumophila* serogroup 1 isolates from patients and the environment in the area of Paris, France, over a 10-year period. *J. Clin. Microbiol.* **37:**2652–2655.

15. **Lück, P. C., J. H. Helbig, H.-J. Hagedorn, and W. Ehret.** 1995. DNA fingerprinting by pulsed-field gel electrophoresis to investigate a nosocomial pneumonia caused by *Legionella bozemanii* serogroup 1. *Appl. Environ. Microbiol.* **61:** 2759–2761.

16. **Rangel-Frausto, M. S., P. Rhomberg, R. J. Hollis, M. A. Pfaller, R. P. Wenzel, C. M. Helms, and L. A. Herwaldt.** 1999. Persistence of *Legionella pneumophila* in a hospital's water system: a 13-year survey. *Infect. Control. Hosp. Epidemiol.* **20:**793–797.

17. **Selander, R. K., R. M. McKinney, T. S. Whittam, W. F. Bibb, D. J. Brenner, F. S. Nolte, and P. E. Pattison.** 1985. Genetic structure of populations of *Legionella pneumophila. J. Bacteriol.* **163:**1021–1037.

18. **van Ketel, R. J., J. ter Schegget, and H. C. Zanen.** 1984. Molecular epidemiology of *Legionella pneumophila* serogroup 1. *J. Clin. Microbiol.* **20:**362–364.

EFFECT OF *LEGIONELLA* TESTING PATTERNS ON THE APPARENT EPIDEMIOLOGY OF LEGIONNAIRES' DISEASE IN AUSTRALIA

Graham Tallis, Agnes Tan, and Norbert Ryan

26

Notification of a legionellosis diagnosis to the health department is required by law in all eight states and territories of Australia. Rates of notification of legionellosis have risen in Australia (population 20 million), from around 0.6 notifications per 100,000 population in 1991, to 1.5 in 2003. In one state, Victoria (population 5 million, or a quarter of the Australian total), the crude incidence rate rose fivefold from 0.4 notifications per 100,000 in 1991 to 2.0 in 2004, so that the rate changed from one of the lowest in Australia to one of the highest (Fig. 1). Also, disproportionately more outbreaks were reported in Victoria compared to other jurisdictions over this period. In the 5-year period 1998 to 2002 21 outbreaks were reported in Victoria, while only 2 were reported elsewhere in Australia. This raises the question of whether these changing notification rates reflect a true rise in incidence in Victoria or a rise in testing, leading to better case and outbreak detection.

It is widely known that *L. longbeachae* appears to be more common in Australia than anywhere else. In a review of national data, 51% of Australian notifications were for *L. pneumophila* infections, and 42% were *L. longbeachae* (4). However, this high proportion of cases due to *L. longbeachae* is not universally seen across Australia, with marked differences between jurisdictions. Between 1998 and 2002, 91% of Victorian notifications were *L. pneumophila* infections. In contrast, Western Australia has reported up to 93% of positive cases being *L. longbeachae* (3).

These data suggest several questions:

- Why are notification rates of legionellosis changing in Victoria?
- Why are rates different from (and higher than) the rest of Australia?
- Why are rates of *L. longbeachae* infections different between Australian jurisdictions?
- Why do so many outbreaks occur in just one jurisdiction—Victoria?

RISING NOTIFICATIONS IN VICTORIA 1991 TO 2005

Fig. 2 illustrates the steady rise in notifications of legionellosis in Victoria from 1991 to 2004. No outbreak-related cases were identified prior to 1996, but a significant proportion thereafter were linked to outbreaks. This coincided with the introduction of the urinary

Graham Tallis Communicable Diseases Section, Department of Human Services, Melbourne, VIC, 3001, Australia. *Agnes Tan* Microbiological Diagnostic Unit Public Health Laboratory, University of Melbourne, VIC, 3010, Australia. *Norbert Ryan* Victorian Infectious Diseases Reference Laboratory, North Melbourne, VIC, 3051, Australia.

Legionella: State of the Art 30 Years after Its Recognition
Edited by Nicholas P. Cianciotto et al.
©2006 ASM Press, Washington, D.C.

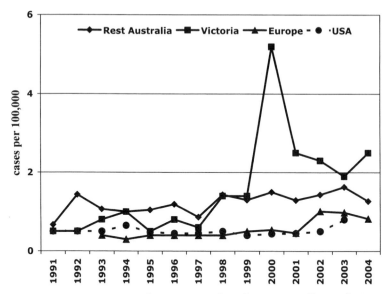

FIGURE 1 International comparison of notification rate of legionellosis.

antigen test in Victoria in 1996. Sporadic and outbreak-related cases were increasingly reported throughout the late 1990s, culminating in the aquarium outbreak in 2000. Thereafter, with tough new regulations covering cooling tower management introduced in the aftermath of the aquarium outbreak, notifications have steadily decreased, with the exception of notifications of *L. longbeachae* infections. *L.* *longbeachae* notifications first increased in the early 1990s and made up nearly half of the total in 1993 and 1994. This increase coincides with the inclusion of *longbeachae* antigens in the in-house immunofluorescent assay developed at the Victorian Infectious Diseases Reference Laboratory (VIDRL) and publicity surrounding the association of the infection with the use of potting mix.

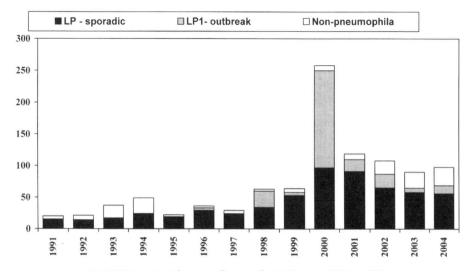

FIGURE 2 Notification of legionellosis, Victoria, 1991 to 2004.

INCREASING USE OF URINARY ANTIGEN TESTING

As elsewhere in the world, diagnosis of *L. pneumophila* serogroup (sg) 1 is increasingly based on a positive urinary antigen test. In Victoria, diagnosis was based solely on urinary antigen testing in 40 to 70% of cases per annum (Surveillance of Notifiable Infectious Diseases in Victoria, 1998 and 2000). While presumably some or many of these cases would have been subsequently detected by another method had the urine test not been available, these data concerning diagnosis underline the importance of the availability and use of this test in case detection and, by extension, outbreak detection.

To explore this further, we obtained test usage data from two sources: VIDRL, from where virtually all Victorian notifications based on urine testing have arisen, and the suppliers of test kits in the Australian market. At VIDRL, average numbers of tests performed grew from 10 per month in 1997, to over 1,000 tests per month in 2000, dropping back to around 500 in 2002. Australian sales of urinary antigen test kits are illustrated by Fig. 3. This shows that Victoria, with only a quarter of the total Australian population, used about three quarters of the test kits. Put another way, by this crude measure, Victorians are about 10 times more likely to have a *Legionella* urinary antigen test than other Australians.

OUTBREAKS

In Victoria, an outbreak is suspected when two or more cases are clustered in time and place, and confirmed if a putative source is identified through environmental testing or epidemiological clustering is convincing (more than two cases, tight physical clustering, etc.) despite no common source being confirmed through environmental testing.

Fig. 2 illustrates the increasing proportion of cases occurring as part of outbreaks. Of the 601 notifications in the 5-year period 1998 to 2002, 91% were *L. pneumophila* sg 1, of which 229 (38% of the total) were outbreak related. Even excluding the Melbourne Aquarium cases, 24% of notifications were outbreak related. In this 5-year period 21 outbreaks were identified, all due to *L. pneumophila* sg 1. Most outbreaks were small (median 4 cases, range 2 to 125), with only two outbreaks having more than 10 cases.

Of the 21 outbreaks confirmed, putative sources were identified in 12. In 10 of these, pulsed field gel electrophoresis showed indistinguishable patterns in clinical and environmental isolates. Of these, eight putative sources were cooling towers, and the other two were spa pools. Only in the aquarium outbreak was a case-control study employed to confirm the observed association. Two other outbreaks were attributed to cooling towers, but no clinical isolate was available for comparison. In the 10 cooling tower–related outbreaks, 7 were in community settings and 3 were in hospital settings.

Median *L. pneumophila* sg 1 counts in these putative sources was 2,400 CFU/ml (when tested according to the Australian/New Zealand Standard method AS/NZS3896), with only three sources having counts less than 1,000 CFU/ml. Of these three outbreaks, two were in hospital settings and the third was in a community setting. The sample from the latter was taken the day after disinfection of the tower and almost certainly did not reflect the tower's count at the time of transmission to cases. This might suggest that finding a count less than 1,000 CFU/ml in an environmental source is probably not indicative of a public health risk, except in hospital settings.

CHANGING RECOGNITION OF RISK PROFILE

The aquarium outbreak presented a unique opportunity to review risk factors for legionellosis. In the case-control study undertaken to study the outbreak, data about traditional risk factors were collected, including data about lung and heart disease, diabetes, immune suppression, alcoholism, and smoking (2). The only risk factor confirmed in the study was smoking, where smokers of more than 70 cigarettes per day were 6.9 times more likely to be ill than nonsmokers ($P < 0.001$). Of the 125 cases, 25% did not require hospitalization, similar to the rate seen in other large outbreaks (1). These data point to a new understanding of the spectrum of Legionnaires'

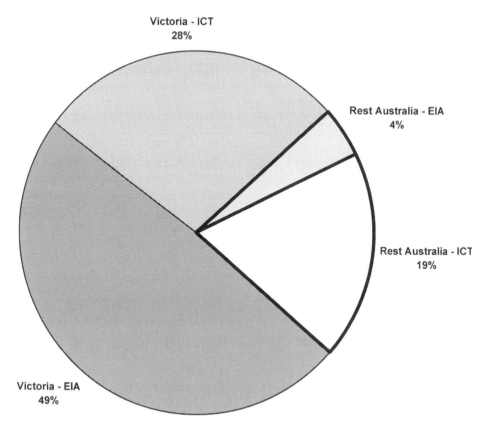

FIGURE 3 Urinary antigen test usage. Victoria versus the rest of Australia, 2002. ICT, immunocromatography; EIA, enzyme immuno-assay.

disease, where the recognized spectrum of illness includes milder cases and where the traditional risk factors are less common than in the classic picture of legionellosis.

In a separate review of risk factors, cases of *L. longbeachae* were more likely to have used potting mix than cases of *L. pneumophila* sg 1 when matched for age, sex, and time of notification (odds ratio, 23.4; 95% confidence interval, 4.2 to 155; $P < 0.001$) (unpublished data). There were no significant differences in having used a spa pool, been exposed to a hospital, or suffering from immune or chronic medical problems. Interestingly, *L. longbeachae* cases were less likely to be smokers (odds ratio 0.1, 95% confidence interval 0.01 to 0.7; $P < 0.05$).

CONCLUSIONS

To return to the original four questions (Why are notification rates of legionellosis changing in Victoria? Why are rates different from [and higher than] the rest of Australia? Why are rates of *L. longbeachae* infections different between Australian jurisdictions? Why are so many outbreaks in just one jurisdiction—Victoria?), it appears that greater (10-fold) use of the urinary antigen test by clinicians is a plausible explanation for all these questions.

Higher testing rates are the most plausible explanation for the rising incidence of legionellosis in Victoria. There is no other factor that could readily explain the fivefold increase in notification rates during the period 1991 to 2004.

These data cannot tell us whether the true incidence of legionellosis was or is higher in Victoria than the rest of Australia to begin with. The regulatory environment is different in each state, and as far as legionellosis incidence is affected by regulation of artificial water sources such as cooling towers and spa pools, it is possible that Victoria had a more risky environment prior to the *Legionella* reforms introduced in 2001. Other ecological factors such as climatic conditions could also cause different disease rates across the Australian continent. In addressing the rising notification rate of legionellosis and outbreaks linked to cooling towers, new regulations were introduced. These regulations mandated registration of cooling towers, mandatory risk management plans for cooling towers, third-party audits of these plans, and an enhanced government technical advisory and outbreak investigation service. The falling number of notifications since 2001 provides some evidence that this strategy is working.

The differences in reported proportions of *Legionella* species across Australia may also be influenced by different testing patterns among clinicians. The higher proportion of Victorian notifications due to *L. pneumophila* sg 1 may reflect higher use of the urinary antigen test rather than a higher true incidence. It is well known that legionellosis is generally under-diagnosed, so it is quite plausible that in Australian jurisdictions with low urinary testing rates, *L. pneumophila* sg 1 cases are missed, leading to a higher proportion of *L. longbeachae* cases, diagnosis of which is mostly based on serology (3). Apart from different uptake of the urinary antigen test, Victorian clinicians may be more attuned, through being sensitized during the multiple outbreaks, to a broader spectrum of illness in legionellosis, and would more readily instigate testing in mild cases of community-acquired pneumonia.

Many would argue that it is not plausible that outbreaks of legionellosis are being routinely missed in other parts of Australia and that the number of outbreaks in Victoria simply indicates a greater risk environment, most clearly exemplified by the aquarium outbreak. This could well be true. However, while it is unlikely that a large explosive outbreak, such as the aquarium, would be missed elsewhere in Australia, the data presented here show that most outbreaks are quite small, and it is plausible that

small clusters consisting of fewer than four cases could be missed. Recognition of more outbreaks in Victoria would be a natural consequence of higher case detection rates. One would have greater confidence that small clusters or outbreaks were not being missed if testing rates were higher in the rest of Australia.

In summary, it would appear that an iterative process has taken place in Victoria, with more testing leading to more case detection, which has led to more outbreaks being detected, raising clinician awareness, leading to more testing, and so on. One implication of the Victorian experience reported here is that any analysis of legionellosis surveillance data, as for many infectious diseases, should be interpreted in the light of testing pattern data. More importantly, there can be no public health confidence where health authorities report low rates of legionellosis notifications unless this is occurring in a context of high testing rates. An international benchmark would be a useful tool in determining the adequacy of *Legionella* surveillance systems.

REFERENCES

1. **Garcia-Felgueiras, A., C. Navarro, D. Fenoll, J. Garcia, P. Gonzalez-Diego, T. Jimenez-Bunuales, M. Rodriguez, R. Lopez, F. Pacheco, J. Ruiz, M. Segovia, B. Balandron, and C. Pelaz.** 2003. Legionnaires' disease outbreak in Murcia, Spain. *Emerg. Infect. Dis.* **9:**915–921.

2. **Greig, J., J. A. Carnie, G. F. Tallis, N. J. Ryan, A. G. Tan, I. R. Gordon, B. Zwolak, J. A. Leydon, C. S. Guest, and W. G. Hart.** 2004. An outbreak of Legionnaires' disease at the Melbourne Aquarium, April 2000: investigation and case-control studies. *Med. J. Aust.* **180:**566–572.

3. **Inglis, T. J. J., F. Haverkort, M. Sears, I. Sampson, and G. Harnett.** 2002. 2002, Epidemiology of legionella infection in Western Australia, p. 353–355. *In* R. Marre, Y. Abu Kwaik, C. Batlett, N. P. Cianciotto, B. S. Fields, M. Frosch, J. Hacker, and P. C. Luck (ed.). *Legionella,* ASM Press, Washington D.C.

4. **Li, J. S., E. D. O'Brien, and C. Guest.** 2002. A review of national legionellosis surveillance in Australia, 1991 to 2000. *Commun. Dis. Intell.* **26:**461–468.

25 YEARS OF SURVEILLANCE FOR LEGIONNAIRES' DISEASE IN ENGLAND AND WALES: WHY NO IMPROVEMENT?

Katherine D. Ricketts, Carol A. Joseph, Timothy G. Harrison, J. V. Lee, and F. C. Naik

27

The Health Protection Agency Centre for Infections has been running a national surveillance scheme for Legionnaires' disease in England and Wales since 1980. Between 1980 and 2004, over 5,000 confirmed and presumptive cases were reported. Over these 25 years advances have been made in diagnostic methods and legislation, aimed at preventing future cases and outbreaks of the disease. With improved control and prevention of cases, earlier diagnosis, and improved surveillance systems allowing timely investigation of cases and clusters over this period, it would not be unreasonable to have expected a decline in the number of cases reported to the center each year. This manuscript discusses some of the reasons why no decrease has yet occurred.

CATEGORY OF CASE

Nosocomial case numbers have been relatively constant over time, with the exception of a peak

Katherine D. Ricketts, Carol A. Joseph, and F. C. Naik Health Protection Agency, Centre for Infections, Respiratory Diseases Department, 61 Colindale Avenue, London NW9 5EQ. *Timothy G. Harrison* Health Protection Agency, Centre for Infections, Respiratory and Systemic Infection Laboratory, Atypical Pneumonia Unit, 61 Colindale Ave., London NW9 5EQ. *J. V. Lee* Health Protection Agency, Centre for Infections, Food Safety Microbiology Laboratory, Water and Environmental Microbiology Reference Unit, 61 Colindale Ave., London NW9 5EQ.

in 1985, when 68 cases were associated with the Staffordshire Hospital outbreak (Fig. 1).

Community cases numbered between 50 and 100 per year between 1980 and 2002, with the exception of 1988 to 1989, when there were 3 substantial community outbreaks in England. England's largest community outbreak to date, in Barrow-in-Furness in 2002, initiated a change and resulted in community case numbers that exceeded 100 in 2003 and 2004, probably due to increased case detection and reporting because of the increased awareness of Legionnaires' disease.

Travel case numbers have been increasing gradually, especially in those who traveled abroad, due in part to a decrease in the costs of travel and to an increased recognition of Legionnaires' disease as a potential diagnosis of atypical pneumonia in a returning traveler. The European Surveillance Scheme for Travel Associated Legionnaires' Disease (EWGLINET) has helped strengthen European national surveillance systems and improve case detection and reporting in travel-associated cases.

OUTBREAK-ASSOCIATED CASES

Over the early part of the past 25 years, the number of outbreaks per year involving 2 or more English or Welsh cases (including outbreaks abroad) remained at 20 or below (Fig. 2).

Legionella: State of the Art 30 Years after Its Recognition
Edited by Nicholas P. Cianciotto et al.
©2006 ASM Press, Washington, D.C.

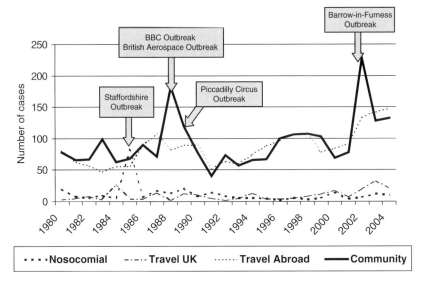

FIGURE 1 Legionnaires' disease cases in England and Wales over 25 years, by category of case.

Since the outbreak in Barrow in 2002, the number of outbreaks per year has not dropped below 29.

The annual proportion of outbreak-associated cases has averaged 25.4% of all reported cases. An improving response to outbreaks should lead to fewer outbreak-associated cases. Although not demonstrating a consistent pattern, the English and Welsh data do suggest a gradual decrease in outbreak-associated cases.

With an increase in the number of outbreaks detected and a decrease in the number of outbreak-associated cases, there would seem to have been both an improvement in the detection of outbreaks and an improved response to outbreaks over the past 25 years.

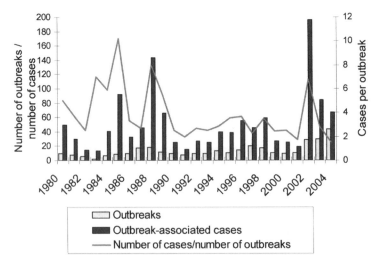

FIGURE 2 English and Welsh outbreak-associated cases and outbreaks involving two or more English or Welsh cases.

GUIDANCE AND LEGISLATION
The guidance and legislation introduced in England and Wales over the past 25 years aims to ensure that all relevant systems are properly installed and maintained and that investigation of these systems (e.g., wet cooling towers) in an outbreak situation is as efficient as possible. These measures are aimed at reducing the absolute number of cases of Legionnaires' disease.

UNDERDIAGNOSIS AND UNDERREPORTING
England and Wales' annual case reports severely underrepresent the true annual incidence of Legionnaires' disease. Using community-acquired pneumonia studies (1, 2), it has been estimated that only 6 to 10% of the expected numbers of cases are being diagnosed.

In contrast, Hospital Episode Statistics show that, of the hospital admissions for pneumonia in patients diagnosed with Legionnaires' disease in 2002 and 2003, 90% were reported to the Health Protection Agency. It therefore seems likely that the England and Wales' national surveillance scheme suffers from a greater degree of underdiagnosis than of underreporting.

URINARY ANTIGEN TEST
The urinary antigen test was introduced in England and Wales in the 1990s and is now the primary diagnostic method for 80% of English and Welsh cases of Legionnaires' disease.

The use of the urinary antigen test has been mirrored by an increase in the number of cases detected in many European countries. The test is quick and easy to perform and might therefore be enabling countries to detect milder cases of the disease that would otherwise have gone undiagnosed (3). However, in England and Wales the number of cases has remained relatively constant despite the increasing use of the test. It therefore appears to be replacing serology as the primary method of diagnosis for cases that would have been detected anyway.

OUTCOME
There have been 635 deaths in cases of Legionnaires' disease reported to the Health Protection Agency—a case fatality rate of 13.2%. The number of deaths per year has increased slightly, reflecting the increase in case numbers, while the case fatality rate has remained relatively constant. This suggests that the increase in cases detected by the centre does not solely consist of less severely ill patients.

CONCLUSION
The simple trend of an increasing number of cases of Legionnaires' disease in England and Wales over the past 25 years masks a more complicated picture. Reporting systems that suffer from underdiagnosis and underreporting are likely to register an increase in case reports (as ascertainment improves) before any actual decrease from improved control and prevention in case numbers is detected.

REFERENCES
1. **Anonymous.** 1987. Community-acquired pneumonia in adults in British hospitals in 1982-1983: a survey of aetiology, mortality, prognostic factors and outcome. The British Thoracic Society and the Public Health Laboratory Service. *Q. J. Med.* **62:**195–220.
2. **Lim, W. S., J. T. Macfarlane, T. C. Boswell, T. G. Harrison, D. Rose, M. Leinonen, and P. Saikku.** 2001. Study of community acquired pneumonia aetiology (SCAPA) in adults admitted to hospital: implications for management guidelines. *Thorax* **56:**296–301.
3. **Joseph, C. A.** 2004. Legionnaires' disease in Europe 2000-2002. *Epidemiol. Infect.* **132:**417–424.

EPIDEMIOLOGICAL SURVEILLANCE OF SEROPOSITIVE LEGIONELLOSIS CASES IN KOREA DURING 1999–2002

Hae Kyung Lee, Soo Jin Baek, Yong In Ju, Jae June Bae, Man Suck Park, and Mi Yeoun Park

28

Since *Legionella pneumophila* was first recognized as the pathogen for Legionnaires' disease (LD) in 1977 (3), epidemiological studies for LD have been reported all over the world (1, 5). Community-acquired and hospital-acquired legionellosis have been occasionally reported in Korea since the recognition of an outbreak of Pontiac fever in 1984 (1, 4, 6). However, there was a lack of epidemiological information on the diagnosed patients that could be used as serological evidence.

We analyzed the epidemiological and clinical data and antibiotic treatment records for the patients diagnosed serologically as having legionellosis in Korea during 1999 to 2002. Fifty-seven patients confirmed with other diseases among 165 seropositive cases with atypical pneumonia were ruled out after a review of the charts, and 108 cases diagnosed as

seropositive for Legionnaires' disease were selected. Although the patients were diagnosed through-out the year, many of seropositive patients were distributed between March and April during the study period (1999 to 2002). Most of patients were in the age group of 50 to 80, and 61.1% of the patients were male (Fig. 1). The common clinical symptoms included fever, cough, sputum, and dyspnea. Underlying disease and risk factors were smoking (33.0%), diabetes mellitus (21.3%), hypertension (13.9%), and cancer therapy (8.4%) (Table 1). Macrolides (49.1%) and quinolones (43.5%) were commonly prescribed. Twenty-eight patients (25.9%) showed reactivity for *Legionella gormanii*, and 14 patients (13.0%) were associated with *Legionella pneumophila* serogroup 1 according to serological tests (Table 2).

Most of the legionellosis cases diagnosed with serological methods in Korea show epidemiological aspects similar to those reported in other countries. In sero-diagnosis of this study, it is remarkable that *L. gormanii* was the most prevalent species in Korea. Therefore, *L. gormanii* infection as well as other *Legionella* species other than *L. pneumophila* should be taken into consideration when serological diagnosis is performed in Korea.

Hae Kyung Lee, Soo Jin Baek, Yong In Ju, Jae June Bae Center for Infectious Disease Research, National Institute of Health, Korea Centers for Disease Control and Prevention, Seoul, 122-701. *Man Suck Park* Center for Disease Investigation and Surveillance, Korea Centers for Disease Control and Prevention, Seoul, 122-701. *Mi Yeoun Park* Center for Immunology and Pathology, National Institute of Health, Korea Centers for Disease Control and Prevention, Seoul, 122-701.

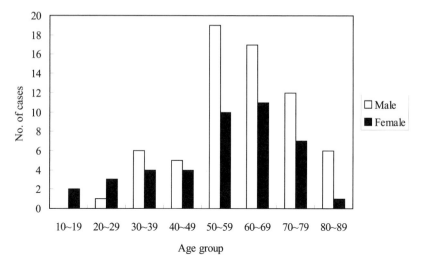

FIGURE 1   Age distribution of 108 patients by sex, 1999 to 2002

TABLE 1   Analysis of risk factors and underlying diseases of 108 patients, 1999 to 2002

Risk factor[a]	No. of cases (%)
Organ transplant	4 (3.7)
Immunosuppressed	11 (10.3)
Alcoholism	19 (18.1)
Smoking	35 (33.0)
Diabetes mellitus	23 (21.3)
Hypertension	15 (13.9)
Cancer therapy	9 (8.4)

[a]Some Patients had more than one risk factor.

TABLE 2   Etiological agents of 108 patients, 1999 to 2002

Etiological agent[a]	No. of cases (%)
L. pneumophila sg 1[b]	14 (13)
L. pneumophila sg 2	0 (0)
L. pneumophila sg 3	6 (5.6)
L. pneumophila sg 4	3 (2.8)
L. pneumophila sg 5	7 (6.5)
L. pneumophila sg 6	7 (6.5)
L. bozemanii	17 (15.7)
L. dumoffii	6 (5.6)
L. micdadei	15 (13.9)
L. gormanii	28 (25.9)
L. feeleii	11 (10.2)
L. longbeachae	18 (16.7)

[a]18 patients showed reactivity with more than one etiological agent.
[b]sg, serogroup.

REFERENCES

1. **Choi, T. Y.** 1998. *Legionella* infection. *Korean J. Clin. Microbiol.* **1:**24–32.
2. **Eldestein, P. H.** 1993. Legionnaries' disease. *Clin. Infect. Dis.* **16:**741–749.
3. **Fraser, D. W., T. R. Tsai, W. Orenstein, W. Parkin, H. J. Beecham, R. G. Sharrar, J. Harris, F. Mallison, S. M. Martin, J. E. Mcdade, C. C. Shepard, P. S Brachman, and the Field Investigation Team.** 1977. Legionnaires' disease: description of an epidemic of pneumonia. *N. Engl. J. Med.* **297:**1189–1197.
4. **Kim, J. S., S. W. Lee, H. S. Shim, D. K. Oh, M. K. Cho, H. B. Oh, J. H. Woo, and Y. S. Chong.** 1985. Outbreak of legionellosis in ICU of K Hospital, Korea. *Korean J. Epidemiol.* **7:**44–58.
5. **Luck, P. C., J. H. Helbig, and M. Schuppler.** 2002. Epidemiology and laboratory diagnosis of *Legionella* infections. *J. Lab. Med.* **26:**127–182.
6. **Seog, W., Y. J. Jung, H. W. Park, H. K. Lee, M. S. Park, M. Y. Park, K S. Park, M. D. Oh, C. Ahn, E. C. Kim, and K. W. Choe.** 1999. A case of community-acquired Legionnaires' disease in a renal transplant recipient. *Infection* **31:**353–357.

PREVALENCE OF LEGIONNAIRES' DISEASE AND INVESTIGATION ON RISK FACTORS: RESULTS OF AN ITALIAN MULTICENTRIC STUDY

Paola Borella, Stefania Boccia, Erica Leoni, Franca Zanetti, Isabella Marchesi, Annalisa Bargellini, Maria Teresa Montagna, Daniela Tatò, Sara Montegrosso, Francesca Pennino, Vincenzo Romano-Spica, Giovanna Stancanelli, and Stefania Scaltriti

29

In Europe, as in Italy, notifications of Legionnaires' disease is increasing year by year (7). It is difficult to say whether this is due to an increase of incidence or to an improvement in the routine diagnostic testing and major attention given by clinicians to recognition and notification of cases (1). Furthermore, the real occurrence of disease is unknown and there is a need for a more accurate assessment of the infection frequency. Due to frequent *Legionella* contamination in domestic and public water systems (2, 3), between 2 and 15% of all cases of community-acquired pneumonia are caused by *Legionella*, while the frequency of nosocomial Legionnaires' disease may be even higher (10 to 40%) (8). Compared to nosocomial cases, community-acquired cases are more numerous but less severe and involve younger, apparently healthy people, and it is generally arduous to find the source of infection. As risk factors, a recent trip or stay in a hotel are often reported, as well as smoking or alcohol consumption, but without a control group it is difficult to establish whether these and other risk factors are involved in the appearance of disease (4). Furthermore, individual susceptibility and resistance to infectious disease can be included among risk factors, and the analysis of genetic polymorphisms may represent a new and promising perspective. Toll-like receptors (TLRs) and chemokine receptors (CCRs) seem to be involved in the immuneresponse to infectious diseases. Indeed, TLR4 has been recently associated with resistance to Legionnaires' disease (5). Both a 32–base pair deletion in the CCR5 gene and a G to A substitution in position 190 of the CCR2 gene seem to be important in protecting the host from infectious agents such as HIV and recurrent respiratory infections (6).

We here present the preliminary results of a large multicentric study carried out in Italy to detect Legionnaires' disease among patients

Paola Borella Department of Public Health Sciences, University of Modena and Reggio Emilia, Via Campi 287, 41100 Modena, Italy. *Stefania Boccia* Institute of Hygiene, Catholic University, Teaching Hospital "A. Gemelli," Largo Francesco Vito 1, 00168 Rome, Italy. *Erica Leoni and Franca Zanetti* Department of Public Medicine and Health, University of Bologna, Via S. Giacomo 12, 40126 Bologna, Italy. *Isabella Marchesi and Annalisa Bargellini* Department of Public Health Sciences, University of Modena and Reggio Emilia, Via Campi 287, 41100 Modena, Italy. *Maria Teresa Montagna and Daniela Tatò* Department of Internal Medicine and Public Health, University of Bari, P.zza Giulio Cesare 11, 70124 Bari, Italy. *Sara Montegrosso and Francesca Pennino* Medicine Institute, University "Federico II" of Naples, Via Pansini 5, 80131 Napoli, Italy. *Vincenzo Romano-Spica* Hygiene and Public Health Unit, Department of Health Sciences, University of Movement Sciences (IUSM), P.zza L. de Bosis 6, 00194 Rome, Italy. *Giovanna Stancanelli and Stefania Scaltriti* S. Raffaele Hospital, Via Stamina d'Ancona 20, 20127 Milan, Italy.

with pneumonia. The general objective was to better evaluate the prevalence of disease within various Italian regions in order to confirm the existing data on the disease frequency and to eventually verify differences related to the geographical area. A case-control study was also designed to investigate behavioral, personal, and environmental risk factors related to the disease. Among studied parameters, the presence of CCR5D32 and CCR264I mutations was investigated to establish their possible role in susceptibility to Legionnaires' disease.

From 2001 to 2004, an active surveillance has been carried out in six hospitals (between 800 and 1,000 beds) located in the north, center and south of Italy. Both community-acquired and suspected nosocomial pneumonia were considered. The inclusion criteria were that patients be adults (>18 years) and affected by pneumonia diagnosed by clinical and/or radiological tests without etiologic identification. Urine, sputum, and serum were collected from each patient at the time of admission. A test for the detection of *Legionella* urinary antigen (Biotest EIA kit) was systematically performed, followed (when possible) by cultural isolation of germ on MWY agar plates (Oxoid) from sputum and by antibody levels (indirect fluorescent-antibody assay method, cutoff titer for positivity 1:256) as confirming tests. A brief questionnaire for collecting personal data and disease characteristics was filled out for each examined patient. Physicians and nurses were informed about the project by initial and periodic meetings to ensure their compliance

with the surveillance protocol. Patients affected by Legionnaires' disease were asked to enter a case-control study, previous informed consensus. As controls, both healthy subjects and patients affected by other kinds of pneumonia were recruited. Case subjects and controls were matched for sex and age (5-year range). A detailed questionnaire on lifestyle, exposure conditions, stress events, and disease data was administered to cases and their controls. Furthermore, a panel of laboratory analyses was requested in both groups: number and percentage of lymphocytes and their subpopulations, chemical-clinical tests (hemogram, proteins electrophoresis, glycaemia, cholesterol, etc.), serum trace elements, and genetic tests. As regards genetic analyses, whole blood was collected and genomic DNA was isolated from peripheral blood cells using standard methodology based on sodium dodecyl sulfate (SDS)/proteinase K lysis and phenol/chloroform extraction. CCR5 and CCR2 genotypes were determined by PCR.

Up to now, 189 cases of Legionnaires' disease from 6,032 examined pneumonia cases were detected; the prevalence was 3.13% but varied between 2.0 and 6.8% depending on the area (Table 1). The case-control design was made up of 130 community-cases and 174 controls (88 with pneumonia and 86 healthy). Concerning diagnostic investigations, the urinary antigen test was positive in 89.6% of cases and in 0 of 83 pneumonia controls. *Legionella* (serogroup 1) was isolated from sputum culture only in 20% of cases. The serologic tests

TABLE 1 Results of active surveillance: number of pneumonia cases surveyed and number/percentage of detected Legionnaires' disease cases according to geographic area.

	2001 to 2002			2003 to 2004		
	Pneumonia	Legionnaires' disease	%	Pneumonia	Legionnaires' disease	%
North	1,468	43	2.9	2,659	53	2.0
Central	309	17	5.5	337	23	6.8
South	679	25	3.7	580	28	4.8
Total	**2,456**	**85**	**3.5**	**3,576**	**104**	**2.9**

were positive in 38.6% of first serum samples and 48% of second serum collected after 10 to 15 days; among negative samples, 8 (22.8%) became positive, while 9 that were already positive showed an increase of titer. Patients affected by Legionnaires' disease compared with both healthy and pneumonia controls showed a higher frequency of chronic diseases (62.0 versus 14.3 and 55.2%, respectively; $P < 0.001$), immune pathologies (22.7 versus 0 and 9.2%, respectively; $P < 0.01$), and immune suppressive therapies (13.9 versus 2.3 and 5.9%, respectively; $P < 0.05$). Modifications in natural killer cell number and trace elements levels (Fe, Zn, and Cu) in patients compared to controls were also detected. Patients with legionellosis were characterized by significantly higher levels of serum copper compared to healthy controls (144.0 ± 25.0 versus 105.2 ± 19.0; $P < 0.001$), most probably as a consequence of an inflammatory process accompanying the acute phase of disease. The CCR5D32 allele did not show major relevant differences after statistical analysis, but a slight association has been observed for the CCR264I allele.

In conclusion, to always perform laboratory tests for Legionnaires' disease appears an effective method to increase the detection of cases. Our active clinical surveillance scheme was able to identify 189 cases, contributing to increased reports in Italy, where in total about 600 cases for year are reported. In our study, the mean prevalence of Legionnaires' disease on recovered pneumonia was approximately 3%, a lower percentage compared to previous investigations (9). Urinary antigen assay was confirmed as the most sensible, specific, and quick test, thus justifying its extensive use for diagnostic purposes (1). On the basis of a mean prevalence of Legionnaires' disease on recovered pneumonia between 2 and 7%, we estimate a cost for the urinary test of between 100 and 300 Euros for each diagnosed patient. This also means that in a large hospital, with about 500 pneumonia cases per year, the additional annual cost for detecting all Legionnaires' disease cases ranges between 500 and 1,500 Euros. In our opinion, this is without doubt a very sustainable cost, taking in consideration that a prompt diagnosis and therapy have the advantage of reducing risk of death, especially for hospital-acquired cases. The preliminary results of the case-control study show that cases have peculiar, although not sufficiently studied, alterations in both immune and biochemical parameters, which deserve further investigations. Furthermore, a more extensive study may support a possible involvement of CCR2 in resistance and susceptibility to Legionnaires' disease.

REFERENCES

1. **Benin, A. L., R. F. Benson, and R. E. Besser.** 2002. Trends in Legionnaires' disease, 1980-1998: declining mortality and new patterns of diagnosis. *Clin. Infect. Dis.* **35**:1039–1046.
2. **Borella, P., M. T. Montagna, S. Stampi, G. Stancanelli, V. Romano-Spica, M. Triassi, I. Marchesi, A. Bargellini, D. Tatò, C. Napoli, F. Zanetti, E. Leoni, M. Moro, S. Scaltriti, G. Ribera d'Alcalà, R. Santarpia, and S. Boccia.** 2005. *Legionella* contamination in hot water of Italian hotels. *Appl. Environ. Microbiol.* **71**:5805–5813.
3. **Borella, P., M. T. Montagna, V. Romano-Spica, S. Stampi, G. Stancanelli, M. Triassi, R. Neglia, I. Marchesi, G. Fantuzzi, D. Tatò, C. Napoli, G. Quaranta, P. Laurenti, E. Leoni, G. De Luca, C. Ossi, M. Moro, and G. Ribera D'Alcalà.** 2004. Legionella infection risk from domestic hot water. *Emerg. Infect. Dis.* **10**:457–464.
4. **Borella, P., M. T. Montagna, V. Romano-Spica, S. Stampi, G. Stancanelli, M. Triassi, A. Bargellini, P. Giacobazzi, F. Vercilli, S. Scaltriti, I. Marchesi, C. Napoli, D. Tatò, G. Spilotros, N. Paglionico, G. Quaranta, M. Branca, M. Tumbarello, P. Laurenti, U. Moscato, E. Capoluongo, G. De Luca, P. P. Legnani, E. Leoni, R. Sacchetti, F. Zanetti, M. Moro, C. Ossi, L. Lopalco, R. Santarpia, V. Conturso, G. Ribera d'Alcalà, and S. Montegrosso.** 2003. Environmental diffusion of *Legionella* spp and legionellosis frequency among patients with pneumonia: preliminary results of a multicentric Italian survey. *Ann. Ig.* **15**:493–503.
5. **Hawn, T. R., A. Verbon, M. Janer, L. P. Zhao, B. Beutler, and A. Aderem.** 2005. Toll-like receptor 4 polymorphisms are associated with resistance to Legionnaires' disease. *Proc. Natl. Acad. Sci. USA.* **102**:2487–2489.
6. **Ianni, A., S. Majore, D. Arzani, I. Carboni, G. M. Corbo, and V. Romano-Spica.** 2001.

CCR2 and CCR5 gene polymorphisms in children with recurrent respiratory infections. *Respir. Med.* **95:**430–432.

7. **Joseph, C. A. and European Working Group for Legionella Infections.** 2004. Legionnaires' disease in Europe 2000-2002. *Epidemiol. Infect.* **132:**417–424.

8. **Seenivasan, M. H., V. L. Yu, and R. R. Muder.** 2005. Legionnaires' disease in long-term care facilities: overview and proposed solutions. *J. Am. Geriatric. Soc.* **53:**875–880.

9. **Vergis, E. N., A. Indorf, T. M. File, J. Phillips, J. Bates, J. Tam, G. A. Sarosi, T. Grayston, J. Summersgill, and V. L. Yu.** 2000. Azytromycin vs cefuroxime plus erythromycin for empirical treatment of community-acquired pneumonia in hospitalized patients. *Arch. Intern. Med.* **160:**1294–1300.

SEROPREVALENCE OF ANTIBODIES TO *LEGIONELLA PNEUMOPHILA* IN NORTHERN ITALY

R. Cosentina, S. Malandrin, P. Valentini, E. Sfreddo,
L. Pirrotta, O. Mercuri, and O. Di Marino

30

All data reported in the scientific literature (1–6) show values of seroprevalence for *Legionella pneumophila* ranging from 1 to 30% or more, according to the different demographic characteristics and job types of the analyzed people (hospital staff, hotel employees, healthy people, and so on). Several surveys show that *Legionella pneumophila* is also present in the animal species living in the natural environment.

This observational epidemiological survey aims, primarily, at evaluating the antibody coverage with particular reference to previous infection (IgG) by looking at predefined populations based on predictable risk versus *Legionella* exposure. Second, some demographic characteristics are evaluated, in particular those that reflect behaviors or situations that seem to facilitate *Legionella* infection.

AIMS

- Evaluating, within homogeneous groups in terms of demographic characteristics or life style, the immune status against Legionella

- Identifying, within homogeneous groups in terms of demographic characteristics or life style, associated or protecting factors with respect to *Legionella* infections
- Hypothesizing individual behavior standards or preventive actions relating to structures potentially subjected to contamination
- Developing a data bank of information regarding antibody concentrations versus time for persons who have been clearly affected by *Legionella* pneumonia.

Participants were recruited from people living in Milan, Italy.

People were selected and divided into the following subpopulations (Fig. 1):

- Group A: University students younger than 35 years old
- Group B: Frequent users (at least 20 times a year over the past 1 year or more) of fitness facilities
- Group C: Hospital staff with 3 years service or more
- Group D: In-patients, 70 years old or more, with at least 5 hospitalizations over the past 3 years
- Group E: Other persons who have been clearly affected in the past by *Legionella* pneumonia

---

R. Cosentina, S. Malandrin, P. Valentini, E. Sfreddo, L. Pirrotta, O. Mercuri, and O. Di Marino Medical Office, Azienda Ospedaliera Fatebenefratelli e Oftalmico, Corso di Porta Nuova, 23–20121 Milan, Italy.

Legionella: State of the Art 30 Years after Its Recognition
Edited by Nicholas P. Cianciotto et al.
©2006 ASM Press, Washington, D.C.
114

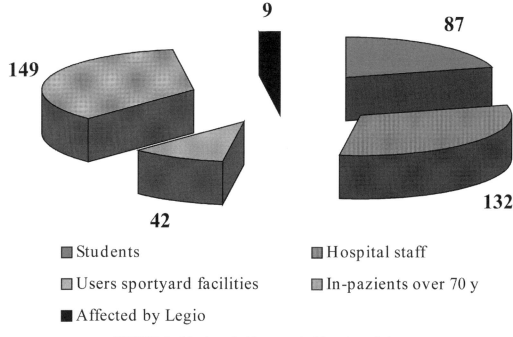

FIGURE 1 Number of subjects recruited, by subpopulation.

RESEARCH METHOD

Information on the project contents was provided by giving subjects a copy of the information text. Participant consent to the project was obtained. A data sheet collecting demographic and lifestyle information was completed. Blood was withdrawn (two samples) to determine the antibody concentration versus *Legionella*. Double-sample tests were used to enable intial separation of positives from negatives; then the positive ones were analyzed to carry out IgG-IgM labeling. The screening test was the Premier Test Elisa Legionella of Meridian Bioscience Inc. Analyses were carried out at Analisi Chimico-cliniche e Microbiologia of Ospedale Fatebenefratelli of Milan.

SURVEY OBJECTIVES

The relative frequencies associated with the confidence limits to 95% were calculated for the following parameters:

- Immunity status versus *Legionella*
- Associated factors to infection
- Protective factors to infection

DATA ANALYSIS

Based on the data collected, in order to devise single behavior models or preventive actions for the facilities potentially affected by contamination, the correlation between the presence or absence of antibodies and the different hypothetic risk factors (translated into a binary conversion code) were evaluated by means of univariate statistical tests. Furthermore, using all variables deemed significant from the univariate analysis, by means of a backward stepwise model, the single impact of all factors was adjusted, taking into account the presence of the others.

PRELIMINARY DATA

Preliminary conclusions show that seroprevalence rate seems to be very low (Fig. 2). Up to September 30th, 2005, 419 subjects had been recruited for blood withdrawal (164 male and 255 female). 17 samples had positive results, corresponding to a serum prevalence of 4.2%. Among the subpopulations studied, frequent users of fitness facilities show the highest share:

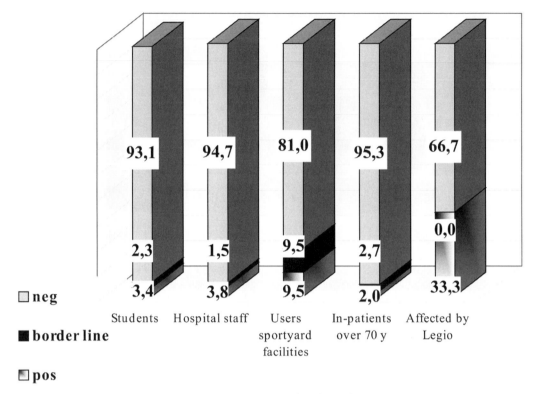

FIGURE 2   Test percentage by subpopulation.

9.5%. Doubtful results came from 2.8% of the samples. The survey will continue with a significant increase in recruitment.

Legionella species are widespread in nature. Disease occurence may be sporadic or epidemic and may occur in the community or in hospitals, and people with compromised immunity are at increased risk. The critical component of the immune system in resistance to legionellosis has not been pinpointed yet, and the role of humoral immunity is not clear; antibodies in all immunoglobulin classes are made after human or experimental infection. Legionellosis can be suspected clinically, but diagnosis can be confirmed only by laboratory testing.

These matters are relevant in confirming the value of studies designed to investigate risks related to lifestyle.

CONCLUSION
This survey, presented at the 20th Annual Meeting of the European Working Group for Le-

gionella Infections, in its present stage, shows total seroprevalence values slightly higher than those generally found in the scientific literature.

Higher serum positivity among frequent users of sport facilities seems to be significant, as does the lower positivity of older in-patients with several hospitalizations. Positive sample confirmation and the evaluation of doubtful samples by means of indirect fluorescent-antibody assay method will help provide a more clear and comprehensive picture.

ACKNOWLEDGMENTS
All costs associated with the laboratory activities for blood tests (kits for analyses of antibody concentration against *Legionella*) were supported by a private company as a pure liberal grant. The costs associated with administrative project management and the insurance coverage for medical activities were funded by Azienda Ospedaliera Fatebenefratelli e Oftalmico di Milano. The project was validated by A.O. Fatebenefratelli e Oftalmico's Ethics Committee on November 24, 2004.

REFERENCES

1. **Borobio, V., C. Martinez, and E. J. Perea.** 1987. Prevalence of anti-Legionella pneumophila antibodies in various groups with different risk factors in Seville (Spain). *Eur. J. Epidemiol.* **3:**436–438.

2. **Franzin, L. and F. Scaramazza.** 1995. Prevalence of Legionella pneumophila serogroup 1 antibodies in blood donors. *Eur. J. Epidemiol.* **11:**475–478.

3. **Nicol, K. L., C. M. Parenti, and J. E. Johnson.** 1991. High prevalence of positive antibodies to Legionella pneumophila among outpatients. *Chest.* **100:**663–666.

4. **Ongut, G., A. Yavuz, D. Ogunc, M. Tuncer, F. Ozturk, D. Mutlu, L. Donmez, D. Colak, F. Ersoy, Yakupoglu, and M. Gultekin.** 2004. Seroprevalence of antibodies to legionella pneumophila in haemodialysis patients. *Transplant. Proc.* **36:**44–46.

5. **Reinthaler, F., F. Mascher, and D. Stunzner.** 1987. Legionella pneumophila: seroepidemiologic studies of dentists and dental personnel in Austria. *Zentralbl. Bakteriol. Mikrobiol. Hyg.* **185:**164–170.

6. **Romano, F., G. Ribera, M. M. D'Errico, G. M. Grasso, and D. Montanaro.** 1989. Levels of anti-Legionella antibodies in a healthy Neapolitan population. *Ann. Ig.* **1:**629–636.

A SEROEPIDEMIOLOGICAL STUDY OF *LEGIONELLA PNEUMOPHILA* ANTIBODIES IN SPANISH PATIENTS: A 13-YEAR RETROSPECTIVE SURVEY

Sebastian Crespi, Albert Torrents, and Miguel A. Castellanos

31

In Spain, where Legionnaires' disease has been a notifiable disease since 1997, the incidence of Legionnaires' disease has increased considerably in the past few years, but a large interregional variability still persists, similar to that existing in other countries. This fact is generally attributed exclusively to changes in the diagnostic and/or declaration rates, but there could be other contributing factors, including changes in exposure patterns, differences in host susceptibility, or the unequal environmental distribution of virulent strains (1). Elucidating the possible existence of other factors, in addition to purely artifactual ones, would be of great interest, not only in improving our understanding of the epidemiology of the disease but also in designing preventive strategies which are more specific and appropriate for each particular situation.

UNDERLYING HYPOTHESES

Seroepidemiological studies of *Legionella pneumophila* have usually been cross-sectional and have, almost always, been utilized to ascertain rates of exposure to the bacteria in apparently healthy populations or in different risk groups (3, 4). In addition, they have been used occasionally in retrospective studies with the objective of evaluating the possible prevalence of Legionnaires' disease in patients with pneumonia (5). We have made this study—a retrospective study of patients with pneumonia—with another objective: to analyze whether or not the different regional incidences observed for Legionnaires' disease in Spain truly reflect different epidemiological patterns.

The underlying hypotheses of our study were as follows:

1. If the regional differences in incidence rates are merely artifactual (that is, are due only to the differences in the epidemiological surveillance system), the seroprevalences obtained in samples of patients with pneumonia—reflecting the infection rate that really occurs—would, however, have to be very similar. If, on the contrary, the regional variations in incidence rates reflect different infection rates, we would be finding a certain correlation between the seroprevalence variations and the incidence variations.

2. On the other hand, if we assume that the improvement in the analytical methods during the study period entailed an evolutive increase (or decrease), it would be logical for

Sebastian Crespi Clinical Laboratory, Policlinica Miramar, Camino de la Vileta 30, 07014 Palma de Mallorca, Spain. *Albert Torrents and Miguel A. Castellanos* Reference Laboratory, Camèlies, 10, 08024 Barcelona, Spain.

Legionella: State of the Art 30 Years after Its Recognition
Edited by Nicholas P. Cianciotto et al.
©2006 ASM Press, Washington, D.C.

this to have had an equal effect on each of the populations being studied. Consequently, it would be possible to statistically compare the evolutive trends of the seroprevalences in each region and to infer possible differences, as appropriate.

SCOPE OF THE SEROEPIDEMIOLOGICAL STUDY

We carried out a retrospective study (1992 to 2004) of the results obtained in the determination of anti–*Legionella pneumophlla* antibodies in 9,143 samples sent to our laboratory, proceeding from inpatients for whom said analysis had been requested. We selected the results of the samples received from three Spanish regions (7,090 samples from Catalonia, 1,247 from the Balearic Islands, and 806 from Andalusia), which were geographically separated and which had presented different incidence rates in the past few years.

Five different commercial tests were used sequentially throughout the period of the study. Four enzyme linked immunosorbent assays and one indirect fluorescent-antibody assay. All the tests detected anti–*Legionella pneumophila* antibodies to serogroups 1 to 6, except one which also incorporated serogroup 7. The results were classified as negative or positive, in accordance with the instructions provided by the manufacturer in each of the tests.

ANALYSIS OF THE RESULTS

The average prevalence of seropositivity to *Legionella pneumophila* antibodies in all the samples analyzed was 7.03% (Table 1). The highest prevalence corresponded to the Balearic Is-

lands (8.66%) and the lowest to Andalusia (5.95%), with Catalonia occupying an intermediate position (6.86%). These data are concordant, although not in a proportional way, with the median incidences of the past 5 years in each region, as issued by the National Epidemiological Institute, which were 4.53 (Balearic Islands), 1.33 (Andalusia), and 4.18 (Catalonia). As for its temporal evolution, in Andalusia, there is an evident rising trend (9.00% in the past 5 years studied compared with 3.27% in the first 5); the same as in Catalonia (10.88% compared with 5.53%) but not in the Balearic Islands, where the difference is much smaller (9.18% compared with 7.54%) (see Fig.1).

Taken together, these data suggest the existence of different epidemiological patterns in each of the three regions. In effect, our study detects different seroprevalences—statistically significant—for each region studied, which suggests that the Spanish geographical variations in the incidence of Legionnaires' disease are not only obedient to merely the differences in the diagnostic and/or declaration rates.

On the other hand, analysis of the temporal evolution of the seroprevalences in each region shows an unforeseen result: whereas in Andalusia and Catalonia there is an evident tendency for them to rise, in the Balearic Islands the annual seroprevalence has remained constant. This result is particularly interesting, as it permits us to discard the increase made in the other regions as being due, at least exclusively, to a greater sensitivity to the last analytical tests utilized. If this is so, the data on the temporal evolution of the seroprevalences

TABLE 1 General data of the seroepidemiological restrospective survey in the three Spanish regions studied

	Samples analyzed	IgG/IgM positive	Seroprevalence (%)	Incidence rate per 100,000
Andalusía	806	48	5.95	1.33
Balearic Islands	1,247	108	8.66	4.53
Catalonia	7,090	487	6.86	4.18
Total	9,143	643	7.03	

Median incidence rates of the past 5 years have been included.

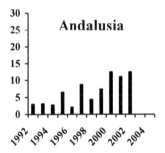 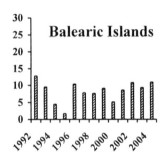 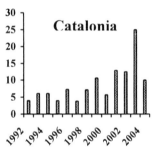

FIGURE 1 Evolution of seroprevalence in Andalusia, Balearic Islands, and Catalonia during the period 1992 to 2004.

suggest that in Andalusia and in Catalonia, the incidence of Legionnaires' disease has shown a tendency to rise independently of, and in addition to, the improvements made in epidemiological surveillance. In the Balearic Islands, on the contrary, the incidence of Legionnaires' disease could have remained relatively constant, with the increase observed being due to the improvements in the surveillance systems introduced in Spain in 1997 and 2000.

Other studies are necessary to confirm these hypotheses. Our study was a retrospective one and we do not have data on the definitive diagnosis of the patients analyzed. In addition, a prospective study would permit the elimination of a possible bias in the selection of patients remitted to the laboratory for analysis. Finally, a study that also analyzes the evolution of the seroprevalence in an apparently healthy population would be useful in extrapolating the results to the general population.

In any case, we believe that the study of the seroprevalences could be a good tool to assist in

demonstrating temporal-spatial variations for Legionnaires' disease, beyond the data provided by the different epidemiological surveillance systems which are, at times, difficult to interpret.

REFERENCES

1. **Bophal, R. S.** 1991. A framework for investigating geographical variation in disease, based on a study of Legionnaire's disease. *J. Public Health Med.* **13:**281–289.
2. **Bophal, R. S.** 1993. Geographical variation of Legionnaire's disease: a critique and guide to future reasearch. *Int. J. Epidemiol.* **22:**1127–1136.
3. **Borobio V., C. Martinez, and E. J. Perea.** 1987. Prevalence of anti-Legionella penumophilla antibodies in various groups with different risk factors in Seville (Spain). *Eur. J. Epidemiol.* **3:**436–438.
4. **Franzin, L., and F. Seramuzza.** 1995. Prevalence of Legionella pneumophilla serogroup 1 antibodies in blood donors. *Eur. J. Epidemiol.* **11:**475–478.
5. **Heltberg I., O. B. Jepsen, S. O. Larsen, and K. Lind.** 1988. Seroepidemiological study of Legionella infection in Denmark: a 28-month restrospective survey. *Dan. Med. Bull.* **35:**95–98.

RISK DIFFERENCES OF LEGIONNAIRES' DISEASE ASSOCIATED WITH TRAVEL IN SPAIN, 1999 TO 2004

Ricardo Casas, Rosa Cano, Carmen Martín, and Salvador Mateo

32

Travel-associated Legionnaires' disease (LD) has caused concern since the early 1980s. As a result, the European Group on *Legionella* Infections (EWGLINET) (2) was set up in 1986 in order to identify travel associated cases and clusters of the disease.

Legionnaires' disease can be severe, especially in some risk groups, members of which travel as much as anyone these days, and it normally causes high alarm in the general population. Spain hosts millions of visitors every year and has one of the strongest tourist industries in the world and consequently a very high density of tourist resorts, especially in coastal areas. Legionnaires' disease at these sites has an international impact, as patients come from all over the world (1, 3). Early detection and notification allow health authorities to undertake control measures at vacation accommodation sites. Now the priority is to develop a fast, effective, and efficient response system to prevent new onsets of this disease. The objective of this work is to measure and compare the risks associated with travel to different destinations in Spain for travelers from several

European countries as well as for Spanish nationals.

We defined incidence rates of legionellosis as cases per 1,000,000 person days. They were calculated for the period 1999 to 2004. Cases notified through EWGLINET and the Spanish reporting system were used as numerators. Denominators were obtained from official data published by the Spanish Census Office (Instituto Nacional de Estadística) on the number of nights spent by tourists in holiday accommodation sites.

A travel-associated case is defined as a person who stayed at least one night in a vacation accommodation during the 10 days before the onset of symptoms. We calculated the incidence rates for Spanish and foreign tourists, as well as for the regions with the major affluence of tourists.

A total of 6,411 cases of LD of any origin (community, nosocomial, and travel associated) were reported trough the Spanish reporting system from 1999 to 2004. Five hundred sixty five cases (9%) were travel associated in Spaniards. However, we only have included 221; those for which we have both full information and that can be assigned to one region. Besides these figures, EWGLINET reported 426 cases in foreign travelers associated with travel to Spain. Eighteen out of 35 countries participating

Ricardo Casas, Rosa Cano, Carmen Martín, and Salvador Mateo Centro Nacional de Epidemiología, Sinesio Delgado 6, Pabellón 12, 28029 Madrid-Spain.

in the EWGLINET scheme reported cases related to Spain during the period, five of them contributing 90% of all notified cases. The total overall incidence rate was slightly higher for foreigners than for Spaniards (0.51 per 10^6 person days versus 0.42 per 10^6 person days). Five out of 19 Spanish regions received the vast majority of tourists. The region which presented the highest incidence rate was C. Valenciana, which had high rates for both foreign and domestic travelers (0.97 per 10^6 person days and 0.73 per 10^6 person days, respectively). The pattern of disease is different in the five regions. In Andalusia, Catalonia, and C. Valenciana incidence rates are high for foreigners, while in the Balearic Islands and the Canary Islands the rates are relatively higher for domestic tourists (Table 1).

For the whole country rates peaked in 2002 and then decreased for both Spanish and foreign travelers (Fig. 1). Major variations exist from year to year in the figures for visitors from individual countries. However, for those countries where figures have been both high and relatively constant, we observed a decrease of incidence rates among Dutch visitors and an increase among British and French tourists. British tourists contribute the highest number of cases, but in relative terms the incidence rates are lead by Dutch, followed by Swedish visitors. The mean days of stay is significantly higher for foreign patients than for Spaniards

TABLE 1 Incidence of legionellosis (cases/1,000,000 person days) in Spain, 1999 to 2004

Year	1999	2000	2001	2002	2003	2004	Cases	Rate	Confidence interval 95%
Spain									
Total	0.23	0.27	0.47	0.78	0.63	0.46	647	0.471	0.43–0.50
Spaniards	0.00	0.00	0.40	0.99	0.67	0.40	221	0.42	0.36–0.47
Foreigners	0.36	0.43	0.52	0.66	0.61	0.50	426	0.51	0.46–0.56
Andalusia									
Total	0.09	0.17	0.23	0.43	0.30	0.15	49	0.23	0.17–0.30
Spaniards	0.00	0.00	0.06	0.30	0.11	0.05	9	0.09	0.04–0.16
Foreigners	0.16	0.31	0.37	0.53	0.48	0.28	40	0.35	0.25–0.48
Balearic Islands									
Total	0.20	0.36	0.34	0.36	0.60	0.29	109	0.35	0.29–0.42
Spaniards	0.00	0.00	0.40	0.59	1.27	0.18	13	0.40	0.23–0.68
Foreigners	0.23	0.40	0.33	0.33	0.51	0.31	96	0.35	0.28–0.42
Canary Islands									
Total	0.10	0.00	0.03	0.14	0.31	0.08	25	0.11	0.07–0.16
Spaniards	0.00	0.00	0.00	0.00	1.12	0.00	8	0.20	0.10–0.39
Foreigners	0.12	0.00	0.03	0.17	0.13	0.10	17	0.09	0.05–0.14
Catalonia									
Total	0.34	0.33	0.33	0.43	0.35	0.40	80	0.36	0.29–0.44
Spaniards	0.00	0.00	0.08	0.24	0.08	0.22	8	0.11	0.05–0.20
Foreigners	0.51	0.49	0.45	0.53	0.50	0.50	72	0.50	0.39–0.62
C. Valenciana									
Total	0.37	0.20	0.68	1.66	1.16	0.87	103	0.83	0.68–1.01
Spaniards	0.00	0.00	0.68	1.71	1.16	0.68	52	0.73	0.56–0.96
Foreigners	0.78	0.44	0.69	1.59	1.15	1.17	51	0.97	0.73–1.23

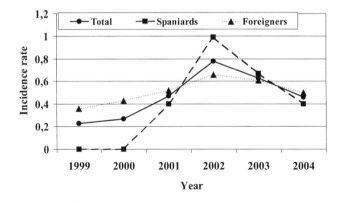

FIGURE 1 Incidence of legionellosis in travelers in Spain, 1999 to 2004.

(mean of 10.0 days and 6.8 days, respectively, $P < 0.05$).

The highest rates are present in the summer, especially during the months September and October. This national pattern coincides with the regional one, although a slight variability is observed in Andalusia and the Canary Islands, and a stronger variability in C. Valenciana, where the rates were highest.

The fact that year after year the highest rates correspond to C. Valenciana, with two exceptions corresponding to Catalonia and the Balearic islands in 2000, may be due to the seasonal closing of the resorts in these regions during the winter months. We have estimated the incidence rate of travel-associated Legionnaires' disease by identifying the denominators as the number of days stayed at accommodation sites. We have used them as an approximation of risk exposure. This is a new approach, as normally calculations correspond to cases per number of tourists. However, differences in national surveillance systems could produce variability in the ascertainment of the number of detected cases and therefore a misinterpretation of the results.

The mean days of stay is significantly higher for foreign patients than for Spaniards from 2002 on. According to data from the Spanish Census Office, the mean stay is globally higher for foreign tourists than for Spanish ones (4.7 and 3.0 days, respectively).

The decrease of global incidence rates for foreigners and Spaniards from 2003 on, coinciding with the entrance of the regulating law for the control of legionellosis (4), permits us to be optimistic in terms of our capacity to prevent this disease in the future.

REFERENCES

1. **Cano R., N. Prieto, C. Martín, C. Pelaz, and S. Mateo.** 2004. Legionnaires' disease clusters associated with travel to Spain during the period January 2001 to July 2003. *Euro. Surveill.* **9:**14–15.
2. **European Working Group for *Legionella* Infections.** 2002. *European Guidelines for Control and Prevention of Travel Associated Legionnaires' Diseases.* P15–20. Public Health Laboratory Service, London, United Kingdom, and http://www.ewgli.org.
3. **Prieto N., R. Cano, C. Martín, and S. Mateo.** 2002. Legionelosis relacionada con viajes a España. Procedimientos y resultados de la red de vigilancia de infecciones por *Legionella* (EWGLINET). *Bol. Epidemiol.* **10:**209–212.
4. **Real Decreto 385/2003.** 2003. Criterios Higiénico-Sanitarios para la Prevención y Control de la Legionelosis. Ministry of Health and Consumers Affaires, Spain.

PERSISTENCE AND GENOTYPIC STABILITY OF *LEGIONELLA* IN A POTABLE-WATER SYSTEM IN A HOTEL OVER A 20-MONTH PERIOD

M. García-Núñez, J. Ferrer, S. Ragull, E. Junyent, E. Sagristà, A. Soler, I. Sanchez, M. L. Pedro-Botet, N. Sopena, and M. Sabriá

33

Legionella pneumophila was first recognized as a cause of Legionnaires' disease in a hotel in Philadelphia in 1976 (4). Since then, few studies have reported the colonization of potable-water supplies in hotels (7, 1).

Chromosomal DNA subtypes responsible for cases of hospital-acquired Legionnaires' disease have been shown to be stable over time (2). Therefore, the identification of chromosomal DNA subtypes of *Legionella* and the demonstration of their persistence within the same aquatic environment (hotel potable-water distribution system) would aid in the epidemiological investigation of travel-associated cases of Legionnaires' disease.

The aims of this study were to determine the prevalence of *Legionella* spp. and to identify the chromosomal DNA subtypes and the genomic stability of these isolates during a 20-month follow-up period in the potable-water distribution system of a hotel.

Marian García-Núñez, Sonia Ragull, Inma Sanchez, Maria Luisa Pedro-Botet, Nieves Sopena, Anna Soler, and M. Sabriá Legionelosis Study Group (GELeg), Infectious Diseases Unit, Hospital Universitario Germans Trias i Pujol (Autonomous University of Barcelona), 08916 Badalona (Barcelona), Spain. *Eva Junyent and Anna Sagristà* Legionelosis Study Group (GELeg), Aqualab SL (Autonomous University of Barcelona), 08203 Sabadell (Barcelona), Spain. *Juan Ferrer* Hosbec Salud, 03500 Benidorm (Alicante) Spain.

From June 2003 to March 2005, 110 water samples were collected from a hotel water supply comprising 11 sample collections of water. The first 10 collections were undertaken the first year, with the last test being made 10 months later. The hot and cold water tanks were analyzed once during the studied period (in the first collection). The same eight distal points were analyzed in each collection, from the same one cold water distal point (distal tap of very infrequent use) and in the same seven hot water distal points. Temperature and free chlorine levels were also measured at each sampling point. Water samples were collected and transported following national guidelines (5).

Samples were processed according to the standard procedure ISO17131:98 (3) for quantitative *Legionella* analysis: all the samples were concentrated by filtration through polycarbonate filters (0.2-mm pore size, Millipore) and resuspended in 5 ml of the distilled water. Aliquots (0.1 ml) of the untreated, acid-buffer-treated and heat-treated (50°C, 30 min) samples were seeded on GVPC (OXOID). The plates were incubated at 36 ± 1°C for 10 days. *Legionella* colonies were confirmed by nutritional requirements and by slide agglutination latex assay (OXOID).

L. pneumophila serogroup (sg.) 1 was isolated in 49 (44.55%) of the 110 samples tested. Cold

TABLE 1 Positive distal points of the hot potable distribution system on each collection round

Date	Positive distal points		CFU/liter *Legionella* spp.	
	n	(%)	Geometric mean	Range[a]
June 2003	9	(85.7)	786	56–55,000
July 2003	7	(100)	187	25–5,000
September 2003	5	(71.4)	192	33–1,815
October 2003	4	(57.1)	269	93–1,680
November 2003	1	(14.3)	25	25
January 2004	5	(71.4)	4,699	3,667–5,000
February 2004	3	(42.8)	82	36–132
March 2004	5	(71.4)	186	38–5,000
May 2004	4	(57.1)	263	150–850
June 2004	5	(71.4)	142	38–330
March 2005	4	(57.1)	183	150–250

[a]Only positive samples

and hot water tanks were negative. The cold water distal point was *L. pneumophila*–positive throughout the whole study period (geometric mean; 6,702 CFU/liter; range, 36 to 245,000 CFU/liter) (Table 1). It should be pointed out that this distal point of cold water was an infrequently used point (roof-top tap) with stagnant water, which may explain the low chlorine levels (range, 0.05 to 0.95 mg of Cl_2 per liter), the heating of water during warmer seasons and thus, colonization by *Legionella* throughout the study period.

On the other hand, *L. pneumophila* was isolated in 39% (38/97) of the hot water samples tested. The potable hot water system remained colonized by *L. pneumophila* sg 1 throughout the study period, with oscillations in the counts (geometric mean: 291 CFU/liter; range, 25 to 55,000 CFU/liter), corresponding to the number of positive points ranging from 14 to 100% in each collection (Table 1). The temperatures in the hot distal points were always >50°C (mean; 54.14°C; range, 51 to 57.3°C). Free chlorine levels remained <0.2 mg/liter (mean, 0.07 mg/liter Cl_2; range, 0.04 to 0.12 mg/liter Cl_2). These results coincide with those reported by other authors and suggest that once a water system is colonized, maintenance of the temperature over 50°C is not sufficient to control colonization by *Le-*

gionella if some corrective measures or actions are not implemented.

Environmental isolates of *Legionella* were genotypically characterized by pulsed-field gel electrophoresis. Genomic DNA was prepared as previously described, with some modifications (6). Briefly, *Legionella* isolates were treated with lysozyme (3 mg/ml [Sigma-Aldrich]) for 24 h at 37°C, with proteinase K (1 mg/ml [Sigma-Aldrich]) for 24 h at 55°C, and were digested with 50 units of *Sfi*I restriction enzyme (New England Biolabs, Beverly, MA) for 18 h at 50°C. Fragments of DNA were separated in a 1% agarose gel prepared and run in 0.5x tris-borate-ethylenediaminete acetate buffer (pH 8.3) in a contour-clamped homogeneous field apparatus (contour-clamped homogeneous electric field DR II system; Bio-Rad, Ivry sur Seine, France) with a constant voltage of 5 V/cm and increasing pulse times to 50.6 s at 14°C for 24 h. A total of 58 isolates were typed by Sfi I-PFGE (source and number of colonies from each sample are listed in Table 2). All the isolates of *L. pneumophila* analyzed presented the same chromosomal DNA subtype A regardless of the period and sample site (Fig. 1).

In this study, we found that *L. pneumophila* serogroup 1 was recovered from multiple sites including showers and outlets. The number of colonized points was uniform throughout the

TABLE 2 Isolates typed by Sfi I pulsed field gel electrophoresis (PFGE)

Date	Source of isolate (distal point)	Isolates typed *n*	PFGE subtype
June 2003	n. 1	1	A
	n. 2	1	A
	n. 3	1	A
	n. 5	1	A
	n. 6	1	A
	n. 7	2	A
	n. 8	1	A
October 2003	n. 2	2	A
	n. 5	2	A
	n. 8	2	A
January 2004	n. 1	2	A
	n. 4	2	A
	n. 5	2	A
	n. 5	2	A
	n. 7	2	A
	n. 8	2	A
June 2004	n. 2	4	A
	n. 3	1	A
	n. 4	4	A
	n. 5	3	A
	n. 5	1	A
	n. 8	2	A
March 2005	n. 1	5	A
	n. 2	4	A
	n. 5	4	A
	n. 8	4	A

Total = 58

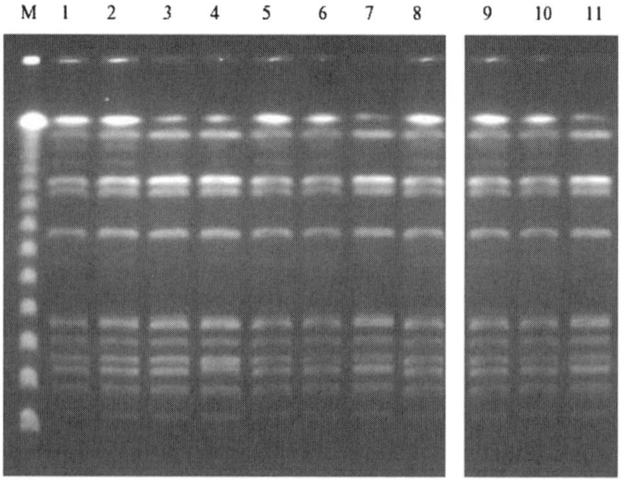

FIGURE 1 Pulsed-field gel electrophoresis. M, Market; 1, point 1 (June 2003); 2, point 2 (October 2003); 3, point 3 (January 2004); 4, point 4 (June 2004); 5, point 5 (June 2003); 6, point 6 (October 2003); 7, point 7 (January 2004); 8, point 8 (June 2004); 9, point 9 (March 2005); 10, point 4 (March 2005); 11, point 11 (March 2005).

20-month study. All *L. pneumophila* isolates analyzed showed the same chromosomal DNA genotype, which indicated the clone stability over the 20-month follow-up period.

ACKNOWLEDGMENTS

This work was supported by ASSOCIACIO IBEMI.

REFERENCES

1. **Borella P., M. T. Montagna, S. Stampi, G. Stancanelli, V. Romano-Spica, M. Triasii, I. Marchesi, A. Bargellini, D. Tato, C. Napoli, F. Zanetti, E. Leoni, M. Moro, S. Scaltriti, G. Ribera D'Alcala, R. Satarpia, and S. Boccia.** 2005. Legionella contamination in hot water of Italian hotels. *Appl. Environ. Microbiol.* **71:**5805–5813.
2. **Garcia-Nuñez, M., S. Ragull, M. L. Pedro-Botet, N. Sopena, E. Reynaga, C. Rey-Joly, J. Morera, and M. Sabria.** 2002. A single environmental clone of Legionella pneumophila serogroup 1 causing hospital-acquired Legionnaires' disease over a 13-year period. 42th Interscience Conference on Antimicrobial Agents and Chemotherapy (ICAAC), San Diego.
3. **ISO 11731:1998(E).** 1998. *Water Quality: Detection and Enumeration of Legionella.*
4. **McDade, J. E., C. C. Shephard, D. W. Fraser, T. R. Tsai, M. A. Resudus, and W. R. Dowdle, and the Laboratory Investigation Team.** 1997. Legionnaires' disease. Isolation of a bacterium and demonstration of its role in other respiratory disease. *N. Engl. J. Med.* **297:**1197–1203.
5. **Real Decreto 865/2003,** de 4 de Julio por el que se establecen los criterios higiénico-sanitarios para la prevención y control de la legionelosis (BOE 17, 18 julio 2003).
6. **Sabria, M., M. Garcia-Núñez, M. L. Pedro Botet, N. Sopena, J. M. Gimeno, E. Reynaga, J. Morera, and C. Rey-Joly.** 2001. Presence and chromosomal subtyping of *Legionella* species in potable water systems in 20 hospitals of Catalonia, Spain. *Infect. Control Hosp. Epidemiol.* **22:**673–676.
7. **Zeybech, Z., and A. Cotuk.** 2002. Relationship between colonization of building water systems by *Legionella pneumophila* and environmental factors, p. 305–308. *In* R. Marre et al. (ed.), *Legionella.* ASM Press, Washington, D.C.

10 YEARS OF *LEGIONELLA* SURVEILLANCE: CHANGE OF *LEGIONELLA* SUBTYPE PRECEDED EPIDEMIC OF NOSOCOMIAL LEGIONNAIRES' DISEASE

Klaus Weist, Christian Brandt, Paul Christian Lück, Jutta Wagner, Tim Eckmanns, and Henning Rüden

34

Although nosocomial *Legionella* infections (nLD) do not occur as frequently as other nosocomial infections (11), there is much concern about their prevention since they may be transmitted by one of the best controlled and most important viands, i.e., drinking water (2). Evidence-based recommendations focus on disease surveillance, monitoring clinical specimens from patients suspected of having atypical pneumonia, and well-kept maintenance of plumbing systems. Best clinical evidence for decontaminating potable water systems remains an unresolved issue (9). No method established guarantees thorough *Legionella* eradication (8), nor is there a proven infective dose of *Legionella* isolates or a valid method to detect virulent strains in advance.

We evaluated the changing patterns of strains causing nLD from a university hospital water system during a decade of active surveillance.

Klaus Weist, Christian Brandt, Tim Eckmanns, and Henning Rüden Institute of Hygiene and Environmental Medicine, Charité—University Medicine Berlin, 12203 Berlin, Germany. *Jutta Wagner* Institute of Microbiology and Infectious Medicine, Charité—University Medicine Berlin, 12203 Berlin, Germany. *Paul Christian Lück* Institute of Medical Microbiology and Hygiene, C-G. Carus University, Dresden, 01307 Dresden, Germany.

10 YEARS OF *LEGIONELLA* SURVEILLANCE IN A UNIVERSITY HOSPITAL WATER SYSTEM

The university hospital studied has 1,100 beds. All medical disciplines are represented, including three intensive care units and oncology wards with one bone marrow transplant unit. Municipal ground water with no prior treatment is provided to the hospital central water supply. Hot water is generated by heating to 70°C for several minutes, followed by cooling down to 53°C by means of a recuperator. The water is stored in three 10,000 liter hot water storage tanks and finally injected to a ring line. Routine water samples were taken at least two times a year from the central water supply as well as from peripheral faucets and shower heads. Additionally, probes were drawn in any potential or proven case of nLD, followed by heating for 2 h up to 65°C in the particular line segment.

DETECTING PATIENTS WITH NOSOCOMIAL *LEGIONELLA* INFECTIONS

All bronchoalveolar lavage specimens were routinely screened for *Legionella*. Bronchial secretions and urine probes (Biotest AG, Dreieich, Germany) from patients with suspicious atypical pneumonia were processed by routine culture methods. One hundred milliliters of environmental water samples were filtered, and 1 ml of the original probe was plated to buffered charcoal yeast extract agar supplemented with 0.1% alpha-ketoglutarate. Typically grown colonies were subcultured on blood agar to show their inability to grow on that agar. *L. pneumophila* serogroups and non-pneumophila species were detected by direct immunofluorescence assays and since 2001 by use of a latex agglutination assay (Oxoid, Wesel, Germany). Isolates of *Legionella* spp. from patients and available corresponding environmental ones were genotyped using an AFLP technique previously described by EWGLI (3) and further subtyped by monoclonal antibody analysis (6).

An MLST-like sequence-based typing technique was used to define allelic profiles (4). *Legionella*-positive clinical specimens were reported to the infection control team, and every case was carefully followed up. Detection of *Legionella* was definied as nLD if the patient was admitted to the university hospital at least 10 days before and no other source of infection could be observed.

CHANGE OF *LEGIONELLA* SUBTYPE PRECEDED EPIDEMIC OF NOSOCOMIAL *LEGIONELLA* INFECTIONS

From January 1995 to 2001, *L. pneumophila* serogroup (sg) 1 could not be detected in the hospital water system. Up to February 2002, no nLD caused by *L. pneumophila* sg 1 had been detected. In 2001, *L. pneumophila* sg 1 was detected in the central water supply for the first time and hence in peripheral water samples, which were positive for *Legionella*. At the end of 2003, single-use filters for water outlets were installed in the bone marrow transplant unit and in oncology wards, where nosocomial cases of nLD had been detected. Up to 30% of peripheral probes showed *L. pneumophila* on average from 1 to 30 CFU/ml, not exceeding 100 CFU/ml. This percentage did not vary during the whole surveillance period, with the

exception of water outlets supplied with filters where all probes were negative for *Legionella*.

From 2002 until 2004, 10 cases of nLD with *L. pneumophila* sg 1 occurred (5 cases in 2002, 4 cases in 2003, and a last case in January 2004). Six of the ten patients' isolates and corresponding peripheral water samples showed indistinguishable AFLP patterns (Fig. 1) and belonged to the subgroup Knoxville, a strain which is monoclonal antibody 3/1 positive (6). The sequence-based typing displayed a unique allelic profile of that strain (Table 1). Until December 2005, no further case of nLD was observed.

CLONAL ORIGIN OF NOSOCOMIAL *LEGIONELLA* INFECTIONS

The switch to *L. pneumophila* sg 1 in environmental water samples was associated with the occurrence of six nLD caused by the epidemic strain *L. pneumophila* sg 1, subtype Knoxville 1. This strain was previously identified to cause nLD in that area (5). We cannot explain the change in *L. pneumophila* patterns and the occurrence of that particular epidemic strain. Notably, neither the treatment of hot water nor the methods used to detect *Legionella* in environmental probes had been changed during

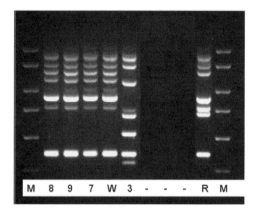

FIGURE 1 AFLP-Patterns of *L. pneumophila* sg 1 isolates from four cases with nLD (Lane numbers correspond with patient no. of Table 1) and 1 isolate detected from a tap water outlet where case patients had been treated. M, marker; R, reference strain.

TABLE 1 Confirmed cases of nosocomial *L. pneumophila* pneumonia in a university hospital during 10 years of surveillance

Pat. no.	Date of diagnosis	Age, gender	Immuno-suppression	Evidence of *L. pneumophila*	Sero-group	Subtype	Outcome
1	Jan. 1999	68 F	Renal transplant recepient	BAL (culture)	8	n.d.[b]	Death
2	March 2001	59 M	None	Immunofluores-cence (lung tissue)	n.d.	n.d.[b]	Death
3	Feb. 2002	40 M	Aplastic anemia	BAL (culture)	1	Bellingham 1	Discharged
4	May 2002	78 M	Chemotherapy (bronchial carcinoma)	BAL (culture)	1	n.d.	Discharged
5	Nov. 2002	74 M	None	Urine (antigen)	1	n.d.	Death
6	Nov. 2002	52 M	High-dose corticosteroids (short term)	BAL (culture)	1	Knoxville 1[a]	Discharged
7	Dec. 2002	62 M	Chemotherapy (autoimmune disease)	Urine (antigen)	1	Knoxville 1[a]	Death
8	May 2003	44 F	High-dose corticosteroids (autoimmune hepatitis)	BAL (culture)	1	Knoxville 1[a]	Discharged
9	June 2003	57 F	Chemotherapy (lymphoma)	BAL (culture)	1	Knoxville 1	Discharged
10	Sept. 2003	68 M	Chemotherapy (bronchial carcinoma)	BAL (culture)	1	Knoxville 1[a]	Death
11	Sept. 2003	62 M	Chemotherapy (bronchial carcinoma)	BAL (culture)	1	Knoxville 1[a]	Death
12	Jan. 2004	53 F	Chemotherapy (bronchial carcinoma)	Urine (antigen)	1	n.d.	Death

[a]Isolates from patients with nLD no. 6, 7, 8, 9, 10, and 11 as well as isolates from the hospital water supply where these patients were treated belonged to the same sequence-based genotype. The allelic profiles of this strain were the following: *flaA:* 3, *pilE:* 4, *asd:* 1, *Mip:* 3, *proA:* 9, *gsp:* 13, *neuA* 11 (4).
[b]n.d., not done

our study. Despite thermal disinfection of potable warm water in the central water supply, about one-third of peripheral water samples displayed *Legionella* spp., although in low concentrations. After the installation of filters, patients were prevented from acquiring further nLD, with the exception of patient 12 (Table 1), who developed nLD during the phase of introducing the filters in certain units.

It has been previously shown that identical genotypes of *L. pneumophila* can persist for years in hospital plumbing systems and affect immunocompromised patients with very low concentrations (<1 CFU/ml) in the water system (7). In the present study there was no association between the occurrence of nLD and the concentration of *Legionella* isolates detected in peripheral water outlets, where the patients suffering from nLD were treated. With the exception of two patients (patients 2 and 5), all had a severe risk factor for acquiring nLD, i.e., immunosuppression.

POTENTIAL IMPACT OF EPIDEMIC STRAINS IN HOSPITAL WATER SYSTEMS

The findings of changing patterns of *L. pneumophila* sg 1 in a hospital water supply and the occurrence of nLD due to a distinct epidemic strain have had two major impacts.

First, high risk patients, the vast majority of the affected patients, should be prevented from any potential contact with *Legionella*. It is believed that mainly the installation of filters in areas of high-risk patients was efficient to prevent further nosocomial cases (10).

Second, screening of water systems for known epidemic strains of *L. pneumophila* sg 1 may be useful for hospitals with units treating high-risk patients. Furthermore, utilization of culture methods to detect nLD in patients should be emphasized. Currently, the majority of *Legionella* infections in Germany are detected by a urine antigen test, and consequently the utilization rate of culture methods to detect *Legionella* has decreased over the past years to 11% in 2004 (1), but only the latter method allows comparison of patient isolates with the ones isolated from the water system.

REFERENCES

1. **Anonymous.** 2005. Legionellose in Deutschland 2004. *Epidemiol. Bull.* **48:**447–451.
2. **Blatt, S. P., M. D. Parkinson, E. Pace, P. Hoffman, D. Dolan, P. Lauderdale, R. A. Zajac, and G. P. Melcher.** 1993. Nosocomial legionnaires' disease: aspiration as a primary mode of disease acquisition. *Am. J. Med.* **95:**16–22.
3. **Fry, N. K., J. M. Bangsborg, A. Bergmans, S. Bernander, J. Etienne, L. Franzin, V. Gaia, P. Hasenberger, B. Baladron Jimenez, D. Jonas, D. Lindsay, S. Mentula, A. Papoutsi, M. Struelens, S. A. Uldum, P. Visca, W. Wannet, and T. G. Harrison.** 2002. Designation of the European Working Group on Legionella Infection (EWGLI) amplified fragment length polymorphism types of Legionella pneumophila serogroup 1 and results of intercentre proficiency testing using a standard protocol. *Eur. J. Clin. Microbiol. Infect. Dis.* **21:**722–728.
4. **Gaia, V., N. K. Fry, B. Afshar, P. C. Lück, H. Meugnier, J. Etienne, R. Peduzzi, and T. G. Harrison.** 2005. Consensus sequence-based scheme for epidemiological typing of clinical and environmental isolates of Legionella pneumophila. *J. Clin. Microbiol.* **43:**2047–2052.
5. **Glaser, S., T. Weitzel, R. Schiller, N. Suttorp, and P. C. Lück.** 2005. Persistent culture-positive Legionella infection in an immunocompetent adult. *Clin. Infect. Dis.* **41:**765–766.
6. **Helbig, J. H., S. Bernander, M. Castellani Pastoris, J. Etienne, V. Gaia, S. Lauwers, D. Lindsay, P. C. Lück, T. Marques, S. Mentula, M. F. Peeters, C. Pelaz, M. Struelens, S. A. Uldum, G. Wewalka, and T. G. Harrison.** 2002. Pan-European study on culture-proven Legionnaires' disease: distribution of Legionella pneumophila serogroups and monoclonal subgroups. *Eur. J. Clin. Microbiol. Infect. Dis.* **21:**710–716.
7. **Mathys, W., M. C. Deng, J. Meyer, and E. Junge-Mathys.** 1999. Fatal nosocomial Legionnaires' disease after heart transplantation: clinical course, epidemiology and prevention strategies for the highly immunocompromised host. *J. Hosp. Infect.* **43:**242–246.
8. **Reichardt, C., M. Martin, H. Ruden, and T. Eckmans.** 2005. Disinfection of hospital water systems and the prevention of *Legionellosis:* what is the evidence?. *In* N. P. Cianciott, et al. (ed.). Legionella: *Proceedings of the 6th Internation Conference.*
9. **Tablan, O. C., L. J. Anderson, R. Besser, C. Bridges, and R. Hajjeh.** 2004. Guidelines for preventing health-care-associated pneumonia, 2003: recommendations of CDC and the Healthcare Infection Control Practices Advisory Committee. *MMWR Recomm. Rep.* **53:**1–36.
10. **Vonberg, R. P., T. Eckmanns, J. Bruderek, H. Ruden, and P. Gastmeier.** 2005. Use of terminal tap water filter systems for prevention of nosocomial legionellosis. *J. Hosp. Infect.* **60:**159–162.
11. **Woodhead, M.** 2002. Community-acquired pneumonia in Europe: causative pathogens and resistance patterns. *Eur. Respir. J. Suppl.* **36:**20s–27s.

REPRESENTATIVE SURVEY OF THE SCOPE OF LEGIONNAIRES' DISEASE AND OF DIAGNOSTIC METHODS AND TRANSMISSION CONTROL PRACTICES IN GERMANY

*Tim Eckmanns, Mona Poorbiazar,
Henning Rüden, and Lüder Fritz*

35

Legionella is the cause of serious hospital-acquired infections, particularly in immunocompromised patients. Few data exist on the incidence of nosocomial Legionnaires' disease (LD) in Germany. LD is a notifiable disease in Germany, but fewer than 500 cases are reported each year. The reported incidence seems to be low in comparison to other European countries (incidence per 1 million inhabitants: Spain, 34.1; Denmark, 19.2; The Netherlands, 17.9; France, 16.9; Germany, 5.0) (1).

There is great uncertainty concerning technical measurements of prevention and screening for *Legionella* contamination in hospital water systems as well as for the diagnosis of LD. The aim of this study was to get a representative overview of the scope of LD and to assess the type and frequency of implemented diagnostic methods and transmission control practices.

DISTRIBUTION OF THE SURVEY QUESTIONNAIRE

We randomly selected 98 out of 2,240 German hospitals. A survey questionnaire of multiple-choice and short-answer questions was com-

posed. The questionnaire was sent to the clinical directors (with two reminders) of each selected hospital. An allowance of 20 Euro was granted to each responder. The anonymous survey included questions about hospital demographics (type of hospital, bed count, presence of risk patients [neonates, oncological, transplanted, HIV positive, and immunsuppressed ICU patients]), episodes of nosocomial and community-acquired LD in the past 5 years, diagnostic testing for LD, environmental sampling practices and the results obtained, as well as questions about prophylactic measurements and maintenance of hospital water systems in general and especially for high-risk patients.

RESULTS OF THE SURVEY QUESTIONNAIRE

Of the 98 surveyed hospitals, 50 (51%) responded with the completed questionnaire to the first mailing, and an additional 10 responded after the second and third mailing remainder, totaling 60 respondents (61% overall response rate). The mean bed count was 294 (range 8 to 1,400) compared to a general mean bed count of 247 in Germany.

Two (3%) hospitals reported definite nosocomial LD in the past 5 years. Nineteen hospitals (32%) had an episode of community-

Tim Eckmanns, Mona Poorbiazar, and Henning Rüden Institute of Hygiene and Environmental Medicine, Charité—University Medicine Berlin, 12203 Berlin, Germany. *Lüder Fritz* Medizinische Klinik Hämatologie/Onkologie, Klinikum Ernst von Bergmann, 14467 Potsdam Germany.

acquired LD. No diagnostic *Legionella* test had been conducted by 18 (32%) of the 56 hospitals that answered that question; 28 (50%) reported a rate of less than one test per month, and 10 (18%) more than one per month. Twenty-one (35%) routinely performed microbiological testing for legionellosis in bronchoalveolar-lavage material among high risk patients. Of the 38 hospitals performing diagnostic tests for *Legionella*, 16 (42%) reported an episode of community-acquired LD in the past 5 years. From the 18 hospitals never using diagnostic tests, 1 (6%) reported an episode of LD. This difference between these two groups is significant ($P = 0.005$).

Peripheral water systems were routinely sampled by 53 (88%) of the hospitals. Three (6%) of those tested their water less than once a year, 20 (38%) once a year, and 30 (57%) more than once a year. Of those 53, 26 (51%) had a positive result, with 3 (6%) of them more than 30% positive peripheral outlets. Out of 52 (88%) hospitals which sampled their central water systems, 14 (29%) detected *Legionella*. In summary, 56 out of 60 (93%) of the hospitals sample their water (peripheral and/or central) for *Legionella*, 31 (55%) with positive results. Of the two hospitals reporting nosocomial LD, only one had positive sampling results, while the other one did not detect *Legionella*.

Measurements for potable water decontamination were performed in 43 (72%) of the hospitals (Table 1). All but one hospital, which used UV-light, implemented thermal measurements for decontamination. Five hospitals used chlorinated water, UV light, or chlorine dioxide in addition. Another five hospitals performed warm water maintenance above 55°C and periodical shock heat above 60°C parallel.

Eighteen hospitals cared specifically for high-risk patients. One hospital treating high-risk patients did not protect these patients from LD at all. Additional measurements for high-risk patients are listed in Table 2.

WHAT HAVE WE LEARNED, AND WHAT NEEDS TO BE DONE?

These are the first representative data on the scope of LD, of diagnostic methods, and transmission infection control practices for LD in Germany. Hospitals responded fairly well to our survey questionnaire (61% response rate). The critical question concerning nosocomial LD was deliberately broad: "Did you have an episode of nosocomial LD in the past 5 years?" Only 2 of the surveyed hospitals reported an episode of nosocomial LD, although in 55% of the hospitals that tested the water (52% of the 60 hospitals), *Legionella* was isolated from the potable water system. Nosocomial LD might be a very rare event, or it is often simply not diagnosed, since 32% of the surveyed hospitals did not perform diagnostic tests for *Legionella* on patients at all. The truth might be in between. In this survey, 9.5 times more hospitals report community-acquired LD than nosocomial LD. Comparing this result to literature reports, which state that in the Unites States 1 of 5 to 1 of 4 of all LD cases are nosocomial (2), might support the thesis that nosocomial LD is

TABLE 1 Measurements for potable water decontamination in 43 hospitals

Measurements	Number* (percentage of 60)
Warm water maintenance temperature >55°C	25 (42%)
Shock heat (periodical) >60°C	23 (38%)
Continuous chlorination	2 (3%)
Ultraviolet light	3 (5%)
Chlorine dioxide	1 (2%)

*Multiple mentioning possible.

TABLE 2 Measurements for high-risk patients in 17 hospitals

Measurements	Number
Point-of-use filters	11
Sterile water in general	1
Sterile water for oropharyngeal care	1
No shower	2
Sterile water for oropharyngeal care and no shower	1
Sterilization of faucet aerator	1

rare in Germany. We cannot explain if this difference is specific for Germany or if nosocomial events are simply underreported in our survey. Of the surveyed hospitals, 32% did not conduct any diagnostic *Legionella* testing, but 93% perform regular environmental testing. Whether the incidence of nosocomial LD is low in reality or due to nonperformance of diagnostic tests—a sign of underdiagnosis—remains an open question. A possible conclusion is that nonperformance of diagnostic tests is a reason for low incidence of LD not only in this study but throughout Germany, too.

Seventy-two percent of the hospitals perform routine measurements for potable water decontamination, while only 3% reported nosocomial cases. An optimistic explanation could be that nosocomial LD is very rare because of the measurements, while the pessimist might say that hospitals spend a lot of money for prophylactic measurements which are perhaps not necessary and neglect the necessity of diagnostic tests on *Legionella*.

Considering the high costs of testing for *Legionella* and for LD prevention, further research is indicated to establish standard procedures concerning low- and high-risk patients.

REFERENCES

1. **Joseph, C. A.** 2004. Legionnaires' disease in Europe 2000-2002. *Epidemiol. Infect.* **132:**417–424.
2. **Marston, B. J., H. B. Lipman, and R. F. Breiman.** 1994. Surveillance for Legionnaires' disease. Risk factors for morbidity and mortality. *Arch. Intern. Med.* **154:**2417–2422.

DISTRIBUTION OF *LEGIONELLA PNEUMOPHILA* GENOTYPES IN PATIENTS AND ENVIRONMENTAL SOURCES

Ed P. F. Yzerman, Jacob P. Bruin, Jeroen W. den Boer, Linda P. Verhoef, and Kim W. van der Zwaluw

36

EPIDEMIOLOGY

Legionnaires' disease is a pneumonia which in The Netherlands is responsible for 4 to 8% of community-acquired pneumonias and for an unknown percentage of hospital-acquired pneumonias (2, 6). Despite this relatively low incidence, a lot of attention is given to Legionnaires' disease because it is considered a preventable disease with a through-the-years stable mortality rate between 5 and 15% (3). Legionellae are part of the microbial community of aquatic ecosystems, natural as well as man-made, which explains why legionellosis occurs worldwide. In many countries Legionnaires' disease is a notifiable disease. Epidemiological study results from the United States show that 90% of Legionnaires' disease cases are caused by *Legionella pneumophila* and, more specifically, 72% by *L. pneumophila* serogroup 1 (5).

A NATIONWIDE STUDY IN THE NETHERLANDS

To compare the Dutch epidemiology with the figures in the United States, we conducted a prospective study that started in August 2002.

Ed P. F. Yzerman and Jacob P. Bruin Regional Public Health Laboratory Kennemerland, Boerhaavelaan 26, 2035 RC Haarlem, The Netherlands. *Jeroen W. den Boer and Linda P. Verhoef* Municipal Health Service Kennemerland, Westergracht 72, 2014 XA Haarlem, The Netherlands. *Kim W. van der Zwaluw* National Institute of Public Health and the Environment, POB 1, 3720 BA Bilthoven, The Netherlands.

This nationwide study consisted of sampling potential sources to which Legionnaires' disease patients had been exposed during their incubation period (2 to 10 days). For that purpose we were given access to national notification data. Both confirmed and probable cases were included in the study. A confirmed case of Legionnaires' disease was defined as a patient suffering from symptoms compatible with pneumonia, with radiological signs of infiltration, and with laboratory evidence of *Legionella* spp. infection. Laboratory evidence included isolation of *Legionella* spp. from respiratory secretions or lung tissue, detection of *L. pneumophila* antigens in urine, and seroconversion or a fourfold or higher rise in antibody titers to *L. pneumophila* in paired acute- and convalescent-phase serum samples. A probable case of Legionnaires' disease was defined as a patient suffering from symptoms compatible with pneumonia, with radiological signs of infiltration, and with laboratory findings suggestive of *Legionella* infection. These findings included a high antibody titer to *L. pneumophila* in a single serum sample, direct fluorescent antibody staining of the organism, and detection of *Legionella* species DNA by polymerase chain reaction in respiratory secretions or lung tissue. Both definitions conform to the criteria of the European Working Group for Legionella Infections (EWGLI) (1).

Patients were excluded from the study if they were abroad during the incubation time.

All 39 municipal health services of The Netherlands participated in the study by incorporating the study procedures into their standard notification protocol. A uniform questionnaire was used by all municipal health services to register the potential sources for each patient. The questionnaire facilitates a structured interview focused at individual exposure to potential sources of infection. Trained personnel from our laboratory subsequently sampled these sources systematically and cultured the water and swab samples according to standard procedures. In short, the water samples were concentrated by filtration, and filtered residues were resuspended in 1 ml sterile water. Of this suspension, 100-μl samples were cultured without dilution and after 10- and 100-fold dilution on buffered charcoal yeast extract agar supplemented with α-ketoglutarate (BCYE-α) agar at 37°C, with increased humidity. In cases of bacterial overgrowth, cultures were repeated after pre- treatment by heating 30 minutes at 50°C. Swab samples were dispersed by immersion in 1 ml sterile water and cultured as described above. Isolates were identified biochemically, and after identification *Legionella* strains were serotyped using commercially available kits containing antiserum against the 14 *L. pneumophila* serogroups and *L. dumoffii*, *L. gormanii*, *L. micdadei*, and *L. bozemanii*. If they belonged to serogroup 1, the strains were genotyped by amplified fragment length polymorphism (AFLP) technique according to EWGLI typing protocols (4). Upon report of a notified culture-proven patient, the medical microbiology laboratory involved was requested to send the patient isolate to our laboratory for sero- and genotyping. Patient and environmental strains were compared, and a potential source was considered to be a true source of infection if a patient isolate and an environmental isolate were indistinguishable by AFLP genotyping. Furthermore, for all isolated strains, we compared the distribution of genotypes causing disease to the distribution of environmental strains in order to investigate

the possibility of targeting preventive measures.

DISTRIBUTION OF *LEGIONELLA* GENOTYPES

In this chapter we present the preliminary results of the distribution of *Legionella* genotypes cultured from patients and environmental sources. Between August 2002 and September 2005, sero- and genotyping of 130 patient isolates and 220 environmental isolates showed that 98% of the patient strains were from the *L. pneumophila* genus. Of these, 87% belonged to *L. pneumophila* serogroup 1, 6% to serogroup 7 to 14, 3% to serogroup 3, 2% to serogroup 2, and 2% to serogroup 6 (Table 1). The most frequently seen EWGLI types were 004 Lyon (21%) and 010 London (10%) (Fig. 1). Another 28% consisted of a mixture of AFLP types that were not yet designated by EWGLI. In sharp contrast, the minority (45%) of environmental strains was of the *L. pneumophila* genus. Of these, the distribution was 57%, 24%, 8%, 6%, 4%, and 1% for serogroups 1, 7 to 14, 5, 3, 2, and 6, respectively. The most frequently seen EWGLI types were 028 Rome (35%), 001 Lugano (13%), 013 London/030 Stockholm (13%), and 003 Glasgow (10%). The not yet designated AFLP patterns represented 17% of the environmental strains.

Comparing the EWGLI types of the *L. pneumophila* serogroup 1 patient strains with the environmental strains, we noticed a distribution that is totally different for both groups of strains. The genotypes 004 Lyon and 010

TABLE 1 Distribution of *L. pneumophila* strains from patients and the environment in serogroups (sg)

	Patient strains number (%)	Environmental strains number (%)
sg 1	112 (87)	56 (57)
sg 2	2 (2)	4 (4)
sg 3	4 (3)	6 (6)
sg 4	0	0
sg 5	0	8 (8)
sg 6	2 (2)	1 (1)
sg 7–14	8 (6)	24 (24)

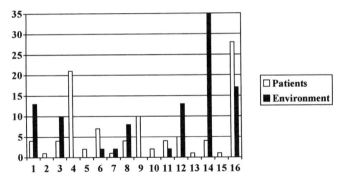

FIGURE 1 Distribution of *L. pneumophila* serogroup 1 isolates from patients and the environment in EWGLI genotypes (%). 1, 001 Lugano; 2, 001 Lugano/028 Rome; 3, 003 Glasgow; 4, 004 Lyon; 5, 005 Rome; 6, 006 Copenhagen; 7, 008 Stockholm; 8, 009 London; 9, 010 London; 10, 015 Dresden; 11, 017 Lugano; 12, 013 London/030 Stockholm; 13, 020 Rome; 14, 028 Rome; 15, 029 London; 16, not yet designated.

London, responsible for almost one-third of all Legionnaires' disease patients in our study period in The Netherlands, were not found in the environmental samples collected from the potential sources that resulted from the structured interviews according to the questionnaires. On the other hand, the genotype 028 Rome is predominantly present in the environmental cultures but was only rarely cultured in patients.

ESTIMATION OF THE RISK OF HUMAN INFECTION

The preliminary results of our study indicate that systematic collection and sampling gives insight to the distribution of the *L. pneumophila* genus and to serotypes and genotypes in humans and in the environment. Our data suggest that aquatic ecosystems, manmade or natural, colonized with *Legionella* strains represent a differentiated risk for causing Legionnaires' disease depending on the presence of *L. pneumophila* versus non–*L. pneumophila* strains and within the pneumophila group depending on the genotypes. Using the distribution as an a priori chance, it may help to estimate the risk for human infection. This implies that genotyping the *L. pneumophila* strains that are isolated from environmental sources can be used as a tool to fine-tune control measures. Based on our findings,

actions in The Netherlands should be more aggressive for the EWGLI genotypes 004 Lyon and 010 London. However, until now we have never isolated these genotypes from environmental samples of potential sources. Legionellae are capable of infecting humans by aerosol inhalation or by drinking and subsequent aspiration of water. For each patient in this study, an inventory of potential sources was drawn up based on this knowledge. However, the results indicate that obviously the inventories were incomplete and the intriguing question arises of which sources were overlooked. The ongoing study may resolve this question in the future.

REFERENCES

1. **Anonymous.** 2002. European Working Group for Legionella Infections. Part 2. Definitions and Procedures for Reporting and Responding to Cases of Travel Associated Legionnaires' Disease, p. 15–20. *In European Guidelines for Control and Prevention of Travel Associated Legionnaires' Disease.* PHLS, London and http://www.ewgli.org.
2. **Bohte, R., R. van Furth, and P. J. van den Broek.** 1995. Aetiology of community-acquired pneumonia: a prospective study among adults requiring admission to hospital. *Thorax* **50:**543–547.
3. **Den Boer, J. W., I. H. Friesema, and J. D. Hooi.** 2002. Reported cases of Legionella pneumonia in the Netherlands, 1987-2000. *Ned. Tijdschr. Geneeskd.* **146:**315–320.

4. **Fry, N. K., J. M. Bangsborg, A. Bergmans, S. Bernander, J. Etienne, L. Franzin, V. Gaia, P. Hasenberger, B. Baladron Jimenez, D. Jonas, D. Lindsay, S. Mentula, A. Papoutsi, M. Struelens, S. A. Uldum, P. Visca, W. Wannet, and T. G. Harrison.** 2002. Designation of the European Working Group on Legionella Infection (EWGLI) amplified fragment length polymorphism types of *Legionella pneumophila* serogroup 1 and results of intercentre proficiency testing using a standard protocol. *Eur. J. Clin. Microbiol. Infect. Dis.* **21:**722–728.

5. **Marston, B. J., H. B. Lipman, and R. F. Breiman.** 1994. Surveillance for Legionnaires' disease. Risk factors for morbidity and mortality. *Arch. Intern. Med.* **154:**2417–2422.

6. **Vegelin, A.L., P. Bissumbhar, J. C. Joore, J. W. Lammers, and I. M. Hoepelman.** 1999. Guidelines for severe community-acquired pneumonia in the western world. *Neth. J. Med.* **55:**110–117.

MOLECULAR COMPARISON OF ISOLATES FROM A RECURRING OUTBREAK OF LEGIONNAIRES' DISEASE SPANNING 22 YEARS

Robert F. Benson, Claressa E. Lucas, Ellen W. Brown, Karen D. Cowgill, and Barry S. Fields

37

During outbreak investigations molecular analytical methods are necessary to discriminate among clinical isolates and to establish a link with environmental isolates. In addition to monoclonal antibodies used as an initial subtyping method, other methods employed have been pulse-field gel electrophoresis, amplified fragment length polymorphosim (AFLP), arbitrarily primed PCR, multilocus variable-number tandem repeat, and sequence-based typing (SBT). Multilocus sequence typing and SBT have been used to study long-term evolutionary relationships among strains of various bacterial genera. In the present study, we used SBT to study the relationships among clinical and environmental isolates of *Legionella pneumophila* isolated from the same hotel during a 22-year period.

MATERIALS AND METHODS

Water samples were collected from the hotel and cultured using standard procedures for the recovery of *Legionella* isolates from environmental sources. An additional isolate was ob-

tained from a patient discovered through a search of the EWGLINET archives that occurred in a Danish visitor to St. Croix in 1998 who spent all his time at the same hotel. Environmental isolates and one patient isolate recovered during the outbreak investigation in 1981 to 1982 were obtained from the Centers for Disease Control and Prevention (CDC) culture collection. To determine the discriminatory power of the SBT procedure, eight isolates that share the same monoclonal antibody (MAb) subtype (1,2,3) were included. These isolates represent four enzyme types based on previous studies using multilocus enzyme electrophoresis (MLEE) (9). All isolates were tested with a panel of MAbs to determine their subtype. Isolates were compared by AFLP according to the procedure described previously (3). Patient isolates and representative environmental isolates from both outbreaks were analyzed by a sequence-based typing scheme (5).

Sequencing was performed on an ABI 3100 using a Big Dye Terminator V1.1 cycle sequencing kit according to the manufacturer's instructions. Forward and reverse sequences were aligned using the SeqMan program of DNA Star. Consensus sequences were compared to the reference sequences obtained from the European Working Group for *Legionella* Infections website (http://www.ewgli.net) using the Meg-

Robert F. Benson, Claressa E. Lucas, Ellen W. Brown, and Barry S. Fields Respiratory Disease Branch, Centers for Disease Control and Prevention, 1600 Clifton Rd., Atlanta, GA 30333. *Karen D. Cowgill* Seattle Biomedical Research Institute, 307 Westlake Ave N, Suite 500, Seattle WA 98109.

TABLE 1 SBT for isolates obtained from St. Croix, U.S.Virgin Islands during 1981 and 2003 epidemic investigations and reference strains

Strain	ET	flaA	pilE	asd	mip	mompS	proA
Concord 4	1	2	10	3	10	9	4
Davenport 1							
Burlington 1	27	3	4	1	1	14	9
Miami Beach 1	25						
Knoxville 1		7	10	17	10	19[a]	4
Lyon Cedex 1	29	2	3	18	15	4	1
Indianapolis 10		6	10	15	28	9	14
Kingston 1	25	3	4	1	1	1	9
D6010-81		3	4	1	1	1	9
D5387-98		11	14	27[a]	16	15	13

[a]New SBT, unofficial designation. All strains were MAb pattern 1,2,3.

Align program of DNA Star. Sequences for all the alleles of each isolate were concatenated, and phylogenetic and molecular analyses were conducted using MEGA version 3.1 (6).

All isolates were subtyped as MAb pattern 1,2,3 (Knoxville 1). The AFLP patterns of environmental and patient isolates recovered during 1981 were similar to each other but different from the pattern obtained for the 2002/2003 isolates and the patient isolate from 1998 which were identical. The results of SBT confirmed that the isolates from 1981 were an identical profile (SBT = 3,4,1,1,1,9); a repre-

sentative pattern is shown in Table 1 for strain D6010-81. The isolates from 2002 and the 1998 patient isolate were different from the previous outbreak but had an identical profile for all isolates (SBT = 11,14,27,16,15,13); a representative SBT pattern is shown in Table 1 for strain D5387-98. Results of SBT analysis for all the MAb 1,2,3 strains examined are also shown in Table 1. An analysis of the strains using MEGA 3.1 showed that the isolates recovered during the 2002 outbreak represent a new clone of MAb 1,2,3 (Fig. 1). A comparison of strains previously analyzed by MLEE

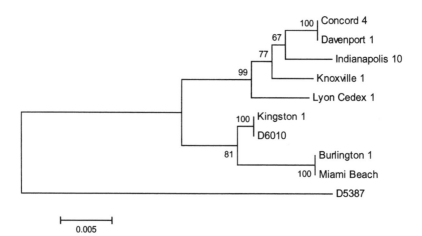

FIGURE 1 Genetic relatedness of St. Croix patient isolates from 1981 and 2003 and reference *Legionella pneumophila* serogroup 1 MAb 1,2,3 pattern.

did not show 100% agreement with the SBT analysis. Two strains, Miami Beach 1 and Kingston 1, which are the same MLEE type (type 25), were different when compared by SBT. Additionally, two strains, Burlington 1 and Miami Beach 1, which shared the same SBT, were different when compared by MLEE types 25 and 27.

CONCLUSIONS

During the 1981/1982 investigation, strains sharing the same MAb pattern as the patient isolate were recovered from several hotels and environmental sources. Monoclonal typing was unable to confirm a source of exposure for these patients. In the subsequent investigation 20 years later, the same monoclonal pattern was again obtained from several environmental sources and an intervening patient isolate in 1998. Using SBT, we were able to show that the strains from each outbreak were different from each other and the colonization of the hotel was the result of a new strain being introduced into the water system, not a clone of the previous isolate. During both outbreak investigations monoclonal subtype 1,2,3 was isolated from several sources, both associated with patients and other nonrelated sources. Therefore, MAb subtyping was not able to determine a definite source of exposure for patients. It is noteworthy that SBT genotyping did not further differentiate the MAb subtypes obtained from these various sources for each outbreak and therefore did not provide any additional information. Aurell et al. (2) they also observed that almost all epidemic isolates had a single sequence type. Darelid et al. (4) reported on the persistence of a single clone in a study of colonization of a municipal and hospital water distribution system over a 12-year period. Aurell et al. (1) reported on the occurrence of an endemic strain in Paris which had persisted in the water distribution system between 1987 and 2002, a 15-year period. Rangel-Frausto et al. (8) also reported on the persistence of a single strain over a 13-year period in a hospital water system. In the present study two different clones were isolated during the 17-year period. The introduction of a new strain into the environment could be due to its ability to multiply at a faster rate than the previous inhabitant. Other studies have shown that endemic clones can cause unrelated cases of Legionnaires' disease. Pruckler et al. (7) reported on the similarity of clinical and environmental isolates from seven outbreaks. In our study the predominant clone was replaced by a totally different clone. Selander et al. (9) have suggested that widespread distribution of Legionella strains could occur by wind and rain transportation. It is possible that the new clone was introduced into the water system after eradication procedures were employed after the 1981 outbreak.

REFERENCES

1. **Aurell, H., J. Etienne, F. Forey, M. Reyrolle, P. Girardo, P. Farge, B. Decludt, C. Campese, F. Vandenesch, and S. Jarraud.** 2003. *Legionella pneumophila* serogroup 1 strain Paris: endemic distribution throughout France. *J. Clin. Microbiol.* **41:**3320–3322.
2. **Aurell, H., P. Farge, H. Meugnier, M. Gouy, F. Foey, G. Lina, F. Vandenesch, J. Etienne, and S. Jarraud.** 2005. Clinical and environmental isolates of *Legionella pneumophila* serogroup 1 cannot be distinguished by sequence analysis of two surface protein genes and three housekeeping genes. *Appl. Environ. Microbiol.* **71:**282–289.
3. **Benson, R. F., B. S. Fields, A. Benin, H. Craddock, and R. E. Besser.** 2002. Application of amplified fragment length polymorphism analysis to subtyping of *Legionella pneumophila* serogroup 6, p. 243–247. *In* R. Marre, Y. Abu Kwaik, C. Bartlett, N. P. Cianciotto, B. S. Fields, M. Frosch, J. Hacker, P. C. Luck (ed.). Legionella. ASM Press, Washington, D.C.
4. **Darelid, J., S. Bernander, K. Jacobson, and S. Lofgren.** 2004. The presence of a specific genotype of Legionella pneumophila serogroup 1 in a hospital and municipal water distribution system over a 12-year period. *Scand. J. Infect. Dis.* **36:**417–423.
5. **Gaia, V., N. K. Fry, B. Afshar, P. C. Luck, et al.** 2005. Consensus sequence-based scheme for epidemiological typing of clinical and environmental isolates of *Legionella pneumophila*. *J. Clin. Microbiol.* **43:**2047–2052.
6. **Kumar, S., K. Tamura, and M. Nei.** 2004. MEGA3: Integrated software for molecular evolutionary genetics analysis and sequence alignment. *Briefings Bioinformatics* **5:**150–163.

7. **Pruckler, J. M., L. A. Mermel, R. F. Benson, C. Giorgio, P. K. Cassiday, R. F. Breiman, C. G. Whitney, and B. S. Fields.** 1995. Comparison of *Legionella pneumophila* isolates by arbitrarily primed PCR and pulsed-field gel electrophoresis: analysis from seven epidemic investigations. *J. Clin. Microbiol.* **33:**2872–2875.

8. **Rangel-Frausto, M. S., P. Rhomberg, R. J. Hollis, M. A. Pfaller, R. P. Wenzel, C. M. Helms, and L. A. Herwaldt.** 1999. Persistence of *Legionella pneumophila* in a hospital's water system: a 13-year survey. *Infect. Control. Hosp. Epidemiol.* **20:**793–797.

9. **Selander, R. K., R. M. McKinney, T. S. Whittam, W. F. Bibb, D. J. Brenner, F. S. Nolte, and P. E. Pattison.** 1985. Genetic structure of populations of *Legionella pneumophila J. Bacteriol.* **163:**1021–1037.

SEQUENCE-BASED TYPING OF *LEGIONELLA PNEUMOPHILA* AS AN AID IN INVESTIGATION OF HOSPITAL-ACQUIRED LEGIONNAIRES' DISEASE

Faiz Fendukly and Sverker Bernander

38

The sequence-based typing (SBT) method was used to determine the allelic profiles of three sporadic clinical isolates as well as seven environmental isolates of *Legionella pneumophila* serogroup 6, isolated at the Bacteriology Laboratory Unit of the Karolinska University Hospital in 2004. The clinical isolates were cultured from patients with suspected nosocomial Legionnaires' disease (LD), while the environmental isolates were cultured from potable water sources in the hospital wards in the vicinity of the three patients being investigated. The six genes which were sequenced for the construction of the SBT profile of the isolates included *flaA, pilE, asd, mip, mompS*, and *proA*, in this predetermined order, and the allelic profile of all the 10 isolates was indistinguishable (3, 13, 1, 28, 14, 9). Four additional geographically unrelated *L. pneumophila* serogroup 6 isolates were included as controls, and their SBT profile was congruent to that mentioned above.

We conclude that the environmental strain isolated from our hospital's drinking water is identical genotypically to the three clinical isolates of *Legionella* and even identical to the control strains isolated outside the Karolinska Hospital. This finding may suggest a smaller number of genotypes belonging to *L. pneumophila* serogroup 6 than was expected. The SBT may prove to be an effective aid for the epidemiological investigation of nosocomial LD.

Legionella species, especially *Legionella pneumophila* is a well-known cause of nosocomially acquired pneumonia, occurring frequently in immunocompromised subjects (7). In order to determine the source of infection, accurate and reliable epidemiological and molecular typing techniques are indispensable for this purpose. Recently the SBT method was introduced by the European Working Group for Legionella Infections (EWGLI) and was shown to be accurate, reproducible, and easy to perform (6). In Sweden three nosocomial outbreaks of LD have been reported during the period from 1990 to 2000 (1–3), but sporadic cases continue to emerge, causing morbidity and mortality (10). In this study the SBT method was used to determine the allelic profiles of three *L. pneumophila* serogroup 6 clinical isolates from suspected cases of nosocomial LD as well as seven environmental *L. pneumophila* serogroup 6 isolates collected from drinking water sources in the vicinity of the

Faiz Fendukly   Department of Clinical Microbiology, L2:02, Karolinska University Hospital, Solna, 171 76 Stockholm, Sweden.   *Sverker Bernander*   Department of Clinical Microbiology, Uppsala University Hospital, 751 85 Uppsala, Sweden.

patients being investigated. The aim of the study was to correlate the nosocomial infections to the environmental strain in the water systems of our hospital.

Three patients with suspected hospital-acquired LD and from whom *L. pneumophila* was isolated were included in this study. The patients were treated in different wards of the Karolinska University Hospital, Stockholm, Sweden. Two young male patients, age 23 and 32, had acute lymphoblastic leukemia and were put on cytotoxic therapy, while the third patient, a 72-year-old woman had chronic vasculitis with glomerulonephritis and was on immunosuppressive therapy. All three patients had long hospital stays ranging from 2 to 3 weeks prior to acquisition of signs and symptoms of LD.

Sputum samples were collected and cultured according to standard microbiological methods (9). Allocation to the serogroup was carried out by the Dresden panel of monoclonal antibodies using an immunofluorescence technique (8).

Subsequent to the isolation of *Legionella* from the clinical specimens, 28 water samples were collected from the tap water sources, showers, and other outlets in the patients' rooms. Water treatment, filtration, culture, and identification of *Legionella* were according to standard procedures applied in our laboratory (9). Allocation to the serogroup was done as mentioned above.

Four additional *L. pneumophila* serogroup 6 isolates geographically unrelated to our hospital were included as controls and were typed by the SBT method as were the clinical and environmental isolates.

The SBT protocol for the epidemiologic typing of *L. pneumophila* was the one proposed by the EWGLI (4). This typing method allows comparison of the allelic profile of an isolate to previously assigned allele numbers of six alleles in this given order of genes: *flaA, pilE, asd, mip, mompS,* and *proA* using the online database at the EWGLI's website. The combination of alleles of the genes in the sequence mentioned above defined the SBT or allelic profile

of the isolate. The procedures for DNA extraction, amplification and sequencing were carried out according to a previously described protocol (4, 6).

Culture of the three sputum samples yielded growth of *L. pneumophila*. Of the 28 water samples collected, 7 showed growth of *L. pneumophila*. All 10 isolates, clinical and environmental, were typed as serogroup 6, and the SBT profile of these isolates as well was the 4 control isolates was indistinguishable (3,13,1,28,14,9). Of the many available phenotypic, immunologic, and molecular typing methods for the characterization and epidemiologic typing of *L. pneumophila*, the amplified fragment length polymorphism (AFLP) and the SBT have mostly gained acceptance by the EWGLI (4, 5). Both methods obviated the need to submit *Legionella* isolates across microbiology laboratories in different countries. The AFLP, on the other hand, suffered a major disadvantage, namely the difficulty in the interpretation of gel images by inexperienced personnel.

Unlike the AFLP, the SBT has proven to be a simple technique, requiring only the ability to amplify DNA fragments by PCR and to sequence the fragments using a DNA sequencer. It has a superior discriminatory capacity than previously described techniques based on gel image interpretation, since it possesses the capability to detect far more variation within a gene locus, resulting in the detection of many more alleles per locus. Another important advantage was the portability of sequence data, allowing results from different laboratories to be compared and shared without the need to exchange strains by conventional mail.

The SBT in this study has confirmed the epidemiological relatedness of the clinical and environmental isolates. Interestingly, the four control isolates collected from other hospitals in Sweden shared the same SBT profile, a finding that can be explained by the possible limited genomic heterogeneity within *L. pneumophila* serogroup 6 compared to serogroup 1. Further typing of a larger number of *L. pneumophila* serogroup 6 isolates is needed to validate this theory.

REFERENCES

1. **Bernander, S., K. Jacobson, J. H. Helbig, P. C. Luck, and M. A. Lundholm.** 2003. Hospital-associated outbreak of Legionnaires' disease caused by *Legionella pneumophila* serogroup 1 is characterized by stable genetic fingerprinting but variable monoclonal antibody patterns. *J. Clin. Microbiol.* **41:**2503–2508.

2. **Bernander, S., K. Jacobson, and M. Lundholm.** 2004. A hospital-associated outbreak of Legionnaires' disease caused by *Legionella pneumophila* serogroups 4 and 10 with a common genetic fingerprinting pattern. *APMIS* **112:**210–217.

3. **Darelid, J., L. Bengtsson, B. Gastrin, H. Hallander, S. Lofgren, B. E. Malmvall, et al.** 1994. An outbreak of Legionnaires' disease in a Swedish hospital. *Scand. J. Infect. Dis.* **26:**417–425.

4. **Gaia, V., F. N., B. Afshar, P. C. Lück, H. Meugnier, J. Etienne, R. Peduzzi, and T. G. Harrison.** 2005. A consensus sequence-based epidemiological typing scheme for clinical and environmental isolates of *Legionella pneumophila, J. Clin. Microbiol.*, in press.

5. **Fry, N .K., J. M. Bangsborg, A. Bergmans, S. Bernander, J. Etienne, L. Franzin, et al.** 2002. Designation of the European Working Group on Legionella Infection (EWGLI) amplified fragment length polymorphism types of *Legionella pneumophila* serogroup 1 and results of intercentre proficiency testing using a standard protocol. *Eur. J. Clin. Microbiol. Infect. Dis.* **21:**722–728.

6. **Gaia, V., N. K. Fry, T. G. Harrison, and R. Peduzzi.** 2003. Sequence-based typing of *Legionella pneumophila* serogroup 1 offers the potential for true portability in legionellosis outbreak investigation. *J. Clin. Microbiol.* **41:**2932–2939.

7. **Gump, D. W., and M. Keegan.** 1986. Pulmonary infections due to Legionella in immunocompromised patients. *Semin. Respir. Infect.* **1:**151–159.

8. **Helbig, J. H., J. B. Kurtz, M. C. Pastoris, C. Pelaz, and P. C. Luck.** 1997. Antigenic lipopolysaccharide components of *Legionella pneumophila* recognized by monoclonal antibodies: possibilities and limitations for division of the species into serogroups. *J. Clin. Microbiol.* **35:**2841–2845.

9. **Isenberg, H. D.** (ed.) *Clinical Microbiology Procedures Handbook.* 1992;1:1. 12.26–30.

10. **Wilczek, H., I. Kallings, B. Nystrom, G. Tyden, and C. G. Groth.** 1985. Experience with 5 cases of Legionnaires' disease at the Department of Transplantation Surgery, Huddinge Hospital, Stockholm, Sweden. *Scand. J. Urol. Nephrol. Suppl.* **92:**67–69.

LEGIONNAIRES' DISEASE ASSOCIATED WITH DEATH AFTER NEAR DROWNING IN LAKE WATER

Jaana Kusnetsov, Satu Pastila, Silja Mentula, and Diane S. J. Lindsay

39

Direct aspiration of water during submersion has caused Legionnaires' disease (LD) cases. We present one such case caused by *Legionella pneumophila* with a novel source of infection.

CASE REPORT

A 53-year-old male was admitted to a Kanta-Häme Central Hospital after submersion for about 2 minutes in shallow water of Lake Vanajavesi, Finland, in August 2002. The patient was healthy but had periodical alcohol abuse. On hospital admission his vital functions were as described in Table 1. He was intubated, and mud, water, and small pieces of decaying wood were aspirated from the trachea. Mechanical ventilation support and intravenous cefuroxime (1.5 g three times a day) were initiated. He developed fever, and the serum cyclic AMP receptor protein CRP concentration increased. Bronchoalveolar lavage revealed purulent secretion with brown mud spots and growth of *Pseudomonas aeruginosa*. Ce-

furoxime was substituted for tobramycin and piperacillin-tazobactam. Treatment with high-dose methylprednisolon was started on day 8 due to suspected acute respiratory distress syndrome. Culture of tracheal secretion yielded *Aspergillus fumigatus*. The patient was continuously febrile, and on day 13, piperacillin-tazobactam was replaced by imipenem, and tracheotomy was performed.

On day 16, *Legionella* urinary antigen was strongly positive by Biotest EIA and levofloxacin was initiated. Serum samples tested contained antibodies against *L. pneumophila* serogroup (sg) 3 (titer, 1/2,048) and *L. pneumophila* sg 4 (1/1,024, in-house IF-method). No samples were available for *Legionella* culture.

On day 20, ventilation improved and CRP concentration decreased to 74 mg/liter, and tobramycin was discontinued. The next day the patient had left-sided convulsions. Imipenem and levofloxacin were replaced by clarithromycin and piperacillin-tazobactam. Bilateral cerebral hypoxic injury of the white matter was detected in the computed tomography scan of the brain. Several tracheal cultures yielded *Aspergillus fumigatus* and *Rhizopus* sp., and there were several thick-walled cavitations surrounded by consolidation in the computed tomography scan of the chest. On day 29, intravenous amphotherisin B at 40 mg four

Jaana Kusnetsov Laboratory of Environmental Microbiology, National Public Health Institute, P.O. Box 95, FI-70701 Kuopio, Finland. *Satu Pastila* Kanta-Häme Central Hospital, Hämeenlinna, Finland. *Silja Mentula* Anaerobe Reference Laboratory, National Public Health Institute, FI-00300 Helsinki, Finland. *Diane S. J. Lindsay* Scottish *Legionella* Reference Laboratory, Stobhill Hospital, Balornock Road, GB-G213UW Glasgow, Scotland, UK.

Legionella: State of the Art 30 Years after Its Recognition
Edited by Nicholas P. Cianciotto et al.
©2006 ASM Press, Washington, D.C.
146

TABLE 1 General vital functions of patient immediately after the near-drowning incident

Vital function	Case	Normal range
pO$_2$ (kPa)	22.1	11.0–14.0
pCO$_2$ (kPa)	9.0	4.5–6.0
pH	7.06	7.36–7.46
Base excess (mmol/liter)	−13.0	−2.5–2.5
Serum potassium (mmol/liter)	3.3	3.3–5.0
Serum sodium (mmol/liter)	142	136–146
Hemoglobin (g/dl)	16.1	13.5–17.0
White cell count (10^9/liter)	3.7	4.0–10.0

times a day was started, and due to a rash, piperacillin-tazobactam was changed to cefepime and metronidazole. The patient developed CO$_2$ retention despite ventilation support. On day 34, a pleural drain was inserted due to a small pneumothorax and some pleural fluid on the left side. Ciprofloxacin was added to the therapy on day 37. After 9 days of amphotericin B treatment, two tracheal samples were still positive for *A. fumigatus*.

After 6 weeks of treatment, the patient again developed a high fever. Amphotericin B was substituted for itraconazole, as azole resistance of the environmental fungi was suspected. On day 50, the patient developed anuric renal failure. His consciousness deteriorated, the blood pressure fell, and he died on day 52. The clinical cause of death was multiorgan failure. The autopsy revealed consolidated lungs with cavitations and purulent foci. The right side of the heart was enlarged, and the lumen of the coronary arteries was moderately narrowed. In the septum of the heart there was an inflammatory focus with some pus.

ENVIRONMENTAL STUDY
Muddy and clear lake water samples were taken from the same point on the beach 40 days after the near-drowning incident. The water temperature then was 10°C, but at the time of the incident it had been 22°C. Water samples were processed according to a standard culture method (ISO 11731). In addition, PCR analysis for legionellae (5s RNA) for the lake water samples were performed (8). The

hospital water system, including samples from four showers and hot and cold tap water from the dialysis unit and haematology department, had been tested for legionellae with the culture method 8 months before the case. After the *Legionella*-positive findings of the patient, two water samples from the intensive care unit were again similarly tested. No *Legionella* was isolated by culture or detected by PCR from the Lake Vanajavesi water samples. The samples taken from the hospital, before and after treating the patient, did not contain culturable *Legionella*.

CONCLUSIONS
The positive urinary antigen result with the Biotest EIA confirms that the patient was suffering from LD. The serology findings suggest that the causative organism of LD was *L. pneumophila* sg 3 and/or sg 4. Appropriate antimicrobial therapy was given once LD was identified, in addition to the antibiotics targeted against other pathogens. The patient suffered from bilateral lung infiltrates, pleural effusion, and cavitation of lung parenchyma, all symptoms of LD. The patient had LD and *P. aeruginosa* and fungal infections. The pulmonary damage led to multiorgan failure, and the secondary or co-infections contributed to his death.

The hospital water system did not contain *Legionella* in any of the samples tested and the case did not fit the criteria for a nosocomial case. No *Legionella* was isolated or detected by PCR either from lake water samples. This is not surprising, as *Legionella* prevails only in low concentrations in a natural nonheated environment. *Legionellae* have been detected by direct fluorescence antibody test in lake water with temperatures ranging from 6 to 63°C (6). The highest water temperature in this Finnish lake was above 20°C during the summer of 2002, which could account for an increase in bacterial counts. The lake samples were negative for *Legionella* when they were taken from colder water (10°C).

The earliest LD case after direct aspiration of water during submersion occurred in 1958 after using underwater breathing apparatus in a freshwater swimming pool (1–3). A lung tissue sample was later inoculated into guinea pigs and yolk sac, and *L. bozemanii* was isolated. Cases of LD have been identified after near drowning incidents in a spa (10) and a hot spring water tub (Yabuuchi, E., L. Wang, M. Arakawa, U. Nonaka, M. Shibako, Abstr. 9th meeting of the European Working Group on Legionella Infections, p. 67–68, 1994). LD cases have also occurred in neonates after water births (7, 9). Immersion in river water has resulted in at least two reported LD cases caused by *L. pneumophila* sg 10 and sg 13 (4, 5). These clinical findings and lack of evidence of a nosocomical case suggest that this LD case resulted from aspiration of lake water and mud. This is the first report of LD caused by *L. pneumophila* associated with a near-drowning incident in lake water.

REFERENCES

1. **Bozeman, F. M., J. W. Humphries, and J. M. Campbell.** 1968. A new group of Rickettsia-like agents recovered from guinea pigs. *Acta Virol.* **12:**87–93.
2. **Brenner, D. J., A. G. Steigerwalt, G. W. Gorman, R. E. Weaver, J. C. Feeley, L. G. Cordes, H. W. Wilkinson, C. Patton, B. M. Thomason, and C. R. Lewallen Sasseville.** 1980. *Legionella bozemanii* sp. nov. and *Legionella dumoffii* sp. nov.: classification of two additional species of *Legionella* associated with human pneumonia. *Curr. Microbiol.* **4:**111–116.
3. **Cordes, L. G., H. W. Wilkinson, G. W. Gorman, B. J. Fikes, and D. W. Fraser.** 1979. Atypical Legionella-like organisms: fastidious water associated bacteria pathogenic for man. *Lancet* **i:**927–930.
4. **Faris, B., C. Faris, M. Schousboe, and C. H. Heath.** 2005. Legionellosis from *Legionella pneumophila* serogroup 13. *Emerg. Infect. Dis.* **11:**1405–1409.
5. **Farrant, J. M., A. E. Drury, and R. P. Thompson.** 1988. Legionnaires' disease following immerson in a river. *Lancet* **i:**460.
6. **Fliermans, C. B., W. B. Cherry, L. H. Orrison, S. J. Smith, D. L. Tison, and D. H. Pope.** 1981. Ecological distribution of *Legionella pneumophila. Appl. Environ. Microbiol.* **41:**9–16.
7. **Franzin, L., C. Scolfaro, D. Cabodi, M. Valera, and P. A. Tovo.** 2001. *Legionella pneumophila* pneumonia in a newborn after water birth: a new mode of transmission. *Clin. Infect. Dis.* **33:**e103–e104.
8. **Lindsay, D. S., W. Abraham, G. Edwards, and R. W. Girdwood.** 2002. Detection of *Legionella*-specific DNA in serum, p. 216–220. *In* R. Marre, Y. Abu Kwaik, C. Bartlett, N. P. Cianciotto, B. S. Fields, M. Frosch, J. Hacker, and P. C. Lück. (ed.). *Legionella.* ASM Press, Washington, D.C.
9. **Nagai, T., H. Sobajima, M. Iwasa, T. Tsuzuki, F. Kura, J. Amemura-Maekawa, and H. Watanabe.** 2003. Neonatal sudden death due to Legionella pneumonia associated with water birth in a domestic spa bath. *J. Clin. Microbiol.* **41:**2227–2229.
10. **Yabuuchi, E., H. Miyamoto, S. Yoshida, R. Shiota, and K. Yamamoto.** 1996. Determination of hot spring spa water as the source of infection in a case of acute pneumonia due to *Legionella pneumophila* serogroup 3, p. 99–101. *In* B. Berdal (ed.). *Legionella Infections and Atypical Pneumonias. Proceedings of the 11th Meeting of the European Working Group on Legionella Infections.* The Norwegian Defence Microbiological Laboratory, Oslo.

IS USE OF POTTING MIX ASSOCIATED WITH *LEGIONELLA LONGBEACHAE* INFECTION? RESULTS FROM A CASE CONTROL STUDY IN SOUTH AUSTRALIA

Bridget O'Connor, Judy Carman, Kerena Eckert, and Graeme Tucker

40

Legionella longbeachae infections in Australia occur more frequently in southern and western Australia, where over 80% of legionellosis can be due to *L. longbeachae* infection (3).

A case control study was performed in South Australia to determine if *L. longbeachae* infection was associated with recent handling of commercial potting mix and to examine possible modes of transmission. Twenty-five laboratory-confirmed case and 75 controls, matched for 5-year age group, sex, and residential postcode, were enrolled between April 1997 and March 1999. Information on underlying illness, smoking history, gardening exposures, and behaviors was obtained by telephone interviews.

The demographics of the cases were similar to controls, but cases had significantly more preexisting cardiac, respiratory, and other medical conditions. Recent use of potting mix was associate with illness (odds ratio [OR] 4.74; 95% confidence interval [95% CI], 1.65 to 13.55; $P = 0.004$) in bivariate analysis. Possible

exposure to waterborne aerosolized *L. longbeachae* by being near hanging pots that were dripping increase the risk of illness (OR, 2.79; 95% CI, 1.05 to 7.47; $P = 0.04$). Other gardening exposures such as gardening frequency, exposure to watering in the garden (as opposed to hanging pots), and exposure to other garden soils were not significantly associated with illness. Being aware of the possible health risk associated with use of potting mix was a significant protective factor against illness (OR, 0.27, 95% CI, 0.10 to 0.73; $P = 0.01$). Of the gardening behaviors that were examined , eating or drinking after gardening without washing one's hands was associated with an increased likelihood of illness (OR, 6.22, 95% CI, 1.95 to 19.77; $P = 0.002$). After adjusting for the effects of other exposure variables (preexisting medical conditions, smoking history, risk awareness, and gardening behaviors) in the model, use of potting mix was no longer significantly associated with illness (Table 1). Better predictors of illness in the multivariate analysis included poor hand-washing practices after gardening, long-term smoking, and being near dripping hanging flower pots. Awareness of a possible health risk with potting mix protected against illness.

This study suggests there are other factors within the gardening environment, as well as

Bridget O'Conner Communicable Disease Control Branch, South Australia Department of Health, Adelaide SA 5000, Australia. *Judy Carman* Department of Public Health, The University of Adelaide SA 5005, Australia. *Kerena Eckert* Division of Health Sciences, University of South Australia. *Graeme Tucker* Epidemiology Branch, South Australia Department of Health.

TABLE 1 Multivariate analysis

Exposure variable	OR	95% CI	P value
Having an underlying cardiac illness	10.78	0.95–122.37	0.055
Having smoked for more than 30 years	19.16	2.25–163.21	0.007[a]
Being near hanging pots that were dripping	8.97	1.41–56.96	0.020[a]
Being aware of possible risk when using potting mix	0.12	0.02–0.70	0.019[a]
Eating or drinking after gardening without washing hands	29.47	1.96–412.14	0.014[a]

[a]Statistically significant at $P < 0.05$ level

intrinsic and behavioral host factors, that are better predictors of *L. longbeachae* infection than recent use of potting mix. Host-related factors such as long-term smoking (a potential marker for underlying respiratory and cardiac disease) and poor gardening hygiene are newly described risk factors for *L. longbeachae* from this study. Previously, cases have been described as being less likely to be current smokers and to have similar rates of chronic medical conditions to the general population (1). However, the finding in this study is consistent with such people having a higher risk of other *Legionella* infections such as *L. pneumophila* (2).

An association between illness and proximity to dripping, hanging pots supports inhalation of contaminated aerosols produced during watering as a possible mode of transmission. Another possible mode of transmission is ingestion of organisms via contaminated hands, which is supported by the association between illness and eating or drinking after gardening before washing hands. These results do not help determine whether one of these methods of transmission is more important than the other, and it is possible that both methods may be important for *L. longbeachae* infection.

Potential limitations in this study result from the sample size and the exposure window used. Retrospective calculations indicated that there was sufficient power to detect an association between potting mix use and *L. longbeachae* infection at the 0.05 level of significance; however the analysis of other gardening exposures and gardening behaviors was hindered by the small sample size. A 4-week exposure window, rather than a time period closer to the 10-day incuba-

tion period, was used as there was less certainty about the incubation period when this study was designed. This may result in misclassification within the exposure variable, as cases or controls reporting exposure between 10 days and 4 weeks before hospitalization or interview may have been misclassified as exposed when they may not have been.

In summary, this study has provided some clarification of the risk factors for *L. longbeachae* infection in Australia. Factors such as exposure to aerosolized organisms, poor gardening hygiene, and long-term smoking may be important predisposing factors to *L. longbeachae* infections. Further information on risk factors for *L. longbeachae* infection that explores the interface between potting mix, other gardening exposures, and behaviors while gardening need to be addressed in a larger study.

Recommendations from this study include:

- Long-term smokers and possibly people with preexisting medical conditions such as respiratory and cardiac illness should be warned about their increased risk of *L. longbeachae* infection. Long-term smokers in particular should be advised to follow good hygiene when gardening and to wash hands before eating, drinking, or smoking.
- Raising people's awareness of a possible health risk when using potting mix should continue in order to protect against *L. longbeachae* infection.

REFERENCES

1. **Cameron, S., D. Roder, C. Walker, and J. Feldheim.** 1991. Epidemiological characteristics of Legionella infection in South Australia: impli-

cations for disease control. *Aust. N. Z. J. Med.* **21:**65–70.

2. **Edelstein, P. H., N. P. Cianciotto.** 2005. *Legionella*, p. 2711–2724, *In* G. Mandell, J. Bennett, R. Dolin (ed.). *Principles and Practice of Infectious Diseases.* Part III. *Infectious Diseases and their Etiologic Agents*. Philadelphia Churchill Livingstone.

3. **Miller, M., P. Roche, K. Yohannes, J. Spencer, M. Bartlett, J. Brotherton, J. Hutchinson, M. Kirk, A. McDonald, and C. Vadjic.** 2005. Australia's notifiable disease status, 2003 Annual Report of the National Notifiable Diseases Surveillance System. *Communicable Dis. Intell.* **29:**1–76.

EPIDEMIOLOGICAL TYPING
OF *LEGIONELLA PNEUMOPHILA*
IN THE ABSENCE OF ISOLATES

Norman K. Fry, Baharak Afshar,
Günther Wewalka, and Timothy G. Harrison,

41

The recently described sequence-based typing (SBT) scheme for epidemiological typing of *Legionella pneumophila* (1) offers many advantages over other methods. However, clinical and environmental isolates are not always available to allow epidemiological typing to be performed. For this reason we have sought to apply the SBT technique directly to clinical and environmental samples where isolates are not available or before they become available. This has the potential to allow more rapid determination of typing data or the possibility of obtaining such data when there is no alternative, i.e., in the absence of isolates.

PROOF OF PRINCIPLE

The first aim was to determine if this direct approach was feasible. This was established using environmental concentrates, together with environmental and clinical isolates, obtained during the investigation of the largest United Kingdom (UK) outbreak to date, in Barrow-in-Furness in 2002. The strategy was to perform a simple genomic DNA extraction method on the environ-

Norman K. Fry, Baharak Afshar, and Timothy G. Harrison Health Protection Agency, Respiratory and Systemic Infection Laboratory, Centre for Infections, London, 61 Colindale Avenue, London NW9 5EQ, UK. *Günther Wewalka* Austrian Agency for Health and Food Protection (AGES), Ges. m. b. H., Institute for Medical Microbiology and Hygiene, Währingerstraße 25a, A-1096 Vienna, Austria.

mental concentrates using the InstaGene Matrix (Bio-Rad) and to use these extracts as a template in the primary PCR amplifications for the three original genes in the SBT method (*flaA, mompS, proA*) (2), purify any products obtained and determine the primary DNA sequence as per the standard protocol. The sequence data obtained would be compared with that obtained from nucleic extraction from both the clinical and environmental isolates.

Successful amplification was achieved from all three genes, and following sequence analysis the allelic profile from the environmental extract was determined as *flaA*(2),*mompS*(2),*proA*(1). This was identical to that obtained from the environmental isolates and the clinical isolates. Thus the proof of principle of obtaining sequence typing data without isolation was successfully established. This led to its employment in a number of real-life investigations of legionellosis using the six genes (*flaA, piLE, asol, mompS, proA*) in the current scheme (4).

EXAMPLE 1: INVESTIGATION OF LEGIONELLOSIS IN A HOLIDAY CAMP IN HAMPSHIRE, UNITED KINGDOM

A holiday complex on Hayling Island, UK, was previously associated with a case of Legionnaires' disease in 1994. The case had antibody levels consistent with infection by *L. pneumophila*, but no isolate was obtained. In June and July 2005 a cluster of three cases was reported from the same complex. Respiratory samples from two suspected cases (cases 1 and 2) were obtained, but *L. pneumophila* was only isolated from one (case 1). PCR analysis of the extract from case 2 was also undertaken using an in-house *L. pneumophila*–specific real-time PCR assay targeting the macrophage infectivity potentiator (*mip*) gene. SBT was performed on DNA extracted from both the isolate (case 1) and the clinical sample (case 2).

EXAMPLE 2: INVESTIGATION OF LEGIONELLOSIS AT AN AUSTRIAN CAMPSITE

A cluster of cases of legionellosis was reported associated with a campsite in Austria. Bronchial secretions were obtained from two cases. Isolation of *L. pneumophila* was attempted, but without success, probably due to prior antibiotic therapy. Environmental investigations resulted in the isolation of *L. pneumophila* serogroup 1, monoclonal antibody (mAb) subgroup Philadelphia from the hot water system of the campsite. The clinical samples were analyzed by *L. pneumophila* PCR and SBT, and the SBT profile of the environmental isolate was also determined.

EXAMPLE 3A: INVESTIGATION OF POSSIBLE NOSOCOMIAL CASE IN A HOSPITAL IN ENGLAND

In February 2004, a 37-year-old male with acute myeloid leukaemia died and four postmortem samples were obtained. The environmental investigation yielded *L. pneumophila* from the hospital. *L. pneumophila* PCR and direct SBT were performed on all clinical samples and compared to the profile obtained from the environmental isolate. All postmortem samples were PCR positive for *L. pneumophila,* and SBT data was obtained before the organism was grown from the samples (see Table 1).

EXAMPLE 3B: INVESTIGATION OF ADDITIONAL NOSOCOMIAL CASE FROM THE SAME HOSPITAL AS EXAMPLE 3A

In September, 2005, another case of legionellosis was reported in a 59-year-old male with chronic myeloid leukaemia. *L. pneumophila* was not isolated from the patient. Since the previous case (in 2004), a strict regime of monitoring had been introduced and ca. 100 samples/month, were tested for the presence of *L. pneumophila*, which was very rarely isolated.

Following this most recent case, a "sporadic" isolate of *L. pneumophila* serogroup 1, mAb subgroup Philadelphia was isolated. *L. pneumophila* PCR and direct SBT were performed on the respiratory sample from the patient and the SBT profile of the environmental isolate was determined. This profile (1,4,3,1,1,1) is the most common profile to date in the UK and worldwide (of those countries for which data are available) but is distinct from the previous case (see Table 1).

CONCLUSIONS AND FURTHER WORK

The proof of principle of direct SBT was established using environmental samples. One concern about testing environmental samples is the presence of mixed *L. pneumophila* populations. However, in typical point-source outbreaks the causative organism is present in very high numbers and thus is preferentially amplified. Direct SBT was also successfully demonstrated using clinical samples including respiratory and postmortem specimens. Reasons for lack of isolation of *L. pneumophila* include antimicrobial therapy and simply not attempting it at the time. Even if isolation is subsequently achieved, direct SBT can offer speed by yielding typing data within 48 hours. Direct SBT can also offer typing data when no alternative, i.e., no *L. pneumophila* isolate is available. However, further optimization of primary amplification from "dirty" clinical and environmental samples and/or samples where a low number of target organisms are present is required. Further work to correlate the crossing point/amount of target DNA using PCR and

TABLE 1 Result of investigation of legionellosis by *L. pneumophila* PCR and SBT in (a) a holiday camp in Hampshire, UK, (b) an Austrian campsite, (c) a nosocomial case in an English hospital, (d) and another case in the same hospital as (c)

		L. pneumophila isolated	*L. pneumophila* PCR	SBT result[b]
a	**Case 1** Male 57 years Seriously ill	Yes sg 1 Allentown	NA	8,10,3,15,18,1
	Case 2 Male 68 years Died	No	Positive	0,10,3,0,18,0 Direct SBT
	Case 3 Female 57 years Mildly ill	No respiratory sample (serology positive)	NA	NA
b	**Hot water system**	Yes sg 1 Philadelphia	NA	2,3,9,10,2,1
	Case 1	No	Positive	2,3,9,10,2,1 Direct SBT
	Case 2	No	Positive[a]	0,0,0,0,0,0 Direct SBT
c	**Post-mortem** Right lung	Yes sg 1 mAb Bellingham	Positive	7,6,17,3,13,11 Direct SBT
	Post-mortem Left lung	Yes sg1 mAb Bellingham	Positive	7,6,17,3,13,11 Direct SBT
	Post-mortem Sputum	Yes sg 1 mAb Bellingham	Positive	7,6,17,3,13,11 Direct SBT
	Post-mortem Tracheal aspirate	Yes sg 1 mAb Bellingham	Positive	7,6,17,3,13,11 Direct SBT
	Environmental investigation	Yes sg 1 mAb Bellingham	Positive	7,6,17,3,13,11
d	**Case 1**	No	Positive	1,4,3,1,1,1 Direct SBT
	Environmental investigation	Yes sg 1 Philadelphia	NA	1,4,3,1,1,1 Direct SBT

[a]The *L. pneumophila* PCR had a late crossing point, i.e., >40 cycles.
[b]Direct SBT, SBT results obtained without isolates; NA, not applicable; sg, serogroup; mAb, monoclonal antibody subgroup

the likelihood of successful SBT from such specimens is planned.

ACKNOWLEDGMENTS

We acknowledge our microbiologist and epidemiologist colleagues in Hampshire, Ms. Kate Ricketts and Dr. Carol Joseph of the Health Protection Agency, Centre for Infections, London, for epidemiological data and Mrs. Petra Hasenberger, Austrian Agency for Health and Food Safety (AGES), Institute for Medical Microbiology and Hygiene, Wien, Austria, for provision of DNA extracts.

REFERENCES

1. **Gaia, V., N. K. Fry, B. Afshar, P. C. Lück, H. Meugnier, J. Etienne, R. Peduzzi, and T. G. Harrison.** 2005. Consensus sequence-based scheme for epidemiological typing of clinical and environmental isolates of *Legionella pneumophila*. *J. Clin. Microbiol.* **43:**2047–2052.
2. **Gaia, V., N. K. Fry, T. G. Harrison, and R. Peduzzi.** 2003. Sequence-based typing of *Legionella pneumophila* serogroup 1 offers the potential for true portability in legionellosis outbreak investigation. *J. Clin. Microbiol.* **41:**2932–2939.

ONLINE IDENTIFICATION OF *LEGIONELLA* SPECIES BY DNA SEQUENCE ANALYSIS: THE MACROPHAGE INFECTIVITY POTENTIATOR GENE AS AN EXAMPLE

Norman K. Fry, Baharak Afshar, William Bellamy, Anthony P. Underwood, Rodney M. Ratcliff, and Timothy G. Harrison

42

Approximately 50 species of *Legionella* have been described and many more are awaiting formal description (1). The identification of the majority of these by classical phenotypic methods is increasingly difficult. The aim of this study was to investigate the feasibility of providing a dedicated database allowing the online identification of putative *Legionella* spp. following standard PCR amplification and sequencing using a previously published identification scheme for *Legionella* (2).

The *Legionella mip* gene sequence database is curated by the Health Protection Agency, Centre for Infections (http://www.hpa.org.uk/cfi/bioinformatics/eugli/legionellamips.htm) and can also be reached via the European Working Group for Legionella Infections (EWGLI) website (http:// www.ewgli.org). A DNA sequence is entered as text or by upload-ing sequencing chromatogram files. The facility to upload chromatogram files was not available for the proficiency panels, so all submitted data for the two panels were from user sequence text files. Two proficiency panels (panel 1 and panel 2) were composed of *Legionella* species from clinical and environmental sources.

A total of 20 institutes from 14 countries participated in these studies, organized on behalf of the EWGLI. The ability of centers to correctly identify *Legionella* species was assessed following the distribution of two proficiency panels, each composed of 10 coded isolates. A standard PCR and sequencing protocol based on a previously published sequence-based identification scheme (2) was used to generate a DNA sequence from the macrophage infectivity potentiator gene (*mip*). A DNA sequence was submitted online to a dedicated web-based database. Identification of user-generated sequences was performed using tools showing: (i) an alignment of all the sequences from the reference alignment, top five database matches, and the user sequence; (ii) a neighbor-joining tree of the alignment, including the reference species, five closest matches from the database, and the user sequence; and (iii) an alignment of the eight sequences from the combined alignment that are

Norman K. Fry, Baharak Afshar, William Bellamy, and Timothy G. Harrison Health Protection Agency, Respiratory and Systemic Infection Laboratory, Centre for Infections, 61 Colindale Avenue, London NW9 5EQ, UK.   *Anthony P. Underwood* Health Protection Agency, Statistics, Modelling, and Bioinformatics Department, Centre for Infections, 61 Colindale Avenue, London, NW9 5EQ, UK.   *Rodney M. Ratcliff* Infectious Disease Laboratories, Institute of Medical and Veterinary Science, PO Box 14, Rundle Mall, Adelaide, SA, 5000, Adelaide, Australia.

TABLE 1   Results of the first EWGLI *Legionella* species *mip* proficiency panel (2003 to 2004) from 10 centers

Study code	Source	*Legionella* species	Consensus identification
A	Environmental	*L. anisa*	89%
B	Clinical	*L. bozemanii*	89%
C	Environmental	*L. dumoffii*	78%
D	Environmental	*L. dumoffii*	88%
E	Environmental	*L. jamestowniensis*	100%
F	Clinical	*L. longbeachae*	90%
G	Environmental	*L. moravica*	100%
H	Clinical	*L. pneumophila*	90%
I	Environmental	*L. quinlivanii*	88%
J	Environmental	*L. sainthelensi*	80%

most similar to the user sequence together with percentage similarity scores.

Results were received from 10 of 16 centers in 9 countries and 17 of 19 centers in 13 countries for the first and second panels, respectively. Four centers submitted results for the second panel after the deadline, so these data were excluded from the primary analysis (see Tables 1 and 2). Percentage similarity scores of <80% were excluded from the analyses. Most centers, 7 of 10, and 11 of 13 (14 of 17) correctly identified all isolates tested (number tested ranged from 5 to 10). Of 20 strains, 18

showed 100% homology to sequences in the database. Average percentage similarities from all centers achieving correct (intended) identification ranged from 94 to 100% (mean, 98%). The most likely explanations for incorrect scores were: (i) strain switching (in one case it appears that an *L. pneumophila* sequence-based typing proficiency panel was inadvertently switched with the *Legionella* spp. panel) and (ii) low-quality sequence used for the identification. Furthermore, training needs were highlighted and addressed, resulting in improvement in sequence quality.

TABLE 2   Results of the second EWGLI *Legionella* species *mip* proficiency panel (2004 to 2005) from 13 centers

Study code	Source	*Legionella* species	Consensus identification
K	Clinical	*L. bozemanii*	92%
L	Clinical	*L. longbeachae*	92%
M	Environmental	*L. jordanis*	92%
N	Environmental	*L. dumoffii*	92%
O	Clinical	*L. bozemanii*	92%
P	Clinical	*L. pneumophila*	92%
Q	Environmental	*L. oakridgensis*	92%
R	Environmental	*L. jamestowniensis*	91%
S	Environmental	*L. pneumophila*	100%
T	Environmental	*L. sainthelensi*	92%

CONCLUSIONS

Genotypic identification of *Legionella* species is an essential requirement for reference laboratories. Standard protocols, dedicated identification libraries, and online tools provide a valuable resource to help achieve this goal. Chromatogram files can now be uploaded directly and an automated tool provides feedback on the quality of the submitted sequence. It is anticipated that additional genes, including those coding for 16S rRNA, RpoB, and RnpB, will be added to the current identification system to aid in the characterization and identification of known and potential novel members of this genus.

ACKNOWLEDGMENTS

This project was supported by funding to Baharak Afshar and William Bellamy from the European Commission. We thank all the centers for their participation in this EWGLI multicenter study.

REFERENCES

1. **Harrison, T. G.** 2005. *Legionella*, p. 1761–1785. *In* S. P. Borriello, P. R. Murray, and G. Funke (ed.). *Topley & Wilson's Microbiology & Microbial Infections*, Vol. 2. *Bacteriology*, 10th ed. Hodder Arnold, London.

2. **Ratcliff, R. M., J. A. Lanser, P. A. Manning, and M. W. Heuzenroeder.** 1998. Sequence-based classification scheme for the genus *Legionella* targeting the *mip* gene. *J. Clin. Microbiol.* **36:**1560–1567.

PULSED-FIELD GEL ELECTROPHORESIS ANALYSIS AND SEQUENCE-BASED TYPING OF *LEGIONELLA PNEUMOPHILA* SEROGROUP 1 ISOLATES FROM JAPAN

Junko Amemura-Maekawa, Fumiaki Kura,
Bin Chang, and Haruo Watanabe

43

In Japan, hot springs and public baths are more probable sources of legionellosis than cooling towers.

Pulsed-field gel electrophoresis (PFGE) is considered to be one of the most discriminative epidemiological methods for subtyping *Legionella pneumophila* strains and for elucidating the sources of infection (2). On the other hand, sequence-based typing (SBT), which was recently developed, is also a powerful epidemiological method. Six genes, *flaA*, *mompS*, *pilE*, *asd*, *mip*, and *proA*, which are probably under selective pressure, were shown to be useful genetic markers for epidemiological typing (4). We used PFGE and SBT to analyze *L. pneumophila* serogroup 1 isolates from Japan and compared the usefulness of these techniques.

All *L. pneumophila* serogroup 1 isolates used in this study were independently obtained from a wide variety of Japanese locations (10 from cooling towers, 10 from public spas and/or hot spring baths, 7 from patients with bath-related infections, and 4 from patients infected at unknown sources (Fig. 1). These isolates were cultivated on a buffered charcoal yeast extract agar (Difco, Detroit, MI) plate for 3 days at 35°C. The cultured colonies were directly used for PFGE and SBT.

Preparations for PFGE typing were carried out as described previously (2). A dendrogram of the PFGE pattern based on Ward's method was constructed using the Fingerprinting II software (Bio-Rad Laboratories).

The 31 epidemiologically unrelated isolates were discriminated into 30 PFGE types (Fig. 2). The dendrogram of the PFGE pattern of isolates from cooling towers formed a distinct genetic cluster. The cluster is quite different from the cluster of isolates derived from public spas and hot spring baths. This result suggests that the PFGE pattern reflects the habitat of isolates rather than their geographic location.

We applied SBT to the same *L. pneumophila* serogroup 1 isolates as those used in the PFGE analysis. The sequences of *flaA*, *pilE*, *asd*, *mip*, *mompS*, and *proA* were determined. The primers used for SBT and the reaction mixture and conditions were the same as those used by Gaia et al. (4, 5; http://www.ewgli.org). The nucleotide sequences obtained were trimmed based on the data of *flaA*, *pilE*, *asd*, *mip*, *mompS*, and *proA* presented by Norman Fry (http://www.ewgli.org). Putative novel variants found in this study were submitted to the curators of the European Working Group

Junko Amemura-Maekawa, Fumiaki Kura, Bin Chang, and Haruo Watanabe   Department of Bacteriology, National Institute of Infectious Diseases, Toyama, Shinjuku-ku, Tokyo, 162-8640, Japan.

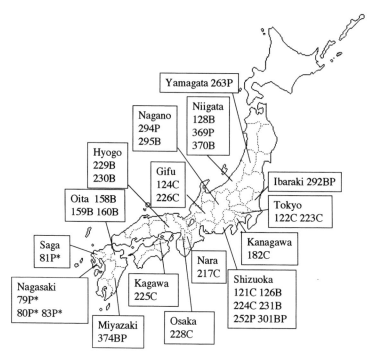

FIGURE 1   Isolates of *L. pneumophila* serogroup 1 in Japan were used in this study. All isolates are epidemiologically unrelated to each other. The names of these isolates are indicated on the map along with the names of the prefectures where they were isolated. C, isolates from cooling tower water; B, isolates from spa and/or hot spring bath water; P, isolates from patients assumed to be infected at spas and/or hot spring baths; P★, isolates from patients infected at unknown sources; BP, representative isolates obtained from both patients and bath water in an outbreak. The P isolates are indicated on the prefectures where the patients visited, and P★ isolates are indicated on the prefectures where the hospitals of patients' admission were located.

for *Legionella* Infections SBT database for verification and assignment of new allelic numbers according to the curators' instructions (http://www.ewgli.org).

The 31 isolates were divided into 8, 9, 11, 8, 11, and 8 types based on the sequences of *flaA*, *pilE*, *asd*, *mip*, *mompS*, and *proA*, respectively (Fig. 2). As a result, the 31 isolates were divided into 16 SB types in total. The isolates with the same SBT demonstrated similar PFGE patterns. A noteworthy finding was that all 10 iso-

lates from cooling towers clustered into a unique type.

Both PFGE and SBT indicated that the spa-bath isolates from Japan were highly diverse. The water used in public spa baths in Japanese resorts is mostly obtained from hot springs. The characteristics of hot spring water, such as chemical substances, pH, and temperature, are highly variable. The genetic diversity of the spa-bath strains may reflect the wide variety of their habitats.

FIGURE 2   Cluster dendrogram by PFGE and sequence-based types of *L. pneumophila* serogroup 1 isolates in Japan. Allele numbers are according to the European Working Group for *Legionella* Infections SBT database for *L. pneumophila* (http://www.ewgli.org), and new alleles that were determined in this research are underlined. Characteristics of hot spring water were quoted from the reference of the hot spring.

Isolate	flaA	pilE	asd	mip	mompS	proA	spring quality
126B	6	16	14	3	21	14	weakly alkaline simple thermal
370B	2	12	3	6	8	14	weakly alkaline hypertonic
229B	6	10	17	6	9	4	chloride
295B	6	10	17	6	9	4	alkaline simple thermal
230B	6	10	21	6	9	4	chloride
252P	10	12	7	3	16	18	
369P	10	12	7	3	16	18	radioactive
128B	6	10	15	28	4	4	
263P	6	10	15	13	4	14	sodium chloride
158B	3	13	1	10	14	9	sulfate
160B	6	6	15	28	4	14	alkaline simple thermal
231B	6	6	15	28	4	14	tap water
292BP	2	3	9	10	2	1	simple thermal (sodium chloride)
374BP	2	3	9	10	2	1	weakly alkaline simple thermal
301BP	6	10	19	3	19	4	sodium hydrogen carbonate
159B	7	6	17	3	14	11	
294P	2	3	18	11	2	10	
79P*	12	8	11	20	5	12	
81P*	4	7	11	3	11	12	
121C	1	4	3	1	1	1	
217C	1	4	3	1	1	1	
225C	1	4	3	1	1	1	
80P*	1	4	3	1	1	1	
83P*	1	4	3	1	1	1	
122C	1	4	3	1	1	1	
182C	1	4	3	1	1	1	
224C	1	4	3	1	1	1	
124C	1	4	3	1	1	1	
228C	1	4	3	1	1	1	
223C	1	4	3	1	1	1	
226C	1	4	3	1	1	1	

Using the methods of PFGE and SBT, it may be possible to determine the environmental source of a strain of unknown origin. In fact, two of four isolates from patients infected at unknown sources were assigned to the cooling tower SB type and considered to derive from cooling towers. The first clinical isolate (no. 79) in Japan (6) was not assigned to the cooling tower SB-type.

Only four SB types were common between Europe and Japan (N. Fry personal communication). One of these was the SB type ($flaA$, $pilE$, asd, mip, $mompS$, $proA$) = (1, 4, 3, 1, 1, 1), which was the cooling tower type identified in our study. This type was most commonly isolated in Europe (see chapter 25). This type has been isolated from water samples from European spa pools, hotels, and hospitals, although it is not known if isolates of this type have been found in European cooling towers.

Although PFGE is the most widely used technique and is generally accepted to be highly effective in discriminating genomic differences, it may have certain drawbacks with regard to interlaboratory reproducibility (3). SBT appears to be less effective at discriminating between strains than PFGE. On the other hand, SBT shows excellent reproducibility, and it is easy to create a database of SBT, because SBT does not require the interpretation of gel images. We have to consider the advantages and limitations of both methods and apply the most suitable method according to the requirements.

This article is partially reprinted from the *Microbiology and Immunology* (1) with permission of the publisher.

ACKNOWLEDGMENTS

We thank Norman Fry and Tim Harrison (Respiratory and Systemic Infection Laboratory, Health Protection Agency) for helpful advice on SBT, providing the sequences of $flaA$, $pilE$, asd, mip, $mompS$, and $proA$ of EUL strains, assigning the alleles newly identified by us the new numbers, and critical reading of the manuscript, and Jun Terajima (Department of Bacteriology, National Institute of Infectious Diseases) for helpful advice on PFGE analysis. We also thank Eiko Yabuuchi (Department of Microbial Bioinformatics, Gifu University School of Medicine), Fumio Gondaira (Denka Seiken Co., Ltd.), Akihito Wada (Department of Bacteriology, National Institute of Infectious Diseases), and the researchers in the prefectural public health institutes in Japan for providing the *L. pneumophila* serogroup 1 isolates.

REFERENCES

1. **Amemura-Maekawa, J., F. Kura, B. Chang, and H. Watanabe.** 2005. *Legionella pneumophila* serogroup 1 isolates from cooling towers in Japan form a distinct genetic cluster. *Microbiol. Immunol.* **49:**1027–1033.
2. **Amemura-Maekawa, J., F. Kura, H. Watanabe, F. Gondaira, and J. Sugiyama.** 2002. Analysis of *Legionella pneumophila* serogroup 1 isolates in Japan by using pulsed-field gel electrophoresis and monoclonal antibodies, p. 302–304. *In* R. Marre, Y. Abu Kwaik, C. Bartlett, N. P. Cianciotto, B. S. Fields, M. Frosch, J. Hacker, and P. C. Lück (ed.). Legionella, ASM Press, Washington, D.C.
3. **De Zoysa, A. S., and T. G. Harrison.** 1999. Molecular typing of *Legionella pneumophila* serogroup 1 by pulsed-field gel electrophoresis with *Sfi*I and comparison of this method with restriction fragment-length polymorphism analysis. *J. Med. Microbiol.* **48:**269–278.
4. **Gaia, V., N. K. Fry, B. Afshar, P. C. Lück, H. Meugnier, J. Etienne, R. Peduzzi, and T. G. Harrison.** 2005. Consensus sequence-based scheme for epidemiological typing of clinical and environmental isolates of *Legionella pneumophila*. *J. Clin. Microbiol.* **43:**2047–2052.
5. **Gaia, V., N. K. Fry, T. G. Harrison, and R. Peduzzi.** 2003. Sequence-based typing of *Legionella pneumophila* serogroup 1 offers the potential for true portability in legionellosis outbreak investigation. *J. Clin. Microbiol.* **41:**2932–2939.
6. **Saito, A., T. Shimoda, M. Nagasawa, H. Tanaka, N. Ito, Y. Shigeno, K. Yamaguchi, M. Hirota, M. Nakatomi, and K. Hara.** 1981. The first case of Legionnaires' disease in Japan. *Kansensyogaku Zasshi* **55:**124–128.

DEVELOPMENT OF AN ONLINE TOOL FOR EUROPEAN WORKING GROUP FOR LEGIONELLA INFECTIONS SEQUENCE-BASED TYPING, INCLUDING AUTOMATIC QUALITY ASSESSMENT AND DATA SUBMISSION

Anthony P. Underwood, William Bellamy, Baharak Afshar,
Norman K. Fry, and Timothy G. Harrison

44

Previous studies have demonstrated the utility of sequence-based typing (SBT) analysis for the epidemiological typing of *Legionella pneumophila* isolates (3, 4, chapter 41). This technique relies on sequence data from six loci per isolate. Ideally each locus should be sequenced in both the forward and reverse directions. A single base change in a locus will result in that locus being assigned a different allele number and is considered enough to differentiate between two isolates or strains. The implication of these features of the typing method is that, with modern robotics and automated sequencing facilities, high throughput is possible and a large amount of data will be generated. In order for efficient data analysis, these data will be required to be stored in a database. High standards of sequence data quality will be essential for the results to be interpreted in an accurate and meaningful way. Assessment of the sequence quality is a skilled task and can often be very subjective. Therefore, the primary aim of the study described in this paper

was to develop novel web-based tools to facilitate: (i) database curation, designation of novel allele types, and allelic profiles; (ii) automated quality assessment of sequence chromatograms.

The outcome of the study is an online tool, which is now accessible via the European Working Group for Legionella Infections website (http://www.ewgli.org) to allow users to assemble sequence traces (from the SBT scheme) and to identify matching preexisting alleles. Putative new types that do not match existing alleles are submitted to the curators for confirmation.

DESCRIPTION OF THE UTILITY OF THE TOOL AND UNDERLYING TECHNOLOGIES

The online tool uses a web-based interface that can be accessed through any standard Internet browser (e.g., Internet Explorer or Firefox), thereby providing an accessible method for both user access and curatorship of a database. The web pages are constructed dynamically using the perl programming language and cgi scripting on a unix-based web server.

The application has been designed so that users upload trace files for individual SBT alleles. These trace files can be in scf or abi formats. The scf file format usually specifies the

Anthony P. Underwood and William Bellamy Health Protection Agency, Statistics, Modelling and Bioinformatics Department, Center for Infections, 61 Colindale Avenue, London NW9 5EQ London, UK. *Baharak Afshar, Norman K. Fry, and Timothy G. Harrison* Health Protection Agency, Respiratory and Systemic Infection Laboratory, Center for Infections, 61 Colindale Avenue, London NW9 5EQ, UK.

dye chemistry that was used in the sequencing reaction. Unknown dye chemistries are assigned a nominal type, which may result in less accurate base-calling. Once uploaded, the sequences are assessed for quality and assembled into a contig.

QUALITY ASSESSMENT AND CONTIG ASSEMBLY

In order to provide a robust quality assessment of DNA sequencing chromatogram files, the industry-standard Phred software was used. This program reads DNA sequencing trace files, calls bases, and assigns a quality value to each called base. The phred quality values have been thoroughly tested for both accuracy and power to discriminate between correct and incorrect base-calls (1, 2). Most SBT types are derived from at least two sequences, typically one forward and one reverse sequence. These require assembly, and the program uses phrap to assemble the sequences and recall the quality in the resulting consensus sequence (5). Extensive sequence analysis and manipulation of the sequences and assembled contig is carried out using the BioPerl toolkit of objects (6). The processes involved include quality-trimming, assessment of the proportion of the contig that is double-stranded, and consensus generation. Significantly, the resulting consensus sequence is a mosaic of the highest quality read segments rather than the frequently used base which is present in the majority at each position.

DATABASE COMPARISON AND RESULT SUBMISSION

The consensus sequence is trimmed to the correct length using start and end markers specific to each SBT locus. These markers will be degenerate since they will consist of all possible combinations of bases present in the current allele database. When attempting to match each marker with the consensus sequence, two mismatches are permitted to allow for novel allelic variants. In order for the consensus sequence to be assigned a locus before checking whether it matches an existing allele of that locus, it must match both the start and end marker sequences for a particular locus. If matching, and consequently trimming, is successful, the trimmed sequence is compared to all allele types in the database, resulting in either an exact match or the closest match. This process allows the user to ascertain whether the type exists in the current database, is a putative new allele, or is not of sufficient quality to be matched.

If the sequence matches an existing type, the overall quality score and detailed quality scores on a base-by-base assessment can be used to provide quantitative confidence in the accuracy of the result. If the sequence is reported to be of sufficiently good quality, but with mismatches to existing types, these can be confirmed by visualizing their positions on the chromatograms directly within the web page. After verification, putative new types can be submitted directly to the curators, again from the web page, after providing contact details. If however, the sequence cannot be trimmed correctly or is reported to be of poor quality, the user has the facility to look at the quality scores of the individual sequences and determine whether resequencing would be beneficial.

If trace files from more than one locus are uploaded, they will be assembled into separate contigs by phrap, and if possible, each locus will be assigned an allelic type. These will be combined to produce an allelic profile. This consists of a string of six numbers, one for each locus. If a locus does not match an existing type, it will be assigned an allele number of -1. The allelic profile will be compared with existing profiles and exact matches reported. A new profile can be submitted to the database curators from within the web page.

These results are all displayed graphically on the web page. Initially a summary is given, but clicking on buttons on the web page displays more detailed information to better inform the user (Fig. 1).

CONCLUSIONS

Reliable typing results are essential for outbreak investigation, and high-quality sequence data are a prerequisite for this. Data analysis

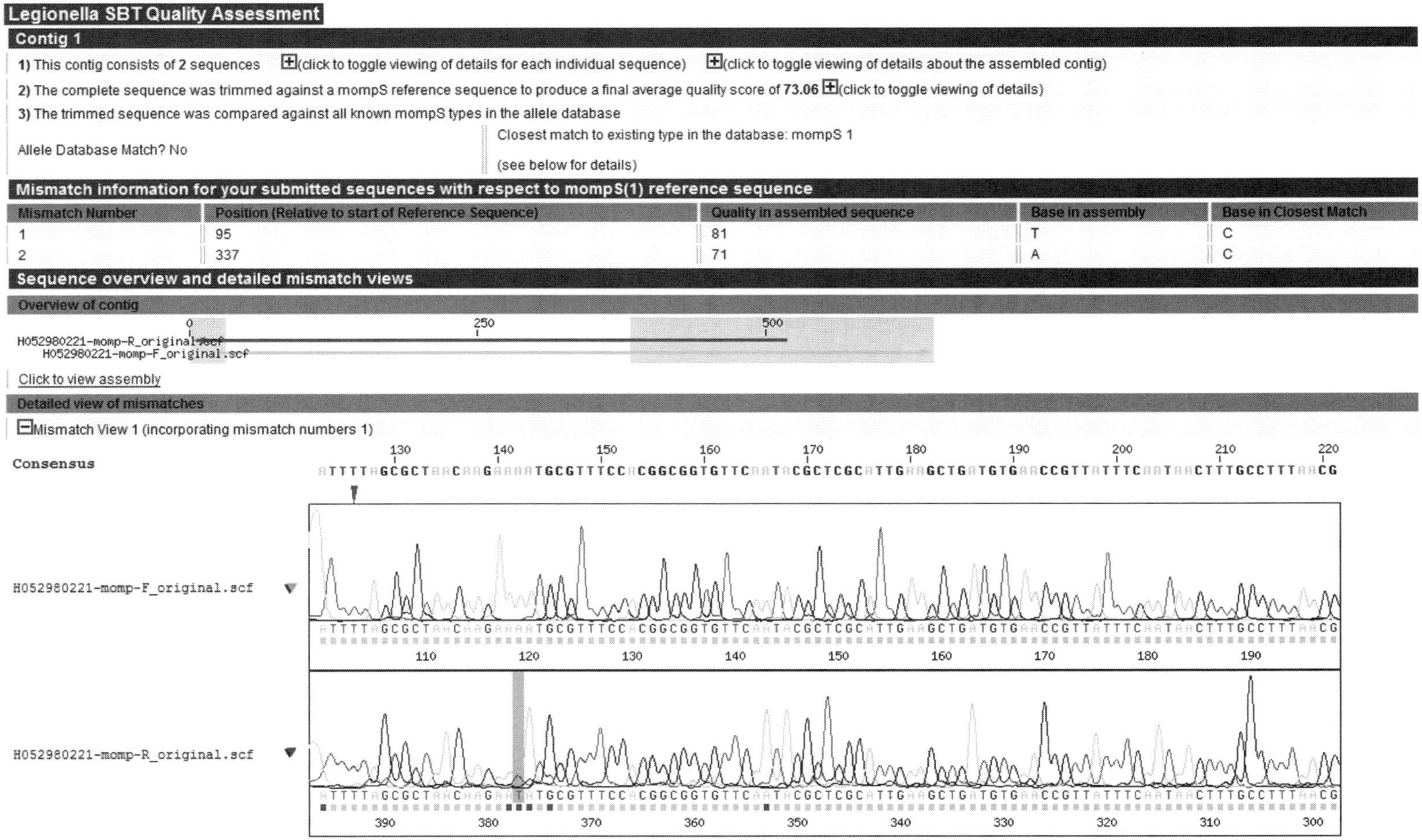

FIGURE 1 The output displayed on the web page after uploading and processing two trace files from one locus of the *L. pneumophila* SBT typing scheme with the sequence quality tool.

from multicenter proficiency SBT analysis has confirmed that misidentification readily occurs when poor-quality trace files are used. This online tool will help systemically check sequence quality and will assist in the determination of minimum standards for a range of sequence quality parameters, ensuring that only high-quality data is accepted for allele and profile identification. Moreover, addition of new alleles to the database is now streamlined.

REFERENCES

1. **Ewing B., and P. Green**. 1998. Base-calling of automated sequencer traces using phred. II. Error probabilities. *Genome Res.* **8:**186–194.
2. **Ewing B., L. Hillier, M. C. Wendl, and P. Green**. 1998. Base-calling of automated sequencer traces using phred. I. Accuracy assessment. *Genome Res.* **8:**175–185.
3. **Gaia V., N. K. Fry, B. Afshar, P. C. Luck, H. Meugnier, J. Etienne, R. Peduzzi, and T. G. Harrison**. 2005. Consensus sequence-based scheme for epidemiological typing of clinical and environmental isolates of *Legionella pneumophila*. *J. Clin. Microbiol.* **43:**2047–2052.
4. **Gaia V., N. K. Fry, T. G. Harrison, and R. Peduzzi**. 2003. Sequence-based typing of *Legionella pneumophila serogroup* 1 offers the potential for true portability in legionellosis outbreak investigation. *J. Clin. Microbiol.* **41:**2932–2939.
5. **Gordon D., C. Abajian, and P. Green**. 1998. Consed: a graphical tool for sequence finishing. *Genome Res.* **8:**195–202.
6. **Stajich J. E., D. Block, K. Boulez, S. E. Brenner, S. A. Chervitz, C. Dagdigian, Fuellen G., J. G. Gilbert, I. Korf, H. Lapp, H. Lehvaslaiho, C. Matsalla, C. J. Mungall, B. I. Osborne, M. R. Pocock, P. Schattner, M. Senger, L. D. Stein, E. Stupka, M. D. Wilkinson, and F. Birney**. 2002. The Bioperl toolkit: perl modules for the life sciences. *Genome Res.* **12:**1611–1618.

MICROBIOLOGY, PATHOGENESIS, IMMUNOLOGY, AND GENETICS

IDENTIFICATION OF TRANSLOCATED SUBSTRATES OF THE *LEGIONELLA PNEUMOPHILA* Dot/Icm SYSTEM WITHOUT THE USE OF EUKARYOTIC HOST CELLS

Ralph R. Isberg and Matthias Machner

45

Survival of *Legionella pneumophila* within lung macrophages is central to its causing pneumonia in humans (25). Within environmental sources the bacteria replicate in amoebae, which act as the major reservoir for *L. pneumophila* prior to contact with the human host (22). Remarkably, most of the steps that occur during intracellular growth of the bacteria in amoebae also appear to take place during growth in macrophages, suggesting that growth within immune cells may simply mimic events that occur in the environment. The bacterium is internalized by phagocytic cells into a membrane-bound compartment that bypasses entry into the endocytic pathway (9). Instead, the *Legionella*-containing vacuole recruits early secretory vesicles trafficking between the endoplasmic reticulum (ER) and Golgi apparatus, allowing intimate interaction of the bacterial compartment with the ER (9). In fact, host GTPases that regulate ER to Golgi traffic are required for intracellular bacterial replication, indicating the importance of a close relationship with this early secretory system (9). Continued replication of *L. pneumophila* appears coordinated with modulation of the vacuolar membrane and the associated ER

throughout its infection cycle within target cells. Therefore, the bacterium must express effector proteins that facilitate this interaction, avoid entry into the endocytic path and promote a connection between replication of the microorganism and growth of the surrounding vacuole.

Intracellular growth and formation of the replication vacuole require the products of 26 *L. pneumophila dot/icm* genes (18, 24). The genes encode components of a secretion complex that translocates macromolecules from the bacterium into host cells (15). Since a number of the encoded proteins show sequence similarity to proteins found in complexes involved in conjugative transfer of DNA, particularly to those related to the antibiotic resistance transfer plasmid R64, the Dot/Icm system is designated as a member of the type IV secretion system (T4SS) family (6). Consistent with a function that is similar to apparatuses involved in conjugative transfer, the Dot/Icm system still retains the ability to transfer DNA (18, 24). The true function of the apparatus, however, is apparently to promote translocation of bacterial effector proteins across host cell membranes, in particular, the plasma membrane during uptake and the vacuolar membrane during intracellular growth of the microorganism (15). The protein substrates of

Ralph R. Isberg and Matthias Machner   Howard Hughes Medical Institute and Dept. of Molecular Biology and Microbiology, Tufts University School of Medicine, 150 Harrison Ave., Boston, MA 02111.

the *L. pneumophila* T4SS are most likely the direct effectors that manipulate host secretory traffic and allow formation of the replication vacuole.

The identities of the translocated substrates of Dot/Icm were at first elusive, as no hunt for bacterial mutants with defective intracellular growth uncovered genes that encode such substrates. Presumably, there is considerable functional redundancy among the various substrates, as more than one protein may have the potential to promote similar steps involved in formation of the replication vacuole. Furthermore, *L. pneumophila* may be able to control several independent intracellular trafficking pathways, each having the ability to promote replication vacuole formation. If this should be the case, then in the absence of one such substrate, other effector proteins may be able to stimulate steps in intracellular growth that compensate for this absence.

As simple hunts for mutants that have lost the ability to grow intracellularly failed to uncover these substrates, alternate strategies had to be developed to identify the Dot/Icm translocated effectors. Foremost among these are bioinformatics approaches and the use of the yeast *Saccharomyces cerevisiae* to recapitulate manipulation of secretory traffic by *L. pneumophila* (4, 19). The former approach has allowed the identification of a translocated effector that promotes guanine nucleotide exchange on small GTPases of the Arf family that regulate ER to Golgi traffic (RalF [15]) as well as proteins involved in egress from amoebae (5). In the case of yeast, a number of *L. pneumophila* proteins have been identified that disrupt vesicular trafficking when they are expressed in this organism, although the mechanisms of action of these proteins are still unclear (4, 19).

In this chapter, we will describe alternate approaches that take advantage of special properties of *L. pneumophila* strains that allowed the formulation of simple screens using toothpicks and petri dishes containing bacteriological medium. Developing readouts in the microorganism that allow screens on solid medium has allowed a number of translocated substrates to

be identified. Interestingly, Dot/Icm-defective mutants were similarly identified by selections on petri dishes, because such mutants had higher viability either on Mueller-Hinton medium (14) or on medium containing sodium chloride (24). Therefore, the behavior of *L. pneumophila* on laboratory medium has allowed workers to make great inroads into the proteins that promote intracellular growth.

SELECTION 1: ISOLATION OF PHENOCOPIES OF THE *dotL* MUTANT

The product of the *dotL* gene is predicted to encode a membrane-associated ATPase that is a member of a large family of proteins found in a number of T4SSs (2, 3). Members of this family are called "coupling proteins" because it is thought that they couple the translocation complex to the translocated substrates (3). In several conjugative DNA apparatuses, biochemical and genetic experiments have indicated that these ATPases bind to the proteins that associate with the transferred DNA and promote DNA mobilization into recipient cells (1, 12, 20). Therefore, it would appear that DotL is a good candidate for a protein that physically associates with some, if not all, of the Dot/Icm substrates. Presumably, any phenotypes associated with loss of the DotL from the bacterium are of great interest, because the absence of proteins that directly interact with DotL may result in similar phenotypes and would allow the identification of translocated substrates.

In fact, the behavior of *dotL* mutants is distinct from most of the other mutants lacking Dot/Icm function. Unlike most other *dot/icm* mutations, loss of DotL function is lethal to the bacterium in the commonly used *L. pneumophila* Philadelphia 1 strain background LP02 (2). This lethality is almost certainly due to a partially assembled T4SS that causes severe misregulation within the bacterium. This was demonstrated by isolating suppressor mutations that allow *dotL* mutants to form colonies. The great majority of these mutations were located in other *dot/icm* genes, indicating that the presence of members of the Dot/Icm complex is responsible for the lethality exhibited in the absence of DotL (7). One model for how such misregulation could occur is that DotL, in conjunction with other proteins, forms a regulatory lid on the Dot/Icm complex, and removal of the lid results in promiscuous movement of molecules through the translocation channel. Such promiscuous translocation had been observed in mutants affecting other secretion systems that promote pathogenesis, such as the *Shigella flexneri* type III secretion system. In the case of *S. flexneri*, the absence of translocated effectors IpaB or IpaD caused unregulated secretion of other substrates, although the resulting strains with such hypersecretion were viable (16). Since it seemed likely that DotL interacted with translocated substrates, and there was precedence for the absence of certain translocated substrates causing misregulation of secretion, we reasoned that the loss of some translocated substrates causes misregulation in the cell in ways that mimic the behavior of *dotL* mutants. Therefore, a mutant hunt was performed to identify these *dotL* "phenocopies" (7). Such mutants should be inviable in an *L. pneumophila* wild-type background, but viable in the presence of a defective T4SS. Interestingly, we found that *dotL* mutants producing low levels of DotA were able to grow on agar plates. This important finding allowed considerable manipulation of phenotypes, because the behavior of mutants that would otherwise be inviable could now be evaluated during intracellular growth.

To identify *dotL* phenocopies, insertion mutations were generated in a *dotA* background (defective T4SS). A plasmid expressing high levels of DotA was then introduced into each of the isolated strains to identify mutants that could not tolerate the presence of high levels of a functional Dot/Icm transporter (7). Each of the identified mutations was then introduced into an *L. pneumophila* strain that produces low levels of DotA, and therefore a limited number of functional Dot/Icm translocators, to detect mutants that behaved as if they were impaired in translocated substrates. Such mutants were determined to be of interest if they had a Dot/Icm

system that appeared to function with similar efficiency to that of wild-type strains. The criterion for demonstrating that the Dot/Icm was functioning properly was based on an assay for the ability of the mutants to cause rapid cytotoxicity to host cells after high multiplicity infection (11). Dot/Icm mutants are known to be defective for killing cells under these conditions, so it was assumed that most mutants defective for intracellular growth and defective for killing target cells under these circumstance were mutations that alter Dot/Icm function, and were not of primary interest. The great majority of mutations, in fact, appeared to directly affect the assembly of the Dot/Icm system or disrupt function of DotL, and many of the impaired proteins are involved in either membrane biogenesis or membrane protein folding (7). However, two mutations did not fall into this class, and they were both in the *lidA* gene, which encodes a 729-amino-acid protein predicted to be rich in coiled-coil structures. Using antibody raised against this bacterial protein, it was demonstrated that LidA is translocated during host cell infection across the vacuolar membrane near the poles of the bacterium. As other T4SSs also appear to translocate macromolecules at the poles of the bacterial rod, the Dot/Icm system conforms with the functional models for other secretion apparatuses (7) (Fig. 1).

Studies of eukaryotic target cells overproducing LidA revealed that this protein manipulates the host cell secretory traffic and has profound effects on either mammalian or yeast cell integrity (8). In mammalian cells, expression of LidA results in disruption of the Golgi and the ER–Golgi intermediate compartment, consistent with LidA binding a protein necessary for vesicle traffic from the ER to the Golgi. In addition, expression of LidA in yeast results in accumulation of the unprocessed form of the secreted yeast invertase. Furthermore, the intracellular membrane morphology of yeast cells is highly disrupted in the presence of LidA (8). Recent work indicates that LidA binds a protein that is directly involved in the movement of vesicles from the ER to the Golgi, indicating that the protein probably

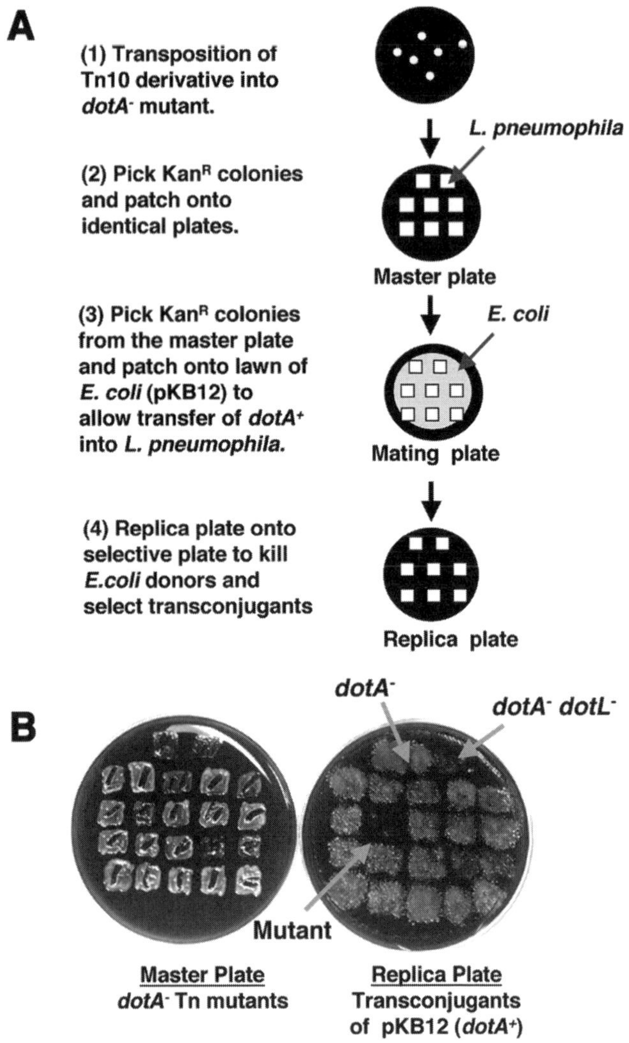

A

(1) Transposition of Tn10 derivative into *dotA⁻* mutant.

(2) Pick Kan^R colonies and patch onto identical plates.

Master plate

(3) Pick Kan^R colonies from the master plate and patch onto lawn of *E. coli* (pKB12) to allow transfer of *dotA⁺* into *L. pneumophila.*

Mating plate

(4) Replica plate onto selective plate to kill *E.coli* donors and select transconjugants

Replica plate

B

FIGURE 1 Isolation of *lidA::*miniTn*10*, a mutation that eliminates a translocated substrate of Dot/Icm (from [7]). (*A*) Insertion mutations were generated in an *L. pneumophila* strain that has a defective Dot/Icm transporter due to a mutation in the *dotA* gene. These strains were then patched onto bacteriological plates and allowed to grow. Strains of high viability were then patched onto a plate seeded with *E. coli* with a mobilizable plasmid that encodes an intact *dotA⁺* gene to allow mating and movement of the *dotA* gene into the mutagenized strains. Such strains will now have an intact Dot/Icm translocator, and should grow like wild-type strains unless the insertion mutation has caused a growth defect when there is a Dot/Icm system present. These mutants with lowered plating efficiency in the presence of an intact Dot/Icm will make a smaller patch on selective media than do strains that can tolerate an intact translocator. (*B*) Charcoal-yeast extract plates showing examples of mutants that fail to tolerate the presence of the Dot/Icm system.

Master Plate
dotA⁻ Tn mutants

Replica Plate
Transconjugants
of pKB12 (*dotA⁺*)

plays a central role in facilitating interaction of the *Legionella*-containing vacuole with the host cell early secretory system.

SELECTION 2: LARGE SCALE SCREENING OF TRANSLOCATED EFFECTORS USING INTERBACTERIAL PROTEIN TRANSFER

The above selection described identification of a subset of substrates that may be important for the proper integrity of the Dot/Icm apparatus. Based on the behavior of other secretion systems (16), it was highly unlikely that mutations in most of the translocated substrates would have the same phenotype as a *lidA* mu-

tation, so a more broad-based screen was desirable, in which translocation of proteins into target cells was directly assayed. The extreme flexibility of the Dot/Icm apparatus allows the formulation of multiple schemes for such screens. Of most interest was the fact that *L. pneumophila* was shown to be able to transfer plasmids into other bacterial strains via the Dot/Icm system (18, 24). We took advantage of this observation to identify a large number of Dot/Icm substrates.

Transfer of plasmids between bacterial strains involves both the membrane components of the T4SS as well as mobilization

(Mob) proteins that bind directly to the transferred DNA. Interestingly in several T4SSs, Mob protein can be transferred into recipient cells in the absence of DNA mobilization, indicating that protein translocation into bacteria can occur in the same fashion that is observed when the pathogen contacts a mammalian cell (21). Therefore, we developed an assay to test if *L. pneumophila* can transfer proteins into bacterial recipient cells and used this system to screen for new translocated substrates. To this end, we adapted a strategy previously used to detect protein transfer via the *Agrobacterium tumefaciens* T4SS into target plant cells (23). In this assay, protein transfer is detected by generating fusion proteins between translocated effectors and the Cre site-specific recombinase. The resulting protein chimera are recognized by the T4SS and can be transferred to bacterial cells, an event determined by detecting Cre activity in the recipient cell.

A number of different readouts have been developed for assaying Cre, which recognizes *lox* recombination sites. In our case, a double selection system was developed in which a recipient bacterium is killed on selective medium unless transfer of the Cre fusion protein occurs. To this end, an *L. pneumophila* recipient strain was constructed in which a DNA cassette containing *sacB*, which confers sucrose sensitivity, and a transcription terminator was flanked by *lox* recombination sites. Upstream

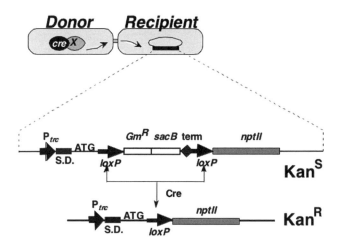

FIGURE 2 Identification of translocated substrates of Dot/Icm by transfer between bacterial strains (from [13]). Fusions of *L. pneumophila* DNA to the gene encoding Cre were constructed and were used to test the ability of the resulting hybrid protein to translocate into a tester *L. pneumophila* strain. Translocation of a hybrid protein from the donor strain can be detected by the generation of kanamycin-resistant (Kan^r) exconjugants. These strains are recombinants that result from removal of a floxed transcriptional terminator located between a promoter and the kan^r gene found on a construction in the recipient bacterial strain. In contrast, in the absence of Cre, bacteria harboring the intact reporter are unable to grow on media containing kanamycin and sucrose. The translocation of Cre hybrid protein into the recipient strain leads to the excision, via recombination at the *loxP* sites, of the DNA fragment that confers sucrose sensitivity (*sacB*) and reconstitution of a functional *loxP-nptII* translational fusion. S.D., Shine-Dalgarno sequence; *nptII*, neomycin phosphotransferase; ◊, transcriptional terminator. Arrows indicate the *trc* promoter and *loxP* sites.

from this cassette is a promoter and a ribosome-binding site, while a promoterless open reading frame (ORF) encoding the kanamycin resistance gene *nptII* is downstream from the cassette (13) (Fig. 2). When this strain is plated on medium containing sucrose and kanamycin, it is unable to form colonies unless the Cre protein is received, which removes the transcription terminator and the *sacB* gene by site-specific recombination between the *lox* sites. Therefore, only if a translocation signal recognized by the Dot/Icm system is fused to Cre, the chimeric protein is transferred from a donor strain expressing such fusion into the recipient strain, leading to loss of the cassette by Cre-mediated recombination. Such recombinants lose *sacB* and generate an active *nptII* gene by fusing it to the promoter and ribosome-binding site, generating a strain that is kanamycin and sucrose resistant. The correct function of this approach was confirmed in control experiments. First, we found that fusions of Cre to the Mob protein encoded by the plasmid RSF1010 were translocated into recipient strains and generated Kan^{r}Suc^r recombinants. Second, when proteins such as LidA and RalF, which were known to be translocated to mammalian cells, were fused to Cre, a similar Kar^{r}Suc^r readout was generated. Therefore, the screen worked and, perhaps

more remarkably, proteins that are normally translocated into mammalian host cells could also be translocated between bacteria.

Although the above screen is technically simple to perform, in that two strains are incubated together and then plated on selective medium, in practice a complete genome survey of fusions would involve analyzing at least 24,000 mating events. To bypass large amounts of screening, a shortcut was taken. We reasoned that a Dot/Icm component must exist that efficiently interacts with effector proteins during translocation. Therefore, we could limit the Cre analysis to binding partners of this component, no matter how weak such binding may appear at first. To this end, we found that DotF was able to bind the previously characterized translocated effector RalF (15). Using a bacterial two-hybrid approach (10), a large number of proteins that showed the potential to bind to DotF were identified, and those that passed this initial screen were then tested in the Cre-Lox translocation assay.

The results of the assay are shown in Table 1. Eleven ORFs were identified using this strategy. Interestingly, many of the ORFs were found to have paralogs in the genome of *L. pneumophila* Philadelphia 1 (17), so in total the analysis identified 26 potential translocated substrates. To demonstrate that the putative

TABLE 1 Translocated substrates identified by the Cre-Lox interbacterial screen

Gene	Protein size (AA)	*L. pneumophila* paralogs	E values or identity of paralogs	Orthologs in other species (E value)
sidA	474	—	—	—
sidB	417	4	7E^{-26}–5E^{-09}	Rtx toxin/lipase (e^{-5})
sidC	913	2	71% identity	—
sidD	471	—	—	—
sidE	1,495	5	0 to 2e^{-05}	—
sidF	912	—	NA	—
sidG	965	—	NA	—
sidH	2,225	3	7e^{-08} to 3e^{-04}	—
sidI	949	—	—	—
sidJ	873	2	56% identity	—
vipD	621	4	7e^{-12} to 3e^{-10}	ExoU/phospholipase(1e^{-7})

translocated effectors were indeed substrates that were transferred into mammalian host cells, antibody raised against one of these proteins, SidC, was used to show that SidC is clearly localized on the cytoplasmic side of *Legionella*-containing vacuoles (13).

CONCLUSIONS

In the work described here, we demonstrated that simple strategies in bacterial genetics, using toothpicks and petri dishes, allowed the identification of proteins that are transferred into mammalian cells during *L. pneumophila* infection. One of the most surprising aspects of this analysis was that although a large number of genes were identified, knockout mutations in most of these genes generate little defect in intracellular bacterial growth. In fact, the few mutations that do alter intracellular growth of *L. pneumophila* do not appear to affect the formation of the replication vacuole or recruitment of ER-derived vesicles, but affect other poorly defined processes (data not shown). Therefore, the factors that promote formation of the replicative compartment appear to be highly redundant. Furthermore, elimination of multiple paralogs of a single gene family does not solve the redundancy issue, suggesting that a similar amino acid sequence does not necessarily mean the proteins have similar functions. Development of functional and biochemical assays for these proteins will be necessary to sort out their roles in intracellular growth.

REFERENCES

1. **Beranek, A., M. Zettl, K. Lorenzoni, A. Schauer, M. Manhart, and G. Koraimann.** 2004. Thirty-eight C-terminal amino acids of the coupling protein TraD of the F-like conjugative resistance plasmid R1 are required and sufficient to confer binding to the substrate selector protein TraM. *J. Bacteriol.* **186:**6999–7006.
2. **Buscher, B. A., G. M. Conover, J. L. Miller, S. A. Vogel, S. N. Meyers, R. R. Isberg, and J. P. Vogel.** 2005. The DotL protein, a member of the TraG-coupling protein family, is essential for viability of Legionella pneumophila strain Lp02. *J. Bacteriol.* **187:**2927–2938.
3. **Cabezon, E., J. I. Sastre, and F. de la Cruz.** 1997. Genetic evidence of a coupling role for the TraG protein family in bacterial conjugation. *Mol. Gen. Genet.* **254:**400–406.
4. **Campodonico, E. M., L. Chesnel, and C. R. Roy.** 2005. A yeast genetic system for the identification and characterization of substrate proteins transferred into host cells by the Legionella pneumophila Dot/Icm system. *Mol. Microbiol.* **56:**918–933.
5. **Chen, J., K. S. de Felipe, M. Clarke, H. Lu, O. R. Anderson, G. Segal, and H. A. Shuman.** 2004. Legionella effectors that promote nonlytic release from protozoa. *Science* **303:**1358–1361.
6. **Christie, P. J., and J. P. Vogel.** 2000. Bacterial type IV secretion: conjugation systems adapted to deliver effector molecules to host cells. *Trends Microbiol.* **8:**354–360.
7. **Conover, G. M., I. Derre, J. P. Vogel, and R. R. Isberg.** 2003. The Legionella pneumophila LidA protein: a translocated substrate of the Dot/Icm system associated with maintenance of bacterial integrity. *Mol. Microbiol.* **48:**305–321.
8. **Derre, I., and R. R. Isberg.** 2005. LidA, a translocated substrate of the Legionella pneumophila type IV secretion system, interferes with the early secretory pathway. *Infect. Immun.* **73:**4370–4380.
9. **Kagan, J. C., and C. R. Roy.** 2002. Legionella phagosomes intercept vesicular traffic from endoplasmic reticulum exit sites. *Nat. Cell. Biol.* **4:**945–954.
10. **Karimova, G., J. Pidoux, A. Ullmann, and D. Ladant.** 1998. A bacterial two-hybrid system based on a reconstituted signal transduction pathway. *Proc. Natl. Acad. Sci. USA* **95:**5752–5756.
11. **Kirby, J. E., J. P. Vogel, H. L. Andrews, and R. R. Isberg.** 1998. Evidence for pore-forming ability by *Legionella pneumophila. Mol. Microbiol.* **27:**323–336.
12. **Lu, J., and L. S. Frost.** 2005. Mutations in the C-terminal region of TraM provide evidence for in vivo TraM-TraD interactions during F-plasmid conjugation. *J. Bacteriol.* **187:**4767–4773.
13. **Luo, Z. Q., and R. R. Isberg.** 2004. Multiple substrates of the Legionella pneumophila Dot/Icm system identified by interbacterial protein transfer. *Proc. Natl. Acad. Sci. USA* **101:**841–846.
14. **Marra, A., S. J. Blander, M. A. Horwitz, and H. A. Shuman.** 1992. Identification of a Legionella pneumophila locus required for intracellular multiplication in human macrophages. *Proc. Natl. Acad. Sci. USA* **89:**9607–9611.
15. **Nagai, H., J. C. Kagan, X. Zhu, R. A. Kahn, and C. R. Roy.** 2002. A bacterial guanine nucleotide exchange factor activates ARF on Legionella phagosomes. *Science* **295:**679–682.
16. **Parsot, C., R. Menard, P. Gounon, and P. J. Sansonetti.** 1995. Enhanced secretion through the Shigella flexneri Mxi-Spa translocon leads to assembly of extracellular proteins into macromo-

lecular structures. *Mol. Microbiol.* **16:**291–300.

17. **Russo, J. J., et al.** 2001. Legionella Genome Project. Department of Microbiology, Columbia University. [Online.]

18. **Segal, G., M. Purcell, and H. A. Shuman.** 1998. Host cell killing and bacterial conjugation require overlapping sets of genes within a 22-kb region of the Legionella pneumophila genome. *Proc. Natl. Acad. Sci. USA* **95:**1669–1674.

19. **Shohdy, N., J. A. Efe, S. D. Emr, and H. A. Shuman.** 2005. Pathogen effector protein screening in yeast identifies Legionella factors that interfere with membrane trafficking. *Proc. Natl. Acad. Sci. USA* **102:**4866–4871.

20. **Szpirer, C. Y., M. Faelen, and M. Couturier.** 2000. Interaction between the RP4 coupling protein TraG and the pBHR1 mobilization protein Mob. *Mol. Microbiol.* **37:**1283–1292.

21. **Szpirer, C. Y., M. Faelen, and M. Couturier.** 2001. Mobilization function of the pBHR1 plasmid, a derivative of the broad-host-range plasmid pBBR1. *J. Bacteriol.* **183:**2101–2110.

22. **Tyndall, R. L., and E. L. Domingue.** 1982. Cocultivation of Legionella pneumophila and free-living amoebae. *Appl. Environ. Microbiol.* **44:** 954–959.

23. **Vergunst, A. C., B. Schrammeijer, A. den Dulk-Ras, C. M. de Vlaam, T. J. Regensburg-Tuink, and P. J. Hooykaas.** 2000. VirB/D4-dependent protein translocation from Agrobacterium into plant cells. *Science* **290:**979–982.

24. **Vogel, J. P., H. L. Andrews, S. K. Wong, and R. R. Isberg.** 1998. Conjugative transfer by the virulence system of Legionella pneumophila. *Science* **279:**873–876.

25. **Winn, W. C., Jr., and R. L. Myerowitz.** 1981. The pathology of the Legionella pneumonias. A review of 74 cases and the literature. *Hum. Pathol.* **12:**401–422.

FUNCTION OF
LEGIONELLA EFFECTORS

Howard A. Shuman, Christopher Pericone, Nadim Shohdy,
Karim Suwwan de Felipe, and Margaret Clarke

46

The fundamental question that makes the biology of *Legionella pneumophila* so fascinating is how this simple prokaryote manages to reprogram tightly regulated organelle trafficking events inside eukaryotic host cells. The ability of *Legionella* to survive the antimicrobial defenses of both mammalian and protozoan phagocytes and to use these cells as sites for replication represents a scaled-down, single-celled version of parasitic relationships that occur in many variations throughout biology. The availability of a variety of genetic tools, a complete genome sequence, and several convenient host systems offers the opportunity to study the molecular mechanisms of the interactions between *Legionella* and its hosts. Examination of the *Legionella* genome sequence revealed the existence of a relatively large number of genes that resemble eukaryotic rather than prokaryotic orthologs (4, 6, 8). The use of the well-characterized soil amoeba *Dictyostelium discoideum* as a host permits the observation of organelle behavior in living cells with a variety

of green fluorescent protein (GFP)-labeled marker proteins (13). Finally, ectopic expression of *Legionella* genes in yeast has provided a convenient genetic system for evaluating the ability of individual gene products to cause derangement of organelle trafficking (20). By taking advantage of these various tools, we have tried to understand how *Legionella* modulates organelle trafficking. The paradigm that has emerged from the work of several laboratories indicates that *Legionella* uses a translocation device called the "Dot/Icm type IV secretion system" to inject effector proteins into host cells; it is these "effectors" that are directly responsible for modulating host cell organelle behavior to the *Legionella's* advantage (16, 19, 22). Since the identification of several effectors, the primary goal has been to understand how the effectors interact with targets in host cells and how these interactions alter organelle trafficking.

LEGIONELLA LIFE INSIDE THE HOST CELL

Whether *Legionella* replicates autonomously in the wild is an open question. However, it certainly parasitizes unicellular amoebae and protozoa and replicates within them until they are destroyed. Because amoebae and other protozoa normally phagocytose and graze on bacteria,

Howard A. Shuman, Christopher Pericone, and Nadim Shohdy Department of Microbiology, Columbia University Medical Center, New York, NY 10032. *Karim Suwwan de Felipe* Integrated Program in Cellular, Molecular and Biophysical Studies, Columbia University Medical Center, New York, NY 10032. *Margaret Clarke* Oklahoma Medical Research Foundation, 825 N.E. 13th Street, Oklahoma City, OK 73104.

Legionella must first avoid destruction following phagocytosis. This is accomplished by preventing the fusion of *Legionella*-containing phagosomes with lysosomes and the maintenance of an intraphagosomal pH near neutrality (10, 11). Once inside the confines of the unfused phagosome or vacuole, the bacteria begin to modulate the behavior of the compartment to suit their needs. The first visual evidence of this activity is the accumulation of small vesicles surrounding the cytosolic face of the phagosomal membrane (9). The origin and function of these structures is not certain, but there is evidence that they are derived from pathways connecting the golgi and the endoplasmic reticulum (ER) (12). The next overt characteristic of the *Legionella*-containing vacuole (LCV) is its association with the ER. The nature of this association is not clear; either the LCV intercepts and captures vesicles containing ER-like membrane or the LCV becomes wrapped with the ER itself. Notwithstanding this ambiguity, at about 2 to 4 h following phagocytosis, the LCV contains large amounts of ER luminal and membrane markers such as BiP or calnexin (13, 21). Although the functional significance of the ER-nature of the LCV is not understood, it is also not restricted to *Legionella*; several other intracellular pathogens replicate in compartments with similar characteristics (17) . Once the LCV acquires ER properties, *Legionella* begins to replicate within the confines of the LCV. As many as several hundred bacteria can be observed inside a single LCV, suggesting that the bacteria can divide several times before exiting the host using either lytic or nonlytic mechanisms (5).

Most of these intracellular trafficking events can be observed with the aid of confocal fluorescence microscopy in real time using living *Dictyostelium discoideum* host cells that express GFP-tagged host proteins that localize to specific intracellular organelles. For example, phagocytosis of *Legionella* that express the *D. discoideum*-RedXpress red fluorescent protein can be monitored using *D. discoideum* that express coronin-GFP. Coronin transiently binds to the actin filaments that are transiently associated

with the phagosome that engulfs the bacterium. Similarly, the VatM subunit of the V-ATPase that is incorporated into the lysosomal membrane can be used to monitor phagosome-lysosome fusion when VatM is fused to GFP. As mentioned above, association of the LCV with ER is easily seen in cells that express calnexin-GFP (13). (Fig. 1).

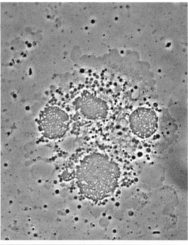

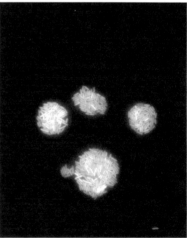

FIGURE 1   This figure shows a single cell of *Dictyostelium discoideum* that was infected by *Legionella pneumophila* Philadelphia-1 strain JR32 expressing green fluorescent protein (GFP). The Dictyostelium cell is no longer intact, yet four distinct *Legionella*-containing vacuoles, each containing numerous bacteria, are clearly visible, either by phase contrast (top) or fluorescence (bottom) imaging.

GENETICS, GENOMICS AND REDUNDANCY

Initially it was thought that genetic screens would yield *Legionella* mutants that could carry out a subset of the events described above. Random screens for *Legionella* mutants that could not kill or replicate within macrophages led to the discovery of the genes encoding the Dot/Icm type IV secretion system (2, 18). However, these mutants cannot inject effectors and are thus completely unable to alter organelle trafficking in host cells. As a consequence, following uptake, the mutants cannot prevent acidification or phagosome-lysosome fusion and are rapidly destroyed within the acidic fused phagolysosome. The same genetic screens did not lead to the identification of the genes encoding the effector Dot/Icm substrates. Alternative approaches based on genome inspection or the use of clever genetic screens for other phenotypes eventually led to the identification of several effectors (7,14,15). In one case, *Legionella* effectors that vaguely resemble components of vesicle fusion proteins (eg. EEA1, interaptin) were shown to play a role in a previously recognized nonlytic release pathway in protozoa that results in packets of bacteria being exocytosed prior to host cell death (5). In most cases, deletion of the gene for a single effector did not result in an observable phenotype. In some cases where several presumably paralogous effector genes were identified, even deletion of several paralogs yielded bacteria with an ostensibly wild-type phenotype (14). These negative results can be interpreted to mean either that most recognized effectors have no role in *Legionella* intracellular life or that the bacteria possess multiple, redundant gene products for functions that are required for intracellular multiplication. If true, such functional redundancy could be the consequence of selection for the ability to grow in a wide variety of protozoa. Indeed, the LepA and LepB proteins mentioned above exhibit such redundancy. Loss of either LepA or LepB does not detectably reduce nonlytic release, yet loss of both proteins results in a severe defect in nonlytic release. This defect can be complemented in trans by restoring the structural gene for either LepA or LepB (5).

BIOINFORMATICS AND ECTOPIC EXPRESSION LEAD THE WAY

Frustrated by the lack of information based on classical genetic approaches, we sought to examine the genome for additional clues using more targeted approaches than the original BLAST-based annotation. We reasoned that because *Legionella* replicates inside eukaryotic hosts, it may have acquired genes from the host and perhaps some of these had been subjected to selection or "sculpted" in such a way that they now contributed to intracellular replication. Using a simple algorithm based on screening functional domains in the Pfam database (available at http://pfam.wustl.edu), we identified a large set of *Legionella pneumophila* Philadelphia-1 genes that were more closely related to eukaryotic genes than to prokaryotic genes (8). Many, but not all, of these genes had also been identified in two other *Legionella pneumophila* isolates (4). These *leg* genes (*Legionella eukaryote-like genes*) could be classified based on the Pfam domain that they encoded or on whether they encoded a product with a predicted specific enzymatic activity. These are listed in Table 1. Comparison of the G+C content of *leg* gene sequences with a collection of 20 "informational" genes (encoding translation and transcription functions) indicated that the G+C content of the *leg* genes was distinct and provides evidence that they were acquired by inter-kingdom horizontal gene transfer. A subset of the *leg* gene products was shown to be Dot/Icm substrates and subject to the same type of growth-phase regulation as other virulence-related functions such as motility. The majority of *leg* gene products appear to contain motifs involved in protein:protein interactions such as ankyrin repeats, leucine-rich repeats, or coiled coils. Remarkably other *leg* gene products are closely related to eukaryotic proteins that are involved in lipid signaling pathways absent in prokaryotes (LegS1 and LegS2), ubiquitination (LegU1, LegU2, LegAU13), or GTP/GDP exchange factors (LegG1, LegG2).

TABLE 1   Summary of *Legionella* eukaryotic-like genes

Gene name	ORF designation	Domain(s)	Best E value	Identification
		Group I: ankyrin repeats		
legA1	lpg2416	Ankyrin repeats	4×10^{-7}	This study (4)
legA2	lpg2215	Ankyrin repeats	5×10^{-4}	This study (4)
legA3	lpg2300	Ankyrin repeats	8×10^{-16}	This study (4)
legAS4	lpg1718	Ankyrin repeats and SET	2×10^{-7}	This study (4)
legA5	lpg2322	Ankyrin repeats	1×10^{-4}	This study (4)
legA6	lpg2131	Ankyrin repeats	3×10^{-6}	This study (4)
legA7	lpg0403	Ankyrin repeats	1×10^{-15}	This study (4)
legA8	lpg0645	Ankyrin repeats	1×10^{-16}	This study (4)
legA9	lpg0402	Ankyrin repeats	5×10^{-20}	This study (4)
legA10	lpg0038	Ankyrin repeats	2×10^{-11}	This study (4)
legA11	lpg0436	Ankyrin repeats	4×10^{-6}	This study (4)
legA12	lpg0483	Ankyrin repeats	2×10^{-4}	This study (4)
legAU13	lpg2144	Ankyrin repeats and FBOX	1×10^{-5}	This study (4)
legA14	lpg2452	Ankyrin repeats	2×10^{-10}	This study (4)
legA15	lpg2456	Ankyrin repeats	6×10^{-9}	This study (4)
		Group II: leucine rich repeats		
legL1	lpg0945	Leucine rich repeats	2×10^{-11}	This study
legL2	lpg1602	Leucine rich repeats	3×10^{-6}	This study
legL3	lpg1660	Leucine rich repeats	9×10^{-6}	This study
legLC4	lpg1948	Leucine rich repeats and coiled coils	6×10^{-8}	This study
legL5	lpg1958	Leucine rich repeats	2×10^{-15}	This study
legL6	lpg2392	Leucine rich repeats	8×10^{-9}	This study
legL7	lpg2400	Leucine rich repeats	3×10^{-9}	This study
legLC8	lpg1890	Leucine rich repeats and coiled coils	4×10^{-8}	This study
		Group III: coiled coils		
legC1	lpg1312	Coiled coils	N/A	This study
legC2/ylfB	lpg1884	Coiled coils	N/A	This study (3)
legC3	lpg1701	Coiled coils	N/A	This study
legC4	lpg1953	Coiled coils	N/A	This study

(Continued)

Two coiled-coil-containing proteins, LegC2/YlfB and LegC7/YlfA, were independently identified in a screen for *Legionella* proteins that are lethal when expressed in yeast (3). The remaining Leg proteins are related to eukaryotic proteins with recognizable enzymatic functions or small molecule-binding domains.

To find out if the *leg* gene products play any role in altering intracellular organelle trafficking, we took advantage of the vacuolar protein sorting (Vps) phenotype of yeast. Wild-type yeast trafficks several degradative proteins to the vacuole, the functional equivalent of the lysosome. When the N-terminal 50 residues of vacuolar carboxypeptidase Y (CPY) is genetically fused to invertase (Inv), a plasma membrane protein, the CPY-Inv hybrid protein is efficiently localized to the vacuole. Because yeast cells cannot transport sucrose across the plasma membrane, the vacuolar CPY-Inv protein is

TABLE 1 *Continued*

Gene name	ORF designation	Domain(s)	Best E value	Identification
		Group III: coiled coils		
legC5	lpg1488	Coiled coils	N/A	This study
legC6	lpg1588	Coiled coils	N/A	This study
legC7/ylfA	lpg2298	Coiled coils	N/A	This study (3)
legC8	lpg2862	Coiled coils	N/A	This study
legLC4	lpg1948	Leucine rich repeats and coiled coils	N/A	This study
legLC8	lpg1890	Leucine rich repeats and coiled coils	N/A	This study
		Group IV: ser/thr kinase		
legK1	lpg1483	Ser/Thr kinase	2×10^{-22}	This study (4)
legK2	lpg2137	Ser/Thr kinase	2×10^{-5}	This study (4)
legK3	lpg2556	Ser/Thr kinase	9×10^{-22}	This study (4)
		Group V: signaling lipid related domains		
legS1	lpg2588	Lipid phosphoesterase	2×10^{-4}	This study
LegS2	lpg2176	Sphingosine 1-P lyase	6×10^{-50}	This study (4)
		Group VI: GDP/GTP exchange		
legG1	lpg1976	RCC-1	7×10^{-9}	This study (4)
legG2	lpg0276	Ras GEF	2×10^{-8}	This study
		Group VII: ubiquitination-related domains		
legAU13	lpg2144	Ankyrin repeats and FBOX	4×10^{-6}	This study (4)
legU1	lpg0171	FBOX	2×10^{-7}	This study (4)
legU2	lpg2830	UBOX	7×10^{-4}	This study (4)
		Group VIII: others		
legAS4	lpg1718	Ankyrin repeats and SET	1×10^{-15}	This study (4)
legN	lpg2720	cNMP binding domain	2×10^{-14}	This study
legY	lpg0422	Amylase	2×10^{-52}	This study (4)
legD1	lpg2694	Phytanoyl-coA dioxygenase	2×10^{-29}	This study (4)
legD2	lpg0515	Phytanoyl-coA dioxygenase	5×10^{-17}	This study (4)
legP	lpg2999	Astacin protease	1×10^{-52}	This study (4)
legT	lpg1328	Thaumatin domain	1×10^{-9}	This study

unable to hydrolyze sucrose and these cells are phenotypically Suc⁻. Perturbations of organelle trafficking result in mislocalization of the CPY-Inv hybrid and enable these yeast cells to hydrolyze extracellular sucrose to glucose and fructose. The Suc⁺ phenotype of these cells is easily visualized with the aid of a colorimetric glucose detection assay (1). We initially identified three *Legionella* genes that produced a Suc⁺ phenotype in yeast cells expressing the CPY-Inv hybrid. These *Legionella vip* (*V*ps inhibition *p*rotein) genes interfere with the trafficking of normal CPY, carboxypeptidase S, and alkaline phosphatase in otherwise wild-type yeast cells. We cloned several of the *leg* genes in a yeast vector to find out if any of them would produce a Vps/Suc⁺ phenotype (20). To ensure that any growth defects produced by the *leg* genes would not go unnoticed, we cloned the *leg* genes in a vector in which the presence of galactose is required for increased gene expression in yeast. We found that LegC7 and LegLC8 both produce a strong Vps/Suc$^+$ phenotype in yeast when induced with galactose. In addition, LegC5 and LegC8 cause a growth defect when expressed at high levels in yeast (de Felipe and Shuman, unpublished data). Previous work indicated that the LegC5, LegLC8, and LegC7 proteins are translocated via the Dot/Icm system to host cells by *Legionella*. These results indicate the utility of the yeast ectopic expression systems for focusing attention on *Legionella* proteins that may play an important role in the ability of *Legionella* to modify host cell organelle trafficking to suit its own needs.

CONCLUSIONS

Standard bacterial genetic analysis provided crucial information about the existence and importance of the Dot/Icm type IV secretion system but has been of little value in identifying and characterizing the molecular basis of *Legionella*'s ability to modify organelle trafficking in host cells. A combination of bioinformatics and novel genetic approaches have resulted in the identification of several interesting translocated effectors. The challenge is now to determine how these effectors alter host cell organelle trafficking. Combining new genetic approaches and in vivo cell biological methods is expected to provide additional information about these mechanisms.

REFERENCES

1. **Bankaitis, V. A., L. M. Johnson, and S. D. Emr.** 1986. Isolation of yeast mutants defective in protein targeting to the vacuole. *Proc. Natl. Acad. Sci. USA* **83:**9075–9079.
2. **Berger, K. H., and R. R. Isberg.** 1993. Two distinct defects in intracellular growth complemented by a single genetic locus in Legionella pneumophila. *Mol. Microbiol.* **7:**7–19.
3. **Campodonico, E. M., L. Chesnel, and C. R. Roy.** 2005. A yeast genetic system for the identification and characterization of substrate proteins transferred into host cells by the Legionella pneumophila Dot/Icm system. *Mol. Microbiol.* **56:**918–933.
4. **Cazalet, C., C. Rusniok, H. Bruggemann, N. Zidane, A. Magnier, L. Ma, M. Tichit, S. Jarraud, C. Bouchier, F. Vandenesch, F. Kunst, J. Etienne, P. Glaser, and C. Buchrieser.** 2004. Evidence in the Legionella pneumophila genome for exploitation of host cell functions and high genome plasticity. *Nat. Genet.* **36:**1165–1173.
5. **Chen, J., K. S. de Felipe, M. Clarke, H. Lu, O. R. Anderson, G. Segal, and H. A. Shuman.** 2004. Legionella effectors that promote nonlytic release from protozoa. *Science* **303:**1358–1361.
6. **Chien, M., I. Morozova, S. Shi, H. Sheng, J. Chen, S. M. Gomez, G. Asamani, K. Hill, J. Nuara, M. Feder, J. Rineer, J. J. Greenberg, V. Steshenko, S. H. Park, B. Zhao, E. Teplitskaya, J. R. Edwards, S. Pampou, A. Georghiou, I. C. Chou, W. Iannuccilli, M. E. Ulz, D. H. Kim, A. Geringer-Sameth, C. Goldsberry, P. Morozov, S. G. Fischer, G. Segal, X. Qu, A. Rzhetsky, P. Zhang, E. Cayanis, P. J. De Jong, J. Ju, S. Kalachikov, H. A. Shuman, and J. J. Russo.** 2004. The genomic sequence of the accidental pathogen Legionella pneumophila. *Science* **305:**1966–1968.
7. **Conover, G. M., I. Derre, J. P. Vogel, and R. R. Isberg.** 2003. The Legionella pneumophila LidA protein: a translocated substrate of the Dot/Icm system associated with maintenance of bacterial integrity. *Mol. Microbiol.* **48:**305-321.
8. **de Felipe, K. S., S. Pampou, O. S. Jovanovic, C. D. Pericone, S. F. Ye, S. Kalachikov, and H. A. Shuman.** 2005. Evidence for acquisition of Legionella type IV secretion substrates via interdomain horizontal gene transfer. *J. Bacteriol.* **187:**7716–7726.
9. **Horwitz, M. A.** 1983. Formation of a novel phagosome by the Legionnaires' disease bacterium (Legionella pneumophila) in human monocytes. *J. Exp. Med.* **158:**1319–1331.
10. **Horwitz, M. A.** 1983. The Legionnaires' disease bacterium (Legionella pneumophila) inhibits phagosome-lysosome fusion in human monocytes. *J. Exp. Med.* **158:**2108–2126.
11. **Horwitz, M. A., and F. R. Maxfield.** 1984. Legionella pneumophila inhibits acidification of its phagosome in human monocytes. *J. Cell. Biol.* **99:**1936–1943.
12. **Kagan, J. C., and C. R. Roy.** 2002. Legionella phagosomes intercept vesicular traffic from endoplasmic reticulum exit sites. *Nat. Cell. Biol.* **4:**945–954.
13. **Lu, H., and M. Clarke.** 2005. Dynamic properties of Legionella-containing phagosomes in Dictyostelium amoebae. *Cell. Microbiol.* **7:**995–1007.
14. **Luo, Z. Q., and R. R. Isberg.** 2004. Multiple substrates of the Legionella pneumophila Dot/Icm

system identified by interbacterial protein transfer. *Proc. Natl. Acad. Sci. USA* **101:**841–846.

15. **Nagai, H., J. C. Kagan, X. Zhu, R. A. Kahn, and C. R. Roy.** 2002. A bacterial guanine nucleotide exchange factor activates ARF on Legionella phagosomes. *Science* **295:**679–682.

16. **Nagai, H., and C. R. Roy.** 2003. Show me the substrates: modulation of host cell function by type IV secretion systems. *Cell. Microbiol.* **5:**373–383.

17. **Roy, C. R.** 2002. Exploitation of the endoplasmic reticulum by bacterial pathogens. *Trends Microbiol.* **10:**418–424.

18. **Sadosky, A. B., L. A. Wiater, and H. A. Shuman.** 1993. Identification of Legionella pneumophila genes required for growth within and killing of human macrophages. *Infect. Immun.* **61:** 5361–5373.

19. **Segal, G., M. Purcell, and H. A. Shuman.** 1998. Host cell killing and bacterial conjugation require overlapping sets of genes within a 22-kb region of the Legionella pneumophila genome. *Proc. Natl. Acad. Sci. USA* **95:**1669–1674.

20. **Shohdy, N., J. A. Efe, S. D. Emr, and H. A. Shuman.** 2005. Pathogen effector protein screening in yeast identifies Legionella factors that interfere with membrane trafficking. *Proc. Natl. Acad. Sci. USA* **102:**4866–4871.

21. **Swanson, M. S., and R. R. Isberg.** 1996. Analysis of the intracellular fate of Legionella pneumophila mutants. *Ann. NY Acad. Sci.* **797:**8–18.

22. **Vogel, J. P., H. L. Andrews, S. K. Wong, and R. R. Isberg.** 1998. Conjugative transfer by the virulence system of Legionella pneumophila. *Science* **279:**873–876.

THE *LEGIONELLA PNEUMOPHILA* Dot/Icm TYPE IV SECRETION SYSTEM

Carr D. Vincent, Kwang Cheol Jeong, Jessica Sexton,
Emily Buford, and Joseph P. Vogel

47

The gram-negative bacterium *Legionella pneumophila* is the causative agent of a potentially fatal form of pneumonia called Legionnaires' disease. There are an estimated 10,000 to 20,000 cases of Legionnaires' disease in the United States every year, with an approximate fatality rate of 10%. Humans become infected by inhaling aerosols generated from contaminated water sources such as cooling towers found in large air conditioning systems used in hospitals and hotels. In the environment, *L. pneumophila* is found in water sources, where it is a professional parasite of freshwater amoebae (12). When humans inhale *L. pneumophila*, the bacteria rapidly encounter ameboid-like cells, alveolar macrophages (17). The highly bactericidal macrophages internalize the bacteria in an attempt to kill them. However, reminiscent of the Trojan horse, *L. pneumophila* is able to not only survive inside the hostile environment of the macrophage but is able to replicate and eventually kill the host cell (17). *L. pneumophila* survives intracellularly by inhibiting acidification of the nascent phagosome and by preventing subsequent phagosome–lysosome fusion (16). Instead of fusing with

lysosomes, phagosomes containing *L. pneumophila* undergo a series of cell-biological maturation steps in which they associate with small vesicles and mitochondria, eventually becoming surrounded by rough endoplasmic reticulum (15). Inside this novel compartment, known as the replicative phagosome, *L. pneumophila* is able to multiply, eventually lysing the host cell and initiating subsequent rounds of infection (15). Although the intracellular life cycle of *L. pneumophila* has been clearly established, the molecular mechanisms used by this organism to survive and replicate inside host cells are not fully understood.

The key to *L. pneumophila*'s virulence appears to be its ability to prevent phagosome–lysosome fusion, as mutants defective in this trait are incapable of replicating inside host cells and thus are unable to cause disease (14, 16). Based on this observation, a variety of approaches have been taken to identify bacterial factors that are necessary for *L. pneumophila* pathogenesis. This chapter will focus on one major class of virulence factors, the *dot/icm* genes. The *dot/icm* genes were discovered in the laboratories of Ralph Isberg at Tufts University and Howard Shuman at Columbia University. They were identified by a variety of selections and screens for *L. pneumophila* mutants that were defective for intracellular growth

Carr D. Vincent, Kwang Cheol Jeong, Jessica Sexton, Emily Buford, and Joseph P. Vogel Department of Molecular Microbiology, Washington University School of Medicine, St. Louis, MO 63110.

and/or defective for macrophage killing. In addition, some of the genes were detected based on proximity to known *dot/icm* genes, i.e., they were transcribed in operons or near operons containing genes known to be required for growth inside host cells. In the Isberg lab, the mutants were called *dot* for defect in organelle trafficking, since all the mutants were unable to alter the endocytic pathway and therefore targeted to the lysosome. In the Shuman lab, the mutants were called *icm* reflecting a deficiency in intracellular multiplication. Since most of the genes were independently identified in two separate labs, they have unfortunately been designated with both a *dot* and an *icm* name. Although multiple screens and selections were done to isolate these genes, none were performed to saturation, and therefore additional *dot/icm* genes may exist.

The initial set of *dot/icm* genes was independently identified via a plate selection and an enrichment strategy in the Shuman and Isberg laboratories, respectively. The Shuman laboratory exploited a previously isolated *L. pneumophila* mutant, 25D, which was isolated by the Horwitz laboratory (14). 25D was isolated by passaging a wild-type strain on a suboptimal medium called supplemented Mueller-Hinton (SMH) agar. Only avirulent *L. pneumophila* strains can grow on SMH agar, whereas both virulent and avirulent strains can replicate on the more commonly used buffered charcoal yeast extract (CYE) plates. Characterization of the 25D mutant by Horwitz and colleagues revealed that it was unable to grow inside macrophages, due to an inability in preventing phagosome-lysosome fusion (14). Subsequently, the Shuman laboratory complemented the 25D mutant for intracellular growth with a cosmid containing chromosomal DNA, thereby identifying an operon initially believed to consist of four open reading frames, *icmW*, *icmX*, *icmY*, and *icmZ* (5). It was later discovered that this locus actually consisted of only three genes, *icmV*, *icmW*, and *icmX*, with *icmV* transcribed divergently from the *icmWX* operon. At approximately the same time, the Isberg laboratory devised a clever approach to enrich

for cells that could survive, but were unable to grow, inside macrophages (3). This method, utilizing a *thyA* mutant of *L. pneumophila*, exploited an enrichment based on thymine-less death, which is similar to the classic penicillin selection. Complementation of these mutants revealed a gene, *dotA*, which was required for proper development of the *L. pneumophila* replicative phagosome inside macrophages (3).

After the identification of these four *dot/icm* genes, both the Isberg and the Shuman laboratories pursued additional strategies to identify more *dot/icm* genes. The Isberg laboratory took three new approaches. First, in order to eliminate *dotA* from their screen, they repeated the thymine-less death enrichment using a wild-type strain containing a second copy of *dotA* (26). Unfortunately this strategy did not discover any additional *dot* genes, but did identify some factors that were required for optimal intracellular growth of *L. pneumophila*. The second strategy employed was similar to the approach used by Horwitz to isolate the 25D mutant (28). However, buffered CYE plates supplemented with 0.65% sodium chloride were used instead of SMH agar. SMH agar contains a low amount of sodium chloride due to the presence of Casamino Acids; the amount of sodium present in SMH agar was later confirmed to cause a decreased plating efficiency of virulent *Legionella* (7). Using this strategy, a collection of salt-resistant *L. pneumophila* mutants was isolated and shown to be defective for intracellular growth and alteration of the endocytic pathway (27). Complementation of these mutants using a genomic library revealed a number of genes that were part of the *dotDCB* and *dotKIHGFEP* operons. The third strategy taken by the Isberg lab to identify new *dot/icm* genes involved screening a collection of ethylmethanesulfonate mutagenized *L. pneumophila* cells for mutants unable to form plaques on an A/J macrophage monolayer (1). Complementation of these mutants identified three genes: *dotH*, *dotI*, and *dotO*. Independently, the Shuman laboratory constructed and screened a large collection of *L. pneumophila* Tn903 insertions for mutants that were unable to kill

HL-60-derived human macrophages. Sequence identification of the sites of insertion revealed a large number of genes that eventually were named *icmB* through *icmT* (22).

Since these original screens, two additional screens have been performed that also isolated *dot/icm* genes. The first was a screen of mini-Tn10 insertions in the *L. pneumophila* Wadsworth strain for mutants that were defective for replication within the human monocytic cell line U937 and the amoebae *Acanthamoeba polyphaga* (13). This identified a number of insertions in known *dot/icm* genes. The second screen was a signature-tagged mutagenesis that isolated genes required for optimal survival and replication within guinea pig lungs (11). This screen re-identified *dotB*, *dotF*, *dotO*, and *icmX* and confirmed that the *dot/icm* genes were required for optimal growth inside a mammalian host (11).

To date, 26 *dot/icm* genes have been identified and are known to be required for optimal intracellular growth of *L. pneumophila*. These genes are clustered primarily in two distinct areas of the *L. pneumophila* chromosome designated region I and region II, suggestive of two or more pathogenicity islands (Figure 1, Figure 2, and Table 1). These regions each consist of approximately 20 kb of DNA and are located

quite distantly from each other on the chromosome. The *L. pneumophila* Philadelphia-1 chromosome is ~3,400 kb in size, and *icmX* in region I is approximately 840 kb from *icmT* in region II (8).

Sequence analysis of the *dot/icm* complementing open reading frames revealed two common characteristics. First, the majority of the genes (19 of 26) code for proteins predicted to be localized to the bacterial membrane. Considering that mutants lacking these genes have almost identical phenotypes, i.e., are unable to grow inside macrophages due to a failure in preventing phagosome–lysosome fusion, the colocalization of Dot/Icm proteins at the bacterial membrane strongly suggests they compose a single large membrane complex. Second, the majority of Dot/Icm proteins have homology to plasmid conjugation proteins of the IncI plasmids ColIb-P9 and R64 (18). Of the 26 known Dot/Icm proteins, 19 have some level of homology to the IncI Tra or Trb proteins. Interestingly, most of the IncI operon structures have been maintained in *L. pneumophila*, although their relative positions have been shuffled.

Based on the homology between the Dot/Icm proteins and proteins involved in the transfer of IncI plasmids, the Isberg and Shuman

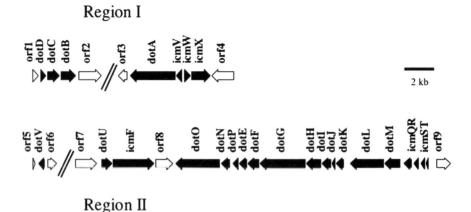

FIGURE 1 The *dot/icm* genes (shown with black arrows) are located in two regions on the *L. pneumophila* chromosome. The white arrows indicate genes that are not required for intracellular growth. The appropriate *dot* or *icm* name is shown above the arrow except in the case where the gene has two names and then only the *dot* name is indicated.

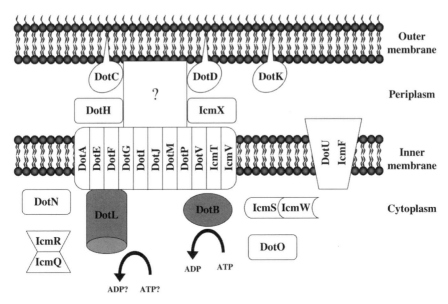

FIGURE 2 Schematic representing predicted locations and possible interactions between the Dot/Icm proteins. The proteins are predicted to localize to the cytoplasm, the inner membrane, the periplasm, or the outer membrane. Known interactions include the following pairs: IcmQ–IcmR, IcmS–IcmW, and DotU–IcmF.

laboratories independently tested to determine if the Dot/Icm apparatus could function as a plasmid transfer system (23, 27). In both labs, the Dot/Icm system was able to mobilize an exogenous DNA substrate, the IncQ plasmid RSF1010, from *L. pneumophila* to another *L. pneumophila* strain or to *Escherichia coli*. In the Isberg lab, transfer of RSF1010 was completely dependent on each *dot/icm* gene examined and occurred in a DNAse-resistant manner, indicative of conjugation and not natural transformation (27). In the Shuman laboratory, transfer was sometimes only partially dependent on the *dot/icm* genes (23). This was later determined to be due to the presence of a second plasmid transfer system in the Shuman laboratory strain background, the *lvhB* operon, which could also mobilize the plasmid substrate.

Based on the homology of the Dot/Icm proteins to conjugal transfer proteins and their ability to transfer a plasmid, the Dot/Icm system was described as a type IV secretion system (T4SS) (23, 27). Type IV secretion systems consist of both plasmid transfer systems and adapted conjugation systems used by patho-

gens to export substrates (9). Most type IV systems are related to the VirB system used by the plant pathogen *Agrobacterium tumefaciens*. These systems have been designated type IVA. In contrast, the *L. pneumophila* Dot/Icm system is more closely related to the transfer apparatus of IncI plasmids and has been designated type IVB. *Coxiella burnetii*, the causative agent of Q fever, also contains a type IVB secretion system that strongly resembles the *L. pneumophila* Dot/Icm T4SS.

It was initially believed, based on the homology to a plasmid transfer system, that the *L. pneumophila* T4SS might transfer a DNA substrate into the host cell similar to the plant pathogen *A. tumefaciens*. However, the *dot/icm* machinery is required immediately upon bacterial uptake and therefore there does not appear to be sufficient time for a DNA effector to be imported, processed, and used inside macrophages as seen with *A. tumefaciens* and plants (21). As a result, it seemed more likely that a protein toxin similar to pertussis toxin might be secreted by the *dot/icm* machinery. In fact, over the past several years, a large number of

TABLE 1 Characteristics of the *L. pneumophila* Dot/Icm proteins

Protein[a]	#aa[b]	Size[c]	IncI[d]	Motifs[e]	Location[f]	Function[g]
DotA	1,048	113	TraY		IM	Structural
DotB	377	42	TraJ	Walker box/VirB11	Cytoplasm	Energy transducer
DotC	303	34	TraI	Lipoprotein	OM	
DotD	163	18	TraH	Lipoprotein	OM	
DotE/IcmC	194	20	TraQ		IM	Structural
DotF/IcmG	269	30	TraP		IM	IM receptor
DotG//IcmE	1,048	108	TraO	VirB10	IM	Energy transducer
DotH/IcmK	360	39	TraN		Periplasm	
DotI/IcmL	212	23	TraM		IM	Structural
DotJ/IcmM	94	11			IM	Structural
DotK/IcmN	189	21	TraL	Lipoprotein	OM	
DotL/IcmO	783	87	TrbC	Walker box/VirD4	IM	Energy transducer
DotM/IcmP	376	43	TrbA		IM	Structural
DotN/IcmJ	208	24	TraT		Cytoplasm	
DotO/IcmB	1,009	112	TraU		Cytoplasm	
DotP/IcmD	132	14	TraR		IM	Structural
DotU	261	30			IM	Chaperone
DotV	180	20			IM	Structural
IcmF	973	111			IM	Chaperone
IcmQ	191	22			Cytoplasm	Pore?
IcmR	120	13			Cytoplasm	Chaperone
IcmS	114	13			Cytoplasm	Chaperone
IcmT	86	10	TraK		IM	Pore?
IcmV	151	18	TraX		IM	Structural
IcmW	151	17			Cytoplasm	Chaperone
IcmX	466	51	TraW		Periplasm	

[a]Proteins are listed with the appropriate Dot or Icm designation.
[b]Number of amino acids.
[c]Predicted size in kilo-Daltons.
[d]Name of the IncI plasmid homologue.
[e]Motifs or homology to the *A. tumefaciens* type IV secretion system.
[f]Location of the protein predicted by a variety of computer algorithms. IM, inner membrane; OM, outer membrane.
[g]Predicted function of the protein.

proteins exported by the Dot/Icm apparatus have been identified. These are covered in detail in other sections of this publication and will not be further discussed here. In summary, *L. pneumophila* appears to have effectively "stolen" the conjugation system of an IncI plasmid and is now using it to inject a large number of protein factors into the host cell in order to alter the endocytic pathway and enable bacterial replication.

The *L. pneumophila* Dot/Icm T4SS has the ability to transfer plasmids between bacterial cells and can export a variety of protein sub-strates into eukaryotic host cells. However, the specific molecular function of most Dot/Icm proteins remains unknown. Over the past several years, a number of laboratories have begun to characterize individual Dot/Icm proteins. We now possess limited information on about half of these proteins. The last part of this chapter will summarize our current knowledge in this area, drawing primarily from the work of the Abu Kwaik, Isberg, Roy, Segal, and Vogel laboratories. Limited individual references are cited (see the review in [22]).

The first Dot protein to be extensively characterized was DotA. This is a large polytopic inner membrane protein of 1,048 amino acids that contains 8 transmembrane domains, 1 large periplasmic domain, and a large carboxy-terminal cytoplasmic domain. Although the specific role of DotA in the type IV secretion apparatus is not known, it has been reported that the DotA protein can be detected in culture supernatants (20). The presence of DotA in the culture supernatants was described as not being due to cell lysis, but no biological significance for the secreted DotA could be detected.

DotU and IcmF were shown to be required for optimal stability of proteins in the *L. pneumophila* type IV secretion apparatus (24). Unlike the other *dot/icm* genes, *dotU* and *icmF* are the only *dot/icm* genes that have homologues in a large number of other pathogens. In these organisms, *dotU* and *icmF* homologues are associated with a large operon that appears to encode a novel secretion system designated IAHP or type VI. Based on their inclusion in another secretion system and their role in stabilizing the Dot/Icm apparatus, it has been proposed that DotU and IcmF may function as macromolecular chaperones (24).

The IcmR protein has been shown to function as a chaperone-like molecule for IcmQ. Although a definitive function for IcmQ remains elusive, the protein has been shown to possess pore-forming activity in vitro (10). The IcmT protein has also been associated with pore formation. One class of *icmT* mutants, called *rib* (release of intracellular bacteria) mutants, is able to replicate intracellularly, similar to a wild-type strain, but are defective in pore formation–mediated cytolysis and egress from mammalian and protozoan cells (4).

Two Dot proteins, DotB and DotL, contain Walker box motifs, suggesting they may function as ATPases. DotB has been purified and shown to form hexamers with ATPase activity. Native DotB localizes mostly to the cell cytoplasm, with a detectable portion associated with the cytoplasmic face of the inner membrane. Similar to the VirB11 protein of *A.*

tumefaciens, DotB has also been shown to play a role in substrate specificity (25). DotL, a second putative T4SS ATPase, has proven to be more refractory to biochemical analysis, and its predicted ATPase activity has not been confirmed. DotL has homology to the type IV coupling protein (T4CP) family, a large class of putative ATPases. T4CPs are reported to function as inner membrane receptors for substrates and may also be involved in providing energy for substrate export across the bacterial envelope. Consistent with its proposed role as a T4CP, DotL was shown to be an inner membrane protein (6). In contrast to most *dot/icm* genes, *dotL*, *dotM*, and *dotN* share the unique property that they are essential for growth of *L. pneumophila* on buffered CYE plates (6). Interestingly, loss of these genes can be suppressed by mutations in other *dot/icm* genes. It has been proposed that lethality due to loss of *dotL/dotM/dotN* may be due to the generation of an unregulated secretion pore, which can then be suppressed by inactivation of components of the unregulated pore (6).

A number of additional Dot/Icm proteins have been proposed to play a direct role in export of substrates. DotF, a small inner membrane protein, was shown to interact with putative T4SS substrates via a bacterial two-hybrid screen (19). Somewhat perplexingly, DotF is not absolutely required for growth of *L. pneumophila* in permissive hosts such as U937 and HL-60 cells. The IcmS protein, reported to resemble type III secretion chaperones, was shown to bind and be required for export of members of the SidE family of T4SS substrates (2).

In summary, the Dot/Icm proteins characterized to date appear to fall into three functional classes: structural, energy transducing, and chaperone-like. A significant amount of work remains to be done in order to fully characterize the Dot/Icm apparatus. First, the predicted localization of the proteins indicated in Table 1 needs to be experimentally confirmed. Second, protein–protein interactions between different Dot/Icm proteins need to be ascertained in order to define functional subcomplexes of Dot/Icm proteins. Third, the location of the *L. pneumophila* Dot/Icm complex needs to be determined. Further characterization of this very complex secretion system will likely reveal insights into how bacteria export proteins and how *L. pneumophila* causes disease.

REFERENCES

1. **Andrews, H. L., J. P. Vogel, and R. R. Isberg.** 1998. Identification of linked *Legionella pneumophila* genes essential for intracellular growth and evasion of the endocytic pathway. *Infect. Immun.* **66:**950–958.
2. **Bardill, J. P., J. L. Miller, and J. P. Vogel.** 2005. IcmS-dependent translocation of SdeA into macrophages by the *Legionella pneumophila* type IV secretion system. *Mol. Microbiol.* **56:**90–103.
3. **Berger, K. H., J. J. Merriam, and R. R. Isberg.** 1994. Altered intracellular targeting properties associated with mutations in the *Legionella pneumophila dotA* gene. *Mol. Microbiol.* **14:**809–822.
4. **Bitar, D. M., M. Molmeret, and Y. A. Kwaik.** 2005. Structure-function analysis of the C-terminus of IcmT of *Legionella pneumophila* in pore formation-mediated egress from macrophages. *FEMS Microbiol. Lett.* **242:**177–184.
5. **Brand, B. C., A. B. Sadosky, and H. A. Shuman.** 1994. The *Legionella pneumophila icm* locus: a set of genes required for intracellular multiplication in human macrophages. *Mol. Microbiol.* **14:**797–808.
6. **Buscher, B. A., G. M. Conover, J. L. Miller, S. A. Vogel, S. N. Meyers, R. R. Isberg, and J. P. Vogel.** 2005. The DotL protein, a member of the TraG-coupling protein family, is essential for viability of *Legionella pneumophila* strain Lp02. *J. Bacteriol.* **187:**2927–2938.
7. **Catrenich, C. E., and W. Johnson.** 1989. Characterization of the selective inhibition of growth of virulent *Legionella pneumophila* by supplemented Mueller-Hinton medium. *Infect. Immun.* **57:**1862–1864.
8. **Chien, M., I. Morozova, S. Shi, H. Sheng, J. Chen, S. M. Gomez, G. Asamani, K. Hill, J. Nuara, M. Feder, J. Rineer, J. J. Greenberg, V. Steshenko, S. H. Park, B. Zhao, E. Teplitskaya, J. R. Edwards, S. Pampou, A. Georghiou, I. C. Chou, W. Iannuccilli, M. E. Ulz, D. H. Kim, A. Geringer-Sameth, C. Goldsberry, P. Morozov, S. G. Fischer, G. Segal, X. Qu, A. Rzhetsky, P. Zhang, E. Cayanis, P. J. De Jong, J. Ju, S. Kalachikov, H. A. Shuman, and J. J. Russo.** 2004. The genomic sequence of the accidental pathogen *Legionella pneumophila*. *Science* **305:**1966–1968.
9. **Christie, P. J., K. Atmakuri, V. Krishnamoorthy, S. Jakubowski, and E. Cascales.** 2005. Biogenesis, architecture, and function of bacterial type IV secretion systems. *Annu. Rev. Microbiol.* **59:**451–485.
10. **Dumenil, G., T. P. Montminy, M. Tang, and R. R. Isberg.** 2004. IcmR-regulated membrane insertion and efflux by the *Legionella pneumophila* IcmQ protein. *J. Biol. Chem.* **279:**4686–4695.
11. **Edelstein, P. H., M. A. Edelstein, F. Higa, and S. Falkow.** 1999. Discovery of virulence genes of *Legionella pneumophila* by using signature tagged mutagenesis in a guinea pig pneumonia model. *Proc. Natl. Acad. Sci. USA* **96:**8190–8195.
12. **Fields, B. S.** 1996. The molecular ecology of Legionellae. *Trends Microbiol.* **4:**286–290.
13. **Gao, L. Y., O. S. Harb, and Y. Abu Kwaik.** 1997. Utilization of similar mechanisms by Legionella pneumophila to parasitize two evolutionarily distant host cells, mammalian macrophages and protozoa. *Infect. Immun.* **65:**4738–4746.
14. **Horwitz, M. A.** 1987. Characterization of avirulent mutant *Legionella pneumophila* that survive but do not multiply within human monocytes. *J. Exp. Med.* **166:**1310–1328.
15. **Horwitz, M. A.** 1983. Formation of a novel phagosome by the Legionnaires' disease bacterium (*Legionella pneumophila*) in human monocytes. *J. Exp. Med.* **158:**1319–1331.
16. **Horwitz, M. A.** 1983. The Legionnaires' disease bacterium (*Legionella pneumophila*) inhibits phagosome-lysosome fusion in human monocytes. *J. Exp. Med.* **158:**2108–2126.
17. **Horwitz, M. A., and S. C. Silverstein.** 1980. Legionnaires' disease bacterium (*Legionella pneumophila*) multiples intracellularly in human monocytes. *J. Clin. Invest.* **66:**441–450.
18. **Komano, T., T. Yoshida, K. Narahara, and N. Furuya.** 2000. The transfer region of IncI1 plasmid R64: similarities between R64 *tra* and *Legionella icm/dot* genes. *Mol. Microbiol.* **35:**1348–1359.
19. **Luo, Z. Q., and R. R. Isberg.** 2004. Multiple substrates of the *Legionella pneumophila* Dot/Icm system identified by interbacterial protein transfer. *Proc. Natl. Acad. Sci. USA* **101:**841–846.
20. **Nagai, H., and C. R. Roy.** 2001. The DotA protein from *Legionella pneumophila* is secreted by a novel process that requires the Dot/Icm transporter. *EMBO J.* **20:**5962–5970.
21. **Roy, C. R., K. H. Berger, and R. R. Isberg.** 1998. *Legionella pneumophila* DotA protein is required for early phagosome trafficking decisions that occur within minutes of bacterial uptake. *Mol. Microbiol.* **28:**663–674.
22. **Segal, G., M. Feldman, and T. Zusman.** 2005. The Icm/Dot type-IV secretion systems of

Legionella pneumophila and *Coxiella burnetii*. *FEMS Microbiol. Rev.* **29:**65–81.

23. **Segal, G., M. Purcell, and H. A. Shuman.** 1998. Host cell killing and bacterial conjugation require overlapping sets of genes within a 22-kb region of the *Legionella pneumophila* genome. *Proc. Natl. Acad. Sci. USA* **95:**1669–1674.

24. **Sexton, J. A., J. L. Miller, A. Yoneda, T. E. Kehl-Fie, and J. P. Vogel.** 2004. *Legionella pneumophila* DotU and IcmF are required for stability of the Dot/Icm complex. *Infect. Immun.* **72:**5983–5992.

25. **Sexton, J. A., H. J. Yeo, and J. P. Vogel.** 2005. Genetic analysis of the *Legionella pneumophila* DotB

ATPase reveals a role in type IV secretion system protein export. *Mol. Microbiol.* **57:**70–84.

26. **Swanson, M. S., and R. R. Isberg.** 1996. Identification of *Legionella pneumophila* mutants that have aberrant intracellular fates. *Infect. Immun.* **64:**2585–2594.

27. **Vogel, J. P., H. L. Andrews, S. K. Wong, and R. R. Isberg.** 1998. Conjugative transfer by the virulence system of *Legionella pneumophila*. *Science* **279:**873–876.

28. **Vogel, J. P., C. Roy, and R. R. Isberg.** 1996. Use of salt to isolate *Legionella pneumophila* mutants unable to replicate in macrophages. *Ann. NY Acad. Sci.* **797:**271–272.

SUBCELLULAR LOCALIZATION OF THE Dot/Icm TYPE IV SECRETION PROTEINS

Kwang Cheol Jeong, Carr D. Vincent,
Emily Buford, and Joseph P. Vogel

48

Legionella pneumophila is the causative agent of Legionnaires' disease, a fatal form of pneumonia in elderly and immunocompromised people. This bacterium replicates inside a variety of freshwater protozoa in the environment, whereas it causes disease in humans by multiplying inside alveolar macrophages. *L. pneumophila* replicates inside host cells by using a type IV secretion system (T4SS) encoded by the *dot/icm* genes (7). This system appears to be an adapted conjugation system that is now used to inject protein substrates into the cytoplasm of an *L. pneumophila* host cell. Most T4SSs are related to the *Agrobacterium tumefaciens* VirB system, including the T4SS of *Bartonella tribocorum, Bordetella pertussis, Brucella abortus, Helicobacter pylori,* and *Rickettsia prowazekii* (2). However, the *L. pneumophila* T4SS is more closely related to the conjugation system found on the IncI plasmids CollB-P9 and R64 (5). In addition to *L. pneumophila,* the pathogen *Coxiella burnetii* contains a T4SS with similarity to the IncI conjugation system (9). Type IV secretion systems that resemble the *A. tumefaciens* VirB system have been classified as type IVA, whereas systems related to the *L. pneumophila* Dot/Icm apparatus are grouped as type IVB.

The *L. pneumophila dot/icm* genes are essential for intracellular multiplication of *L. pneumophila* (8, 10). In contrast to wild-type cells, *dot/icm* mutants cannot alter the endocytic pathway of their host, and therefore the nascent phagosome rapidly acquires endocytic markers such as LAMP-1 soon after formation. The failure of *dot/icm* mutants to alter phagosome trafficking prevents intracellular replication of the bacterium and renders the strain avirulent. The Dot/Icm T4SS exports a large number of substrates into host cells (1,6). We previously reported that one subset of these proteins, the SidE family of T4SS substrates, localize to the cytoplasmic face of the phagosomal membrane adjacent to the poles of the bacterial cell after injection into the host cell (1). This appears to be a shared trait of many *L. pneumophila* T4SS substrates, since at least two others, LidA and SidC, display a similar localization pattern (3, 6).

To determine how these T4SS substrates localize in a polar fashion, we examined the subcellular location of this secretory apparatus using an immunofluorescence assay. Twenty microliters of an *L. pneumophila* strain Lp02 stationary phase culture were briefly fixed in methanol, resuspended in phosphate-buffered saline (PBS), and allowed to adhere to poly-L-lysine-coated microscope slides. Lysozyme was used to permeabilize the cells, which were then

Kwang Cheol Jeong, Carr D. Vincent, Emily Buford, and Joseph P. Vogel Department of Molecular Microbiology, Washington University School of Medicine, St. Louis, MO 63110.

washed with PBS and incubated with various primary antibodies. After incubation, cells were washed with PBS decorated with Oregon green-conjugated goat anti-rabbit IgG, and stained with 4′, 6′-diamidino-2-phenylindole (DAPI) to detect DNA. Fluorescence antifade reagent was added, and the immunostained cells were observed with a fluorescence microscope.

We initiated our studies with a series of controls by probing cells with antibodies against well-studied *L. pneumophila* proteins or components known to localize to distinct cellular compartments. Control antibodies recognized were the cytoplasmic protein isocitrate dehydrogenase (lpg0816), an inner membrane protease with homology to *E. coli* YaeL that we have called RipA (lpg0505), the periplasmic chaperone Mip (lpg0791), and the cell surface molecule lipopolysaccharide. Antibodies were generated by the Vogel laboratory (RipA) or were obtained from Linc Sonenshein (ICDH), J.H. Helbig (Mip), and ViroStat lipopolysaccharide. Using control antibodies, in conjunction with DAPI staining, we were able to identify *L. pneumophila* cytoplasmic, inner membrane, periplasmic, and outer membrane compartments

via immunofluorescence microscopy (data not shown). Notably, none of the control antibodies localized to a specific cellular subdomain such as a bacterial pole.

To determine the localization of the *L. pneumophila* T4SS apparatus, we initially investigated the distribution of one Dot protein, DotF. DotF, predicted to localize to the inner membrane by hydrophilicity analysis, has been shown to interact with secreted substrates via a bacterial two-hybrid screen (6). Immunofluorescence microscopy using a polyclonal antibody that recognizes DotF, a most generous gift of Ralph Isberg (Tufts University), revealed that this protein was specifically localized to the bacterial poles in the wild-type *L. pneumophila* strain Lp02 (Fig. 1). Quantitative analysis showed that more than 95% of cells had at least one focus of fluorescence at a pole. The majority of the cells showed bipolar DotF staining. The polar staining was specific to DotF because it was not apparent in a strain lacking the *dotF* gene. Analysis of DotF staining in a collection of nonpolar *dot/icm* mutants indicated that one or more additional Dot/Icm proteins are required for proper DotF localization. A

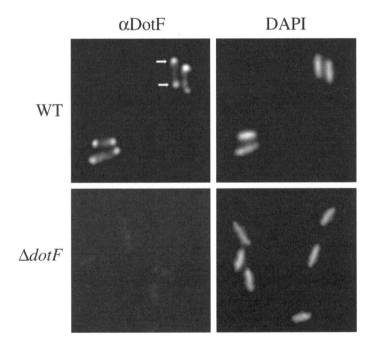

FIGURE 1 Intracellular localization of DotF in wild type (WT) and a Δ*dotF* mutant. Cells were grown to stationary phase and analyzed by fluorescence microscopy. Cells were detected by staining with the DNA stain DAPI. Foci of DotF proteins were detected using an antibody that specifically recognizes DotF.

number of additional Dot/Icm proteins were also localized to the poles (data not shown). Therefore, based on these results, it seems most likely that the localization pattern of the *L. pneumophila* T4SS substrates inside host cells is due to the polar localization of the secretion apparatus itself.

Similar to this, it has previously been shown that the T4SS of *A. tumefaciens* also localizes to a bacterial subdomain (Judd et al., 2005). Using a similar immunofluorescence assay, the Das laboratory was able to determine that a number of the VirB proteins localize to a bacterial pole (Judd et al., 2005). Moreover, the authors determined that one VirB protein, VirB8, appeared to play a key role in establishing VirB localization. Interestingly, VirB8 appears to be able to localize to a cell pole by itself and then functions as an anchor or target to recruit other VirB proteins to the cell pole, presumably through protein-protein interactions. The polar localization of the *A. tumefaciens* VirB T4SS has been proposed to provide a selective advantage since this plant pathogen attaches to its host cell in a polar manner, thereby allowing intimate contact between the secretion system and the plant cell. Likewise, it possible that polar localization of the *L. pneumophila* T4SS may be advantageous to the pathogen in allowing contact between the bacterium and the membrane of the newly formed phagosome.

Based on our observations, and the work on the *A. tumefaciens* VirB system, it is possible that polar localization is a conserved feature of all adapted type IV secretion systems. In addition, it is worth noting that a large number of important bacterial proteins and processes are localized to the poles. This includes the flagellar apparatus, methyl-accepting chemotaxis proteins, competence proteins, and a number of virulence factors such as the *Shigella flexneri* IcsA protein (4). There are several possible explanations for why proteins may be specifically localized to a bacterial pole. First, it is possible that these proteins simply need to be localized

to any specific subcellular structure in order to increase local concentration, and the bacterial pole represents a specific subcellular domain. Alternatively, as in the case of adapted conjugation systems and IcsA, there may be a selective advantage or molecular requirement for the proteins to be localized to the bacterial pole.

REFERENCES

1. **Bardill, J. P., J. L. Miller, and J. P. Vogel.** 2005. IcmS-dependent translocation of SdeA into macrophages by the *Legionella pneumophila* type IV secretion system. *Mol. Microbiol.* **56**:90–103.

2. **Christie, P. J., K. Atmakuri, V. Krishnamoorthy, S. Jakubowski, and E. Cascales.** 2005. Biogenesis, architecture, and function of bacterial type IV secretion systems. *Annu. Rev. Microbiol.* **59**:451–485.

3. **Conover, G. M., I. Derre, J. P. Vogel, and R. R. Isberg.** 2003. The *Legionella pneumophila* LidA protein: a translocated substrate of the Dot/Icm system associated with maintenance of bacterial integrity. *Mol. Microbiol.* **48**:305–321.

4. **Janakiraman, A., and M. B. Goldberg.** 2004. Evidence for polar positional information independent of cell division and nucleoid occlusion. *Proc. Natl. Acad. Sci. USA* **101**:835–840.

5. **Komano, T., T. Yoshida, K. Narahara, and N. Furuya.** 2000. The transfer region of IncI1 plasmid R64: similarities between R64 *tra* and *Legionella icm/dot* genes. *Mol. Microbiol.* **35**:1348–1359.

6. **Luo, Z. Q., and R. R. Isberg.** 2004. Multiple substrates of the *Legionella pneumophila* Dot/Icm system identified by interbacterial protein transfer. *Proc. Natl. Acad. Sci. USA* **101**:841–846.

7. **Segal, G., M. Feldman, and T. Zusman.** 2005. The Icm/Dot type-IV secretion systems of *Legionella pneumophila* and *Coxiella burnetii*. *FEMS Microbiol. Rev.* **29**:65–81.

8. **Segal, G., M. Purcell, and H. A. Shuman.** 1998. Host cell killing and bacterial conjugation require overlapping sets of genes within a 22-kb region of the *Legionella pneumophila* genome. *Proc. Natl. Acad. Sci. USA* **95**:1669–1674.

9. **Sexton, J. A., and J. P. Vogel.** 2002. Type IVB secretion by intracellular pathogens. *Traffic* **3**:178–185.

10. **Vogel, J. P., H. L. Andrews, S. K. Wong, and R. R. Isberg.** 1998. Conjugative transfer by the virulence system of *Legionella pneumophila*. *Science* **279**:873–876.

DEFINING THE TRANSLOCATION PATHWAY OF THE *LEGIONELLA PNEUMOPHILA* TYPE IV SECRETION SYSTEM

Carr D. Vincent, Jonathan R. Friedman, and Joseph P. Vogel

49

Legionella pneumophila is a respiratory pathogen that causes a severe pneumonia known as Legionnaires' disease. The ability of *L. pneumophila* to cause disease is primarily due to its ability to survive inside alveolar macrophages, immune cells capable of destroying most bacteria. The bacteria accomplish this by delaying phagosome-lysosome fusion and establishing a protected intracellular compartment in the macrophage, referred to as the replicative phagosome. The bacteria replicate inside the host cell, eventually lysing the cell and spreading to other macrophages.

Twenty-six genes involved in delaying phagosome-lysosome fusion have been identified and are collectively termed *dot* (defective in organelle trafficking) or *icm* (intracellular multiplication) genes (7). The Dot/Icm proteins make up a type IVB secretion system that is required for delivery of multiple protein substrates into the host cell cytoplasm, where they are believed to alter the endocytic pathway and allow the bacterial phagosome to avoid fusion with lysosomal markers. Type IV secretion systems (T4SSs) are used by many pathogens to deliver substrates to host cells, in-

cluding *Helicobacter pylori*, *Bordetella pertussis*, and *Brucella abortus*. The secretion systems used by these pathogens are ancestrally related to plasmid transfer systems and are thus referred to as adapted conjugation systems. This evolutionary relationship is clear in the case of the *L. pneumophila* Dot/Icm T4SS, as this secretion system retains the ability to transfer the broad host range RSF1010 plasmid to a bacterial recipient (3).

Our goal is to characterize the molecular details of how the Dot/Icm secretion complex assembles and functions to export substrates. Our approaches to understanding these processes are based on techniques that have proven successful in characterizing the canonical T4SS, the VirB/D4 system from the plant pathogen *Agrobacterium tumefaciens*. In this system, understanding which components interact has been useful in determining functional subcomplexes of the secretion apparatus. Biochemical and genetic analyses have revealed that the *A. tumefaciens* secretion complex is composed of three subcomplexes: an energy-transducing subcomplex, the T pilus subcomplex, and a transmembrane "core" subcomplex (2). The energy-transducing subcomplex consists of three proteins, VirB4, VirB11, and VirD4, all of which contain a Walker box nucleoside triphosphate binding motif that is required for substrate

Carr D. Vincent, Jonathan R. Friedman, and Joseph P. Vogel Department of Molecular Microbiology, Washington University Medical School, St. Louis, MO 63110.

secretion. The T pilus complex consists primarily of the major pilin subunit VirB2, with VirB7 and VirB5 as minor components of the T pilus. The core subcomplex consists of VirB6, VirB7, VirB8, VirB9, and VirB10. VirB6, VirB8, and VirB10 localize to the bacterial inner membrane, whereas the VirB7 lipoprotein is found in the outer membrane linked by a disulfide bridge to VirB9. The interactions between these five subunits have been extensively characterized via coimmunoprecipitation experiments, yeast and bacterial two-hybrid assays, and copurification, all of which strongly support the idea that these proteins form a transmembrane subcomplex that is the core structure of the secretion apparatus.

In contrast to the VirB/D4 T4SS, relatively little is known about the structure of the Dot/Icm T4SS, and limited homology to the VirB/D4 system makes inferences about protein functions difficult. The likely components of an energy-transducing complex are DotB and DotL, the only Dot/Icm components that have clear Walker box motifs. These two proteins are homologs of VirB11 and VirD4, respectively. The DotB protein has been characterized in some detail, and it is known that ATP hydrolysis by this protein is required for substrate secretion and may have a role in determining substrate specificity (8, 9). DotL is an essential protein that has been shown to be the type IV coupling protein for the Dot/Icm system and is thought to serve as the inner membrane receptor for secreted substrates (1). At this time, it is unknown whether the Dot/Icm complex assembles a pilus, and no clear homologs exist for the T pilus subcomplex proteins. Of the VirB/D4 core subcomplex components, only VirB10 has homology to a Dot/Icm component; DotG. Three Dot/Icm proteins, DotC, DotD, and DotK, contain lipobox motifs, suggesting that they may localize and function similarly to the lipoproteins VirB9 and VirB7 (10).

Using the approaches successfully taken to identify functional subcomplexes in the *A. tumefaciens* T4SS, we have begun to characterize the structure of the Dot/Icm complex by localizing the Dot/Icm proteins and analyzing their stability in the absence of other components. First, to localize the Dot/Icm proteins, we have utilized differential solubilization of inner and outer membrane proteins using the detergent Triton X-100. Wild-type *L. pneumophila* strain Lp02 cells grown to early stationary phase were lysed in a French pressure cell, followed by low-speed centrifugation to remove unlysed cells. The cleared lysate was then subjected to ultracentrifugation (100,000 × *g* for 1 h) to separate soluble and membrane proteins. The membrane pellet was resuspended in buffer containing 1% Triton X-100 followed by incubation on ice for 30 min. The suspension was ultracentrifuged (100,000 × *g* for 1 h) to pellet insoluble proteins, and fractions of the supernatant (containing inner membrane proteins) and the resuspended pellet (containing outer membrane proteins) were saved for analysis by sodium dodecyl sulfate-polyacrylimide gel electrophoresis followed by Western blotting with specific antibodies. Fig. 1 shows fractionation of the control proteins MomP (outer membrane), signal peptidase LepB (inner membrane), and isocitrate dehydrogenase (soluble), demonstrating that the protein fractions obtained by this procedure are relatively free of contaminants from other cell compartments and that there is not significant degradation of proteins during the fractionation procedure. Analysis of these protein fractions with anti-Dot/Icm antibodies has allowed us to assign the subcellular localization of many of the Dot/Icm proteins; shown in Fig. 1 are DotB (a homolog of VirB11) and DotL (the type IV coupling protein). Using this fractionation technique, DotB localizes primarily to the soluble fraction. This result is consistent with previous findings and with the proposed role of DotB in determining substrate specificity; cytoplasmic localization would allow DotB to interact with substrates and with the cytoplasmic face of the secretion complex (8). DotL was shown to localize to the inner membrane, again consistent with previous findings and with its role as the type IV coupling protein for the Dot/Icm complex (1). All

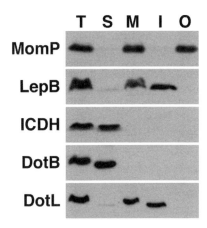

FIGURE 1 Subcellular localization of Dot/Icm proteins using Triton X-100 solubility. Cells were grown to early stationary phase and lysed by French press, and membrane proteins were separated from soluble proteins by ultracentrifugation. Inner membrane protein fractions were then extracted with Triton X-100, followed by ultracentrifugation to separate insoluble outer membrane proteins. Shown are Western blots performed with antibodies specific to the proteins listed to the left of each blot. T, total protein; S, soluble protein; M, total membrane protein; I, inner membrane protein; O, outer membrane protein.

known type IV coupling proteins localize to the inner membrane, where they function as the inner membrane receptors for secreted substrates. Because detergent solubility can occasionally be misleading as an indicator of protein localization, the localizations of the Dot/Icm proteins will be confirmed using sucrose density gradients. Information about the subcellular localization of proteins in the secretion complex will be useful not only in determining potential protein interactions and subcomplexes within the machinery, but it will also provide clues about the potential roles of the Dot/Icm proteins in substrate export.

To identify functional subcomplexes of the Dot/Icm T4SS, we are currently examining protein interactions between Dot/Icm components. One approach we are taking to accomplish this is to determine effects on protein stability caused by deletions of other *dot/icm* genes. This approach is based on the fact that proteins that interact in a complex often require their interaction partners for stability. For example, it has been shown that loss of the *A. tumefaciens* T4SS component VirB6 affects the stability of VirB3 and VirB5 (6). To identify potential protein-protein interactions between Dot/Icm proteins, we assayed the effects of *dot/icm* deletions on the stability of other proteins in the secretion complex. We generated in-frame deletions of all of the known *dot/icm* genes and analyzed stationary-phase cell lysates by Western blot with antibodies specific to various Dot/Icm proteins. Our results confirm previous reports of interactions between IcmQ and its chaperone IcmR and between the secretion chaperones IcmS and IcmW (4, 5). In addition, we have identified additional *dot/icm* mutants that exhibit stability defects on other components of the Dot/Icm complex (Vincent el al., in preparation), suggesting potential protein interactions and subcomplexes within the Dot/icm secretion apparatus. Future work will include confirmation of these interactions and biochemical characterization of the protein subcomplexes. Ideally, these studies will result in the identification of Dot/Icm subcomplexes with discrete functions, similar to those described for the *A. tumefaciens* T4SS, with the long-term goal of understanding the role of each of the individual Dot/Icm proteins in the secretion apparatus.

REFERENCES

1. **Buscher, B. A., G. M. Conover, J. L. Miller, S. A. Vogel, S. N. Meyers, R. R. Isberg, and J. P. Vogel.** 2005. The DotL protein, a member of the TraG-coupling protein family, is essential for viability of *Legionella pneumophila* strain Lp02. *J. Bacteriol.* **187:**2927–2938.
2. **Christie, P. J., and E. Cascales.** 2005. Structural and dynamic properties of bacterial type IV secretion systems (review). *Mol. Membr. Biol.* **22:** 51–61.
3. **Christie, P. J., and J. P. Vogel.** 2000. Bacterial type IV secretion: conjugation systems adapted to deliver effector molecules to host cells. *Trends Microbiol.* **8:**354–360.
4. **Coers, J., J. C. Kagan, M. Matthews, H. Nagai, D. M. Zuckman, and C. R. Roy.** 2000. Identification of Icm protein complexes that play distinct roles in the biogenesis of an organelle permissive for *Legionella pneumophila* intracellular growth. *Mol. Microbiol.* **38:**719–736.

5. **Dumenil, G., and R. R. Isberg.** 2001. The *Legionella pneumophila* IcmR protein exhibits chaperone activity for IcmQ by preventing its participation in high-molecular-weight complexes. *Mol. Microbiol.* **40:**1113–1127.

6. **Hapfelmeier, S., N. Domke, P. C. Zambryski, and C. Baron.** 2000. VirB6 is required for stabilization of VirB5 and VirB3 and formation of VirB7 homodimers in *Agrobacterium tumefaciens. J. Bacteriol.* **182:**4505–4511.

7. **Segal, G., M. Feldman, and T. Zusman.** 2005. The Icm/Dot type-IV secretion systems of *Legionella pneumophila* and *Coxiella burnetii. FEMS Microbiol. Rev.* **29:**65–81.

8. **Sexton, J. A., J. S. Pinkner, R. Roth, J. E. Heuser, S. J. Hultgren, and J. P. Vogel.** 2004. The *Legionella pneumophila* PilT homologue DotB exhibits ATPase activity that is critical for intracellular growth. *J. Bacteriol.* **186:**1658–1666.

9. **Sexton, J. A., H. J. Yeo, and J. P. Vogel.** 2005. Genetic analysis of the *Legionella pneumophila* DotB ATPase reveals a role in type IV secretion system protein export. *Mol. Microbiol.* **57:**70–84.

10. **Yerushalmi, G., T. Zusman, and G. Segal.** 2005. Additive effect on intracellular growth by *Legionella pneumophila* Icm/Dot proteins containing a lipobox motif. *Infect. Immun.* **73:**7578–7587.

LOSS OFF PATATIN-LIKE PHOSPHOLIPASE A CAUSES REDUCED INFECTIVITY OF *LEGIONELLA PNEUMOPHILA* IN AMOEBA AND MACROPHAGE INFECTION MODELS

Philipp Aurass, Sangeeta Banerji, and Antje Flieger

50

L. pneumophila possesses a large variety of lipolytic enzyme activities that affect phospholipids. In addition to the phospholipid-degrading phospholipase A (PLA) and lysophospholipase A (LPLA) activities, glycerophospholipid: cholesterol acyltransferase (GCAT) activity has been described (1, 4, 5). Remarkably, those activities have been assigned only partially to the secreted enzymes PlaA (LPLA) and PlaC (PLA/LPLA/GCAT) as well as the cell-associated PLA/LPLA, PlaB. (1, 6, 7). Therefore, one of our research goals is the search for new *L. pneumophila* PLA and LPLA activities. A previous study showed the wide distribution of genes coding for patatin-like proteins (PLPs) among bacterial genomes (2). Patatin is known as the main potato storage glycoprotein possessing lipid acylhydrolase activity (3). Bacterial PLPs typically possess four conserved regions of protein homology (blocks I to IV), in which blocks II and IV harbor the putative active-site serine and aspartic acid residues of the catalytical dyad, respectively (3). The first published bacterial PLP was the type III secreted *Pseudomonas aeruginosa* exoenzyme U

(ExoU). ExoU is known for its cytotoxicity toward eucaryotic cells and causes acute lung injury in animal models. Furthermore, it possesses PLA and LPLA activities which could be linked to its strong cytotoxicity (8, 9). Considering that *L. pneumophila* still possesses uncharacterized secreted and cell-associated PLA/LPLA activities and since putative lipolytic PLPs occur in many bacteria, we screened the genome of *L. pneumophila* strain Philadelphia-1 for PLPs with help of the text search tool of the pedant database by using the keyword "patatin" (http://pedant.gsf.de). We found that the genome of *L. pneumophila* Philadelphia-1 coded for 11 PLPs, designated PatA to PatK. All of them harbored the members of the characteristic putative active-site dyad, serine and aspartate. One of them, PatA, had already been described by Shohdy and colleagues (10) to be secreted by the *L. pneumophila* Dot/Icm type IVB secretion system and was designated VipD. VipD was further characterized as a protein with impact on the vacuolar trafficking in yeast when expressed in these cells. Next, we were interested in whether some of the 11 *L. pneumophila* PLPs showed protein homology to *P. aeruginosa* ExoU. One PLP, *L. pneumophila* PatA/VipD, was most closely related to *P. aeruginosa* ExoU (27% identity, expect value: 1×10^{-16}). This prompted us to investigate whether

Philipp Aurass, Sangeeta Banerji, and Antje Flieger   Research Group NG 5, Robert Koch-Institut, Nordufer 20, 13353 Berlin, Germany.

PatA/VipD might be a new virulence factor of *L. pneumophila*.

L. PNEUMOPHILA PLP GENES ARE EXPRESSED DURING EXTRACELLULAR GROWTH

First, we were interested in whether all of the 11 PLP genes are expressed in *L. pneumophila* Philadelphia-1 during growth in laboratory media. In order to assess expression of the PLP genes, mRNA was isolated at four growth phases (early logarithmic, mid-logarithmic, late logarithmic, and early stationary) in standard laboratory media and subsequently used for reverse transcriptase PCR with *L. pneumophila* Philadelphia-1 PLP gene-specific primers. Control reactions to check for DNA contamination and presence of RNA were done and are not shown here. We found that all 11 *L. pneumophila* PLP genes were expressed at the 4 considered growth phases, indicating that all of them were expressed during extracelluar growth (Fig. 1 and data not shown).

L. PNEUMOPHILA PatA/VipD POSSESSES LPLA ACTIVITY WHEN EXPRESSED IN *E. COLI*

P. aeruginosa ExoU shows LPLA and PLA activity when activated by an unknown eukaryotic factor (8, 9). We now aimed to assess the lipolytic activity of PatA/VipD. To this purpose, we cloned the *patA/vipD* gene with its putative promoter region into *E. coli* DH5α and incubated the culture supernatant and cell lysates of clones harboring plasmid pPA3 (pBCKS + *patA/vipD*) as well as the empty pBCKS vector with PLA substrates dipalmitoylphosphatidylglycerol and dipalmitoylphosphatidylcholine, LPLA substrates monopalmitoyllysophosphatidylglycerol, monopalmitoyllysophosphatidylcholine, and monopalmitoyllysophosphatidylethanolamine, and a lipase substrate 1-monopalmitoylglycerol. Additionally, in order to assess whether recombinant PatA/VipD was likewise activated by a eukaryotic factor, bacterial cell lysates were incubated with lysates of U937 monocyte cells prior to lipid incubation. Cell lysates of *E. coli* carrying plasmid pKH192 (pBCKS + *plaB*), containing the gene for the cell-associated *L. pneumophila* PLA/LPLA, PlaB, served as a positive control for our enzyme assay (7). Our results show that the cell lysate of the *E. coli* clone carrying *patA/vipD* in *trans* displayed high LPLA activity, as it possessed an increased capacity to release free fatty acids from the three employed LPLA substrates compared to the clone carrying the empty vector (Fig. 2). No increased PLA or lipase activ-

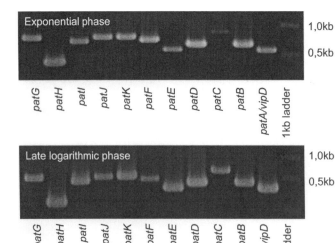

FIGURE 1 Expression of *L. pneumophila* PLP genes during growth in laboratory media. Expression of the 11 *L. pneumophila* Philadelphia-1 PLP genes, *patA/vipD* to *patK*, was examined during exponential and late logarithmic growth phase in BYE broth. RNA was isolated and reverse transcriptase PCR was performed. The results are representative of two independent experiments.

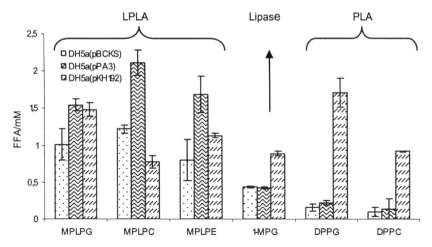

FIGURE 2 *L. pneumophila patA/vipD* shows LPLA activity when expressed in *E. coli*. Release of free fatty acids from monopalmitoyllysophosphatidylglycerol, monopalmitoyllysophosphatidylcholine, monopalmitoyllysophosphatidylethanolamine, 1-monopalmitoylglycerol, dipalmitoylphosphatidylglycerol, and dipalmitoylphosphatidylcholine by three times concentrated cell lysates of *E. coli* DH5α containing the pBCKS vector in comparison to *E. coli* DH5α containing its derivative pPA3 with *L. pneumophila patA/vipD*. *E. coli* carrying pKH192, harboring the *L. pneumophila* PLA and LPLA gene *plaB* (7), was used as a positive control. Free fatty acids were quantified after 24 h incubation at 37°C. Data are means +/− standard deviation of three experiments and are shown as the difference between fatty acids released by bacterial lysate and by negative control (40 mM Tris-HCl buffer).

ity could be detected in the clones harboring pPA3, implying that recombinant *L. pneumophila* PatA/VipD is an LPLA but does not possess PLA and lipase activities. Preincubation with monocyte lysates did not lead to enhanced lipolysis of the employed substrates or to a changed substrate specificity of recombinant PatA/VipD, suggesting that this enzyme cannot be activated by a eukaryotic factor (data not shown). In summary, our data show that *patA/vipD* expressed in *E. coli* possesses LPLA activity and incubation with U937 monocyte cell lysates does not lead to further enzyme activation.

L. PNEUMOPHILA *patA/vipD* MUTANTS SHOW DECREASED LIPOLYTIC ACTIVITY

To assess lipolytic activities of *L. pneumophila* lacking a functional *patA/vipD* gene, knockout mutants were generated by introducing a Kn[r] cassette into the chromosomal *patA/vipD* gene of *L. pneumophila* Philadelphia-1. *L. pneumophila* wild type and the *patA/vipD* mutants were grown to late exponential phase, and cell lysates and culture supernatants of the bacteria were tested for PLA, LPLA, and lipase activities (7). Culture supernatants of the *L. pneumophila patA/vipD* mutants showed a significantly reduced ability to hydrolyze LPLA substrates. Remarkably, hydrolysis of PLA substrates and the lipase substrate was decreased as well. Therefore, in addition to LPLA activity, PatA/VipD likely shows PLA and lipase activities, and we speculate that this difference compared to recombinant PatA/VipD might be due to activation of the enzyme by a *Legionella*-specific factor.

LOSS OF THE FUNCTIONAL L. PNEUMOPHILA *patA/vipD* GENE LEADS TO AVIRULENCE IN AMOEBA AND MACROPHAGE INFECTION MODELS

We examined the multiplication of *L. pneumophila patA/vipD* mutants within its potential natural host *Acanthamoeba castellanii* as well as U937 macrophages. We inoculated multiplicities of infection of 0.1 and monitored the multiplication of the bacteria over 96 h. Bacteria were counted by determining CFU in 24-h intervals. In the *Acanthamoebae* infection model, we found that the *L. pneumophila patA/vipD* mutant strains were unable to replicate, whereas the wild type showed a multiplication rate of more than 100-fold after 96 h (data not shown). Inside U937 macrophages, the mutants also showed no multiplication (data not shown).

In summary, our study shows that all 11 PLP genes were expressed during growth of *L. pneumophila* in laboratory media. Furthermore, expression of *patA/vipD* in *E. coli* revealed increased LPLA activity, and *L. pneumophila patA/vipD* mutants showed reduced secreted LPLA and PLA activities. Importantly, in coinfection assays with amoebae and macrophages, the *L. pneumophila patA/vipD* mutant strains were severely impaired for intracellular replication. Thus, PatA/VipD is a new type IVB secreted phospholipase of *L. pneumophila* with essential importance during host cell infection.

REFERENCES

1. **Banerji, S., M. Bewersdorff, B. Hermes, N. P. Cianciotto, and A. Flieger.** 2005. Characterization of the major secreted zinc metalloprotease-dependent glycerophospholipid:cholesterol acyltransferase, PlaC, of *Legionella pneumophila*. *Infect. Immun.* **73**:2899–2909.
2. **Banerji, S., and Flieger, A.** 2004 Patatin-like proteins: a new family of lipolytic enzymes present in bacteria? *Microbiology* **150**:522–525.
3. **Finck-Barbancon, V., J. Goranson, L. Zhu, T. Sawa, J. P. Wiener-Kronish, S. M. Fleiszig, C. Wu, L. Mende-Mueller, and D. W. Frank.** 1997. ExoU expression by *Pseudomonas aeruginosa* correlates with acute cytotoxicity and epithelial injury. *Mol. Microbiol.* **25**:547–557.
4. **Flieger, A., S. Gong, M. Faigle, M. Deeg, P. Bartmann, and B. Neumeister.** 2000. Novel phospholipase A activity secreted by *Legionella* species. *J. Bacteriol.* **182**:1321–1327.
5. **Flieger, A., S. Gong, M. Faigle, S. Stevanovic, N. P. Cianciotto, and B. Neumeister.** 2001. Novel lysophospholipase A secreted by *Legionella pneumophila*. *J. Bacteriol.* **183**:2121–2124.
6. **Flieger, A., B. Neumeister, and N. P. Cianciotto.** 2002. Characterization of the gene encoding the major secreted lysophospholipase A of *Legionella pneumophila* and its role in detoxification of lysophosphatidylcholine. *Infect. Immun.* **70**: 6094–6106.
7. **Flieger, A., K. Rydzewski, S. Banerji, M. Broich, and K. Heuner.** 2004. Cloning and characterization of the gene encoding the major cell-associated phospholipase A of *Legionella pneumophila*, *plaB*, exhibiting hemolytic activity. *Infect. Immun.* **72**:2648–2658.
8. **Phillips, R. M., D. A. Six, E. A. Dennis, and P. Ghosh.** 2003. In vivo phospholipase activity of the *Pseudomonas aeruginosa* cytotoxin ExoU and protection of mammalian cells with phospholipase A2 inhibitors. *J. Biol. Chem.* **278**:41326–41332.
9. **Sato, H., D. W. Frank, C. J. Hillard, et al.** 2003. The mechanism of action of the *Pseudomonas aeruginosa*-encoded type III cytotoxin, ExoU. *EMBO J.* **22**:2959–2969.
10. **Shohdy, N., J. A. Efe, S. D. Emr, and H. A. Shuman.** 2005. Pathogen effector protein screening in yeast identifies *Legionella* factors that interfere with membrane trafficking. *Proc. Natl. Acad. Sci. USA* **102**:4866–4871.

IDENTIFICATION OF A CYTOTOXIC *LEGIONELLA PNEUMOPHILA LpxB* PARALOGUE IN A MULTICOPY SUPPRESSOR SCREEN USING *ACANTHAMOEBA CASTELLANII* AS A SELECTIVE HOST

Urs Albers, Katrin Reus, and Hubert Hilbi

51

Environmental amoebae, including *Acanthamoeba castellanii* and the social amoeba *Dictyostelium discoideum*, are free-living primordial phagocytes that feed on bacteria. Some bacteria evolved to prevent digestion by amoebae and instead persist and replicate within the phagocytes. *Legionella pneumophila* is a parasite of freshwater and soil amoebae (3, 12) including *Dictyostelium* (6, 14). Upon encountering an amoeba, *L. pneumophila* triggers its uptake via pathogen-directed phagocytosis (7) and replicates within a specific vacuole until the host cell is ultimately killed. These processes are dependent on the bacterial Icm/Dot type IV secretion system (T4SS) (11, 13, 14). The Icm/Dot T4SS translocates proteins into the host cell, thus subverting phagocytosis and vesicle trafficking. Many protein substrates of the Icm/Dot T4SS have been identified, yet for most of them the effects on host cell functions are poorly defined.

THE AMOEBAE PLATE TEST: GROWTH OF *L. PNEUMOPHILA* ON AGAR PLATES IN THE PRESENCE OF *A. CASTELLANII*

In order to establish a robust plate test that allows genetic screens for *L. pneumophila* factors playing a role in the interaction with phago-

cytes, we established the amoebae plate test (APT) (1). In the APT, *L. pneumophila* wild-type strain JR32 or *icm/dot* mutants are spotted in serial dilutions on charcoal-yeast extract (CYE) agar plates impregnated with 4×10^6 *A. castellanii* and incubated for 5 to 7 days at 30°C. Under these conditions, wild-type *L. pneumophila* formed robust colonies even at high dilutions (Fig. 1). However, an *icmT* mutant strain (and most other *icm/dot* mutants) did not grow at all, while *icmG* and *icmS* mutant strains showed an intermediate phenotype. The growth defects could be complemented by supplying the corresponding genes on a plasmid. In the absence of amoebae, all strains grew equally well on the CYE agar plates.

ISOLATION OF CYTOTOXIC *Icm/Dot* SUPPRESSOR PLASMIDS USING THE AMOEBAE PLATE TEST

The IcmG and IcmS proteins are not constituents of the membrane-spanning part of the Icm/Dot T4SS; rather, these proteins bind to secreted effector proteins, thus possibly facilitating their translocation (2, 5, 9, 10). Perhaps this specific function accounts for the only partial phenotype in the APT of *L. pneumophila* strains lacking IcmG or IcmS. We used *icmS* and *icmG* mutants to screen by the APT an *L. pneumophila* genomic library for multicopy

Urs Albers, Katrin Reus, and Hubert Hilbi   Institute of Microbiology, Swiss Federal Institute of Technology (ETH), 8093 Zürich, Switzerland.

Legionella: State of the Art 30 Years after Its Recognition
Edited by Nicholas P. Cianciotto et al.
©2006 ASM Press, Washington, D.C.

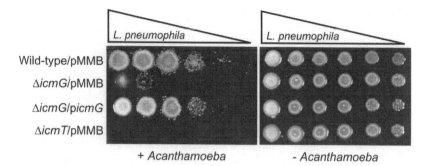

FIGURE 1   Growth of *L. pneumophila* on agar plates in the presence of amoebae: the
APT. *L. pneumophila* wild-type and *icm/dot* mutant strains were spotted in serial dilutions onto CYE agar plates in presence or absence of 4×10^6 *A. castellanii* and incubated for 5 days at 30°C. The strains harbored either an empty vector (pMMB) or a
complementing plasmid (p*icm*G). While wild-type and the complemented *icm*G mutant strain grew robustly even at high dilutions, the *icm*G and *icm*T mutant strains
showed a partial or complete growth defect in presence of amoebae. In the absence of
amoebae, all bacterial strains grew equally well.

suppressors. Thus, we identified a suppressor
plasmid complementing the *icm*S mutant and
several suppressor plasmids which allowed the
*icm*G mutant host strain to grow more robustly
in the presence of *A. castellanii*.

The suppressor plasmids were analyzed for
effects on intracellular replication and cytotoxicity. Cytotoxicity against *A. castellanii* was determined 2 days postinfection (30°C) by uptake of the fluorescent dye propidium iodide
(Fig. 2). Under these conditions, an *icm*G mutant strain was barely cytotoxic, as the amoebae
remained spread out on the surface of the culture dish and excluded propidium iodide.
However, if the *icm*G mutant was complemented with a plasmid expressing the *icm*G
gene, almost all amoebae rounded up and were
killed, i.e., accumulated the dye. While none of
the suppressor plasmids identified increased
the efficiency of intracellular replication, some
of the plasmids rendered the *icm*G mutant host
strain more cytotoxic for *A. castellanii*. Suppressor genes on APT-selected plasmids were identified by deletion of individual open reading
frames. Thus, the *lcs* (*Legionella* cytotoxic suppressor) genes *A*, –*B*, –*C* and –*D* were identified. The corresponding proteins show sequence similarity to hydrolases, NlpD-related
metalloproteases, lipid A disaccharide syn-
thases, and ABC transporters, respectively.

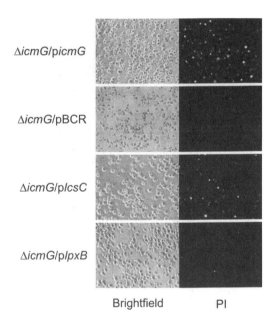

FIGURE 2   *L. pneumophila* *lcs*C but not *lpx*B expressed in an *icm*G mutant strain is cytotoxic for *A.
castellanii*. *A. castellanii* was infected at a multiplicity of
infection of 50 with *L. pneumophila* *icm*G mutant strains
(Δ*icm*G) harboring a complementing plasmid (p*icm*G),
an empty plasmid (pBCR), or plasmids expressing *lcs*C
(p*lcs*C) or *lpx*B (p*lpx*B) from the P$_{tac}$ promotor. Cytotoxicity was quantified 2 days postinfection (30°C) by
uptake of the fluorescent dye propidium iodide (PI).
Bright field and fluorescence micrographs are shown.

IDENTIFICATION AND CHARACTERIZATION OF *LcsC*, A CYTOTOXIC PARALOGUE OF THE LIPID A DISACCHARIDE SYNTHASE *LpxB*

The cytotoxic suppressor gene *lcsC*, a homologue of lipid A disaccharide synthase (*lpxB*), was investigated in more detail by deletion as well as overexpression analysis. Overexpression of *lcsC* but not its paralogue *lpxB* in an *L. pneumophila icmG* mutant strain was cytotoxic for *A. castellanii* (Fig. 2), without enhancing intracellular replication of the bacteria within the amoebae. Toxicity of *lcsC* was observed only in an *icmG* mutant background, but not in other *L. pneumophila* mutants (*icmS*, *icmF*, *rpoS*), which are also partially impaired for intracellular replication within phagocytes. Furthermore, Icm/Dot-dependent efficient phagocytosis by *Dictyostelium* (16) was abolished by expression of *lcsC* in wild-type *L. pneumophila* or in an *icmG* mutant strain, while uptake of an *icmT* mutant or viability of the strains apparently was not affected.

The physiological function and the mechanism of toxicity of LcsC are currently not known. *L. pneumophila* LcsC and its paralogue LpxB share 31% identity on an amino acid level and are 30% or 42% identical, respectively, to the LpxB orthologue from *E. coli*. LpxB is the lipid A disaccharide synthase and in the course of lipid A biosynthesis catalyzes the glycosyl transferase reaction between UDP-2,3-diacylglucosamine and UDP-2,3-diacylglucosamine-1-phosphate to form the tetra-acylated lipid IV.

In the genome of *L. pneumophila* Philadelphia-1 (4), *lpxB* (lpg2945) is clustered with the lipid A biosynthesis genes *lpxD* (lpg2944), *lpxA* (lpg2943), and *lpxL* (lpg2940), suggesting that LpxB is a lipid A disaccharide synthase involved in lipopolysaccharide biosynthesis. On the other hand, *lcsC* (lpg1371) is clustered with genes involved in the maintenance of cell shape and peptidoglycan metabolism, such as *rodA* (lpg1374), encoding the rod shape–determining protein RodA, and *pbpA* (lpg1375), encoding the penicillin-binding protein 2. The vicinity of these genes might suggest that LcsC catalyzes a glycosyl transferase reaction during peptidoglycan metabolism. However, in the *L. pneumophila* genome, *lcsC* is located immediately adjacent to and probably forms an operon with a gene annotated as an NADH-dependent myo-inositol-2-dehydrogenase (lpg1372). A closer inspection revealed that this oxidoreductase is 39% identical to the GnnA protein of *Acidithiobacillus ferrooxidans*, which together with the GnnB protein catalyzes the conversion of UDP-N-acetylglucosamine (UDP-GlcNAc) to UDP-GlcNAcN3 (15). The latter compound contains an amino instead of a hydroxyl group at position 3 of the GlcNAc moiety. A gene similar to *A. ferrooxidans gnnB* (39% identity on an amino acid level) is also present in the *L. pneumophila* genome (lpg1424) but not located in the immediate vicinity of *gnnA* (lpg1372). UDP-GlcNAcN3 is acylated by LpxA and serves as a starter molecule for lipid A biosynthesis by *A. ferrooxidans* and presumably also by *L. pneumophila* (8). These bioinformatic findings suggest that *L. pneumophila lcsC* and *lpxB* both participate in lipid A biosynthesis. Finally, not only the glycosyl transferase gene *lpxB* but also the acyl transferase genes *lpxA* and *lpxD* are present in at least two copies in the *L. pneumophila* genome. Common to these genes is that they alter the acylation state of lipid A precursors in the lipopolysaccharide biosynthesis pathway. Different transferase paralogues might have different acyl substrate specificities and thus determine the variable length of the lipid A acyl chains identified in *L. pneumophila* (8).

CONCLUSIONS

The Icm/Dot T4SS is not only required for the upregulation of phagocytosis and intracellular replication of *L. pneumophila*, but also for growth of *L. pneumophila* in the presence of *A. castellanii* on agar plates. Using partially growth defective *icm/dot* mutant strains, we screened an *L. pneumophila* chromosomal library for suppressors and identified *lcsC*, a cytotoxic *lpxB* paralogue. Based on in silico analysis, *lpxB* as well as *lcsC* might participate in lipid A biosynthesis. The APT established here may prove useful to discover other bacterial factors relevant for interactions with amoeba.

ACKNOWLEDGMENTS

We would like to thank Howard A. Shuman, in whose laboratory this study was initiated, for support and encouragement. The work was funded by grants from the Swiss National Science Foundation (631-065952) and the Swiss Federal Institute of Technology, ETH (TH 17/02-3).

REFERENCES

1. **Albers, U., K. Reus, H. A. Shuman, and H. Hilbi.** 2005. The amoebae plate test implicates a paralogue of *lpxB* in the interaction of *Legionella pneumophila* with *Acanthamoeba castellanii*. *Microbiology* **151**:167–182.

2. **Bardill, J. P., J. L. Miller, and J. P. Vogel.** 2005. IcmS-dependent translocation of SdeA into macrophages by the *Legionella pneumophila* type IV secretion system. *Mol. Microbiol.* **56**:90–103.

3. **Bozue, J. A., and W. Johnson.** 1996. Interaction of *Legionella pneumophila* with *Acanthamoeba castellanii*: uptake by coiling phagocytosis and inhibition of phagosome-lysosome fusion. *Infect. Immun.* **64**:668–673.

4. **Chien, M., I. Morozova, S. Shi, H. Sheng, J. Chen, S. M. Gomez, G. Asamani, K. Hill, J. Nuara, M. Feder, J. Rineer, J. J. Greenberg, V. Steshenko, S. H. Park, B. Zhao, E. Teplitskaya, J. R. Edwards, S. Pampou, A. Georghiou, I. C. Chou, W. Iannuccilli, M. E. Ulz, D. H. Kim, A. Geringer-Sameth, C. Goldsberry, P. Morozov, S. G. Fischer, G. Segal, X. Qu, A. Rzhetsky, P. Zhang, E. Cayanis, P. J. De Jong, J. Ju, S. Kalachikov, H. A. Shuman, and J. J. Russo.** 2004. The genomic sequence of the accidental pathogen *Legionella pneumophila*. *Science* **305**:1966–1968.

5. **Coers, J., J. C. Kagan, M. Matthews, H. Nagai, D. M. Zuckman, and C. R. Roy.** 2000. Identification of Icm protein complexes that play distinct roles in the biogenesis of an organelle permissive for *Legionella pneumophila* intracellular growth. *Mol. Microbiol.* **38**:719–736.

6. **Hagele, S., R. Kohler, H. Merkert, M. Schleicher, J. Hacker, and M. Steinert.** 2000. *Dictyostelium discoideum*: a new host model system for

intracellular pathogens of the genus *Legionella*. *Cell. Microbiol.* **2**:165–171.

7. **Hilbi, H., G. Segal, and H. A. Shuman.** 2001. Icm/Dot-dependent upregulation of phagocytosis by *Legionella pneumophila*. *Mol. Microbiol.* **42**:603–617.

8. **Kooistra, O., Y. A. Knirel, E. Lüneberg, M. Frosch, and U. Zahringer.** 2002. Phase variation in *Legionella pneumophila* serogroup 1, subgroup OLDA, strain RC1 influences lipid A structure, p. 68–73. *In* R. Marre, Y. Abu Kwaik, C. Bartlett, N. P. Cianciotto, B. S. Fields, M. Frosch, J. Hacker, and P. C. Luck (ed.). Legionella. ASM Press, Washington, D.C.

9. **Luo, Z. Q., and R. R. Isberg.** 2004. Multiple substrates of the *Legionella pneumophila* Dot/Icm system identified by interbacterial protein transfer. *Proc. Natl. Acad. Sci. USA* **101**:841–846.

10. **Ninio, S., D. M. Zuckman-Cholon, E. D. Cambronne, and C. R. Roy.** 2005. The *Legionella* IcmS-IcmW protein complex is important for Dot/Icm-mediated protein translocation. *Mol. Microbiol.* **55**:912–926.

11. **Otto, G. P., M. Y. Wu, M. Clarke, H. Lu, O. R. Anderson, H. Hilbi, H. A. Shuman, and R. H. Kessin.** 2004. Macroautophagy is dispensable for intracellular replication of *Legionella pneumophila* in *Dictyostelium discoideum. Mol. Microbiol.* **51**:63–72.

12. **Rowbotham, T. J.** 1980. Preliminary report on the pathogenicity of *Legionella pneumophila* for freshwater and soil amoebae. *J. Clin. Pathol.* **33**:1179–1183.

13. **Segal, G., and H. A. Shuman.** 1999. *Legionella pneumophila* utilizes the same genes to multiply within *Acanthamoeba castellanii* and human macrophages. *Infect. Immun.* **67**:2117–2124.

14. **Solomon, J. M., A. Rupper, J. A. Cardelli, and R. R. Isberg.** 2000. Intracellular growth of *Legionella pneumophila* in *Dictyostelium discoideum*, a system for genetic analysis of host-pathogen interactions. *Infect. Immun.* **68**:2939–2947.

15. **Sweet, C. R., A. A. Ribeiro, and C. R. Raetz.** 2004. Oxidation and transamination of the 3″-position of UDP-N-acetylglucosamine by enzymes from *Acidithiobacillus ferrooxidans*. Role in the formation of lipid A molecules with four amide-linked acyl chains. *J. Biol. Chem.* **279**:25400–25410.

16. **Weber, S., C. Ragaz, K. Reus, and H. Hilbi.** 2006. *Legionella pneumophila* exploits phosphatidylinositol-4 phosphate to anchor secreted effector proteins to the replicative vacuole. PLoS Pathog **2**:e46.

TYPE II PROTEIN SECRETION AND TWIN-ARGININE TRANSLOCATION PROMOTE THE PATHOGENESIS OF *LEGIONELLA PNEUMOPHILA*

Ombeline Rossier and Nicholas P. Cianciotto

52

Translocation of fully or partially folded proteins across bacterial membranes is a remarkable property of the type II secretion (T2S) system and the twin-arginine translocation (Tat) pathway. In gram-negative bacteria, the T2S system spans the bacterial envelope and secretes proteins from the periplasm to the exterior milieu. Type II exoproteins are exported from the cytoplasm in two steps: most proteins first use the general secretion (Sec) pathway for their translocation across the inner membrane and once in the periplasm they recruit the T2S machinery to be transferred across the outer membrane (Fig. 1) (5). In *Legionella pneumophila*, the T2S system is encoded by the *lsp* (*Legionella* secretion pathway) and *pilD* genes, which are located in five separate loci, i.e., *lspC*, *lspDE*, *lspFGHIJK*, *lspLM*, and *pilD* (18). Translocation of folded proteins across the cytoplasmic membrane is achieved by the Tat pathway (Fig. 2) (13). Tat substrates have a signal peptide that harbors two consecutive arginine residues (13). In *Lpn* the inner membrane-bound components of the Tat machinery are encoded in two loci, i.e., *tatAB* and *tatC* (16). In the following paragraphs, we will summarize our current knowledge of the significance of T2S and Tat for *L. pneumophila*.

DEFINING THE Lsp SECRETOME

Our mutational analysis showed that the Lsp pathway promotes the secretion of at least 11 degradative enzymes, including acid phosphatases, chitinase, zinc-metalloprotease, ribonuclease, mono-, di-, triacylglycerol lipases, phospholipase A (PLA), lysophospholipases A (LPLA) and phospholipases C (PLC) (Table 1). The major secreted protein in culture supernatants is the zinc-metalloprotease encoded by *proA/mspA* (8, 21). This protein, which exhibits hemolytic and cytotoxic activities in vitro (15), facilitates tissue damage in vivo (12). In addition, ProA promotes the activation of PlaC, an Lsp-secreted protein with LPLA and glycerophospholipid:cholesterol acyltransferase activities (4). Since ProA appears to facilitate the presence of lipase and PLA activity in culture supernatants (Table 1), a similar mechanism of enzyme activation may also occur for other secreted lipolytic enzymes. It is also conceivable that some *proA*-dependent activities are secreted independently of Lsp but activated in the supernatant, and this possibility needs further investigation. Identification and characterization of Lsp exoenzymes were carried out using several different strategies. The zinc-metalloprotease ProA

Ombeline Rossier and Nicholas P. Cianciotto Department of Microbiology and Immunology, Northwestern University Medical School, Chicago, IL 60611.

Legionella: State of the Art 30 Years after Its Recognition
Edited by Nicholas P. Cianciotto et al.
©2006 ASM Press, Washington, D.C.

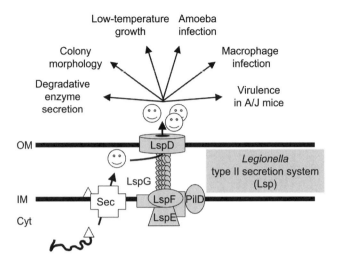

FIGURE 1 Model for T2S by the Lsp system and its role in *Legionella pneumophila*. Most type II exoproteins are synthesized with an N-terminal signal peptide that targets them to the general secretion pathway (Sec). Following translocation across the inner membrane, their signal peptide (triangle) is cleaved, and they can acquire some secondary structure in the periplasm. Proteins can then recruit the T2S, which spans the bacterial envelope, to cross the outer membrane. In *L. pneumophila*, the T2S machinery is called Lsp for *Legionella* secretion pathway. It includes the ATPase LspE, the inner membrane protein LspF, the pseudopilin LspG, the prepilin peptidase PilD, and the outer membrane secretin LspD (5). In *L. pneumophila*, our mutagenesis studies have demonstrated that Lsp facilitates the secretion of degradative enzymes. In addition to influencing the morphology of colonies on agar plates, it also potentiates growth at low temperature. Finally, the Lsp pathway promotes intracellular infection in amoebae and human macrophages as well as virulence in the A/J mouse model of Legionnaires' disease. OM, outer membrane; IM, inner membrane; Cyt, cytoplasm.

FIGURE 2 Model for Tat and its role in *L. pneumophila*. Unlike the substrates of the Sec pathway, Tat substrates are folded prior to export across the cytoplasmic membrane. They are targeted to the Tat pathway by their amino-terminal signal peptide, which harbors twin arginines. The Tat machinery includes the inner membrane components TatA, TatB, and TatC. In *L. pneumophila*, our mutagenesis studies have demonstrated that Tat facilitates the secretion of phospholipase C activity, cytochrome *c*–dependent respiration, growth under low-iron conditions, and intracellular infection. Moreover it appears to promote biofilm formation (7). OM, outer membrane; IM, inner membrane; Cyt, cytoplasm.

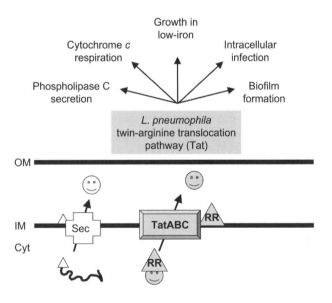

TABLE 1 Secreted enzymatic activities promoted by Lsp

Activity in culture supernatants and corresponding substrates tested	Genes or loci promoting the presence of activity in supernatants	
	Lsp secretion genes or loci tested	Other genes tested
Acid phosphatase		
Tartrate-sensitive *p*-nitrophenol (*p*-NP) phosphate hydrolase	*lspDE, lspF, lspG, pilD* (2, 17, 18)	*map* (1)
Tartrate-resistant *p*-NP phosphate hydrolase	*lspDE, lspF, lspG, pilD* (18)	
Chitinase		
p-NP triacetyl chitotriose	*lspF* (Soderberg and Cianciotto, unpublished)	
Protease		
Casein hydrolase	*lspDE, lspF, lspG, pilD* (2, 11, 17, 18)	*proA* (11, 18, 21)
Ribonuclease		
Baker's yeast type III RNA hydrolase	*lspDE, lspF, pilD* (2, 18)	
Acyltransferase		
Glycerophospholipid:cholesterol acyltransferase	*lspDE* (4)	*plaC, proA* (4)
Lipases		
p-NP caprylate hydrolase	*lspDE, lspF, lspG, pilD* (2, 17, 18)	*lipA, proA* (3, 18)
p-NP palmitate hydrolase	*lspDE, lspF, lspG, pilD* (2, 17, 18)	*lipA, proA* (3, 18)
Monoacylglycerol lipase	*lspDE, lspF, lspG, pilD* (2, 3, 17, 18)	*lipA, plaA, plaC, proA* (3, 4, 10, 18)
1,2-DG hydrolase	*lspDE, pilD* (3)	
Tricaprylin hydrolase	*lspDE, pilD* (3)	*lipA, lipB* (3)
Phospholipase A (PLA)		
Phosphatidylcholine hydrolase	*lspDE, lspG, pilD* (2, 17, 18)	*plaA, plaC, proA* (4, 10)
Lysophospholipase A (LPLA)		
Lysophosphatidylcholine hydrolase	*lspDE, lspG, lspF, pilD* (9, 17, 18)	*plaA, plaC* (4, 10)
Phospholipase C (PLC)		
p-NP phosphorylcholine hydrolase	*lspDE, lspG, lspF, pilD* (2, 3, 17, 18)	*plcA, mip* (3) (DebRoy and Cianciotto, unpublished)

and the major LPLA PlaA were purified from culture supernatants (8, 9). Screening for loss of acid phosphatase activity in *L. pneumophila* supernatants led to mutants in the major acid phosphatase (*map*) (1). Finally, examination of the genome of strain Philadelphia-1 (then unfinished) and subsequent targeted mutagenesis allowed identification of genes encoding the lipases LipA and LipB, the PLC PlcA and the LPLA/acyltransferase PlaC (3, 4). Cloning of the structural genes for zinc-metalloprotease (*proA*), LPLA (*plaA*), major acid phosphatase

(*map*) and LPLA/acyltransferase (*plaC*) was confirmed by the acquisition of the corresponding enzymatic activity by *Escherichia coli* (1, 4, 10, 15, 21).

Although all these strategies have been successful, they all have their limitation in defining the complete set of proteins secreted by the Lsp pathway. First, the genes encoding the secreted chitinase, ribonuclease, and tartrate-resistant acid phosphatase have not yet been identified. Second, genetic analysis indicates that some identified exoenzymes do not account for all the corresponding Lsp-dependent activity; e.g., supernatants of the *plcA*-negative strain retain 50% of the PLC activity of the wild type, indicating that there is more than one Lsp-secreted PLC (3). Finally, there are probably some additional Lsp-secreted proteins whose function has not been tested for. Therefore, our laboratory is currently analyzing the Lsp secretome by two-dimensional electrophoresis and mass spectrometry. Using this proteomics approach, we have confirmed that ProA, Map, and PlaA are present in wild-type culture supernatants. Moreover, we have identified two proteins with homology to chitinase and ribonuclease (DebRoy and Cianciotto, unpublished results). Finally, most of the other secreted proteins identified so far bear no homology with proteins of known function, indicating that this analysis is likely to expand the repertoire of type II exoproteins. In summary, the Lsp pathway promotes the secretion of various degradative enzymes, some of which could facilitate nutrient acquisition. In addition, our mutational analysis described below indicates that some of the secreted proteins play a role in growth at low-temperature, intracellular infection, and virulence.

ROLE OF Lsp FOR EXTRACELLULAR GROWTH BY *L. PNEUMOPHILA*

Among the phenotypes associated with loss of the Lsp pathway is a change in colony morphology on buffered yeast extract agar plates, suggesting that some Lsp-dependent secreted or outer membrane proteins play a role in modifying bacterial surface properties (17). Although exhibiting normal growth in buffered yeast extract at 37°C, *lsp* mutants are impaired in their growth at temperatures below 30°C (19), indicating that the Lsp may play a role in survival of *L. pneumophila* in the environment (see chapter 53).

Lsp PROMOTES INTRACELLULAR GROWTH OF *L . PNEUMOPHILA*

When tested in the amoebal hosts *Hartmannella vermiformis* and *Acanthamoeba castellanii*, the Lsp-negative strains exhibit a severe replication defect, i.e., 100- to 1000-fold CFU recovery defect at 48 h postinfection (hpi) compared to wild type (11, 14, 17, 18), indicating that the Lsp pathway is required for intracellular replication of *L. pneumophila* in amoebae. The contribution of Lsp secretion to human cell infection appears more modest, although it depends on the type of cells tested. Indeed, in U937 human macrophage-like cells, the *lsp*-negative strains exhibit a 5- to 10-fold reduction in CFU recovery compared to wild type at 24 hpi (17, 18). In contrast, the replication defect is more severe in peripheral blood derived–monocytes, with 100-fold less CFUs recovered from cells infected with the *lsp* mutant than for the wild-type strain (14). Therefore, the Lsp pathway promotes intracellular infection of amoebae and human cells by *L. pneumophila*. To our knowledge, the *L. pneumophila* Lsp pathway is the only T2S system to be implicated in intracellular parasitism.

The nature of the Lsp exoprotein(s) important for intracellular replication has so far remained elusive. Mutational analysis of the genes encoding all the Lsp effectors identified to date (i.e., *lipA*, *lipB*, *map*, *plaA*, *plaC*, *plcA*, and *proA*) indicates that the corresponding secreted proteins are dispensable for intracellular growth in amoebae and human cells (1, 3, 4, 10, 12, 21). On the one hand, this result could suggest that the secreted factor(s) critical for intracellular parasitism has (have) not yet been identified. On the other hand, it is possible that some of the secreted factors have redundant functions. In the latter case, only a mutational analysis inactivating several genes in the same strain might uncover the role of the exoproteins in pathogenesis. We have initiated such a study by generating a strain defective in the lipase genes *lipA* and *lipB*. This strain exhibited normal growth within host cells, indicating that the combined inactivation of two of the five known lipolytic enzymes does not affect the parasitism potential of *L. pneumophila* (3). Generation of strains carrying multiple targeted mutations is under way.

Lsp PROMOTES VIRULENCE OF *L. PNEUMOPHILA* IN THE A/J MOUSE MODEL OF LEGIONNAIRE'S DISEASE

The ability to replicate intracellularly in human cells has been linked to *L. pneumophila* virulence in vivo. Therefore, as a next step in understanding the significance of Lsp secretion for *L. pneumophila*, we used the A/J mouse model of Legionnaire's disease. In a first approach, bacterial strains were evaluated in a competition assay. Most significantly, 72 h after intratracheal infection of a mixture of equal amounts of wild-type and *lsp*-negative strains, there were approximately 15-fold more CFUs of the wild-type strain than the *lsp* mutants in the lungs, indicating that Lsp promotes multiplication of *L. pneumophila* in vivo (18). In a second approach, strains were assessed in a growth kinetics assay. Bacteria were injected in the trachea of distinct groups of A/J mice. Whereas the wild type replicated in the first 48 h following the infection, the numbers of *lsp*-negative strains declined, yielding 50-fold fewer CFUs compared to the wild type at 48 hpi (18). Thus, Lsp-dependent secretion is critical for multiplication and/or survival of *L. pneumophila* in the lungs of A/J mice following intratracheal infection. Compared to the modest intracellular replication defect observed in U937 cells in vitro, the defect seen in vivo is much more severe. This observation could reflect a role of T2S beyond macrophage intracellular infection, such as infection of alveolar epithelial cells, host tissue destruction, resistance to antimicrobial compounds (including the lysophospholipids present in the lung surfactant) or modulation of the immune system. The nature of the Lsp exoproteins important for virulence is not yet known, since strains disrupted in genes encoding the known Lsp effectors remain to be evaluated in A/J mice. However, using sera from mice infected with the wild type, we could show that some Lsp-secreted proteins are expressed in vivo (18). In summary, our results demonstrate that Lsp secretion promotes virulence of *L. pneumophila* in A/J mice.

THE Tat PATHWAY PROMOTES SECRETION OF PHOSPHOLIPASE C ACTIVITY

In our analysis of factors that influence Lsp secretion, we investigated the significance of the Tat pathway for *L. pneumophila*. In strains 130b and Lens, PlcA, a Lsp-secreted phospholipase C, carries a consensus motif for Tat substrates (3, 16). Accordingly, supernatants of a *tat* mutant show a reduction in phospholipase C activity, while maintaining normal levels of other secreted activities (16). Therefore, the Tat system promotes the secretion of phospholipase C activity by *L. pneumophila*, and further experiments will test whether PlcA is a Tat substrate in *L. pneumophila*.

Tat POTENTIATES CYTOCHROME *c*−DEPENDENT RESPIRATION

In an in silico screen for Tat substrates in the genome of *L. pneumophila*, we identified 32 additional putative substrates, based on the presence of 2 consecutive arginines in their signal peptide (16). Among them is PetA, a protein homologous to the iron-sulfur subunit of the cytochrome *c* reductase complex. In accordance with our screen's prediction, translocation of the epitope-tagged PetA to the periplasm was *tat*-dependent (16). In addition, we demonstrated that Tat promotes cytochrome *c*−dependent respiration by *L. pneumophila* (16). Hence, the Tat pathway appears to be important for the assembly of a functional respiratory complex in *L. pneumophila*.

Tat FACILITATES GROWTH UNDER LOW-IRON CONDITIONS

Under standard conditions, a *tat*-negative strain exhibits normal extracellular growth. However, addition of the ferrous iron chelator 2, 2′-dipyridyl (DIP) to the growth medium leads to slower replication of the *tat* mutant, indicating that the Tat pathway facilitates growth under low-iron conditions (16). In addition, Tat-dependent mechanisms of iron acquisition appear relevant for intracellular replication in the amoebal hosts. Indeed, although a *tat*-defective strain grows normally in the amoebal host *H. vermiformis* under standard conditions, addition of 25 μM DIP to the co-culture, a concentration that does not affect the wild type, is associated with a 20-fold decrease in CFUs of the *tat*-defective strains at 72 hpi (16). Therefore, the Tat pathway promotes both extracellular as well as intracellular growth under low-iron conditions. Since a *tat*-defective strain produces normal level of legiobactin, the *L. pneumophila* siderophore, it appears that iron acquisition processes other than legiobactin production are *tat*-dependent (16). Further work aims to identify the nature of the Tat substrate(s) promoting growth in iron-limiting conditions.

Tat PROMOTES INTRACELLULAR INFECTION INDEPENDENTLY OF TYPE II SECRETION

As described above, the Tat pathway does not appear to promote intracellular growth within *H. vermiformis* under standard conditions (16). However, intracellular replication of *tat* mutants was recently reported to be defective in the amoebal model *A. castellanii*, with a 100-fold recovery defect compared to the wild type at 72 hpi (7). These results suggest that the Tat pathway is more important for replication within *Acanthamoeba* than within *Hartmannella*. When tested for growth within the human macrophage-like U937 cells, *tat* mutants exhibit a 10-fold reduction in CFUs recovered from the monolayers 48 hpi compared to wild type (7, 16). Since the Tat pathway promotes the secretion of at least one type II–dependent enzymatic activity in *L. pneumophila*, we examined whether the intracellular growth defect of the *tatB* mutants in macrophages was due to the influence of Tat on type II secreted proteins. Strains disrupted in genes essential for Tat and Lsp secretion had a greater replication defect than strains lacking either secretion system alone; i.e., at 48 hpi there was a ca. 40-fold reduction in CFU recovery for the *tatB lspF* and *tatB pilD* mutants, whereas the reductions for the *lspF*-negative strain and the *tatB*-negative strain were only ca. 9-fold (16). These data suggest that the *L. pneumophila* pathway promotes intracellular infection of U937 cells by

processes that are independent of type II secretion. Most significantly, our in vitro infection data demonstrate, for the first time, the importance of Tat for optimal intracellular infection of human host cells.

IS THERE A CROSS-TALK BETWEEN THE DIFFERENT SECRETION SYSTEMS IN *L. PNEUMOPHILA*?

The genome of *L. pneumophila* now appears to encode several secretion systems that are important for intracellular parasitism, including T2S, Tat, and the Dot/Icm type IV secretion system (16, 18, 20). It now appears clear that the Tat pathway influences the secretion of phospholipase C, a Lsp-dependent activity, thereby linking Lsp and Tat secretion systems for secretion in vitro. However, the importance of Tat for intracellular growth of *L. pneumophila* is independent of *lsp* genes, suggesting that both pathways facilitate macrophage infection independently. A link between Lsp and Dot/Icm secretion was suggested by the isolation of a mutation in the Lsp pathway in a screen for suppressors of a *dot/icm*-dependent phenotype (6). However, Lsp-defective strains do not display phenotypes normally associated with a defective Dot/Icm pathway, such as salt sensitivity, loss of contact-mediated cell cytotoxicity, defect in induction of host cell apoptosis, and phagosome association with late endosomal marker LAMP-1 (14, 22; Rossier and Cianciotto unpublished). Therefore, there is so far no clear link between Lsp and Dot/Icm. Finally, a cross-talk between the Tat and Dot/Icm pathways has not yet been investigated.

In conclusion, the Lsp system promotes the secretion of degradative enzymes, growth at low temperature, intracellular replication within amoebae and macrophages, as well as virulence in A/J mice. Moreover, the Tat pathway facilitates secretion of phospholipase C, cytochrome *c*–dependent respiration, growth in iron-limiting conditions, as well as intracellular replication. Further identification and characterization of Lsp and Tat substrates should therefore enhance our understanding of the pathogenesis of *L. pneumophila*.

REFERENCES

1. **Aragon, V., S. Kurtz, and N. P. Cianciotto.** 2001. The *Legionella pneumophila* major acid phosphatase and its role in intracellular infection. *Infect. Immun.* **69:**177–185.

2. **Aragon, V., S. Kurtz, A. Flieger, B. Neumeister, and N. P. Cianciotto.** 2000. Secreted enzymatic activities of wild-type and *pilD*-deficient *Legionella pneumophila*. *Infect. Immun.* **68:**1855–1863.

3. **Aragon, V., O. Rossier, and N. P. Cianciotto.** 2002. *Legionella pneumophila* genes that encode lipase and phospholipase C activities. *Microbiology* **148:**2223–2231.

4. **Banerji, S., M. Bewersdorff, B. Hermes, N. P. Cianciotto, and A. Flieger.** 2005. Characterization of the major secreted zinc metalloprotease-dependent glycerophospholipid:cholesterol acyltransferase, PlaC, of *Legionella pneumophila*. *Infect. Immun.* **73:**2899–2909.

5. **Cianciotto, N. P.** 2005. Type II secretion: a protein secretion system for all seasons. *Trends Microbiol.* **13:**581–588.

6. **Conover, G. M., I. Derre, J. P. Vogel, and R. R. Isberg.** 2003. The Legionella pneumophila LidA protein: a translocated substrate of the Dot/Icm system associated with maintenance of bacterial integrity. *Mol. Microbiol.* **48:**305–321.

7. **De Buck, E., L. Maes, E. Meyen, L. Van Mellaert, N. Geukens, J. Anne, and E. Lammertyn.** 2005. *Legionella pneumophila* Philadelphia-1 *tatB* and *tatC* affect intracellular replication and biofilm formation. *Biochem. Biophys. Res. Commun.* **331:**1413–1420.

8. **Dreyfus, L. A., and B. H. Iglewski.** 1986. Purification and characterization of an extracellular protease of *Legionella pneumophila*. *Infect. Immun.* **51:**736–743.

9. **Flieger, A., S. Gong, M. Faigle, S. Stevanovic, N. P. Cianciotto, and B. Neumeister.** 2001. Novel lysophospholipase A secreted by *Legionella pneumophila*. *J. Bacteriol.* **183:**2121–2124.

10. **Flieger, A., B. Neumeister, and N. P. Cianciotto.** 2002. Characterization of the gene encoding the major secreted lysophospholipase A of *Legionella pneumophila* and its role in detoxification of lysophosphatidylcholine. *Infect. Immun.* **70:**6094–6106.

11. **Hales, L. M., and H. A. Shuman.** 1999. *Legionella pneumophila* contains a type II general secretion pathway required for growth in amoebae

as well as for secretion of the Msp protease. *Infect. Immun.* **67:**3662–3666.

12. **Moffat, J. F., P. H. Edelstein, D. P. Regula, Jr., J. D. Cirillo, and L. S. Tompkins.** 1994. Effects of an isogenic Zn-metalloprotease-deficient mutant of *Legionella pneumophila* in a guinea-pig pneumonia model. *Mol. Microbiol.* **12:**693–705.

13. **Palmer, T., and B. C. Berks.** 2003. Moving folded proteins across the bacterial cell membrane. *Microbiol.* **149:**547–556.

14. **Polesky, A. H., J. T. Ross, S. Falkow, and L. S. Tompkins.** 2001. Identification of *Legionella pneumophila* genes important for infection of amoebas by signature-tagged mutagenesis. *Infect. Immun.* **69:** 977–987.

15. **Quinn, F. D., and L. S. Tompkins.** 1989. Analysis of a cloned sequence of *Legionella pneumophila* encoding a 38 kD metalloprotease possessing haemolytic and cytotoxic activities. *Mol. Microbiol.* **3:**797–805.

16. **Rossier, O., and N. P. Cianciotto.** 2005. The *Legionella pneumophila tatB* gene facilitates secretion of phospholipase C, growth under iron-limiting conditions, and intracellular infection. *Infect. Immun.* **73:**2020–2032.

17. **Rossier, O., and N. P. Cianciotto.** 2001. Type II protein secretion is a subset of the PilD-dependent processes that facilitate intracellular infection by *Legionella pneumophila*. *Infect. Immun.* **69:**2092–2098.

18. **Rossier, O., S. R. Starkenburg, and N. P. Cianciotto.** 2004. *Legionella pneumophila* type II protein secretion promotes virulence in the A/J mouse model of Legionnaires' disease pneumonia. *Infect. Immun.* **72:**310–321.

19. **Soderberg, M. A., O. Rossier, and N. P. Cianciotto.** 2004. The type II protein secretion system of *Legionella pneumophila* promotes growth at low temperatures. *J. Bacteriol.* **186:**3712–3720.

20. **Swanson, M. S., and B. K. Hammer.** 2000. *Legionella pneumophila* pathogenesis: a fateful journey from amoebae to macrophages. *Annu. Rev. Microbiol.* **54:**567–613.

21. **Szeto, L., and H. A. Shuman.** 1990. The *Legionella pneumophila* major secretory protein, a protease, is not required for intracellular growth or cell killing. *Infect. Immun.* **58:**2585–2592.

22. **Zink, S. D., L. Pedersen, N. P. Cianciotto, and Y. Abu-Kwaik.** 2002. The Dot/Icm type IV secretion system of *Legionella pneumophila* is essential for the induction of apoptosis in human macrophages. *Infect. Immun.* **70:**1657–1663.

THE TYPE II PROTEIN SECRETION SYSTEM OF *LEGIONELLA PNEUMOPHILA* IS IMPORTANT FOR GROWTH IN IRON-RICH MEDIA AND SURVIVAL IN TAP WATER AT LOW TEMPERATURES

Maria A. Söderberg and Nicholas P. Cianciotto

53

The type II protein secretion system of *Legionella pneumophila* is important for bacterial survival in protozoa, macrophages, and the lungs of A/J mice (2). Type II secretion is one of five protein secretion systems that can mediate the export of proteins across the gram-negative outer membrane into the extracellular milieu and/or into target cells (1). Presently, *L. pneumophila* is known to express both type II and type IV secretion systems. Type II secretion (*lsp*) mutants of *L. pneumophila* grow normally in bacteriologic media at 30 to 37°C. Interestingly, however, they have severe growth defects at 12 to 25°C when cultured in buffered yeast extract (BYE) broth and on buffered charcoal yeast extract (BCYE) agar, indicating that type II secretion has an important function in low-temperature survival (3). This growth defect is reversed by the reintroduction of the intact *lsp* gene into the mutant, confirming the role of type II secretion in low-temperature growth (3). The importance of Lsp for low-temperature growth is without precedent (1) and thus has been the subject of further investigation.

The low-temperature defect of *lsp* mutants of *L. pneumophila* serogroup 1 strain 130b was partially due to the presence of 0.25 g/liter ferric pyrophosphate in the growth media, since bacterial growth on agar plates at 25°C improved significantly when that supplement was omitted (Fig. 1). Indeed, the growth of an *lspDE* mutant (2) improved at least 100-fold, going from an efficiency of plating of 0.05% for plates with ferric pyrophosphate to an efficiency of plating of 5% on plates without the compound. The inhibitory effect was due to the supplemental iron and not the pyrophosphate, since the *lsp* mutant was as defective for low-temperature growth on BCYE containing other iron sources as it was on BCYE containing ferric pyrophosphate. The inhibitory effect of iron was specific to the *lsp* mutant and to low-temperature growth, since the growth of wild type at 25°C did not change with variations in the amount of supplemental iron, and the growth of the *lsp* mutant was not negatively influenced by iron supplementation when incubated at 37°C. How can iron impact the growth of an *lsp* mutant? Since iron reacts with hydrogen peroxide in media through the Fenton reaction to create damaging free radicals (4), one possible scenario is that a factor needed for oxidative defense is secreted through the type II secretion system. Alternatively, the Lsp system secretes a protein that binds iron, sequestering it and thereby minimizing toxic

Maria A. Söderberg and Nicholas P. Cianciotto   Department of Microbiology and Immunology, Northwestern University, Chicago, IL 60611.

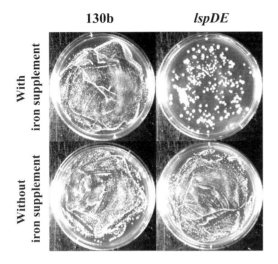

130b   *lspDE*

With iron supplement

Without iron supplement

FIGURE 1 The effect of iron on the low-temperature growth of *L. pneumophila* wild type and *lsp* mutants. Wild-type 130b and an *lspDE* mutant were grown on standard BCYE agar for 3 days at 37°C, and then the bacteria were resuspended to the same optical density, serially diluted, and plated onto BCYE agar containing or not containing the traditional 0.25 g/liter ferric pyrophosphate supplement. The inoculated plates were then incubated at 37 or 25°C, and growth was observed over the next 12 days. The presented image is the bacterial growth that resulted from plating the 10^{-3} dilution on agar incubated at 25°C. These results are representative of two independent experiments.

A.

37°C

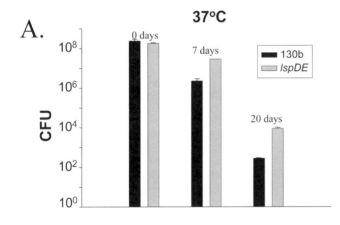

B.

17°C

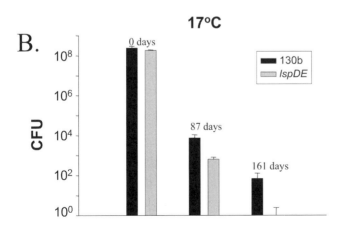

FIGURE 2 The survival of wild type and *lsp* mutant *L. pneumophila* in tap water at different temperatures. Wild-type 130b and *lspDE* mutant bacteria were grown in BYE broth to log phase at 37°C and then centrifuged. The resultant bacterial pellets were resuspended in filter-sterilized tap water, recentrifuged, and again resuspended in tap water in order to remove traces of BYE broth. Then, equal numbers of wild type and *lspDE* mutant bacteria were inoculated into flasks containing 50 ml of filtered sterilized tap water, and the cultures were incubated at 37°C (A) or 17°C (B) with shaking. Aliquots were removed at various times, and CFU determinations were made by plating serial dilutions on BCYE agar. The results presented are the means and standard deviations from duplicate samples and are representative of two independent experiments.

radical formation. That this *lsp* mutant defect is only seen at lower temperatures could be due to the increase in oxygen solubility that occurs with drops in temperature and/or temperature regulation of the type II effectors.

In the environment, *L. pneumophila* is found in freshwater and man-made water systems. To begin to better determine if type II secretion is important for survival at lower temperatures in the environment, wild-type 130b and an *lspDE* mutant were compared for their ability to persist in filtered sterilized tap water incubated at 37 and 17°C (Fig. 2). As in BYE or BCYE, the *lsp* mutant was not defective in water at 37°C (Fig. 2A). In fact, the mutant persisted better than wild type over the 20-day course of the experiment. In contrast, when in 17°C tap water, the *lspDE* mutant exhibited a marked defect in survival (Fig. 2B). For example, after 87 days, it displayed a 12-fold reduced recoverability. Furthermore, whereas the wild type was still in evidence after 161 days, the *lsp* mutant was not, implying at least a 100-fold reduction in recoverability. These data suggest that *L. pneumophila* type II protein secretion is important for low-temperature survival not only in rich media but also in water samples. Thus, the *Legionella* type II secretion system would appear to have a very important role in the ability of *L. pneumophila* to persist for long periods of time in those water sources, including those that are the sources of human infection.

Since nearby wild-type colonies or supernatants can promote the growth of an *lsp* mutant on BCYE agar at room temperature (3, unpublished results), we suspect that a secreted protein is required for the low-temperature survival of *L. pneumophila*. Since the growth of an *lsp* mutant on BCYE agar lacking the iron supplement improved significantly but not to wild-type levels, more than one protein is likely required for low-temperature adaptation. The identity of type II secreted proteins and those that are expressed and perhaps hyperexpressed at low temperatures is being determined by examining proteins contained in different culture supernatants using two-dimensional polycrylamide gel electrophoresis analysis. In sum, the results presented indicate that the *L. pneumophila* type II secretion system is not only important for growth at 35 to 37°C in host cells but is also critical for extracellular growth and survival at lower temperatures, including iron-rich bacteriological media and environmental water samples. Thus, type II secretion appears to play a central and multifaceted role in the natural history of Legionnaires' disease.

REFERENCES

1. **Cianciotto, N. P.** 2005. Type II secretion: a protein secretion system for all seasons. *Trends Microbiol.* **13**:581–588.
2. **Rossier, O., S. R. Starkenburg, and N. P. Cianciotto.** 2004. *Legionella pneumophila* type II protein secretion promotes virulence in the A/J mouse model of Legionnaires' disease pneumonia. *Infect. Immun.* **72**:310–321.
3. **Söderberg, M. A., O. Rossier, and N. P. Cianciotto.** 2004. The type II protein secretion system of *Legionella pneumophila* promotes growth at low temperatures. *J. Bacteriol.* **186**:3712–3720.
4. **Winterbourn, C. C.** 1995. Toxicity of iron and hydrogen peroxide: the Fenton reaction. *Toxicol. Lett.* **82/83**:969–974.

IDENTIFICATION OF PUTATIVE SUBSTRATES OF THE *LEGIONELLA PNEUMOPHILA* Tat SECRETION PATHWAY VIA TWO-DIMENSIONAL PROTEIN GEL ELECTROPHORESIS

*E. De Buck, L. Maes, J. Robben, J.-P. Noben,
J. Anné, and E. Lammertyn*

54

Different secretion pathways have been shown to play a role in the virulence of *Legionella pneumophila* (4). Recently we showed the presence of the twin-arginine translocation (Tat) pathway in *L. pneumophila* Philadelphia-1 (2) and its importance in intracellular replication and biofilm formation (3). The Tat pathway translocates folded proteins across the cytoplasmic membrane. Proteins transported through this secretion route typically carry two arginine residues or a lysine-arginine pair in their signal peptide. It is known that this secretion pathway plays a role in the virulence of different human and plant pathogens (1). In order to study the importance of the Tat pathway in the virulence of *L. pneumophila*, we initiated the identification of Tat substrates and their possible involvement in virulence. Since some Tat substrates might be transported across the outer membrane following Tat-dependent transport across the cytoplasmic membrane, we looked for differential spots in culture media of the wild-type strain and two Tat secretion mutants (*tatB* and *tatC* mutant) by two-dimensional protein gel electrophoresis analysis. Therefore, 150 μg of protein was separated in the first dimension using IPG strips with pH 4 to 7, followed by separation in the second dimension on 12.5% acrylamide gels. Protein patterns, visualized by silver staining, were compared (Fig. 1) and differential proteins were identified by tandem mass spectrometry (nanoLC-ESI-MS/MS) of the tryptic peptides.

Three proteins were found to be absent from the culture medium of the Tat secretion mutants: LvrE, Lpg1962 (a peptidyl-prolyl cis-trans isomerase), and Lpg2320 (a hypothetical protein). They all have a predicted signal peptide with a motif aberrant from the expected twin-arginine motif which is one arginine for LvrE, a double lysine motif for Lpg1962, and a lysine-arginine motif for Lpg2320. Another protein, IcmX, was present in a higher amount in the culture media of the Tat secretion mutants. IcmX is a protein required for biogenesis of the *L. pneumophila* replicative organelle (5) and may be present in a higher amount in the culture media of the *tatB* mutant and *tatC* mutant due to the absence of another phagosome biogenesis factor that is Tat dependent.

Genes encoding LvrE, Lpg1962, and Lpg2320 were cloned and overexpressed in *L. pneumophila* wild type and *tat* mutants. Tat dependence could be confirmed for LvrE. When

E. De Buck, L. Maes, J. Anné and E. Lammertyn Laboratory of Bacteriology, Rega Institute for Medical Research, Katholieke Universiteit Leuven, Minderbroedersstraat 10, B-3000 Leuven, Belgium. *J. Robben and J.-P. Noben* Biomedical Research Institute, Hasselt University, Agoralaan, Building A, B-3590 Diepenbeek, Belgium.

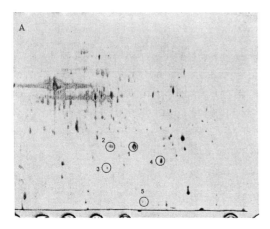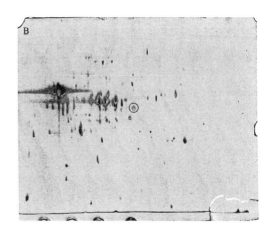

FIGURE 1 Two-dimensional gel electrophoresis (12.5% acrylamide; pI 4 to 7) on the supernatant of *L. pneumophila* Philadelphia-1 wild type (A) and *tatC* mutant (B). 1, LvrE; 2, 3, LvrE fragments; 4, Lpg1962; 5, Lpg2320; 6, IcmX.

LvrE-specific antibodies were used, no protein was observed in the supernatant of the *tatB* mutant and the *tatC* mutant, while a clear protein band was visible for the wild-type sample. On the other hand, no difference in secretion pattern was found between wild type and *tat* mutants for Lpg1962 when the FLAG tag and anti-FLAG antibodies were used. For Lpg2320, experiments are still in progress.

The LvrE protein was further studied in detail. This protein, with a predicted molecular weight of 28.8 kDa and a pI of 6.16, shows no sequence similarity to any known sequence found in Genbank. The *lvrE* gene is situated in between the *lvh* (*Legionella vir* homologues) genes on the *L. pneumophila* Philadelphia-1 genome that encode the Lvh type IV secretion system. This region is a DNA island with a higher GC content compared to the remainder of the chromosome. In addition to the 11 *lvh* genes, 5 other open reading frames were found and were named *lvrA-E* (*Legionella vir* region) but were shown not to play a role in type IV secretion (6).

On investigating *lvrE* transcription, reverse transcriptase PCR analysis using *lvrE* specific primers showed that transcription also takes place in the *tat* mutants, indicating that loss of LvrE in the supernatant of the mutants is not due to an effect at the transcription level. To get an idea about the possible function of LvrE, the *lvrE* gene was replaced by a kanamycine resistance gene using double homologous recombination. To this order, the *lvrE* upstream flanking sequence was cloned as a *XhoI/HindIII* fragment at one site of the kanamycine resistance gene in the plasmid pBSKan (3). In a next step the *lvrE* downstream flanking sequence was cloned as a *PstI/BamHI* fragment at the other site of the resistance gene. The entire resistance cassette was then introduced in the wild-type strain by natural competence (7). The resulting *lvrE* mutation was confirmed: at mRNA level with reverse transcriptase-PCR (with *lvrE* specific primers), at DNA level with Southern blot hybridization (with a *lvrE* specific DNA probe) and at protein level using LvrE specific antibodies. A complemented mutant was constructed by introducing a plasmid containing the *lvrE* gene downstream of a *tac* promoter in the plasmid pMMBN. The latter plasmid is a derivative of pMMB207, with an *NdeI* restriction site introduced at the ATG codon immediately downstream of the *tac* promoter and part of the *mobA* gene deleted (*AgeI* digestion).

Next, intracellular replication of the *lvrE* mutant was studied. Infection assays were performed with *Acanthamoeba castellanii* and in the monocytic U937 cell line, differentiated into

A

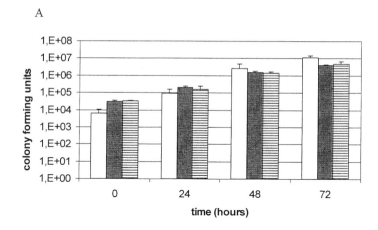

B

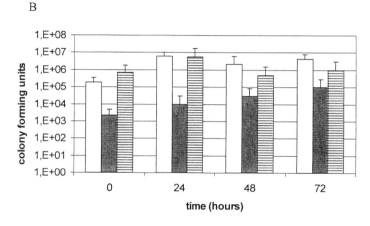

FIGURE 2 Replication of wild type (white), *lvrE* mutant (grey) and complemented mutant (striped) in *A. castellanii* (A) and differentiated U937 cells (B).

macrophage-like cells by treatment with phorbol 12-myristate 13-acetate 72 h prior to infection. Infection assays were performed during 72 h with a multiplicity of infection (MOI) of 2. In amoebae the *lvrE* mutant showed a small decrease in replication after 48 h, but this could not be complemented. In U937 cells the mutant was clearly impaired in infection and replication, while the complemented strain almost reached wild-type values (Fig. 2).

Based on two-dimensional analysis, three proteins were found to be absent in the culture media of the *L. pneumophila tatB* and *tatC* mutant. These proteins all contain a signal peptide without two arginine residues but with a lysine-arginine motif or even a more aberrant motif. For one of these proteins, LvrE, Tat dependence was confirmed using specific antibodies. It was

shown that an *L. pneumophila lvrE* mutant is clearly impaired for replication in differentiated U937 cells so this protein could be responsible for the decrease in intracellular replication of the Tat secretion mutants. Studies on the possible function of LvrE are ongoing.

REFERENCES
1. **Berks, B. C., T. Palmer, and F. Sargent.** 2003. The Tat protein translocation pathway and its role in microbial physiology. *Adv. Microb. Physiol.* **47:** 187–254.
2. **De Buck, E., I. Lebeau, L. Maes, N. Geukens, E. Meyen, L. Van Mellaert, J. Anné, and E. Lammertyn.** 2004. A putative twin-arginine translocation pathway in *Legionella pneumophila. Biochem. Biophys. Res. Commun.* **317:**654–661.
3. **De Buck, E., L. Maes, E. Meyen, L. Van Mellaert, N. Geukens, J. Anné, and E. Lammertyn.** 2005. *Legionella pneumophila* Philadelphia-1

tatB and *tatC* affect intracellular replication and biofilm formation. *Biochem. Biophys. Res. Commun.* **331:**1413–1420.

4. **Lammertyn, E., and J. Anné.** 2004. Protein secretion in *Legionella pneumophila* and its relation to virulence. *FEMS Microbiol. Lett.* **238:**273–279.

5. **Matthews, M., and C. R. Roy.** 2000. Identification and subcellular localization of the *Legionella pneumophila* IcmX protein: a factor essential for establishment of a replicative organelle in eukaryotic host cells. *Infect. Immun.* **68:**3971–3982.

6. **Segal, G., J. J. Russo, and H. A. Shuman.** 1999. Relationships between a new type IV secretion system and the *icm/dot* virulence system of *Legionella pneumophila. Mol. Microbiol.* **34:**799–809.

7. **Sexton, J. A., and J. P. Vogel.** 2004. Regulation of hypercompetence in *Legionella pneumophila. J. Bacteriol.* **186:**3814–3825.

IDENTIFICATION OF TARGET PROTEINS OF THE Lss SECRETION SYSTEM OF *LEGIONELLA PNEUMOPHILA* CORBY

Christiane Albert, Sebastian Jacobi, Emmy De Buck, Elke Lammertyn, and Klaus Heuner

55

Protein secretion plays an important role for the virulence of pathogenic bacteria. *Legionella pneumophila* exhibits a type II secretion system, encoded by the *lsp* genes. This type II secretion system is PilD-dependent and seems to be required for full virulence of *L. pneumophila*. In addition, two type IV secretion systems have been described: the *dot/icm* encoded and the *lvh* encoded type IV secretion systems. The type IV secretion system encoded by the *dot/icm* loci has been shown to be absolutely required for the infection process of *L. pneumophila*. The Lvh system was found to play a role in host cell infection by *L. pneumophila* at 30°C. A short description of protein secretion systems described for *L. pneumophila* was recently published (7).

We have described a first putative type I secretion system (Lss) of *L. pneumophila* which is encoded by the *lssXYZABD* locus (Fig. 1) (5). Comparison of the *lssXYZABD* genes of the virulent strains *L. pneumophila* Paris, *L. pneumophila* Corby, and *L. pneumophila* Philadelphia I results in high identities of 94 to 99%. In order to identify substrates of this Lss secretion system,

we performed comparative two-dimensional polyacrylamide gel electrophoresis (2D-PAGE) analysis of extracellular proteins of *L. pneumophila* Corby wild type, Corby *lssB* mutant, and complemented *lssB* mutant strain. Proteins of interest have been analyzed by mass spectrometry and N-terminal sequencing.

Protein p13 was not found in the supernatant of the *lssB* mutant, and secretion of protein p14 was highly reduced (Fig. 2A). All differences observed in the mutant strain could be complemented by an intact *lssB* gene in trans, showing that the obtained differneces are a result of the inactivation of *lssB* (data not shown). Our results indicate that the proteins p13 and p14 are secreted in *L. pneumophila* Corby in an *lssB*-dependent manner. Both proteins exhibit a functional signal sequence (Fig. 2B), and 2D-PAGE analysis of *L. pneumophila* 130b wild type and 130b *lspG* mutant strain elucidated that secretion of protein p13 is *lspG*-dependent in strain 130b (data not shown). Both proteins were further analyzed by MS/MS. The obtained peptide sequences of protein p13 revealed that the protein is a member of the VirK family (4). The gene *lpg1832*, which encodes a VirK homolog in strain Philadelphia, was recently reported as a putative candidate virulence gene of *Legionella* (1). The *virK* gene of *L. pneumophila* encodes a mature protein with

Christiane Albert, Sebastian Jacobi, and Klaus Heuner Institut für Molekulare Infektionsbiologie, Julius-Maximilians Universität Würzburg, Röntgenring 11, D-97070 Würzburg, Germany. *Emmy De Buck and Elke Lammertyn* Rega Institute for Medical Research, Katholieke Universiteit Leuven, Minderbroedersstraat 10, B-3000 Louvain, Belgium.

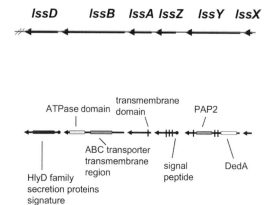

FIGURE 1 Genomic organization and putative protein function of the *lss* locus of *L. pneumophila*. The first line of arrows indicates genes present in the *lss* locus. Putative proteins and protein motifs are shown below. DedA, DedA–related protein family; PAP2, type 2 phosphatidic acid phosphatase superfamily.

a theoretical mass of 13.1 kDa and a pI of 5.8, which is in good agreement with the data retrieved by two–dimensional gel analysis (data not shown). Protein p14 was found to be a homolog of a protein encoded by the gene *lpl1897* in *L. pneumophila* Lens. The function

of this protein is not known. Since LssB and LssZ were identified in silico as putative Tat secretion system substrates (2, 3), we examined a *tatC* mutant of *L. pneumophila* Corby for VirK secretion by 2D-PAGE analysis. However, the experiments revealed that secretion of proteins p13 and p14 is not *tatC*-dependent (data not shown).

CONCLUSIONS

In this study we showed that the identified VirK protein of *L. pneumophila* is secreted in an LssB- but not a TatC-dependent manner. Referring to our results, we imply that the Lss secretion system is also linked to the Lsp type II secretion system in *L. pneumophila*. We speculate that VirK and p14 are secreted via a concerted action of the Lss and the Lsp system which includes cleavage of the signal sequences during secretion. A similar interaction between different protein secretion systems was recently described by Nishi et al. (8). In *Agrobacterium* and *Rhizobium* VirK is associated with the *vir*-regulon, and it was postulated that VirK should have a function in the interaction of the bacteria with the host

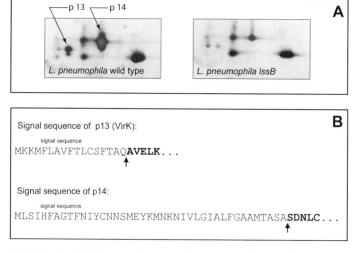

FIGURE 2 (A) Cutout of comparative analysis of extracellular proteins of *L. pneumophila* Corby wild type and Corby *lssB* mutant. Proteins of interest were analyzed by mass spectrometry and N-terminal sequencing. (B) Signal sequences of p13 and p14. Arrows mark the putative sites of processing. Amino acids identified by N-terminal sequencing are in bold.

cells (6). Therefore, we started to generate a *virK* mutant strain of *L. pneumophila* Corby and an antibody against VirK to further characterize the expression, secretion, and function of VirK of *L. pneumophila*.

REFERENCES

1. **Chien, M., I. Morozova, S. Shi, H. Sheng, J. Chen, S. M. Gomez, G. Asamani, K. Hill, J. Nuara, M. Feder, J. Rineer, J. J. Greenberg, V. Steshenko, S. H. Park, B. Zhao, E. Teplitskaya, J. R. Edwards, S. Pampou, A. Georghiou, I. C. Chou, W. Iannuccilli, M. E. Ulz, D. H. Kim, A. Geringer-Sameth, C. Goldsberry, P. Morozov, S. G. Fischer, G. Segal, X. Qu, A. Rzhetsky, P. Zhang, E. Cayanis, P. J. De Jong, J. Ju, S. Kalachikov, H. A. Shuman, and J. J. Russo.** 2004. The genomic sequence of the accidental pathogen *Legionella pneumophila*. *Science* **305:**1966–1968.
2. **Cianciotto, N. P., and O. Rossier.** 2005. The *Legionella pneumophila tatB* gene facilitates secretion of phospholipase, growth under iron-limiting conditions, and intracellular infection. *Infect. Immun.* **73:**2020–2032.
3. **De Buck, E., I. Lebeau, L. Maes, N. Geukens, E. Meyen, L. Van Mellaert, J. Anné, and E. Lammertyn.** 2004. A putative twin-arginine translocation pathway in *Legionella pneumophila*. *Biochem. Biophys. Res. Commun.* **317:**654–661.
4. **Hattori, Y., K. Iwata, K. Suraji, N. Ohta, A. Katoh, and K. Yoshida.** 2001. Sequence characterization of the vir region of a nopaline type Ti plasmid, pTi-SAKURA. *Genes Genet. Syst.* **76:** 121–130.
5. **Jacobi, S., and K. Heuner.** 2003. Description of a putative type I secretion system in *Legionella pneumophila*. *Int. J. Med. Microbiol.* **293:**349–358.
6. **Kalogeraki, V. S., and S. C. Winans.** 1998. Wound-released chemical signals may elicit multiple responses from an *Agrobacterium tumefaciens* strain containing an octopine-type Ti plasmid. *J. Bacteriol.* **180:**5660–5667.
7. **Lammertyn, E., and J. Anné.** 2004. Protein secretion in *Legionella pneumophila* and its relation to virulence. *FEMS Microbiol. Lett.* **238:**273–279.
8. **Nishi, J., J. Sheikh, K. Mizuguchi, B. Luisi, V. Burland, A. Boutin, D. J. Rose, F. R. Blattner, and J. P. Nataro.** 2003. The export of coat protein from enteroaggregative *Escherichia coli* by a specific ATP-binding cassette transporter system. *J. Biol. Chem.* **278:**45680–45689.

LEGIONELLA PNEUMOPHILA Mip: NEW FUNCTION FOR AN OLD PROTEIN?

Sruti DebRoy and Nicholas P. Cianciotto

56

The macrophage infectivity potentiator (Mip) protein of *Legionella pneumophila* is one of the most studied *Legionella* proteins and has long been known to promote virulence (3). The reduced infectivity of *mip* mutants has demonstrated that Mip is required for the optimal establishment and replication of *L. pneumophila* in macrophage cell lines, alveolar macrophages, blood monocytes, lung epithelial cells, amoebae, ciliates, and guinea pigs (3, 4, 12). In support of this, other studies have shown that Mip is expressed within infected macrophages and amoebae (7), and antibodies directed against Mip have been observed within sera from patients with Legionnaires' disease. The *mip* gene is present and expressed in other species of *Legionella* (9), and *mip* mutants of *L. micdadei* and *L. longbeachae* are also defective for intracellular infection (5).

Mip is a positively charged, 24-kDa protein that is expressed on the bacterial surface (7). The protein has been purified and shown to possess peptidyl-prolyl *cis/trans* isomerase (PPIase) activity (6). PPIases catalyze the slow *cis/trans* isomerization of a peptidyl-prolyl bond and have been implicated in protein folding and maturation. Mip belongs to the family of

FK506-binding proteins, and like its eukaryotic counterparts, has a PPIase activity that is inhibited by the immunosuppressant drug FK506 (6). Comparison of primary structures and crystallographic studies shows that Mip proteins have an N-terminal and a C-terminal domain. The N-terminal domain is responsible for dimerization, and cross-linking studies have shown that Mip forms homodimers in solution as well as on the surface of the bacteria (11). The C-terminal domain binds FK506 and has the PPIase active site. The crystal structure also reveals that the N-terminal and C-terminal domains of Mip are connected by a long α-helix, the longest freestanding helix seen so far in any protein structure (10). Site-directed mutagenesis studies have identified residues in the PPIase active fold that, when altered, drastically reduce the PPIase activity of Mip (12). Whereas the N-terminal domain of Mip and the dimerization of Mip are required for both intracellular infectivity and virulence, the PPIase activity of Mip is needed for full virulence in guinea pigs but not for infection of amoebae and human phagocytes in vitro (8, 12). Despite these many studies, the substrate(s) and molecular target of Mip have remained unknown.

In the course of studying type II protein secretion, we have uncovered a possible target

Sruti DebRoy and Nicholas P. Cianciotto   Department of Microbiology Immunology, Northwestern University Medical School, Chicago, IL 60611.

Legionella: State of the Art 30 Years after Its Recognition
Edited by Nicholas P. Cianciotto et al.
©2006 ASM Press, Washington, D.C.

for Mip. Type II secretion is one of five protein secretion systems that can mediate the export of proteins across the gram-negative outer membrane into the extracellular milieu and/or into target cells (2), and we have shown that the type II protein secretion system of *L. pneumophila* is required for optimal replication in macrophages, amoebae, and mice. Using the type II pathway, *L. pneumophila* secretes numerous enzymes whose activities can be detected in culture supernatants. Among these activities is a *p*-nitrophenyl phosphorylcholine (*p*-NPPC) hydrolase activity that is often associated with phospholipase C enzymes (1). A screen designed to isolate transposon mutants that were defective for secretion of lipolytic enzymes yielded several mutants that displayed reduced *p*-NPPC hydrolase activity in their culture supernatants (1). Interestingly, one of these mutants (previously designated NU247) proved to contain a single transposon insertion in the *mip* gene, suggesting, for the first time, that Mip might influence protein secretion. The strain lacked Mip as observed on immunoblots of cell lysates probed with a polyclonal Mip antibody. NU247 reproducibly showed a 40 to 70% reduction in *p*-NPPC hydrolase activity in its culture supernatants in comparison to the wild-type strain 130b (Table 1). In preparation for complementation studies, a comparable reduction in secreted enzyme activity was observed for NU247 containing the vector pMMB2002 (Table 1). Introduction of p*mip*, a plasmid representing (only) full-length *mip* cloned into pMMB2002, into NU247 restored the secreted *p*-NPPC hydrolase activity of the mutant to wild-type levels (Table 1), confirming that Mip is needed for optimal secreted activity in culture supernatants. Since NU247 is not defective for secreted protease, esterase, and acid phosphatase activities (1), Mip does not appear to influence the secretion of all type II–dependent exoenzymes.

This is the first evidence of a potential target for Mip action and also the first instance of a PPIase being implicated in secretion of proteins across the outer membrane or in their activity. It is tempting to speculate that Mip might also have a chaperone-like function similar to that shown for other members of the FK506-binding protein family. Given its location on the bacterial surface, two simple hypotheses can be proposed to account for our observations (Fig. 1). On the one hand, Mip might be involved in the extracellular release of an active enzyme that has *p*-NPPC hydrolase activity (Fig. 1A). On the other hand, Mip might interact with a newly secreted protein and, by virtue of its PPIase and/or possible chaperone activities, cause structural changes that render the protein enzymatically active (Fig. 1B). In summary, these data indicate that Mip promotes the presence of enzyme activity in the culture supernatants of *L. pneumophila*. Further work is needed to determine how Mip mediates this event and if this function explains the

TABLE 1 Levels of *p*-NPPC hydrolase activity in culture supernatants of wild type 130b and the *mip* mutant NU247, carrying the vector pMMB2002 or full-length *mip* cloned into pMMB2002 (p*mip*)[a].

Strain	*p*-NPPC hydrolysis units		
	Expt. 1	Expt. 2	Expt. 3
130b	0.224 ± 0.015	ND	ND
NU247	0.117 ± 0.023	ND	ND
130b (pMMB2002)	ND	0.232 ± 0.054	0.149 ± 0.038
NU247 (pMMB2002)	ND	0.1115 ± 0.019	0.087 ± 0.008
NU247 (p*mip*)	ND	0.1935 ± 0.026	0.16 ± 0.024

[a]Each data point represents the mean and standard deviation of triplicate cultures. The differences between 130b and NU247, with or without the vector, are significant ($P < 0.05$; Student's *t* test)

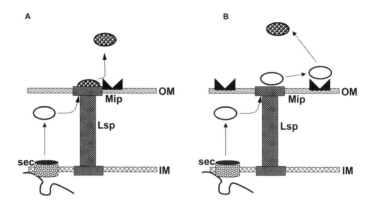

FIGURE 1 Models to explain the impact of Mip on a secreted *p*-NPPC hydrolase activity. Initially, like most type II-secreted proteins, the unfolded *p*-NPPC hydrolase interacts with components of the Sec system (2). After the signal peptide is cleaved, the protein is translocated across the inner membrane into the periplasm, where it is folded, possibly with the help of periplasmic chaperones. The protein then interacts with the type II secretion apparatus (Lsp), in order to transit across the outer membrane. At this point, surface-localized Mip may be involved in the final release of active enzyme into the extracellular milieu (*A*). Alternately, Mip might act on newly released protein and, through additional folding (e.g., PPIase) reactions, render it an active enzyme (*B*). Although not depicted here, it is possible that the *p*-NPPC hydrolase substrate is translocated across the inner membrane via the twin-arginine translocation (Tat) system (2).

requirement for Mip in intracellular infection and virulence. Since Mip is highly conserved in the *Legionella* genus, and surface and secreted Mip-like proteins are present in other pathogenic microorganisms (8), Mip and Mip-like proteins might promote the secretion of other important effectors.

REFERENCES

1. **Aragon, V., S. Kurtz, A. Flieger, B. Neumeister, and N. P. Cianciotto.** 2000. Secreted enzymatic activities of wild-type and *pilD*-deficient *Legionella pneumophila. Infect. Immun.* **68:**1855–1863.
2. **Cianciotto, N. P.** 2005. Type II secretion: a protein secretion system for all seasons. *Trends Microbiol.* **13:**581–588.
3. **Cianciotto, N. P., B. I. Eisenstein, C. H. Mody, G. B. Toews, and N. C. Engleberg.** 1989. A *Legionella pneumophila* gene encoding a species-specific surface protein potentiates initiation of intracellular infection. *Infect. Immun.* **57:**1255–1262.
4. **Cianciotto, N. P., J. Kim Stamos, and D. W. Kamp.** 1995. Infectivity of *Legionella pneumophila mip* mutant for alveolar epithelial cells. *Curr. Microbiol.* **30:**247–250.
5. **Doyle, R. M., T. W. Steele, A. M. McLennan, I. H. Parkinson, P. A. Manning, and M. W. Heuzenroeder.** 1998. Sequence analysis of the *mip* gene of the soilborne pathogen *Legionella longbeachae. Infect. Immun.* **66:**1492–1499.
6. **Fischer, G., H. Bang, B. Ludwig, K. Mann, and J. Hacker.** 1992. Mip protein of *Legionella pneumophila* exhibits peptidyl-prolyl-cis/trans isomerase (PPIase) activity. *Mol. Microbiol.* **6:**1375–1383.
7. **Helbig, J. H., P. C. Luck, M. Steinert, E. Jacobs, and M. Witt.** 2001. Immunolocalization of the Mip protein of intracellularly and extracellularly grown *Legionella pneumophila. Lett. Appl. Microbiol.* **32:**83–88.
8. **Kohler, R., J. Fanghanel, B. Konig, E. Luneberg, M. Frosch, J. U. Rahfeld, R. Hilgenfeld, G. Fischer, J. Hacker, and M. Steinert.** 2003. Biochemical and functional analyses of the Mip protein: influence of the N-terminal half and of peptidylprolyl isomerase activity on the virulence of Legionella pneumophila. *Infect. Immun.* **71:**4389–4397.
9. **Ratcliff, R. M., S. C. Donnellan, J. A. Lanser, P. A. Manning, and M. W. Heuzenroeder.** 1997. Interspecies sequence differences in the Mip protein from the genus *Legionella*: implications for

function and evolutionary relatedness. *Mol. Microbiol.* **25**:1149–1158.

10. **Riboldi-Tunnicliffe, A., B. Konig, S. Jessen, M. S. Weiss, J. Rahfeld, J. Hacker, G. Fischer, and R. Hilgenfeld.** 2001. Crystal structure of Mip, a prolylisomerase from *Legionella pneumophila. Nat. Struct. Biol.* **8**:779–783.

11. **Schmidt, B., S. Konig, D. Svergun, V. Volkov, G. Fischer, and M. H. Koch.** 1995. Small-angle X-ray solution scattering study on the dimerization of the FKBP25mem from *Legionella pneumophila. FEBS Lett.* **372**:169–172.

12. **Wintermeyer, E., B. Ludwig, M. Steinert, B. Schmidt, G. Fischer, and J. Hacker.** 1995. Influence of site specifically altered Mip proteins on intracellular survival of *Legionella pneumophila* in eukaryotic cells. *Infect. Immun.* **63**:4576–4583.

PHOSPHOLIPASES A OF *LEGIONELLA PNEUMOPHILA:* VIRULENCE FACTORS BY DIVERSITY?

Antje Flieger

57

Legionella pneumophila possesses several virulence-mediating factors; for example, it has two protein export systems: the type II (Lsp) and the type IVB (Dot/Icm) secretion systems. Both systems transport effector molecules, including lipolytic proteins such as phospholipases. Bacterial phospholipases in particular have been shown to contribute to pathogenesis in many ways, e.g., (i) by hydrolyzing cell membrane phospholipids (e.g., exit of the bacterium from the phagosome or host cell and cytotoxicity), (ii) by generating second messengers, such as 1,2-diacylglycerol or lysophosphatidylcholine (e.g., influence on host cell signalling), (iii) generation of lysophospholipids (e.g., eliciting apoptosis or acting as a chemo attractant, causing inflammation), and (iv) destructing lung surfactant phospholipids (e.g., impairment of lung function). Those enzymes may therefore be important tools of the bacterium with respect to intracellular survival, bacterial spread, host cell modification or damage, and development of the disease.

L. *pneumophila* possesses two major phospholipase activities, phospholipase A (PLA) and lysophospholipase A (LPLA), both capable

Antje Flieger Research Group NG 5—Pathogenesis of *Legionella* Infection, Robert Koch-Institut, 13353 Berlin, Germany.

of hydrolyzing phospholipids and generating the reaction products: free fatty acids (PLA and LPLA), cytotoxic lysophospholipids (via the action of PLA), and water-soluble phosphodiesters, like glycerophosphorylcholine (via the subsequent action of LPLA on lysophospholipids) (3, 4). LPLAs comprise a subclass of PLAs; in contrast to "true" PLAs, they preferentially cleave the single remaining fatty acid from a lysophospholipid molecule. A PLA, however, releases one of the two fatty acids present in diacylphospholipids and thereby generates the substrate lysophospholipid of an LPLA.

GDSL-HYDROLASE FAMILY AND PlaB

So far, four exported proteins, PlaA, PlaB, PlaC, and PlaD, have been found to contribute to the L. *pneumophila* PLA/LPLA activities (Fig. 1, Table 1). PlaA is the major secreted LPLA and is important for the detoxification of highly cytotoxic lysophospholipids (5). PlaB, on the other hand, remains associated with the outer shell of the bacterial cell, possesses both PLA and LPLA activities, and lyses red blood cells (6). PlaC, another secreted activity, comprises acyltransferase activity in addition to its PLA and LPLA activities. PlaC is therefore characterized by its capacity not only

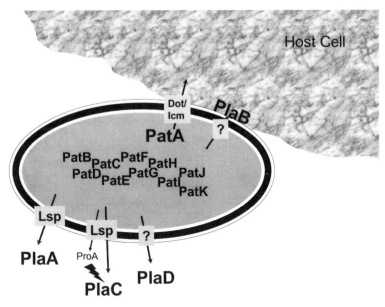

FIGURE 1 Overview of the secreted and cell-associated PLA, LPLA, and acyl-transferase activities of *L. pneumophila*. The protein with only LPLA activity is PlaA. PlaB, PlaD, and PatA/VipD possess PLA and LPLA activities. In addition to PLA and LPLA activity, PlaC confers an acyltransferase activity and is directly or indirectly activated by the zinc-metalloprotease ProA. PatB to PatK represent so far uncharacterized putative lipolytic proteins. Furthermore, the dependency on specific secretion systems (Lsp = type II secretion, Dot/Icm = type IVB secretion) is shown or unknown (designated by ?). Proteins whose activity or expression depends on host cell contact are shown in proximity to the host cell.

to cleave fatty acids from lipids but to transfer them to cholesterol, which is a major component of eukaryotic but not prokaryotic membranes (1). PlaD shares protein homology with PlaA and PlaC and has been recently found to contribute to *L. pneumophila* secreted PLA as well as LPLA activity (see chapter 59). All three proteins belong to the GDSL family of lipolytic enzymes, composed of members with PLA, LPLA, acyltransferase, lipase, and hemolytic activities (10). PlaB, on the other hand, seems not to belong to one of the known classes of lipolytic serine hydrolases (6). PlaA and PlaC harbor a Sec-signal peptide within their amino acid sequence and are secreted via the type II Lsp machinery (1, 7). In contrast, the mode of PlaB and PlaD secretion is not known but is likely to be Lsp independent, because these proteins do not possess a Sec-signal sequence. Single *L. pneumophila* knockout mutants in the *plaA*, *plaB*,

plaC, and *plaD* genes are not attenuated in the amoeba or macrophage host cell models, probably showing that similar enzymatic activities, especially within one enzyme family, might complement for each other when one activity is missing. Therefore, the construction of mutants harboring several knockouts would be a future goal.

THE PATATIN-LIKE PROTEIN FAMILY
Furthermore, a new family of potential PLAs, the family of patatin-like proteins (PLP), with homology to the human cytosolic phospholipase A as well as to the PLA/LPLA, cytotoxin, and type III secreted virulence factor ExoU of *Pseudomonas aeruginosa*, has been described in *L. pneumophila* and in many other bacteria (2, 8) (see chapter 50). Patatins are a group of plant storage glycoproteins that show lipid acyl hydrolase activity. In animal models and hu-

TABLE 1 Overview of the PLA activities of *L. pneumophila*, their biochemical activities, localization, secretion type, and enzyme classification type

Protein	Activity	Localization	Secretion type	Enzyme type	Reference
PlaA	LPLA	Secreted	II (Lsp)	GDSL	4, 5, 7
PlaB	PLA, LPLA, hemolytic	Cell-associated but surface exposed	Unknown	Not classified	6
PlaC	PLA, LPLA, acyltransferase	Secreted	II (Lsp)	GDSL	1
PlaD	PLA, LPLA	Secreted	Unknown	GDSL	See chapter 59
PatA/VipD	PLA, LPLA	Secreted, injected in host cell	IVB (Dot/Icm)	Patatin-like	See chapter 50
PatB to PatK	Unknown	Unknown	Unknown	Patatin-like	See chapter 50

man infections, the toxicity of ExoU has been linked to the development of lung injury, sepsis, and bacterial dissemination (8). Thus, ExoU is both an important virulence factor of *P. aeruginosa* and the first patatin-like bacterial hydrolase characterized. Interestingly, *L. pneumophila* possesses the high number of 11 members of this group (PatA to PatK; Fig. 1, Table 1; see chapter 50). One of the corresponding proteins, PatA, was found to be an LPLA when expressed in *Escherichia coli*. The *patA* to *patK* genes of *L. pneumophila* have been determined to be expressed during bacterial growth in laboratory media. Furthermore, by the construction of knockout mutants, it was shown that *patA* contributes to secreted bacterial PLA and LPLA activities and is essential for intracellular replication of *L. pneumophila*, during both amoeba and macrophage infection, which implies an important role for lipolytic enzymes with respect to the pathogenic behavior of the bacterium (see chapter 50). Recently, Shohdy et al. identified several type IVB (Dot/Icm) secreted effector proteins of *L. pneumophila* by screening an *L. pneumophila* library for vesicle transport interference in a yeast model. Among the effectors identified was PatA, which was designated VipD and was additionally found to be injected into the host cell by *L. pneumophila* only minutes after host cell infection (9). Therefore, PatA/VipD is the first type IVB secreted effector PLA/LPLA so far analyzed which additionally has a major impact on host cell infection. In the future, it would be of great interest to examine the substrate specificities and secretion modes of the other 10 PLPs present in *L. pneumophila* as well as to characterize their impact on *L. pneumophila* host cell infection.

In short, *L. pneumophila* possesses more than 10 PLAs which are type II (PlaA and PlaC) and type IVB secretion system (PatA/VipD) dependently exported by the bacterium or exported in a yet unknown manner (PatB to PatK, PlaB, PlaD). The high number of lipolytic enzymes present in *L. pneumophila* implies that lipids may be an important source of nutrients for the bacterium, and/or lipid hydrolysis is an essential step in bacterial establishment, especially within the host cell. All of these lipolytic enzymes belong to serine hydrolase families which have recently drawn attention (PLP and GDSL) or remain unclassified, such as *L. pneumophila* PlaB. Therefore, future work on those mostly uncharacterized enzymes will address the impact of the manifold phospholipases on Legionnaires disease pathogenesis, their biochemical properties as well as their activating factors and regulation cascades.

REFERENCES

1. **Banerji, S., M. Bewersdorff, B. Hermes, N. P. Cianciotto, and A. Flieger.** 2005. Characterization of the major secreted zinc metalloprotease-dependent glycerophospholipid:cholesterol acyltransferase, PlaC, of *Legionella pneumophila*. *Infect. Immun.* **73**:2899–2909.

2. **Banerji, S. and A. Flieger.** 2004. Patatin-like proteins: a new family of lipolytic enzymes present in bacteria? *Microbiology* **150:**522–525.

3. **Flieger, A., S. Gong, M. Faigle, M. Deeg, P. Bartmann, and B. Neumeister.** 2000. Novel phospholipase A activity secreted by *Legionella* species. *J. Bacteriol.* **182:**1321–1327.

4. **Flieger, A., S. Gong, M. Faigle, S. Stevanovic, N. P. Cianciotto, and B. Neumeister.** 2001. Novel lysophospholipase A secreted by *Legionella pneumophila. J. Bacteriol.* **183:**2121–2124.

5. **Flieger, A., B. Neumeister, and N. P. Cianciotto.** 2002. Characterization of the gene encoding the major secreted lysophospholipase A of *Legionella pneumophila* and its role in detoxification of lysophosphatidylcholine. *Infect. Immun.* **70:**6094–6106.

6. **Flieger, A., K. Rydzewski, S. Banerji, M. Broich, and K. Heuner.** 2004. Cloning and characterization of the gene encoding the major cell-associated phospholipase A of *Legionella pneu-mophila, plaB,* exhibiting hemolytic activity. *Infect. Immun.* **72:**2648–2658.

7. **Rossier, O., and N. P. Cianciotto.** 2001. Type II protein secretion is a subset of the PilD-dependent processes that facilitate intracellular infection by *Legionella pneumophila. Infect. Immun.* **69:**2092–2098.

8. **Sato, H., D. W. Frank, C. J. Hillard, J. B. Feix, R. R. Pankhaniya, K. Moriyama, V. Finck-Barbancon, A. Buchaklian, M. Lei, R. M. Long, J. Wiener-Kronish, and T. Sawa.** 2003. The mechanism of action of the *Pseudomonas aeruginosa*-encoded type III cytotoxin, ExoU. *EMBO J.* **22:**2959–2969.

9. **Shohdy, N., J. A. Efe, S. D. Emr, and H. A. Shuman.** 2005. Pathogen effector protein screening in yeast identifies *Legionella* factors that interfere with membrane trafficking. *Proc. Natl. Acad. Sci. USA* **102:**4866–4871.

10. **Upton, C., and J. T. Buckley.** 1995. A new family of lipolytic enzymes? Trends *Biochem. Sci.* **20:**178–179.

IDENTIFICATION AND CHARACTERIZATION OF *LEGIONELLA PNEUMOPHILA* PHOSPHOLIPASES A

Sangeeta Banerji, Margret Müller,
Stefan Stevanovic, and Antje Flieger

58

Legionella pneumophila possesses secreted and cell-associated phospholipase A (PLA) activities which may play a role during the establishment of the severe pneumonia as they are capable of destroying lung surfactant, a phospholipid mono-layer in the alveoli that is essential for the stability of the lung (3). Furthermore, PLAs could be involved in the remodeling and lysis of the phagosomal membrane and might create reaction products capable of interfering with the signal transduction of the host (8, 10). So far, we have characterized three *L. pneumophila* PLAs: a secreted lysophospholipase A (LPLA), designated PlaA; a secreted acyltransferase with additional PLA/LPLA activities, PlaC; and a cell-associated PLA/LPLA, PlaB (1, 4, 5). Nevertheless, *L. pneumophila* still possesses PLA/LPLA activities which have not been assigned to any specific protein and which we aim to identify.

THREE PLA-CANDIDATE PROTEINS WERE IDENTIFIED IN THE *L. PNEUMOPHILA* CULTURE SUPERNATANT.

For the isolation of secreted PLAs, concentrated culture supernatant was fractionated by anion exchange chromatography. The major proteins of the anion-exchange fractions with PLA activity were N-terminally sequenced. We found three PLA-candidate proteins. The first protein, designated Aas (gi52840834), possessed conserved amino acid domains which were characteristic for a group of acyltransferases involved in phospholipid biosynthesis. The second protein is encoded by an open reading frame which is located within a region holding components of the *L. pneumophila* Lvh type IVA secretion system and therefore has been named *lvrE* (gi52841476), for *Legionella vir* region gene E (9). The LvrE sequence displayed homology to the protein phosphatase 2C-like protein (gi68550832) of *Pelodictyon phaeoclathratiforme*. The third protein, designated Unk1 (gi52840444), showed some homology to a triacylglycerol lipase/esterase (gi1674078) of *Mycoplasma pneumoniae*. In order to investigate the contribution of the newly identified proteins to the PLA activity of *L. pneumophila* and their importance for the infection of host cells, knockout mutants of the relevant protein genes were constructed in *L. pneumophila* and characterized with respect to their lipolytic activities and intracellular replication in amoeba and macrophage host cells.

L. PNEUMOPHILA AAS INFLUENCES THE BACTERIAL MEMBRANE LIPID COMPOSITION AND DETOXIFIES CYTOLYTIC LYSOPHOSPHOLIPIDS.

The best-characterized homolog of *L. pneumophila* Aas is the Aas protein of *Escherichia*

Sangeeta Banerji and Antje Flieger Research Group NG 5—Pathogenesis of *Legionella* Infection, Robert Koch-Institut, 13353 Berlin, Germany. *Margret Müller and Stefan Stevanovic* Interfakultäres Institut für Zellbiologie, Universität Tübingen, 72076 Tübingen, Germany.

Legionella: State of the Art 30 Years after Its Recognition
Edited by Nicholas P. Cianciotto et al.
©2006 ASM Press, Washington, D.C.

coli. E. coli Aas is an inner-membrane protein with acyltransferase activity, shown to be involved in the acylation of endogenous and exogenous lysophosphatidylethanolamine and in the incorporation of the resulting diacylphosphatidylethanolamine into the bacterial membrane (7). Accordingly, *E. coli aas* mutants accumulate lysophosphatidylethanolamine in their membrane (6). Although *L. pneumophila* Aas was sequenced from the bacterial culture supernatant, we found that the corresponding 130b mutants possessed slightly reduced PLA and LPLA activities in their cell lysates but not in the culture supernatant (data not shown), indicating that Aas might also be present in association with the cell and might fulfil a function comparable to *E. coli* Aas. We therefore analyzed the bacterial membrane lipid composition of the *L. pneumophila aas* mutants and the 130b wild type by thin-layer chromatography (TLC). Indeed, the *L. pneumophila aas* mutants displayed reduced levels of dia-

cylphosphatidylcholine and diacylphosphatidylethanolamine and an elevated level of lysophosphatidylcholine compared to the wild type, indicating that Aas might be involved in the transfer of fatty acids to lysophospholipids and incorporation of diacylphospholipids into the bacterial membrane (Fig. 1A). Since *L. pneumophila* Aas can transfer fatty acids to lysophospholipids, it could also serve to neutralize cytolytic lysophospholipids by converting them to diacylphospholipids.

To address this issue, we monitored the growth of wild type and *aas* mutants in the presence of lysophosphatidylcholine (MPLPC) and found that in contrast to the wild type, the growth of the *aas* mutants was impaired in the presence of the lysophospholipid (Fig. 1B). This result shows that Aas is necessary to counteract the toxic effects of lysophospholipids, probably by converting them to phospholipids which might then be incorporated into the bacterial membrane. Finally, the role

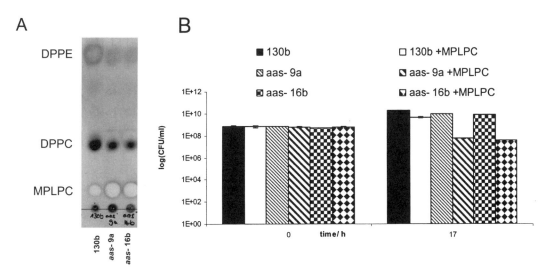

FIGURE 1 Comparative TLC analysis of cell lipids from *L. pneumophila* 130b wild type and *aas* mutants and the effect of cytolytic MPLPC on bacterial viability. (*A*) To analyze the cell lipids, bacteria were grown to mid-logarithmic phase and lipids were directly extracted from the cell lysates and separated by TLC (number of experiments: 2). (*B*) To assess the effect of cytolytic MPLPC on the viability of *L. pneumophila*, the wild type and two independent *aas* mutants were grown to mid-logarithmic phase, 0.2 mM MPLPC was added, cultures were grown for another 16 h, and the CFU were determined (number of experiments: 3). aas-9a and aas-16b represent two independent *L. pneumophila* 130b *aas* mutants. DPPE, dipalmitoylphosphatidylethanolamine; DPPC, dipalmitoylphosphatidylcholine; MPLPC, monopalmitoylphosphatidylcholine.

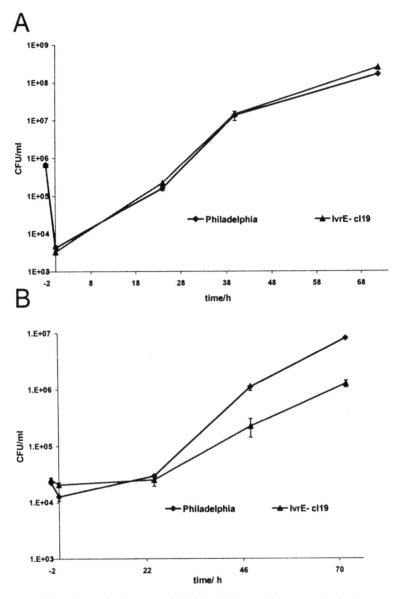

FIGURE 2 Intracellular infection by *L. pneumophila* Philadelphia-1 wild type and an *lvrE* mutant as well as 130b wild type and *unk1* mutants. Strains Philadelphia-1 and an *lvrE* mutant as well as strains 130b and the *unk1* mutants unk1-cl.1 and unk1-cl.2 were used to infect cultures of U937 macrophages (*A*, *C*) or monolayers of *A. castellanii* amoebae (*B*, *D*) at a multiplicity of infection of 1 or 0.01, respectively. At various time points postinoculation, the number of bacteria were quantified by plating aliquots on buffered charcoal yeast extract agar. Results represent the means and standard deviations of triplicate samples and are representative of three (*A*, *B*, *D*) or two (*B*) independent experiments.

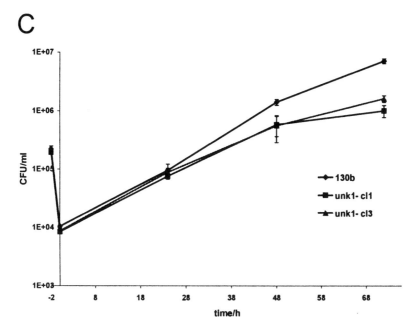

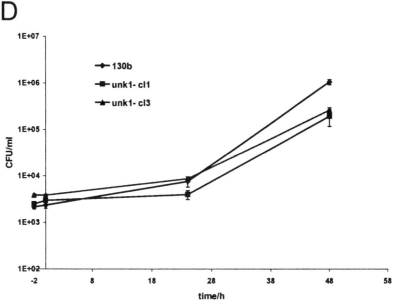

FIGURE 2 *Continued*

of Aas during *L. pneumophila* intracellular infection was assessed in the three host models, U937 macrophages, A549 epithelial cells, and *A. castellanii* amoebae, and Aas was found to be dispensable, because *L. pneumophila aas* mutants showed the same increase in CFU in all three hosts during 72 h of infection as the wild type (data not shown). In summary, our data show that *L. pneumophila* Aas contributes to the incorporation of phospholipids into the bacterial membrane, and in addition to *L. pneumophila* PlaA, is another enzyme involved in the detoxification of lysophospholipids (4).

L. PNEUMOPHILA LVRE MUTANTS SHOW REDUCED PLA ACTIVITY AND ARE ATTENUATED IN AMOEBAL INFECTIONS.

The LvrE sequence displayed homology to a protein phosphatase 2C-like protein (gi68550832) of *Pelodictyon phaeoclathratiforme* (expect value, 2×10^{-11}; identity, 30%; similarity, 44%) and low homology to the Omp36 porin (gi13384123) of *Enterobacter aerogenes* (expect value 0.14; identity, 34%; similarity, 48%). We found that the culture supernatant of an early logarithmic *L. pneumophila* Philadelphia-1 *lvrE* mutant hydrolyzed less phosphatidylglycerol (DPPG), lysophosphatidylglycerol (MPLPG), lysophosphatidylcholine (MPLPC), and 1-monopalmitoylglycerol (1-MPG) than the wild-type culture supernatant, whereas the hydrolysis of phosphatidylcholine (DPPC) was comparable to the wild type (data not shown). The cell-associated PLA activity of the mutant did not differ from the wild-type activity (data not shown). With the aim of assessing the importance of LvrE for infection, *L. pneumophila* Philadelphia-1 wild type and an *lvrE* mutant were used for the infection of U937 macrophages and *Acanthamoeba castellanii*. Interestingly, while the replication of the *lvrE* mutant in the macrophage host was comparable to the wild type, it was attenuated in the amoebal host (Fig. 2A, B).

L. PNEUMOPHILA UNK1 DOES NOT POSSESS LIPOLYTIC PROPERTIES BUT IS ESSENTIAL FOR INTRACELLULAR INFECTION.

The third protein, Unk1, sequenced from PLA-active anion exchange-fractions, displayed weak homology to a triacylglycerol lipase/esterase (gi1674078), Lip3, of *Mycoplasma pneumoniae* (expect value, 1; identity, 25%; similarity, 37%). However, an alignment of the two proteins showed that in spite of good overall homology of Unk1 with Lip3 of *M. pneumoniae*, our protein lacked the lipase motif GXSXG, which encloses the catalytic active serine of many lipolytic enzymes. We examined the corresponding *L. pneumophila* 130b *unk1* mutants for hydrolysis of diacylphospholipids, monoacylphospholipids, and 1-MPG and found that Unk1 did not contribute to the secreted PLA and LPLA activities of *L. pneumophila* (data not shown). Nevertheless, we were interested in its role during intracellular infection. To this purpose, we infected U937 cells with *L. pneumophila* 130b wild type and *L. pneumophila unk1* mutants. Interestingly, the CFU of the *L. pneumophila unk1* mutants was approximately one logarithmic unit behind the CFU of the wild type from 48 h after infection (Fig. 2C). Similar results were obtained for the infection of *A. castellanii* amoebae (Fig. 2D). In summary, our data indicate an essential role for Unk1 in infections of amoebae and macrophages by *L. pneumophila*.

We have identified three proteins, Aas, LvrE, and Unk1, present in the *L. pneumophila* culture supernatant. Aas contributes to the cell membrane integrity of *L. pneumophila*, but is dispensable for the infection of host cells. However, it is likely that the enzymatic functions of Aas might have more impact in vivo, because an unbalanced lipid composition of the bacterial membrane can render the bacterium more susceptible to host defence mechanisms. The second protein, LvrE, contributes to the secreted PLA/LPLA activities of *L. pneumophila* and is essential in amoebal but not macrophage infections. Since LvrE is a substrate of the type

II secretion system, this phenotype is consistent with the more severe defect of type II secretion mutants in amoebal than in macrophage infections (2). The enzymatic function of Unk1 is so far unknown. It shows, in spite of weak homology to lipases, no lipolytic activity, but is essential for the infection of host cells by *L. pneumophila*. Since bacterial lipolytic enzymes are important bacterial tools for surviving both inside and outside of hosts, their characterization promotes our understanding of the life cycle of *L. pneumophila*.

REFERENCES

1. **Banerji S., M. Bewersdorff, B. Hermes, N. P. Cianciotto, and A. Flieger.** 2005. Characterization of the major secreted zinc metalloprotease-dependent glycerophospholipid:cholesterol acyltransferase, PlaC, of *Legionella pneumophila*. *Infect. Immun.* **73:**2899–2909.
2. **De Buck E., L. Maes, E. Meyen, L. Van Mellaert, N. Geukens, J. Anne, and E. Lammertyn.** 2005. *Legionella pneumophila* Philadelphia-1 *tatB* and *tatC* affect intracellular replication and biofilm formation. *Biochem. Biophys. Res. Commun.* **331:**1413–1420.
3. **Flieger A., S. Gong, M. Faigle, H. A. Mayer, U. Kehrer, J. Mussotter, P. Bartmann, and B. Neumeister.** 2000. Phospholipase A secreted by *Legionella pneumophila* destroys alveolar surfactant phospholipids. *FEMS Microbiol. Lett.* **188:**129–133.
4. **Flieger A., B. Neumeister, and N. P. Cianciotto.** 2002. Characterization of the gene encoding the major secreted lysophospholipase A of *Legionella pneumophila* and its role in detoxification of lysophosphatidylcholine. *Infect. Immun.* **70:**6094–6106.
5. **Flieger A., K. Rydzewski, S. Banerji, M. Broich, and K. Heuner.** 2004. Cloning and characterization of the gene encoding the major cell-associated phospholipase A of *Legionella pneumophila, plaB*, exhibiting hemolytic activity. *Infect. Immun.* **72:**2648–2658.
6. **Hsu L., S. Jackowski, and C. O. Rock.** 1991. Isolation and characterization of *Escherichia coli* K-12 mutants lacking both 2-acyl-glycerophosphoethanolamine acyltransferase and acyl-acyl carrier protein synthetase activity. *J. Biol. Chem.* **266:**13783–13788.
7. **Jackowski S., P. D. Jackson, and C.O. Rock.** 1994. Sequence and function of the *aas* gene in *Escherichia coli*. *J. Biol. Chem.* **269:**2921–2928.
8. **Molmeret M., D. M. Bitar, L. Han, and Y. Abu Kwaik.** 2004. Disruption of the phagosomal membrane and egress of *Legionella pneumophila* into the cytoplasm during the last stages of intracellular infection of macrophages and *Acanthamoeba poly-phaga*. *Infect. Immun.* **72:**4040–4051.
9. **Segal G., J. J. Russo, and H. A. Shuman.** 1999. Relationships between a new type IV secretion system and the *icm/dot* virulence system of *Legionella pneumophila*. *Mol. Microbiol.* **34:**799–809.
10. **Tilney L. G., O. S. Harb, P. S. Connelly, C. G. Robinson, and C. R. Roy.** 2001. How the parasitic bacterium *Legionella pneumophila* modifies its phagosome and transforms it into rough ER: implications for conversion of plasma membrane to the ER membrane. *J. Cell. Sci.* **114:**4637–4650.

CHARACTERIZATION OF GDSL-HYDROLASES OF THE LUNG PATHOGEN *LEGIONELLA PNEUMOPHILA*

Sangeeta Banerji, Elena Rastew, Björn Hermes, and Antje Flieger

59

Legionella pneumophila possesses secreted phospholipase A (PLA) and lysophospholipase A (LPLA) activities which might be involved in bacterial virulence, because they represent tools for modification and lysis of host cell membranes and for the interference with the host signal transduction pathways (4). The *L. pneumophila* genomes Philadelphia-1, Paris, and Lens code for three members of the GDSL family of lipolytic proteins, namely PlaA, PlaC, and PlaD (2, 3, 5). The first two have already been characterized, but the enzymatic function of PlaD is still unknown (1, 5). PlaA is the major secreted LPLA of *L. pneumophila* (5). PlaC is a secreted glycerophospholipid:cholesterol acyltransferase (GCAT) with additional PLA and LPLA activities which needs activation by a factor present in the *L. pneumophila* culture supernatant (1). Activation of PlaC is directly or indirectly dependent on the zinc metallo protease ProA (1). We aimed to identify the factor which directly activates PlaC. Furthermore, we sought to characterize PlaD, the third *L. pneumophila* GDSL enzyme, and to compare the lipolytic properties as well as the capacity of *L. pneumophila plaD* mutants to multiply intracel-lularly with *L. pneumophila* wild type and *plaA* and *plaC* mutants.

THE ZINC METALLOPROTEASE ProA DIRECTLY ACTIVATES PlaC

The PlaC-activating factor, characterized by its ability to enhance GCAT activity of PlaC expressed in *Escherichia coli*, was biochemically purified from the culture supernatant of an *L. pneumophila* 130b *plaC* mutant by anion exchange chromatography (AEX) (1). The main protein in the activating fractions consistently was the zinc metalloprotease ProA. Since the amount of protease activity present in the fractions also correlated with the strength of enhancing the GCAT activity of PlaC, we concluded that ProA directly activated PlaC (Fig. 1). In order to confirm that *L. pneumophila* ProA can directly activate PlaC, the cell lysates of *E. coli* clones harboring *plaC* in *trans* were incubated with periplasmic fractions of *E. coli* clones harboring *proA* in *trans* which showed protease activity and were then assayed for GCAT activity by thin-layer chromatography. Indeed, we found that the GCAT activity of cell lysates from *E. coli* clones carrying *plaC* was enhanced by coincubation with ProA-containing *E. coli* fractions (data not shown). So far, host cell proteins such as human interleukin-2 and human T lymphocyte CD4 have

Sangeeta Banerji, Elena Rastew, Björn Hermes, and Antje Flieger Research Group NG 5—Pathogenesis of *Legionella* Infection, Robert Koch-Institut, 13353 Berlin, Germany.

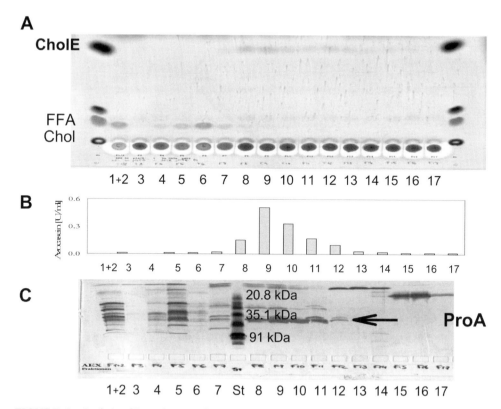

FIGURE 1   Analysis of AEX fractions from culture supernatant of an *L. pneumophila* 130b *plaC* mutant with respect to PlaC activating properties and protease activity. Culture supernatant from an *L. pneumophila* 130b *plaC* mutant was purified by AEX, and the fractions (1 to 17) were assayed for induction of GCAT activity detected by the formation of cholesterol ester by *E. coli* harboring *plaC* (*A*) and for protease activity (*B*). Fractions were also analyzed by reducing sodium dodecylsulfate-polyacrylamide gel electrophoresis (*C*). Similar results were obtained on one more occasion. AEX, anion exchange chromatography; Chol, cholesterol; CholE, cholesterol ester; FFA, free fatty acid; St, protein mass standard.

been shown to be substrates degraded by ProA, but PlaC is the first identified *L. pneumophila* substrate where cleavage by ProA enhances its activity (8).

L. PNEUMOPHILA PlaD IS A SECRETED PHOSPHOLIPASE A AND LYSOPHOSPHOLIPASE A

Since the two *L. pneumophila* paralogs of PlaD (PlaA, PlaC) possess PLA and/or LPLA activities, we examined *L. pneumophila* Corby *plaD* mutants for lipolytic activities (1, 5). Cell lysates and culture supernatants of *L. pneumophila* Corby wild type as well as of *plaD* mutants were incubated with diacylphospholipids (DPPC and

DPPG) to check for PLA activity, monoacylphospholipids (lysophospholipids) (MPLPC and MPLPG) to check for LPLA activity, and the lipid 1-monopalmitoylglycerol (1-MPG) to study lipase activity; the amount of released fatty acids was also determined. The cell lysates of the Corby wild type and the *plaD* mutants released comparable amounts of fatty acids from all employed substrates (data not shown). In contrast, compared to wild type, the culture supernatants of the *plaD* mutants displayed a 50% reduction in the hydrolysis of the diacylphospholipids DPPG and DPPC and a 30% reduction in the hydrolysis of the monoacylphospholipid MPLPG and the lipid 1-MPG,

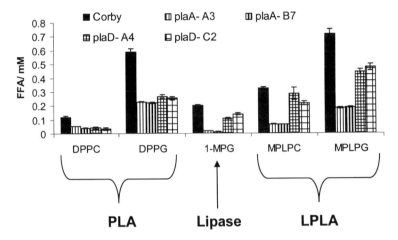

FIGURE 2 Lipolytic activities of *L. pneumophila* Corby culture supernatants from wild type and *plaD* mutants. *L. peumophila* Corby culture supernatants from wild type, *plaD*, and *plaA* knockout mutants were obtained at late logarithmic growth phase and incubated with dipalmitoylphosphatidylcholine (DPPC), dipalmitoyllysophosphatidylglycerol (DPPG), 1-monopalmitoylglycerol (1-MPG), monopalmitoyllysophosphatidylcholine (MPLPC), and monopalmitoyllysophosphatidylglycerol (MPLPG) at 37°C. Subsequently, the release of fatty acids was quantified. Similar results were obtained on two more occasions. plaA-A3 and plaA-B7 represent two independent *L. pneumophila* Corby *plaA* mutants. Likewise, plaD-A4 and plaD-C2 represent two independent Corby *plaD* mutants.

(Figure 2). Our data therefore show that PlaD is a PLA and a LPLA which is secreted to the *L. pneumophila* culture supernatant.

To investigate whether *L. pneumophila* Corby *plaD* could confer PLA and LPLA activities to *E. coli*, cell lysates as well as the culture supernatants from *E. coli* clones carrying *plaD* in *trans* were examined for hydrolysis of PLA, LPLA, and lipase substrates. The culture supernatants from the *E. coli* clones harboring *plaD* released the same amount of fatty acids from all employed substrates as the clones carrying the pBCKS vector control (data not shown). The cell lysates of the *E. coli* clones harboring *plaD* in *trans* showed the same hydrolysis of DPPG, DPPC, and the lipid 1,2 dipalmitoylglycerol but an increased hydrolysis of the monoacylphospholipids MPLPC and MPLPG compared to the clone carrying the empty vector. Thus, our results show that *L. pneumophila* PlaD displays LPLA activity when expressed in *E. coli*. We did not detect PLA activity of PlaD

in *E. coli*, although the *L. pneumophila* Corby *plaD* mutants showed a high reduction in their PLA activity. Maybe the overexpression of the *plaD* gene led to imperfect folding of the protein or some additional factor, as is the case for PlaC, which was missing in *E. coli*, resulting in restricted enzymatic activity of PlaD.

To assess the importance of the GDSL enzyme PlaD during the infection of amoebae, *L. pneumophila* Corby wild type and Corby *plaD* mutants were used to infect *Acanthamoeba castellanii*. The *plaD* mutants showed the same increase in CFU as the wild type, namely an increase of approximately three logarithmic units during 48 h of infection (data not shown). Moreover, an infection of U937 macrophages demonstrated that *plaD* was likewise dispensable for macrophage infection (data not shown). These results are in accordance with the findings that *L. pneumophila plaA* and *plaC* mutants do not display a growth defect in the two studied infection models, either (1, 5). However,

GDSL hydrolases of other bacteria, e.g. *Aeromonas hydrophila* SatA and *Salmonella enterica* SseJ, both of which probably possess similar enzymatic activities as the *L. pneumophila* GDSL enzymes, have been identified as virulence determinants (6, 7). Our findings demonstrate that the three *L. pneumophila* GDSL enzymes have overlapping enzymatic functions (especially in their LPLA activity), which might enable the bacterium to compensate for the loss of one. Therefore, investigating the virulence of *L. pneumophila* mutants lacking two or all three GDSL enzymes is an important future issue. The presence of three GDSL family enzymes with overlapping functions furthermore points to the high level of redundancy in the *L. pneumophila* genome with respect to possible virulence factors.

REFERENCES

1. **Banerji S., M. Bewersdorff, B. Hermes, N. P. Cianciotto, and A. Flieger.** 2005. Characterization of the major secreted zinc metalloprotease-dependent glycerophospholipid:cholesterol acyltransferase, PlaC, of *Legionella pneumophila*. *Infect. Immun.* **73:**2899–2909.
2. **Cazalet, C., C. Rusniok, H. Bruggemann, N. Zidane, A. Magnier, L. Ma, M. Tichit, S. Jarraud, C. Bouchier, F. Vandenesch, F. Kunst, J. Etienne, P. Glaser, and C. Buchrieser.** 2004. Evidence in the *Legionella pneumophila* genome for exploitation of host cell functions and high genome plasticity. *Nat. Genet.* **36:**1165–1173.
3. **Chien M., I. Morozova, S. Shi, H. Sheng, J. Chen, S. M. Gomez, G. Asamani, K. Hill, J. Nuara, M. Feder, J. Rineer, J. J. Greenberg, V. Steshenko, S. H. Park, B. Zhao, E. Teplitskaya, J. R. Edwards, S. Pampou, A. Georghiou, I. C. Chou, W. Iannuccilli, M. E. Ulz, D. H. Kim, A. Geringer-Sameth, C. Goldsberry, P. Morozov, S. G. Fischer, G. Segal, X. Qu, A. Rzhetsky, P. Zhang, E. Cayanis, P. J. De Jong, J. Ju, S. Kalachikov, H. A. Shuman, and J. J. Russo.** 2004. The genomic sequence of the accidental pathogen *Legionella pneumophila*. *Science* **305:**1966–1968.
4. **Flieger A., S. Gong, M. Faigle, M. Deeg, P. Bartmann, and B. Neumeister.** 2000. Novel phospholipase A activity secreted by Legionella species. *J. Bacteriol.* **182:**1321–1327.
5. **Flieger A., B. Neumeister, and N. P. Cianciotto.** 2002. Characterization of the gene encoding the major secreted lysophospholipase A of *Legionella pneumophila* and its role in detoxification of lysophosphatidylcholine. *Infect. Immun.* **70:**6094–6106.
6. **Lee K. K., and A. E. Ellis.** 1990. Glycerophospholipid:cholesterol acyltransferase complexed with lipopolysaccharide (LPS) is a major lethal exotoxin and cytolysin of *Aeromonas salmonicida*: LPS stabilizes and enhances toxicity of the enzyme. *J. Bacteriol.* **172:**5382–5393.
7. **Ohlson M. B., K. Fluhr, C. L. Birmingham, J. H. Brumell, and S. I. Miller.** 2005. SseJ deacylase activity by *Salmonella enterica* serovar Typhimurium promotes virulence in mice. *Infect. Immun.* **73:**6249–6259.
8. **Mintz C. S., R. D. Miller, N. S. Gutgsell, and T. Malek.** 1993. *Legionella pneumophila* protease inactivates interleukin-2 and cleaves CD4 on human T cells. *Infect. Immun.* **61:**3416–3421.

GENETIC AND STRUCTURAL
EXAMINATION OF THE
LEGIOBACTIN SIDEROPHORE

Kimberly A. Allard, Domenic Castignetti, David Crumrine,
Prakash Sanjeevaiah, and Nicholas P. Cianciotto

60

IRON AND *LEGIONELLA PNEUMOPHILA*

The importance of iron for *L. pneumophila* growth has long been appreciated, based upon minimal growth studies in chemically defined medium (CDM). Supporting data confirm that *L. pneumophila* assimilates significant amounts of iron upon incubation with $^{55}Fe^{2+}$ (8). Many studies have shown a role for iron in pathogenesis; e.g., macrophages treated with iron chelators such as desferrioxamine mesylate inhibit *L. pneumophila* growth (1, 2, 6, 8). Thus, *L. pneumophila* has a nonnegotiable requirement for iron and must therefore encode specific determinant(s) for acquiring this metal that are relevant to its intra- and extracellular lifestyles.

SIDEROPHORES BIND IRON AND FACILITATE ITS UTILIZATION

Bacteria have evolved numerous ways to acquire iron from the environment, but one of the most common mechanisms is the produc-

tion of siderophores. Siderophores are electronegative, iron-regulated, low-molecular-weight compounds produced by bacteria and fungi that bind ferric iron and facilitate its internalization via specific receptors. Siderophores are secondary metabolites mainly synthesized by nonribosomal peptide synthetases that are similar to those used for antibiotic synthesis (3). The genes responsible for siderophore biosynthesis are iron regulated and often clustered with genes required for transport. Once the siderophore is synthesized, it must be exported out of the bacterial cell, and in some cases, the siderophore exit route is generated by proteins homologous with permeases of the major facilitator superfamily of proton-motive force-dependent membrane efflux pumps (9). Iron-bound siderophores are usually recognized at the cell surface by specific receptors in both gram-positive and gram-negative bacteria that internalize the ferrisiderophore. Typically, the internalized, siderophore-bound Fe^{3+} is then reduced to free Fe^{2+} in the cytoplasm by ferric reductases and is subsequently utilized in metabolic processes. Siderophores are extremely effective in binding the Fe^{3+} ion because they contain the most effective iron binding ligands in nature, consisting of hydroxamate, catecholate, and α-hydroxycarboxylate ligands that form

Kimberly A. Allard and Nicholas P. Cianciotto Department of Microbiology-Immunology, Northwestern University Medical School, Chicago, IL 60611. *Domenic Castignetti* Department of Biology, Loyola University of Chicago, Chicago, IL 60626. *David Crumrine and Prakash Sanjeevaiah* Department of Chemistry, Loyola University of Chicago, Chicago, IL 60626.

hexadentate Fe^{3+} complexes, thus satisfying the six coordination sites on ferric ions (10).

LEGIOBACTIN, THE *L. PNEUMOPHILA* SIDEROPHORE

For many years, *L. pneumophila* was thought not to produce siderophores based upon negative results obtained from Arnow and Csáky assays, which identify catecholate and hydroxamate structures, as well as the Chrome Azurol S (CAS) assay, which detects iron chelators independently of structure (5, 7). However, we showed that *L. pneumophila* could produce a high-affinity iron-chelator (4). When grown at 37°C in a low-iron CDM, *L. pneumophila* secretes a low-molecular-weight substance that is reactive in the CAS assay. The siderophore activity is iron-repressed and only observed when the CDM is inoculated with legionellae that had been grown to log or early-stationary phase (4). We designated the

iron-chelating activity in *L. pneumophila* supernatants as legiobactin.

GENETICS OF LEGIOBACTIN PRODUCTION

lbtA, an *L. pneumophila* gene involved in legiobactin production, encodes a protein that is related to several siderophore synthetases, showing 23% identity and 40% similarity to IucA and 26% identity and 44% similarity to IucC. Using reverse transcriptase PCR, we observed that *lbtA* mRNA levels are greater in *L. pneumophila* 130b grown in deferrated CDM than in bacteria grown in iron-supplemented CDM, suggesting that *lbtA* is iron regulated. Multiple mutants containing an antibiotic-resistance insertion in *lbtA* consistently displayed 40 to 70% less CAS-reactive material than wild type when grown in iron-deplete CDM (Fig. 1; compare wild type and *lbtA* mutant containing vector pMMB2002). Since wild-type supernatants

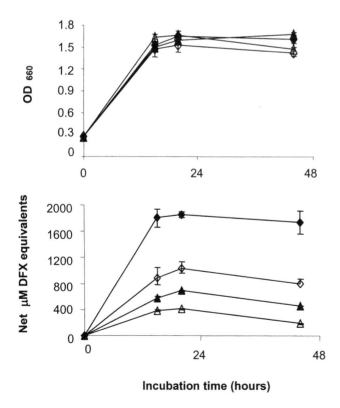

FIGURE 1 Siderophore production by *L. pneumophila* wildtype and an *lbtA* mutant. Wildtype 130b with pMMB2002 (◇) or plbtA (◆) and *lbtA* mutant with pMMB2002 (Δ) or plbtA (▲) were grown in buffered yeast extract to an optical density at 660 nm (OD_{660}) of 1.0; inoculated into deferrated CDM to an OD_{660} of 0.3; and then incubated at 37°C. At various time points, the growth of the cultures was monitored spectrophotometrically (top), and the CAS reactivity of culture supernatants was examined (bottom). The values presented represent the means and standard deviations from duplicate cultures. The CAS reactivity of the mutant's cultures was significantly different from that of the wildtype and complemented mutant cultures, at all times of incubation ($P < 0.05$; Student's *t*- test). The results presented are representative of at least four independent experiments.

derived from iron-deplete CDM support growth of wild-type legionellae in buffered-charcoal yeast extract (BCYE) agar containing an inhibitory concentration of 2,2'-dipyridyl (DIP) and an *feoB* mutant on BCYE agar lacking the iron supplement, we investigated the growth-promoting ability of *lbtA* mutant supernatants. Unlike CAS-positive, wild-type supernatants, *lbtA* mutant supernatants do not stimulate growth of the wild type or the *feoB* mutant. Thus, the *lbtA* mutant was lacking in its ability to elaborate a biologically active iron chelator. Introduction of a plasmid (plbtA)– containing *lbtA* increased CAS reactivity of the mutant (Fig. 1), but plbtA did not restore reactivity to wild-type levels. However, supernatants from the complemented mutant were capable of rescuing growth of the *feoB* mutant on BCYE that lacked the iron supplement and 130b in DIP-containing BCYE agar.

These data suggest that *lbtA* is required for the expression of legiobactin, but the incomplete complementation observed suggested that downstream gene(s) might also be involved in legiobactin production. Indeed, an *lbtA* mutant containing a nonpolar deletion mutation was fully complemented by introduction of plbtA. According to the completed genomes of *L. pneumophila* strains Philadelphia-1, Paris, and Lens, *lbtA* is the first gene in a three-gene operon. The two genes downstream of *lbtA* are predicted to encode members of the MFS class of proton motive force-dependent membrane efflux pumps. The second gene in the operon, which we designate *lbtB*, encodes a protein with 12 transmembrane spanning domains that is 23% identical and 44% similar to the *E. coli* bicyclomycin resistance protein Bcr, and 21% identical and 39% similar to the *E. coli* tetracycline efflux

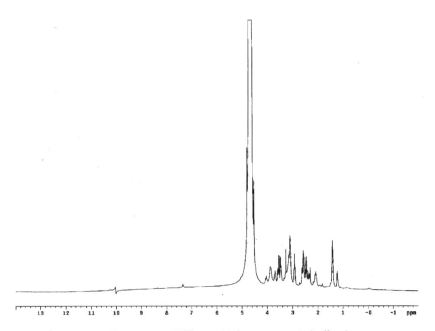

FIGURE 2 Legiobactin H-NMR spectrum. Wild-type 130b was grown in buffered yeast extract to an OD_{660} of 1.0; inoculated into deferrated CDM to an OD_{660} of 0.3; and then incubated at 37°C. After 24 h, the CAS-reactive supernatants were harvested and filtered through 3-kDa cut-off ultrafiltration filters. Low-molecular-weight supernatants were then concentrated 16-fold by rotary evaporation and subjected to anion-exchange high-pressure liquid chromatography. Purified legiobactin was then concentrated and desalted by ultrafiltration using 500-kDa cut-off filters and dissolved ion D_2O. 1H-NMR spectra were recorded on a Varian 300-Mhz NMR spectrometer.

pump, TetA. Several *lbtB* mutants consistently produced 40 to 70% less CAS-reactive material than wild type and did not promote growth of the *feoB* mutant in BCYE agar lacking the iron supplement. All *lbtB* mutants were fully complemented by addition of *lbtB* in *trans*. Thus, like *lbtA*, *lbtB* is important for legiobactin production. The last gene in the operon is not required for legiobactin production. Although *lbtA* is expressed intracellularly, the *lbtA* mutant infects U937 macrophages and *Hartmannella vermiformis* amoebae as wild type does, even when the host cell was treated with the iron chelators DIP or desferrioxamine mesylate. Thus, *lbtA* and *lbtB* are necessary for optimal legiobactin production and, although not required for intracellular growth, are likely important for growth of the bacteria in the environment.

STRUCTURAL CHARACTERISTICS OF LEGIOBACTIN

The CAS-reactive compound produced by *L. pneumophila* has the characteristics of a siderophore, as the activity is iron repressed, less than 1 kDa in size, resistant to heat and protease treatment, and promotes the growth of iron-starved legionellae. The CAS-reactive substance does not react in the Csáky and Arnow assay, suggesting that legiobactin is a hydroxy-carboxylate siderophore. Indeed, legiobactin does not extract into common solvents that are used to extract catecholate and hydroxamate siderophores. Since legiobactin is anionic at neutral pH, anion-exchange high-pressure liquid chromatography is used to purify the compound to homogeneity. The legiobactin peak is the portion of *L. pneumophila* supernatants that promotes growth of iron-starved legionellae and is absent in the *lbtA* mutant supernatants. As was the case using ^{13}C nuclear magnetic resonance (NMR) analysis, proton NMR analysis of purified legiobactin demonstrates that the siderophore contains only aliphatic residues (Fig. 2). Currently we are working toward determining the structure of legiobactin based on two-dimensional-NMR, elemental analysis, and Maldi experiments.

REFERENCES

1. **Byrd, T. F., and M. A. Horwitz.** 1989. Interferon gamma-activated human monocytes downregulate transferrin receptors and inhibit the intracellular multiplication of *Legionella pneumophila* by limiting the availability of iron. *J. Clin. Invest.* **83:**1457–1465.
2. **Byrd, T. F., and M. A. Horwitz.** 1991. Lactoferrin inhibits or promotes *Legionella pneumophila* intracellular multiplication in nonactivated and interferon gamma-activated human monocytes depending upon its degree of iron saturation. Iron-lactoferrin and nonphysiologic iron chelates reverse monocyte activation against *Legionella pneumophila*. *J. Clin. Invest.* **88:**1103–1112.
3. **Crosa, J. H., and C. T. Walsh.** 2002. Genetics and assembly line enzymology of siderophore biosynthesis in bacteria. *Microbiol. Mol. Biol. Rev.* **66:**223–249.
4. **Liles, M. R., T. Aber Scheel, and N. P. Cianciotto.** 2000. Discovery of a nonclassical siderophore, legiobactin, produced by strains of *Legionella pneumophila*. *J. Bacteriol.* **182:**749–757.
5. **Liles, M. R., and N. P. Cianciotto.** 1996. Absence of siderophore-like activity in *Legionella pneumophila* supernatants. *Infect. Immun.* **64:**1873–1875.
6. **Pope, C. D., W. O'Connell, and N. P. Cianciotto.** 1996. *Legionella pneumophila* mutants that are defective for iron acquisition and assimilation and intracellular infection. *Infect. Immun.* **64:**629–636.
7. **Reeves, M. W., L. Pine, J. B. Neilands, and A. Balows.** 1983. Absence of siderophore activity in *Legionella* species grown in iron-deficient media. *J. Bacteriol.* **154:**324–329.
8. **Robey, M., and N. P. Cianciotto.** 2002. *Legionella pneumophila feoAB* promotes ferrous iron uptake and intracellular infection. *Infect. Immun.* **70:**5659–5669.
9. **Wandersman, C., and P. Delepelaire.** 2004. Bacterial iron sources: from siderophores to hemophores. *Annu. Rev. Microbiol.* **58:**611–647.
10. **Winkelmann, G.** 2002. Microbial siderophore-mediated transport. *Biochem. Soc. Trans.* **30:**691–696.

EUKARYOTIC-LIKE PROTEINS OF *LEGIONELLA PNEUMOPHILA* AS POTENTIAL VIRULENCE FACTORS

Fiona M. Sansom, Hayley J. Newton and Elizabeth L. Hartland

61

Our bioinformatic analysis of the *Legionella pneumophila* Philadelphia 1 genome revealed four novel open reading frames (ORFs) that have no known prokaryotic homologues but are predicted to encode products with significant domain homology to eukaryotic proteins (3). All four ORFs were also found in the genome of the Lens and Paris strains of *L. pneumophila* (2). These ORFs were predicted to encode two putative eukaryotic-like ecto-nucleosidases (*lpg1905* and *lpg0971*), a putative protein with a collagen-like domain (*lpg-2644*), and another with low homology to surface antigens of *Plasmodium* (*lpg1488*; Table 1). We postulated that the gene products of these eukaryotic-like ORFs may interact with eukaryotic host cell proteins and thus play a role in the subversion of host cellular trafficking pathways by *L. pneumophila* and the establishment of the replicative niche inside host cells.

GENE TRANSCRIPTION

We initially performed reverse transcriptase (RT)-PCR to determine if the ORFs were transcribed and thus if in vitro characterization was possible. RNA was extracted from station-ary phase broths of *L. pneumophila* 130b using the Epicentre Masterpure RNA Purification Kit (MCR85102). cDNA was synthesized, and specific primers for each gene were used to amplify 200 to 1,100 base pair portions of each ORF. This revealed that all four ORFs were transcribed in stationary phase (Fig. 1).

CONTRIBUTION OF INDIVIDUAL GENES TO HOST CELL INTERACTIONS

To investigate the potential role that each eukaryotic-like gene under investigation may play in interactions with host cells, insertional mutants were constructed for each ORF. A portion of each gene was cloned into either pCRScript or pGEM-T Easy and a kanamycin cassette inserted into the coding region. The inactivated ORF was then introduced into *L. pneumophila* 130b by natural transformation, and the wild-type gene was replaced by allelic exchange as described previously (6). Briefly, bacteria were grown stationary at 30°C with ~30 μg of plasmid DNA carrying the mutated gene to an optical density at 660 of ~1.6. Bacteria were plated onto charcoal-yeast extract (CYE) plates supplemented with 25 μg of kanamycin per ml and grown at 37°C for 3 days. Genomic DNA was extracted from kanamycin-resistant colonies, and primers

Fiona M. Sansom, Hayley J. Newton, and Elizabeth L. Hartland Australian Bacterial Pathogenesis Program, Department of Microbiology, Monash University, Victoria 3800, Australia.

TABLE 1 Characteristics of the *L. pneumophila* eukaryotic-like ORFs.

ORF	Predicted size	Motifs	Homology
lpg2644	492 amino acids	N-terminal signal peptide 170-aa long region of internal repeats	Domain homology to type VI collagen
lpg1488	865 amino acids	Predicted coiled-coil region (628 to 724 aa)	Domain homology to surface antigens of *Plasmodium falciparum*
lpg1905	393 amino acids	Hydrophobic region (1 to 35 aa). Five apyrase conserved regions	Homologous to eukaryotic GDA 1/CD39 NTPDase family
lpg0971	381 amino acids	Signal peptide (1 to 21 aa) five apyrase conserved regions	Homologous to eukaryotic GDA 1/CD39 NTPDase family

external to the mutated region were used to confirm replacement of the wild-type gene with the insertion mutation as evidenced by an increase in size of the PCR product equivalent to the size of the kanamycin cassette.

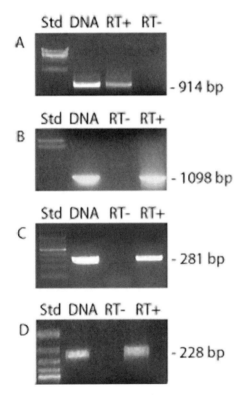

FIGURE 1 Reverse transcription (RT)-PCR analysis of putative genes predicted to encode products with homology to eukaryotic proteins. (*A*) *lpg1905*, (*B*) *lpg1488*, (*C*) *lpg0971*, (*D*) *lpg2644*. DNA, genomic DNA control; RT+, reverse transcription PCR; RT-, control PCR reaction without reverse transcription step.

Mutant strains were then assayed for entry and replication defects in THP-1 cells, a human monocyte cell line that can be chemically differentiated into macrophage-like cells, and A549 cells, a human carcinoma cell line with characteristics of type II alveolar epithelial cells. THP-1 cells were infected with bacteria at a multiplicity of infection of 5 for 2 h before gentamicin treatment (100 µg/ml) for 1 h to kill extracellular bacteria. Cells were then lysed with 0.1% digitonin and serial dilutions plated onto CYE plates at time 2, 24, 48, and 72 to determine viable counts. Similar assays were performed using A549 cells with the exception that a multiplicity of infection of 100 was used.

Preliminary results suggested there was no significant difference to wild-type *L. pneumophila* 130b either invasion into or replication within THP-1 cells for any of the mutant strains. In A549 cells, however, a significant reduction in entry was observed for the Δ*lpg1905* mutant compared to the wild-type strain (Fig. 2A). To investigate this result further, additional assays were carried out in which the gentamicin step was omitted and cells were only washed to remove unattached bacteria before lysis. This enabled us to calculate the total number of bacteria in association with the cells. Parallel assays were carried out to measure the percentage of these cell-associated bacteria that became intracellular to discover if an attachment or true entry defect was responsible for the overall decrease in invasion observed. For the Δ*lpg1905* mutant, a significant decrease in both host cell attachment and host cell invasion was observed. (Fig. 2B, C).

A

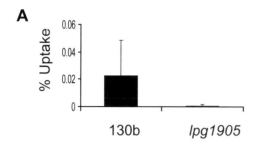

B

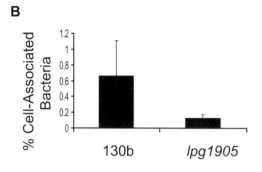

C

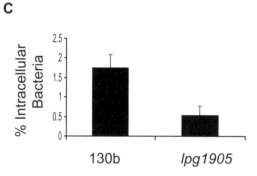

FIGURE 2 Interaction of *L. pneumophila* 130b and *L. pneumophila* Δ*lpg1905* with A549 epithelial cells. (*A*) Percentage of the inoculum that is intracellular after a 2-h infection period (*P* < 0.01, unpaired two tailed t-test) (*B*) Percentage of the inoculum that associates with cells after a 2-h infection period (*P* < 0.02, unpaired two tailed t-test). (*C*) Percentage of the total cell-associated bacteria that become intracellular after a 2-h infection period (*P* < 0.008, unpaired two tailed t-test).

LOCALIZATION OF THE PROTEIN PRODUCTS OF EUKARYOTIC-LIKE ORFs

To further characterize the products of *lpg1905* and *lpg2644*, a combination of epitope-tagging and specific polyclonal antibodies were used to localize the proteins within the bacteria. The promoter for *mip*, a constitutively expressed gene in *L. pneumophila*, was cloned into pMMB207 to create an expression vector, and the specific genes of interest were amplified by PCR to include a C-terminal FLAG epitope tag and cloned downstream of the *mip* promoter in pMMB207. Constructs were then electroporated into *L. pneumophila* 130b, and bacteria containing the plasmids were selected by incubation on CYE plates containing 6 μg of chloramphenicol per ml.

Bacteria carrying the FLAG constructs were then lysed by sonication, and lysates were separated into soluble and insoluble fractions by high-speed centrifugation. At the same time culture supernatants were precipitated with 10% (wt/vol) TCA. Fractions were then analyzed by Western blot using monoclonal antibodies to the FLAG epitope to detect protein. Epitope tagging and Western blot analysis was successfully carried out for *lpg1488*, revealing the presence of Lpg1488 in both soluble and insoluble fractions (Fig. 3A), and for *lpg2644*, demonstrating the presence of Lpg2644 in the soluble, insoluble, and secreted protein fractions, demonstrating that this protein is exported by *L. pneumophila*.

In addition, *lpg2644* and *lpg1905* were cloned into expression vectors pET28a and pRSET-B and expressed in *E. coli* BL21 with a 6xHis N-terminal tag. Proteins were purified using Ni-agarose columns, and antibodies to the purified proteins were raised in rabbits. Derivatives of *L. pneumophila* 130b were then fractionated as described above, and Western blotting was carried out using the specific polyclonal antibodies. This confirmed that Lpg2644 was secreted by wild-type *L. pneumophila* 130b. To determine if Lpg2644 was exported via the Dot/Icm type IV secretion system or the Lsp type II secretion system, precipitated culture supernatants of a Δ*dotA* and a Δ*lspDE* mutant were also analyzed by Western blot. This demonstrated that neither secretion pathway was involved in the export of Lpg2644 (Fig. 3B).

In addition, Western blot analysis of bacterial fractions using the Lpg1905 polyclonal

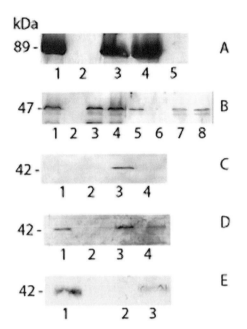

FIGURE 3 (*A*) Western blot analysis of derivatives of *L. pneumophila* 130b detected with anti-FLAG monoclonal antibodies. Lane 1,130b (pMIP:*lpg2644*-FLAG), whole cell lysate (WCL); lane 2, 130b (pM-MB207) negative control (WCL); lane 3, 130b (pMIP:*lpg2644*-FLAG) soluble fraction (S); lane 4, 130b (pMIP:*lpg2644*-FLAG), insoluble fraction (I); lane 5, 130b (pMIP:*lpg2644*-FLAG), supernatant fraction (SN). (*B*) Derivatives of 130b detected with mono-specific antibodies to recombinant Lpg2644. Lane 1, 130b WCL; lane 2; Δ*lpg2644* WCL; lane 3, Δ*lspDE* WCL; lane 4 Δ*dotA* WCL; lane 5, 130b SN; lane 6, Δ*lpg2644* SN; lane 7, Δ*lspDE* SN; lane 8, Δ*dotA* SN. (*C*) Derivatives of 130b detected with monospecific antibodies to recombinant Lpg1905. Lane 1, 130b WCL; lane 2, Δ*lpg1905* WCL; lane 3, 130b SN; lane 4, Δ*lpg1905* SN. (*D*) Derivatives of 130b detected with monospecific antibodies to recombinant Lpg1905. Lane 1, 130b SN; lane 2, Δ*lpg1905* SN; lane 3, Δ*proA* SN; lane 4, Δ*dotA* SN. (*E*) Derivatives of 130b detected with monospecific antibodies to recombinant Lpg1905. Lane 1; 130b SN, lane 2, Δ*lpg1905* SN; lane 3, Δ*lspDE* SN.

antisera detected the protein only in the supernatant fraction of cultures (Fig. 3C, D), suggesting that the majority of the protein is secreted by the bacteria. Lpg1905 was also detected in the supernatant of a Δ*dotA* mutant, demonstrating that the Dot/Icm type IV secretion system is not involved in export of Lpg1905. Western blot analysis of the supernatant fraction of an *lspDE* mutant using the Lpg1905 polyclonal antisera detected only a very faint band compared to 130b supernatant. This was despite an increased level of contamination by cytoplasmic proteins (due to cell breakdown) in the supernatant fraction, evident on the silver stained gel, for the *lspDE* mutant strain compared to 130b. Although this result requires confirmation, it suggests that the Lsp Type II secretion system is involved in the export of Lpg1905 (Fig. 3E).

CONCLUSION

L. pneumophila 130b possesses four ORFs that are predicted to encode products with homology to eukaryotic but not prokaryotic proteins. Here we show that all are expressed in stationary phase and that the product of *lpg1905* is secreted and appears to play a role in host cell attachment and entry. Export of Lpg1905 does not require the Dot/Icm system and may involve the Lsp type II secretion system. Lpg1905 shares significant similarity with eukaryotic ecto-NTP/NDPases and possesses five apyrase conserved regions, suggesting that Lpg1905 may also be able to perform this function. The role of NTPDases in eukaryotes varies but includes regulation of platelet aggregation by vascular NTPDases in humans (4) and control of protease glycosylation in *Caenorhabditis elegans* (5). Interestingly, two NTPDases are also found in the intracellular protozoan parasite *Toxoplasma gondii*, only one of which is associated with virulence (1).

Another of the eukaryotic-like proteins in this study, Lpg2644, was also secreted by *L. pneumophila* 130b in a process apparently independent of both the Dot/Icm secretion system and the Lsp secretion system, suggesting that other secretion systems present in *L. pneumophila* are likely to be involved. To date, the role of Lpg2644 in pathogenesis is unknown, as no defect for an Δ*lpg2644* mutant could be observed in THP-1 or A549 cell infections. However, the role of *lpg2644* and *lpg1905* may be further elucidated by virulence studies in amoebae and in the A/J mouse model of infection.

REFERENCES

1. **Asai, T., S. Miura, D. Sibley, H. Okabayashi, and T. Takeuchi.** 1995. Biochemical and molecular characterization of nucleoside triphosphate hydrolase isoenzymes from the parasitic protozoan *Toxoplasma gondii. J. Biochem. Chem.* **270:**11391–11397.

2. **Cazalet, C., C. Rusniok, H. Bruggemann, N. Zidane, A. Magnier, L. Ma, M. Tichit, S. Jarraud, C. Bouchier, F. Vandenesch, F. Kunst, J. Etienne, P. Glaser, and C. Buchrieser.** 2004. Evidence in the Legionella pneumophila genome for exploitation of host cell functions and high genome plasticity. *Nat. Genet.* **36:**1165–1173.

3. **Chien, M., I. Morozova, S. Shi, H. Sheng, J. Chen, S. M. Gomez, G. Asamani, K. Hill, J. Nuara, M. Feder, J. Rineer, J. J. Greenberg, V. Steshenko, S. H. Park, B. Zhao, E. Teplitskaya, J. R. Edwards, S. Pampou, A. Georghiou, I. C. Chou, W. Iannuccilli, M. E. Ulz, D. H. Kim, A. Geringer-Sameth, C. Goldsberry, P. Morozov, S. G. Fischer, G. Segal, X. Qu, A. Rzhetsky, P. Zhang, E. Cayanis, P. J. De Jong, J. Ju, S. Kalachikov, H. A. Shuman, and J. J. Russo.** 2004. The genomic sequence of the accidental pathogen *Legionella pneumophila. Science* **305:**1966–1968.

4. **Kaczmarek, E., K. Koziak, J. Sevigny, J. B. Siegel, J. Anrather, A. R. Beaudoin, F. H. Bach, and S. C. Robson.** 1996. Identification and characterization of CD39/vascular ATP diphosphohydrolase. *J. Biol. Chem.* **271:**33116–33122.

5. **Nishiwaki, N., Y. Kubota, Y. Chigira, S. K. Roy, M. Suzuki, M. Schvarzstein, Y. Jigami, N. Hisamoto, and K. Matsumoto.** 2004. An NDPase links ADAM protease glycosylation with organ morphogenesis in *C. elegans. Nat. Cell Biol.* **6:**31–37.

6. **Stone, B. J., and Y. A. Kwaik.** 1999. Natural competence for DNA transformation by *Legionella pneumophila* and its association with expression of type IV pili. *J. Bacteriol.* **181:**1395–1402.

ROLE OF *LEGIONELLA PNEUMOPHILA*-SPECIFIC GENES IN PATHOGENESIS

*Hayley J. Newton, Fiona M. Sansom,
Vicki Bennett-Wood, and Elizabeth L. Hartland*

62

There are more than 40 named species of *Legionella,* but 80 to 90% of Legionnaires' disease is caused by *L. pneumophila* serogroup 1 (11). *L. micdadei* and *L. longbeachae* are the next most common etiological agents of Legionnaires' disease and together account for approximately 2 to 5% of disease worldwide (11). Interestingly this trend is not universal, as approximately 30% of Legionnaires' disease in Australia and New Zealand is attributed to *L. longbeachae* (11). These epidemiological differences suggest that *L. pneumophila*, in particular serogroup 1, is more virulent than other *Legionella* species.

Several studies have compared the virulence traits of different *Legionella* spp., yet little is known about the genetic basis of these phenotypic differences. Unlike *L. pneumophila*, *L. micdadei* appears to replicate within a vacuole that does not recruit host rough endoplasmic reticulum (4, 6). *L. longbeachae* and *L. micdadei* replicate within human macrophage cell lines to a level comparable to *L. pneumophila*, but they show varying abilities to replicate within pro-

tozoan hosts (4, 9). In addition, non-*pneumophila* species of *Legionella* are less cytotoxic to a range of host cells (4, 6).

Southern hybridization analysis shows that strains of *L. micdadei* and *L. longbeachae* possess dot/icm homologues; however, the completeness, functionality, and expression of the type IV secretion system has not been thoroughly examined in either species (6, 8).

To investigate genetic differences between *L. pneumophila* and *L. micdadei*, we performed a low stringency genomic subtractive hybridization between a serogroup 1 isolate of *L. pneumophila* and a clinical isolate of *L. micdadei*. Initial characterization of the strains included electron microscopy which confirmed that *L. pneumophila* strain 02/41 establishes a characteristic replicative vacuole surrounded by rough endoplasmic reticulum within THP-1 cells, while *L. micdadei* 02/42 does not (data not shown).

Subtractive hybridization revealed 151 open reading frames (ORFs) present in *L. pneumophila* 02/41 and absent in *L. micdadei* 02/42. Sequence analysis of these ORFs showed that 36% corresponded to hypothetical proteins and ORFs with no significant homologues and no known function. The remaining clones had a range of predicted functions and included 16 ORFs with known

Hayley J. Newton, Fiona M. Sansom, and Elizabeth L. Hartland Australian Bacterial Pathogenesis Program, Department of Microbiology, Monash University, Victoria 3800, Australia. *Vicki Bennett-Wood* Australian Bacterial Pathogenesis Program, Department of Microbiology and Immunology, University of Melbourne, Victoria 3010, Australia.

or putative roles in virulence, 10 putative regulators, and 13 genes putatively involved in cell wall and lipopolysaccharide biosynthesis.

From this cohort of *L. pneumophila* ORFs, 40 were selected for an investigation of their prevalence among *Legionella* species using low stringency Southern hybridization. DNA from *L. pneumophila* isolates of various serogroups and strains of *L. longbeachae*, *L. micdadei*, *L. gormanii*, and *L. jordanis* were examined, and this showed that 12 of the 40 ORFs investigated were present only in *L. pneumophila*. Among these *L. pneumophila*-specific ORFs were a hypothetical protein, *lpg2222*; 2 genes identified previously, *ladC* and *lepB* (2, 10), two putative transcriptional regulators, *pleD* and

lpg1357; two putative zinc-dependent proteases, *lpg1176* and *lpg2977*, the latter sharing homology with virulence-associated proteases, and ORFs predicted to be involved in cell wall and lipopolyssacharide synthesis, including *orf3*, *orf21*, and *orf22* of the known LPS biosynthesis gene cluster (7); a putative lipid A biosynthesis acyltransferase (*waaM*); and a putative N-acetylmuramoyl-L-alanine amidase (*lpg2699*). Two of these ORFs, *ladC* and *lpg2222*, were inactivated through insertion of a kanamycin cassette, and the subsequent mutants were characterized using mammalian cell culture models of *L. pneumophila* infection.

LadC is predicted to be an adenylate cyclase, as it possesses an intact HAMP signal

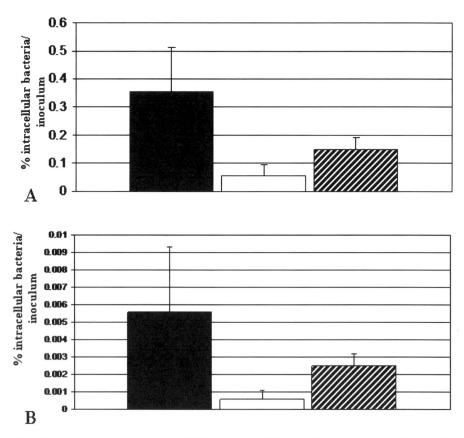

FIGURE 1   *L. pneumophila* entry into THP-1 (*A*) and A549 cells (*B*). *L. pneumophila* 130b (black), *L. pneumophila ladC::km* (white), and *L. pneumophila ladC::km* (pMIP:*ladC*) (stripes) are shown. *L. pneumophila ladC::km* shows statistically significant reduced entry in comparison to *L. pneumophila* 130b and *L. pneumophila ladC::km* (pMIP:*ladC*), *P* < 0.05.

transduction region and the catalytic domain for adenylate cyclases. Furthermore, this 55-kDa predicted protein has a putative N-terminal signal sequence and N-terminal transmembrane domain. An *L. pneumophila ladC::km* mutant replicated in THP-1 and A549 cells, although bacteria were recovered at lower numbers (data not shown). Closer examination revealed that this reflected a reduced level of entry into THP-1 cells that is also observed in A549 cells (Fig. 1). There are several mechanisms by which LadC, as a putative adenylate cyclase, could influence the invasion of host cells. LadC, through control of cAMP levels within the bacterial cell, may influence the regulation of *L. pneumophila* virulence determinants or it may play a direct role in host-pathogen interactions. Further investigation is required to elucidate the exact mechanism which LadC contributes to the relationship between *L. pneumophila* and eukaryotic cells.

Lpg2222 is a hypothetical 41-kDa protein that shares similarity with EnhC. The protein is predicted to contain an N-terminal signal sequence and eight tetratricopeptide repeats (TPRs). Interestingly, an *L. pneumophila lpg-2222::km* mutant also showed reduced entry into THP-1 and A549 cells (Fig. 2). We have termed this protein LpnE, for *L. pneumophila* entry protein. The finding that LpnE is required for full entry of *L. pneumophila* into host cells correlates with its homology to EnhC, which also enhances the ability of *L. pneumophila* to invade host cells. These phenotypes are presumably related to the presence of multiple TPRs (3). TPRs are associated with a range of functions in eukaryotic cells through their ability to mediate protein-protein interactions (5). We predict that these regions may be important for LpnE and EnhC-mediated interaction of *L. pneumophila* with host cells either via direct interaction with host cell proteins or through interactions with other *L.*

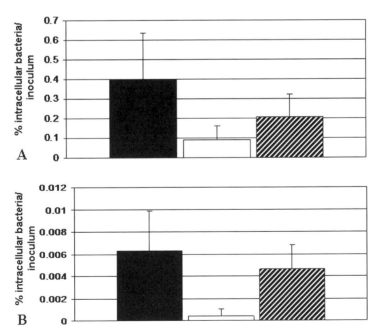

FIGURE 2 *L. pneumophila* entry into THP-1 (*A*) and A549 cells (*B*). *L. pneumophila* 130b (black), *L. pneumophila lpnE::km* (white), and *L. pneumophila lpnE::km* (pMIP:*lpnE*) (stripes) are shown. *L. pneumophila lpnE::km* shows statistically significant reduced entry in comparison to *L. pneumophila* 130b and *L. pneumophila lpnE::km* (pMIP:*lpnE*), $P < 0.05$.

pneumophila virulence determinants. Furthermore, the genome sequences have revealed that *L. pneumophila* possesses another three putative proteins with TPRs, including LidL, which may also contribute to the pathogen's interaction with eukaryotic cells (1).

Both *L. pneumophila ladC::km* and *L. pneumophila lpnE::km* demonstrate levels of association to THP-1 and A549 cells comparable to wild type, confirming that these virulence determinants are directly involved in *L. pneumophila* entering these host cells rather than initial attachment.

In summary, this investigation has demonstrated that *L. pneumophila* possesses virulence determinants that are absent in other *Legionella* species. Specifically, LadC, a putative adenylate cyclase, and LpnE, a protein with eight predicted TPRs, are both important for the entry of *L. pneumophila* into THP-1 and A549 cells.

REFERENCES

1. Cazalet, C., C. Rusniok, H. Bruggemann, N. Zidane, A. Magnier, L. Ma, M. Tichit, S. Jarraud, C. Bouchier, F. Vandenesch, F. Kunst, J. Etienne, P. Glaser, and C. Buchrieser. 2004. Evidence in the *Legionella pneumophila* genome for exploitation of host cell functions and high genome plasticity. *Nature Genet.* **36**:1165–1173.
2. Chen, J., K. S. de Felipe, M. Clarke, H. Lu, O. R. Anderson, G. Segal, and H. A. Shuman. 2004. *Legionella* effectors that promote nonlytic release from protozoa. *Science* **303**:1358–1361.
3. Cirillo, S. L., L. J., and J. D. Cirillo. 2000. Identification of novel loci involved in entry by *Legionella pneumophila*. *Microbiology* **146**:1345–1359.
4. Gao, L. Y., M. Susa, B. Ticac, and Y. Abu-Kwaik. 1999. Heterogeneity in intracellular replication and cytopathogenicity of *Legionella pneumophila* and *Legionella micdadei* in mammalian and protozoan cells. *Microb. Pathog.* **27**:273–287.
5. Goebl, M., and M. Yanagida. 1991. The TPR snap helix: a novel protein repeat motif from mitosis to transcription. *Trends Biochem. Sci.* **16**:173–177.
6. Joshi, A. D., and M. S. Swanson. 1999. Comparative analysis of *Legionella pneumophila* and *Legionella micdadei* virulence traits. *Infect. Immun.* **67**:4134–4142.
7. Luneberg, E., N. Zetzmann, D. Alber, Y. A. Knirel, O. Kooistra, U. Zahringer, and M. Frosch. 2000. Cloning and functional characterization of a 30 kb gene locus required for lipopolysaccharide biosynthesis in *Legionella pneumophila*. *Int. J. Med. Microbiol.* **290**:37–49.
8. Morozova, I., X. Qu, S. Shi, G. Asamani, J. E. Greenberg, H. A. Shuman, and J. J. Russo. 2004. Comparative sequence analysis of the *icm/dot* genes in *Legionella*. *Plasmid* **51**:127–47.
9. Neumeister, B., M. Schoniger, M. Faigle, M. Eichner, and K. Deitz. 1997. Multiplication of different *Legionella* species in Mac 6 cells and in *Acanthamoebae castellanii*. *Appl. Environ. Microbiol.* **63**:1219–1224.
10. Rankin, S., Z. Li, and R. R. Isberg. 2002. Macrophage-Induced Genes of *Legionella pneumophila*: Protection from Reactive Intermediates and Solute Imbalance during Intracellular Growth. *Infect. Immun.* **70**:3637–3648.
11. Yu, V. L., J. F. Plouffe, M. C. Pastoris, J. E. Stout, M. Schousboe, A. Widmer, J. Summersgill, T. File, C. M. Heath, D. L. Paterson, and A. Chereshsky. 2002. Distribution of *Legionella* species and serogroups isolated by culture in patients with sporadic community-acquired pneumonia: an international collaborative study. *J. Infect. Dis.* **186**:127–128.

THE Hsp60 CHAPERONIN OF *LEGIONELLA PNEUMOPHILA*: AN INTRIGUING PLAYER IN INFECTION OF HOST CELLS

Audrey Chong, Angela Riveroll, David S. Allan,
Elizabeth Garduño, and Rafael A. Garduño

63

L. pneumophila is an intracellular bacterial pathogen that alters the vesicular and organellar traffic of its host cell, leading to the establishment of a specialized vacuole where it replicates (7). *L. pneumophila* effectors responsible for cell entry and for the alteration of organelle trafficking have just begun to be described and thus remain poorly understood. However, it is known that a functional Dot/Icm type IV secretion system is required for *L. pneumophila* to alter the vesicular and organellar traffic of the host cell (1, 2). Likely, this is because the *L. pneumophila* Dot/Icm system is required for the secretion of virulence effector proteins to the bacterial cell surface, or for the translocation of such effector proteins across the plasma membrane (or the vacuolar membrane) of the host cell (10, 11).

The 60-kDa chaperonins (also known as Hsp60 proteins) are a family of highly conserved proteins, present in all cellular forms of life, whose main function is to help other proteins fold properly (14). In bacteria, the protein-folding molecular machine consists of a barrel-shaped, Hsp60 14-mer, associated with a 7-mer ring of the cochaperonin Hsp10. The large size of this molecular machine and the lack of known protein secretion signals in all Hsp60 proteins characterized to date suggest that chaperonins are cytoplasmic proteins. However, we have found that 30 to 50% of the *L. pneumophila* Hsp60 appears to be extracytoplasmic (4) and may even be found on the bacterial cell surface (5) and free in the lumen of the vacuole where *L. pneumophila* replicates in HeLa cells (4, 6), although not necessarily as a multimer. By exploring potential virulence functions of the surface-exposed (extracellular) Hsp60, we previously determined that it mediates invasion of HeLa cells by *L. pneumophila* (5). Here we report that through genetic and functional studies in yeast and mammalian cells, we have also demonstrated that *L. pneumophila* Hsp60 alters signaling cascades in eukaryotic cells, modifies the actin cytoskeleton of mammalian cells, and alters the trafficking of mitochondria in mammalian cells, making Hsp60 a potential effector in *L. pneumophila*'s intracellular establishment. For clarity, we will subsequently refer to the *L. pneumophila* Hsp60 as HtpB, to distinguish it from the *E. coli* Hsp60 known as GroEL. This distinction is necessary

Audrey Chong, Angela Riveroll, David S. Allan, and Elizabeth Garduño Department of Microbiology and Immunology, Sir Charles Tupper Medical Building, Dalhousie University, Halifax, NS, Canada, B3H-1X5. *Rafael A. Garduño* Department of Microbiology and Immunology, and Division of Infectious Diseases—Department of Medicine, Sir Charles Tupper Medical Building, Dalhousie University, Halifax, NS, Canada, B3H-1X5.

due to the functional differences between these two chaperonins, as detailed below.

HOW DOES THE *L. PNEUMOPHILA* Hsp60 (HtpB) GET TO THE BACTERIAL CELL SURFACE?

In looking for possible mechanisms by which HtpB can be secreted, we first explored whether known secretion systems present in *L. pneumophila* were involved. The Philadelphia-1 *L. pneumophila* strain Lp02 possesses at least two type IV secretion systems (1, 3), one of which is the Dot/Icm system described above as being essential for virulence. We thus decided to determine whether defined, loss-of-function mutations in *dot* genes would have an impact on the subcellular localization of HtpB. Through the use of a protease sensitivity assay that assesses exposure of HtpB on the surface of *L. pneumophila* (5), we determined that *dotA* and *dotB* mutant derivatives of Lp02 had virtually no protease-accessible HtpB; i.e., these mutants displayed no surface-exposed HtpB (Fig. 1). In addition, immunogold electron microscopy, performed as previously described with an HtpB-specific rabbit antiserum and a secondary gold-conjugated antibody against rabbit immunoglobulin (4), showed that the Lp02 *dot* mutants, particularly the *dotB* mutant, had an increased amount of periplasmic gold particles (Fig. 2). This interesting result suggested that a nonfunctional Dot/Icm system results in the accumulation of HtpB in the periplasm. Our interest derived from the additional implication that HtpB appears to be translocated to the *L. pneumophila* cell surface by a Dot/Icm-related mechanism via the periplasm. Therefore, we set out to determine whether HtpB was found in the periplasm of the wild-type Lp02 strain and identify its derivative *dotA* and *dotB* mutants. Immunoblot analysis of concentrates containing proteins released from the periplasm by osmotic shock (12) confirmed that approximately 1% of the total cell-associated HtpB is free in the periplasm of *L. pneumophila* strain Lp02. In addition, osmotic shock concentrates from the *dotA* mutant appeared to have about two-fold more

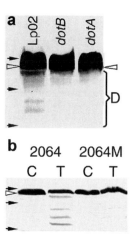

FIGURE 1 Immunoblots of sodium dodecylsulfate-polyacrylamide gel eclectrophoresis-resolved proteins from whole bacterial cells treated with trypsin (250 µg trypsin/10^9 bacteria) and immunostained with an HtpB-specific monoclonal antibody and alkaline phosphatase-conjugated rabbit antimouse IgG. (*a*) Band patterns produced in the wild-type strain Lp02 and its two derivatives JV303 (*dotB* mutant) and JV309 (*dotA* mutant). The bracket labeled with a D marks the region where HtpB degradation products appear, and the hollow arrowheads point at full-size nondegraded HtpB. (*b*) Band patterns showing a clear degradation of trypsin-accessible HtpB in the virulent strain 2064 (used as a positive control for the assay [5]) and the lack of trypsin-accessible HtpB in the 2064M salt-tolerant mutant (used as a negative control for the assay [5]). C, control, trypsin-free sample; T, trypsin-treated sample. The hollow arrowhead points at full-size nondegraded HtpB. Black arrows at the left side of both panels point to the position of three of the broad-range prestained protein size markers (New England Biolabs): from top to bottom, 62-, 47-, and 37.5-kDa, respectively.

periplasmic HtpB than its parent strain, Lp02, as determined by densitometric analysis of the HtpB protein band immunostained with an HtpB-specific monoclonal antibody. The *dotB* mutant had between two- and four-fold more periplasmic HtpB than its parent strain, Lp02. In summary, a small proportion of the total HtpB present in *L. pneumophila* cells seems to access the periplasm (by a still unknown mechanism) and from there reach the surface by a Dot/Icm system-related mechanism.

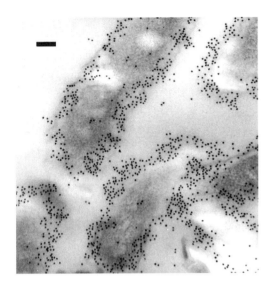

FIGURE 2 Immunoelectron microscopy of ultra-thin sections of *L. pneumophila* Lp02 derivative JV303 (*dotB* mutant). Sectioned bacteria were labeled with HtpB-specific polyclonal antibody and antirabbit goat IgG conjugated with gold spheres. Gold spheres are ∼10 nm in diameter and appear as well-defined black dots on the micrograph. Notice the abundant clustering-banding of gold particles at the periphery of the bacterial sections, corresponding to the bacterial cell envelope and periplasmic regions. Labeling of the Lp02 parent strain results in a single row of gold particles lining the outer membrane (not shown), but not in clustering of abundant gold particles. The size bar represents 100 nm.

BESIDES MEDIATING INVASION OF HELA CELLS BY *L. PNEUMOPHILA*, WHAT OTHER VIRULENCE-RELATED FUNCTIONS CAN EXTRACELLULAR HtpB PERFORM?

Since HtpB is an essential protein, we have been unable to genetically delete the *htpB* gene to study its involvement in *L. pneumophila* virulence functions. However, we have developed several functional models to allow us to identify and subsequently characterize novel HtpB functions. Here, we describe some results derived from the application of three functional models: (i) the expression of recombinant HtpB in the cytoplasm of the genetically tractable eukaryote *Saccharomyces cerevisiae*, (ii) the expression of recombinant HtpB in the cytoplasm of the Chinese hamster ovary cell line (CHO), and (iii) the production of HtpB-coated mi-

crospheres to study the "from without" effects of HtpB.

HtpB Induces Pseudohyphal Growth in the Budding Yeast *S. cerevisiae*

Recombinant HtpB (expressed in yeast from the pEMBLyex4 plasmid carrying the *htpB* gene under the control of its galactose-inducible promoter) caused yeast cells to elongate, bud in a unipolar fashion to form filaments, and penetrate solid medium, a phenotype collectively termed pseudohyphal growth. Recombinant GroEL (the Hsp60 of *E. coli*) did not induce pseudohyphal growth when expressed in the cytoplasm of *S. cerevisiae*, suggesting the uniqueness of HtpB. *S. cerevisiae* tightly controls the induction of pseudohyphal growth, which naturally occurs during mating (sexual reproduction) and in response to nitrogen starvation (9). That a chaperonin would trigger this tightly regulated physiological process in yeast is indeed mechanistically appealing. Therefore, we set out to determine whether HtpB induces pseudohyphal growth in *S. cerevisiae* through the signaling pathways known to control this process in response to natural signals.

In *S. cerevisiae*, the Ras2 global regulator controls two separate signaling pathways involved in pseudohyphal growth: the Ste MAP-kinase and the cAMP/PKA pathways (9). Yeast mutants with single deletions in members of these known Ras2p-activated pathways could still elongate and bud in a unipolar fashion but could not form long pseudohyphae or invade solid medium upon expression of recombinant HtpB. However, when HtpB was expressed in a *ras2* deletion mutant, no pseudohyphal growth was observed at all. These data indicate for the first time that HtpB has the capacity to specifically interfere with eukaryotic signaling pathways and show the unique ability of HtpB to initiate a developmental change in *S. cerevisiae* by acting upstream or at the level of the Ras2 global regulator. The potential meaning of this interaction for the pathogenesis of *L. pneumophila* remains to be established. In the meantime, the yeast expression system remains as a useful functional model to study the unique biochemistry of HtpB in relation to eukaryotic cell signaling.

HtpB Alters the Organization of the Actin Cytoskeleton in Mammalian Cells

Recombinant HtpB (expressed in CHO cells stably transfected after integration of the pTRE-2hyg plasmid carrying the *htpB* gene under the control of the Tet controlled transactivator) caused CHO cells to lose their stress fibers (Fig. 3a and b). In addition, HtpB-expressing cells appeared to have an increased amount of cortical polymerized actin showing a "framing" effect (Fig. 3b).

To further study this intriguing effect of HtpB in mammalian cells, we used the yeast two-hybrid method, aimed at identifying potential molecular partners with which HtpB may interact. A commercially available HeLa cell cDNA library randomly fused to the Gal4 activation domain, and kept in the *S. cerevisiae* yeast strain Y187 (Matchmaker, BD Biosciences, Palo Alto, Calif.), was screened with the "bait" plasmid carrying a fusion between *htpB* and the Gal4 DNA-binding domain. Positive interactions between HtpB and putative HeLa cell proteins were selected through the expression of four reporter genes: *HIS3, ADE2* (which allows growth in media without adenine and histidine), *MEL1*, and *lacZ* (which allows the breakdown of the chromogenic substratum X-alpha-gal). Clones that formed blue colonies on media containing X-alpha-gal, but lacking adenine and histidine, were considered to express at least one HeLa cell protein that physically interacted with HtpB. Interestingly, one of the double-screened positive clones carried the cDNA for Merlin-associated protein. Merlin-associated protein binds to Merlin and may play a cooperative role in Merlin-mediated functions (8). Merlin, in turn, is a protein closely related to ezrin, radixin, and moesin, which are active actin reorganizers, particularly of cortical actin (13, 15). As exciting as these results are, the interaction of HtpB with actin-reorganizing proteins requires further confirmation through various biochemical, cell biological, functional, and genetic approaches.

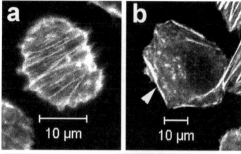

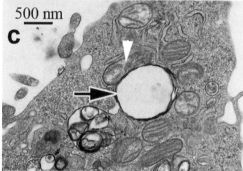

FIGURE 3 Effects of HtpB on mammalian cells as evaluated through fluorescence microscopy (*a* and *b*) or transmission electron microscopy (*c*). Panels a and b are images of CHO cells stained with fluorescently labeled phalloidin (which binds to polymerized actin), taken in a Zeiss confocal microscope. CHO cells normally show numerous stress fibers (clearly seen as bright lines going across the CHO cell in panel a. When expressed as a recombinant protein in the cytoplasm of CHO cells, HtpB causes the disappearance of actin stress fibers as shown in panel b. In addition, HtpB causes the appearance of peripheral bundles of polymerized actin (e.g., white arrowhead in panel b), an effect that we have called "framing." (c) Electron micrograph of a sectioned CHO cell showing an internalized HtpB-coated microsphere (black arrow) surrounded by numerous mitochondria. Notice that some mitochondria (e.g., white arrowhead) come in close apposition to the membrane surrounding the internalized microsphere.

HtpB-Coated Microspheres Alter Organellar and Vesicular Trafficking in Host Cells

The two series of experiments described above rely on the expression of recombinant HtpB in the cytoplasm of eukaryotic cells. However, it remains to be determined whether or not HtpB is translocated into the cytoplasm of the host cell during *L. pneumophila* infection. To better mimic the way a host cell would encounter HtpB during *L. pneumophila* infection, we used HtpB-coated microspheres. That is, microspheres would display HtpB in a fashion similar to bacterial cells expressing HtpB on their cell surfaces. We surmised that this experimental model could allow us to study (or mimic) some of the effects that HtpB may exert upon mammalian cells during the early stages of *L. pneumophila* infection. Because we have already identified HtpB as a mediator of invasion of nonphagocytic cells by *L. pneumophila* (5), here we have focused on postinternalization events.

The intracellular trafficking of fluorescent microspheres was examined by both electron and fluorescence microscopy. Transmission electron microscopy clearly showed that vacuoles containing internalized HtpB-coated microspheres were associated with numerous mitochondria (Fig. 3c). In contrast, vacuoles containing control BSA-coated microspheres did not recruit mitochondria as frequently, or in the same numbers, as HtpB-coated microspheres. These observations were confirmed by fluorescence microscopic analysis, in which fluorescent blue HtpB-coated microspheres were clearly surrounded by red mitochondria stained with Mitotracker red (Molecular Probes). At 1 h after adding the microspheres to our CHO cells, 53 ± 3% of the HtpB-coated microspheres colocalized with, or were surrounded by, mitochondria, whereas 30 ± 5% of the BSA-coated microspheres were surrounded by mitochondria. It should be noted here that the scoring of positive events was based only on the presence of surrounding mitochondria, regardless of the number of mitochondria present. Interestingly, GroEL was not able to produce the same effect as HtpB since only 38 ± 5% of the GroEL-coated microspheres were surrounded by mitochondria. This effect was not due to the presence of too many HtpB-coated microspheres in the cytoplasm (so that microspheres would be forced by space constraints to be around mitochondria), since the number of microspheres per cell was similar among the differently coated microspheres.

To determine whether the coated microspheres would be delivered to lysosomes, we analyzed the colocalization of microspheres with lysosomes preloaded with the fluorescent complex Texas red-Ovalbumin (or TrOv). At 1 h after addition of the microspheres, only 23 ± 2% of the HtpB-coated microspheres colocalized with TrOv, whereas 32 ± 3% of the BSA-coated microspheres colocalized with TrOv. Again, GroEL was unable to prevent fusion with lysosomes to the same extent as HtpB did, since 29 ± 1% of the GroEL-coated microspheres colocalized with TrOv. These results were confirmed in bone marrow-derived macrophages, where the HtpB-coated microspheres showed the lowest levels of colocalization with TrOv, in relation to BSA- and GroEL-coated microspheres. In summary, internalized HtpB-coated microspheres were able to recruit mitochondria and avoid, to some extent, their delivery to lysosomes. These two postinternalization HtpB-mediated events (recruitment of mitochondria and abrogation of endosome-lysosome fusion) are likely relevant to the intracellular establishment of *L. pneumophila* in host cells, since these events constitute a hallmark of infection by wild type strains of *L. pneumophila* (7).

CONCLUSION

Previous results from our laboratory indicate that a fraction of the *L. pneumophila* chaperonin, HtpB, is surface-exposed and mediates invasion of HeLa cells by *L. pneumophila* (4, 5). In addition, the fact that *L. pneumophila* copiously releases HtpB into the lumen of the replicative vacuole, as determined by immunofluorescence microscopy (6) or immunogold electron microscopy (4), suggests that HtpB may have other virulence-related functions. HtpB is an essential protein and, therefore, its encoding gene cannot be deleted from the *L. pneumophila* genome to study its involvement in *L. pneumophila* pathogenesis. However, we have developed several functional models that have allowed us to study some of the activities that HtpB may mediate in relation to the intracellular establishment of *L. pneumophila*. If further experimentation confirms the role of Dot/Icm in the translocation of HtpB to the cell surface, HtpB will join the group of Dot/Icm-secreted effectors that may contribute to the intracellular establishment of *L. pneumophila*. The abilities of HtpB to specifically alter eukaryotic signaling pathways, cytoskeletal organization, and organellar traffic are indeed functional characteristics that fit well into HtpB's potential role as an *L. pneumophila* virulence effector.

REFERENCES

1. **Andrews, H. L., J. P. Vogel, and R. R. Isberg.** 1998. Identification of linked *Legionella pneumophila* genes essential for intracellular growth and evasion of the endocytic pathway. *Infect. Immun.* **66:**950–958.

2. **Brand, B. C., A. B. Sadosky, and H. A. Shuman.** 1994. The *Legionella pneumophila icm* locus: a set of genes required for intracellular multiplication in human macrophages. *Mol. Microbiol.* **14:** 797–808.

3. **Brassinga, A. K. C., M. F. Hiltz, G. R. Sisson, M. G. Morash, N. Hill, E. Garduño, P. H. Edelstein, R. A. Garduño, and P. S. Hoffman.** 2003. A 65-kilobase pathogenicity island is unique to Philadelphia-1 strains of *Legionella pneumophila*. *J. Bacteriol.* **185:**4630–4637.

4. **Garduño, R. A., G. Faulkner, M. A. Trevors, N. Vats, and P. S. Hoffman.** 1998. Immunolocalization of Hsp60 in *Legionella pneumophila*. *J. Bacteriol.* **180:**505–513.

5. **Garduño, R.A., E. Garduño, and P. S. Hoffman.** 1998. Surface-associated Hsp60 chaperonin of *Legionella pneumophila* mediates invasion in a HeLa cell model. *Infect. Immun.* **66:**4602–4610.

6. **Hoffman, P. S., L. Houston, and C. A. Butler.** 1990. *Legionella pneumophila htpAB* heat shock operon: nucleotide sequence and expression of the 60-kilodalton antigen in *L. pneumophila*-infected HeLa cells. *Infect. Immun.* **58:**3380–3387.

7. **Horwitz, M. A.** 1983. Formation of a novel phagosome by the Legionnaires' disease bacterium (*Legionella pneumophila*) in human monocytes. *J. Exp. Med.* **158:**1319–1331.

8. **Lee, I. K., K. S. Kim, H. Kim, J. Y. Lee, C. H. Ryu, H. J. Chun, K. U. Lee, Y. Lim, Y. H. Kim, P. W. Huh, et. al.** 2004. MAP, a protein interacting with a tumor suppressor, merlin, through the run domain. *Biochem. Biophys. Res. Commun.* **325:**774–783.

9. **Lengeler, K. B., R. C. Davidson, C. D'Souza, T. Harashima, W. C. Shen, P. Wang, X. Pan, M. Waugh, and J. Heitman.** 2000. Signal transduction cascades regulating fungal development and virulence. *Microbiol. Mol. Biol. Rev.* **64:**746–785.

10. **Luo, Z. -Q., and R. R. Isberg.** 2004. Multiple substrates of the *Legionella pneumophila* Dot/Icm system identified by interbacterial protein transfer. *Proc. Natl. Acad. Sci. USA* **101:**841–846.

11. **Nagai, H., J. C. Kagan, X. Zhu, R. A. Kahn, and C. R. Roy.** 2002. A bacterial guanine nucleotide exchange factor activates ARF on *Legionella* phagosomes. *Science* **295:**679–682.

12. **Nossal, N. G., and L. A. Heppel.** 1966. The release of enzymes by osmotic shock from *Escherichia coli* in exponential phase. *J. Biol. Chem.* **241:** 3055–3062.

13. **Rouleau, G. A., P. Merel, M. Lutchman, M. Sanson, J. Zucman, C. Marineau, K. Hoang-Xuan, S. Demczuk, C. Desmaze, B. Plougastel, et al.** 1993. Alteration in a new gene encoding a putative membrane-organizing protein causes neuro-fibromatosis type 2. *Nature* **363:**515–521.

14. **Sigler, P. B., Z. Xu, H. S. Rye, S. G. Burston, W. A. Fenton, and A. L. Horwich.** 1998. Structure and function in GroEL-mediated protein folding. *Annu. Rev. Biochem.* **67:**581–608.

15. **Takeuchi, K., A. Kawashima, A. Nagafuchi, and S. Tsukita.** 1994. Structural diversity of band 4.1 superfamily members. *J. Cell Sci.* **107:**1921–1928.

LIPOPOLYSACCHARIDE ARCHITECTURE OF *LEGIONELLA PNEUMOPHILA* GROWN IN BROTH AND HOST CELLS

*Jürgen H. Helbig, Esteban Fernandez-Moreira,
Enno Jacobs, Paul Christian Lück, and Martin Witt*

64

The involvement of lipopolysaccharide (LPS) in *Legionella pneumophila* pathogenesis has been under consideration since a phase-variable expression was found in a strain belonging to *L. pneumophila* serogroup 1, monoclonal subgroup OLDA (5, 6). For this individual strain, the phenotypic LPS variations were caused by chromosomal insertion and excision of a 30 kb unstable genetic element (4). Phase variable LPS architecture, however, was not described for the wild-type strains mostly used in pathogenic studies. These strains belong to *L. pneumophila* serogroup 1, monoclonal subgroup Philadelphia (Lp02, Wadsworth 130b, AM 511), or monoclonal subgroup Knoxville (Corby) and carry the so-called virulence-associated LPS component which can be recognized by the monoclonal antibody (MAb) 3/1 of the Dresden panel (1). All mutants described to influence pathogenicity possess nearly the same genomic equipment for LPS synthesis. Therefore, it cannot be ruled out that some of the gene loci characterized as being essential for intracellular replication can also code for vectors of LPS impact on host cell modulation. Regarding the LPS involvement in pathogenicity, a second point to mention is the inability to find rough mutants that confirm an essential role for the complete LPS. Due to the absence of mutants with major LPS changes, current investigations of the LPS contribution to virulence are limited mainly to phenotypic methods. However, pathogenic studies have rarely focused on morphological changes of the bacterial surface and its immediate surroundings, although these structures initiate the first step of interaction between bacteria and host cells. We demonstrate here for the first time to what extent the LPS architecture is involved in extracellular structures of *L. pneumophila* bacteria. Furthermore, the intraphogosomal/intracellular shedding of LPS components is investigated.

DotA AND *letA* MUTANTS GROWN IN BROTH POSSESS CHANGED LPS STRUCTURES

The *L. pneumophila* strain Lp02 and its *dotA* and *letA* mutants were grown in broth until different growth phases. Slide preparations were made using the Cytospin technique (Shandon). After fixation with Leucoperm (Serotec, Düsseldorf, Germany) LPS was labeled with

Jürgen H. Helbig, Enno Jacobs, and Paul Christian Lück   Medical Faculty TU Dresden, Institute of Medical Microbiology and Hygiene, D-01307 Dresden, Germany.   *Esteban Fernandez-Moreira*   University of Michigan Medical School, Department of Microbiology and Immunology, Ann Arbor, MI, 48109.   *Martin Witt*   Medical Faculty TU Dresden, Institute of Anatomy, D-01307 Dresden, Germany.

MAb 3/1 followed by antimouse IgG FITC (Sigma, Munich, Germany). Fig. 1A to 1D illustrate continuous-tone graphs of the LPS architecture of Lp02, *dotA*, and *letA*. For the exponential growth phase, it is typical that small LPS vesicles of strain Lp02 are still connected to the cell wall, but released lamellar LPS structures were also observed (Fig. 1A). In the postexponential growth phase vesicles were massively released (Fig. 1B). Nevertheless, lamellar structures of LPS are partly associated with the bacterial cell wall. The *DotA* mutant also shed LPS structures intensively in the postexponential growth phase, but caused by defects in the secretion apparatus, its architecture is different when compared with the wild type (Fig. 1C). The vesicles are distinctly smaller, the releasing is focused on particular outer membrane segments, and released LPS components may build circular structures. In the exponential growth phase the

letA mutant was already slightly stringy. In the postexponential phase it developed filamentous and capsule-like LPS-containing structures (Fig. 1D).

Our investigation of the surface architecture of strain Lp02 and its *dotA* and *letA* mutants substantiate that the LPS equipment depends at least on both genes. Therefore, the valuation of the influence of *dotA* and *letA* on pathogenicity also has to consider the phenotypic changes of the LPS surface structures, because they initiate the first step of interaction between bacteria and host cells.

L. PNEUMOPHILA SHED LPS DURING THE EXPONENTIAL GROWTH PHASE IN MOLECULAR FORM

In order to investigate in which form LPS is shed in the exponential growth phase, aliquots of broth cultures from *L. pneumophila* strains

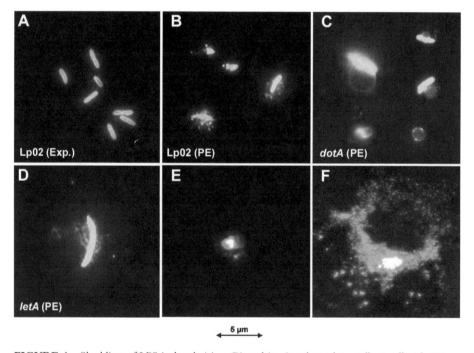

FIGURE 1 Shedding of LPS in broth (*A* to *D*) and in *Acanthamoeba castellanii* cells 6 h (*E*) or 19 h (*F*) after infection with *L. pneumophila* strain Corby. LPS structures were labeled with MAb 3/1 (*A* to *D*) or double-labeled with MAb 59/1 plus MAb 3/1 (*E* and *F*). Objects positive for both antibodies appear white; only MAb 3/1-positive structures appear grey. MAb 59/1-positive only was not obtained. Exp, exponential growth phase; PE, postexponential growth phase. For details, see text.

Lp02 and Corby were collected after diafiltration through Vivaspin 300 kDa (Vivascience, Hannover, Germany) or after sterile filtration (0.2 μm and 0.45 μm). The amounts of LPS were quantified by enzyme-linked immunosorbent assay. LPS was immobilized in microtiter wells and detected by MAb 3/1 followed by antimouse-IgG horseradish peroxidase. Approximately 75% of the total released LPS were found in the fraction obtained by Vivaspin, with an exclusion size of 300 kDa. This indicates that LPS is shed mainly in molecular form during the exponential phase and not as part of complex outer membrane vesicles. Interestingly, a second LPS component recognized by our MAb 59/1 could not be detected in any fraction. MAb 3/1-positive strains possess this component only as short O-chain LPS (data not shown). Moreover, until the postexponential growth phase in broth cultures only approximately 5% of bacteria of strains Corby and Lp02 can be labeled by MAb 59/1. With initiation of the browning phenomenon, the clouds of bacteria render MAb 59/1 positive (data not shown).

The intracellular LPS shedding was investigated after infection of *Acanthamoeba castellanii* and human phagocytic U937 cells with strain Corby. Fig. 1E shows a typical amoebal phagosome 8 h after infection. The white cluster of the continuous-tone graphs represents the intraphagosomal bacteria being positive by both MAb 3/1 (fluorescein isothiocyanta stained) and MAb 59/1 (Cy3 stained). Inside the phagosome, the bacteria are surrounded by LPS only carrying the epitope of MAb 3/1. After lysis of host cells we observed that legionellae were released either as bacterial clusters positive for both MAbs or as single MAb 3/1-positive bacteria being MAb 59/1-positive or negative (Fig. 1F).

The shedding of LPS in infected U937 cells was substantiated by electron microscopy. Fourteen hours after infection with strain Corby, bacteria containing phagosomes were detected by transmission microscopy (Fig. 2A). For localization of LPS components, paraformaldehyde-fixed cells were cryosubstituted and embedded in Lowicryl HM20. The presence of MAb 3/1-positive structures was detected by immunogold labeling. Fig. 2A and 2C demonstrate that the phagosomal membrane (large arrows) was not destroyed. In this stage, shed LPS is not only detectable inside the phagosome (Fig. 2B/C) but also when crossing the phagosomal membrane (ellipse) and even in the host cell cytoplasm (circle). The immunogold technique does not allow reliable differentiation between vesicles and soluble LPS. But keeping in mind that purified vesicles have an average diameter of more than 200 nm (Fernandez-Moreira, unpublished data), it can be assumed that during this stage of infection

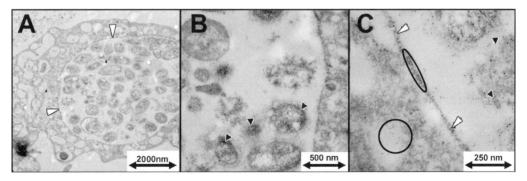

FIGURE 2 Transmission electron micrograph (A) and immunogold labeling (B, C) of U937 cells infected by *L. pneumophila* strain Corby (14 h postinfection). The *Legionella* LPS was labeled with MAb 3/1. Distinguishing, large white arrows; phagosome membrane (*A* and *C*), small black arrows; LPS on bacterial surface (*B*), ellipse; shed LPS crossing the phagosomal membrane (*C*), circle; shed LPS inside the cytoplasm (*C*).

the shed LPSs inside the cytoplasm are soluble components that have crossed the as-far-as-possible intact phagosomal membrane and take part in the complex host cell modulation or only in degradation of the cytoskeleton. Therefore, further investigation should be directed at finding the subcellular host cell targets in order to characterize the involvement of LPS in pathogenesis. We assume that wrapping up of LPS in vesicles and its transport through egress pores is the second step for initiating the necrotic process.

The ability to modify and shed LPS allows *L. pneumophila* to build lamellar structures and vesicles and to release soluble LPS that may modulate host cells, enhance survival in aerosols, or stabilize biofilms. At least the two last processes were triggered by high hydrophobic distinctive features caused by the *lag1* gene (7). This gene codes for an acetyltransferase conducted by the virulence-associated *O*-acetylation of long-chain *Legionella* LPS. Strains recognized by MAb 3/1 are responsible for nearly 80% of community-acquired or travel-associated legionellosis (2). Of course, the presence or absence of this LPS component alone cannot explain the high virulence of MAb 3/1-positive strains. In this study we have demonstrated that the whole LPS architecture is influenced by at least two systems, namely the type IV secretion apparatus and *letA*/*letS* regulatory elements.

ACKNOWLEDGMENTS

This work was supported in part by a grant from the Deutsche Forschungsgemeinschaft to J. H. (HE 2160/6-1).

The authors are grateful to Sigrid Gäbler, Martina Rossmann, Ines Wolf, and Volker Bellmann for excellent technical assistance.

REFERENCES

1. **Helbig, J. H., P. C. Lück, Y. A. Knirel, W. Witzleb, and U. Zähringer.** 1995. Molecular characterization of a virulence-associated epitope on the lipopolysaccharide of *Legionella pneumophila* serogroup 1. *Epidemiol. Infect.* **115:**71–78.
2. **Helbig, J. H., S. Bernander, M. Castellani Pastoris, J. Etienne, V. Gaia, S. Lauwers, D. Lindsay, P. C. Lück, T. Marques, S. Mentula, M. F. Peeters, C. Pelaz, M. Struelens, S. A. Uldum, G. Wewalka, and T. G. Harrison.** 2002. Pan-European study on culture-proven Legionnaires' disease: distribution of *Legionella pneumophila* serogroups and monoclonal subgroups. *Eur. J. Clin. Microbiol Infect. Dis.* **21:**710–716.
3. **Lüneberg, E., B. Meyer, N. Daryab, O. Kooistra, U. Zähringer, M. Rhode, J. Swanson, and M. Frosch.** 2001. Chromosomal insertion and excision of a 30 kb unstable genetic element is responsible for phase variation of lipopolysaccharide and other virulence determinants in *Legionella pneumophila. Mol. Microbiol.* **39:**1259–1271.
4. **Lüneberg, E., U. Zähringer, Y. A. Knirel, D. Steinmann, M. Hartmann, I. Steinmetz, M. Rohde, J. Köhl, and M. Frosch.** 1998. Phase-variable expression of lipopolysaccharide contributes to virulence of *Legionella pneumophila. J. Exp. Med.* **188:**49–60.
5. **Swanson, M. S., and E. Fernandez-Moreia.** 2002. A microbial strategy to multiply in macrophages: the pregnant pause. *Traffic* **3:**170–77.
6. **Zou, C. H., Y. A. Knirel, J. H. Helbig, U. Zähringer, and C. S. Mintz.** 1999. Molecular cloning and characterization of a locus responsible for *O*-acetylation of the O polysaccharide of *Legionella pneumophila* serogroup 1 lipopolysaccharide. *J. Bacteriol.* **181:**4137–4141.

Lag-1 ACETYLATION OF LIPOPOLYSACCHARIDE

Natalie N. Whitfield and Michele S. Swanson

65

The premier virulence trait of *Legionella pneumophila* is its ability to avoid lysosomal degradation by amoebae and macrophages. When engulfed by phagocytic cells, transmissive *L. pneumophila* cells establish a vacuole that is separated from the endosomal network and instead interacts with the secretory pathway (reviewed in references 6 and 9). In mouse macrophages, the pathogen persists for at least 6 h in this autophagosome-like compartment; subsequently, the vacuole contains replicative *L. pneumophila* and acquires lysosomal characteristics, including an acidic pH (8). Here our major interest is to determine whether developmental changes in the lipopolysaccharide (LPS) affect the ability of *L. pneumophila* to alter phagosome maturation.

LPS is the formidable barrier of gram-negative bacteria, and its role in virulence is well documented. Modifications of the LPS of *L. pneumophila* serogroup 1 strains have been correlated to virulence and serum sensitivity (5). The "virulence-associated" LPS epitope recognized by monoclonal antibody (MAb) 3/1 is prevalent among clinical isolates of *L. pneumophila* serogroup 1 (4). The epitope of MAb 3/1 is dependent upon a functional *lag-1* (lipopolysaccharide-associated gene) gene, which encodes the O-acetyl transferase that modifies the eighth carbon position on the legionaminic acid monomers of the *L. pneumophila* O-antigen. Spontaneous mutants lacking the *lag-1* gene lose reactivity with MAb 3/1, and expression of the gene in *trans* restores MAb 3/1 reactivity as well as acetyltransferase activity (3, 4, 10). In addition to their role in specific antibody recognition, these O-acetyl groups are a major contributor to the hydrophobicity of the LPS (3).

To test whether acetylation of the O-antigen affects phagosome maturation, we constructed a *lag-1* mutant and compared its fate in murine macrophages to a wild-type serogroup 1 *L. pneumophila* strain (Lp02). The Dot/Icm type IV secretion system of *L. pneumophila* is essential for establishment of the replication vacuole, since *dot* mutants fail to alter endocytic trafficking and reside in a compartment with late endosomal markers (reviewed in references 2 and 7). Thus, we examined any *dot*-specific effects by also analyzing a *lag-1 dotA* double mutant in addition to a *dotA* single mutant.

First, we investigated whether the LPS of the *lag-1* mutant differed from wild type by using immunofluorescence, Western blot, and bacterial adherence to hydrocarbon (BATH) assays. *L. pneumophila* appeared to acetylate

Natalie N. Whitfield and Michele S. Swanson   Department of Microbiology and Immunology, University of Michigan, Ann Arbor, MI 48109.

their LPS specifically during the replicative phase, since exponential phase bacteria react strongly with MAb 3/1, but postexponential phase cells did not (Table 1). As expected, *lag-1* mutants lacked the acetylations required for recognition by MAb 3/1, as determined by both immunofluorescence and Western analysis. Bacterial adherence to the hydrocarbon *n*-hexadecane, which reflects surface hydrophobicity or charge, is a developmentally regulated trait of *L. pneumophila*, dependent on the growth phase of the broth cultures from which the cells are isolated (2). Unlike the wild type, replicative phase *L. pneumophila lag-1* mutants did not bind to *n*-hexadecane (Table 1). Thus, *L. pneumophila* binding to hexadecane correlated to reactivity to MAb 3/1; both activities required the *Lag-1* acetyltransferase and were maximal in the replicative phase. Confident that the *lag-1* mutants lacked LPS-specific acetylations, we proceeded to examine their virulence phenotypes by using several well-characterized assays (1).

Cytotoxicity was determined by analyzing macrophage viability after a 1–h incubation with bacteria at a high multiplicity of infection. In addition, the ability of *lag-1* mutants to enter and replicate in macrophages was assessed by incubating bacteria with primary murine bone marrow-derived macrophages for 2 to 72 h, and then determining the fraction of the inoculum that was intracellular and viable at 2 h as well as the total yield at subsequent times. An acetylated LPS was not required for *L. pneumophila* to be cytotoxic to macrophages or to replicate within macrophages (Table 1). Nevertheless, LPS acetylation did alter trafficking of *L. pneumophila* in mouse macrophages.

When macrophages were infected with *L. pneumophila* lacking an acetylated LPS, ~60% of replicative phase bacteria avoided lysosomal degradation, whereas only 10% of wild-type replicative cells evaded degradation (Table 1). A large percentage of those replicative *lag-1* mutant bacteria that avoided degradation nevertheless colocalized with the late endosomal and lysosomal protein LAMP-1, suggesting they are stalled before reaching the degradative lysosomes. This aberrant trafficking pattern is also typical of postexponential phase *dotA* mutants and formalin-killed postexponential phase wild-type *L. pneumophila* (reviewed in reference 2). Intact rods that colocalized with LAMP-1 were also evident within macropages infected with replicative *lag-1dotA* double mutant bacteria (Fig. 1), indicating that a significant number of E phase *lag-1* bacteria arrest maturation of their vacuole by a Dot/Icm-independent mechanism. Thus, lack of acetylation of LPS by *L. pneumophila* correlated with residence and survival in a vacuole with features of late endosomes.

TABLE 1 Effect of acetylation on growth phase-regulated phenotypes of *L. pneumophila*

Strain	MAb 3/1[a]	Attachment to hexadecane[b]	Cytotoxicity[c]	Intracellular Replication[c]	Lamp1 Evasion[d]	Lysosome Evasion[c]
Lp02 PE	−	−	+	+	+	+
Lp02 E	+	+	−	+	−	−
lag-1 PE	−	−	+	+	+	+
lag-1 E	−	−	ND	+	+/−	+
dotA PE	+	+	−	−	−	+
lag-1dotA PE	−	−	−	−	−	+
lag-1dotA E	−	−	ND	−	−	+

[a]Determined using immunofluorescence as described (8) and by Western analysis of LPS preparations.
[b]Bacterial cultures were incubated for 10 min with *n*-hexadecane at 37°C and then incubated at room temperature to allow phase separation. The *n*-hexadecane phase was removed by aspiration and the optical density at 600 (OD_{600}) of aqueous phase was determined. Attachment was expressed by the equation $[(OD_{Before} - OD_{After})/OD_{Before}] \times 100$.
[c]Binding and entry, cytotoxicity, and intracellular degradation were assayed as previously described (1).
[d]Lamp-1 colocalization was determined as previously described (1), where + indicates >60%, +/− indicates >30%, and − indicates <30% colocalization after 2 h incubation at a multiplicity of infection of 1.

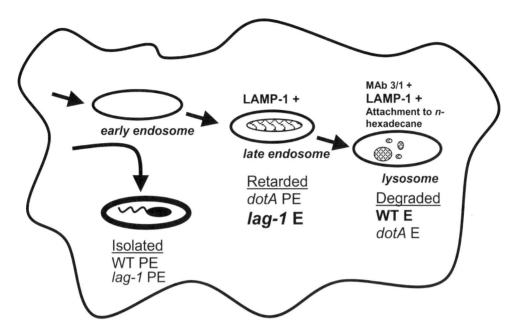

FIGURE 1 Model for role of LPS acetylation of *L. pneumophila* in lysosomal evasion. E, exponential growth phase; PE, postexponential growth phase.

Recently Esteban Fernandez-Moreira in our laboratory found that *L. pneumophila* shed developmentally regulated glycoconjugates that inhibit phagosome-lysosome fusion (2). We have extended these observations by implicating acetylation of LPS as one determinant of the intracellular fate of *L. pneumophila* in mouse macrophages. In the replicative phase, the *lag-1* mutant does not attach to *n*-hexadecane or react with MAb 3/1, traits that correlate with the ability of wild-type *L. pneumophila* to evade degradation in lysosomes (Fig. 1). Accordingly, our working model is that, by acetylating its LPS during the replicative phase, intracellular bacteria release the inhibition to fusion with endosomal vacuoles while increasing their resistance to lysosomal enzymes (8). As a consequence, the pathogen can exploit the host endosomal pathway as a source of not only nutrients but also the membrane needed to expand its replication niche. Questions remaining to be answered include (i) Is there a difference in adherence and uptake of LPS acetylation mutants by macrophages? (ii) Do replicative *lag-1* mutants exhibit the same de-

lay before multiplying intracellularly that wild-type replicative bacteria do? (iii) What compartment do *lag-1* bacteria replicate within? Although its exact contribution to pathogenesis remains to be determined, it is evident that *L. pneumophila* LPS is a dynamic molecule. By modifying the pathogen's surface according to growth phase, the LPS armor can alternately contribute both to transmission and to replication within key defenders of the human immune system.

REFERENCES

1. **Byrne, B., and M. S. Swanson.** 1998. Expression of *Legionella pneumophila* virulence traits in response to growth conditions. *Infect. Immun.* **66:** 3029–3034.
2. **Fernandez-Moreira, E., J. H. Helbig, and M. S. Swanson.** Membrane vesicles shed by *Legionella pneumophila* inhibit fusion of phagosomes with lysosomes. *Infect. Immun.* **74:**3285–3295.
3. **Kooistra, O., E. Luneberg, B. Lindner, Y. A. Knirel, M. Frosch, and U. Zahringer.** 2001. Complex O-acetylation in *Legionella pneumophila* serogroup 1 lipopolysaccharide. Evidence for two genes involved in 8-O-acetylation of legionaminic acid. *Biochemistry* **40:**7630–7640.

4. **Luck, P. C., T. Freier, C. Steudel, Y. A. Knirel, E. Luneberg, U. Zahringer, and J. H. Helbig.** 2001. A point mutation in the active site of *Legionella pneumophila* O-acetyltransferase results in modified lipopolysaccharide but does not influence virulence. *Int. J. Med. Microbiol.* **291:**345–352.

5. **Luneberg, E., U. Zahringer, Y. A. Knirel, D. Steinmann, M. Hartmann, I. Steinmetz, M. Rohde, J. Kohl, and M. Frosch.** 1998. Phase-variable expression of lipopolysaccharide contributes to the virulence of *Legionella pneumophila*. *J. Exp. Med.* **188:**49–60.

6. **Molofsky, A. B., and M. S. Swanson.** 2004. Differentiate to thrive: lessons from the *Legionella pneumophila* life cycle. *Mol. Microbiol.* **53:**29–40.

7. **Sexton, J. A., and J. P. Vogel.** 2002. Type IVB secretion by intracellular pathogens. *Traffic* **3:**178–185.

8. **Sturgill-Koszycki, S., and M. S. Swanson.** 2000. *Legionella pneumophila* replication vacuoles mature into acidic, endocytic organelles. *J. Exp. Med.* **192:**1261–1272.

9. **Swanson, M. S., and E. Fernandez-Moreira.** 2002. A microbial strategy to multiply in macrophages: the pregnant pause. *Traffic* **3:**170–177.

10. **Zou, C. H., Y. A. Knirel, J. H. Helbig, U. Zahringer, and C. S. Mintz.** 1999. Molecular cloning and characterization of a locus responsible for O acetylation of the O polysaccharide of *Legionella pneumophila* serogroup 1 lipopolysaccharide. *J. Bacteriol.* **181:**4137–4141.

IMMUNOCHEMICAL ANALYSIS OF *LEGIONELLA PNEUMOPHILA* OUTER MEMBRANE VESICLES

Jürgen H. Helbig, Esteban Fernandez-Moreira, Paul Christian Lück, Enno Jacobs, and Michele S. Swanson

66

For more than 25 years it is been known, that *Legionella pneumophila* forms evaginations at its outer membrane (3). More recent studies demonstrated that *L. pneumophila* releases these outer membrane (OM) vesicles in both macrophages (1) and broth (2). Furthermore, when immobilized via lipopolysaccharide (LPS)-specific antibodies onto beads, vesicles of the virulent *L. pneumophila* strain Lp02 arrest phagosome-lysosome fusion (2). Accordingly, the highly hydrophobic LPS of *L. pneumophila* may modulate host cell membrane traffic. The strain Lp02 belongs to serogroup 1, monoclonal subgroup Philadelphia. This subgroup is characterized by carrying the virulence-associated LPS epitope recognized by the Dresden Panel monoclonal antibody (MAb) 3/1 (8). Nearly 80% of community-acquired and travel-associated legionellosis is caused by MAb 3/1-positive strains (5). To begin to investigate how vesicles contribute to pathogenesis, the composition of OM vesicles was analyzed by immunochemistry. Here we present the first description of

the components of OM vesicles produced by *L. pneumophila*.

Using MAbs that recognize different LPS epitopes or protein components (Table 1), we first analyzed vesicles from strain Lp02. Next these results were compared with those of three serologically different MAb 3/1-negative patient isolates of our collection. In particular, we investigated strain Görlitz 6543 (serogroup 1, monoclonal subgroup Bellingham), and strains L270 and L138, which were typed as serogroup 3 and serogroup 6, respectively. All three strains were cultured in charcoal-yeast extract broth to the postexponential phase, and then released vesicles were collected by ammonium sulfate precipitation and Optiprep gradient ultracentrifugation as described elsewhere (2). To investigate whether their composition was developmentally regulated, vesicles of the MAb 3/1-positive strain Lp02 were also prepared at two additional stages of the postexponential growth phase, determined by optical density of the broth (see Fig. 1). Concentrations of Lp02 vesicles were equilibrated quantitatively according to their 2-keto-3-deoxyoctulosonic acid (KDO), a component of lipid A, since the length of the O-chain on their LPS may differ. The composition of purified vesicles was compared by immumoblot analysis and enzyme-linked immunosorbent assay using the panel of MAbs listed in Table 1.

Jürgen H. Helbig, Paul Christian Lück, and Enno Jacobs Medical Faculty TU Dresden, Institute of Medical Microbiology and Hygiene, D-01307 Dresden, Germany. *Esteban Fernandez-Moreira and Michele S. Swanson* University of Michigan Medical School, Department of Microbiology and Immunology, Ann Arbor, MI 48109.

TABLE 1 MAbs used for immunochemical analysis of *L. pneumophila* OM vesicles

MAb	Recognized component	Epitope description/localization (reference)
Monofluo (BIO-RAD)	OmpM	Species-specific epitope of the major outer membrane protein
MAb 80/1	p13	Species-specific epitope of a not-yet identified protein (this study)
MAb 22/1	Mip	Genus-specific epitope localized in the active site of Mip PPIase (6)
MAb 31/1	Hsp60	Unknown epitope on *Legionella* and other gram-negative bacteria (this study)
MAb 85/5	Flagella apparatus	Heat-labile conformational epitope expressed by *flaA* (this study)
MAb 61/1	Aup	Unknown epitope of amoebal uptake protein (10)
MAb 8/5	LPS	Serogroup 1-specific epitope of core region (8)
MAb 3/1	LPS	Virulence-associated epitope associated with 8-*O*-acetylgroup of legioaminic acid residues located on O-chain of different MAb subgroups of serogroup 1 (8)
MAb 10/6	LPS	MAb subgroup Bellingham-specific epitope of O-chain (5)
MAb 59/1	LPS	Epitope of short chain LPS molecules of different MAb subgroups of serogroup 1 (this study)
MAb 4/7	LPS	Serogroup 3-specific epitope of core region/O-chain (7)
MAb 9/2	LPS	Serogroup 6-specific epitope of core region/O-chain (7)
MAb 32/3	LPS	Serogroup-cross-reactive epitope core region/O-chain (7)

DETECTION OF SELECTED PROTEIN COMPONENTS

The outer membrane vesicles of all *Legionella* strains contain as major proteins OmpM, detected by Monofluo anti-*Legionella* reagent (BioRad, Munich, Germany) and a 13-kDa *L. pneumophila*-specific protein recognized by the MAb 80/1 that up to now has not been characterized (Fig. 1 and 2). In comparison to OmpM, the 13-kDa protein is only a minor component of whole cell lysates (data not shown). MAb 80/1 was raised against vesicles of an Lp02 *letA* mutant which is defective for expression of transmissive phase traits. As seen in Fig. 1A and 2A, the epitope recognized by MAb 80/1 is also detectable on a protein of approximately 32 kDa which is suspected to be a trimeric porin of the 13-kDa monomer, because porins are known to migrate slowly during sodium dodecyl sulfate gel electrophoresis. Additional components of the vesicles included the chaperone Hsp60 (4), the amoebal uptake protein Aup (10), and the Mip PPIase protein which is necessary for full virulence (6) (data not shown). Also identified by our MAb 48/5 using enzyme-linked immunosorbent assay was a heat-labile epitope of the flagella apparatus that was not expressed either on the surface of *flaA* flagellin mutant bacteria or on outer membrane vesicles of *letA* regulatory mutants (data not shown).

CHARACTERIZATION OF LPS COMPONENTS ENRICHED IN OM VESICLES

To investigate in more detail whether *L. pneumophila* alters the composition of OM vesicles during its life cycle, Lp02 vesicles obtained from cultures at three stages of postexponential growth phase were analyzed by immunoblot (Fig. 1). Fig. 1A and Fig. 1B demonstrate the reactivity of the LPS-specific MAb 8/5 and MAb 3/1 with vesicles that have the same KDO concentration. In contrast, vesicle samples prepared from a later phase of postexponential growth and containing eight times more lipid A were less reactive to MAbs that recognize the O-chain and/or the core region (Fig. 1C). Moreover, these late postexponential phase vesicles contain LPS of shorter lengths,

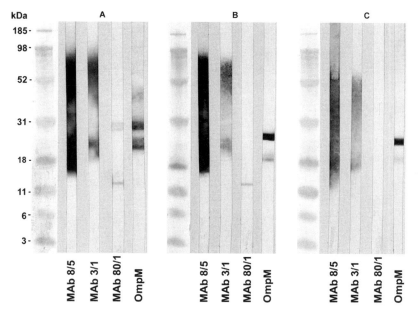

FIGURE 1 Immunoblot analysis of Lp02 vesicles obtained at different growth phases. (*A* and *B*) vesicles released at postexponential phase culture densities of optical density at 600 nm of (OD$_{600nm}$) 3.5 (*A*) and 4.2 (*B*). (*C*) Vesicles prepared from later postexponential phase cultures characterized by their decrease from the peak OD$_{600nm}$. The KDO equivalent of the vesicle solution was adjusted to 0.2 mg of KDO per 1 ml of solution for *A* and *B*. The concentration in *C* amounted to 1.5 mg of KDO per 1 ml of solution.

indicating that *L. pneumophila* developmentally regulates its LPS structure. We assume that lipid A remains anchored to intravesical structural proteins, whereas carbohydrate moieties of LPS are altered, either before or after vesicle formation.

To begin to determine whether the composition of OM vesicles varies between *L. pneumophila* strains, the LPS components of vesicles obtained from three MAb 3/1–negative patient isolates were analyzed. For the serogroup 1 Görlitz 6543 strain, the serogroup 1-specific MAb 8/5 and three other Dresden Panel MAbs that recognize the core or O-chain of LPS were applied. In each case, when compared to Lp02 (Fig. 1A), shorter-chain LPS molecules were detected (Fig. 2A). Shorter forms of LPS were also observed when two strains of other serogroups were probed with either serogroup-specific MAbs (MAb 4/7 or

MAb 9/2) or the serogroup-cross-reacting MAb 32/3 (Fig. 2B and 2C).

LPS MODIFICATIONS MAY ALTER THE FUNCTION OF OM VESICLES DURING LIFE CYCLE

Because outer membrane vesicles of *L. pneumophila* are dynamic and complex structures, measuring quantitative changes in their composition by immunochemical analysis is difficult. Nevertheless, the current data allow preliminary conclusions about individual components. The LPS-enriched vesicles are composed of several structural as well as virulence-associated proteins of *L. pneumophila*, e.g., Mip, Aup, and Hsp60. The organelles also contain components of the flagella apparatus, a preferred target of Toll-like receptors that trigger innate immune responses. Therefore, we postulate that *L. pneumophila* modulates host cell biology

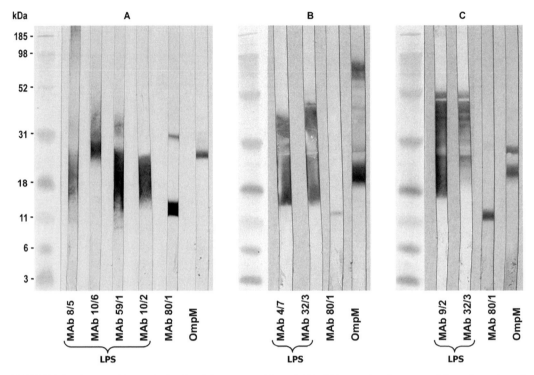

FIGURE 2 Immunoblot analysis of vesicles prepared from MAb 3/1-negative *L. pneumophila* strains cultured to the postexponential growth phase indicated by OD_{600nm} of 3.5. (*A*) Strain Görlitz 6543 (serogroup 1, monoclonal subgroup Bellingham); (*B*) strain L270 (serogroup 3); (*C*) strain L138 (serogroup 6). The LPS-specific MAbs are indicated.

during formation and maturation of its phagosome using complex pathogen-associated molecular patterns; a major constituent of its armature is LPS, which may affect the phagosome maturation machinery directly due to its hydrophobicity or indirectly by forming vehicles that deliver virulence factors to the phagosomal membrane or both. Consistent with this view, Fernandez-Moreiera et al. (2) demonstrated that, when compared to those of virulent strain Lp02, vesicles of the *letA* mutant are not only significantly impaired in their ability to inhibit phagosome-lysome fusion but also contain modified LPS species. The LetA/LetS two-component regulatory system is known to be involved in *dot/icm* transduction and also contributes to the stationary phase stress response (9). In the present study we extend this paradigm by demonstrating that *L. pneumophila* alters the protein and LPS components of OM

vesicles during its life cycle. The detection of particular *dot/icm* gene products inside the OM vesicles would be one approach to investigate how OM vesicles contribute to *L. pneumophila* pathogenesis during phagosome formation and maturation, as well as during biofilm establishment and persistence. For example, after lysis of infected amoeba, vesicles produced by intracellular *L. pneumophila* could stabilize the biofilm structure.

In addition to growth phase-dependent regulation, we also demonstrated that the LPS content of OM vesicles varies between *L. pneumophila* serotypes. In comparison to the other serotypes of *L. pneumophila* examined, the virulence-associated LPS structure recognized by MAb 3/1 correlated with vesicles enriched in long-chain LPS. Because the corresponding epitope is formed by 8-*O*-acetylation of the legioaminic acid residues of the LPS O-chain (11), the hydrophobicity of such vesicles is very high, a feature likely to enhance survival of bacteria within aerosols. The hydrophobic LPS structure of MAb 3/1-positive vesicles may also be critical for *L. pneumophila* to perturb phagosome maturation. These and other provocative hypotheses can be tested directly by comparing, for example, the capacity of MAb 3/1-negative and -positive OM vesicles to arrest delivery of phagocytosed beads to the lysosomes (2).

ACKNOWLEDGMENTS

This work was supported in part by grants from the Deutsche Forschungsgemeinschaft to J. H. (HE 2160/6-1) and the National Institutes of Health to M. S. (2R01 AI040694).

The authors are grateful to Sigrid Gäbler, Ines Wolf, and Volker Bellmann for excellent technical assistance.

REFERENCES

1. **Derre, I., and R. R. Isberg.** 2004. *Legionella pneumophila* replication vacuole formation involves rapid recruitment of proteins of the early secretory system. *Infect. Immun.* **72:**3048–3053.

2. **Fernandez-Moreira, E., J. H. Helbig, and M. S. Swanson.** 2006. Membrane vesicles shed by *Legionella pneumophila* inhibit fusion of phagosomes with lysosomes. *Infect. Immun.* **74** (In press).

3. **Flesher, A. R., S. Ito, B. J. Mansheim, and D. L. Kasper.** 1979. The cell envelope of the Legionnaires' disease bacterium. *Ann. Intern. Med.* **90:** 628–630.

4. **Garduno, R. A., E. Garduno, and P. S. Hoffman.** 1998. Surface-associated Hsp60 chaperonin of *Legionella pneumophila* mediates invasion in HeLa cell model. *Infect. Immun.* **66:**4602–4610.

5. **Helbig, J. H., S. Bernander, M. Castellani Pastoris, J. Etienne, V. Gaia, S. Lauwers, D. Lindsay, P. C. Lück, T. Marques, S. Mentula, M. F. Peeters, C. Pelaz, M. Struelens, S. A. Uldum, G. Wewalka, and T. G. Harrison.** 2002. Pan-European study on culture-proven Legionnaires' disease: distribution of *Legionella pneumophila* serogroups and monoclonal subgroups. *Eur. J. Clin. Microbiol Infect. Dis.* **21:**710–716.

6. **Helbig, J. H., B. König, H. Knospe, B. Bubert, C. Yu, P. C. Lück, A. Riboldi-Tunnicliffe, R. Hilgenfeld, E. Jacobs, J. Hacker, and G. Fischer.** 2003. The PPIase active site of *Legionella pneumophila* Mip protein is involved in the infection of eukaryotic host cells. *Biol. Chem.* **384:** 125–137.

7. **Helbig, J. H., J. B. Kurtz, M. Castellani Pastroris, C. Pelaz, and P. C. Lück.** 1997. The antigenic lipopolysaccharide components of *Legionella pneumophila* recognized by monoclonal antibodies: possibilities and limitations for division of the species into serogroups. *J. Clin. Microbiol.* **35:**2841–2845.

8. **Helbig, J. H., P. C. Lück, Y. A. Knirel, W. Witzleb, and U. Zähringer.** 1995. Molecular characterization of a virulence-associated epitope on the lipopolysaccharide of *Legionella pneumophila* serogroup 1. *Epidemiol. Infect.* **115:**71–78.

9. **Lynch, D., N. Fieser, K. Glöggler, V. Forsbach-Birk, and R. Marre.** 2003. The response regulator LetA regulates the stationary-phase stress response in *Legionella pneumophila* and is required for efficient infection of *Acanthamoeba castellanii. FEMS Microbiol. Lett.* **219:**241–248.

10. **Steudel, C., J. H. Helbig, and P. C. Lück.** 2002. Characterization of a 16 kilodalton species-specific protein of *Legionella pneumophila* promoting uptake in amoebae, p. 165–169. *In* R. Marre (ed.), *Legionella.* American Society for Microbiology, Washington, D.C.

11. **Zou, C. H., Y. A. Knirel, J. H. Helbig, U. Zähringer, and C. S. Mintz.** 1999. Molecular cloning and characterization of a locus responsible for *O*-acetylation of the O polysaccharide of *Legionella pneumophila* serogroup 1 lipopolysaccharide. *J. Bacteriol.* **181:**4137–4141.

CONTRIBUTION OF *LEGIONELLA*'S SURFACE TO THE PREGNANT PAUSE VIRULENCE STRATEGY

Esteban Fernandez-Moreira, Jürgen H. Helbig, and Michele S. Swanson

67

THE PREGNANT PAUSE VIRULENCE STRATEGY OF *L. PNEUMOPHILA*

Legionella pneumophila has a complex life cycle composed of at least three developmental states, each characterized by a different cell type: the replicative, transmissive, and mature intracellular forms (6). Once engulfed by a phagocytic cell, transmissive *L. pneumophila* inhibit phagosome maturation for several hours, a pattern also observed for the intracellular pathogens *Coxiella* and *Leishmania* (9). How long *L. pneumophila* remains isolated from the endosomal pathway is determined in part by the host cell (7). In mouse macrophages, after 8 to 10 h *L. pneumophila* differentiates to the replicative form, which no longer inhibits phagosome-lysosome fusion. Furthermore, once *L. pneumophila* have differentiated to the replicative form, phagosome-lysosome fusion stimulates rather than inhibits bacterial growth (8). We aim to understand how during its life cycle in mouse macrophages *L. pneumophila* first delays and then exploits phagosome-lysosome fusion, a virulence strategy we have named "the pregnant pause" (9).

Although more than 15 serogroups of *L. pneumophila* have been defined according to

lipopolysaccharide structure, serogroup 1 isolates account for ~80% of cases of Legionnaires' disease (3). Surface properties of transmissive *L. pneumophila* have also been implicated as a mechanism to arrest phagosome maturation, a critical step in pathogenesis (5, 9). It is noteworthy that this mechanism is independent of the Dot/Icm type-IV secretion apparatus, the most extensively studied virulence tool of *L. pneumophila* and one that is common to strains of different serogroups. Amrita Joshi in our laboratory found that both formalin-killed transmissive *L. pneumophila* and viable transmissive *dotA* mutants persist for days in late endosomal-like vacuoles that are stalled short of the digestive lysosomes (5). Like other gram-negative bacteria, *L. pneumophila* form small vesicles on their outer membranes (2). Here we tested whether these vesicles retain the same surface properties and activities as whole microbes. Specifically, we tested the hypothesis that *L. pneumophila* releases LPS-rich vesicles that arrest phagosome maturation while also remodeling the bacterial surface during differentiation to the replicative form.

MEMBRANE VESICLES: A VEHICLE OF THE PREGNANT PAUSE

L. pneumophila cultured in broth displays many attributes of the replicative and transmissive phases observed inside macrophages (9).

Esteban Fernandez-Moreira and Michele S. Swanson University of Michigan Medical School, Department of Microbiology and Immunology, Ann Arbor, MI, 48109. *Jürgen H. Helbig* Medical Faculty TU Dresden, Institute of Medical Microbiology and Hygiene, D-01307 Dresden, Germany.

Therefore, as transmissive forms, we analyzed postexponential (PE) wild type (WT) and *dotA* bacteria (optical density at 600 [OD_{600}] 3.0 to 4.0); replicative forms were either exponential (E) WT (OD_{600} 0.2 to 2.0) or PE *letA* cells, regulatory mutants that are locked in the replicative phase. Since the yield of vesicles from exponential phase cultures is prohibitively low, *letA* mutants permit us to collect sufficient amounts of replicative vesicles from cultures of high cell density. As a control for a microbe that does not survive inside macrophages, we also analyzed vesicles produced by *Escherichia coli*. Vesicles shed by *L. pneumophila* or *E. coli* into broth were purified by gradient ultracentrifugation (4).

Membrane vesicles shed by transmissive *L. pneumophila* exhibited the same lectin-binding properties as whole bacteria. Compared to those produced by PE *letA* mutants, membrane vesicles collected from WT PE *L. pneumophila* have higher affinity for the sialic acid-specific *Limulus polyphemus* lectin (1). As expected, this lectin binding was inhibited by 5 mM sialic acid.

Since lectin-binding is coordinately expressed with the capacity to inhibit phagosome maturation (1), we next tested whether vesicles blocked delivery of beads to the lysosomes. Vesicles were attached to protein G-beads with an LPS-specific antibody (Fig. 1A), and the coupling reaction was evaluated by sodium dodecyl sulfate-polyacrylamide gel electrophoresis, silver staining, and the purpald assay, a colorimetric reagent that reports the 2-keto-3-deoxyoctulosonic acid component of *Legionella* LPS (1). Next the beads were incubated with macrophages whose lysosomes contained fluorescein-dextran, and then the fate of the beads was scored by fluorescence microscopy.

Only 28% of beads coated with WT vesicles colocalized with lysosomal probes (Fig. 1B), whereas ~70% of beads were lysosomal when they were decorated with vesicles shed by the differentiation mutant *letA* or by *E. coli* (Fig. 1C). Furthermore, the inhibitory activity was independent of the Dot/Icm machinery, as membrane vesicles purified from the *dotA* mutant

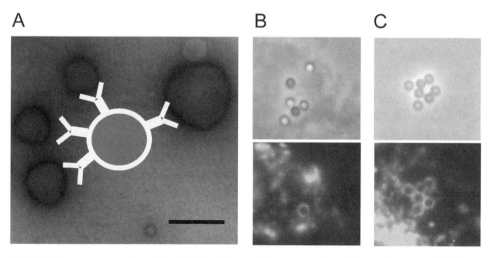

FIGURE 1 *L. pneumophila* vesicles inhibit delivery of phagocytosed particles to lysosomes. (A) Membrane vesicles shed by WT *L. pneumophila* broth cultures were isolated by their buoyant density on an Optiprep gradient and visualized by negative staining. Bar = 100 nm. The diagram (not to scale) illustrates that vesicles were attached to 1-μm polystyrene beads (white) before their delivery to macrophages. (B) When coated with vesicles isolated from WT or *dotA* mutant cultures, most of the beads are in arrested phagosomes, as they do not colocalize with lysosomal FDx. (C) When polystyrene beads are bound to *letA* mutant or *E. coli* K-12 vesicles, they typically colocalize with lysosomal FDx.

performed like WT vesicles. WT vesicles inhibited phagosome maturation in macrophages derived from the bone marrow of either permissive *lgn1* mutant A/J mice or restrictive *lgn1*[+] C57Bl/6J mice. Thus, phagosome arrest by vesicles occurs independently of the ability of *L. pneumophila* to replicate in its host. Inhibition of phagosome maturation was temporary, as judged by the results of a series of pulse-chase experiments. Macrophages treated for 30 min with WT vesicles were incubated in fresh media for 1, 5, or 10 h before being fed with beads: 33, 52, and 71% of these, respectively, colocalized with the lysosomal marker (1).

THE ROLE OF MEMBRANE VESICLES IN *L. PNEUMOPHILA* PATHOGENESIS

As it differentiates to the transmissive form, *L. pneumophila* gains the capacity to inhibit phagosome maturation and also alters the composition of the glycoconjugates on its surface. Like *Salmonella*, *L. pneumophila* fills its phagosome with membrane vesicles (Fig. 2A). Presumably, once intercalated into host membranes, LPS alters the biophysical properties of the phagosome, a virulence strategy proposed for *Leishmania* and *Mycobacteria* (reviewed in reference 9). Although glycoconjugates isolated from virulent *Mycobacteria* and *Leishmania* are sufficient to arrest phagosome maturation, LPS purified from either transmissive or replicative *L. pneumophila* failed to arrest phagosome-lysosome fusion (data not shown). Therefore, we cannot rule out contributions of components other than LPS to the vesicles' inhibitory activity. However, we favor the alternative interpretation that to inhibit maturation of the surrounding phagosome, LPS must be properly oriented.

From broth cultures, we can obtain two forms of *L. pneumophila* that resemble those observed in macrophages (Fig. 2B). As has been documented for many microorganisms, *L. pneumophila* modifies the composition of the LPS on its surface during its life cycle. Compared to replicative cells, transmissive bacteria express LPS species of higher molecular weight (1). Furthermore, only vesicles produced in the transmissive phase (PE WT and PE *dotA* cul-

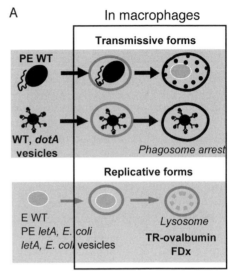

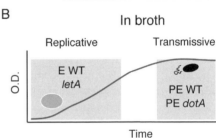

FIGURE 2 The ability of membrane vesicles to inhibit phagosome maturation is developmentally regulated. (A) Vesicles purified from PE WT or *dotA* cultures inhibit phagosome maturation as well as transmissive phase cells, but vesicles from PE *letA* or *E. coli* cultures do not. In addition to inhibiting phagosome maturation, vesicle release would also remodel the *L. pneumophila* surface as it differentiates from the transmissive to the replicative form. (B) At the transition from E phase (OD$_{600}$ < 2.0) to PE phase (OD$_{600}$ 3.0 to 4.0), WT and *dotA* mutant *L. pneumophila* differentiate to the transmissive form, but *letA* regulatory mutants do not.

ture), not the replicative phase (PE *letA*), inhibit phagosome maturation. The O-chain of *L. pneumophila* LPS is a homopolymer of legionaminic acid, a unique sugar structurally similar to sialic acid whose striking features are the lack of free hydroxyl groups and a high degree of O-acetylation. Therefore, changes in acetylation of legionaminic acid could modulate the inhibitory activity of vesicles and affect binding to hexadecane and to lectin. Perhaps by modifying the pattern of LPS acetylation in response to its nutrient supply, *L. pneumophila* can first inhibit and later exploit phagosome-lysosome fusion. By this model, transmissive bacteria release vesicles that stall phagosome maturation. During this pregnant pause, *L. pneumophila* alters the composition of its LPS not only to release the block to fusion but also to tolerate the harsh conditions of the lysosomal compartment, which the progeny exploit as a source of nutrients and membrane to support proliferation.

ACKNOWLEDGMENTS

We thank Dotty Sorensen for electron microscopy assistance and Meta Kuehn for her vesicle purification protocol.

This work was supported by the National Institute of Allergy and Infectious Diseases of the National Institute of Health, 2 R01 AI040694 and a Ministerio de Educación, Ciencia y Deportes/Fullbright-Spain postdoctoral fellowship to E. F.-M.

REFERENCES

1. **Fernandez-Moreira, E., J. H. Helbib, and M. S. Swanson.** 2006. Membrane vesicles shed by *Legionella pneumophila* inhibit fusion of phagosomes with lysosomes. *Infect. Immun.* **74:**3285–3295.
2. **Flesher, A. R., S. Ito, B. J. Mansheim, and D. L. Kasper.** 1979. The cell envelope of the Legionnaires' disease bacterium. Morphologic and biochemical characteristics. *Ann. Intern. Med.* **90:** 628–630.
3. **Helbig, J. H., S. Bernander, M. Castellani Pastoris, J. Etienne, V. Gaia, S. Lauwers, D. Lindsay, P. C. Luck, T. Marques, S. Mentula, M. F. Peeters, C. Pelaz, M. Struelens, S. A. Uldum, G. Wewalka, and T. G. Harrison.** 2002. Pan-European study on culture-proven Legionnaires' disease: distribution of *Legionella pneumophila* serogroups and monoclonal subgroups. *Eur. J. Clin. Microbiol. Infect. Dis.* **21:**710–716.
4. **Horstman, A. L., and M. J. Kuehn.** 2002. Bacterial surface association of heat-labile enterotoxin through lipopolysaccharide after secretion via the general secretory pathway. *J. Biol. Chem.* **277:** 32538–32545.
5. **Joshi, A. D., S. Sturgill-Koszycki, and M. S. Swanson.** 2001. Evidence that Dot-dependent and –independent factors isolate the *Legionella pneumophila* phagosome from the endocytic network in mouse macrophages. *Cell. Microbiol.* **3:** 99–114.
6. **Molofsky, A. B., and M. S. Swanson.** 2004. Differentiate to thrive: lessons from the *Legionella pneumophila* life cycle. *Mol. Microbiol.* **53:**29–40.
7. **Sauer, J. D., J. G. Shannon, D. Howe, S. F. Hayes, M. S. Swanson, and R. A. Heinzen.** 2005. Specificity of *Legionella pneumophila* and *Coxiella burnetii* vacuoles and versatility of *Legionella pneumophila* revealed by coinfection. *Infect. Immun.* **73:**4494–4504.
8. **Sturgill-Koszycki, S., and M. S. Swanson.** 2000. *Legionella pneumophila* replication vacuoles mature into acidic, endocytic organelles. *J. Exp. Med.* **192:**1261–1272.
9. **Swanson, M. S., and E. Fernandez-Moreira.** 2002. A microbial strategy to multiply in macrophages: the pregnant pause. *Traffic* **3:**170–177.

NEW INSIGHTS INTO PATHOGENESIS OF *LEGIONELLA PNEUMOPHILA* INFECTION: FROM BEDSIDE FINDINGS TO ANIMAL MODELS

Kazuhiro Tateda, Soichiro Kimura,
Etsu T. Fuse, and Keizo Yamaguchi

68

Legionella pneumophila is a gram-negative intracellular pathogen that often causes a serious and life-threatening pneumonia (7, 12). Potentially lethal complications, including acute lung injury and acute respiratory distress syndrome, are frequent consequences in these patients, although the exact pathogenesis of lung injury in *Legionella* disease is still poorly understood (14). Despite aggressive supportive care, including antibiotic therapy and oxygen supplementation, high mortality rates reaching 50% have been reported, especially in immunocompromised patients (3, 11).

In lung tissue, bacteria multiply in several types of host cells, including macrophages, monocytes, and alveolar epithelial cells (5, 8). Cytopathogenicity of this bacterium to host cells has been well demonstrated although incompletely understood. Accumulating data indicates that *L. pneumophila* can induce apoptosis in macrophages and alveolar epithelial cells in vitro (4, 9).

Oxygen supplementation is commonly given to patients with severe pneumonia, including *Legionella* disease. This is a critical supportive therapy, especially for patients demonstrating severe hypoxemia. However, prolonged administration or even transient supplementation of oxygen can promote lung damage. Cells at risk for hyperoxia-induced injury include alveolar epithelial cells and lung microvascular endothelial cells (2, 15). Although mechanisms of oxygen toxicity to the lungs have not been carefully defined, it is likely that apoptosis plays a role in hyperoxia-associated lung injury (1, 15). Moreover, the effect of hyperoxia as a cofactor in the development of acute lung injury in bacterial pneumonia is unknown.

In this paper, we present topics concerning the pathogenesis of *Legionella* infections, such as exaggeration of acute lung injury by hyperoxia and a role of *Pseudomonas* quorum-sensing molecules in the ecological niche of *Legionella* organisms.

HYPEROXIA EXAGGERATES ACUTE LUNG INJURY BY *L. PNEUMOPHILA* PNEUMONIA

L. pneumophila is a major cause of life-threatening pneumonia, which is characterized by a high incidence of acute lung injury and resultant severe hypoxemia. Mechanical ventilation using high oxygen concentrations is often required in the treatment of patients with *L. pneumophila* pneumonia. Unfortunately, oxygen

Kazuhiro Tateda, Soichiro Kimura, Etsu T. Fuse, and Keizo Yamaguchi Department of Microbiology and Infectious Diseases, Toho University School of Medicine, Tokyo 143-8540, Japan.

itself may propagate various forms of tissue damage, including acute lung injury. The effect of hyperoxia as a cofactor in the course of *L. pneumophila* pneumonia is poorly understood.

We first examined the effects of hyperoxia on survival in permissive A/J and nonpermissive C57BL/6 mice with *Legionella* pneumonia (10, 13). After i.t. administration of bacteria, one group of animals was kept in a hyperoxic conditions for 60 h, while another group was placed in room air during the observation period. Interestingly, in both strains of mice, drastic decreases in survival were observed in the group of animals exposed to hyperoxic conditions. Specifically, in C57BL/6 mice, only 10% of control mice died, whereas more than 90% of mice exposed to hyperoxia died by 7 days post-infectious challenge (Fig. 1). A similar increase in lethality was observed in infected A/J mice under hyperoxic conditions as compared to infected animals breathing room air. Importantly, hyperoxia treatment alone (90 to 94% oxygen for 60 h), without infection, induced no death of mice in either strain. To determine whether the hyperoxia-associated increase in lethality was specific for *Legionella* pulmonary infection, we assessed survival of mice with *Klebsiella pneumoniae* pneumonia in normal and hyperoxic conditions. In contrast to *Legionella* pneumonia, we did not observe any detrimental effects of hyperoxia on survival of animals with *Klebsiella* pneumonia. These data clearly demonstrate that hyperoxia increases the lethality of *L. pneumophila* pneumonia, but not *K. pneumoniae* pneumonia. Moreover, an increase of lethality was observed in both permissive A/J and nonpermissive C57BL/6 mice, suggesting that intracellular growth of *Legionella* is not required for hyperoxia-induced lethality.

The enhanced lethality was associated with an increase in lung permeability, but not changes in either lung bacterial burden or leukocyte accumulation. To investigate the cause of accelerated lung injury in hyperoxia, we examined quantitative markers of apoptosis, including histone-associated DNA fragments and caspase-3, in the lungs of mice 2 days after *Legionella* challenge (Fig. 2). Histone-associated DNA fragments are a marker for DNA fragmentation, one of the main characteristics of apoptosis, whereas caspase-3 is an essential protease mediating apoptosis. Hyperoxia alone did not induce evidence of apoptosis. However, *Legionella* infection induced a clear increase in the presence of both markers in the lungs. The

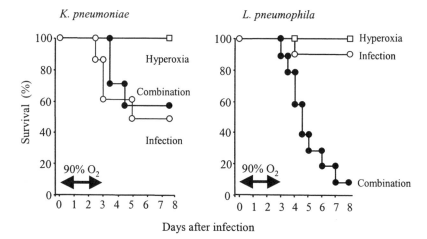

FIGURE 1 Effects of hyperoxia on survival of mice with pneumonia. Mice were infected with *L. pneumophila* and *K. pneumoniae* and then kept in room air or hyperoxia (90% O$_2$) for 60 h. Survival of mice was observed for 8 days after infection. Reprinted from reference 13 with permission.

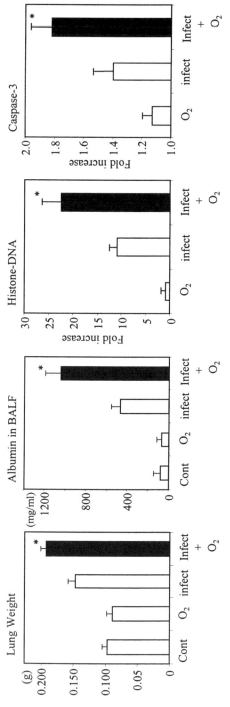

FIGURE 2 Changes of lung weight and albumin in bronchoalveolar lavage fluid (BALF) of mice with *Legionella* pneumonia in the setting of hyperoxia. Mice were intratracheally infected with *L. pneumophila* and then kept in room air or hyperoxia. Lung weight, albumin in BALF, histone-DNA, and caspase-3 in the lung were examined on day 2 (n = 5). ★ $P < 0.05$. Reprinted from reference 13 with permission.

increase in DNA fragments and caspase-3 activity was most marked in the lungs of infected animals exposed to hyperoxia, with 22- and 1.8-fold increases, respectively, compared to control uninfected animals ($P < 0.05$). In addition, evidence of DNA laddering was observed in agarose gel analysis after extraction of chromosomal DNA from infected lungs, which was more pronounced in the lungs of hyperoxic animals (data not shown). These studies indicate that hyperoxia treatment exaggerates apoptosis in the lungs of mice infected with *L. pneumophila*. In addition, we did not observe an increase of histone-associated DNA levels in *K. pneumoniae*-infected lungs in the setting of hyperoxia. These results are well correlated with the survival data and further suggest that the degree of apoptosis correlates with lung injury. Terminal deoxynucleotidyltransferase-mediated dUTP-biotin nick end labeling staining of infected lung sections demonstrated increased apoptosis in hyperoxic mice, predominantly in macrophages (alveolar and interstitial) and alveolar epithelial cells. Fas-deficient mice demonstrated partial resistance to the lethal effects of *Legionella* infection induced by hyperoxia. These results demonstrate that hy-

peroxia serves as an important cofactor for the development of acute lung injury and lethality in *L. pneumophila* pneumonia. Exaggerated apoptosis, in part through Fas receptor signaling, may accelerate hyperoxia-induced acute lung injury in *Legionella* pneumonia.

INTERSPECIES COMMUNICATION BETWEEN *LEGIONELLA* AND *PSEUDOMONAS*

Bacteria commonly communicate each other by a cell-to-cell signaling mechanism referred to as quorum-sensing. Recent studies have shown that *Las*-quorum-sensing autoinducer N-3-oxododecanoyl-L-homoserine lactone (3-oxo-C_{12}-HSL) of *Pseudomonas aeruginosa* possesses a variety of functions, not only intraspecies signaling, but also interspecies and interkingdom communications. We report effects of *Pseudomonas* 3-oxo-C_{12}-HSL on growth and virulence factor expression of other species of bacteria which frequently colocalize with *P. aeruginosa* in nature (6). Type strains and clinical isolates of *Legionella pneumophila*, *Serratia marcescens*, *Proteus mirabilis*, *Ecsherichia coli*, *Alcaligenes faecalis*, and *Stenotrophomonas maltophilia* were used in this study. The bacteria were incubated in liquid

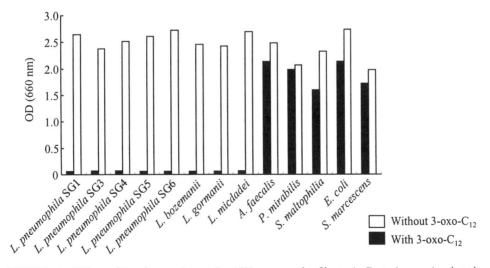

FIGURE 3 Effects of *Pseudomonas* 3-oxo-C_{12}-HSL on growth of bacteria. Bacteria were incubated in liquid medium for 24 h with 50 μM of 3-oxo-C_{12}-HSL, and optical densities were examined. Reprinted from reference 6 with permission.

medium with or without synthetic 3-oxo-C_{12}-IISL or its analogues (5 to 50 μM), and 24 h later their growths were examined by plating. Effect of filter-sterilized culture supernatants of *P. aeruginosa* PAO1 and its *Las*-quorum-sensing–deficient mutant was examined. Expressions of virulence factor genes (*dotA*, *rtxA*, or *lvh*) were explored in *L. pneumophila* by reverse transcriptase-PCR in the presence of 3-oxo-C_{12}-HSL. 3-oxo-C_{12}-HSL suppressed growth of all *Legionella* species examined, but not other organisms such as *Escherichia coli* and *Proteus mirabilis* (Fig. 3). We also observed growth suppression of *L. pneumophila* by culture supernatant of *P. aeruginosa* PAO-1, but not that of *Las*-quorum-sensing mutant. The growth of *L. pneumophila* was completely inhibited by 50 μM of 3-oxo-C_{12}-HSL, and interestingly, significant suppressions of virulence factor genes were demonstrated in *L. pneumophila* exposed to *Pseudomonas* 3-oxo-C_{12}-HSL. Our results suggest that *Pseudomonas* quorum-sensing autoinducer 3-oxo-C_{12}-HSL possesses bacteriostatic and virulence factor-suppressing activity against *L. pneumophila*. These data identify a novel activity of quorum-sensing autoinducer in bacterial interspecies communication.

REFERENCES

1. **Barazzone, C., S. Horowitz, Y. R. Donati, I. Rodriguez, and P. F. Piguet.** 1998. Oxygen toxicity in mouse lung: pathways to cell death. *Am. J. Respir. Cell. Mol. Biol.* **19**:573–581.
2. **Barazzone, C., and C. W. White.** 2000. Mechanisms of cell injury and death in hyperoxia: role of cytokines and Bcl-2 family proteins. *Am. J. Respir. Cell. Mol. Biol.* **22**:517–519.
3. **el-Ebiary, M., X. Sarmiento, A. Torres, S. Nogue, E. Mesalles, M. Bodi, and J. Almirall.** 1997. Prognostic factors of severe *Legionella* pneumonia requiring admission to ICU. *Am. J. Respir. Crit. Care. Med.* **156**:1467–1472.
4. **Gao, L. Y., and Y. Abu Kwaik.** 1999. Apoptosis in macrophages and alveolar epithelial cells during early stages of infection by *Legionella pneumophila* and its role in cytopathogenicity. *Infect. Immun.* **67**:862–870.
5. **Horwitz, M. A., and S. C. Silverstein.** 1980. Legionnaires' disease bacterium (*Legionella pneumophila*) multiples intracellularly in human monocytes. *J. Clin. Invest.* **66**:441–450.
6. **Kimura, S., K. Tateda, Y. Ishii, E. T. Fuse, and K. Yamaguchi.** *Pseudomonas aeruginosa* Las-quorum-sensing autoinducer suppresses growth of and virulence factors in *Legionella pneumophila*. Submitted.
7. **Marston, B. J., H. B. Lipman, and R. F. Breiman.** 1994. Surveillance for Legionnaires' disease. Risk factors for morbidity and mortality. *Arch. Intern. Med.* **154**:2417-2422.
8. **Mody, C. H., R. Paine, M. S. Shahrabadi, R. H. Simon, E. Pearlman, B. I. Eisenstein, and G. B. Toews.** 1993. *Legionella pneumophila* replicates within rat alveolar epithelial cells. *J. Infect. Dis.* **167**:1138–1145.
9. **Muller, A., J. Hacker, and B. C. Brand.** 1996. Evidence for apoptosis of human macrophage-like HL-60 cells by *Legionella pneumophila* infection. *Infect. Immun.* **64**:4900–4906.
10. **Nara, C., K. Tateda, T. Matsumoto, A. Ohara, S. Miyazaki, T. J. Standiford, and K. Yamaguchi.** 2004. *Legionella*-induced acute lung injury in the setting of hyperoxia: protective role of tumour necrosis factor-alpha. *J. Med. Microbiol.* **53**:727–733.
11. **Pedro-Botet, M. L., M. Sabria-Leal, N. Sopena, J. M. Manterola, J. Morera, R. Blavia, E. Padilla, L. Matas, and J. M. Gimeno.** 1998. Role of immunosuppression in the evolution of Legionnaires' disease. *Clin. Infect. Dis.* **26**:14–19.
12. **Reingold, A. L.** 1988. Role of legionellae in acute infections of the lower respiratory tract. *Rev. Infect. Dis.* **10**:1018–1028.
13. **Tateda, K., J. C. Deng, T. A. Moore, M. W. Newstead, R. Paine, 3rd, N. Kobayashi, K. Yamaguchi, and T. J. Standiford.** 2003. Hyperoxia mediates acute lung injury and increased lethality in murine *Legionella* pneumonia: the role of apoptosis. *J. Immunol.* **170**:4209–4216.
14. **Tkatch, L. S., S. Kusne, W. D. Irish, S. Krystofiak, and E. Wing.** 1998. Epidemiology of legionella pneumonia and factors associated with legionella-related mortality at a tertiary care center. *Clin. Infect. Dis.* **27**:1479–1486.
15. **Waxman, A. B., O. Einarsson, T. Seres, R. G. Knickelbein, J. B. Warshaw, R. Johnston, R. J. Homer, and J. A. Elias.** 1998. Targeted lung expression of interleukin-11 enhances murine tolerance of 100% oxygen and diminishes hyperoxia-induced DNA fragmentation. *J. Clin. Invest.* **101**:1970–1982.

INDUCTION OF APOPTOSIS DURING INTRACELLULAR REPLICATION OF *LEGIONELLA PNEUMOPHILA* IN THE LUNGS OF MICE

M. Santic, M. Molmeret, S. Jones, R. Asare,
A. Abu-Zant, M. Doric, and Y. Abu Kwaik

69

Legionella pneumophila has been shown to induce apoptosis within macrophages, monocytic cell lines, and alveolar epithelial cells. The mechanisms and significance of *L. pneumophila*-associated apoptosis are not well understood.

It has been shown that *L. pneumophila* induces Dot/Icm-dependent activation of caspase-3 in the host cell, which results in apoptosis. Although *L. pneumophila* induces activation of caspase-3 during the early stages of infection, apoptosis is not triggered until the late stages of intracellular replication. The induction of caspase-3 activation apoptosis by *L. pneumophila* is very well documented in vitro. It is not known whether it occurs in vivo during Legionnaires' disease. It is also not known whether genetically susceptible inbred mice strains, but not resistant ones, are susceptible to the *L. pneumophila* Dot/Icm-mediated induction of apoptosis.

The experiment is designed to address the Dot/Icm-mediated induction of apoptosis in genetically susceptible A/J mice during *L. pneumophila* infection. In our experiment A/J and BALB/c mice were intratracheally inocu-lated with 10^6 cfu of *L. pneumophila* AA100-GFP or the *dotA*-GFP mutant. At different time points after infection (2, 24, 48, and 72 h) mice were sacrificed and the lungs were removed. The number of colony-forming units of *L. pneumophila* AA100 or the *dotA* mutant strain in the lungs was determined by the plate dilution method using buffered charcoal yeast extract agar. After 3 days of incubation at 37°C the colonies were enumerated and the results were expressed as the number of colony-forming units per lung. For histopathological analysis, the excised lungs were inflated and fixed in formalin for 24 h, dehydrated, and embedded in paraffin. Sections were cut and stained with In Situ Cell Death Detection Kit, POD, POD (Roche Diagnostic, Germany) according to their protocol. An analysis of the histology of the lung tissue in the presence of intracellular green fluorescent protein (GFP)-expressing bacteria and apoptotic cells (terminal deoxynucleotidyl transferase-mediated dUTP-biotin nick end labeling [TUNEL] positive) was carried out using laser-scanning confocal microscopy.

Consistant with previous report, we showed that *L. pneumophila* AA100 replicated in the lungs of A/J mice with the peak of infection at 48 h (Fig. 1A) (8). The *dotA* mutant did not replicate in the lungs of A/J mice, and the mutant was gradually and slowly cleared from the

M. Santic and M. Doric   Department of Microbiology, Medical Faculty, University of Rijeka, Rijeka 51000, Croatia. *M. Molmeret, S. Jones, R. Asare, A. Abu-Zant, and Y. Abu Kwaik* Department of Microbiology and Immunology, Room 316, University of Louisville College of Medicine, Louisville, KY 40202.

lungs (Fig. 1A). Our data showed for the first time that the Dot/Icm type IV secretion system is essential for intrapulmonary replication of *L. pneumophila* in susceptible mice. These in vivo data are consistent with the established role of the Dot/Icm system in the intracellular infection in tissue culture systems in vitro. *L. pneumophila* AA100 did not replicate in the lungs of BALB/c mice (Fig. 1B), in contrast to its robust intrapulmonary replication in A/J mice.

We have recently shown that robust activation of caspase-3 is exhibited throughout the early and exponential intracellular replication of *L. pneumophila*, but apoptosis is delayed and is not triggered in the infected cells until the late stages of infection, concomitant with termination of intracellular replication (1). Despite the numerous studies that documented induction of apoptosis by *L. pneumophila* in several in vitro tissue culture systems, only one study analyzed cell death during the infection in a mouse model of disease (10). Therefore, we used in situ cell analyses to examine apoptosis in the lung tissue of A/J and BALB/c mice infected with *L. pneumophila* AA100-GFP and the *dotA*-GFP mutant by using laser scanning confocal microscopy. Any nuclei

stained black by TUNEL were considered apoptotic, regardless of the intensity of the staining.

At 2 h after infection of A/J mice with *L. pneumophila* AA100-GFP, most of the cells had normal morphology without any sign of apoptosis (Fig. 2A). However, at 24 and 48 h after infection of A/J mice with *L. pneumophila* AA100-GFP, most of the cells became apoptotic (Fig. 2A). *L. pneumophila* replicated intracellulary and was mainly localized in apoptotic cells. In the lungs of A/J mice infected with the *dotA*-GFP mutant, histopathological changes were insignificant without any apoptotic process (Fig. 2B). We did not observe any apoptotic process in the lungs of BALB/c mice infected with *L. pneumophilla* AA100-GFP (Fig. 3).

Pathogens have evolved distinct mechanisms to modulate apoptosis, and this modulation plays a central role in the complex balance between an invading pathogen and host defense, which may be beneficial or detrimental to survival and proliferation of the pathogen (5). The Dot/Icm type IV secretion system is very well documented to be essential for the ability of *L. pneumophila* to modulate phagosome biogenesis and to replicate intracellularly within macrophages in vitro (6). Our data show

A

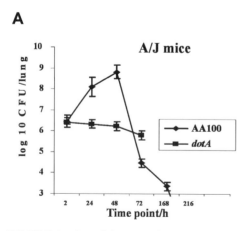

B

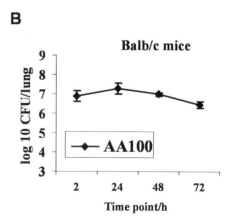

FIGURE 1 Growth kinetics of *L. pneumophila* AA100 and its isogenic *dotA* mutant in the lungs of A/J (*A*) and BALB/c (*B*) mice. Animals were inoculated with *L. pneumophila* AA100 or the *dotA* mutant. At specific time points, the mice were sacrificed and the number of bacteria was determined in the lungs at 2, 24, 48, and 72 h after infection. The error bars represent standard deviations of five mice per group, and results shown are representative of three independent experiments.

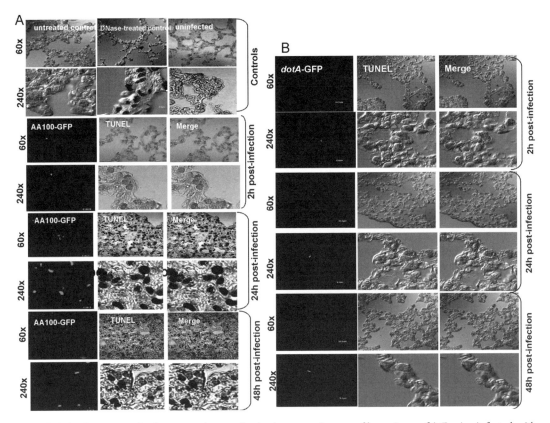

FIGURE 2 Representative laser scanning confocal microscopy images of lung tissue of A/J mice infected with *L. pneumophila* AA100 (*A*) and *dotA* mutant (*B*). A/J mice were infected with the *L. pneumophila* AA100-GFP and the *dotA*-GFP mutant with 10^6 cfu/mouse. At 2, 24, and 48 h after infection lungs were processed to be sectioned and labeled for TUNEL. Lung tissue of uninfected A/J mice, DNase-treated, and untreated lung tissues were used as controls. Apoptotic cells were labeled using TUNEL (black), and the bacteria are shown in gray (GFP). The experiments were done in triplicate using 5 mice for each time point, and the images are representative of 20 microscopic fields from each animal. The results are representative of three independent experiments.

that the Dot/Icm type IV secretion system is essential for intrapulmonary replication and manifestation of disease in the A/J mice model of Legionnaires' disease. Our in vivo studies confirm the in vitro findings on the crucial role of the Dot/Icm system in the pathogenesis of *L. pneumophila*.

Although the induction of apoptosis during the late stages of infection of macrophages by *L. pneumophila* is well documented in vitro (7), it has never been examined whether it also occurs in vivo during experimental Legionnaires' disease in animal models. Our data show a dramatic induction of pulmonary apoptosis during the late stages of infection of the permissive A/J

mice by *L. pneumophila*, similar to infection of macrophages in vitro (9).

L. pneumophila can multiply within different cell types. Most inbred mouse strains are nonpermissive for intracellular replication of *L. pneumophila*. The genetic permissiveness of mice to *L. pneumophila* infection (2) is attributed to allelic polymorphism of the *naip5/ bircle* gene (4, 11). It has been shown that upon infection of C57BL/6 (nonpermissive) mice macrophages in vitro, ~20% of infected macrophages become TUNEL positive at 14 h after infection (3). However, our study did not reveal any apoptotic cells in the lungs of resistant BALB/C mice infected with *L. pneumophila*

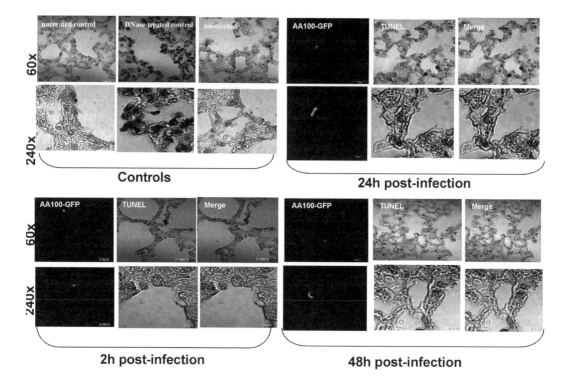

FIGURE 3 Failure of *L. pneumophila* to induce pulmonary apoptosis in resistant BALB/c mice. Representative laser scanning confocal microscopy images of lung tissue of BALB/c mice infected with the *L. pneumophila* AA100. BALB/c mice were infected with *L. pneumophila* AA100-GFP with 10^6 cfu/mouse. At 2, 24, and 48 h after infection, lungs were processed to be sectioned and labeled for TUNEL. Lung tissue of uninfected BALB/c mice, DNase-treated, and untreated lung tissues were used as controls. Apoptotic cells were labeled using TUNEL (black), and the bacteria are shown in gray (GFP). The experiments were done in triplicate using 5 mice for each time point, and the images are representative of 20 microscopic fields from each animal. The results are representative of three independent experiments.

strain AA100. It is important to note that our data do not exclude the possibility that intracellular replication is essential for manifestation of apoptosis. However, activation of caspase-3, which eventually leads to apoptosis, is essential for intracellular replication of *L. pneumophila* (12). Therefore, there is a correlation between the ability to trigger caspase-3 activation and to replicate intracellularly, which culminates in the induction of apoptosis.

In summary, *L. pneumophila* AA100 induces apoptosis in vivo during the late stages of Legionnaires' disease in experimental animals. Genetically susceptible A/J mice are susceptible to the *L. pneumophila* Dot/Icm-mediated induction of apoptosis. Apoptosis is triggered at the late stages of infection, concomitant

with termination of intracellular replication. BALB/c mice are resistant to infection of *L. pneumophila* AA100 as well as induction of apoptosis.

REFERENCES:

1. **Abu-Zant, A., M. Santic, M. Molmeret, S. Jones, J. Helbig, and Y. Abu Kwaik.** 2005. Incomplete activation of macrophage apoptosis during intracellular replication of *Legionella pneumophila*. *Infect. Immun.* **73**:5339–5349.

2. **Beckers, M. C., S. Yoshida, K. Morgan, E. Skamene, and P. Gros.** 1995. Natural resistance to infection with *Legionella pneumophila*: chromosomal localization of the Lgn1 susceptibility gene. *Mamm. Genome* **6**:540–545.

3. **Derre, I., and R. R. Isberg.** 2004. Macrophages from mice with the restrictive Lgn1 allele exhibit

multifactorial resistance to Legionella pneumophila. *Infect. Immun.* **72:**6221-6229.

4. **Diez, E., S. H. Lee, S. Gauthier, Z. Yaraghi, M. Tremblay, S. Vidal, and P. Gros.** 2003. Birc1e is the gene within the Lgn1 locus associated with resistance to *Legionella pneumophila. Nat. Genet.* **33:**55–60.

5. **Gao, L.-Y., and Y. Abu Kwaik.** 2000. Hijacking the apoptotic pathways of the host cell by bacterial pathogens. *Microb. Infect..* **2:**1705–1719.

6. **Molmeret, M., S. D. Zink, L. Han, A. Abu-Zant, R. Asari, D. M. Bitar, and Y. Abu Kwaik.** 2004. Activation of caspase-3 by the Dot/Icm virulence system is essential for arrested biogenesis of the Legionella-containing phagosome. *Cell. Microbiol.* **6:**33–48.

7. **Neumeister, B., M. Faigle, K. Lauber, H. Northoff, and S. Wesselborg.** 2002. *Legionella pneumophila* induces apoptosis via the mitochondrial death pathway. *Microbiology* **148:**3639–3650.

8. **Pedersen, L. L., M. Radulic, M. Doric, and Y. Abu Kwaik.** 2001. HtrA homologue of Legionella pneumophila: an indispensable element for intracellular infection of mammalian but not protozoan cells. *Infect. Immun.* **69:**2569–2579.

9. **Rajan, S., G. Cacalano, R. Bryan, A. J. Ratner, C. U. Sontich, A. van Heerckeren, P. Davis, and A. Prince.** 2000. Pseudomonas aeruginosa induction of apoptosis in respiratory epithelial cells: analysis of the effects of cystic fibrosis transmembrane conductance regulator dysfunction and bacterial virulence factors. *Am. J. Respir. Cell Mol. Biol.* **23:**304–312.

10. **Tateda, K., J. C. Deng, T. A. Moore, M. W. Newstead, R. Paine, 3rd, N. Kobayashi, K. Yamaguchi, and T. J. Standiford.** 2003. Hyperoxia mediates acute lung injury and increased lethality in murine *Legionella pneumonia*: the role of apoptosis. *J. Immunol.* **170:**4209–4216.

11. **Wright, E. K., S. A. Goodart, J. D. Growney, V. Hadinoto, M. G. Endrizzi, E. M. Long, K. Sadigh, A. L. Abney, I. Bernstein-Hanley, and W. F. Dietrich.** 2003. Naip5 affects host susceptibility to the intracellular pathogen *Legionella pneumophila. Curr. Biol.* **13:**27–36.

12. **Zink, S. D., L. Pedersen, N. P. Cianciotto, and Y. Abu Kwaik.** 2002. The Dot/Icm type IV secretion system of *Legionella pneumophila* is essential for the induction of apoptosis in human macrophages. *Infect. Immun.* **70:**1657–1663.

THE ROLE OF THE PHAGOSOMAL TRANSPORTER (Pht) FAMILY OF PROTEINS IN *LEGIONELLA PNEUMOPHILA* PATHOGENESIS

John-Demian Sauer and Michele S. Swanson

70

LEGIONELLA PNEUMOPHILA DIFFERENTIATION AND NUTRIENT STARVATION

A central aspect of the pathogenesis of *Legionella pneumophila* is its ability to differentiate in response to nutrient availability. In the transmissive form, the bacterium efficiently infects host cells, amoebas, or macrophages without being killed; however, it is unable to replicate. In the replicative form, bacteria replicate freely but rarely survive ingestion by phagocytic host cells. The underlying mechanisms of the differentiation process have been extensively studied using broth cultures (2). One predominant signal controlling induction of the transmissive state is nutrient availability, and components of the classic stringent response to amino acid starvation have been implicated in this regulatory pathway. While the complete regulatory cascade has not been elucidated, many of the integral players have been identified, including the ppGpp synthetase *relA*; the sigma factors *rpoS, rpoN,* and *fliA*; the two-component signaling system *letA/S*; and the posttranscriptional regulator *csrA* (1, 3, 4, 6, 7). However, much remains to be learned about what environmental signals trigger differentiation of intracellular transmissive bacteria to the replicative form. Furthermore, the specific nutrients *L. pneumophila* acquires from its host to exploit its intracellular vacuole as a replication niche are not known. Recently, we have identified a family of proteins, the phagosomal transporters (Pht), that we believe equip *L. pneumophila* not only to acquire nutrients from the host phagosome, but also to trigger bacterial differentiation and support their subsequent replication.

THE Pht PROTEIN FAMILY AND *phtA* MUTANT ANALYSIS

The Pht proteins are members of the major facilitator superfamily of proteins which are ATP-independent transporters that perform diverse functions in both prokaryotic and eukaryotic cells (9). One of these transporters, PhtA, is required by intracellular *L. pneumophila* for threonine acquisition, differentiation, and growth (10). Accordingly, we have developed a model in which Pht proteins are responsible for acquisition of nutrients in the nascent phagosome; we postulate that this process is absolutely required both for *L. pneumophila* differentiation and growth within host macrophages (Fig. 1). By identifying the substrates of each of the Pht family members, we seek to reveal the

John-Demian Sauer and Michele S. Swanson Department of Microbiology and Immunology, University of Michigan Medical School, Ann Arbor, MI 48109.

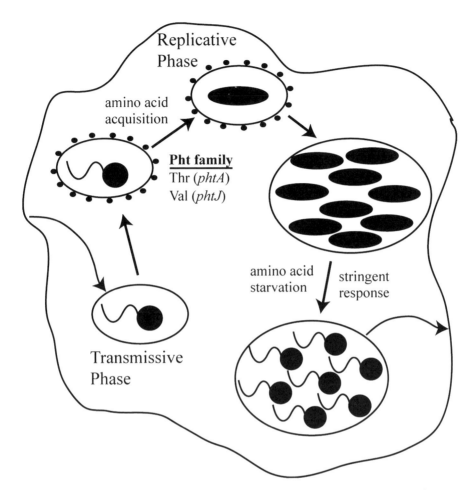

FIGURE 1 Model of Pht function in intracellular differentiation and replication. In the proposed intracellular model of differentiation, we believe that the Pht family of transporters plays a central role in the differentiation from the transmissive form to the replicative form following internalization of *L. pneumophila*. Recognition of rich intracellular stores of nutrients via the Pht transporters allows release of the stringent response and the subsequent differentiation to the replicative form as a prerequisite for intracellular growth. Upon exhaustion of nutrients, the stringent response triggers differentiation back to the transmissive form.

intracellular growth requirements of this fastidious bacterium.

PhtA was initially determined to be required for growth within macrophages by analyzing intracellular growth of *phtA* null mutants. Mutant bacteria could be rescued from a dormant state by inducing a plasmid-born wild-type *phtA* gene even 72 h postinfection, indicating that their growth defect was clearly different from that caused by loss of type IV secretion. Subsequent immunofluorescence microscopy revealed that *phtA* is not required for proper intracellular trafficking, since *phtA* null mutants appropriately avoided the endocytic pathway following internalization, as determined by lack of colocalization with the late endosomal/lysosomal protein LAMP-1. Additionally, *phtA* mutants established a replication-competent vacuole, marked by the endoplasmic reticulum resident protein calnexin, where the bacteria resided throughout their static infection of macrophages (10).

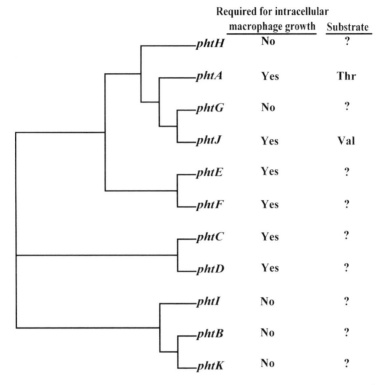

FIGURE 2 *L. pneumophila* Pht family phylogeny. Clustal W alignment of the protein sequences of the *L. pneumophila* Philadelphia 1 strain indicates the relation of the Pht family of proteins to one another. Analysis of growth of *pht* mutants shows that some Phts are required for growth within A/J mouse bone marrow-derived macrophages while others are not. PhtA and PhtJ are most likely threonine and valine transporters, respectively.

As members of the major facilitator superfamily, Pht proteins are predicted to be transporters. This observation led to the hypothesis that PhtA was responsible for nutrient acquisition within the host cell. To test this prediction, we developed a minimal medium to mimic the intracellular growth phenotypes of wild type and the *phtA* mutant, then used this medium to screen for supplements that could rescue the mutant growth defect. Using Biolog Phenotype Microarray technology, we determined that peptides containing threonine bypassed the requirement for PhtA in minimal medium. Subsequent analysis showed that addition of excess threonine in either amino acid or peptide form also rescued the intracellular growth defect of a *phtA* mutant. Therefore, we predict that PhtA is a threonine transporter that enables *L. pneu-*

mophila to acquire threonine in the stringent environment of the replication vacuole (10).

Knowing that nutrient starvation in broth triggers differentiation of replicative cells to the transmissive form, we tested if threonine limitation, caused by mutation of the *phtA* locus, impaired differentiation of intracellular bacteria. Using a phenotypic assay, NaCl sensitivity, we showed that bacteria purified from macrophages 1 h after infection remain in the transmissive state, regardless of PhtA activity. By 10 h after infection, however, wild-type bacteria differentiate to the NaCl-resistant replicative phase, while *phtA* mutants do not. We confirmed that *phtA* mutants have a differentiation defect by reverse transcriptase-PCR analysis of transcript levels of *flaA* and *rpoB*, indicators of the transmissive phase and replicative phase, respectively. Supplementation of macrophage cultures with excess threonine prior to infection also restored differentiation of *phtA* mutants to the replicative form; accordingly, threonine acquisition appears to be a critical hurdle for *L. pneumophila* to surmount during intracellular infection of phagocytes (10).

ADDITIONAL *pht* MUTANT ANALYSIS AND FUTURE IMPLICATIONS

As indicated in Fig. 2, there are 10 additional Pht family members in the *L. pneumophila* Philadelphia 1 genome. Analysis of mutants that lack one of each of the other members of the Pht family indicated that some of the putative transporters are required for growth within mouse macrophages, but others are not. In particular, the *phtB*, *phtG*, *phtH*, *phtIK* genes are not essential for growth within macrophages, while *phtC*, *phtD*, *phtE*, *phtF*, and *phtJ* are required for replication. Further examination of Pht function indicates that *phtJ*, formerly described by Harb and Abu Kwaik as *milA* (5), equips *L. pneumophila* to obtain valine, since supplementation of macrophage cultures with valine rescues the *phtJ* mutant intracellular growth defect.

Building on knowledge of signals that trigger differentiation of replicative cells to the transmissive form, we postulate that the Pht family may be critical to releasing transmissive cells from the stringent response, thereby allowing intracellular replication. Until the bacterium obtains each of the amino acids it requires, uncharged tRNAs may continue to activate RelA and generate the alarmone ppGpp. As a consequence, the stringent response machinery maintains the transmissive state. By this reasoning, the Pht family represents the major mechanism by which *L. pneumophila* surveys its surroundings to gauge whether or not a particular environment is permissive for replication.

It is interesting to note that the *pht* gene family is not only conserved within *L. pneumophila* (11 members in Philadelphia 1 and Lens strains, 9 in Paris), but 9 family members are also present in the genome of *Coxiella burnetii* and 4 to 6 *pht* genes are encoded by *Francisella tularensis*. These three related intracellular pathogens each exhibit biphasic life cycles within amoebas in soil and water (8). Accordingly, we postulate that the Pht family represents a conserved microbial mechanism that evolved to facilitate acquisition of nutrients from host amoebas and other single-celled protozoan hosts in the environment; subsequently, Pht transporters enabled these bacteria to become opportunistic pathogens of higher organisms.

REFERENCES

1. **Bachman, M. A., and M. S. Swanson.** 2001. RpoS co-operates with other factors to induce *Legionella pneumophila* virulence in the stationary phase. *Mol. Microbiol.* **40:**1201–1214.
2. **Byrne, B., and M. S. Swanson.** 1998. Expression of *Legionella pneumophila* virulence traits in response to growth conditions. *Infect. Immun.* **66:** 3029–3034.
3. **Hammer, B. K., and M. S. Swanson.** 1999. Co-ordination of *Legionella pneumophila* virulence with entry into stationary phase by ppGpp. *Mol. Microbiol.* **33:**721–731.
4. **Hammer, B. K., E. S. Tateda, and M. S. Swanson.** 2002. A two-component regulator induces the transmission phenotype of stationary-phase *Legionella pneumophila. Mol. Microbiol.* **44:**107–118.
5. **Harb, O. S., and Y. Abu Kwaik.** 2000. Characterization of a macrophage-specific infectivity locus (milA) of *Legionella pneumophila. Infect. Immun.* **68:**368–376.
6. **Jacobi, S., R. Schade, and K. Heuner.** 2004. Characterization of the alternative sigma factor sigma54 and the transcriptional regulator FleQ of *Legionella pneumophila*, which are both involved in the regulation cascade of flagellar gene expression. *J. Bacteriol.* **186:**2540–2547.
7. **Molofsky, A. B., and M. S. Swanson.** 2003. *Legionella pneumophila* CsrA is a pivotal repressor of transmission traits and activator of replication. *Mol. Microbiol.* **50:**445–461.
8. **Oyston, P. C., A. Sjostedt, and R. W. Titball.** 2004. Tularaemia: bioterrorism defence renews interest in *Francisella tularensis. Nat. Rev. Microbiol.* **2:**967–978.
9. **Pao, S. S., I. T. Paulsen, and M. H. Saier, Jr.** 1998. Major facilitator superfamily. *Microbiol. Mol. Biol. Rev.* **62:**1–34.
10. **Sauer, J. D., M. A. Bachman, and M. S. Swanson.** 2005. The phagosomal transporter A couples threonine acquisition to differentiation and replication of *Legionella pneumophila* in macrophages. *Proc. Natl. Acad. Sci. USA* **102:**9924–9929.

A ROLE FOR PHOSPHOINOSITIDE METABOLISM IN PHAGOCYTOSIS AND INTRACELLULAR REPLICATION OF *LEGIONELLA PNEUMOPHILA*

Stefan S. Weber, Curdin Ragaz, Katrin Reus, and Hubert Hilbi

71

Legionella pneumophila is a facultative intracellular bacterium which replicates within amoebae and macrophages by a similar mechanism. Intracellular replication within both protozoa and mammalian cells requires the bacterial *icm/dot* genes, (13) encoding a conjugation apparatus related to type IV secretion systems (T4SSs). Notably, the haploid social amoeba *Dictyostelium discoideum* sustains intracellular replication of *L. pneumophila* in an Icm/Dot-dependent manner (4, 12, 14).

More recently, it has become clear that the Icm/Dot T4SS not only functions within host cells but also determines the initial interactions of *L. pneumophila* with phagocytic and non-phagocytic host cells. The Icm/Dot apparatus modulates phagocytosis by amoebae as well as by primary macrophages and macrophage-like cell lines (5, 15) and translocates presynthesized effector proteins into eukaryotic host cells before bacterial internalization occurs (11).

Dependent on the Icm/Dot T4SS, endocytic maturation of the *Legionella*-containing vacuole (LCV) is blocked shortly after phagocytosis, and the phagosome neither fuses with lysosomes nor acidifies. Instead, the LCV is transformed into a replication-permissive vacuole by interception with the early secretory pathway at endoplasmic reticulum (ER) exit sites (6). Formation of the LCV requires the function of small GTPases (Sar1, Arf1) and results in the acquisition of the ER marker calnexin, the v-SNARE Sec22b, and the small GTPases Arf1 and Rab1 (2, 7).

To date, more than 30 Icm/Dot-secreted proteins have been identified, many of which form families of paralogues. Yet, for most of these proteins it is not known whether and how they contribute to the formation of the LCV. To understand how *L. pneumophila* forms its replicative vacuole, the biochemical functions as well as the spatial and temporal organization of the Icm/Dot-secreted proteins need to be elucidated.

PHOSPHOINOSITIDE METABOLISM REGULATES VESICLE TRAFFICKING

Phosphoinositides (PIs) are lipid second messengers that play a pivotal role in membrane dynamics during phagocytosis, endocytosis, exocytosis, and other vesicle trafficking pathways (1, 3). The inositol carbohydrate moiety of PIs is phosphorylated at positions 3, 4, and/or 5 by specific kinases or dephosphorylated by phosphatases, respectively. Thus, distinct PIs are produced in a time- and organelle-specific manner

Stefan S. Weber, Curdin Ragaz, Katrin Reus, and Hubert Hilbi Institute of Microbiology, Swiss Federal Institute of Technology (ETH), 8093 Zürich, Switzerland.

Legionella: State of the Art 30 Years after Its Recognition
Edited by Nicholas P. Cianciotto et al.
©2006 ASM Press, Washington, D.C.

and serve as membrane anchors for effector proteins during membrane trafficking. $PI(3,4,5)P_3$ and $PI(3)P$ are generated by class I and class III phosphatidylinositol-3 kinases (PI3Ks) and mediate the closure of the phagocytic cup during phagocytosis or fusion of the phagosome with early endosomes during endocytosis, respectively. $PI(4)P$ is a marker of the Golgi apparatus, which regulates trafficking from the *trans* Golgi network to the plasma membrane.

PHOSPHATIDYLINOSITOL-3 KINASES ARE DISPENSABLE FOR PHAGOCYTOSIS BUT INVOLVED IN INTRACELLULAR REPLICATION OF *L. PNEUMOPHILA*

To analyze whether PI metabolism plays a role in phagocytosis, trafficking, and intracellular replication of *L. pneumophila*, we used a *Dictyostelium* strain lacking the class I PI3K-1 and -2 ($\Delta PI3K1/2$ [18]). The $\Delta PI3K1/2$ mutant shows a number of phenotypes likely related to vesicle trafficking. The mutant is impaired for (i) vegetative growth in axenic medium and on bacterial lawns, (ii) phagocytosis of live or autoclaved bacteria, (iii) pinocytosis of fluid markers, (iv) maturation of phagosomes to "spacious" phagosomes via homotypic fusion, and (v) possibly exocytosis. Also, in the $\Delta PI3K1/2$ mutant the PI profile is altered; as the levels of $PI(3,4)P_2$ and $PI(3,4,5)P_3$ are reduced, the level of $PI(4)P$ is elevated, while $PI(3)P$ and $PI(4,5)P_2$ remain unchanged (17).

Phagocytosis of *L. pneumophila* by *Dictyostelium* was determined by a gentamicin protection assay (Fig. 1) and by flow cytometry (16). Using these assays, wild-type *L. pneumophila* was found to be phagocytosed about

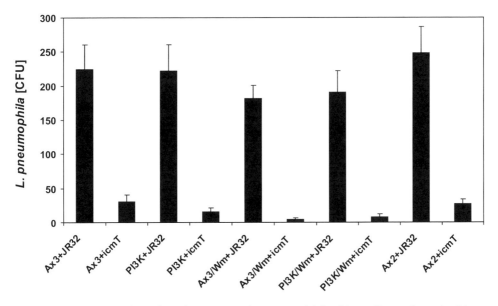

FIGURE 1 Icm/Dot-dependent phagocytosis of *L. pneumophila* by *Dictyostelium* as determined by a gentamicin protection assay. 5×10^4 *Dictyostelium* wild type (Ax3, Ax2) and $\Delta PI3K1/2$ (PI3K; a A×3 derivative lacking the phosphatidylinositol-3 kinases 1 and -2) were infected at a multiplicity of infection of 10 in a 96-well plate with wild-type *L. pneumophila* or *icmT* mutant strain grown for 3 days on charcoal-yeast extract agar plates. Where indicated, the *Dictyostelium* strains were treated with the PI3K inhibitor wortmannin (Wm; 0.1 μM, 1 h). The infection was synchronized by centrifugation (10 min, 880 × g), and after 10 min of incubation, extracellular bacteria were killed by the addition of medium containing gentamicin (0.1 mg/ml). After 1 h at 25°C, the medium was aspirated, and the infected amoebae were lysed in medium containing saponin (0.1%, 15 min). The number of *L. pneumophila* protected from gentamicin within the amoebae was determined by plating 10 μl of the lysate.

one order of magnitude more efficiently than an *icmT* mutant lacking a functional Icm/Dot T4SS (Δ*icmT*). Icm/Dot-dependent phagocytosis was observed for different *Dictyostelium* wild-type strains, including Ax3 and Ax2, and occurred at a multiplicity of infection of 1 to 100. While phagocytosis of wild-type *L. pneumophila* was only slightly affected in Δ*PI3K1/2* (−1%) or by addition of the PI3K inhibitors wortmannin (Wm, −19%) or LY294002 (data not shown) (16), uptake of Δ*icmT* was substantially reduced under these conditions. (−50 to 86%).

Wild-type *L. pneumophila* (but not Δ*icmT*) replicated more efficiently in Δ*PI3K1/2* or in wild-type *Dictyostelium* treated with PI3K inhibitors (16). Within 6 to 8 days, about 100 times more wild-type bacteria were released from *Dictyostelium* lacking functional PI3Ks compared to wild-type amoebae. Moreover, Δ*icmT* was degraded more slowly by Δ*PI3K1/2*, indicating that PI3Ks play a role in the degradative endocytic pathway. To analyze whether PI3Ks affect trafficking of wild-type *L. pneumophila* in *Dictyostelium*, we used red fluorescent bacteria expressing the protein DsRed-Express and calnexin-GFP-labeled amoebae. PI3Ks were found to play a role in trafficking of wild-type *L. pneumophila*, as the transition from "tight" to "spacious" LCVs (8, 9) was impaired in Δ*PI3K1/2* or in wild-type *Dictyostelium* treated with LY294002 (Fig. 2) (16). Since *L. pneumophila* replicated more efficiently in absence of PI3Ks, the transition from tight to spacious vacuoles does not seem to be required for the formation of the LCV.

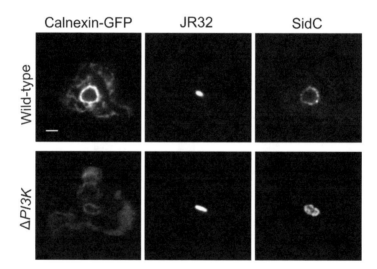

FIGURE 2 Phosphatidylinositol-3 kinases modulate trafficking of *L. pneumophila* within *Dictyostelium* and affect the amount of the Icm/Dot-secreted protein SidC on the *Legionella*-containing vacuole. Confocal laser scanning micrographs are shown of LCVs in calnexin-green fluorescent protein-labeled *Dictyostelium* wild-type strain Ax3 (upper panel) or Δ*PI3K1/2* lacking the phosphatidylinositol-3 kinases-1 and -2 (lower panel). The amoebae were infected with DsRed-expressing wild-type *L. pneumophila* (JR32) for 1 h and immuno-labeled for the Icm/Dot-secreted protein SidC with an affinity-purified antibody and Cy5-conjugated secondary antibody. In wild-type *Dictyostelium*, spacious LCVs were formed preferentially, which bound lower amounts of SidC. However, in Δ*PI3K1/2*, a higher number of tight LCVs were observed, which accumulated on average 1.5 times more SidC. Fluorescence intensity was quantified by subtracting the averaged intensity within background areas from the intensity of a sample area. Bar, 2 μm.

BINDING OF ICM/DOT-SECRETED PROTEINS TO PI(4) PHOSPHATE LINKS TYPE IV SECRETION AND PHOSPHOINOSITIDE METABOLISM

Some Icm/Dot-secreted proteins localize to the membrane of the LCV and therefore are attractive candidates to test the hypothesis that *L. pneumophila* effector proteins anchor to the LCV via PIs. To investigate whether Icm/Dot-secreted proteins bind in vitro to PIs immobilized on nitrocellulose membranes, we performed a lipid protein overlay assay using glutathione S-transferase fusion proteins and an anti-glutathione S-transferase antibody. Thus, SidC (10) and its paralogue SdcA were found to specifically bind to PI(4) phosphate but not to other PIs or lipids (16). The result was confirmed in a pull-down assay, where we employed phospholipid vesicles containing either PI(4)P or other PIs. Using red fluorescent bacteria and calnexin-green fluorescent protein-labeled *Dictyostelium*, we also established that in *L. pneumophila*-infected amoebae more SidC bound to the LCV in absence of PI3Ks (Fig. 2) (16). This result indicates that the PI status of the host cell determines the amount of SidC bound to the LCV and is in agreement with the observation that in *Dictyostelium* Δ*PI3K1/2* the PI3K substrate PI(4)P is present in higher amounts compared to the complemented strain (17).

CONCLUSIONS

The results described here establish that *L. pneumophila* triggers a phagocytic pathway bypassing a requirement for PI3Ks but replicates more efficiently in absence of PI3Ks. The availability of defined mutants and the ease of genetic manipulation renders *Dictyostelium* a good model to further study the role of PI metabolism during phagocytosis and intracellular replication of *L. pneumophila*. Moreover, the finding that the Icm/Dot-secreted protein SidC and its paralogue SdcA specifically bind PI(4) phosphate identifies PI(4) phosphate as a lipid marker of the LCV and provides a mechanistic link between the Icm/Dot-dependent subversion of host cell trafficking by *L. pneumophila* and host cell PI metabolism.

ACKNOWLEDGMENTS

We would like to thank Richard A. Firtel (University of California, San Diego), Hans Faix (Hannover Medical School, Hannover, Germany), Annette Müller-Taubenberger, and Günther Gerisch (MPI for Biochemistry, Martinsried, Germany), for supplying *Dictyostelium* strains and the calnexin-green fluorescent protein expression plasmid.

This work was supported by grants from the Swiss National Science Foundation (631-065952), the Swiss Federal Institute of Technology (TH 17/02-3), and the Velux Foundation.

REFERENCES

1. **De Matteis, M. A., and A. Godi.** 2004. PI-loting membrane traffic. *Nat. Cell. Biol.* **6:**487–492.
2. **Derre, I., and R. R. Isberg.** 2004. *Legionella pneumophila* replication vacuole formation involves rapid recruitment of proteins of the early secretory system. *Infect. Immun.* **72:**3048–3053.
3. **Gillooly, D. J., A. Simonsen, and H. Stenmark.** 2001. Phosphoinositides and phagocytosis. *J. Cell Biol.* **155:**15–18.
4. **Hagele, S., R. Kohler, H. Merkert, M. Schleicher, J. Hacker, and M. Steinert.** 2000. *Dictyostelium discoideum*: a new host model system for intracellular pathogens of the genus *Legionella*. *Cell. Microbiol.* **2:**165–171.
5. **Hilbi, H., G. Segal, and H. A. Shuman.** 2001. Icm/Dot-dependent upregulation of phagocytosis by *Legionella pneumophila*. *Mol. Microbiol.* **42:**603–617.
6. **Kagan, J. C., and C. R. Roy.** 2002. *Legionella* phagosomes intercept vesicular traffic from endoplasmic reticulum exit sites. *Nat. Cell. Biol.* **4:**945–954.
7. **Kagan, J. C., M. P. Stein, M. Pypaert, and C. R. Roy.** 2004. *Legionella* subvert the functions of rab1 and sec22b to create a replicative organelle. *J. Exp. Med.* **199:**1201–1211.
8. **Li, Z., J. M. Solomon, and R. R. Isberg.** 2005. *Dictyostelium discoideum* strains lacking the RtoA protein are defective for maturation of the *Legionella pneumophila* replication vacuole. *Cell. Microbiol.* **7:**431–442.
9. **Lu, H., and M. Clarke.** 2005. Dynamic properties of *Legionella*-containing phagosomes in *Dictyostelium* amoebae. *Cell. Microbiol.* **7:**995–1007.
10. **Luo, Z. Q., and R. R. Isberg.** 2004. Multiple substrates of the *Legionella pneumophila* Dot/Icm system identified by interbacterial protein transfer. *Proc. Natl. Acad. Sci. USA* **101:**841–846.
11. **Nagai, H., E. D. Cambronne, J. C. Kagan, J. C. Amor, R. A. Kahn, and C. R. Roy.** 2005. A C-terminal translocation signal required for Dot/Icm-dependent delivery of the *Legionella* RalF protein to host cells. *Proc. Natl. Acad. Sci.*

USA **102:**826–831.

12. **Otto, G. P., M. Y. Wu, M. Clarke, H. Lu, O. R. Anderson, H. Hilbi, H. A. Shuman, and R. H. Kessin.** 2004. Macroautophagy is dispensable for intracellular replication of *Legionella pneumophila* in *Dictyostelium discoideum. Mol. Microbiol.* **51:**63–72.

13. **Segal, G., and H. A. Shuman.** 1999. *Legionella pneumophila* utilizes the same genes to multiply within *Acanthamoeba castellanii* and human macrophages. *Infect. Immun.* **67:**2117–2124.

14. **Solomon, J. M., A. Rupper, J. A. Cardelli, and R. R. Isberg.** 2000. Intracellular growth of *Legionella pneumophila* in *Dictyostelium discoideum*, a system for genetic analysis of host-pathogen interactions. *Infect. Immun.* **68:**2939–2947.

15. **Watarai, M., I. Derre, J. Kirby, J. D. Growney, W. F. Dietrich, and R. R. Isberg.** 2001. *Legionella pneumophila* is internalized by a macro-pinocytotic uptake pathway controlled by the Dot/Icm system and the mouse *lgn1* locus. *J. Exp. Med.* **194:**1081–1096.

16. **Weber, S., C. Ragaz, K. Reus, and H. Hilbi.** 2006. *Legionella pneumophila* exploits phosphatidylinositol-4 phosphate to anchor secreted effector proteins to the replicative vacuole. PLoS Pathog:**2:**e46.

17. **Zhou, K., S. Pandol, G. Bokoch, and A. E. Traynor-Kaplan.** 1998. Disruption of *Dictyostelium* PI3K genes reduces [^{32}P]phosphatidylinositol 3,4 bisphosphate and [^{32}P]phosphatidylinositol trisphosphate levels, alters F-actin distribution and impairs pinocytosis. *J. Cell Sci.* **111:**283–294.

18. **Zhou, K., K. Takegawa, S. D. Emr, and R. A. Firtel.** 1995. A phosphatidylinositol (PI) kinase gene family in *Dictyostelium discoideum*: biological roles of putative mammalian p110 and yeast Vps34p PI3-kinase homologs during growth and development. *Mol. Cell. Biol.* **15:**5645–5656.

INTERACTION WITH THE CILIATE *TETRAHYMENA* MAY PREDISPOSE *LEGIONELLA PNEUMOPHILA* TO INFECT HUMAN CELLS

Elizabeth Garduño, Gary Faulkner, Marco A. Ortiz-Jimenez, Sharon G. Berk, and Rafael A. Garduño

72

Legionella pneumophila, the causative agent of Legionnaires' disease, is transmitted from the environment to humans. Legionnaires' disease is not transmitted from person to person and therefore is considered an environmental disease. Although patients with Legionnaires' disease may expell *L. pneumophila*-ladden droplets through coughing, the inhalation of such droplets by bystanders does not result in Legionnaires' disease. We have hypothesized that specific events absent in the human host environment, but exclusively associated with *L. pneumophila*'s natural environment (e.g., the interaction with freshwater protozoa) must predispose *L. pneumophila* to infect humans.

L. pneumophila follows a biphasic developmental cycle that alternates between two main forms; a noninfectious, replicative or exponential phase form (2, 4, 6) and a highly infectious mature intracellular form (MIF) (3, 6). Observed by electron microscopy, the replicative form shows a typical gram-negative envelope structure, a cytoplasm rich in ribosomes, and the characteristic features of cell division (4, 6). In contrast, the MIF displays a complex envelope architecture, an electron-dense cytoplasm with large inclusions, and no morphological indicators of cell division (4, 6). We have inferred that MIFs, which are abundantly produced in infected amoebae (1, 7, 9) and naturally predominate in freshwater, must be the transmissible infectious form of *L. pneumophila*. Therefore, the fact that MIFs are inconspicuous in human cultured macrophages (6) may begin to explain why Legionnaires' disease is not spread from person to person. Macrophages constitute the main target cells during *L. pneumophila* human infection.

Our objective here was to determine whether the interaction of *L. pneumophila* with freshwater ciliates provides additional insights into the factors that may predispose *L. pneumophila* to infect human cells.

CILIATES MAY PREDISPOSE *L. PNEUMOPHILA* TO INFECT HUMAN CELLS BY PACKAGING LEGIONELLAE INTO RESPIRABLE-SIZE INFECTIOUS UNITS CONTAINING HUNDREDS OF BACTERIA

The ciliate *Tetrahymena* sp. (isolated from a water cooling tower in Tennessee) has been previously reported to abundantly produce

Elizabeth Garduño and Gary Faulkner Department of Microbiology and Immunology, Dalhousie University, 5850 College Street, Halifax, Nova Scotia, Canada B3H-1X5. *Marco A. Ortiz-Jimenez* Instituto de Investigaciones Biomedicas, Apartado Postal 70228, Ciudad Universitaria, 04510, Mexico, D.F., Mexico. *Sharon G. Berk* Center for the Management, Utilization and Protection of Water Resources, Tennessee Technological University, Cookeville, TN 38505. *Rafael A. Garduño* Department of Microbiology and Immunology and Division of Infectious Diseases-Department of Medicine, Dalhousie University, 5850 College Street, Halifax, Nova Scotia, Canada B3H-1X5.

legionellae-laden vesicles (8). Our ultrastructural characterization of these vesicles (by standard transmission electron microscopy techniques, as described in reference 4) indicated that they are not always surrounded by a membrane. Furthermore, immunogold labeling with a rabbit antiserum directed against the major outer membrane of *L. pneumophila* suggested that most of the membrane material found in these vesicles appears to be of bacterial origin, rather than of host origin. Therefore, in this report we decided to use the term *pellets* instead of the term *vesicles*.

When ciliates feed on *L. pneumophila* Philadelphia-1 strains Lp1-SVir and Lp02, the rate of pellet formation varied proportionally to the bacteria-to-ciliate ratio. For instance, at a bacteria to ciliate ratio of 10^2, ciliates produced ~1 pellet/hour, whereas at a bacteria to ciliate ratio of 10^4, ciliates produced ~5 pellets/hour. The average pellet had a diameter of 4.2 μm and was estimated to contain ~100 legionellae cells. Although all pellets contained live legionellae, the proportion of live to dead bacterial cells varied from pellet to pellet. Routinely, pellets were isolated from live ciliates mixed with free bacteria, by repeated centrifugation at $700 \times g$ for 15 min (to sediment ciliates and pellets but not free bacteria) followed by a 10-min period during which live ciliates were allowed to swim back into suspension. By transmission electron microscopy we clearly determined that both human and mouse macrophages were capable of ingesting whole isolated pellets. Moreover, isolated pellets were infectious in a plaque assay with L929 mouse cells (5). The plaquing efficiency of isolated pellets was higher than that of free legionellae grown in vitro (Table 1), an observation that could be accounted for (at least in part) by the fact that each pellet (equivalent to one infectious unit) was actually formed by ~100 bacteria. Therefore, we concluded that the *Tetrahymena*-mediated packaging of free legionellae into infectious units with a payload of ~100 bacteria may be an important predisposing factor in the infection of human cells.

CILIATES MAY PREDISPOSE *L. PNEUMOPHILA* TO INFECT HUMAN CELLS BY PROMOTING THE DIFFERENTIATION OF LEGIONELLAE INTO MIFs

Routinely, we fed *Tetrahymena* cells with stationary phase *L. pneumophila* grown in vitro (2). It should be noted that stationary phase *L. pneumophila* cells are partially differentiated forms that do not have the full maturation traits of MIFs (6). However, the cells contained in the expelled pellets showed all the previously described (4) ultrastructural characteristics of MIFs (Fig. 1). Interestingly, based

TABLE 1 *Tetrahymena* sp.-expelled pellets are 10- to 100-fold more infectious than stationary phase (agar-grown) free legionellae[a]

Experiment no.	Efficiency of plaque formation %[b]	
	Lp1-SVir in pellets	Agar-grown Lp1-SVir
1	8.9	0.6
2	1.5	0.02
3	4.0	
4	7.4	

[a]Infectivity was measured through a plaque assay on L929 cell monolayers following the method reported by Fernandez et al. (5). The higher the percent efficiency of plaque formation, the more infectious the legionellae are.

[b]Efficiency of plaque formation is reported as a percent, based on the following formula: percent efficiency = (number of plaques formed/number of colony-forming units added) × 100.

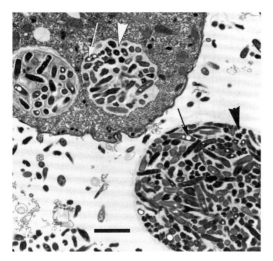

FIGURE 1 Transmission electron micrograph of a sectioned specimen of *Tetrahymena* sp., which had been feeding on *L. pneumophila* strain Lp02 for 20 h. A partial view of a *Tetrahymena* cell containing two legionellae-laden food vacuoles is seen on the top-left corner of the micrograph, whereas a partial view of an expelled pellet is seen on the bottom-right corner of the micrograph. Arrowheads point at MIFs with an irregular shape and dark cytoplasm. Arrows point at MIFs with large cytoplasmic inclusions. Size bar represents 2 μm.

on total counts of colony forming units, direct light microscopy counts, and the short time (~1 h) that expires before the first pellets to emerge after feeding the ciliates, we confidently concluded that *L. pneumophila* cells did not multiply to any obvious extent while in transit through the ciliate. This implies that the *L. pneumophila* stationary phase forms were capable of differentiating into MIFs in the absence of bacterial cell replication. Therefore, we surmised that when *Tetrahymena* sp. ingests immature legionellae, which sometimes may be found free in freshwater, *L. pneumophila* would be induced to rapidly differentiate into MIFs. This way, ciliates may be predisposing *L. pneumophila* cells to infect humans by promoting the formation of highly infectious MIFs. It should be remembered here that the intracellular replication of *L. pneumophila* in cultured human macrophages

is not conducive to the effective formation of MIFs (6).

CILIATES MAY PREDISPOSE *L. PNEUMOPHILA* TO INFECT HUMAN CELLS BY PROVIDING FITNESS ADVANTAGES TO THE PELLETED LEGIONELLAE

We specifically determined whether pelleted legionellae were more resistant to desiccation than free legionellae. Serial dilutions of isolated pellets or free stationary phase legionellae were dispensed into wells of a 24-well cell culture plate, and allowed to dry for 2 weeks in a desiccator. Then, the wells containing the dried legionellae (free or in pellets) were seeded with L929 cells in minimal essential medium. L929 cells were allowed to adhere and grow for 24 h to form a monolayer, after which a layer of agarose-solidified minimal essential medium was added, as to perform a plaque assay (5). Dried free legionellae produced no plaques, even at the highest concentration, suggesting that the dried legionellae either had entered the viable but nonculturable state or were no longer viable. In contrast, we observed numerous plaques in the wells containing the dried pellets, suggesting that the pelleted legionellae remained infectious and withstood desiccation in a viable state.

When a similar experiment was conducted using amoebae (*Acanthamoeba*) instead of L929 cells, we observed infected amoebae only in the samples containing dried pellets. Because amoebae are known to rescue viable but nonculturable forms of *L. pneumophila* (10), we concluded that the dried free legionellae had actually lost their viability. These results indicate that the *Tetrahymena* sp.-mediated packaging of legionellae into pellets protects *L. pneumophila* from desiccation (and perhaps from other environmental stresses), allowing *L. pneumophila* to remain infectious for extended periods. This may be a predisposing factor that may explain (at least in part) the ability of *L. pneumophila* to survive long-distance, airborne travel.

CONCLUSION

In summary, the interaction with ciliates may predispose *L. pneumophila* to infect humans by packaging legionellae into respirable-size infectious units containing hundreds of MIFs, a process also associated with some environmental fitness advantages. Ciliates thus may play a previously unrecognized, yet important, role in the transmission of Legionnaires' disease.

REFERENCES

1. **Berk, S. G., R. S. Ting, G. W. Turner, and R. J. Ashburn.** 1998. Production of respirable vesicles containing live *Legionella pneumophila* cells by two *Acanthamoeba* spp. *Appl. Environ. Microbiol.* **64:**279–286.

2. **Byrne, B., and M. S. Swanson.** 1998. Expression of *Legionella pneumophila* virulence traits in response to growth conditions. *Infect. Immun.* **66:** 3029–3034.

3. **Cirillo, J. D., S. L. Cirillo, L. Yan, L. E. Bermudez, S. Falkow, and L. S. Tompkins.** 1999. Intracellular growth in *Acanthamoeba castellanii* affects monocyte entry mechanisms and enhances virulence of *Legionella pneumophila*. *Infect. Immun.* **67:**4427–4434.

4. **Faulkner, G., and R. A. Garduño.** 2002. Ultrastructural analysis of differentiation in *Legionella pneumophila*. *J. Bacteriol.* **184:**7025–7041.

5. **Fernandez, R. C., S. H. S. Lee, D. Haldane, R. Sumarah, and K. R. Rozee.** 1989. Plaque assay for virulent *Legionella pneumophila*. *J. Clin. Microbiol.* **27:**1961–1964.

6. **Garduño, R. A., E. Garduño, M. Hiltz, and P. S. Hoffman.** 2002. Intracellular growth of *Legionella pneumophila* gives rise to a differentiated form dissimilar to stationary phase forms. *Infect. Immun.* **70:**6273–6283.

7. **Greub, G., and D. Raoult.** 2003. Morphology of *Legionella pneumophila* according to their location within *Hartmanella vermiformis*. *Res. Microbiol.* **154:**619–621.

8. **McNealy, T., A. L. Newsome, R. A. Johnson, and S. G. Berk.** 2002. Impact of amoebae, bacteria and *Tetrahymena* on *Legionella pneumophila* multiplication and distribution in an aquatic environment, p. 170–175. *In* R. Marre, Y. AbuKwaik, C. Bartlett, N. Cianciotto, B. S. Fields, M. Frosch, J. Hacker, and P. C. Lück (ed.). *Legionella*. ASM Press, Washington, D.C.

9. **Rowbotham, T. J.** 1986. Current views on the relationships between amoebae, legionellae and man. *Isr. J. Med. Sci.* **22:**678–689.

10. **Steinert, M., L. Emödy, R. Amann, and J. Hacker.** 1997. Resuscitation of viable but nonculturable *Legionella pneumophila* Philadelphia JR32 by *Acanthamoeba castellani*. *Appl. Environ. Microbiol.* **63:**2047–2053.

GENETICS OF MOUSE MACROPHAGE RESISTANCE TO *LEGIONELLA PNEUMOPHILA*

Russell E. Vance, Tao Ren, Dario S. Zamboni,
Craig R. Roy, and William F. Dietrich

73

Genetics can be a powerful approach with which to dissect the biological relationships between pathogens and their hosts. Genetic approaches are most successful when used to analyze a biological system with large phenotypic differences that are influenced by relatively few genes and alleles. The growth of *Legionella pneumophila* in mouse macrophages has proven to be such a such a system. Macrophages from C57BL/6 mice permit little if any replication of *Legionella*, whereas macrophages from A/J mice permit three to four logs of bacterial growth over 4 to 5 days. Genetic studies have established that polymorphisms in *Naip5* (neuronal apoptosis inhibitory protein 5) appear to explain the entire difference in permissiveness of B6 and A/J macrophages to *Legionella* replication (5, 20).

Naip5 is located within a cluster of tandemly repeated *Naip*-paralogs that vary in number and organization from strain to strain (8). The *Naip* genes were so named because initial observations suggested a connection to neuronal apoptosis (18), but current evidence suggests

the expression and function of the *Naip* genes is not solely within the nervous system. Accordingly, the *Naip5* gene has recently been renamed *Birc1e* (or baculoviral IAP repeat-containing 1e), though the former name is still more commonly used. Evidence that *Naip5* rather than another *Naip* gene is responsible for resistance to *Legionella* came from experiments in which a B6-derived BAC transgene containing only a complete *Naip5* gene conferred resistance to *Legionella* in A/J macrophages (5, 20). It should be noted that the A/J genome also contains a *Naip5* gene; however, the A/J *Naip5* allele appears to be expressed at lower levels and encodes 14 amino acid polymorphisms as compared to the B6 allele. The precise polymorphism(s) within *Naip5* that accounts for the *Legionella* phenotype has yet to be identified.

Macrophages from (B6 × A/J)F_1 hybrid mice exhibit a completely resistant phenotype, indicating that resistance is dominant. This has been interpreted to mean that B6 expresses a functional allele of *Naip5* that acts to provide resistance to *Legionella* and that the A/J allele lacks this function. An alternative model, in which the A/J *Naip5* allele is a functional allele that promotes permissiveness to *Legionella*, would require that the A/J *Naip5* allele is haploinsufficient (nonfunctional when present in

Russell E. Vance, Tao Ren, and William F. Dietrich Genetics Department, Harvard Medical School, 77 Avenue Louis Pasteur, Rm. 358, Boston, MA 02115. *Dario S. Zamboni and Craig R. Roy* Section of Microbial Pathogenesis, Yale University School of Medicine, Boyer Center for Molecular Medicine, 295 Congress Avenue, New Haven, CT 06536.

only one copy), so as to explain the dominance of the B6 phenotype. Such a scenario seems unlikely, given that transgenic experiments (5, 20) indicate that the B6 allele of *Naip5* is indeed functional. Nevertheless, it remains possible that the A/J *Naip5* protein is functional or partially functional: for example, it may provide only a low level of resistance to *Legionella* or higher levels of resistance to other pathogens.

The *Naip5* protein contains three recognizable domains (Fig. 1): (i) three N-terminal baculovirus inhibitor of apoptosis repeats (BIRs); (ii) a central nucleotide-binding domain (NBD), sometimes also called a NACHT or NOD domain, that is believed to mediate protein-protein oligomerization; and (iii) C-terminal leucine-rich repeats (LRRs). LRRs are also found in the extracellular domain of transmembrane Toll-like receptors, where they are believed to mediate the (direct or indirect) detection of microbe-associated molecules such as lipopolysaccharide, flagellin, double-stranded RNA, etc. Interestingly, the slightly more specific combination of

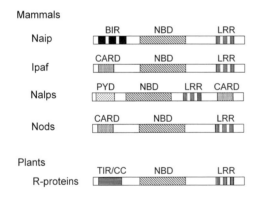

FIGURE 1 NBD-LRR proteins. The domain structure of several NBD-LRR proteins is shown (not to scale). NBD, nucleotide binding domain (thought to mediate protein-protein oligomerization); LRR, leucine rich repeats (thought to mediate sensing of microbial products); BIR, baculovirus inhibitor of apoptosis repeat; CARD, caspase recruitment domain; PYD, pyrin domain; TIR, Toll/IL-1 receptor signaling domain; CC, coiled coil domain; Naip, neuronal apoptosis inhibitory protein; Ipaf, ICE-protease activating factor (ICE is another name for caspase-1); Nalp, NACHT-, LRR-, and PYD-containing protein (NACHT is another name for the NBD).

LRRs with a central NBD is also found in a large number of other proteins in plants and animals, and this superfamily of NBD-LRR proteins have been given various names including Nod-like or CATERPILLAR (9, 19). Unlike the Toll-like receptors, NBD-LRR proteins do not contain transmembrane domains and are believed to localize within the cytoplasm.

Plant NBD-LRR proteins have long been recognized to play important roles in resistance to pathogens (14). More recent work has now established that many mammalian NBD-LRR proteins are also involved in pathogen detection. Best characterized are the Nod1 and Nod2 proteins, which have been shown to participate (directly or indirectly) in the recognition of bacterial peptidoglycan fragments (such as muramyl dipeptide or iE-DAP) (2, 6, 7, 10). Nod2-deficient mice exhibit defects in defense against orally delivered *Listeria monocytogenes* (11). Other NBD-LRR family members in the mouse include the *Nalps* and *Ipaf*. A recent study from the Dietrich laboratory has shown that a mouse *Nalp1* gene controls susceptibility to anthrax lethal toxin (1). In addition, targeted mutation of *Ipaf* prevented caspase-1 activation in response to *Salmonella* infection (13). Thus, *Naip5* exhibits homology to an interesting class of intracellular proteins that appear to play key roles in detecting and responding to pathogens. It must be emphasized, however, that in no case has any of these mammalian proteins been shown to bind directly to microbial products. It should also be emphasized that there are NBD-LRR proteins (e.g., CIITA) that have no known function in pathogen detection.

In the cases that have been studied in detail, it appears that NBD-LRR proteins are normally maintained in a nonoligomerized form that does not initiate downstream signaling. Truncation of the C-terminal LRR usually results in a constitutively active (oligomerized) form of the protein, so it has been proposed that the LRR functions as an autoinhibitory domain. It is believed that recognition of a ligand by the LRR relieves the autoinhibition, resulting in oligomerization (mediated by the central NBD) and downstream signaling (mediated by the N-terminal signaling domain). In many NBD-LRR proteins, the N-terminal signaling domain is a caspase recruitment domain (CARD) that recruits caspases or other CARD-containing proteins such as the RIP2 (RICK, CARDIAK) kinase (Fig. 1). Naip proteins are unique among NBD-LRR proteins in containing N-terminal BIR domains and no CARDs. Some BIR-containing proteins such as XIAP have been found to regulate caspase activation and apoptosis. Indeed, there is evidence from in vitro assays that the isolated BIR repeats of human NAIP interact with and inhibit caspases-3, -7, and -9 (3, 12). The biological significance of these observations remains to be understood (see below).

Based on the homology of *Naip5* to other pathogen-detector proteins, our working model was that Naip5 recognized *Legionella* (directly or indirectly), leading to oligomerization of Naip5 and downstream signaling mediated by the BIR motifs. The signaling pathways activated by the BIR motifs in Naip5 were unknown, so initially we tested some likely candidates. We found that overexpression of *Naip5* (or ΔLRR *Naip5* variants that are presumably constitutively active) did not stimulate NF-κB or IRF-dependent reporter constructs in 293T cells, though these reporters were strongly activated by positive control proteins such as NOD2 and TBK1 (data not shown). We also examined whether permissiveness of mouse macrophages to *Legionella* growth was affected by inhibitors of the mitogen activated protein (MAP) kinases, another important class of signaling molecules in the immune system. Again, however, we found that inhibitors of p38, ERK1/2, or JNK MAP kinases did not significantly affect resistance or permissiveness of B6 or A/J macrophages (data not shown). Therefore, we concluded that classic immune signaling pathways, including NF-κB, IRFs and (MAP) kinases, were not likely to be downstream of Naip5.

Based on studies with human NAIP (mentioned above), we considered the possibility that mouse Naip5 might interact with (and in-

hibit) caspases-3 and -7. Indeed, work in Abu Kwaik's laboratory over the past several years has indicated that caspase-3 is activated in human macrophages in response to *Legionella* infection, and that caspase-3/7 inhibitors can reverse the permissiveness of human macrophages to *Legionella* growth (15). Interestingly, we observed *Legionella*-induced activation of caspases-3/7 occurred preferentially in A/J-derived (susceptible) macrophages (Fig. 2A), but whether caspase activation *causes* permissiveness or viceversa was not established. In fact, we found that treatment of macrophages with the cell-permeable caspase-3/7 inhibitor Z-DEVD-FMK had no effect on the permissiveness of A/J-derived macrophages or on the resistance of B6-derived macrophages (Fig. 2B; 20). In addition, work from Craig Roy's lab has shown that B6-backcrossed caspase-3 knockout macrophages were as resistant to *Legionella* as wild-type B6 macrophages, and A/J-backcrossed caspase-3 knockout macrophages were as permissive to *Legionella* as wild-type A/J macrophages (21). Thus, activation of caspase-3/7 appears to be a consequence and not a cause of permissiveness to *Legionella* replication. The contrast between our results and those from the Abu Kwaik laboratory has not been explained but may represent a mouse-human difference.

A clue to the nature of B6 macrophage resistance to *Legionella* growth was obtained when another caspase inhibitor, Z-VAD-FMK, was found to partially reverse the resistance of B6 macrophages (21). Z-VAD-FMK is a relatively broad-spectrum caspase inhibitor that is also toxic to macrophages at high concentrations. Importantly, therefore, it was found that macrophages from B6-backcrossed caspase-1 knockout mice were also more permissive for *Legionella* growth than were wild-type B6 macrophages (21). Furthermore, caspase-1 was found to be preferentially activated in B6 macrophages upon *Legionella* infection (21). These results were immediately of interest because other pathogens, notably *Salmonella*, initiate an unusual form of macrophage death that depends on caspase-1 (but not caspase-3) and is distinguishable from

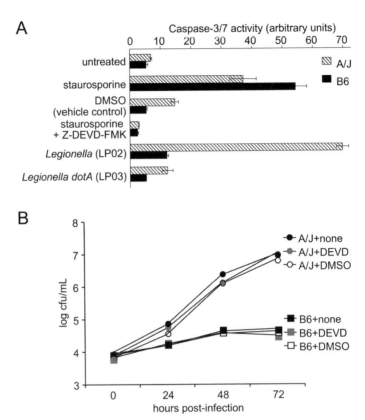

FIGURE 2 *Legionella* activates caspase-3/7 in permissive mouse macrophages. (*A*) Caspase-3/7 activity was measured with the CaspaseGlo assay from Promega. Bone marrow-derived macrophages were obtained from either A/J or B6 macrophages and were infected at multiplicity of infection of 2 for 15 h before assaying caspase activity. As a positive control, macrophages were treated with staurosporine (50 nM) or with DMSO, as a vehicle control. Z-DEVD-FMK is a caspase-3/7 inhibitor and was used at 50 μM. (*B*) Bone marrow macrophages were treated with 50 μM caspase-3/7 inhibitor (Z-DEVD-FMK), a concentration confirmed to be functional (see panel A), or with dimethyl sulfoxide as a vehicle control. The macrophages were then infected with *Legionella* (LP02) and harvested at daily intervals to determine colony-forming units (cfu). The figure shows that although A/J macrophages activate caspase-3 in response to *Legionella* infection, activation of caspase-3 is not required for permissiveness to *Legionella* growth. Similar growth of *Legionella* in Z-DEVD-FMK-treated macrophages was observed previously (20).

apoptosis by its extreme rapidity (<4 h) and necrosis-like characteristics. The idea that *Legionella* might activate caspase-1 was also consistent with previous observations in which *Legionella* was found to induce a rapid cell death preferentially in B6 macrophages (4). It should be noted that caspase-1 is unlikely to be the sole effector of macrophage death, or of macrophage resistance to *Legionella,* since caspase-1 knockout macrophages sometimes exhibited only partial or slight permissiveness as compared to wild-type B6 macrophages (data not shown). Other caspases may have overlapping functions with caspase-1.

The foregoing observations can be assembled with other data from several laboratories

(16) into a model of how B6 macrophages restrict *Legionella* growth. In this model, a single *Legionella* bacterium infecting a macrophage initiates the formation of its replicative vacuole over the course of several hours. During this time, B6 macrophages sense the presence of *Legionella* and initiate the activation of caspase-1 in a manner that appears to depend on Naip5. Caspase-1 activation results in a rapid macrophage death that, in a large proportion of B6 macrophages, occurs before extensive bacterial replication can take place. Caspase-1-mediated death also occurs prior to full activation of caspase-3 or -7, explaining why B6 macrophages appear to exhibit minimal activation of these apoptotic mediators (Figure 2). In contrast, A/J macrophages activate caspase-1 to a much lesser degree and generally survive long enough to allow several rounds of bacterial replication. Eventually, in A/J macrophages, caspase-3 and -7 are activated, and the macrophage undergoes apoptosis or simply lyses due to the large number of intracellular bacteria. The bacteria that are released from A/J macrophages have not only increased in number, but have also had ample opportunity to reexpress virulence genes (e.g., dot/icm genes) required for infection of a new host cell. These genes are believed to be downregulated during the early intracellular replication phases; thus, bacteria released prematurely from B6 macrophages are not only fewer in number but may also exhibit reduced infectivity for new host macrophages. Together, these effects may mediate the strong caspase-1- and *Naip5*-dependent restriction of bacterial growth observed in B6 macrophages.

There are many unresolved issues. For example, there is no clear molecular explanation for how Naip5's BIR domains, classically considered caspase-*inhibition* domains, are involved in caspase-1 *activation*. Future studies will undoubtedly shed more light on these mechanistic details. It is nevertheless clear, however, that caspase-1 activation is likely to be complex and that other proteins are likely to work in concert with Naip5. For example, the *Naip5* homolog *Ipaf* is required for caspase-1 activation

by *Salmonella* (13) and is also required for restriction of *Legionella* growth (21).

If our model of Naip5 activation in response to *Legionella* is correct, a final key question is how Naip5 senses *Legionella*. We reasoned that a *Legionella* mutant that did not produce the putative ligand for Naip5 would possibly be able to evade the growth restriction normally imposed by B6 macrophages. Thus, we screened a large library of *Legionella* transposon mutants for strains capable of growth in B6 macrophages. Our screen was close to saturating, and remarkably, the only gene we identified was *flaA* (17), encoding flagellin, the primary structural component of the flagellum. We did not hit genes encoding other essential components of the flagellum, suggesting that it is loss of flagellin itself, rather than loss of motility or a flagellum, that permits evasion of *Naip5*-mediated growth restriction. Indeed, a *fliI* mutant, which still expresses flagellin but is nonmotile and nonflagellated, was growth restricted in B6 macrophages (17). An unmarked deletion mutant of *flaA* exhibited the same robust growth phenotype in B6 macrophages as the transposon mutants, arguing that polar or second-site mutations are unlikely to explain our results (17). Finally, fitting with the idea that caspase-1-mediated macrophage death is required for *Naip5*-dependent growth restriction, the flagellin-deficient mutants failed to trigger the rapid caspase-1 dependent cell death of B6 macrophages (17). Our results are entirely consistent with independent studies from Molofsky, Swanson and colleagues (see chapter 76).

Our current model is that *Legionella* flagellin is detected by B6 macrophages in a manner dependent on Naip (and possibly Ipaf), leading to caspase-1 activation, rapid cell death, and nonpermissiveness for *Legionella* growth. Whether Naip or Ipaf is a direct intracellular receptor for flagellin will likely be difficult to establish convincingly, as even the much more thoroughly characterized Toll and Nod proteins have not been unequivocally demonstrated to bind directly to their putative ligands. It has nevertheless been extremely satisfying to see how the concerted application of mouse and bacterial genetics has led to several insights into the nature of innate macrophage resistance to *Legionella*.

REFERENCES

1. **Boyden, E., and W. F. Dietrich.** 2006. Nalp1b controls mouse macrophage susceptibility to anthrax lethal toxin. *Nat. Genet.* **38:**240–4.
2. **Chamaillard, M., M. Hashimoto, Y. Horie, J. Masumoto, S. Qiu, L. Saab, Y. Ogura, A. Kawasaki, K. Fukase, S. Kusumoto, M. A. Valvano, S. J. Foster, T. W. Mak, G. Nunez, and N. Inohara.** 2003. An essential role for NOD1 in host recognition of bacterial peptidoglycan containing diaminopimelic acid. *Nat. Immunol.* **4:**702–707.
3. **Davoodi, J., L. Lin, J. Kelly, P. Liston, and A. E. MacKenzie.** 2004. Neuronal apoptosis-inhibitory protein does not interact with Smac and requires ATP to bind caspase-9. *J. Biol. Chem.* **279:**40622–40628.
4. **Derre, I., and R. R. Isberg.** 2004. Macrophages from mice with the restrictive Lgn1 allele exhibit multifactorial resistance to Legionella pneumophila. *Infect. Immun.* **72:**6221–6229.
5. **Diez, E., S. H. Lee, S. Gauthier, Z. Yaraghi, M. Tremblay, S. Vidal, and P. Gros.** 2003. Birc1e is the gene within the Lgn1 locus associated with resistance to Legionella pneumophila. *Nat. Genet.* **33:**55–60.
6. **Girardin, S. E., I. G. Boneca, L. A. Carneiro, A. Antignac, M. Jehanno, J. Viala, K. Tedin, M. K. Taha, A. Labigne, U. Zahringer, A. J. Coyle, P. S. DiStefano, J. Bertin, P. J. Sansonetti, and D. J. Philpott.** 2003. Nod1 detects a unique muropeptide from gram-negative bacterial peptidoglycan. *Science* **300:**1584–1587.
7. **Girardin, S. E., I. G. Boneca, J. Viala, M. Chamaillard, A. Labigne, G. Thomas, D. J. Philpott, and P. J. Sansonetti.** 2003. Nod2 is a general sensor of peptidoglycan through muramyl dipeptide (MDP) detection. *J. Biol. Chem.* **278:**8869–8872.
8. **Growney, J. D., J. M. Scharf, L. M. Kunkel, and W. F. Dietrich.** 2000. Evolutionary divergence of the mouse and human Lgn1/SMA repeat structures. *Genomics* **64:**62–81.
9. **Inohara, N., M. Chamaillard, C. McDonald, and G. Nunez.** 2005. NOD-LRR Proteins: role in host-microbial interactions and inflammatory disease. *Annu. Rev. Biochem.* **74:**355–383.
10. **Inohara, N., Y. Ogura, A. Fontalba, O. Gutierrez, F. Pons, J. Crespo, K. Fukase, S. Inamura, S. Kusumoto, M. Hashimoto, S. J. Foster, A. P. Moran, J. L. Fernandez-Luna, and G. Nunez.** 2003. Host recognition of bacterial muramyl dipeptide mediated through NOD2. Implications for Crohn's disease. *J. Biol. Chem.* **278:**5509–5512.

11. **Kobayashi, K. S., M. Chamaillard, Y. Ogura, O. Henegariu, N. Inohara, G. Nunez, and R. A. Flavell.** 2005. Nod2-dependent regulation of innate and adaptive immunity in the intestinal tract. *Science* **307:**731–734.
12. **Maier, J. K., Z. Lahoua, N. H. Gendron, R. Fetni, A. Johnston, J. Davoodi, D. Rasper, S. Roy, R. S. Slack, D. W. Nicholson, and A. E. MacKenzie.** 2002. The neuronal apoptosis inhibitory protein is a direct inhibitor of caspases 3 and 7. *J. Neurosci.* **22:**2035–2043.
13. **Mariathasan, S., K. Newton, D. M. Monack, D. Vucic, D. M. French, W. P. Lee, M. Roose-Girma, S. Erickson, and V. M. Dixit.** 2004. Differential activation of the inflammasome by caspase-1 adaptors ASC and Ipaf. *Nature* **430:**213–218.
14. **Martin, G. B., A. J. Bogdanove, and G. Sessa.** 2003. Understanding the functions of plant disease resistance proteins. *Annu. Rev. Plant. Biol.* **54:**23–61.
15. **Molmeret, M., S. D. Zink, L. Han, A. Abu-Zant, R. Asari, D. M. Bitar, and Y. Abu Kwaik.** 2004. Activation of caspase-3 by the Dot/Icm virulence system is essential for arrested biogenesis of the Legionella-containing phagosome. *Cell. Microbiol.* **6:**33–48.
16. **Molofsky, A. B., and M. S. Swanson.** 2004. Differentiate to thrive: lessons from the Legionella pneumophila life cycle. *Mol. Microbiol.* **53:**29–40.
17. **Ren T., D. S. Zamboni, C. R. Roy, W. F. Dietrich, R. E. Vance.** 2006. Flagellin-deficient *Legionella* mutants evade caspase-1- and Naip5-mediated macrophange immunity. *Plos Pathogens.* **2:**175–183
18. **Roy, N., M. S. Mahadevan, M. McLean, G. Shutler, Z. Yaraghi, R. Farahani, S. Baird, A. Besner-Johnston, C. Lefebvre, X. Kang, et al.** 1995. The gene for neuronal apoptosis inhibitory protein is partially deleted in individuals with spinal muscular atrophy. *Cell* **80:**167–178.
19. **Ting, J. P., and B. K. Davis.** 2005. CATERPILLER: a novel gene family important in immunity, cell death, and diseases. *Annu. Rev. Immunol.* **23:**387–414.
20. **Wright, E. K., S. A. Goodart, J. D. Growney, V. Hadinoto, M. G. Endrizzi, E. M. Long, K. Sadigh, A. L. Abney, I. Bernstein-Hanley, and W. F. Dietrich.** 2003. Naip5 affects host susceptibility to the intracellular pathogen Legionella pneumophila. *Curr. Biol.* **13:**27–36.
21. **Zamboni, D. S., K. S. Kobayashi, T. Kohlsdorf, Y. Ogura, E. M. Long, R. E. Vance, K. Kuida, S. Mariathasan, V. M. Dixit, R. A. Flavell, W. F. Dietrich, and C. R. Roy.** Birc1e control over caspase-1 function restricts intracellular replication of microbial pathogens. *Nat. Immunol.* **7:**318–325.

Birc1e/Naip5 IN MACROPHAGE FUNCTION AND SUSCEPTIBILITY TO INFECTION WITH *LEGIONELLA PNEUMOPHILA*

Anne Fortier and Philippe Gros

74

Genetic analysis in the mouse has been used to identify host genes and proteins that play important roles in natural defences against a broad range of infectious diseases. In mouse, susceptibility to infection with *Legionella pneumophila* is genetically controlled. While peritoneal macrophages from most inbred strains including C57BL/6J are not permissive to intracellular replication of *L. pneumophila* Philadelphia-1, A/J macrophages allow uncontrolled bacterial proliferation. Segregation analysis in informative crosses derived from A/J and C57BL/6 mice has shown that differential intracellular replication of *Legionella* in this model is controlled by a single gene or group of closely linked genes, with resistance being inherited in a dominant fashion (14). This locus, designated *Lgn1*, was subsequently mapped to the distal portion of chromosome 13 (1, 3). High-resolution linkage mapping, together with physical mapping studies established that the minimal physical interval of the *Lgn1* locus contains a cluster of *Naip* genes (6) initially identified as positional candidates for human spinal muscular atrophy. Functional complementation studies in vivo in transgenic animals harboring individual BAC (bacterial artificial chromosomes) clones cov-

ering the *Lgn1* region was used to determine which of the *Birc1/Naip* gene copies is responsible for the *Lgn1* effect. Independent BAC clones derived from *Legionella*-resistant strains (C57BL/6J, 129X1) were microinjected into fertilized FVB eggs. F_0 animals were crossed to the *Legionella*-susceptible A/J strain ($Lgn1^s$) to generate [$F_0 \times$ A/J] F_1 mice, followed by backcrossing to A/J to transfer individual transgenic BAC clones onto the $Lgn1^s$ genetic background. Such animals were then tested for complementation of the *Legionella*-susceptibility phenotype. Positive complementation was observed for two BAC clones. The only full-length coding sequence contained within the region of overlap between these two BAC clones was the *Birc1e/Naip5* gene copy.

These results have suggested that overexpression of a functional *Birc1e/Naip5* polypeptide corrects *Legionella*-susceptibility in A/J macrophages (4). Additionally, strain-specific amino acid sequence polymorphisms have been identified in the *Birc1e/Naip5* protein from A/J mice, and suppression of Birc1e/Naip5 protein expression by antisense RNA was shown to enhance *L. pneumophila* replication (13). Those studies have clearly established that intact *Birc1e/Naip5* function is essential to control intracellular replication of *L. pneumophila* in mouse macrophages. However, the molecular

Anne Fortier and Philippe Gros   Department of Biochemistry, McGill University, Montreal, Qc, H3G 1Y6, Canada.

Legionella: State of the Art 30 Years after Its Recognition
Edited by Nicholas P. Cianciotto et al.
©2006 ASM Press, Washington, D.C.

mechanism by which Birc1e/Naip5 contributes to macrophage defenses against infections has remained poorly understood.

Intracellular pathogens have evolved different strategies to inhibit or escape the bacteriostatic or bactericidal defense mechanisms of host macrophages. For instance, some microorganisms, such as *Coxiella burnetii*, have developed protective mechanisms that allow them to replicate within acidified and fully mature phagolysosome; others, such as *Listeria monocytogenes* and *Shigella flexneri*, dissolve the phagosomal membrane and multiply in the cytoplasm; *Mycobacterium tuberculosis* and *Chlamydia trachomatis* inhibit maturation of the phagosome including fusion with lysosomes and multiply within a vacuole that retains endosomal markers. Finally, bacteria such as *Brucella abortus* and *L. pneumophila* modulate phagosome maturation to create a replicative niche that resembles the endoplasmic reticulum (ER). *Legionella* modulates the maturation of its vacuole in such a way that the *Legionella*-containing phagosome (LCP) does not acidify and avoids fusion with lysosomes; instead, LCPs become surrounded by vesicles and mitochondria (7). Electron microscopy studies indicated that LCP morphology diverges from classic phagosomes: within the first 5 min of infection, LCPs become surrounded by host vesicles; 15 min postinfection, the thickness of the phagosomal membrane resembles that of the ER, and 6 h after infection, ribosomes are found attached to the cytoplasmic face of LCPs (12). Although endocytic markers, such as Lamp1, Lamp2 and the lysosomal acid protease Cathepsin D, are excluded (2, 11), proteins residing in secretory vesicles cycling between the ER and Golgi apparatus are associated with most LCPs within 30 min, and resident ER proteins such as Calnexin are found associated with the LCP within 1 to 2 h (8). A recent study demonstrated the presence of the resident ER protein glucose-6-phosphatase (G6Pase), protein disulfide isomerase, and proteins with the ER-retention signal KDEL at LCPs, indicating ER fusion with the phagosome (10). It is this ER-derived organelle, called the replicative organelle, which

supports *Legionella* growth. This behavior is specific for virulent *Legionella* isolates and is not seen with avirulent species, which typically reside in vacuoles that interact sequentially with endocytic vesicles and subsequently with lysosomes.

To get better insight into the function and molecular mechanism of action of *Birc1e/Naip5* in controlling intracellular replication of *L. pneumophila*, we assessed the effect of resistant and susceptible *Birc1e/Naip5* genotypes on intracellular survival of *Legionella*, including phagosome maturation. We performed morphological and biochemical characterization of LCPs formed in macrophages from control A/J-susceptible (nontransgenic) and from transgenic mice that carry the resistant allele of *Birc1e/Naip5* using two-color immunofluorescence microscopy and electron microscopy. Nontransgenic permissive A/J peritoneal and bone marrow-derived macrophages allowed massive *L. pneumophila* replication over a 72-h infection period (2.8 ± 0.3 and 1.6 ± 0.4 log CFU respectively) compared to macrophages from *Birc1e/Naip5*-resistant transgenic animals (0.74 ± 0.18 and -0.52 ± 0.26 log CFU respectively). Differences in colocalization with lysosomal and ER markers were observed between LCPs formed in transgenic and nontransgenic macrophages; for example, $75 \pm 8\%$ of LCPs formed in *Birc1e/Naip5*-resistant transgenic macrophages acquired cathepsin D by 2 h postinfection compared to only $35 \pm 7\%$ in permissive macrophages and only $17 \pm 3\%$ of LCPs formed in *Birc1e/Naip5*-resistant transgenic macrophages were calnexin-positive 24 h postinfection compared to $46 \pm 4\%$ in permissive macrophages. Of LCPs formed in *Birc1e/Naip5*-resistant transgenic macrophages, within 1 h, $60 \pm 6\%$ acquired internalized endocytic marker (HRP) compared to $27 \pm 8\%$ in control macrophages. Ultrastructural studies showed the presence of ribosomes at the phagosomal membrane in $56 \pm 6\%$ of the LCPs formed in permissive macrophages and in only $21 \pm 4\%$ of the LCPs formed in *Birc1e/Naip5*-resistant transgenic macrophages. Furthermore, G6Pase staining can only be observed in the lumen of LCPs in nontransgenic macrophages 6 h postinfection.

Our results show that expression of the resistant allele of Birc1e/Naip5 impairs the ability of *Legionella* to escape the endocytic pathway and to modulate the formation of an ER-derived replicative vacuole. It has been proposed that, like other members of the NLR family, Birc1e/Naip5 may function as an intracellular sensor of *L. pneumophila* products and may trigger immune responses through inflammatory caspases (5, 9, 15).

REFERENCES

1. **Beckers, M. C., E. Ernst, E. Diez, C. Morissette, F. Gervais, K. Hunter, D. Housman, S. Yoshida, E. Skamene, and P. Gros.** 1997. High-resolution linkage map of mouse chromosome 13 in the vicinity of the host resistance locus *Lgn1. Genomics* **39:**245–263.

2. **Clemens, D. L., and M. A. Horwitz.** 1995. Characterization of the *Mycobacterium tuberculosis* phagosome and evidence that phagosomal maturation is inhibited. *J. Exp. Med.* **181:**257–270.

3. **Dietrich, W. F., D. M. Damron, R. R. Isberg, E. S. Lander, and M. S. Swanson.** 1995. *Lgn1,* a gene that determines susceptibility to *Legionella pneumophila,* maps to mouse chromosome 13. *Genomics* **26:**443–450.

4. **Diez, E., S. H. Lee, S. Gauthier, Z. Yaraghi, M. Tremblay, S. Vidal, and P. Gros.** 2003. *Birc1e* is the gene within the *Lgn1* locus associated with resistance to *Legionella pneumophila. Nat. Genet.* **33:**55–60.

5. **Fortier, A., E. Diez, and P. Gros.** 2005. *Naip5/Birc1e* and Susceptibility to *Legionella pneumophila. Trends Microbiol.* **13:**328–335.

6. **Growney, J. D., and W. F. Dietrich.** 2000. High-resolution genetic and physical map of the *Lgn1* interval in C57BL/6J implicates *Naip2* or *Naip5* in *Legionella pneumophila* pathogenesis. *Genome Res.* **10:**1158–1171.

7. **Horwitz, M. A.** 1983. The Legionnaires' disease bacterium (*Legionella pneumophila*) inhibits phagosome-lysosome fusion in human monocytes. *J. Exp. Med.* **158:**2108–2126.

8. **Kagan, J. C., and C. R. Roy.** 2002. *Legionella* phagosomes intercept vesicular traffic from endoplasmic reticulum exit sites. *Nat. Cell. Biol.* **4:**945–954.

9. **Martinon, F., and J. Tschopp.** 2005. NLRs join TLRs as innate sensors of pathogens. *Trends Immunol.* **26:**447–454.

10. **Robinson, C. G., and C. R. Roy.** 2005. Attachment and fusion of endoplasmic reticulum with vacuoles containing Legionella pneumophila. *Cell. Microbiol.* **Online Early**.

11. **Swanson, M. S., and R. R. Isberg.** 1996. Identification of Legionella pneumophila mutants that have aberrant intracellular fates. *Infect. Immun.* **64:**2585–2594.

12. **Tilney, L. G., O. S. Harb, P. S. Connelly, C. G. Robinson, and C. R. Roy.** 2001. How the parasitic bacterium *Legionella pneumophila* modifies its phagosome and transforms it into rough ER: implications for conversion of plasma membrane to the ER membrane. *J. Cell Sci.* **114:**4637–4650.

13. **Wright, E. K., S. A. Goodart, J. D. Growney, V. Hadinoto, M. G. Endrizzi, E. M. Long, K. Sadigh, A. L. Abney, I. Bernstein-Hanley, and W. F. Dietrich.** 2003. *Naip5* affects host susceptibility to the intracellular pathogen *Legionella pneumophila. Curr. Biol.* **13:**27–36.

14. **Yoshida, S., Y. Goto, Y. Mizuguchi, K. Nomoto, and E. Skamene.** 1991. Genetic control of natural resistance in mouse macrophages regulating intracellular *Legionella pneumophila* multiplication in vitro. *Infect. Immun.* **59:**428–432.

15. **Zamboni, D., K. S. Kobayashi, T. Kohlsdorf, Y. Ogura, E. M. Long, R. E. Vance, K. Kuida, S. Mariathasan, V. M. Dixit, R. A. Flavell, W. F. Dietrich, and C. R. Roy.** 2006. The Birc1e cytosolic pattern-recognition receptor contributes to the detection and control of *Legionella pneumophila* infection. *Nat. Immunol.* **7:**318–325.

LOCUS ON CHROMOSOME 13 IN MICE INVOLVED IN CLEARANCE OF *LEGIONELLA PNEUMOPHILA* FROM THE LUNGS

*Seiji Kobayashi, Fumiaki Kura, Junko Amemura-Maekawa,
Bin Chang, Naoki Yamamoto, and Haruo Watanabe*

75

Legionella pneumophila is a causative agent of pneumonia in humans. Until now, the mechanisms that control the *L. pneumophila* infection in the lungs have not yet been thoroughly assessed. We found that the mouse strain B10.A/SgSnSlcNiid (B10.A/Niid) was highly susceptible to intranasal infection by *L. pneumophila* 80-045 serogroup 1, while the mouse strain C57BL/10 was resistant to it (Fig. 1). The purpose of the present study is to identify the loci responsible for regulating *L. pneumophila* replication in the lungs, by comparing susceptible B10.A/Niid with other resistant inbred strains.

First, the susceptibility of B10.A/Niid mice might be a general response against the *L. pneumophila* species. Alternatively, the response could be strain dependent. Therefore, we tested the susceptibility to five kinds of laboratory and clinical strains of *L. pneumophila* serogroup 1. We observed that B10.A/Niid mice were susceptible to all the *L. pneumophila* serogroup 1 strains used in the experiment, whereas the

Seiji Kobayashi, Fumiaki Kura, Junko Amemura-Maekawa, Bin Chang, and Haruo Watanabe Department of Bacteriology, National Institute of Infectious Diseases, Toyama 1-23-1, Shinjuku-ku, Tokyo, 162-8640, Japan. *Naoki Yamamoto* Department of Molecular Virology, Graduate School of Medicine, Tokyo Medical and Dental University, Yushima 1-5-45, Bunkyo-ku, Tokyo, 113-8519, Japan.

C57BL/10 mice were not. In contrast, B10.A/Niid mice showed similar clearance of the *L. pneumophila* Chicago (serogroup 6) and *L. bozemanii* WIGA strains that was similar to the clearance observed in C57BL/10 mice. Thus, the susceptibility of B10.A/Niid mice could be partly attributed to their inability to sufficiently clear *L. pneumophila* serogroup 1 strains.

Second, in order to identify the locus that is responsible for the susceptibility phenotype on the major histocompatibility complex (MHC) region of the H-2^a haplotype, we analyzed the influence of the MHC haplotypes on *L. pneumophila* infection by using three types of B10.A intra-MHC-congenic mice, i.e., B10.A (2R), B10.A (3R), and B10.A (5R), as well as the parental mice, B10.A/SgSnJ, obtained from the Jackson laboratory. Surprisingly, *L. pneumophila* could not replicate in the lungs of any of these mice, indicating that the H-2 region might not be involved in the susceptibility of B10.A/Niid mice to *L. pneumophila* and that B10.A/Niid mice might possess spontaneous defects. Furthermore, a series of mating experiments were performed to characterize the genetic basis of the B10.A/Niid susceptibility phenotype. B10.A/Niid mice were mated with C57BL/10 mice, and the susceptibility of the F$_1$ progeny was assessed. All the (B10.A/Niid × C57BL/10) F$_1$ animals were resistant to *L. pneumophila*

Legionella: State of the Art 30 Years after Its Recognition
Edited by Nicholas P. Cianciotto et al.
©2006 ASM Press, Washington, D.C.

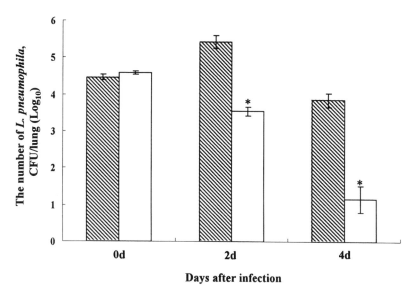

FIGURE 1 The mouse strain B10.A/Niid is susceptible to pulmonary *L. pneumophila* infection. B10.A/Niid (striped bar) and C57BL/10 strains (white bar) were infected with 3×10^4 CFU of *L. pneumophila* 80-045 (serogroup 1) per lung by intranasal infection. Data were mean ± standard deviation of 6 to 8 mice. The data shown are the combined data from two experiments. The asterisk (★) indicates a significant difference ($P < 0.05$) between B10.A/Niid and C57BL/10 strains.

infection in the lungs. The ratio of resistant individuals to susceptible individuals was 17:18 for B10.A/Niid × F_1 and F_1 × B10.A/Niid backcross progeny. Thus, the susceptibility of B10.A/Niid mice appears to be due to a major recessive gene mutation, although the susceptibility pattern of the F_2 progeny showed a somewhat continuous distribution between the resistant C57BL/10 and susceptible B10.A/Niid parents.

Third, to identify the responsible locus, we performed a whole-genome single-point linkage analysis on 280 (AKR/N × B10.A/Niid)F_2 mice using 81 simple sequence-length markers that are polymorphic for B10.A/Niid and AKR/N. On the basis of this analysis, an *L. pneumophila* susceptibility locus with significant linkage was mapped between *D13Die26* and *D13Mit287* (mapped position 54 to 57 cM) on chromosome 13 ($P < 0.00001$) (Fig. 2); no additional significant linkages were observed. The mapped locus was near the *Lgn1* locus that determines the ability of macrophages to

restrict *L. pneumophila* growth (1, 2). A/J mouse macrophages are permissive for *L. pneumophila* growth (*Lgn1^s*).

Last, we compared the replication in thioglycolate-elicited macrophages obtained from B10.A/Niid mice and the B10.A/Niid-*Lgn1^s* congenic mouse strain that was constructed by introgressively backcrossing (B10.A/Niid × A/J) F_1 mice to B10.A/Niid mice. We found that thioglycolate-elicited macrophages from B10.A/Niid mice could restrict *L. pneumophila* growth, while those from B10.A/Niid-*Lgn1^s* mice could not. This result suggests that the susceptibility locus in B10.A/Niid mice is different from *Lgn1*.

In conclusion, our present study indicates that B10.A/Niid mice have a genetic defect that affects *L. pneumophila* serogroup 1 growth in the lungs. Furthermore, we also found that the responsible locus was located near *Lgn1* but was phenotypically distinct from it. We propose to designate this locus as *Lgn2*.

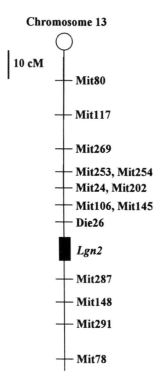

Chromosome 13

10 cM

Mit80

Mit117

Mit269

Mit253, Mit254

Mit24, Mit202

Mit106, Mit145

Die26

Lgn2

Mit287

Mit148

Mit291

Mit78

FIGURE 2 Schematic representation of chromosome 13. Markers shown were used to genotype mice of the F₂ cross. Approximate location of the markers was obtained from the mouse genome database. The centromere is indicated by a large circle.

REFERENCES

1. Diez, E., S. H. Lee, S. Gauthier, Z. Yaraghi, M. Tremblay, S. Vidal, and P. Gros. 2003. *Birc1e* is the gene within the *Lgn1* locus associated with resistance to *Legionella pneumophila. Nat. Genet.* **33**:55–60.

2. Wright, E. K., S. A. Goodart, J. D. Growney, V. Hadinoto, M. G. Endrizzi, E. M. Long, K. Sadigh, A. L. Abney, I. Bernstein-Hanley, and W. F. Dietrich. 2003. *Naip5* affects host susceptibility to the intracellular pathogen *Legionella pneumophila. Curr. Biol.* **13**:27–36.

INFLAMMATORY IMMUNE RESPONSE TO CYTOSOLIC FLAGELLIN PROTECTS MICE FROM *LEGIONELLA PNEUMOPHILA* INFECTION

Michele S. Swanson, Brenda G. Byrne, Natalie W. Whitfield,
Etsu T. Fuse, Kazuhiro Tateda, and Ari B. Molofsky

76

Legionella pneumophila has earned a reputation as a public scourge, and rightly so. This bacterium continues to cause dramatic epidemics of pneumonia when unsuspecting people inhale aerosols of contaminated water. However, specialists know that the attack rate in community outbreaks is actually relatively low. Typically, only ~5% of those exposed to *L. pneumophila* become ill, and most vulnerable people have preexisting conditions that weaken the natural defenses of their airways or immune systems. Thus, *L. pneumophila* is a classic opportunistic pathogen that most healthy people eliminate without overt symptoms. How the innate immune system effectively recognizes and responds to *L. pneumophila* within lungs is the focus of this study. By extending two key observations made by mouse and bacterial geneticists, we develop a model in which contamination of the macrophage cytoplasm by flagellin triggers a proinflammatory cell death that not only denies the pathogen its safehaven for replication but also recruits immune cells to join the fight.

HISTOPATHOLOGY OF LEGIONNAIRES' DISEASE

By analyzing 53 clinical cases of Legionnaires' disease, Winn and Myerowitz provided the first comprehensive picture of the lung pathology caused by *L. pneumophila* (33). In this typically bilateral disease, the lungs display acute bronchiolitis and alveolitis. The afflicted tissues are bathed in fluid rich in neutrophils, macrophages, and red blood cells mixed with copious fibrin and other proteinaceous material. Extensive lysis of white blood cells within the alveoli that contain *L. pneumophila* is another hallmark of the human infection.

With the goal of identifying the molecular basis of *L. pneumophila* pathogenesis, investigators quickly turned to animal models. Baskerville, Winn, and colleagues demonstrated that guinea pig infections mimic several aspects of human disease, including the acute inflammatory response in which alveoli accumulate fluid that contains white blood cells and fibrin (4, 5, 32). Likewise, a variety of other mammals were found to be susceptible to infection by *L. pneumophila,* including rhesus monkeys, marmosets, and rats (5, 32).

Despite the versatility of this promiscuous microbe, which replicates not only in mammals and their phagocytes, but also within a variety of amoebae and protozoa, mice are

Michele S. Swanson, Brenda G. Byrne, Natalie W. Whitfield, and Ari B. Molofsky Department of Microbiology and Immunology, University of Michigan Medical School, Ann Arbor, MI 48109-0620. *Etsu T. Fuse and Kazuhiro Tateda* Department of Microbiology and Infectious Diseases, Toho University School of Medicine, 5-21-16 Ohmorinishi, Ohtaku, Tokyo 143-8540, Japan.

naturally resistant to infection. By comparing *L. pneumophila* replication in macrophages isolated from six strains of mice, Yamamoto and colleagues identified one exception: the A/J mouse (34). Brieland and colleagues verified that, in addition to bacterial replication within macrophages, key histological and immunological features of the A/J mouse response to *L. pneumophila* mimicked that of humans and guinea pigs (8). Inflammation, accumulation of neutrophils and macrophages, and death of host cells characterize the lungs a few days after infection by the respiratory route. By exploiting the observation that susceptibility to *L. pneumophila* behaves as a single Mendelian trait, the Dietrich and Gross laboratories identified *naip5* as a critical determinant of mouse resistance to this pathogen (reviewed in reference 12). Based on these and other pioneering pathology and immunology studies in the 1980s and 1990s, workers in the field today continue to depend on small animal models to investigate the host-pathogen interactions that govern the outcome of *L. pneumophila* infection.

L. PNEUMOPHILA TOXICITY TO MACROPHAGES REQUIRES BOTH FLAGELLIN AND TYPE IV SECRETION

Seeking the bacterial factors responsible for the inflammation and cell death that are clinical hallmarks of Legionnaires' disease, Conlan and coworkers focused on proteases secreted by *L. pneumophila*. When instilled into the airways of guinea pigs, one of its proteases causes the tissue injury and inflammation typical of infection by live microbes (3). Nevertheless, additional microbial factors likely damage host tissue, since the pathology caused by *L. pneumophila* that lack major secretory protease is indistinguishable from that produced by the wild-type strain (6).

In addition to secreted proteases, *L. pneumophila* expresses another toxic activity—one that requires an intimate association with the target cell (19). Husmann and Johnson demonstrated that when contacted directly by large numbers of *L. pneumophila*, guinea pig and mouse macrophages rapidly die. Using lysis of red and white blood cells to quantify this toxic activity, Kirby and colleagues discovered that *L. pneumophila* utilize type IV secretion to insert pores of <3 nm diameter into eukaryotic membranes (21).

Although a type IV secretion system is necessary for *L. pneumophila* to intoxicate macrophages, it is not sufficient. To lyse macrophages, *L. pneumophila* also requires flagellin (25). First, it was noted that *L. pneumophila* simultaneously becomes highly cytotoxic and motile as the replication period ends during culture in either broth or macrophages (1, 9). A closer analysis revealed that *L. pneumophila* that lack flagella are not toxic to macrophages, yet they still insert pores into red blood cells by type IV secretion (25). By engineering a panel of mutants with defects in synthesis, assembly, or activity of flagella, Molofsky and coworkers established that motility was dispensable for macrophage intoxication, whereas flagellin protein was essential (Fig. 1). Thus, by a mechanism that requires the pore-forming type IV secretion system, *L. pneumophila* flagellin triggers lysis of macrophages.

BACTERIAL FLAGELLINS STIMULATE INFLAMMATION IN MAMMALS

Flagellar-based motility is mediated by a sophisticated organelle whose components are strictly regulated and whose assembly is carefully orchestrated. As expected for a complex multicomponent structure, the proteins of the flagellum are highly conserved among bacterial species. In the bacterial world, flagella contribute to many essential functions, including motility, chemotaxis, biofilm formation, and adherence to inert surfaces or host tissues (reviewed in reference 29). Among the pathogenic microbes, flagellar-based motility is especially critical for colonization of mucosal tissues, such as those that line the stomach and the gut. Likewise, flagella are critical for *L. pneumophila* fitness. A survey of 54 strains of *Legionella* conducted by the Fields laboratory revealed that 40 express flagellin, as judged by reactivity to a specific antibody (7). Furthermore, the majority

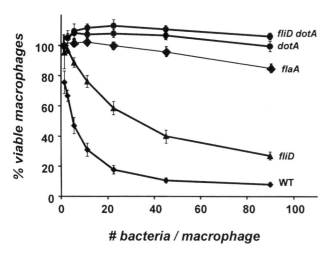

FIGURE 1 *L. pneumophila* requires flagellin but not motility to induce macrophage death. After centrifugation with two-fold dilutions of the strains indicated, A/J mouse macrophages and microbes were incubated for 1 h, and then viability was determined by Alamar Blue reduction. Shown are mean percent of viable macrophages ± standard error pooled from three or more experiments in multiplicity of infection bins of two-fold dilutions; the middle value for each bin is indicated. WT, wild-type *L. pneumophila* strain Lp02; *dotA* mutants lack type IV secretion but are fully motile; *flaA* mutants lack the structural protein flagellin; *fliD* mutants secrete but do not assemble flagellin; *fliD dotA* mutants lack motility and type IV secretion but secrete flagellin protein.

of the nonflagellate strains failed to replicate in the amoebal host, *Hartmannella vermiformis*. Using a genetic approach, the Fields laboratory extended and strengthened the link between motility and virulence. Of 10 transposon mutants isolated on the basis of their failure to express flagellin protein, 7 were also defective for replication in amoebae and macrophages (28). Subsequently, other workers found that mutants that lack the flagellin gene *flaA* infect host cells poorly and are not toxic to macrophages (10, 25). Furthermore, *L. pneumophila* that lack FliA, a sigma factor known to be critical for transcription of flageller genes, also produce excess pigment, evade macrophage lysosomes poorly, and fail to replicate in the amoebae *Dicytostelium discoideum* (15, 18, 25). Thus, to thrive in its natural environment and in tissue culture models of infection, *L. pneumophila* relies on its flagellar regulon to coordinate expression of multiple transmission traits.

On the other hand, its conservation, extracellular location, and critical role in colonization together make flagella a prized target of the immune system (reviewed in reference 29). Thus, flagellin is classified as a PAMP, or pathogen–associated molecular pattern, an early warning of infection that is readily detected by the innate immune system. Extracellular flagellin is recognized by a receptor on the surface of eukaryotic cells, Toll-like receptor 5 (TLR-5; 17). The TLR adaptor protein MyD88 then activates the NF-κB and the mitogen-activated protein kinase signal transduction pathways, which coordinate production of secreted factors that mount the innate immune response, including tumor necrosis factor alpha interleukin 6 (IL-6), IL-8, and nitric oxide (reviewed in reference 29). As a consequence, inflammation ensues: Professional phagocytes are recruited to the site and activated to combat the infection. That flagellin itself triggers an inflammatory response has been well established in a variety of experimental models of respiratory and gastrointestinal infections, including those caused by *Escherichia coli*, *Salmonella enterica* serovars, and *Pseudomonas aeruginosa* (reviewed in reference 29). Likewise, *L. pneumophila* flagellin is recognized by TLR5 on human cells (16), and its ability to trigger release of proinflammatory cytokines has been demonstrated in a mouse model of disease (30).

MOUSE MACROPHAGE SUSCEPTIBILITY TO FLAGELLIN DEPENDS ON Naip5 AND CASPASE 1

Knowing that bacterial flagellins trigger a rapid, proinflammatory innate immune response, we postulated that lysis of macrophages by motile *L. pneumophila* was not a pathogen tactic, but rather a host defense. Indeed, the intracellular pathogens *Shigella flexneri*, *Salmonella enterica*, and *Francisella tularenesis* each stimulate a caspase 1-dependent cell death in which macrophages rapidly release mature IL-1β and IL-18 (reviewed in references 11, 22). Therefore, we investigated whether *L. pneumophila* triggers a similar cell death in macrophages, one that Cookson and colleagues have named "pyroptosis" from the Greek *pyro,* to invoke fire and fever, and *ptosis,* or "falling," to emphasize that the death is programmed (11).

For this purpose, we analyzed a panel of *L. pneumophila* flagellar regulon and type IV secretion mutants in a series of cell biological assays. Together, the results show that in mouse macrophages, *L. pneumophila* induces a rapid proinflammatory death that is distinct from classic apoptosis, as judged by multiple criteria: nuclear morphology, plasma membrane permeability, secretion of IL-1β, caspase-3 activation, sensitivity to caspase inhibitors, and the speed of the response. After exposure for 1 h to Flagellin[+] type IV secretion[+] *L. pneumophila*, A/J mouse macrophages were permeable and had condensed nuclei. As expected for pyroptosis, macrophages are protected from flagellin-dependent toxicity by Ac-YVAD-CHO, a peptide inhibitor of caspase-1. In contrast, a caspase-3 inhibitor has little effect on macrophage intoxication or viability, as judged by lactate dehydrogenase release and by reduction of Alamar Blue, respectively. Furthermore, within 1 h of exposure to *L. pneumophila*, macrophages released

IL-1β (Fig. 2). Liberation of active IL-1β required not only *L. pneumophila* type IV secretion and flagellin, but also macrophage caspase-1 activity.

Because the protein Naip5 confers resistance of mice to *L. pneumophila*, we investigated whether it contributes to the host response to Flagellin[+] microbes. Naip5 is composed of three modules: N-terminal baculoviral inhibitor-of-apoptosis repeats, a central nucleotide-binding oligomerization (NOD) domain, and C-terminal leucine-rich repeats (reviewed in reference 12). Analogous to Toll-like receptors, so called NOD proteins are thought to detect extracellular microbial products in the cytosol. The biochemical activity of murine Naip5 is not yet known, but the closely related human protein Naip/Birc1 inhibits apoptosis by binding effector caspases through its baculoviral inhibitor-of-apoptosis repeat domain. Other NOD family members, including IPAF and the NALPs, interact with the inflammasome, a caspase-1-containing complex that can be triggered by microbial products (reviewed in reference 23). Thus, in response to the intracellular pathogens *S. enterica, S. flexneri,* and *F. tularensis,* the inflammasome initiates a proinflammatory cell death, or pyroptosis (11, 22).

Indeed, restrictive Naip5[+] C57Bl/6 macrophages are more sensitive than permissive *naip5* mutant A/J cells to Flagellin[+] *L. pneu-*

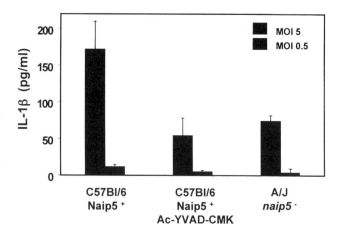

FIGURE 2 Secretion of IL-1β by macrophages in response to *L. pneumophila* is controlled by Naip5 and caspase-1. After infecting the macrophages shown for 1 h with wild-type *L. pneumophila* at the multiplicity of infection indicated, secreted interleukin-1β was quantified. Where indicated, macrophages were treated for 1 h before and during the infection with 100 μM Ac-YVAD-cmk, an inhibitor of caspase-1.

TABLE 1 Recognition of flagellin by mouse macrophages restricts *L. pneumophila* infection

Trait	Resistant C57Bl/6 mice	Permissive A/J mice
Naip5 protein	+	−/+
Susceptibility of macrophages to intoxication by cytosolic flagellin	+	+/−
IL-1β secretion induced by flagellin	+	+/−
Restriction of replication in macrophages		
Wild-type *L. pneumophila*	+	−
flaA mutant *L. pneumophila*	−	−
Restriction of replication in lung		
Wild-type *L. pneumophila*	+	−/+
flaA mutant *L. pneumophila*	−/+	−/+

mophila equipped for type IV secretion, as judged by the quantities of lactate dehydrogenase and IL-1β the macrophages released 1 or 6 h after exposure to *L. pneumophila* (Fig. 2, Table 1). Compared to resistant cells, A/J macrophages express reduced levels of a Naip5 mutant protein that harbors multiple amino acid substitutions (reviewed in reference 12). When infected by type IV secretion-competent, Flagellin$^+$ *L. pneumophila*, murine macrophages initiated a Naip5- and caspase-1-dependent proinflammatory death program (Fig. 2).

L. pneumophila can induce apoptosis in numerous cell types, including human peripheral blood monocytes, human monocyte, epithelial and T cell lines, and mouse alveolar macrophages and epithelial cells (13, 14, 26, 27). Therefore, we tested more rigorously whether flagellin triggers classic apoptosis in mouse macrophages. After a 5-h treatment with staurosporine, an inducer of apoptosis, 30% of the cells contained activated caspase-3, as judged by immunofluorescence microscopy, and a majority showed morphological hallmarks of an apoptotic response including chromatin condensation and nuclear blebbing, intact plasma membranes, and viability. A different pattern was observed for macrophages incubated with a high multiplicity of infection of wild-type *L. pneumophila*: <5% contained appreciable activated caspase-3 even 5 h after infection, and by 1 h, ~75% of the cells were permeable and had condensed nuclei. Thus, in mouse

macrophages, flagellate *L. pneumophila* induces a rapid proinflammatory death best described as pyroptosis.

L. PNEUMOPHILA THAT LACK FLAGELLIN ESCAPE DEFENSES OF MOUSE MACROPHAGES AND LUNGS

If the macrophage response to flagellin is indeed a host mechanism to combat infection, we postulated that *L. pneumophila* that lack flagellin would escape the Naip5. As expected, Naip5$^+$ C57Bl/6 macrophages restricted replication of wild-type *L. pneumophila* and other strains that encode flagellin (wild type, *fliD, flaA pFlaA, flhB, fliI, motAB*) to the level observed for type IV secretion mutants. In stark contrast, flagellin-null mutants replicated freely in restrictive Naip5$^+$ C57Bl/6 or BALB/c macrophages: Their yield increased more than 100-fold during a 3-day infection, comparable to that achieved by wild-type *L. pneumophila* in permissive *naip5* mutant A/J macrophages (Table 1).

Even more striking, flagellin also has a dramatic impact on the outcome of *L. pneumophila* infection in the mouse model of disease. As documented previously, within the lungs of restrictive Naip5$^+$ mice, wild-type *L. pneumophila* failed to replicate; by the third day, the number of colony-forming units had decreased ~50 fold. In striking contrast, *L. pneumophila* that lack flagellin replicated in Naip5$^+$ C57BL/6 mice: The number of colony-forming units gradually increased for 2 days, then rapidly declined, reproducing the pattern of both wild-type and *flaΛ* mutant *L. pneumophila* within the lungs of permissive mice (Table 1). Therefore, detection of flagellin is one critical component of the robust murine innate immune response that confers resistance to *L. pneumophila* infection.

WHAT DOES THE MOUSE MODEL OF *L. PNEUMOPHILA* INFECTION TEACH US ABOUT HUMAN DISEASE?

By considering the results of these and published laboratory experiments in the context of the histopathology of clinical infection, the following model can be put forth (Fig. 3). During phagocytosis, the *L. pneumophila* type IV secretion system inserts pores into the macrophage membrane to deliver virulence effectors that perturb phagosome maturation. Flagellin protein that diffuses through these pores is detected by Naip5, either directly via its leucine-rich repeat region, or indirectly by hetero-oligomerization with another NOD protein that itself binds flagellin. Consequently, Naip5 activates the inflammasome, either directly or by cooperating with adaptor proteins that regulate caspase-1, such as ASC, Ipaf, or a NALP protein (reviewed in reference 23). The activated inflammasome then coordinates matura-

tion and secretion of proinflammatory cytokines to combat the infection. In particular, we postulate that release of IL-1β and IL-18 coordinates the inflammatory response, including recruitment of neutrophils and development of the Th1 phenotype and subsequent production of gamma interferon, hallmarks of a successful response to *L. pneumophila* infection (31). Having identified key bacterial and host factors that govern the innate immune response of cultured macrophages, we are now in a position to test key aspects of this model using a more realistic experimental system, respiratory infection of mice.

Although *L. pneumophila* is cleared efficiently by most mouse strains, other *Legionella* species replicate in a variety of mice. For example, unlike *L. pneumophila* serogroup, 1 strains of *L. dumoffii*, *L. micdadei*, *L. bozemanae*, and *L. feeleii* replicated freely not only in macrophages from A/J mice, but also those of C57BL/6 and DBA/2 mice (20, 24). Therefore, additional insight into the innate immune response to *L. pneumophila* can be obtained by verifying that other species of *Legionella* are also motile and form pores in mammalian membranes. If so, it would be valuable to learn if their flagellin protein contains the motif that presumably in-

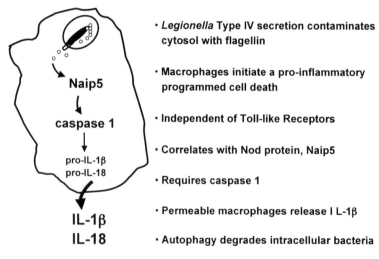

FIGURE 3 Model for the innate immune response of mice to *L. pneumophila* infection. See text for details.

teracts directly with a NOD cytosolic receptor protein. In this regard, it is noteworthy that many species of microbes encode a flagellin that lacks the amino acid sequence that makes the microbe vulnerable to detection by TLR5 (2).

Another line of inquiry that warrants attention is a direct comparison of the response to cytosolic flagellin by human and mouse macrophages. Although *L. pneumophila* infections in A/J mouse lungs exhibit many key features of human disease, the molecular mechanisms that control the host response may vary. For example, humans encode a single *naip* gene, whereas mice carry a tandem repeat of seven copies of a homologous gene. Therefore, the respective contributions of human and mouse TLR and NOD pathways to detection of PAMPs, including flagellin, remains to be investigated.

REFERENCES

1. **Alli, O. A., L. Y. Gao, L. L. Pedersen, S. Zink, M. Radulic, M. Doric, and Y. Abu Kwaik.** 2000. Temporal pore formation-mediated egress from macrophages and alveolar epithelial cells by *Legionella pneumophila. Infect. Immun.* **68:**6431–6440.
2. **Andersen-Nissen, E., K. D. Smith, K. L. Strobe, S. L. Barrett, B. T. Cookson, S. M. Logan, and A. Aderem.** 2005. Evasion of Toll-like receptor 5 by flagellated bacteria. *Proc. Natl. Acad. Sci. USA* **102:**9247–9252.
3. **Baskerville, A., J. W. Conlan, L. A. Ashworth, and A. B. Dowsett.** 1986. Pulmonary damage caused by a protease from *Legionella pneumophila. Br. J. Exp. Pathol.* **67:**527–536.
4. **Baskerville, A., A. B. Dowsett, R. B. Fitzgeorge, P. Hambleton, and M. Broster.** 1983. Ultrastructure of pulmonary alveoli and macrophages in experimental Legionnaires' disease. *J. Pathol.* **140:**77–90.
5. **Baskerville, A., R. B. Fitzgeorge, M. Broster, and P. Hambleton.** 1983. Histopathology of experimental Legionnaires' disease in guinea pigs, rhesus monkeys and marmosets. *J. Pathol.* **139:** 349–362.
6. **Blander, S. J., L. Szeto, H. A. Shuman, and M. A. Horwitz.** 1990. An immunoprotective molecule, the major secretory protein of *Legionella pneumophila*, is not a virulence factor in a guinea pig model of Legionnaires' disease. *J. Clin. Invest.* **86:**817–824.
7. **Bosshardt, S. C., R. F. Benson, and B. S. Fields.** 1997. Flagella are a positive predictor for virulence in *Legionella. Microb. Pathogen.* **23:**107–

112.
8. **Brieland, J., P. Freeman, R. Kunkel, C. Chrisp, M. Hurley, J. Fantone, and C. Engleberg.** 1994. Replicative *Legionella pneumophila* lung infection in intratracheally inoculated A/J mice. *Am. J. Pathol.* **145:**1537–1546.
9. **Byrne, B., and M. S. Swanson.** 1998. Expression of *Legionella pneumophila* virulence traits in response to growth conditions. *Infect. Immun.* **66:** 3029–3034.
10. **Dietrich, C., K. Heuner, B. C. Brand, J. Hacker, and M. Steinert.** 2001. Flagellum of *Legionella pneumophila* positively affects the early phase of infection of eukaryotic host cells. *Infect. Immun.* **69:**2116–2122.
11. **Fink, S. L., and B. T. Cookson.** 2005. Apoptosis, pyroptosis, and necrosis: mechanistic description of dead and dying eukaryotic cells. *Infect. Immun.* **73:**1907–1916.
12. **Fortier, A., E. Diez, and P. Gros.** 2005. Naip5/Birc1e and susceptibility to *Legionella pneumophila. Trends Microbiol.* **13:**328–335.
13. **Gao, L.-Y., and Y. Abu Kwaik.** 1999. Apoptosis in macrophages and alveolar epithelial cells during early stages of infection by *Legionella pneumophila* and its role in cytopathogenicity. *Infect. Immun.* **67:**862–870.
14. **Hagele, S., J. Hacker, and B. Brand.** 1998. *Legionella pneumophila* kills human phagocytes but not protozoan host cells by inducing apoptotic death. *FEMS Microbiol. Lett.* **169:**51–58.
15. **Hammer, B. K., E. S. Tateda, and M. S. Swanson.** 2002. A two-component regulator induces the transmission phenotype of stationary-phase *Legionella pneumophila. Mol. Microbiol.* **44:**107–118.
16. **Hawn, T. R., A. Verbon, K. D. Lettinga, L. P. Zhao, S. S. Li, R. J. Laws, S. J. Skerrett, B. Beutler, L. Schroeder, A. Nachman, A. Ozinsky, K. D. Smith, and A. Aderem.** 2003. A common dominant TLR5 stop codon polymorphism abolishes flagellin signaling and is associated with susceptibility to Legionnaires' disease. *J. Exp. Med.* **198:**1563–1572.
17. **Hayashi, F., K. D. Smith, A. Ozinsky, T. R. Hawn, E. C. Yi, D. R. Goodlett, J. K. Eng, S. Akira, D. M. Underhill, and A. Aderem.** 2001. The innate immune response to bacterial flagellin is mediated by Toll-like receptor 5. *Nature* **410:** 1099–1103.
18. **Heuner, K., C. Dietrich, C. Skriwan, M. Steinert, and J. Hacker.** 2002. Influence of the alternative sigma(28) factor on virulence and flagellum expression of *Legionella pneumophila. Infect. Immun.* **70:**1604–1608.
19. **Husmann, L. K., and W. Johnson.** 1994. Cytotoxicity of extracellular *Legionella pneumophila*.

Infect. Immun. **62:**2111–2114.

20. **Izu, K., S. Yoshida, H. Miyamoto, B. Chang, M. Ogawa, H. Yamamoto, Y. Goto, and H. Taniguchi.** 1999. Grouping of 20 reference strains of Legionella species by the growth ability within mouse and guinea pig macrophages. *FEMS Immunol. Med. Microbiol.* **26:**61–68.

21. **Kirby, J. E., J. P. Vogel, H. L. Andrews, and R. R. Isberg.** 1998. Evidence of pore-forming ability by *Legionella pneumophila. Mol. Microbiol.* **27:**323–336.

22. **Mariathasan, S., D. Weiss, V. Dixit, and D. Monack.** 2005. Innate immunity against *Francisella tularensis* is dependent on the ASC/caspase-1 axis. *J. Exp. Med.* **202:**1043–1049.

23. **Martinon, F., K. Burns, and J. Tschopp.** 2002. The inflammasome: a molecular platform triggering activation of inflammatory caspases and processing of proIL-beta. *Mol. Cell.* **10:**417–426.

24. **Miyamoto, H., K. Maruta, M. Ogawa, M. C. Beckers, P. Gros, and S. Yoshida.** 1996. Spectrum of Legionella species whose intracellular multiplication in murine macrophages is genetically controlled by Lgn1. *Infect. Immun.* **64:**1842–1845.

25. **Molofsky, A. B., L. M. Shetron-Rama, and M. S. Swanson.** 2005. Components of the *Legionella pneumophila* flagellar regulon contribute to multiple virulence traits, including lysosome avoidance and macrophage death. *Infect. Immun.* **73:**5720–5734.

26. **Muller, A., J. Hacker, and B. Brand.** 1996. Evidence for apoptosis of human macrophage-like HL60 cells by *Legionella pneumophila* infection. *Infect. Immun.* **64:**4900–4906.

27. **Neumeister, B., M. Faigle, K. Lauber, H. Northoff, and S. Wesselborg.** 2002. *Legionella pneumophila* induces apoptosis via the mitochondrial death pathway. *Microbiology* **148:**3639–3650.

28. **Pruckler, J. M., R. F. Benson, M. Moyenuddin, W. T. Martin, and B. S. Fields.** 1995. Association of flagellum expression and intracellular growth of *Legionella pneumophila. Infect. Immun.* **63:**4928–4932.

29. **Ramos, H. C., M. Rumbo, and J. C. Sirard.** 2004. Bacterial flagellins: mediators of pathogenicity and host immune responses in mucosa. *Trends Microbiol.* **12:**509–517.

30. **Ricci, M. L., A. Torosantucci, M. Scaturro, P. Chiani, L. Baldassarri, and M. C. Pastoris.** 2005. Induction of protective immunity by *Legionella pneumophila* flagellum in an A/J mouse model. *Vaccine* **23:**4811–4820.

31. **Tateda, K., T. A. Moore, J. C. Deng, M. W. Newstead, X. Zeng, A. Matsukawa, M. S. Swanson, K. Yamaguchi, and T. J. Standiford.** 2001. Early recruitment of neutrophils determines subsequent T1/T2 host responses in a murine model of *Legionella pneumophila* pneumonia. *J. Immunol.* **166:**3355–3361.

32. **Winn, W. C.** 1982. Legionnaire's pneumonia after intratracheal inoculation of guinea pigs and rats. *Lab Invest.* **47:**568–578.

33. **Winn, W. C., and Myerowitz, R. L.** 1981. The pathology of the legionella pneumonias. *Hum. Path.* **12:**401–422.

34. **Yamamoto, Y., T. W. Klein, C. A. Newton, R. Widen, and H. Friedman.** 1988. Growth of *Legionella pneumophila* in thioglycolate-elicited peritoneal macrophages from A/J mice. *Infect. Immun.* **56:**370–375.

A PEPTIDOGLYCAN-ASSOCIATED LIPOPROTEIN OF *LEGIONELLA PNEUMOPHILA* ACTIVATES TOLL-LIKE RECEPTOR 2 IN MURINE MACROPHAGES

Mi Jeong Kim, In Kyeong Lee, Jin-Ah Yang, Hee-Sun Sim, Jang Wook Sohn, and Min Ja Kim

77

The innate immune system has an important function in activation and shaping of the adaptive immune response through the induction of costimulatory molecules and cytokines. The Toll-like receptors (TLRs), which have recently been characterized as receptors of innate immunity (1), are one of the most important pattern recognition receptor families. TLRs recognize various classes of pathogen-associated molecular pattern (2, 3).

Legionnaires' disease is an important cause of epidemic and sporadic pneumonia in humans. Although the mechanisms are not incompletely understood, cell-mediated immunity against *Legionella pneumophila* plays an important role in inhibition of bacterial growth in a susceptible host, thereby facilitating resolution of *Legionella* infections. We have previously shown that a 19-kDa *Legionella* peptidoglycan-associated lipoprotein (PAL) is a species-common antigen that successfully induces the PAL-specific immune

responses in mice delivered with plasmid DNA3-PAL (4). The purpose of this study was to determine whether PAL can activate TLRs involved in innate immunity that might be linked to development of specific immune responses.

Recombinant-PAL (rPAL) of *L. pneumophila* serogroup 1 was prepared as described previously (4). Purified rPAL was checked for lipopolysaccharide (LPS) contamination by the kinetic assay. Interleukin 6 (IL-6) and tumor necrosis factor alpha (TNF-α) production were measured in peritoneal macrophages from BALB/c mice by enzyme-linked immunosorbent assay. Mice macrophages were preincubated with TLR2- or TLR4-blocking monoclonal antibody (MAb) (20 μg/ml) for 30 min and then subjected to stimulation with 10 to 1000 ng of r-PAL per ml for 18 h. Compared with cytokine production by r-PAL in the absence of the blocking antibodies, both TNF-α and IL-6 production were inhibited with TLR2- or TLR4-blocking MAb. Inhibition of cytokine production was significantly higher with TLR2-blocking antibody than TLR4 antibody ($P < 0.001$), depending on the dose of rPAL. The degree of inhibition by TLR2 antibody ranges from 88 to 97% for IL-6, and 66.7 to 81.8% for TNF-α, whereas inhibition by

Mi Jeong Kim, In Kyeong Lee, Jin Ah-Yang, and Hee-Sun Sim Research Institute of Emerging Infectious Disease, Korea University, Seoul, 136-705, Republic of Korea. *Jang Wook Sohn and Min Ja Kim* Research Institute of Emerging Infectious Disease, and Division of Infectious Disease, Department of Internal Medicine, Seoul 136-705, Republic of Korea.

TLR4 antibody ranges from 16.4 to 51% for IL-6, and 23.8 to 66.7% for TNF-α. To exclude the effect of possible *Escherichia coli* LPS contamination in the rPAL preparation on the cytokine production, the neutralization experiment with TLR2-blocking antibody was performed in macrophages from the LPS-hyporesposive C3H/HeJ mice. It also showed that r-PAL-induced cytokine production was inhibited significantly in the presence of TLR2-blocking antibody. These results indicate that the *Legionella* PAL is likely to activate TLR2-mediated signaling in murine macrophages, and they have important implications for the development of immune responses by *Legionella* surface antigens.

REFERENCES

1. **Akira, S., K. Takeda, and T. Kaisho.** 2001. Toll like receptors: critical proteins linking innate and acquired immunity. *Nat. Immunol.* 675–680.
2. **Brightbill, H. D., D. H. Libraty, and S. R. Krutzik.** 1999. Host defense mechanisms triggered by microbial lipoproteins through Toll like receptor. *Science* **285:**732–736.
3. **Takeuchi, O., K. Hoshino, T. Kawai, H. Sanjo, H. Takada, T. Ogawa, K. Takeda, and S. Akira.** 1999. Differential roles of TLR2 and TLR4 in recognition of gram-negative and gram-positive bacterial cell wall components. *Immunity* **11:**443–451.
4. **Yoon, W. S., S. H. Park, Y. K. Park, S. C. Park, J. I. Sin, and M. J. Kim.** 2002. Comparison of responses elicited by immunization with a Legionella species common lipoprotein delivered as naked DNA or recombinant protein. *DNA Cell Biol.* **21:**99–107.

LEGIONELLA INFECTION OF BONE MARROW DENDRITIC CELLS INDUCES MODULATION BY CATECHINS

James Rogers, Izabella Perkins, Alberto van Olphen,
Nicholas Burdash, Thomas W. Klein, and Herman Friedman

78

Green tea polyphenols include (-)-epigallocatechin gallate (EGCG), (-)-epigallocatechin (EGC), (-)-epicatechin gallate (ECG), and (-)-epicatechin (EC). EGCG is the major polyphenol present in green tea and is commonly known as tea catechin, accounting for more than 40% of polyphenols in green tea (7).

Previous studies from this laboratory showed that this catechin has marked modulating effects on cytokine production by immune cells. For example, reports from this laboratory on the continuous murine alveolar macrophage cell line (MH-S cells) showed that EGCG induced selective upregulation of Th1 helper cell cytokines important for antimicrobial cell-mediated immunity (interleukin 12 [IL-12] and tumor necrosis factor alpha [TNF-α]) and simultaneous downregulation of IL-10, an interleukin associated with biasing toward Th2 helper cells important in humoral antibody-based immunity (5). Other studies from this laboratory demonstrated upregulation of macrophage-produced gamma interferon, including mRNA for gamma interferon, in macrophage cultures treated with EGCG (6).

In the present study, primary murine (BALB/c) bone marrow-derived dendritic cells (DCs), a phagocytic monocytic cell essential for innate immunity to intracellular microorganisms like *Legionella*, were infected in vitro with the bacteria. Production of the Th1 helper cell-activating cytokine IL-12 and the proinflammatory cytokine TNF-α produced mainly by phagocytic cells and important for antimicrobial immunity was determined in cell cultures by enzyme-linked immunosorbent assay (ELISA).

Treatment of the cells with EGCG enhanced production of TNF-α in a dose-dependent manner in the DC cultures (Fig. 1). In contrast, EGCG inhibited, in a dose-dependent manner, production of IL-12 (Fig. 2).

In related studies, it was recently reported that EGCG inhibited IL-12 production by lipopolysaccharide-treated DCs (1). In contrast, topical application of EGCG before UVB exposure has been reported to increase IL-12 production in draining lymph nodes of C3H/HeN mice (3).

EGCG has also been reported to decrease LPS-induced TNF-α production in a dose-dependent manner in the murine macrophage cell line RAW 264.7 and to similarly inhibit LPS-induced TNF-α production in elicited BALB/c mouse peritoneal macrophages, effects

James Rogers, Izabella Perkins, Alberto van Olphen, Nicholas Burdash, Thomas W. Klein, and Herman Friedman Department of Medical Microbiology and Immunology, University of South Florida 12901 Bruce B. Downs Boulevard, Tampa, FL 33612.

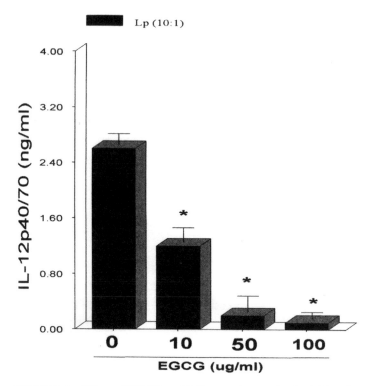

FIGURE 1 Effects of EGCG on TNF-α production by dendritic cells infected for 24 h with *L. pneumophila* (Lp). TNF-α levels in culture supernatants were determined by enzyme-linked immunosorbent assay and results are expressed as a mean value in nanograms per milliliter ± standard error of the mean from three independent experiments. The asterisk indicates statistically significant differences of $P < 0.05$ from values obtained with non-EGCG-treated *L. pneumophila*-infected DCs.

attributed in part through blocking NF-κB activation (10). However, previous studies from this laboratory have shown that EGCG diminished nicotine-induced inhibition of TNF-α production by alveolar macrophages infected with *Legionella pneumophila* as well as reversed cigarette smoke condensate downregulation of TNF-α in response to *L. pneumophila* infection (6). In cultured human peripheral blood mononuclear cells, EGCG was also reported to stimulate production of TNF-α (9). All of these reports suggest that cytokine modulatory effects of catechins have varied effects depending upon the immune cell subpopulation studied.

Assessment of the physiological relevance of the findings presented here must take into account maximum achievable EGCG concen-

trations attainable in vivo. For example, a 1,600-mg oral dose of EGCG under fasting conditions has been reported to achieve a maximum human plasma level of 3.4 μg/ml (8). This level is three times less than the lowest concentration used in this in vitro study (10 μg), equivalent to eight times the highest reported daily intake from tea (8), making it likely that only pharmaceutical prepared formulations of green tea would reach the plasma catechin levels used in this study. However, it is likely that concentrations of catechins at tissue sites are higher than in the blood. For example, concentrations of EGCG in the oral cavity 400 to 1,000 times greater than that in plasma have been obtained when green tea solution (1.2 g green tea solids per 200 ml of water) is

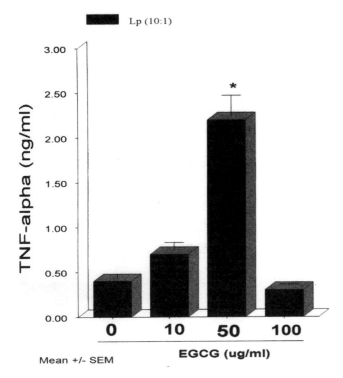

Lp (10:1)

FIGURE 2 Effects of EGCG on IL-12 p40/p70 production by DCs infected for 24 h with *L. pneumophila* (Lp). Results are expressed as a mean value in nanograms per milliliter ± standard error of the mean from three independent experiments. The asterisks indicate statistically significant differences of $P < 0.05$ from values obtained with non-EGCG-treated *L. pneumophila*-infected DCs.

held in the mouth without swallowing (4). It is also important to note that EGCG has a relatively short half-life in vivo, ranging from 1.87 to 4.58 h for a 50- to 1,600-mg dose (8), whereas DCs used in the present study were cultured with EGCG for a period of 24 h. The short half-life of catechins in vivo might be overcome, however, by repeated administration, which is feasible given the reported low toxicity of this catechin and given that even high doses, such as 1,600 mg, are well tolerated by human subjects (8).

The seemingly opposed effects of EGCG on IL-12 and TNF-α production are not without precedent. Ho et al. (2) reported a similar effect, downregulation of IL-12 and upregulation of TNF-α, on human dendritic cells treated with aspirin.

In conclusion, we present evidence that EGCG can inhibit IL-12 production while enhancing TNF-α production in bone marrow-derived DCs infected with *Legionella*. Elucidation of the molecular mechanisms for the divergent effects of EGCG on IL-12 and

TNF-α production, as well as how such a dichotomy may be important to *Legionella* infection, will require further study.

ACKNOWLEDGMENTS

Supported in part by grants from the NIH (AI 4618), the Florida HighTech Program, and the Florida Veterinary Specialists and Cancer Treatment Center, Tampa, FL.

REFERENCES

1. **Ahn, S. C., G. Y. Kim, J. H. Kim, S. W. Baik, M. K. Han, H. J. Lee, D. O. Moon, C. M. Lee, J. H. Kang, B. H. Kim, Y. H. Oh, and Y. M. Park.** 2004. Epigallocatechin-3-gallate, constituent of green tea, suppresses the LPS-induced phenotypic and functional maturation of murine dendritic cells through inhibition of mitogen-activated protein kinases and NF-κB. *Biochem. Biophys. Res. Commun.* **313:**148–155.

2. **Ho, L., D. Chang, H. Shiau, C. Chen, T. Hsieh, Y. Hsu, and C. Wong.** 2001. L J. Aspirin differentially regulates endotoxin-induced IL-12 and TNFα production in human dendritic cells. *Scand. J. Rheumatol.* **30:**346–352.

3. **Katiyar, S. K., A. Challa, T. S. McCormick, K. D. Cooper, and H. Mukhtar.** 1999. Prevention

of UVB-induced immunosuppression in mice by the green tea polyphenol (-)-epigallocatechin e gallate may be associated with alterations in IL-10 and IL-12 production. *Carcinogenesis* **20:**2117–2124.

4. **Lambert, J. D., and C. S. Yang.** 2003. Mechanisms of cancer prevention by tea constituents. *J. Nutr.* 3262S–3267S.

5. **Matsunaga, K., T. W. Klein, H. Friedman, and Y. Yamamoto.** 2001. *Legionella pneumophila* replication in macrophages inhibited by selective immunomodulatory effects on cytokine formation by epigallocatechin gallate, a major form of tea catechins. *Infect. Immun.* **69:**3947–3953.

6. **Matsunaga, K., T. W. Klein, H. Friedman, and Y. Yamamoto.** 2002. Epigallocatechin gallate, a potential immunomodulatory agent of tea components, diminishes cigarette smoke condensate-induced suppression of anti-*Legionella pneumophila* activity and cytokine responses of alveolar macrophages. *Clin. Diagn. Lab. Immunol.* **9:**864–871.

7. **Nakagawa, K., and T. Miyazawa.** 1997. Chemiluminescence: high-performance liquid chromatographic determination of tea catechin, (-)-epigallocatechin 3-gallate, at picomole levels in rat and human plasma. *Anal. Biochem.* **248:**41–49.

8. **Ullman, U., J. Haller, J. P. DeCourt, N. Girault, J. Girault, A. S. Richard-Caudron, B. Pineau, and P. Weber.** 2003. A single ascending dose study of epigallocatechin gallate in healthy volunteers. *JIMR.* **31:**88–101.

9. **Sakagami, H., M. Takeda, K. Sugaya, T. Omata, H. Takahashi, M. Tamamura, Y. Hara, and T. Shimamura.** 1995. Stimulation by epigallocatechin gallate of interleukin-1 production by human peripheral blood mononuclear cells. *Anticancer Res.* **15:**971–974.

10. **Yang, F., W. J. Villiers, C. J. McClain, and G. W. Varilek.** 1998. Green tea polyphenols block endotoxin-induced tumor necrosis factor-production and lethality in a murine model. *J. Nutr.* **128:**2334–2340.

GENE EXPRESSION AND VIRULENCE IN *LEGIONELLA*: THE FLAGELLAR REGULON

Klaus Heuner, Sebastian Jacobi, Christiane Albert,
Michael Steinert, Holger Brüggemann, and Carmen Buchrieser

79

In *Legionella pneumophila* the expression of the transmissive phenotype is genetically linked to the expression of the flagellum (6, 12). Early after the first description of *L. pneumophila,* flagellated bacteria were identified in the human lung. Then it was shown that the expression of the flagellum is linked to the virulent phenotype of *Legionella* (12, 15–17). Experiments indicated that the flagellum is a positive predictor for the virulence of *Legionella* but is not needed for intracellular replication. On the other hand, motility might be an important factor that allows *L. pneumophila* to survive in its aquatic habitat. Motility could be involved in colonization, e.g., biofilm formation, or in reaching a new host for replication. It is still proposed that in the natural environment *Legionella* replicates inside of protozoa. *L. pneumophila* exhibits a single monopolar flagellum which is composed of the flagellin subunit FlaA (8). A homolog of *flaA* was found in nearly all *Legionella* strains tested so far. It was shown that the expression of the flagellum is modulated by different environmental factors and that the flagellum is not expressed during replication or in the presence of high concentrations of nutrients. For a review see reference 13.

THE FLAGELLAR REGULON OF *L. PNEUMOPHILA*

In general, the flagellar system is very complex, and in most bacteria it is composed of more then 50 genes (Fig. 1A). Since the genomes of three *Legionella* species are sequenced, it is possible to identify all genes which seem to be members of the flagellar regulon of *L. pneumophila* (Fig. 1B). The operon structure is very similar to that of *Pseudomonas* or *Salmonella* (13). We have recently identified homologs of the *flgAMN* genes, which we have missed in a recent study (13), as well as a homolog of MotY, which indicates that the flagellum of *L. pneumophila* exhibits a sodium-driven motor (MotAB) (2). However, we did not detect homologs of various chemotaxis genes (*cheWA*, *cheZYBR*, *cheB*). This probably should explain the absence of swarming or chemotaxis in *Legionella*. On the other hand, sometimes specific conditions are needed to demonstrate chemotaxis or swarming activity of a microorganism. Thus, we may not have yet identified the conditions required for inducing these activities in *L. pneumophila*.

Klaus Heuner, Sebastian Jacobi, Christiane Albert, and Michael Steinert Institut für Molekulare Infektionsbiologie, Julius-Maximilians Universität Würzburg, Röntgenring 11, D- 97070 Würzburg, Germany. *Holger Brüggemann and Carmen Buchrieser* Unité de Génomique des Microorganismes Pathogènes, Institut Pasteur, 25 rue du Dr. Roux, 75724 Paris Cedex15, France.

A.

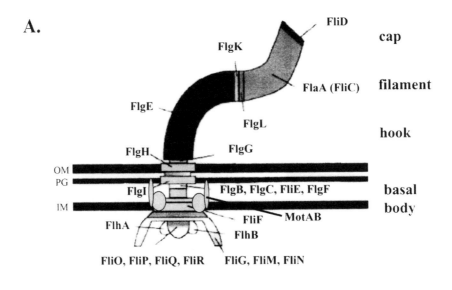

B.

FIGURE 1   (*A*) Schematic drawing of a flagellum and (*B*) the flagellar operon structure of *L. pneumophila* Paris (modified from reference 13). Underlined genes are found together in one chromosomal region. Bars represent the direction of putative operons in *L. pneumophila*. I to IV, classes of the regulatory cascade (see also Fig. 2); black dot, σ^{54} promoter element; white dot, σ^{28} promoter element; grey dot, σ^{70} promoter element; grey square, class III promoter, putatively FleR/RpoN-dependant; IM, inner membrane; OM, outer membrane; PG, peptidoglycan layer.

As mentioned above, *Legionella* is commonly found in aquatic habitats, but the bacteria are also able to infect the human lung. Therefore, *Legionella* has to modulate gene expression to be able to respond to and to survive in such different environments. *L. pneumophila* exhibits a biphasic life cycle, and the transmissive, phase of *L. pneumophila* is associated with the expression of the flagellum (15, 17). Furthermore, the expression of the *flaA* gene is modulated by temperature, growth phase, osmolarity, viscosity, and the nutrient stage (8, 10).

We aimed to characterize the cascade of *flaA* regulation to obtain more information about mechanisms of gene regulation and about virulence of *L. pneumophila*. To analyze the cascade of flagellar gene regulation we identified, cloned and characterized the regulators which may be involved in *flaA* regulation. The regulatory genes were identified experimentally or by screening the flagellar operons for putative promoter consensus sequences in silico (see Fig. 1B). Specific mutants of the identified regulators were generated and anlayzed for flagellar gene expression (9, 11,

12, 14). We demonstrated that the alternative sigma factor FliA (σ^{28}) directly regulates the expression of the flagellin. We also showed, that the sigma-54 activator protein FleQ is the master regulator of the flagellar regulon (14). Together with the σ^{54} factor (RpoN), FleQ regulates the expression of most of the flagellar operons (class 2 genes; see Fig. 2) (2, 14). However, the expression of FliA is not RpoN or FleQ dependent (14). In addition, a member of the LysR family of transcriptional regulators (FlaR) also seems to be involved in *flaA* expression (11).

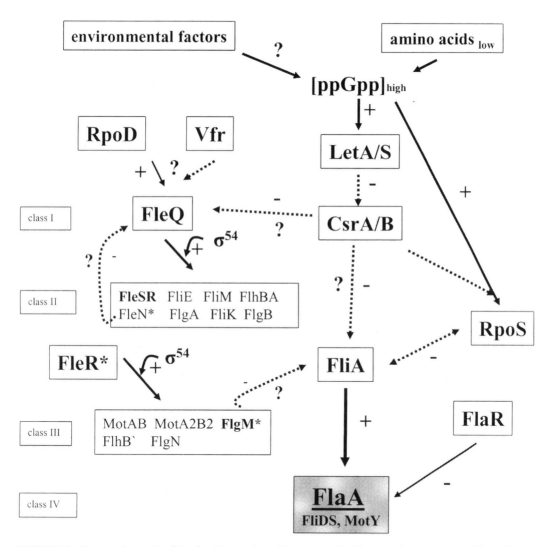

FIGURE 2 Proposed cascade of the flagellar regulon of *L. pneumophila*. The cascade was generated from data retrieved by analyzing flagellar mutant strains and by transcriptome analysis (see text). FleQ is the master regulator of the flagellar genes. FliA seems to be transcribed independently of FleQ and is the direct regulator of the class IV genes. Dotted bars represent an unknown mode of regulation (direct or indirect). ★, indicates regulation activity has to be approved by experiments using specific mutant strains; +, positive regulation; −, negative regulation; CsrA, carbon storage regulator; FlaA, flagellin; FlaR, transcriptional regulator (LysR family); FleQ and FleR, σ^{54} activator protein; FliA, alternative σ^{28} factor; LetA and LetS, two-component system; RpoN, alternative σ^{54} factor; RpoS, alternative sigma factor. (modified from references 2 and 13).

Other groups demonstrated that RelA, RpoS, LetAS, and CsrA are also involved in *flaA* expression in a yet unknown manner (1, 4, 6, 15). Results are summarized in Fig. 2.

Recently, we developed the ability to analyze flagellar gene expression by transcriptome analysis using wild-type *L. pneumophila* strain Paris and the isogenic *fliA* mutant strain (2). From the results obtained, we were able to generate a more detailed cascade of the flagellar regulon of *L. pneumophila* (see Fig. 2). The results support our earlier findings that FleQ is the master regulator of the flagellar regulon. Furthermore, we showed that the activator protein FleR together with RpoN may be responsible for expression of the class 3 genes (2). In a recent study we demonstrated that *flaA* expression is reduced in FleQ and RpoN mutant strains (14). Since FliA is the direct regulator of *flaA* expression, and *fliA* expression, seems to be σ^{70}-dependent (Heuner, unpublished data), it is still unclear at which level of the cascade there is a link between *flaA* and basal body gene expression. It is likely that we will identify LetA or CsrA as regulators responsible for the observed connection.

THE LINK BETWEEN THE FLAGELLAR REGULON AND THE VIRULENCE OF *L. PNEUMOPHILA*

As mentioned above, the regulation of the flagellum is linked to the expression of the virulent phenotype of *Legionella*. Phenotypic characteristics which are associated with the virulence of *L. pneumophila* are the ability to lyse human erythrocytes, to infect host cells, and to replicate inside host cells. Therefore, we analyzed different flagellar mutants by using invasion, coculture, and hemolysis assays. We demonstrated that the flagellin is needed for the invasion capacity of *L. pneumophila,* but intracellular replication was not significantly reduced in amoebae (3). In addition, intracellular replication of a *fleQ* mutant strain was also not significantly reduced (14). On the other hand, whereas a *flaA* mutant was still able to replicate in *Dictyostelium discoideum*, the *fliA* mutant was not (12). Recently it was also shown that the

alternative sigma facor FliA is involved in the expression of the cytotoxic phenotype (6). These results indicate that on the level of FliA there is a link between the regulation of flagellum expression and the expression of factors involved in the virulence and fitness of *L. pneumophila*. We characterized the FliA-regulon to identify the genes responsible for the observed reduced virulence of the *fliA* mutant strain in host cells. For this purpose we performed a comparative study analyzing the transcriptomes of the *L. pneumophila* wild-type strain Paris and the isogenic *fliA* mutant strain during intracellular replication in *A. castellanii* (2). Ten *fliA*-dependant operons were identified. Only three operons belong to the flagellar regulon, whereas the other seven operons do not. These operons encode a putative regulator protein exhibiting a GGDEF/EAL domain, three proteins possessing a signal sequence, and some proteins with yet unknown functions.

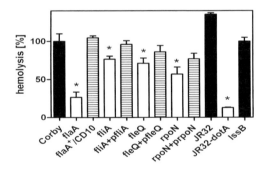

FIGURE 3 Hemolysis assay of sheep red blood cells with different *L. pneumophila* flagellar mutant strains at postexponential phase. The contact-dependent hemolysis assay was carried out at different time points during growth as stated recently (5), but the incubation time was reduced to 4 h. Hemolysis values of different mutants are expressed as a percentage of hemolysis by wild-type *L. pneumophila* Corby (defined as 100%). *L. pneumophila* Corby and *L. pneumophila* JR32 wild-type strains and *lssB* mutant strain (black columns); flagellar mutant strains *flaA*, *fliA*, *fleQ*, *rpoN*, and *dotA* mutant (white columns); complemented strains with the corresponding gene (striped columns). A minimum of three independent experiments in triplicate were performed. Error bars represent standard deviation. Asterisks represent values for mutant strains statistically different from wild-type hemolysis with a value of $P < 0.05$ as determined with the two-sided Student's *t*-test.

The identified putative virulence factors are now under further investigation to identify the effectors responsible for the observed reduced cytotoxicity and fitness of the *fliA* mutant strain of *L. pneumophila*.

In addition, we investigated our mutant strains for their contact-dependent hemolytic activity using a liquid hemolysis assay. Results are shown in Fig. 3. The regulatory mutants (*fleQ*, *rpoN*, and *fliA*) showed a reduced hemolytic activity, which could be complemented by an intact gene in *trans*. We were surprised that our *flaA* mutant strain also showed a reduction in the hemolytic activity. Experiments are under way to characterize the mechanisms responsible for this effect and to identify the target protein(s) involved in the flagellar-associated hemolytic activity of *L. pneumophila*.

CONCLUSIONS

Many virulence genes have been cloned, but still, little is known about their regulation. By analyzing the flagellar regulon of *L. pneumophila,* we identified regulators, such as sigma factors, antisigma factors, sigma factor activator proteins, and further transcriptional regulators, and structural genes belonging to the flagellar regulon. As a result of these investigations, a cascade of the flagellar regulon was proposed and several aspects of the link between the flagellar regulon and the expression of the virulent phenotype of *L. pneumophila* are better understood. Thus the analysis of the flagellar regulon and of genes influencing *flaA* expression is a useful means for identifying factors involved in the virulence of *L. pneumophila*. Interestingly, at this meeting Michele Swanson's group demonstrated that flagellin mediates cell death of mouse macrophages by activating the cytosolic Naip5 (Birc1e) receptor. Similar results were also presented by R. Vance, who was searching for the effectors of the Lgn1 locus associated with the permissiveness of A/J mice to *L. pneumophila* infection. In addition, it was shown recently that mutations in Toll-like receptor 5, which mediates the flagellin-dependant recognition of bacteria by the innate immune system in humans, is associated with the incidence of *L. pneumophila* infection. (7). These results together indicate that various phenotypes associated with the flagellar regulon are of importance for virulence of *L. pneumophila*, demonstrating the importance of research into the flagellar regulon.

REFERENCES

1. **Bachman, M. A., and M. S. Swanson.** 2001. RpoS co-operates with other factors to induce *Legionella pneumophila* virulence in the stationary phase. *Mol. Microbiol.* **40:**1201–1214.
2. **Brüggemann H., A. Hagman, M. Jules, O. Sismeiro, M.-A. Dillies, C. Gouyette, F. Kunst, M. Steinert, K. Heuner, J.-Y. Coppée, and C. Buchrieser.** Virulence strategies for infecting phagocytes deduced from the transcriptional program of *Legionella pneumophila*, submitted.
3. **Dietrich, C., K. Heuner, B. C. Brand, J. Hacker, and M. Steinert.** 2001. Flagellum of *Legionella pneumophila* positively affects the early phase of infection of eukaryotic host cells. *Infect. Immun.* **64:**2116–2122.
4. **Fettes, P. S., V. Forsbach-Birk, D. Lynch, and R. Marre.** 2001. Overexpresssion of a *Legionella pneumophila* homologue of the *E. coli* regulator *csrA* affects cell size, flagellation, and pigmentation. *Int. J. Med. Microbiol.* **291:**353–360.
5. **Flieger, A., K. Rydzewski, S. Banerji, M. Broich, and K. Heuner.** 2004. Cloning and characterization of the gene encoding the major cell-associated phospholipase A of *Legionella pneumophila*, *plaB*, exhibiting hemolytic activity. *Infect. Immun.* **72:**2648–2658.
6. **Hammer, B. K., E. S. Tateda, and M. S. Swanson.** 2002. A two-component regulator induces the transmission phenotype of stationary-phase *Legionella pneumophila*. *Mol. Microbiol.* **44:**107–118.
7. **Hawn, T. R., A. Verbon, K. D. Lettinga, L. P. Zhao, S. S. Li, R. J. Laws, S. J. Skerrett, B. Beutler, L. Schroeder, A. Nachman, A. Ozinsky, K. D. Smith, and A. Aderem.** 2003. A common dominant TLR5 stop codon polymorphism abolishes flagellin signaling and is associated with susceptibility to Legionnaires' disease. *J. Exp. Med.* **198:**1563–1572.
8. **Heuner, K., L. Bender-Beck, B. C. Brand, P. C. Lück, K.-H. Mann, R. Marre, M. Ott, and J. Hacker.** 1995. Cloning and genetic characterization of the flagellum subunit gene (*flaA*) of *Legionella pneumophila* serogroup 1. *Infect. Immun.* **63:**2499–2507.
9. **Heuner, K., J. Hacker, and B. C. Brand.** 1997. The alternative sigma factor σ²⁸ of *Legionella*

pneumophila restores flagellation and motility to an *Escherichia coli fliA* mutant. *J. Bacteriol.* **179:**17–23.

10. **Heuner K, B. C. Brand, and J. Hacker.** 1999. The expression of the flagellum of *Legionella pneumophila* is modulated by different environmental factors. *FEMS Microbiol. Lett.* **175:**69–77.

11. **Heuner, K., C. Dietrich, M. Steinert, U. B. Göbel, and J. Hacker.** 2000. Cloning and characterization of a *Legionella pneumophila* specific gene encoding a member of the LysR family of transcriptional regulators. *Mol. Gen. Genet.* **264:**204–211.

12. **Heuner, K., C. Dietrich, C. Skriwan, M. Steinert, and J. Hacker.** 2002. Influence of the alternative σ^{28} factor on virulence and flagellum expression of *L. pneumophila*. *Infect. Immun.* **70:**1604-1608.

13. **Heuner, K., and M. Steinert.** 2003. The flagellum of *Legionella pneumophila* and its link to the expression of the virulent phenotype. *Int. J. Med. Microbiol.* **293:**133–145.

14. **Jacobi, S., J. Hacker, R. Schade, and K. Heuner.** 2004. Cloning and characterization of the alternative sigma factor σ^{54} and the transcriptional regulator FleQ of *Legionella pneumophila*, both involved in the regulation cascade of flagellar gene expression. *J. Bacteriol.* **186:**2540–2547.

15. **Molofsky, A. B., and M. S. Swanson.** 2004. Differentiate to thrive: lessons from the *Legionella pneumophila* life cycle. *Mol. Microbiol.* **53:**29–40.

16. **Pruckler, J. M., R. F. Benson, M. Moyenuddin, W. T. Martin , and B. S. Fields.** 1995. Association of flagellum expression and intracellular growth of *Legionella pneumophila*. *Infect. Immun.* **63:**4928–4932.

17. **Rowbotham, T. J.** 1986. Current views on the relationships between amoebae, legionellae and man. *Isr. J. Med. Sci.* **22:**678–689.

IDENTIFICATION OF *LEGIONELLA PNEUMOPHILA* GENES UNDER TRANSCRIPTIONAL CONTROL OF LpnR REGULATORY PROTEINS

E. Lammertyn, L. Maes, E. De Buck, I. Lebeau, and J. Anné

80

In a search for *Legionella pneumophila* Philadelphia proteins with sequence similarity to members of the LuxR family of transcriptional regulators, three novel proteins were identified. These were designated LpnR1 (lpg 2557), LpnR2 (lpg 1946), and LpnR3 (lpg 1448), and although these proteins were not quorum-sensing related, they act as transcriptional regulators. To get more insight into the function of these proteins, isogenic deletion mutants were constructed. Analysis of these mutants revealed that LpnR2 and LpnR3 positively affect flagellin expression. Furthermore, LpnR2 proved to be necessary for efficient invasion of *Acanthamoeba castellanii*, and LpnR3 for efficient intracellular replication in this protozoan host (2).

The isogenic *lpnR3* deletion mutant was further characterized by comparative proteomic analysis using two-dimensional gel electrophoresis in combination with mass spectrometric analysis. In addition, LpnR3-dependent expression of some of the identified proteins was analyzed by semi-quantitative reverse transcriptase (RT)-PCR analysis and by using an in-house-developed two-plasmid-based system in *E. coli*.

Preliminary comparative analysis of protein patterns of *L. pneumophila* wild type and Δ*lpnR3* generated by two-dimensional gel electrophoresis, performed as described by Lebeau et al. (3), showed that 23 proteins are differentially expressed. Thirteen proteins seem to have a lower expression in the mutant strain. Proteins with a lower expression of at least a factor of 10 in the absence of the LpnR3 protein are acetoacetate decarboxylase (lpg 2971), malate dehydrogenase/oxidoreductase (lpg 0672), and triacylglycerol lipase (lpg 1889). Other spots that have a clearly down-regulated expression could not be identified by mass spectrometry and are thus not included in this analysis.

In this work, for the malate dehydrogenase gene (*maeA*), a semiquantitative RT-PCR analysis was performed to check transcriptional control by LpnR3. Total RNA was isolated from *L. pneumophila,* and *L. pneumophila* Δ*lpnR3* was grown to exponential and stationary growth phase in culture broth. RT-PCR with *maeA*-specific primers (5′-atggatccggcaatcaaagtaac-3′ and 5′-atggatcctagaacatatggttc-3′) was performed on 100 ng of total RNA (Fig. 1). Control reactions on the 16S rRNA resulted in a signal of equal intensity for wild-type and Δ*lpnR3* samples (data not shown). Controls for

E. Lammertyn, L. Maes, E. De Buck, I. Lebeau, and J. Anné Laboratory of Bacteriology, Rega Institute for Medical Research, Katholieke Universiteit Leuven, Minderbroedersstraat 10, B-3000 Leuven, Belgium.

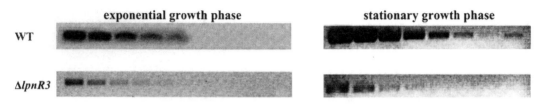

FIGURE 1 RT-PCR analysis of the *maeA* gene. RNA was isolated from *L. pneumophila* wild type and *L. pneumophila* Δ*lpnR3*, and RT-PCR with *maeA*-specific primers was performed on 100 ng total RNA. Amplification fragments were analyzed after 30 cycli and loaded on agarose gel in twofold dilutions. Control reactions on the 16S rRNA resulted in a signal of equal intensity for wild-type and Δ*lpnR3* samples (data not shown). Controls for the presence of traces of DNA in the samples were negative.

the presence of traces of DNA in the samples were negative.

This showed that *maeA* expression is clearly decreased in the absence of LpnR3, during both exponential and stationary growth, indicating that the *maeA* gene is under transcriptional control of LpnR3.

To investigate whether the regulation of *maeA* expression by LpnR3 occurs in a direct or indirect manner or whether additional factors are required besides LpnR3 to activate transcription, binding of the LpnR3 protein to the putative *maeA*-promoter region upstream of the gene was investigated using a two-plasmid-based system in *E. coli* as outlined in Fig. 2. By means of PCR, the *L. pneumophila lpnR3* gene was amplified from chromosomal DNA as a template using the primer pair LPR3A (5′-tacatatgcctaattcatcaaatg-3′) and LPR3B (5′-tagcggccgcttatcttgtaattattcc-3′). The resulting 1-kb fragment was cloned in pGEM-T Easy, generating pGEMlpnR3, and subsequently

transferred as an *Nde*I/*Not*I restriction fragment to the vector pEX50 (1) upstream of the IPTG-inducible *tac* promoter. This resulted in the vector pEXlpnR3.

In a second cloning strategy, the *lacZα* gene fragment encoding part of β-galactosidase was isolated from pUC19 by PCR using the primer pair lacZF (5′-tagatatcacaggaaacacctatgacc-3′) and lacZR (5′-tatacgtactatgcggcatcagagc-3′). The amplified fragment, 300 bp in length, was cloned in pGEM-T Easy, generating pGEMlacZ, and subsequently transferred to pACYC184 as an *Eco*RV/*Not*I restriction fragment in order to generate the vector pACYClacZ. In this way, *lacZ* is preceded by a suitable ribosome-binding site but lacks an upstream promoter. Subsequently, the *maeA* promoter region was isolated by PCR (using the primers 5′-atgatatccataatcgatttaagg-3′ and 5′-atgatatctttgttactttgattgc-3′) using chromosomal *L. pneumophila* DNA as a template and cloned upstream of the *lacZ* gene in pACYClacZ,

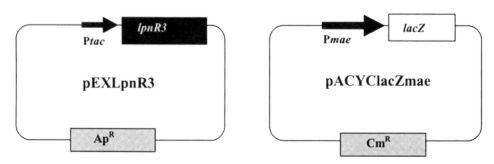

FIGURE 2 Schematic presentation of the two-plasmid system used. *E. coli* cells containing both plasmids were grown in the presence or absence of IPTG.

resulting in pACYClacZmaeA. Both plasmids were introduced into competent *E. coli* cells, and cells were grown in the presence or absence of the inducer isopropyl-β-O-thiogalactopyranoside (IPTG) on Luria broth plates containing X-Gal (5-bromo-4-chloro-3-indolyl-β-D-galactopyranoside. This enabled us to analyze transcriptional activity from the *maeA* promoter in the absence or presence of LpnR3 by adding IPTG. In the absence of IPTG, and thus in the absence of *lpnR3* transcription, no blue color was formed, indicating that the *maeA* promoter has no activity in *E. coli* as such. The appearance of blue color when IPTG was added to the medium, and LpnR3 was thus expressed, indicated that the *maeA*-promoter can be activated by LpnR3, indicating a direct regulation of *maeA* transcription by LpnR3 without the need for any additional *L. pneumophila*-specific factor, although one should keep in mind that the *E. coli* background is different from that of *L. pneumophila*.

To conclude, we can say that it is possible to analyze differential expression in *L. pneumophila* wild type and *L. pneumophila* Δ*lpnR3* using two-dimensional analysis. Furthermore, we found that the malate dehydrogenase gene can be transcriptionally activated by LpNR3 in a direct way without the need for additional factors. Activation of transcription of two other genes, i.e., the triacyl glycerol lipase and the acetoacetate decarboxylase genes, on the other hand, is dependent on LpnR3, but rather indirectly, or requires an additional factor. While initially a role of LpnR3 in the virulence process was suggested, these results indicate an additional role in bacterial cell metabolism.

REFERENCES

1. **Geukens, N., E. Lammertyn, L. Van Mellaert, S. Schacht, K. Schaerlaekens, V. Parro, S. Bron, Y. Engelborghs, R. P. Mellado, and J. Anné.** 2001. Membrane topology of the *Streptomyces lividans* type I signal peptidases. *J. Bacteriol.* **183:**4752–4760.

2. **Lebeau, I., E. Lammertyn, E. De Buck, L. Maes, N. Geukens, L. Van Mellaert, and J. Anné.** 2004. Novel transcriptional regulators of *Legionella pneumophila* that affect replication in *Acanthamoeba castellanii*. *Arch. Microbiol.* **181:**362–370.

3. **Lebeau, I., E. Lammertyn, E. De Buck, L. Maes, N. Geukens, L. Van Mellaert, L. Arckens, J. Annè, and S. Clerens.** 2005. First proteomic analysis of *Legionella pneumophila* based on its developing genome sequence. *Res. Microbiol.* **156:**119–129.

MODULATION OF *rpoH* EXPRESSION USING AN ANTISENSE STRATEGY

Fanny Ewann and Paul S. Hoffman

81

Legionella pneumophila and related species reside in aquatic environments as intracellular parasites of protozoa. When transmitted by aerosols to susceptible human hosts, *L. pneumophila* infects alveolar macrophages, often producing an atypical and sometimes fatal pneumonia known as Legionnaires' disease. One of the major proteins produced under in vitro and in vivo conditions by *L. pneumophila* is the highly conserved 60-kDa heat-shock protein (Hsp60) encoded by the essential gene *htpB* (6). In contrast to most bacteria, the *L. pneumophila* Hsp60 is located in the periplasm and translocation to the bacterial surface is both necessary for infectivity and dependent upon a functional Dot/Icm type IV secretion system (3). Hsp60 mediates invasion of HeLa cells, blocks organelle trafficking, and is the most abundant protein secreted into the phagosomes of host cells throughout the course of intracellular multiplication (4).

In *Escherichia coli* the heat shock response is controlled by sigma factor 32 (RpoH) encoded by *rpoH*. RpoH is highly conserved through

Fanny Ewann and Paul S. Hoffman Division of Infectious Diseases, University of Virginia, Charlotteville, VA 22903 and Departments of Microbiology and Immunology, and Medicine, Division of Infectious Diseases, Faculty of Medicine, Dalhousie University, Halifax, Nova Scotia, Canada B3H 4H7.

evolution (e.g., 75% similarity between *E. coli* and *L. pneumophila*). The *htpAB* operon (Hsp10 and Hsp60) of *L. pneumophila* contains a consensus heat shock promoter (TNtCNCcCTTGAA-13 to15 nt-CCCCATtTa), and expression of *htpAB* in *E. coli* requires RpoH (2). The relative level of Hsp60 in *L. pneumophila* is nearly 50-fold higher than levels of GroEL in *E. coli*. We do not know if the high level of Hsp60 produced in *L. pneumophila* is necessary for or in excess of that required for biological function. Moreover, we do not know to what extent RpoH is either expressed or required for gene expression.

The objective of our study is to control levels of Hsp60 production by modulating expression of *rpoH* in order to evaluate the role of Hsp60 and the heat shock response in pathogenesis. We developed three strategies to alter *rpoH* expression: (i) construct an *rpoH* knockout mutant by allelic replacement; (ii) replace the endogenous *rpoH* promoter with an isopropyl-β-D-thiogalactopyranoside (IPTG)-inducible promoter; and (iii) to knock down levels of RpoH and HtpAB with antisense RNA. All of our attempts to knock out *rpoH* have proven unsuccessful, suggesting that RpoH function is essential for viability. Moreover, attempts to complement chromosomal deletions of *rpoH* by transcomplementation or by promoter swapping (replacement of the *rpoH* endogenous

FIGURE 1 RNA level of *htpB* in recombinant *L. pneumophila* Philadelphia-1 strains. The recombinant strains containing the *htpB* half length and *htpAB*, *rpoH* half length or full length antisense constructs (lanes 2 to 5) were subjected to a 24-h induction with 5 mM IPTG. RNA was extracted and subjected to RT-PCR with primers specific for the *htpA* (★) and *rplJ* (internal control) genes. The signal was compared to the one obtained for the parental strain containing either the empty vector (lane 1) or the same vector expressing *gfp* (lane 6).

promoter with a controllable promoter) have also proven unsuccessful. Either the *rpoH* gene is located in a dead zone for recombination or more likely the introduction of strong promoters into this locus causes constitutive expression of downstream genes associated with cell division (*ftsYEX*) (1). Alternatively, the *rpoH* locus might contain *cis*-acting regulatory elements that control cell division genes.

Antisense technology is a simple and often underutilized strategy that relies on the antisense mRNA (complementary to sense mRNA) to form a duplex with sense mRNA and thereby ablate translation. The relative effi-

ciency of antisense to knockdown protein levels is a function of the nucleotide sequence chosen and the degree of mRNA secondary structure (natural duplexes in mRNA). Antisense constructs of *rpoH* and *htpAB* were produced by cloning the genes (or various length sequences) in reverse under an IPTG-inducible promoter (5). The resulting plasmids were then introduced into *L. pneumophila* Philadelphia-1 and subsequently induced with different concentrations of IPTG. The expression of *htpB* was determined by reverse transcriptase PCR, and the production of Hsp60 by various antisense constructs and wild-type strains were assessed by sodium dodecyl-sulfate polyacrylamide gel electrophoresis and immunoblot developed with monospecific anti-Hsp60 serum.

As shown in Fig. 1, expression levels of *htpAB* were decreased relative to controls (*rplJ*) in strains expressing *htpAB*, *htpB* antisense constructs (lane 2, 3), as well as in the strain expressing the full-length *rpoH* antisense RNA (lane 5), following IPTG induction. The latter results demonstrate that the *htpAB* operon is under control of RpoH in *L. pneumophila*. The immunoblot analysis of the IPTG-induced strains (Fig. 2) showed a slight reduction in Hsp60 levels in strains expressing either the *htpB*, *htpAB*, or full-length *rpoH* antisense (lanes 2, 3, 5). However, the variation in protein

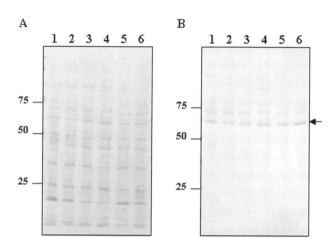

FIGURE 2 Hsp60 production in the antisense-containing strains. *L. pneumophila* strains Philadelphia-1 recombinant containing the empty vector (lane 1), the same vector expressing *gfp* (lane 6), the *htpB* half length, *htpAB*, and *rpoH* half length or full length antisense constructs (lanes 2 to 5) were subjected to a 24-h induction with 5mM IPTG. The protein extract obtained from the different strains were run on sodium dodecyl-sulfate-polyacrylamide gel electrophoresis and transferred on a nitrocellulose membrane. After Ponceau Red staining (*A*), the production of Hsp60 was assessed by immunoblotting with anti-Hsp60 antibodies (*B*).

levels did not correlate with the variation in transcript levels, suggesting that Hsp60 expression might be under posttranscriptional regulation. In summary, the preliminary data presented in this report suggest that antisense can be used to knock down protein expression levels in *L. pneumophila*.

REFERENCES

1. **Chien, M., I. Morozova, S. Shi, H. Sheng, J. Chen, S. M. Gomez, G. Asamani, K. Hill, J. Nuara, M. Feder, J. Rineer, J. J. Greenberg, V. Steshenko, S. H. Park, B. Zhao, E. Teplitskaya, J. R. Edwards, S. Pampou, A. Georghiou, I. C. Chou, W. Iannuccilli, M. E. Ulz, D. H. Kim, A. Geringer-Sameth, C. Goldsberry, P. Morozov, S. G. Fischer, G. Segal, X. Qu, A. Rzhetsky, P. Zhang, E. Cayanis, P. J. De Jong, J. Ju, S. Kalachikov, H. A. Shuman, and J. J. Russo.** 2004. The genomic sequence of the accidental pathogen *Legionella pneumophila*. *Science* **305:**1966–1968.

2. **Cowing, D. W., J. C. Bardwell, E. A. Craig, C. Woolford, R. W. Hendrix, and C. A. Gross.** 1985. Consensus sequence for Escherichia coli heat shock gene promoters. *Proc. Natl. Acad. Sci. USA* **82:**2679–2683.

3. **Garduno, R. A., G. Faulkner, M. A. Trevors, N. Vats, and P. S. Hoffman.** 1998. Immunolocalization of Hsp60 in *Legionella pneumophila*. *J. Bacteriol.* **180:**505–513.

4. **Garduño, R. A., E. Garduno, and P. S. Hoffman.** 1998. Surface-associated hsp60 chaperonin of *Legionella pneumophila* mediates invasion in a HeLa cell model. *Infect. Immun.* **66:**4602–4610.

5. **Morales, V. M., A. Backman, and M. Bagdasarian.** 1991. A series of wide-host-range low-copy-number vectors that allow direct screening for recombinants. *Gene* **97:**39–47.

6. **Weeratna, R., D. A. Stamler, P. H. Edelstein, M. Ripley, T. Marrie, D. Hoskin, and P. S. Hoffman.** 1994. Human and guinea pig immune responses to *Legionella pneumophila* protein antigens OmpS and Hsp60. *Infect. Immun.* **62:**3454–3462.

NOVEL USE OF *HELICOBACTER PYLORI* NITROREDUCTASE (*rdxA*) AS A COUNTERSELECTABLE MARKER IN ALLELIC VECTOR EXCHANGE TO CREATE *LEGIONELLA PNEUMOPHILA* PHILADELPHIA-1 MUTANTS

Ann Karen C. Brassinga, Matthew A. Croxen, Charles J. Shoemaker, Michael G. Morash, Jason J. LeBlanc, and Paul S. Hoffman,

82

The majority of Legionnaires' disease cases reported annually in the United States is caused by serogroup 1 strains of *Legionella pneumophila* (4). The availability of three genome sequences for serogroup 1 strains (2, 3) provides a unique opportunity to study both the genetic diversity among isolates and the core genes associated with environmental fitness and pathogenesis. Both random transposon-based mutagenesis and directed allelic replacement mutagenesis strategies have been used to identify or assess virulence-associated genes in *L. pneumophila*. One of the more common approaches to create gene knockout mutants involves introducing mutant constructs into *L. pneumophila* on suicide plasmids containing the *sacB* counterselectable marker. One such vector is pBOC20, which contains the multiple cloning site from pHXK cloned into pEA75, which contains a chloramphenicol resistance gene and a counterselectable levansucrase *sacB* gene (5, 6). The vector was utilized successfully to disrupt the

Paul S. Hoffman, Ann Karen C. Brassinga, and Charles J. Shoemaker Department of Internal Medicine, Division of Infectious Diseases and International Health and Department of Microbiology and Immunology, University of Virginia School of Medicine, Charlottesville, VA 22908. *Matthew A. Croxen, Michael G. Morash, and Jason J. LeBlanc* Departments of Microbiology and Immunology, and Medicine, Division of Infectious Diseases, Faculty of Medicine, Dalhousie University, Halifax, Nova Scotia, Canada B3H 4H7.

hemin-binding protein (*hbp*) in *L. pneumophila* Wadsworth 130b (6). However, previous experience in our laboratory shows that utilizing pBOC20 has proven somewhat inefficient, as indicated by the high number of false-positive gene knockout colonies scored. In this report, we test the feasibility of replacing the *sacB* gene with another suicide gene, *rdxA*, from *Helicobacter pylori*, which encodes an NADPH oxygen insensitive nitroreductase. Here we show that the improved suicide vector pKBOXR (*rdxA*) and use of metronidazole in counterselection is superior to the *sacB* in *L. pneumophila*.

H. pylori rdxA encodes an oxygen insensitive NADPH nitroreductase which exhibits high substrate specificity for the prodrug metronidazole, which is rapidly reduced to DNA-damaging adducts of hydroxylamine, a bactericidal agent (7), thereby acting as a potential and novel counterselectable marker. To create the novel *rdxA* counterselectable allelic exchange vector pKBOXR, pBOC20 was sequenced (DalGEN Sequencing Facility, Halifax, Nova Scotia), providing a guideline for the removal of *sacB* via high fidelity PCR (Roche) amplification of the vector by primers (PF VecHpaI GCGATAGTTAACAAACGCAAA-AGAAAATGCCGA and PR VecNdeI GCG-ATACATATGTCATGTCTCCTTTTTTAT-GTA) annealing to the flanking regions of *sacB*

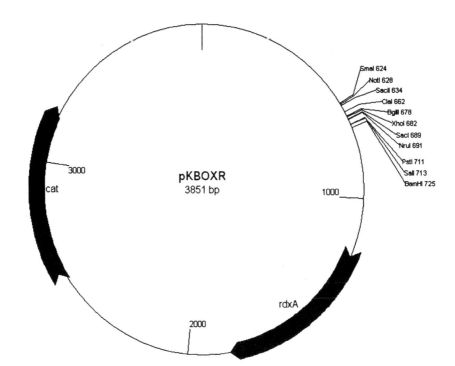

FIGURE 1 Vector map of pKBOXR with unique multiple cloning restriction sites.

and the ligation/insertion of PCR-amplified *rdxA* (PF RdxANdeI GCGATACATATGA-AATTTTTGGATCA and PR RdxAHpaI GC-GATAGTTAACTCACAACCAAGT) (Hot-Start Taq, Qiagen) from *H. pylori* SS1 (Fig. 1). For comparative purposes, Lp02 electroporated with pKBOXR and pBOC20 was plated on buffered charcoal yeast extract (BCYE) supplemented with and without metronidazole (20 μg) (Sigma) to determine the sensitivity or resistance, if any, to metronidazole. As shown in Table 1, Lp02 pKBOXR and Lp02 pBOC20 are susceptible and resistant, respectively, to metronidazole, therefore ensuring the effectiveness of the novel use of *rdxA* for counterselection.

To create a gene knockout construct, flanking regions of ~500 bp of the targeted gene were individually PCR amplified and ligated into the multiple cloning site of an appropriate high-copy cloning vector (i.e., pBluescript, pUC19), after which an antibiotic cassette (kanamycin or gentamicin) was inserted be-

tween the two flanking regions. Finally, the construct is then excised from the high-copy vector and ligated (sticky or blunt) into the appropriate restriction sites of pKBOXR to create the allelic gene knockout plasmid construct (Fig. 2). To obtain potential gene knockout transformants, ~3 μg of DNA was electroporated into electrocompetent Lp02 cells and plated on BCYE medium supplemented with

TABLE 1 Efficiency of plating of Lp02 strains with and without metronizadole (20 μg/ml)

Strain	Efficiency of plating[a]	
	Without metronidazole	With metronidazole
Lp02 pBOC20	1.0	1.24 ± 1.8
Lp02 pKBOXR	1.0	0

[a]Efficiency of plating was calculated based on colony counts for each strain on BCYE agar without supplementation with metronidazole, and each count was normalized to 1.0. Bacterial counts on BCYE medium with metronidazole were then compared to counts on BCYE medium without metronidazole.

i) Insertion of one PCR amplified ~500 bp region flanking targeted gene into multiple cloning region of high copy cloning vector.

ii) Insertion of the other PCR amplified ~500 bp region flanking targeted gene into multiple cloning region of high copy cloning vector joining the first insert.

iii) Insertion of appropriate antibiotic cassette into BamHI site created by the joining of the first two inserts into multiple cloning region of high copy cloning vector.

iv) Excision of knock-out gene construct from high copy cloning vector for insertion into the corresponding unique restriction sites in the lower copy pKBOXR vector which is then electroporated into Lp02. Resultant metronidazole resistant transformants are then confirmed via PCR and RT-PCR for sucessful chromosomal gene replacement.

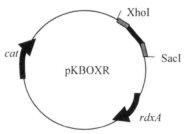

FIGURE 2   Target gene knockout strategy. Drawings are not to scale, and restriction sites are used as examples.

streptomycin (100 μg/ml), thymidine (100 μg/ml), kanamycin (40μg/ml), or gentamicin (10 μg/ml) and grown at 37°C for 3 to 4 days. Transformants were then replica plated on media supplemented with and without metronidazole (20 μg/ml) for counter-selection. Transformants that were resistant to metronidazole were considered to be positive knockout colonies and were further validated for absence of the targeted chromosomal gene via PCR and

reverse transcriptase PCR as well as for loss of the plasmid by resistance to chloramphenicol.

To date, utilization of the novel allelic exchange vector pKBOXR in our laboratory has generated several successful chromosomal gene knockouts of *ahpC1*, *ahpC2D*, *ihfA*, and *ihfB* in *L. pneumophila* Lp02. The pKBOXR vector appears to be more efficient than pBOC20 in generating knockout gene mutant strains, with the additional benefit that *L. pneumophila* is not subjected to osmotic stress during the counter-selectable process. This is the first instance, to our knowledge, of the novel use of metronidazole as a counter-selection reagent.

ACKNOWLEDGMENTS

We thank Michele Swanson and Nicholas Cianciotto for their kind gifts of *L. pneumophila* Lp02 and pBOC20, respectively.

This study was supported by the CIHR grant (MT14443) and the University of Virginia start-up funds to P.S.H. and CIHR and University of Virginia postdoctoral research funds to A.K.C.B.

REFERENCES

1. **Brassinga, A. K., M. F. Hiltz, G. R. Sisson, M. G. Morash, N. Hill, E. Garduno, P. H. Edelstein, R. A. Garduno, and P. S. Hoffman.** 2003. A 65-kilobase pathogenicity island is unique to Philadelphia-1 strains of *Legionella pneumophila. J. Bacteriol.* **185:**4630–4637.

2. **Cazalet, C., C. Rusniok, H. Bruggemann, N. Zidane, A. Magnier, L. Ma, M. Tichit, S. Jarraud, C. Bouchier, F. Vandenesch, F. Kunst, J. Etienne, P. Glaser, and C. Buchrieser.** 2004. Evidence in the *Legionella pneumophila* genome for exploitation of host cell functions and high genome plasticity. *Nat. Genet.* **36:**1165–1173.

3. **Chien, M., I. Morozova, S. Shi, H. Sheng, J. Chen, S. M. Gomez, G. Asamani, K. Hill, J. Nuara, M. Feder, J. Rineer, J. J. Greenberg, V. Steshenko, S. H. Park, B. Zhao, E. Teplitskaya, J. R. Edwards, S. Pampou, A. Georghiou, I. C. Chou, W. Iannuccilli, M. E. Ulz, D. H. Kim, A. Geringer-Sameth, C. Goldsberry, P. Morozov, S. G. Fischer, G. Segal, X. Qu, A. Rzhetsky, P. Zhang, E. Cayanis, P. J. De Jong, J. Ju, S. Kalachikov, H. A. Shuman, and J. J. Russo.** 2004. The genomic sequence of the accidental pathogen *Legionella pneumophila. Science* **305:**1966–1968.

4. **Fields, B. S., R. F. Benson, and R. E. Besser.** 2002. *Legionella* and Legionnaires' disease: 25 years of investigation. *Clin. Microbiol. Rev.* **15:**506–526.

5. **O'Connell, W. A., J. M. Bangsborg, and N. P. Cianciotto.** 1995. Characterization of a *Legionella micdadei mip* mutant. *Infect. Immun.* **63:**2840–2845.

6. **O'Connell, W. A., E. K. Hickey, and N. P. Cianciotto.** 1996. A *Legionella pneumophila* gene that promotes hemin binding. *Infect. Immun.* **64:** 842–848.

7. **Sisson, G., J. Y. Jeong, A. Goodwin, L. Bryden, N. Rossler, S. Lim-Morrison, A. Raudonikiene, D. E. Berg, and P. S. Hoffman.** 2000. Metronidazole activation is mutagenic and causes DNA fragmentation in *Helicobacter pylori* and in *Escherichia coli* containing a cloned *H. pylori* RdxA(+) (Nitroreductase) gene. *J. Bacteriol.* **182:**5091–5096.

ANALYSIS OF GENE EXPRESSION IN *LEGIONELLA* DURING AXENIC GROWTH AND INFECTION

Sergey Pampou, Irina Morozova, David Hilbert,
Karim Suwwan de Felipe, Pavel Morozov, James J. Russo, Howard
A. Shuman, and Sergey Kalachikov

83

To understand the genetic basis of the ability of *Legionella* to adapt to a wide variety of different environmental settings, we have initiated a long-term genome-wide gene expression profiling study of *Legionella pneumophila* wild-type strains Philadelphia 1 and JR32 and several *Legionella* mutants during their growth in rich axenic media as well as at different stages of *Acanthamoeba* infection using whole-genome microarrays.

The availability of the 12X-deep sequenced genomic library generated by the *L. pneumophila* sequencing project (1) allowed us to begin collecting open reading frame-containing shotgun clones for construction of a genomic microarray at the early stages of the sequencing project. PCR-amplified products of over 3,200 shotgun clone inserts representing all genes

identified in the *Legionella* genome and covering it completely were spotted at equal concentrations on aminosilane glass slides. Total RNA extracts isolated from axenically grown *Legionella* or *Legionella*-infected *Acanthamoeba* cultures were fluorescently labeled by reverse transcription and hybridized to the arrays. Expression profiles for differentially expressed genes (Bonferroni corrected *P* value, <0.05) were analyzed by hierarchical clustering and other methods. The gene expression measurements for key operons were independently confirmed by real-time PCR.

Gene expression profiles of wild type and *rpoS* mutants were compared during growth in rich axenic media. Given the fact that *rpoS*-null mutants of *L. pneumophila* are unable to replicate in protozoa or explanted macrophages, it is of considerable interest to identify genes that are not expressed in the absence of RpoS and to find out which ones are required for intracellular replication. In order to identify such genes, we compared the expression profiles of wild-type *L. pneumophila* strain JR32 and an isogenic *rpoS* null strain, LM1376. Each strain was grown in rich liquid medium (AYE), and samples were removed for RNA isolation during log and stationary phases.

Successive time points were compared to one another using a hierarchical clustering

Sergey Pampou, James J. Russo, Irina Morozova, and Sergey Kalachikov Columbia Genome Center, Columbia University, Russ Berrie Pavilion Rm 406, 1150 St Nicholas Ave, New York, NY 10032. *David Hilbert* Department of Anatomy and Cell Biology, College of Physicians and Surgeons, Columbia University, 1216 Hammer Health Science Center, 701 W 168 St, New York, NY 10032. *Karim Suwwan de Felipe and Howard A. Shuman* Department of Microbiology, College of Physicians and Surgeons, 1216 Hammer Health Science Center, 701 W 168 St, New York, NY 10032. *Pavel Morozov* Joint Centers for Systems Biology, Columbia University, Irving Cancer Research Center, 1130 St Nicholas Ave, New York, NY 10032.

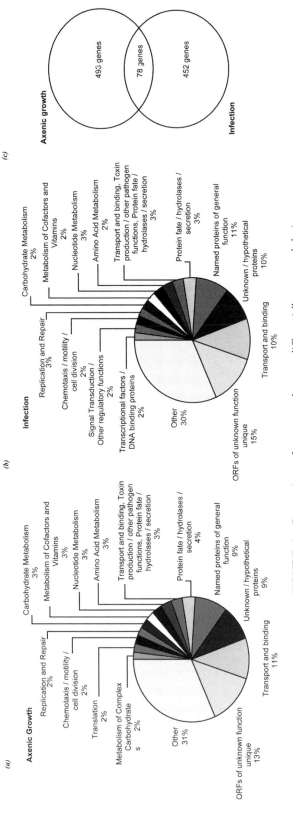

FIGURE 1 Comparison of gene complements differentially expressed during axenic growth and infection. While GO functional categories are represented in relatively equal proportions between axenic growth (a) and earlier stages of the *Acanthamoeba* infection (b), the actual number of common genes between the two groups does not exceed 20% (c).

algorithm. About 30% of the bacterial genome was found to be expressed under any single condition tested in the axenic growth experiments. Expression of about 200 genes was shown to be effected in the *rpoS* mutant. Of particular interest is the stationary phase-dependent expression of the genes encoding the group of heavy-metal efflux pumps discovered initially by McClain et al. (3) and later by Rankin et al. (4). These genes belong to an approximately 120-kb contiguous region containing ~100 genes, most found to be under the control of RpoS σ-factor during the stationary growth phase. Presumably of phage or transposon origin, the region carries a large cassette of metal efflux genes. A putative promoter consensus sequence for RpoS-dependent genes was derived from this analysis.

This region has a higher G+C content (41%) than the overall genome (38%) and is surrounded by genes encoding transposases and phage-related genes. A gene encoding tRNA^{thr} is located at one boundary; tRNAs are known to be preferential sites for mobile element insertions. These properties suggest that the entire region may have been acquired by horizontal gene transfer and represents a pathogenesis-related island. Within this region, 15 predicted transcriptional units (multigene operons) are up-regulated in an RpoS-dependent manner (the results were independently confirmed by real-time PCR). Although there are genes within this region that are not regulated in this manner, 57 out of the 96 predicted genes are shown to be affected by RpoS (at the present time no data are available for 11 of the genes); many of these are predicted to encode efflux pumps for heavy metals and other toxic substances. Indeed, a portion of this region corresponds to the 40-kb region identified initially by McClain et al. (3) and expanded by Rankin et al. (4) as being induced during growth within macrophages.

Gene expression profiles of wild type and *rpoS* mutants were also compared during *Acanthamoeba* infection. Wild-type JR32 and *rpoS*-null mutants were grown in AYE and resus-pended in AC buffer. *A. castellanii* trophozooites (4×10^6) were seeded in microtiter wells and infected with bacteria grown in AYE at a multiplicity of infection of 10. Infection was initiated by centrifugation of the bacteria onto the adherent amoebae. Uningested bacteria were washed away and the monolayer of infected amoebae was cultured at 37°C. At 2, 3, and 5 h postinfection the cells were harvested and the total RNA was isolated, labeled, and hybridized to the arrays. Again, about 30% of the bacterial genome was found to be expressed under any single condition. About 500 genes were found to be expressed during the internalization and early stages of infection; 78 of them were also expressed during growth in liquid culture. A comparison of expressed gene complements between axenic growth and infection and some gene ontology (GO) functional gene categories (http://www.geneontology.org/) (2) expressed during the *Acanthamoeba* infection are shown in Fig. 1. A sharp change in the gene expression profiles occurred between 1 and 2 h postinfection, after which changes were more gradual. Differentially expressed genes included operons of *enh*, *dot/icm*, *lvh*, and a large number of genes encoding metal and other transporters and signal transduction proteins, among which are several homologs of eukaryotic proteins. Further studies should provide a more detailed picture of coordinated gene expression during *Legionella* infection as well as the molecular events that accompany it.

REFERENCES

1. Chien, M., I. Morozova, S. Shi, H. Sheng, J. Chen, S. M. Gomez, G. Asamani, K. Hill, J. Nuara, M. Feder, J. Rineer, J. J. Greenberg, V. Steshenko, S. H. Park, B. Zhao, E. Teplitskaya, J. R. Edwards, S. Pampou, A. Georghiou, I. C. Chou, W. Iannuccilli, M. E. Ulz, D. H. Kim, A. Geringer-Sameth, P. Goldsberry C, Morozov, S. G. Fischer, G. Segal, X. Qu, A. Rzhetsky, P. Zhang, E. Cayanis, P. J. De Jong, J. Ju, S. Kalachikov, H. A. Shuman, and J. J. Russo. 2004. The genomic sequence of the accidental pathogen Legionella pneumophila. *Science* **305:**1966–1968.

2. **Gene Ontology Consortium.** 2000. Gene ontology: tool for the unification of biology. *Nat. Genet.* **25:**25–29.

3. **McClain, M. S., M. C. Hurley, J. K. Brieland, and N. C. Engleberg.** 1996. The *Legionella pneumophila* hel locus encodes intracellularly induced homologs of heavy-metal ion transporters of Alcaligenes spp. *Infect. Immun.* **64:**1532–1540.

4. **Rankin, S., Z. Li, and R. R. Isberg.** 2002. Macrophage-induced genes of *Legionella pneumophila:* protection from reactive intermediates and solute imbalance during intracellular growth. *Infect. Immun.* **70:**3637–3648.

BRINGING THE GENOME OF
LEGIONELLA PNEUMOPHILA TO LIFE:
THE TRANSCRIPTIONAL PROGRAM
DURING INFECTION OF
ACANTHAMOEBA CASTELLANII

H. Brüggemann, A. Hagman, O. Sismeiro,
M.-A. Dillies, K. Heuner, M. Steinert,
C. Gouyette, J.-Y. Coppée, and C. Buchrieser

84

A major step forward in *Legionella* research was the determination of the genome sequences of three clinical serogroup 1 isolates of *Legionella pneumophila*: strains Paris and Lens (1) and strain Philadelphia 1 (3). The genomes are 3.34 to 3.50 Mbp in size and contain ~3,000 genes, revealing a variety of astonishing features unique to *Legionella*, such as the extended array of eukaryotic-like proteins, hot candidates involved in interfering in phagocytic functions. However, little is known about the transcriptional program of *L. pneumophila* during the infectious process. Gene expression experiments have so far only been carried out on some dozen single genes, mostly under in vitro conditions. To bring the entire genome sequence "to life" we recorded the genome-wide transcriptome to detect which genes are expressed at the different stages of the life cycle of *L. pneumophila* during the infection of eukaryotic host cells. We applied microarray technology, a powerful approach to assign the *L. pneumophila* genes to distinct steps in the course of host cell exploitation and manipulation, to reveal novel virulence determinants and regulatory cascades governing virulence formation, as well as metabolic traits allowing intracellular growth of *L. pneumophila*.

THE MICROARRAY DESIGN AND THE CHIP EXPERIMENTS

A whole-genome microarray was constructed which contained 3,823 gene-specific oligonucleotides representing every gene predicted in the *L. pneumophila* strain Paris genome plus each loci specific to strains Lens (302 genes) and Philadelphia (285 genes). This allowed the analyses of the transcriptional profiles of all three sequenced *L. pneumophila* strains. In this short article, some results of the in vivo transcriptional program of the three *L. pneumophila* strains during their life cycles in the natural eukaryotic host *Acanthamoeba castellanii* will be presented.

PHASE-DEPENDENT GENE EXPRESSION PATTERNS

All three *L. pneumophila* strains switched from replicative, nonmotile cells, observed in the early infection period, to highly motile cells in the end of the infection cycle before the lysis

H. Brüggemann, A. Hagman, and C. Buchrieser Unité de Génomique des Microorganismes Pathogènes, Institut Pasteur, 28, rue du Dr. Roux, 75724 Paris Cedex 15, France. *M.-A. Dillies, O. Sismeiro, and J.-Y. Coppée* Institut Pasteur, Plate-forme puces à ADN, 28, rue du Dr. Roux, 75724 Paris Cedex 15, France. *K. Heuner and M. Steinert* Institut für Molekulare Infektionsbiologie, Universität Würzburg, Röntgenring 11, 97070, Würzburg, Germany. *C. Gouyette* Institut Pasteur, Plate-forme Synthèse d'Oligonicléotides, 28, rue du Dr. Roux, 75724 Paris Cedex 15, France.

of the amoebal host, as determined by microscopic observation and bacterial growth quantifications. For gene expression profiling, three time points for RNA extraction have been chosen, which correspond to exponential, postexponential, and stationary phase *Legionella*. A biphasic life cycle is clearly evident from the transcriptional data, with sets of replicative phase (RP) genes upregulated at the earlier time point and transmissive phase (TP) genes upregulated at the later time points. The latter phase showed an elevated variation in gene expression. The global transcriptional programs of all three *L. pneumophila* strains were very similar, indicating that common regulatory mechanisms govern their life cycles.

GENE EXPRESSION IN THE REPLICATIVE PHASE

The RP is characterized by a strong upregulation of the core metabolism. All components of the respiratory chain (complex I-IV) are strongly transcribed in the RP. The terminal oxidases, cytochrome *c* (*coxC/ctgA/coxA/coxB*), cytochrome *o* (*cyoABC*), and cytochrome *d* oxidase (*cydA/cydB*) were upregulated, suggesting that during oxidative phosphorylation within the replicative vacuole, different oxygen partial pressures can be accommodated. An alternative low aeration cytochrome *d* oxidase (*qxtA/qxtB*) was induced in the TP, indicating that oxygen may become limiting at the end of the bacterial proliferation period. During the replication period, a strong oxidative stress response was observed, as indicated by the upregulation of components homologous to alkyl hydroperoxide reductases, catalase-peroxidase and an iron/manganese-dependent superoxide dismutase.

What are the preferred growth substrates? Based on studies of broth cultures, *L. pneumophila* is thought to utilize only amino acids, mainly serine. Indeed, during the RP in amoebae, genes encoding components of amino acid uptake and degradation pathways are upregulated, in particular, for the catabolism of serine, threonine, glycine, tyrosine, alanine, and histidine. *L. pneumophila* also appears to catabolize carbohydrates in the RP, as judged by the up-

regulation of a gene cluster encoding the Entner-Doudoroff pathway, and a putative glucokinase and sugar transporter. Expression of a eukaryotic-like glucoamylase may allow intracellular *L. pneumophila* to utilize starch or glycogen.

MOTILITY AND VIRULENCE TRAITS ARE STRONGLY INDUCED IN THE TRANSMISSIVE PHASE

When switching from a fast-growing bacterium to a motile, highly infectious microbe, the pathogen fundamentally alters its transcriptional program. Core metabolic functions as well as the transcriptional and translational machinery are repressed in the late phase of the infectious cycle. Instead, about 402 genes are commonly induced in all three *L. pneumophila* strains in the TP. Among the strongest induced genes are those belonging to the flagellar operons. Interestingly, these genes can be clustered in four classes, based on their time-dependent manner; the four classes are expressed in a successively cascading pattern. The strongest induced gene in the TP was *flaA*, encoding the structural flagellum filament protein flagellin.

In addition, most of the so far known virulence factors were upregulated in the late phase of the intracellular life cycle, for example, proteins secreted by the Dot/Icm type IV secretion system that are predicted to inhibit phagosome maturation and alter its trafficking (6, 8). The genes of 14 type IV secretion system substrates and their homologs were strongly upregulated in the TP, averaging more than five-fold more RNA than observed in the RP: *ralF*; *sidB*-paralog *sdbB*; *sidC* and paralog *sdcA*; *sidG*; *sidE*-paralogs *sdeA*, *sdeB*, *sdeC*, *lpp1615*, and *lpp1453*; *sidH*-paralogs *sdhB* and *lpp2886*; and the recently identified effectors *vipD* and *vipE* (10). However, not all substrates of the Dot/Icm system exhibited the same pattern: the effectors LepA and LepB, which contribute to the exit from the amoebal phagosome (2), and others, such as paralogs of *sidE* (*sdeD, lpp2062, lpp2093*) and paralogs of *sidB* (*sdbA, sdbC, lpp2657*), showed no significant gene expression changes during the intracellular life cycle in *A. castellanii*.

Also strongly induced in the TP were other virulence factors that are not dependent on type IV secretion. One example is the enhanced entry proteins EnhABC, which contain multiple copies of the protein-protein interaction domain Sel-1 and are thought to affect the initial encounter with a eukaryotic cell (4). Strong induction in the TP also occurred for three homologs of EnhA and two of EnhC. Other putative virulence factors that showed a similar expression pattern include *rtxA*, seven of the ankyrin repeat proteins, and nine proteins containing eukaryotic domains predicted to be involved in host cell modulation. In addition, the TP was characterized by the upregulation of over 90 genes that lack obvious homology to database entries. One exception to the general rule of TP expression of known virulence factors was the gene encoding the macrophage infectivity potentiator (Mip). This surface-associated protein has peptidyl prolyl cis/trans isomerase activity and is needed for full virulence of *L. pneumophila*; its RNA was upregulated in the RP, pointing to a role in the maintenance of a functional replication vacuole.

SIGMA FACTORS, TRANSCRIPTIONAL REGULATORS AND GGDEF/EAL PROTEINS LIKELY GOVERN THE BIPHASIC LIFE CYCLE

Timely differentiation of RP *L. pneumophila* into the TP form must be precise to ensure coordinated expression of functions critical for motility, contact-dependent cytotoxicity, proper phagosome trafficking, and other virulence traits (5, 7). Accordingly, many regulatory genes were differentially expressed in RP and TP. Previously characterized regulatory factors whose expression is strongly induced in the TP are, for instance, the transmission trait enhancer LetE and the alternative sigma factor FliA (σ^{28}). Among the most significantly upregulated genes in the TP, we identified four two-component systems, six LysR-family members, four additional transcriptional regulators

[CpxR, PaiA-, MoxR and TetR-like], and two integration host factors.

A specific group of regulatory proteins that have not yet been examined in *L. pneumophila* is presumably implicated in the control of the late phase of transmission: Nine genes encoding GGDEF and/or EAL domain proteins were upregulated eightfold or more. Moreover, all five adenylate/guanylate cyclases and three of five cyclic AMP/cyclic GMP binding proteins that are most likely connected to the group of GGDEF/EAL domain proteins were similarly upregulated. Proteins containing these domains modulate the intracellular level of bis-(3'-5')-cyclic dimeric guanosine monophosphate (c-di-GMP), a recently discovered second messenger in bacteria (9).

CONCLUSIONS

Our analysis of the transcriptome of three *L. pneumophila* strains during the life cycle in amoebae provides new insights into the pathobiology of this unique host-parasite interaction. The in vivo analysis allowed validation of the life cycle model predicted mainly from broth-grown cells and allowed a more subtle differentiation of life cycle phases. The findings provide a valuable resource and conceptual framework to increase knowledge of the *Legionella* life cycle and the host functions exploited during infection of amoebae and macrophages.

ACKNOWLEDGMENTS

Financial support was received from the Institut Pasteur, the Centre National de la Recherche (CNRS), the Consortium National de Recherche en Génomique (CNRG) du Réseau National des Genopoles (RNG), and the Centre National de Référence des *Legionella*. H. Brüggemann is holder of a fellowship of the German Academy of Natural Scientists Leopoldina (funding number BMBF-LPD9901/8-101), and A. Hagman received a grant from the Network of Excellence "Europathogenomics" LSHB-CT-2005-512061.

REFERENCES

1. Cazalet, C., C. Rusniok, H. Bruggemann, N. Zidane, A. Magnier, L. Ma, M. Tichit, S. Jarraud, C. Bouchier, F. Vandenesch, F. Kunst, J. Etienne, P. Glaser, and C. Buchrieser. 2004.

Evidence in the *Legionella pneumophila* genome for exploitation of host cell functions and high genome plasticity. *Nat. Genet.* **36:**1165–1173.

2. **Chen, J., K. S. de Felipe, M. Clarke, H. Lu, O. R. Anderson, G. Segal, and H. A. Shuman.** 2004. *Legionella* effectors that promote nonlytic release from protozoa. *Science* **303:**1358–1361.

3. **Chien, M., I. Morozova, S. Shi, H. Sheng, J. Chen, S. M. Gomez, G. Asamani, K. Hill, J. Nuara, M. Feder, J. Rineer, J. J. Greenberg, V. Steshenko, S. H. Park, B. Zhao, E. Teplits-kaya, J. R. Edwards, S. Pampou, A. Georg-hiou, I. C. Chou, W. Iannuccilli, M. E. Ulz, D. H. Kim, A. Geringer-Sameth, C. Golds-berry, P. Morozov, S. G. Fischer, G. Segal, X. Qu, A. Rzhetsky, P. Zhang, E. Cayanis, P. J. De Jong, J. Ju, S. Kalachikov, H. A. Shuman, and J. J. Russo.** 2004. The genomic sequence of the accidental pathogen *Legionella pneumophila*. *Science* **305:**1966–1968.

4. **Cirillo, S. L., J. Lum, and J. D. Cirillo.** 2000. Identification of novel loci involved in entry by *Legionella pneumophila*. *Microbiology* **146:**1345–1359.

5. **Heuner, K., and M. Steinert.** 2003. The flagellum of *Legionella pneumophila* and its link to the expression of the virulent phenotype. *Int. J. Med. Microbiol.* **293:**133–143.

6. **Luo, Z. Q., and R. R. Isberg.** 2004. Multiple substrates of the *Legionella pneumophila* Dot/Icm system identified by interbacterial protein transfer. *Proc. Natl. Acad. Sci. USA* **101:**841–846.

7. **Molofsky, A. B., and M. S. Swanson.** 2004. Differentiate to thrive: lessons from the *Legionella pneumophila* life cycle. *Mol. Microbiol.* **Online publication date: 4-Jun-2004**.

8. **Nagai, H., and C. R. Roy.** 2003. Show me the substrates: modulation of host cell function by type IV secretion systems. *Cell. Microbiol.* **5:**373–383.

9. **Romling, U., M. Gomelsky, and M. Y. Galperin.** 2005. C-di-GMP: the dawning of a novel bacterial signalling system. *Mol. Microbiol.* **57:**629–639.

10. **Shohdy, N., J. A. Efe, S. D. Emr, and H. A. Shuman.** 2005. Pathogen effector protein screening in yeast identifies *Legionella* factors that interfere with membrane trafficking. *Proc. Natl. Acad. Sci. USA* **102:**4866–4871.

GENOME REARRANGEMENTS AND HORIZONTAL GENE TRANSFER IN *LEGIONELLA PNEUMOPHILA*

Irina Morozova, Pavel Morozov, Sergey Pampou,
Karim Suwwan de Felipe, Sergey Kalachikov,
Howard A. Shuman, and James J. Russo

85

Legionella pneumophila is found in diverse ecological niches such as axenic cultures, biofilms with other microbes, and intracellular vacuoles of protozoa and human cells. This may have provided ample opportunities for the organisms to rearrange their genomes and to accumulate genes that help them survive in these different environments. *Legionella*'s ability to survive in harsh conditions such as in plumbing systems treated with potent biocides may require additional genes necessary for detoxification. The needed gene pool can be assembled via gene family expansions and horizontal gene transfer from other organisms. Comparative genome sequence analysis of different *L. pneumophila* strains has indeed indicated that such genome rearrangements of different scales have occurred in the past: (i) large-scale relocations of genomic regions, most apparent from dot plots (Fig. 1) and graphs of gas chromatography (GC) skew; (ii) probable transpositions of gene cassettes, such as that for an extensive efflux region, by nearby mobile elements; and (iii) complicated history of acquisition of individual genes within certain gene sets, such as those of the *icm/dot* locus that are critical for the pathogenicity of *Legionella* and genes of probable eukaryotic origin acquired by horizontal gene transfer events. Examples of the third category of rearrangements have already been reported (3–6), and in the case of the *icm/dot* "pathogenicity island" indicate a probable composite outcome of gain of genes by horizontal transfer, perhaps at more than one point in time, gene loss within the locus, and more traditional vertical evolution events.

Fig. 1 shows genome comparisons of the Philadelphia 1 strain (2) and those of Paris and Lens (1), indicating positions of the large-scale genome rearrangements. These include an inversion at the replication terminus in the Lens strain relative to the other two alternative chromosomal insertion sites among the three strains for the *lvr/lvh* gene-containing plasmid pLP45, which can exist in both chromosomal and episomal forms, as supported by PCR and BAC sequencing, and numerous smaller breaks in the plot at the positions of IS and other mobile elements. Some of these points of genome rearrangements, for example, the translocation

Irina Morozova, Sergey Pampou, Sergey Kalachikov, and James J. Russo Columbia Genome Center, Columbia University, Russ Berrie Pavilion Rm 408, 1150 St. Nicholas Ave, New York, NY 10032. *Pavel Morozov* Joint Centers for Systems Biology, Columbia University, Irving Cancer Research Center, 1130 St. Nicholas Ave, New York, NY 10032. *Karim Suwwan de Felipe and Howard A. Shuman* Department of Microbiology, College of Physicians and Surgeons, 1216 Hammer Health Science Center, 701 W 168 St., New York, NY 10032.

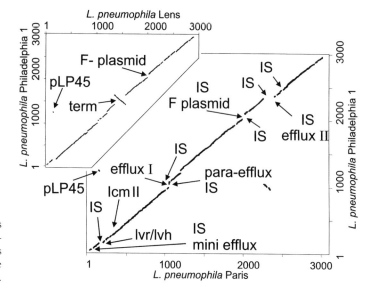

FIGURE 1 Genome alignments (dot plots) of *L. pneumophila* Philadelphia 1 against Lens and Paris strains show the location of major genome rearrangement and acquisition events.

position of the *lvh/lvr* gene region (corresponding to the pLP45 plasmid in the Philadelphia 1 strain), are visible on the Paris strain GC-skew plot (not shown), which cannot be explained merely by the higher than average G+C content of the region, indicating that the translocation is probably a relatively recent event. It is noteworthy that these mobile elements, and possibly the genes of the plasmid itself, are duplicated or exist in multiple copies within individual genomes; presumably, this is a vestige of their past evolutionary history or an indication of their current lifestyle.

Evidence for past mobilization of gene cassettes by transposable elements is exemplified by an extensive locus of ~100 kb in the genome of *L. pneumophila* that contains a very high proportion of its genes encoding membrane transporters. Given its diverse environmental niches and relatively broad host range, the bacterium would be expected to harbor an equally broad range of proteins for uptake of the available nutrients and disposal of toxic molecules. This is in fact borne out by consideration of the annotated genomes, which revealed over 350 binding proteins and permeases for 62 distinct classes of substrates and in which multidrug and heavy metal transporters are particularly highly represented compared

to other bacteria (2), a number of which are found within the 100-kb efflux region. Microarray expression data showed many of these to be RpoS dependent, both by comparison of their response to growth in wild-type versus RpoS-deleted strains and by the identification of upstream σ^S-dependent promoter consensus motifs (see chapter 83). This efflux gene cassette is located at different positions in the genomes of the three sequenced strains (Fig. 2a) and is in the vicinity of a nearby mobile element (designated "para-efflux" in Fig. 2b) that may have participated in the efflux cassette mobilization. Other typical features of a mobile element are also found in the immediate vicinity, including tRNAs, DNA repair genes, and numerous IS elements. Genomic landscape analysis revealed several other potential mobile elements in *L. pneumophila* with similar gene organization. Phylogenetic and gene composition analysis indicate that different *L. pneumophila* strains display distinct acquisition histories for these gene subsets, suggesting gene rearrangements as well as repeated horizontal gene transfer events. Several dozen eukaryotic gene homologs with distinct phenotypes which appear to be acquired by gene transfer were identified (3).

Overall, the genome of *L. pneumophila* reveals features supporting genome rearrangement

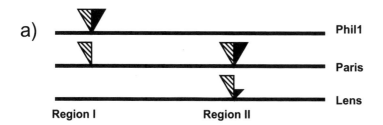

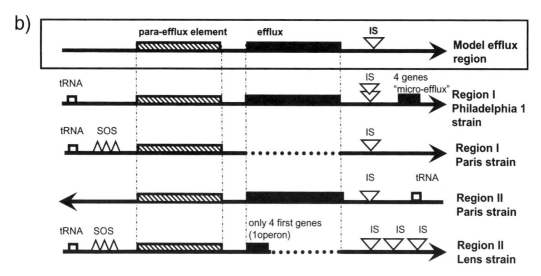

FIGURE 2 Gene content of 100-kb efflux regions in three strains. (a) Efflux elements can be in two different locations in the analyzed genomes. In all three genomes the efflux gene cassette is located next to the para-efflux element which constitutes a remnant of a presumably mobile element.(b) Structure comparisons of the four regions shown in panel a. The insertion of the mobile element most likely occured at the tRNA gene site, and the regions are surrounded by numerous IS elements. In two cases, the error-prone DNA repair SOS system operon is found in the vicinity.

events of different scales. About 5% of the Philadelphia 1 genome encode phage-derived, plasmid-derived, and IS elements, including *icm/dot*, *lvh/lvr*, *tra/trb*, and other gene sets, many of which are dispersed but some of which occur in clusters. *L. pneumophila* strains harbor pathogenicity islands along with several plasmids that can exist in chromosomal and episomal forms. Given its intracellular existence, it is not surprising that examination of the genome sequence offers evidence that *L. pneumophila* can be an active participant in horizontal gene transfer. When one adds on genes of presumably eukaryotic origin, up to 7% of the genome may display evidence of horizontal

transfer since its emergence as a distinct species. As more strains of *L. pneumophila* are sequenced, along with other species of *Legionella*, much more will be learned about their evolutionary origins, lifestyle, and potential for pathogenicity.

REFERENCES

1. **Cazalet. C., C. Rusniok, H. Brüggemann, N. Zidane, A. Magnier, L. Ma, M. Tichit, S. Jarraud, C. Bouchier, F. Vandenesch, F. Kunst, J. Etienne, P. Glaser, and C. Buchrieser** . 2004. Evidence in the *Legionella pneumophila* genome for exploitation of host cell functions and high genome plasticity. *Nat. Genet.* **36:**1165–1173.

2. **Chien. M., I. Morozova, S. Shi, H. Sheng, J. Chen, S. M. Gomez, G. Asamani, K. Hill, J. Nuara, M. Feder, J. Rineer, J. J. Greenberg, V. Steshenko, S. H. Park, B. Zhao, E. Teplitskaya, J. R. Edwards, S. Pampou, A. Georghiou, I. C. Chou, W. Iannuccilli, M. E. Ulz, D. H. Kim, A. Geringer-Sameth, C. Goldsberry, P. Morozov, S. G. Fischer, G. Segal, X. Qu, A. Rzhetsky, P. Zhang, E. Cayanis, P. J. De Jong, J. Ju, S. Kalachikov, H. A. Shuman, and J. J. Russo.** 2004. The genomic sequence of the accidental pathogen *Legionella pneumophila.* *Science* **305:**1966–1968.

3. **de Felipe, K. S., S. Pampou, O. S. Jovanovic, C. D. Pericone, S. F. Ye, S. Kalachikov, and H. A. Shuman.** 2005. Evidence for acquisition of *Legionella* type IV secretion substrates via interdomain horizontal gene transfer. *J. Bacteriol.* **187:** 7716–7726.

4. **Feldman, M., and G. Segal.** 2004. A specific genomic location within the icm/dot pathogenesis region of different *Legionella* species encodes functionally similar but nonhomologous virulence proteins. *Infect. Immun.* **72:**4503–4511.

5. **Ko, K. S., H. K. Lee, M. Y. Park, and Y. H. Kook.** 2003. Mosaic structure of pathogenicity islands in *Legionella pneumophila. J. Mol. Evol.* **57:** 63–72.

6. **Morozova, I., X. Qu, S. Shi, G. Asamani, J. E. Greenberg, H. A. Shuman, and J. J. Russo.** 2004. Comparative sequence analysis of the icm/dot genes in *Legionella. Plasmid.* **51:**127–147.

GENETIC DIVERSITY OF
LEGIONELLA PNEUMOPHILA

C. Cazalet, S. Jarraud, Y. Ghavi-Helm, J. Allignet, F. Kunst,
J. Etienne, P. Glaser, and C. Buchrieser

86

Among the 46 *Legionella* species described to date, *Legionella pneumophila* seems to have a higher capacity to cause disease in humans, as it is responsible for 90% of the diagnosed legionellosis cases. Furthermore, among the 15 serogroups described within the species *L. pneumophila*, serogroup 1 is responsible for over 80% of legionellosis cases. Complete genome sequences of three *L. pneumophila* serogroup 1 strains (Paris, Lens, and Philadelphia) were determined and published in 2004 (3, 4). Whole-genome comparisons revealed a marked plasticity, as three different plasmids are present and about 10% of the genes of each strain are strain specific. In order to explore this diversity in a large number of strains, a *Legionella* biodiversity array was designed containing probes (internal PCR products) specific for all those genes that are variable among the three strains as well as of conserved ones that have known or putative implication in virulence. On the basis of these

criteria, 1,338 probes were spotted in duplicate on the *Legionella* biodiversity array.

The gene content of 180 *Legionella* isolates, 165 *L. pneumophila* strains, and 25 representatives of *Legionella* species other than *L. pneumophila* isolated from the environment during epidemics or from sporadic cases were analyzed by DNA-DNA hybridization using the *Legionella* biodiversity array. The method used identifies genes as present when the percentage of the DNA identity between the genomic DNA tested and the probe present on the array is at least 80%. The results will be discussed here with respect to three different questions: (i) What is the distribution of virulence factors and genes coding for proteins putatively implicated in virulence? (ii) What is the degree of plasticity and what are the evolutionary relationships among *Legionella* genomes? (iii) Can this tool be used for epidemiological studies?

DISTRIBUTION OF VIRULENCE FACTORS AMONG THE DIFFERENT STRAINS

As described briefly above, *L. pneumophila* serogroup 1 is largely overrepresented among human legionellosis cases, suggesting that virulence differences exist among different *L. pneumophila* serogroups and other *Legionella* species. In order to address this question and to

C. Cazalet, Yad Ghavi-Helm, F. Kunst, P. Glaser, and C. Buchrieser Unité de Génomique des Microorganismes Pathogènes, Institut Pasteur, 28, rue du Dr. Roux, 75724 Paris Cedex 15, France. *J. Allignet* Institut Pasteur, Plate-forme puces à ADN, CNR/Sante publique, 28, rue du Dr. Roux, 75724 Paris Cedex 15, France. *S. Jarraud and J. Etienne* Centre National de Référence des *Legionella*, Laboratoire de Bactériologie INSERM E-0230, Faculté de Médecine, IFR 62, 69372, Lyon Cedex 08.

try to find a possible link between virulence differences and a specific pattern of virulence genes, we analyzed the distribution of these factors among 180 strains. As deduced from hybridization results, the known virulence genes (e.g., *dot/icm* gene cluster, *enhABC*, *lss* type II secretion system, etc.) are conserved among all clinical and environmental isolates of *L. pneumophila*. The only exception is *ralF* coding for a Dot/Icm substrate, which is absent from five of the *L. pneumophila* strains tested. RalF perturbs vesicle trafficking by functioning as an exchange factor for the ADP ribosylation factor (ARF) family of guanosine triphosphatases (GTPases). The RalF protein was described as required for the localization of ARF on phagosomes containing. *L. pneumophila* (5). Probably another factor, fulfilling the same function, is present in these strains. This study also reports for the first time the distribution of the eukaryotic-like genes identified during sequence analysis among *L. pneumophila*. Except for a few exceptions, all of these genes are conserved throughout *L. pneumophila*, further strengthening the hypothesis that they are required for intracellular growth and that they might play a role in virulence.

PLASTICITY AND EVOLUTION OF THE *LEGIONELLA* GENOMES

In order to better understand plasticity and evolution in *L. pneumophila* we used hierarchical clustering to analyze the hybridization patterns obtained. This approach allowed us to define five subgroups among the *L. pneumophila* isolates tested. Furthermore, the macroarray allowed us to distinguish *L. pneumophila* strains from *Legionella* strains of other species as they clustered separately. However, no correlation of gene content and epidemiological characteristics, which could explain virulence differences, was observed.

With respect to biodiversity, the hybridization results further underlined the results from complete genome sequence comparison. *L. pneumophila* and other *Legionella* species are highly diverse, as many of the strain-specific genes of strain Paris are conserved mainly only within the subgroup of different Paris strains, the strain-specific genes of strain Lens mainly among Lens strains, and those of strain Philadelphia in the Philadelphia subgroup but are rarely present in strains belonging to other subgroups. As reported in the literature, the DNA identity among certain genes within the different species of *Legionella* can often be less than 80%. This is in agreement with our observation that DNA of non-*L. pneumophila* isolates hybridizes with very few probes. In contrast, some mobile genetic elements such as plasmids, the efflux pump cluster, and the 65-kb pathogenicity island of strain Philadelphia (2) may be highly conserved, as probes specific for these elements were also detected in isolates from other *Legionella* species. They show a heterogeneous distribution in *L. pneumophila*, suggesting that they are transferred horizontally. The cluster encoding the *lvh* type IV secretion and the 130-kb island containing several efflux pumps have in addition a mosaic structure with respect to different isolates, further underlining the high genome plasticity of *Legionella*. For example, the *lvh* region is highly conserved in all three sequenced genomes but the flanking DNA regions are specific to each strain and different from each other (Fig. 1). The size of these flanking regions is also different in each strain, as it is flanked by 11 kb and 22 kb in strain Paris, by 5.8 kb and 30 kb in strain Lens and by 14 kb and 30 kb in strain Philadelphia 1. Other strains analyzed in this study which carry the *lvh* type IV secretion system seem to have different rearrangements in the specific parts flanking the *lvh* region, as the hybridizing regions differ.

Despite the high genome plasticity observed, hierarchical clustering of the hybridization results of the genes selected for the *Legionella* biodiversity array correlates with results of *rpoB* sequence typing and therefore with phylogeny. However, the groups obtained do not correlate with serogroups, thus suggesting that lipopolysaccharide genes determining the serogroup are horizontally exchangeable.

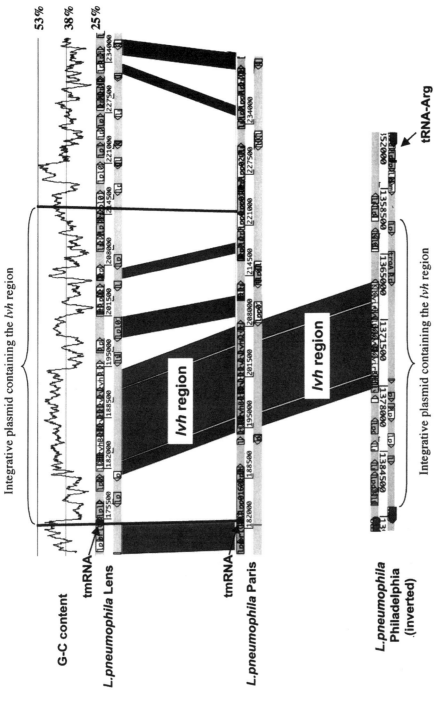

FIGURE 1 Schematic representation of the integrative multicopy plasmid carrying the *lvh* type IV system and comparison of this region in the three strains. The scale on the right indicates the G+C content values in this region. The black curve indicates variations in G+C content along this region in strain Lens. Dark grey bars, homologous regions in the three genomes; light grey arrow, genes predicted in this region. Arrows and numbers indicate the insertion position on the chromosome of each strain.

UTILITY OF THE *LEGIONELLA* BIODIVERSITY ARRAY FOR EPIDEMIOLOGY

Finally, in order to evaluate a possible use of the *Legionella* biodiversity array as an epidemiological tool, we focused on the results obtained for the group of Paris strains investigated. The *L. pneumophila* Paris strains are responsible for 30% of legionellosis cases in Paris and about 12% in France and are widespread in the environment but are indistinguishable with classical typing tools such as pulsed-field gel electrophoresis or multilocus sequence typing (1). The *Legionella* biodiversity array allowed subdivision of the otherwise indistinguishable group of Paris strains. Variations between different Paris strains were observed in some large genomic islands such as the presence/absence of the 65kb pathogenicity island which was identified in the Philadelphia strain (2), variations in the 130kb cluster encoding efflux pumps (6) or the 130-kb plasmid which can be absent or present in part. Variations among specific genes such as *rtxA* and genes encoding restriction/modification proteins were also observed. Furthermore traits specific to the lineage of Paris strains were identified: an autotransporter, an additional cytochrome oxidase, an islet of 17 genes, 14 of which have weak similarity to eukaryotic genes, and 23 additional genes. These results should allow establishment of a multiplex PCR test for rapid identification and subtyping of Paris strains.

CONCLUDING REMARKS

Here we report for the first time comparative genomic data at the genome level for *L. pneumophila* and the genus *Legionella*. Our results emphasize the high diversity among *L. pneumophila* strains and other *Legionella* species and highlight specific plasticity regions. Considering the diversity defined on the basis of results obtained with the biodiversity array, additional studies using whole genome-microarrays are under way, to further define the core genome of *Legionellae*. Together with these results, the *Legionella* biodiversity array will be a promising typing tool for *L. pneumophila*.

ACKNOWLEDGMENTS

This work received financial support from the Institut Pasteur, the Centre National de la Recherche (CNRS), and from the Microarray Plateform, NRC/Public Health, and the Agence Française de Sécurité Sanitaire de l'Environnement et du Travail (AFSSET), France contract ARCL-2005-00. C. Cazalet is holder of a fellowship jointly financed by the CNRS (France) and Veolia Water–Anjou Recherche.

REFERENCES

1. **Aurell, H., J. Etienne, F. Forey, M. Reyrolle, P. Girardo, P. Farge, B. Decludt, F. Vandenesch, and S. Jarraud.** 2003. *Legionella pneumophila* serogroup 1 strain Paris: endemic distribution throughout France. *J. Clin. Microbiol.* **41:**3320–3322.
2. **Brassinga, A. K., M. F. Hiltz, G. R. Sisson, M. G. Morash, N. Hill, E. Garduno, P. H. Edelstein, R. A. Garduno, and P. S. Hoffman.** 2003. A 65-kilobase pathogenicity island is unique to Philadelphia-1 strains of *Legionella pneumophila. J. Bacteriol.* **185:**4630–3637.
3. **Cazalet, C., C. Rusniok, H. Bruggemann, N. Zidane, A. Magnier, L. Ma, M. Tichit, S. Jarraud, C. Bouchier, F. Vandenesch, F. Kunst, J. Etienne, P. Glaser, and C. Buchrieser.** 2004. Evidence in the *Legionella pneumophila* genome for exploitation of host cell functions and high genome plasticity. *Nat. Genet.* **36:**1165–1173.
4. **Chien, M., I. Morozova, S. Shi, H. Sheng, J. Chen, S. M. Gomez, G. Asamani, K. Hill, J. Nuara, M. Feder, J. Rineer, J. J. Greenberg, V. Steshenko, S. H. Park, B. Zhao, E. Teplitskaya, J. R. Edwards, S. Pampou, A. Georghiou, I. C. Chou, W. Iannuccilli, M. E. Ulz, D. H. Kim, A. Geringer-Sameth, C. Goldsberry, P. Morozov, S. G. Fischer, G. Segal, X. Qu, A. Rzhetsky, P. Zhang, E. Cayanis, P. J. De Jong, J. Ju, S. Kalachikov, H. A. Shuman, and J. J. Russo.** 2004. The genomic sequence of the accidental pathogen *Legionella pneumophila. Science* **305:**1966–1968.
5. **Nagai, H., J. C. Kagan, X. Zhu, R. A. Kahn, and C. R. Roy.** 2002. A bacterial guanine nucleotide exchange factor activates ARF on *Legionella* phagosomes. *Science* **295:**679–682.
6. **Rankin, S., Z. Li, and R. R. Isberg.** 2002. Macrophage-induced genes of *Legionella pneumophila*: protection from reactive intermediates and solute imbalance during intracellular growth. *Infect. Immun.* **70:**3637–3648.

THE PROBLEM OF COMPLEXITY

Rodney M. Ratcliff

87

The genus *Legionella* is now recognized to contain 50 species and one genomospecies. While *Legionella pneumophila* is the predominant *Legionella* species associated with infection in humans and accounts for >90% of all reported cases worldwide (23), nearly half of the species have been associated with disease, and in some regions the association can be frequent, such as for *L. longbeachae* in Australia (9). In addition, strains of many species are isolated during the environmental screening which often follows an outbreak and during routine surveillance of at-risk water sources such as cooling towers, spars, and institutional potable water supplies. Given the unreactivity of legionellae to the commonly employed biochemical tests (21), the identification of such isolates is problematic, necessitating complex identification methods.

Early publications of *Legionella* species relied on the few discriminating phenotypic characteristics as well as DNA relatedness, antigenicity as measured by immunofluorescence and/or agglutination, and the characterization of the cellular, especially branched-chained, fatty acids and isoprenoid quinones (2). While a number of these may still be relevant to defin-

ing a strain as belonging to the genus *Legionella*, with the exception of DNA relatedness, none are now capable of resolving all of the currently recognized strains. Fig. 1 presents detailed examples of serological cross-reactivity using a quantitative microagglutination checkerboard titration method incorporating absorbed polyvalent antisera. Uniquely resolving monovalent antisera for all species do not exist. While such serological tests still have a use in assisting in the rapid identification of strains from commonly isolated species, the results can no longer be considered definitive. Similarly, the characterization of the cellular fatty acids and isoprenoid quinones showed early promise, but the detection and characterization of additional species has demonstrated that some species share identical profiles for these cellular components (3, 22). Furthermore, culture conditions and the age of the culture from which an extract is prepared, as well as the extraction method, equipment components, and assay conditions employed to characterize the fatty acid and isoprenoid quinones, all influence the result, making comparative analysis between laboratories difficult.

While DNA relatedness (also termed DNA-DNA hybridization) remains discriminating and continues to be the final arbiter of novelness, the arbitrariness of the empirical 70%

Rodney M. Ratcliff   Infectious Diseases Laboratories, Institute of Medical and Veterinary Science, Adelaide, Australia.

Type of cross-reactivity
- broad cross-reactivity with poor titration endpoint
    e.g. *L. parisiensis, L. anisa, L. bozemanii* (PAB group)

- identical titers
    e.g. *L. spiritensis - L. taurinensis, L. worsleiensis - L. quateirensis*

- high titer cross-reactivity to as many as three to six other species
    e.g. *L. micdadei, L. cincinnatiensis, L. jordanis, L. santicrusis, L. quinlivanii,*
    L. sainthelensi sg 1&2, *L. brunensis* sg 2, FM-1-679

- Some serogroups of *L. pneumophila* cross-react with other species
    e.g. *L. pn* sg 5 & 8 with *L. gormanii, L. dumoffii, L. worsleiensis, L. quateirensis*

- some species can only be recognised by their cross-reactivity spectra to antisera to other
  species
    e.g. *L. oakridgensis, L. sainthelensi*

FIGURE 1   Examples of serological cross-reactivity using a quantitative microagglutination
checkerboard titration method (V. Drasar, personal communication).

relatedness cut-off used to define a new species is being questioned about lacking a theoretical basis (14) and about its ability to always group bacteria into true ecological units (6, 20). Alternative models such as the community phylogeny approach, which define ecotypes based directly on the genetic sequence, are being proposed as a more rational approach (6, 14 and chapter 88). DNA relatedness is also subject to technical issues such as reproducibility and operator interpretation. In addition, each potentially novel strain needs to be tested against every other recognized species (50 for *Legionella*), a prohibitively complex, time consuming and expensive process beyond the resources of most laboratories, and which is unavailable for the classification of currently nonculturable prokaryotes.

Not surprisingly given these difficulties, the growing availability of gene sequence-based phylogenetic schemes has increasingly been relied on to assist in the characterization of *Legionella* species. The essentially digital nature of the data results in less ambiguity when determining the sequence leading to high reproducibility between laboratories and is inherently compatible with computer analysis and database storage. In addition, the sequence is intrinsically several orders of magnitude more informative and thus more resolving than most

other phenotypic methods. For example, a 500 base-pair sequence contains 2,000 potential character states (500 base positions multiplied by four possible nucleotides at each position), 10- to 100-fold more than for phenotypic methods, although in reality the number of informative character states is less because not every base position in the sequence is variable and thus informative with respect to the sequence from another strain or species. Phylogenetic schemes for *Legionella* have been published targeting the following genes; 16S rRNA gene (5, 8), the macrophage infectivity potentiator gene (*mip*) (17), the RNA polymerase β-subunit (*rpoB*) (10, 13), the RNase P RNA gene (*rnpB*) (19) and the DNA gyrase A subunit (*gyrA*) (4) although only the first three have been used to formally publish novel species. Other molecular methods, such as randomly amplified polymorphic DNA (12), ribotyping (7) and 16S-23S interspacer region PCR (18) have also been used, and while these are often inexpensive, interlaboratory reproducibility and/or the restricted number of species currently analyzed limit their use.

The current trend in publishing new *Legionella* species, in spite of the difficulties mentioned above, is to utilize culture and phenotypic characteristics, serology, cellular fatty acids and isoprenoid quinone profiles, DNA relatedness, and a molecular phylogenetic analysis targeting usually more than one gene. However, more recent publications of *Legionella*-like amoebic pathogens (legionellae which currently can only be grown in coculture with amoebae) as *Legionella* species have by necessity precluded some of these analyses (1, 11, 15). The prohibitive cost of performing all of these analyses when many are no longer discriminating has resulted in the accumulation of uncharacterized *Legionella* strains, which may represent potential novel species, in laboratory collections around the world. As a consequence, the awareness of these potential novel species is limited, frequently excluding them from comparative analyses of *Legionella* species.

To investigate the greater use of sequence-based phylogenetic analyses in lieu of serology, fatty acid and isoprenoid quinone profiles, and even perhaps DNA relatedness, 56 strains representing 33 potentially novel species were phylogenetically assessed with 6 gene targets. These strains had been isolated from several laboratories in the United States, Europe, Japan, and Australia and identified as potentially novel by one of three criteria. First, following a phylogenetic analysis based on sequence from the *mip* gene, 49 strains representing 30 potentially novel species were selected because their genetic distance from recognized species was greater than that between recognized species. Thirty-seven of these comprised thirteen potentially novel species, each with more than one strain per novel species. The remaining potentially novel species were represented by a single isolate. Note that this definition of novelness is arbitrary without a theoretical basis, but adequate for the purposes of this study. Second, DNA studies on two strains indicated that they were novel (R. Benson, personal communication), even though their *mip* sequence-derived genetic distance was closer to a recognized species than that between recognized species. Last, the remaining five strains were selected because they were genetically related to either of these latter two strains, based on their *mip* sequence-derived genetic distance.

The six gene targets used in this analysis

were the five listed above, together with the zinc metallo-protease (*proA*, also known as *mspA*) gene. Published primers and amplification conditions were used for *mip* (17), *rpoB* (10, 13), *rnpB* (19) and *gyrA* (4). The following primers were used for the additional two gene targets: 16S rRNA, Leg-16S-F (5′-GGCTCA-GATTGAACGCTGGCGG-3′) and Leg-16S-R (5′-ACCCACTCCCATGGTGTGACGG-3′); *proA*, Leg-proA-F (5′-TGCATTRTAYG-CNGGNTAYGTNATHAARCAYATGTA-3′) and Leg-proA-R (5′-GTGTCCARTART-CCATRTTNGCYTTNACCAT-3′). Typical PCR reagents and cycling conditions (annealing temperature of 55°C for 30 sec) were used for both. Amplification products were sequenced in both directions using dye terminator chemistry primed with the amplification primers. The 16S rRNA-derived amplification products were additionally sequenced with the following internal primers to ensure accurate sequence determination of the entire product: Leg-16S-498-F (5′-CAGCAGCCGCGGTA-ATACG-3′), Leg-16S-1043-F (5′-TGTCGT-CAGCTCGTGTCGTG-3′), and Leg-535-R (5′-TTTACGCCCAGTAATTCCG-3′). Each amplification reaction was performed at least twice, once for each of the sense and antisense sequencing reactions, to minimize the possibility of sequencing errors. A consensus of the sense and antisense sequences was constructed for each gene target from each strain, from which the primer sequences had been excised, and any ambiguities resolved with repeat sequencing before each set of gene target sequences was phylogenetically compared with that published (or ascertained as part of this project for *proA* and *gyrA*) for the type strains of each species. Table 1 presents the size and degree of sequence variability of the amplified products for each gene target, a measure of the comparative informativeness of each gene target. Fig. 2 graphically presents the results of the phylogenetic analysis using a simple UPGMA algorithm to measure sequence similarity.

For a gene to be an ideal target for phylogenetic analysis, it must be present in all species, not be subject to hyper-mutation pressures as

TABLE 1 Sequence statistics for the six gene targets

Gene target	Product size[a] (no. of nucleotides)	Informative base sites (%)	Maximum pairwise variation (%)
16S rRNA	1362–1364	23	10
mip	616–659	56	31
rpoB	312–331	46	32
gyrA	380–384	49 (42)	40 (27)
rnpB	291–330	48	23
proA	514–523	64	36

[a]The number of nucleotides in the product varies depending on the presence of insertions and deletions. Numbers in parentheses (for gyrA) are the percentages recalculated with strain LC1863, and extreme outlier, absent.

may occur for those which encode antigenic determinants, and be sufficiently phylogenetically informative as to resolve all species. In addition, each gene target should associate given strain with the same species or strain cluster (i.e., not be subject to high rates of homologous recombination), and, if present in multiple copies within the genome (e.g., ribosomal genes), the sequence homology between the copies must be high and not contain frequent base variations, deletions, or insertions with respect to the other genomic copies of the same gene. Finally, the primers ought to universally amplify the gene target product from all strains of the genus, a challenge for primers targeting protein-encoding genes where third base redundancy in particular allows considerable sequence variation even in conserved regions. Typically, such primers are initially designed from an alignment of relatively few sequences from known species and may not contain sufficient base redundancy to prime a PCR reaction targeting the genome of novel wild strains.

In reality, it is unlikely that these conditions are all met for any gene target for a genus as diverse as *Legionella*. In this analysis, only the primers targeting the 16S rRNA gene were universally able to amplify product from all type and wild strains, with, in particular, those targeting *rpoB* and *gyrA* failing for a number of strains. Furthermore, the sequences from the *gyrA* target appeared unable to resolve all species, with those from *L. jamestowniensis* and *L. brunensis*, and *L. adelaidensis* and *L. birming-*

hamensis, being identical, respectively. The coalescence theory of species formation (14) suggests that sequence identity between species should only occur if there has been relatively recent horizontal gene transfer between species, the result of a homologous recombination event, so this detected sequence identity needs to be confirmed to ensure that it is not the consequence of laboratory contamination. Given that some of the strains tested here are genetically among the most divergent of all *Legionella* species, even trace amounts of contamination with DNA from other *Legionella* species can preferentially amplify if the primer homology to the DNA of the strain being investigated is low. Such a contamination event probably accounts for the published sequence identity of *rnpB* for *L. maceachernii* and *L. micdadei* (19), as a unique sequence, different from that published for *L. maceachernii*, was ascertained during this analysis. A similar event probably also accounts for erroneous sequences deposited in GenBank for the *mip* gene for other *Legionella* species (U60163 and U60164). A single base deletion or insertion was additionally detected between the multiple 16S rRNA genes of strain 89-2071. This is easily discernable in the sequence chromatogram, where clean dye peaks at each base position are followed after the insertion/deletion site by a mixture of two dye peaks at each base position, because the population of amplified product from one of the gene copies is now asynchronous with that from the other gene copies. The antisense sequence will simi-

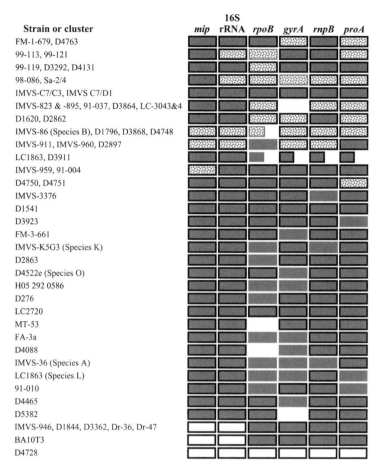

Strain or cluster	mip	16S rRNA	rpoB	gyrA	rnpB	proA
FM-1-679, D4763						
99-113, 99-121						
99-119, D3292, D4131						
98-086, Sa-2/4						
IMVS-C7/C3, IMVS C7/D1						
IMVS-823 & -895, 91-037, D3864, LC-3043&4						
D1620, D2862						
IMVS-86 (Species B), D1796, D3868, D4748						
IMVS-911, IMVS-960, D2897						
LC1863, D3911						
IMVS-959, 91-004						
D4750, D4751						
IMVS-3376						
D1541						
D3923						
FM-3-661						
IMVS-K5G3 (Species K)						
D2863						
D4522e (Species O)						
H05 292 0586						
D276						
LC2720						
MT-53						
FA-3a						
D4088						
IMVS-36 (Species A)						
LC1863 (Species L)						
91-010						
D4465						
D5382						
IMVS-946, D1844, D3362, Dr-36, Dr-47						
BA10T3						
D4728						

FIGURE 2   Graphical representation of strain relatedness for each of the six gene targets. The presence of a rectangle indicates that a sequence (or sequences if there are more than one strain in the clade) was ascertained for that gene target. A partial rectangle indicates that a sequence could not be ascertained for one or more of the strains within a multistrain clade. A grey infill indicates that the genetic distance for that gene target and strain/cluster from a recognized species is greater than that between recognized species, whereas no infill indicates that the genetic distance from a recognized species is less than that between recognized species. For those clades containing more than one strain, a solid infill indicates sequence identity between strains, while a dot pattern indicates sequence variability within the clade. A black border to the rectangle indicates approximate congruence of the location of the clade with the most common topology of the phylogenetic tree.

larly demonstrate the same transition from clean to mixed dye peaks after the insertion/deletion site, but in the opposite orientation when the sequence chromatograms are aligned.

Nevertheless, just as one cannot rely on the sequence from a single gene target to adequately characterize a strain, the potential availability of up to six gene targets will provide sufficient data to achieve a definitive speciation, even if some of the targets do not comply with all the above suitability criteria. Of the six gene targets assessed in this analysis, while all were informative for most strains, *gyrA* was the most problematic in being able to generate product, the sequence of which was capable of uniquely resolving all species. Feddersen et al. included sequences from only 11 species in their analysis (4), none of which have been deposited in a public database. This necessitated that the *gyrA* sequence be reamplified and sequenced from the type strains for this analysis, but the published primers are not sufficiently broadly reactive and failed to generate product from six type strains and three wild strains.

Of the 49 strains from 30 potentially novel species whose *mip* sequence-derived genetic distance to recognized species was greater than that between recognized species (first selection criteria), the 37 strains comprising 13 potentially novel species (i.e., those for which there were more than one strain) were similarly grouped by all six gene targets, with the exception of D1620 and D2862, and IMVS-86, D1796, D3868, and D4748, where some strains were located in separate regions of the phylogenetic tree but were still variant from any type strain. This disparity is probably the result of a past recombination event. The level of sequence variation within each clade representing a potentially novel species was not identical for all gene targets, probably the result of variable functional restraint on the mutation rates within each gene target. With the exception of D4465 and D5382, all 49 strains representing 30 potential novel species were confirmed by all six gene targets, in that their genetic distance from recognized species was greater than that for currently recognized strains. For the two exceptions, only DNA was available for testing, which may have been contaminated, as the ascertained sequence was identical to known common species, so the analyses of these two strains needs to be repeated with reextracted DNA from single colonies. Not all strains were located in the same region of the phylogenetic tree by all gene targets, especially *rpoB* and *gyrA*. Noncongruence in the inferred topology of evolutionary relationships between different species by different gene targets is common, especially for large genera such as *Legionella* (16). This may relate to the different evolutionary pressures on the different genes but is most likely artifactual, the result of the nucleotide substitution saturation, where more recent mutational substitutions "overwrite" earlier substitutions which might inform about the bifurcation of the more ancient lineages.

Of the remaining seven strains representing three potentially novel species based on the DNA relatedness of the two strains (second selection criteria), combined with their inferred relationships with each other based on the *mip* gene sequence (third selection criteria; last three strains/clusters, Fig. 2), all but *mip* and 16S rRNA grouped IMVS-946, D1844, D3362, Dr-36 and Dr-47, and BA10T3 into a single clade sufficiently distant from recognized species to infer species novelness. All gene targets located BA10T3 sufficiently close to the IMVS-946–Dr-47 clade to infer it belongs to the same species. D4748 was inferred to be closely related to *L. longbeachae* by all gene targets, in contrast to the DNA relatedness.

Such an analysis as presented here is complex, comparing over 650 sequences from 6 genes from recognized and potential species of *Legionella*. Some of the primers used, especially those for *gyrA* and *rpoB*, need additional redundancy to enable the amplification of all *Legionella* strains. In addition, some of the strains need to be reanalyzed to ensure that the ascertained sequence is accurate and free from contamination errors. In spite of these difficulties, this analysis demonstrates the robustness and potential of using multiple gene targets to accurately identify novel *Legionella* species and predicts that the *Legionella* genus contains over 80 species. Such a molecular approach, especially when structured on a scientifically rational system such as the community phylogeny approach (chapter 88) to defining ecotypes which are genuinely ecologically distinct strain clusters, will provide a sound scientific basis to speciation, because it is based on the same criteria which give rise to species. Moreover it will obviate the need to rely on serology, and fatty acid and isoprenoid quinone profiles, to formally speciate novel strains. How many sequence targets are adequate, and if DNA relatedness studies remain a requisite component of speciation, requires additional study. In the interim, the use of a combined approach using the sequenced-based phylogeny to identify "nearest neighbor" species will greatly simplify the use of total DNA relatedness studies.

ACKNOWLEDGMENTS

The following colleagues are acknowledged for their generosity in supplying strains used in this analysis. Norma Sangster, Institute of Medical and Veterinary Science, Adelaide, Australia; Robert Benson, National Center for Infectious Diseases, Center for Disease Control and Prevention, USA; Timothy Harrison and Norman Fry, Health Protection Agency, London, UK; Vladimir Drasar, National *Legionella* Reference Laboratory, Public Health Institute, Vyškov, Czech Republic; Jürgen Helbig, Institut für Mikrobiologie und Hygiene, Universitätsklinikum Dresden, Dresden, Germany; Atsushi Saito and Michio Koide, First Department of Internal Medicine, Nishihara, Okinawa, Japan; Maddalena Castellani-Pastoris, Department of Bacteriology, Instituto Superiore di Sanita, Rome, Italy; Jerome Etienne, Laboratoire de Microbiologie, Hôpital Edouard Herriot, Lyon, France; Richard Birtles, Unite des Rickettsies, Faculté de Médecine, Université de la Mediterrannée, Marseille, France.

REFERENCES

1. **Adeleke, A. A., B. S. Fields, R. F. Benson, M. I. Daneshvar, J. M. Pruckler, R. M. Ratcliff, T. G. Harrison, R. S. Weyant, R. J. Birtles, D. Raoult, and M. A. Halablab.** 2001. *Legionella drozanskii* sp. nov., *Legionella rowbothamii* sp. nov. and *Legionella fallonii* sp. nov.: three unusual new *Legionella* species. *Int. J. Syst. Evol. Microbiol.* **51:** 1151–1160.
2. **Brenner, D. J.** 1987. Classification of the *Legionellae. Semin. Respir. Infect.* **2:**190–205.
3. **Diogo, A., A. Verissimo, M. F. Nobre, and M. S. da Costa.** 1999. Usefulness of fatty acid composition for differentiation of *Legionella* species. *J. Clin. Microbiol.* **37:**2248–2254.
4. **Feddersen, A., H. G. Meyer, P. Matthes, S. Bhakdi, and M. Husmann.** 2000. *GyrA* sequence-based typing of *Legionella. Med. Microbiol. Immunol.* **189:**7–11.
5. **Fry, N. K., S. Warwick, N. A. Saunders, and T. M. Embley.** 1991. The use of 16S ribosomal RNA analyses to investigate the phylogeny of the family *Legionellaceae. J. Gen. Microbiol.* **137:**1215–1222.
6. **Gevers, D., F. M. Cohan, J. G. Lawrence, B. G. Spratt, T. Coenye, E. J. Feil, E. Stackebrandt, Y. Van de Peer, P. Vandamme, F. L. Thompson, and J. Swings.** 2005. Opinion: re-evaluating prokaryotic species. *Nat. Rev. Microbiol.* **3:**733–739.
7. **Grimont, F., M. Lefèvre, E. Ageron, and P. A. Grimont.** 1989. rRNA gene restriction patterns of *Legionella* species: a molecular identification system. *Res. Microbiol.* **140:**615–626.
8. **Hookey, J. V., N. A. Saunders, N. K. Fry, R. J. Birtles, and T. G. Harrison.** 1996. Phylogeny of *Legionellaceae* base on small-subunit ribosomal DNA sequences and proposal of *Legionella lytica* comb. nov. for *Legionella*-like amoebal pathogens. *Int. J. Syst. Bacteriol.* **46:**526–531.
9. **Inglis, T. J. J., F. Haverkort, M. Sears, I. Sampson, and G. Harnett.** 2002. Epidemiology of *Legionella* infection in Western Australia, p. 353–355. *In* R. Marre, Y. Abu Kwaik, C. Bartlett, N. P. Cianciotto, B. S. Fields, M. Frosch, J. Hacker, and P. C. Lück (ed.), *Legionella.* ASM Press, Washington, D.C.
10. **Ko, K. S., H. K. Lee, M. Y. Park, K. H. Lee, Y. J. Yun, S. Y. Woo, H. Miyamoto, and Y. H. Kook.** 2002. Application of RNA polymerase beta-subunit gene (*rpoB*) sequences for the molecular differentiation of *Legionella* species. *J. Clin. Microbiol.* **40:**2653–2658.
11. **La Scola, B., R. J. Birtles, G. Greub, T. J. Harrison, R. M. Ratcliff, and D. Raoult.** 2004. *Legionella drancourtii* sp. nov., a strictly intracellular amoebal pathogen. *Int. J. Syst. Evol. Microbiol.* **54:** 699–703.
12. **Lo Presti, F., S. Riffard, F. Vandenesch, and J. Etienne.** 1998. Identification of *Legionella* species by random amplified polymorphic DNA profiles. *J. Clin. Microbiol.* **36:**3193–3197.
13. **Nielsen, K., P. Hindersson, N. Hoiby, and J. M. Bangsborg.** 2000. Sequencing of the *rpoB* gene in *Legionella pneumophila* and characterization of mutations associated with rifampin resistance in the *Legionellaceae. Antimicrob. Agents Chemother.* **44:** 2679–2683.
14. **Palys, T., L. K. Nakamura, and F. M. Cohan.** 1997. Discovery and classification of ecological diversity in the bacterial world: the role of DNA sequence data. *Int. J. Syst. Bacteriol.* **47:**1145–1156.
15. **Park, M., S. T. Yun, M. S. Kim, J. Chun, and T. I. Ahn.** 2004. Phylogenetic characterization of *Legionella*-like endosymbiotic X-bacteria in *Amoeba proteus*: a proposal for 'Candidatus Legionella jeonii' sp. nov. *Environ. Microbiol.* **6:**1252–1263.
16. **Ratcliff, R. M., S. C. Donnellan, J. A. Lanser, P. A. Manning, and M. W. Heuzenroeder.** 1997. Interspecies sequence differences in the Mip

protein from the genus *Legionella*: implications for function and evolutionary relatedness. *Mol. Microbiol.* **25:**1149–1158.

17. **Ratcliff, R. M., J. A. Lanser, P. A. Manning, and M. W. Heuzenroeder.** 1998. Sequence-based classification scheme for the genus *Legionella* targeting the *mip* gene. *J. Clin. Microbiol.* **36:**1560–1567.

18. **Riffard, S., F. Lo Presti, P. Normand, F. Forey, M. Reyrolle, J. Etienne, and F. Vandenesch.** 1998. Species identification of *Legionella* via intergenic 16S-23S ribosomal spacer PCR analysis. *Int. J. Syst. Bacteriol.* **48:**723–730.

19. **Rubin, C. J., M. Thollesson, L. A. Kirsebom, and B. Herrmann.** 2005. Phylogenetic relationships and species differentiation of 39 *Legionella* species by sequence determination of the RNase P RNA gene *rnpB. Int. J. Syst. Evol. Microbiol.* **55:**2039–2049.

20. **Vandamme, P., B. Pot, M. Gillis, P. de Vos, K. Kersters, and J. Swings.** 1996. Polyphasic taxonomy, a consensus approach to bacterial systematics. *Microbiol. Rev.* **60:**407–438.

21. **Wilkinson, H. W.** 1988. Reactions of *Legionella* species on biochemical tests, p. 28. *In* H. W. Wilkinson (ed.). *Hospital-Laboratory Diagnosis of Legionella Infections.* Centers for Disease Control and Prevention, Atlanta, GA.

22. **Wilkinson, I. J., N. Sangster, R. M. Ratcliff, P. A. Mugg, D. E. Davos, and J. A. Lanser.** 1990. Problems associated with identification of *Legionella* species from the environment and isolation of six possible new species. *Appl. Environ. Microbiol.* **56:**796–802.

23. **Yu, V. L., J. F. Plouffe, M. C. Pastoris, J. E. Stout, M. Schousboe, A. Widmer, J. Summersgill, T. File, C. M. Heath, D. L. Paterson, and A. Chereshsky.** 2002. Distribution of *Legionella* species and serogroups isolated by culture in patients with sporadic community-acquired legionellosis: an international collaborative survey. *J. Infect. Dis.* **186:**127–128.

SEQUENCE-BASED DISCOVERY OF ECOLOGICAL DIVERSITY WITHIN *LEGIONELLA*

Frederick M. Cohan, Alexander Koeppel, and Daniel Krizanc

88

Bacteria of the genus *Legionella* are principally pathogens of freshwater amoebae and other protozoa, but many *Legionella* species can also opportunistically cause pneumonia in humans (9, 19). Such infection is particularly likely when water facilities harbor amoebae and their *Legionella* pathogens and produce mist that is inhaled (5, 9). The *Legionella* species that cause human pneumonia are ecologically diverse in their natural and newly staked artificial reservoirs, in the amoebae that they can infect, and in the diseases they cause in humans (2, 3, 14, 16, 17). Even very close relatives, for example within the species *Legionella pneumophila*, can differ in their ecological properties (4, 6, 7; see chapter 84).

Public health measures to prevent and treat *Legionella*-based disease should benefit from efforts to discover all of the ecologically distinct populations within the *Legionella* species known to cause pneumonia. This information may aid in the development of vaccines against the diversity of *Legionella* pathogens and to treat the diversity of diseases caused by close relatives. Understanding the ecological diversity within *Legionella* may help in identifying all environmental reservoirs that can sustain the different potential *Legionella* pathogens.

Approaches based on DNA sequence diversity have proved useful for identifying ecologically distinct populations among close relatives of bacteria. Sequence-based methods provide a general protocol for discovering different populations, even before we know what is ecologically distinct about them. The rationale is that populations playing different ecological roles frequently correspond to discrete DNA sequence clusters for every gene in the genome. One limitation of this approach, however, is that any bacterial taxon is likely to have a nested hierarchy of sequence clusters, with subclusters within clusters and so on, and it is not usually clear which level of sequence cluster is the one that corresponds to ecologically distinct populations. Nevertheless, the recently developed "community phylogeny" approach provides a way to choose the level of cluster most likely to correspond to ecotypes (F. M. Cohan and D. Krizanc, in preparation).

Here we apply the community phylogeny method to demarcate the ecologically distinct populations within the genus *Legionella*, based on the sequence of the macrophage infectivity potentiator (*mip*) gene from 496 *Legionella* isolates, obtained from Rodney M. Ratcliff (personal communication) (16), representing all

Frederick M. Cohan and Alexander Koeppel Department of Biology, Wesleyan University, Middletown, CT 06459. *Daniel Krizanc* Department of Mathematics and Computer Science, Wesleyan University, Middletown, CT 06459.

characterized species within *Legionella*, as well as many uncharacterized groups (A. Koeppel, F. M. Cohan, and R. M. Ratcliff, in preparation).

CLADE SEQUENCE DIVERSITY

Martin (13) developed a method for describing the sequence diversity within a taxon (as applied to *Legionella* in Fig. 1) for the purpose of comparing speciation rates over time and over clades. The key is to interpret the number of sequence clusters (or bins) defined at a particular level of sequence identity as the number of lineages at some point in the past that have survived to the present (e.g., bins defined at low sequence-identity levels correspond to lineages from the distant past). Martin showed

that a log-linear increase in the number of bins with increasing sequence identity is consistent with a constant net rate of speciation. An adaptive radiation early in the history of a clade is indicated by a steeper slope relating bins and sequence identity at low levels of sequence identity. This is observed in the clade sequence diversity of *Legionella* (Fig. 1), which rapidly increased in diversity from one bin at 60% identity to 23 bins at 80%, and then maintained a constant, lower rate of speciation until ~99% identity. Perhaps the early years of the *Legionella* lifestyle offered many opportunities, with a great diversity of amoebae awaiting ecological specialization in *Legionella*. Because the community phylogeny analysis depends on

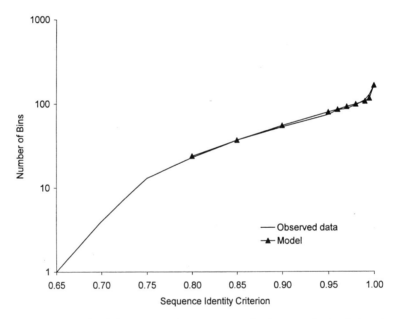

FIGURE 1 The clade sequence diversity pattern for the genus *Legionella*, based on the *mip* gene. The observed pattern and the pattern predicted by the model's optimal rates of ecotype formation and periodic selection, as well as the optimal number of extant ecotypes, are indicated. The number of bins (or sequence clusters) at a given level of sequence identity indicate the number of lineages, surviving to this day that existed at some point in the past. The steep rise in the number of bins, from 60 to 80% identity, suggests an adaptive radiation early in the history of *Legionella*. The log-linear increase from 80% until 99% indicates a constant rate of ecotype formation over the time period represented. Because the community phylogeny analysis assumes a constant rate of ecotype formation, the analysis was applied only for sequence identity levels ≥80%. The sharp increase in the number of bins at 99% identity reflects the facile sequence divergence within ecotypes.

a constant rate of net speciation, we analyzed only the part of the clade sequence diversity with ≥80% sequence identity.

The flair of increased diversity at 98 to 100% identity is typically seen in clade sequence diversity patterns for bacteria (F. M. Cohan and D. Krizanc, in preparation) (1); this flair is also seen in eukaryotes and viruses (12). The inflection point (e.g., at ~99% in Legionella) is caused by a difference in the cause of divergence to the left and right of the inflection (1). The more divergent lineages to the left of the inflection can survive to the present only if the lineages represent different ecotypes; therefore, the increase in number of bins left of 99% depends on the (low) rate of net ecotype formation. However, to the right, recently diverged lineages represent the facile sequence divergence *within ecotypes*. The rate of ecologically meaningless sequence divergence within an ecotype is expected to be much greater than that requiring ecotype formation—hence the flair of diversity at >99% sequence identity.

The community phylogeny approach adds to the structure of analysis laid out by Martin (13) and Acinas (1) by explicitly estimating the rate parameters associated with the history of the group, as well as the number of ecotypes present among the strains sampled.

ESTIMATING THE PARAMETERS OF *LEGIONELLA* EVOLUTION

The community phylogeny approach assumes the stable ecotype model of bacterial evolution (10, 11, 18), while accounting for other possible models. In the stable ecotype model, each ecologically distinct population (an ecotype) endures many "periodic selection" events during its long lifetime; in each such event, natural selection favoring an adaptive mutant brings about a purging of diversity in all genes, as the progeny of that adaptive mutant outcompete all other strains occupying the same niche. In this model, new ecotypes are formed at a rate much lower than that of periodic selection, either through mutation or horizontal transfer.

The community phylogeny analysis evaluates different sets of parameter values for their likelihood of yielding a phylogeny consistent with the observed clade sequence diversity of the group (Fig. 1) (F. M. Cohan and D. Krizanc, in preparation). The model's parameters are: Ω, the net rate of ecotype formation, taking into account the rate of ecotype extinction (per ecotype per unit time); σ, the rate of periodic selection (per ecotype per unit time); and n, the number of ecotypes within the sample of strains. The unit of time is the expected time to substitution in the *mip* sequence.

Here is a broad overview of the maximum likelihood approach for finding the trio of parameter values best fitting the observed curve of clade sequence diversity for *Legionella* (Fig. 1), given the number of sequences sampled (F. M. Cohan & D. Krizanc, in preparation). For a given trio of parameter values, the community phylogeny program simulated the history of the genus, allowing for ecotype formation and periodic selection to occur stochastically at the rates Ω and σ, respectively, leading to the n ecotypes at present. Each of at least 100 replicate simulations of a parameter trio was evaluated for its fit to the observed clade sequence diversity in Fig. 1. The fraction of replicate simulations of a trio yielding a sufficiently good fit to the observed data was our estimate of the likelihood associated with the trio. We searched for the trio of parameter values yielding the observed data with maximum likelihood.

This approach yielded a solution closely fitting the observed clade sequence diversity pattern for *Legionella* (Fig. 1) (A. Koeppel, F. M. Cohan, and R. M. Ratcliff, in preparation). In the maximum likelihood solution, the rate of periodic selection ($\sigma = 7.28$) dwarfs the net rate of ecotype formation ($\Omega = 0.0264$), with 276 periodic selection events on average between net ecotype-formation events (i.e., those ecotype-formation events that yield descendants still surviving and represented in our sample of strains). That periodic selection recurs many times between net ecotype-forming events is consistent with the stable ecotype model, which requires a relatively low rate of ecotype formation, with each ecotype enduring many periodic selection events during its lifetime.

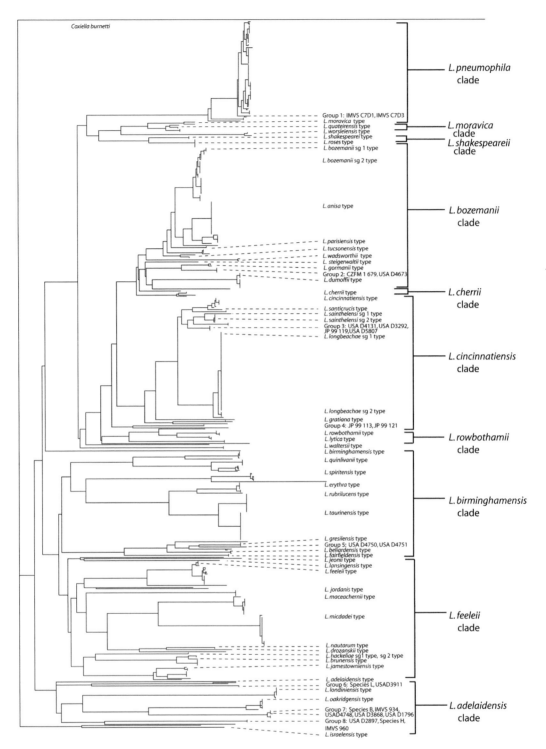

FIGURE 2 Neighbor joining tree for the genus *Legionella*, with *Coxiella* as an outgroup. The genus was divided into 10 clades, plus some additional strains that did not fit clearly within these clades. Each of the clades was tested for whether the Ω and σ rates estimated for the whole genus apply to the clade. A sample of eight orphan groups that are not closely related to any characterized species are labeled. Each of these groups was tested for consistency with being a single ecotype and for distinctness as a separate ecotype from the most closely related, characterized species.

We note, however, that the Ω of our model is the rate of *net* ecotype-formation events, including only the formation of those lineages that are still present today. So, Ω is the difference between the rate of ecotype formation and the rate of ecotype extinction. There could be extremely high rates of both ecotype formation and extinction, with little difference between them (the difference being Ω). This is the scenario laid out by the species-less model of bacterial diversity (10, 11, 18). If this model is correct, then the sequence clusters demarcated as ecotypes are likely to contain multiple, young, ecologically distinct populations that have not yet had time to diverge into separate sequence clusters. We shall provide evidence of such young populations within *L. pneumophila*.

The model estimated 114 ecotypes within *Legionella*, with a 95% confidence interval of 97 to 135 (A. Koeppel, F. M. Cohan, and R. M. Ratcliff, in preparation), far beyond the 50 species described.

HETEROGENEITY OF RATES OF ECOTYPE FORMATION AND PERIODIC SELECTION

We tested whether the rates Ω and σ determined for the whole genus apply as constants for each of the clades delineated in Fig. 2. In only the case of the *L. pneumophila* clade was the clade-specific solution ($\Omega = 0.194$; $\sigma = 6.03$) a significant improvement over the whole-genus solution (likelihood ratio = 1290, 3 degrees of freedom, $P < 0.005$) (A. Koeppel, F. M. Cohan, and R. M. Ratcliff, in preparation). The net rate of ecotype formation was 7.46 times greater for *L. pneumophila* than for the whole genus. Perhaps early in its history, *L. pneumophila* was preadapted to specialize on a great diversity of amoebae or perhaps to specialize to a great diversity of environmental conditions (e.g., temperature).

DEMARCATION OF ECOTYPES WITHIN *L. PNEUMOPHILA*

We next demarcated the individual ecotypes within *L. pneumophila*, assuming the *pneumophila*-specific values of Ω and σ (Fig. 3). Our ap-

proach was to first describe the sequence diversity within a given clade (e.g., group E in Fig. 4A) and then to determine the number of ecotypes that yields the greatest likelihood of a fit to the observed clade sequence diversity (Fig. 4B). We then found the largest clades that were each consistent with a single ecotype (i.e., the confidence interval for *n* included one ecotype). In this process, we checked to see if a given group could be expanded to include ever larger clades. For example, when strains of the most closely related group, F, were added to group E, the confidence interval no longer included one ecotype, so we concluded that these groups were different ecotypes.

We demarcated 11 putative ecotypes within *L. pneumophila*; these are the groups labeled A to K. We note that many of the most common pathogens within the species are included within putative ecotype E.

Consider next whether the sequence clusters identified as putative ecotypes are likely to correspond 1:1 to ecotypes, as predicted under the stable ecotype model. We previously developed alternative models in which a single ecotype can correspond to multiple sequence clusters (10, 11, 18). For example, in the geotype-plus-Boeing model, limited dispersal may result in a single ecotype diverging into a different sequence cluster for each geographic region; then, increased human transport in recent times could result in the reunion of formerly geographical isolates. This model yields, for at least the near future, multiple sequence clusters per ecotype. Thus, to take into account models yielding a 1:many relationship between ecotypes and sequence clusters, ecologists must determine that each putative ecotype is indeed ecologically distinct from others.

There is an indication that at least some of the putative ecotypes demarcated in Fig. 3 are ecologically distinct in the hosts and the environmental conditions where they thrive. For example, putative ecotype J, represented by the type strain of sg5, appears to differ in host specificity from ecotype E (6). Whereas the strains of putative ecotype E (including strains

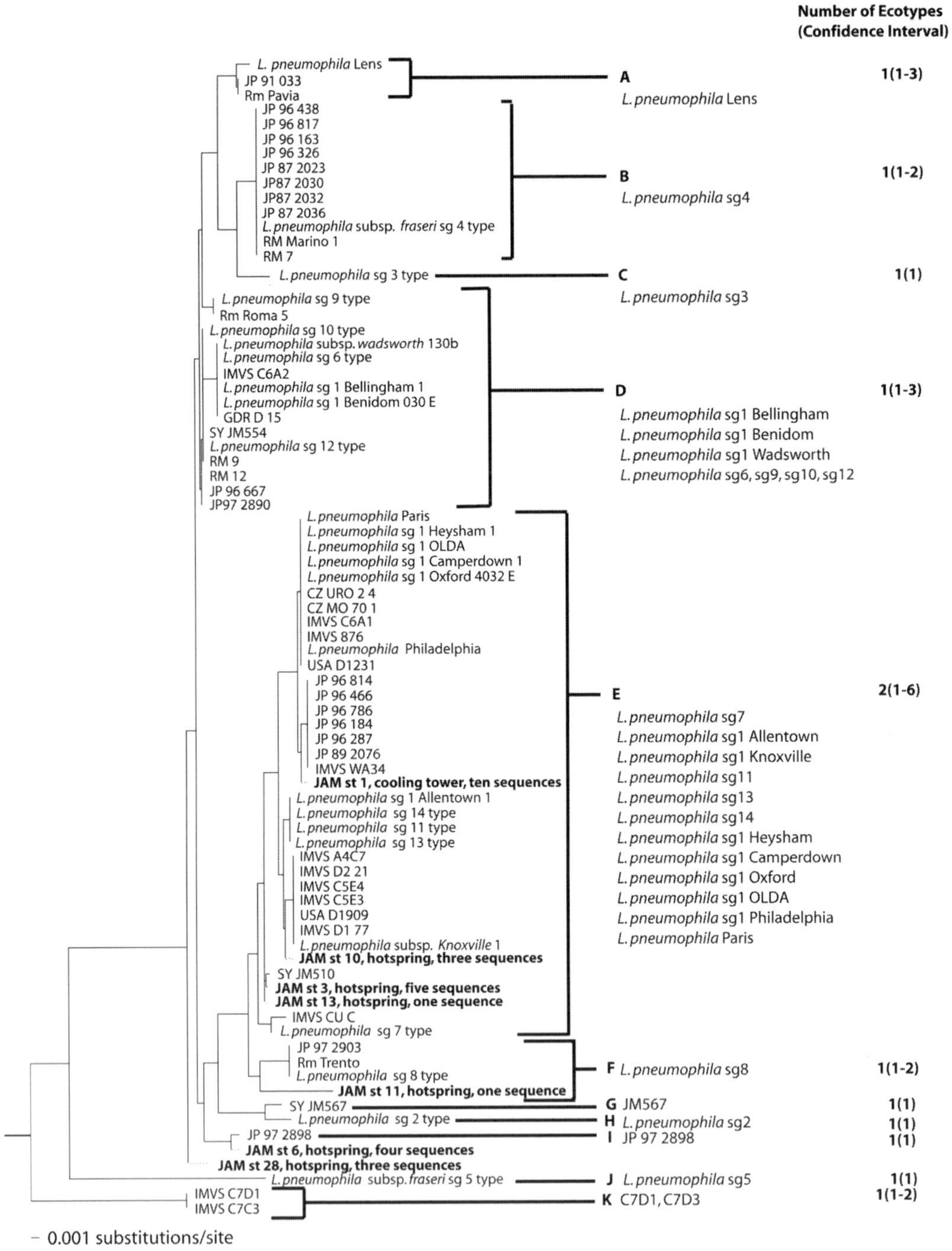

FIGURE 3 Neighbor joining tree of *L. pneumophila*. Eleven putative ecotypes, labeled A to K, were demarcated by the community phylogeny method. The sequences in bold, supplied by J. Amemura-Maekawa, were not included in the community phylogeny analysis because they are shorter. Two strains previously identified as members of *L. pneumophila* subsp. *fraseri* are labeled.

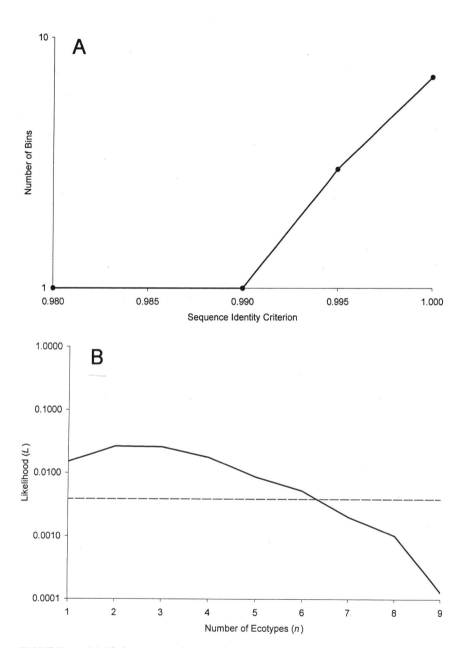

FIGURE 4 (A) Clade sequence diversity for group E within *L. pneumophila*. (B) The likelihood values for various numbers of ecotypes (*n*), using the *pneumophila*-specific rates of Ω and σ. The confidence interval is based on the threshold line indicated (where the likelihood ratio is 6.83). In this case, n = 2 yields the maximum likelihood of a fit to the observed clade sequence diversity for group E, but the confidence interval ranges from 1 to 6.

Philadelphia and Paris) can grow on the amocbac *Acanthamoeba polyphaga* and *A. castellanii*, the sg5 type strain cannot. Also, ecotypes A and E appear to differ in the time courses of gene expression during infection of a eukaryotic host cell, as seen in differences between the Lens strain of A versus the Paris and Philadelphia strains of E (see chapter 84). Given the confirmation of ecotypes so far, the *Legionella* community might expand its intensive study of *Legionella* ecology and physiology beyond Philadelphia, Paris, and Lens to include other putative ecotypes.

We have developed models allowing a many:1 relationship between ecotypes and sequence clusters (10, 11, 18). For example, in the species-less model, ecotypes are formed and extinguished at a high rate, such that many young, closely related ecotypes cannot be distinguished by the sequences of most genes. Therefore, confirmation that a sequence-based putative ecotype is indeed an ecotype will require tests that the group is ecologically homogeneous.

A recent study by Amemura-Maekawa and colleagues suggests possible ecological heterogeneity within putative ecotype E (4). Cooling tower and spa baths in Japan were found to support nonoverlapping sets of sequence types, but all falling within putative ecotype E (Fig. 3). The one sequence type consistently found in cooling towers is in a subclade distinct from the spa bath sequence types, which suggests that this ecological divergence might represent a very new invention of an ecotype, too young to be distinguished as an ecotype by community phylogeny. We note that putative ecotype E is the one labeled group more likely to contain two ecotypes than one (Fig. 3), and that this group might contain up to six ecotypes. These considerations increase the likelihood that the spa bath and cooling tower isolates might represent different, very young ecotypes within putative ecotype E.

THE TAXONOMIC STATUS OF UNCHARACTERIZED CLADES

Within the genus are various groups that are not closely related to any described species (chapter 87). A sample of eight such groups is labeled in Fig. 2. In seven of these eight cases, the group was found to be consistent with a single ecotype; in group 5, the two strains of the group were hypothesized to represent two ecotypes. We then tested whether adding the group's most closely related named species would make the pool of strains inconsistent with a single ecotype. Each group was shown to be an ecotype (or two ecotypes) distinct from the most closely related named species. We can now offer two explanations for why there are so many more putative ecotypes (estimated at 114) than species already characterized (about 50): the existence of ecotypes outside of characterized species and multiple ecotypes within at least one species (*L. pneumophila*) and possibly within others.

TOWARD A NATURAL TAXONOMY OF *LEGIONELLA*

We have previously proposed a theory-based concept of bacterial species, which takes into account multiple models of bacterial evolution, while allowing for operational demarcation of species in everyday systematics (10, 11). We suggested that species be demarcated as the smallest groups with both a history of coexistence and a prognosis for future coexistence. A history of coexistence can be revealed by the groups falling into different sequence clusters, and a prognosis for future coexistence is suggested by the groups being ecologically distinct in nature. The community phylogeny analysis provides an objective and theory-based method for identifying sequence clusters that are likely to correspond to ecotypes with a long history of coexistence.

Consider first the *Legionella* groups demarcated as ecotypes outside of any uncharacterized species. When such a group is found to be ecologically distinct from the groups most closely related to it, the group will have earned species status. Because these groups are not already within named species, their description as new species should be straightforward.

Application of this approach is more involved for existing, named species that appear to contain multiple ecotypes (e.g., *L. pneumophila*).

When a putative ecotype identified by community phylogeny is shown to be ecologically distinct in nature, it will have demonstrated the attributes of species. However, to avoid taxonomic confusion, we have suggested that infraspecific groups with the qualities of species should be given a trinomial name with an "ecovar" label, e.g., *L. pneumophila* ecovar *pneumophila* (10, 11). It is our hope that a more precise taxonomy, giving names to all the long-coexisting and ecologically distinct groups within a named species, will allow ecologists and epidemiologists to better predict the properties of newly isolated strains (11).

ACKNOWLEDGMENTS

F.M.C. thanks Rodney M. Ratcliff for many interesting discussions on the ecology and systematics of *Legionella* and for providing the sequence data on which this project was based. We thank Junko Amemura-Maekawa for providing sequence data from Japanese strains from spa baths and cooling towers.

This work was funded by NSF grant EF-0328698 and by grants from Wesleyan University.

REFERENCES

1. **Acinas, S. G., V. Klepac-Ceraj, D. E. Hunt, C. Pharino, I. Ceraj, D. L. Distel, and M. F. Polz.** 2004. Fine-scale phylogenetic architecture of a complex bacterial community. *Nature* **430:**551–554.

2. **Adeleke, A., J. Pruckler, R. Benson, T. Rowbotham, M. Halablab, and B. Fields.** 1996. *Legionella*-like amebal pathogens: phylogenetic status and possible role in respiratory disease. *Emerg. Infect. Dis.* **2:**225–230.

3. **Adeleke, A. A., B. S. Fields, R. F. Benson, M. I. Daneshvar, J. M. Pruckler, R. M. Ratcliff, T. G. Harrison, R. S. Weyant, R. J. Birtles, D. Raoult, and M. A. Halablab.** 2001. *Legionella drozanskii* sp. nov., *Legionella rowbothamii* sp. nov. and *Legionella fallonii* sp. nov.: three unusual new *Legionella* species. *Int. J. Syst. Evol. Microbiol.* **51:**1151–1160.

4. **Amemura-Maekawa, J., F. Kura, B. Chang, and H. Watanabe.** 2005. *Legionella pneumophila* Serogroup 1 isolates from cooling towers in Japan form a distinct genetic cluster. *Microbiol. Immunol.* **49:**1027–1033.

5. **Arnow, P. M., T. Chou, D. Weil, E. N. Shapiro, and C. Kretzschmar.** 1982. Nosocomial Legionnaires' disease caused by aerosolized tap water from respiratory devices. *J. Infect. Dis.* **146:**460–467.

6. **Berk, S. G., R. S. Ting, G. W. Turner, and R. J. Ashburn.** 1998. Production of respirable vesicles containing live *Legionella pneumophila* cells by two *Acanthamoeba* spp. *Appl. Environ. Microbiol.* **64:**279–286.

7. **Brüggemann, H., C. Cazalet, and C. Buchrieser.** 2006. Adaptation of *Legionella pneumophila* to the host environment: role of protein secretion, effectors and eukaryotic-like proteins. *Curr. Opin. Microbiol.* **9:**86–94.

8. **Cazalet, C., C. Rusniok, H. Brüggemann, N. Zidane, A. Magnier, L. Ma, M. Tichit, S. Jarraud, C. Bouchier, F. Vandenesch, F. Kunst, J. Etienne, P. Glaser, and C. Buchrieser.** 2004. Evidence in the *Legionella pneumophila* genome for exploitation of host cell functions and high genome plasticity. *Nat. Genet.* **36:**1165–1173.

9. **Fields, B. S., R. F. Benson, and R. E. Besser.** 2002. *Legionella* and Legionnaires' disease: 25 years of investigation. *Clin. Microbiol. Rev.* **15:**506–526.

10. **Gevers, D., F. M. Cohan, J. G. Lawrence, B. G. Spratt, T. Coenye, E. J. Feil, E. Stackebrandt, Y. Van de Peer, P. Vandamme, F. L. Thompson, and J. Swings.** 2005. Opinion: re-evaluating prokaryotic species. *Nat. Rev. Microbiol.* **3:**733–739.

11. **Godreuil, S., F. Cohan, H. Shah, and M. Tibayrenc.** 2005. Which species concept for pathogenic bacteria? An E-Debate. *Infect. Genet. Evol.* **5:**375–387.

12. **Ho, S. Y., M. J. Phillips, A. Cooper, and A. J. Drummond.** 2005. Time dependency of molecular rate estimates and systematic overestimation of recent divergence times. *Mol. Biol. Evol.* **22:**1561–1568.

13. **Martin, A. P.** 2002. Phylogenetic approaches for describing and comparing the diversity of microbial communities. *Appl. Environ. Microbiol.* **68:**3673–3682.

14. **Newsome, A. L., T. M. Scott, R. F. Benson, and B. S. Fields.** 1998. Isolation of an amoeba naturally harboring a distinctive *Legionella* species. *Appl. Environ. Microbiol.* **64:**1688–1693.

15. **Ratcliff, R. M., J. A. Lanser, P. A. Manning, and M. W. Heuzenroeder.** 1998. Sequence-based classification scheme for the genus *Legionella* targeting the mip gene. *J. Clin. Microbiol.* **36:**1560–1567.

16. **Rowbotham, T. J.** 1993. *Legionella*-like amoebal pathogens, p. 137–140. *In* J. M. Barbaree, R. F. Breiman, and A. P. Dufour (ed.). *Legionella: Current Status and Emerging Perspectives.* American Society for Microbiology, Washington, D.C.

17. **Steele, T. W., J. Lanser, and N. Sangster.** 1990. Isolation of *Legionella longbeachae* serogroup 1 from potting mixes. *Appl. Environ. Microbiol.* **56:**49–53.

18. **Ward, D. M., and F. M. Cohan.** 2005. Microbial diversity in hot spring cyanobacterial mats:

pattern and prediction. *In* W. P. Inskeep and T. McDermott (ed.). *Geothermal Biology and Geochemistry in Yellowstone National Park.* Thermal Biology Institute, Bozeman, MT.

19. **Yu, V. L., J. F. Plouffe, M. C. Pastoris, J. E. Stout, M. Schousboe, A. Widmer, J. Summersgill, T. File, C. M. Heath, D. L. Paterson, and A. Chereshsky.** 2002. Distribution of *Legionella* species and serogroups isolated by culture in patients with sporadic community-acquired legionellosis: an international collaborative survey. *J. Infect. Dis.* **186:**127–128.

GENOME SEQUENCING
AND GENOMICS

Carmen Buchrieser, Paul S. Hoffman,
James J. Russo, and Joseph P. Vogel

89

The purpose of this panel discussion was to explore issues related to *Legionella* genomics. A critical breakthrough in this area recently occurred with the determination of the genome sequences of three clinical, serogroup 1 isolates of *Legionella pneumophila* subsp. *pneumophila*. The three strains are *L. pneumophila* Paris, Lens (4) and Philadelphia-1 (5). Strain Paris is an abundant, endemic strain in France and Europe that is associated with nosocomial and community-acquired disease (1). Strain Lens is an epidemic strain that caused a large outbreak of Legionnaires' disease in France from November 2003 to January 2004 (9). This outbreak consisted of 86 cases, including 17 deaths, suggesting strain Lens is particularly virulent (9). Strain Philadelphia-1 was derived from the original reported outbreak of Legionnaires' disease that occurred at the American Legion convention held in Philadelphia in 1976 (7). The genome sequences obtained from these three strains have already proven to be incredi-

bly useful as a genetic tool and they have provided numerous new and valuable insights into the biology of *L. pneumophila* (3–5).

Analysis of the completed genome sequences of *L. pneumophila* strains Paris, Lens and Philadelphia-1 revealed that each consists of 3.3 to 3.5 million base pairs, encodes approximately 3,000 genes, and has an average G+C content of 38%. Similar to other bacterial genomes, they are composed of a conserved core and a flexible gene pool. The conserved core encodes for many functions including the biosynthesis of proteins, lipids, and the cell envelope and for proteins involved in major metabolic pathways. Intriguingly, about one quarter of the conserved genes were present only in *Legionella* species, suggesting that they might code for proteins with novel *Legionella*-specific functions. The flexible gene pool consists largely of plasmids, IS-elements, transposases, and genomic islands. About 2,500 genes make up the conserved core while about 300 genes are strain specific and thus constitute the flexible gene pool. Since the three analyzed strains belong to the same species and same serogroup, this level of diversity is remarkable.

The previous analysis compared the genomes of the three *L. pneumophila* species. To extend these observations, the three genomes can be compared with the genome of a genetically

Carmen Buchrieser Unité de Genominque des Microorganismes Pathogenes, Institut Pasteur, Paris, France. *Paul S. Hoffman* Division of Infectious Diseases and International Health, University of Virginia Health System, Charlottesville, VA 22908. *James J. Russo* Columbia Genome Center, Columbia University, New York, NY 10032. *Joseph P. Vogel* Department of Molecular Microbiology, Washington University School of Medicine, St. Louis, MO 63110.

related intracellular pathogen, *Coxiella burnetii* (11), and the genome of a nonpathogenic strain, the *Escherichia coli* K-12. Applying reductionist strategies, including the use of internet-based programs (e.g., Neurogadgets.com), provides an independent method to identify common core genes. This technique involves first identifying common genes that are orthologous matches (e.g., homology score of e^{-15} to e^{-30}), then subtracting presumed housekeeping genes, defined as those genes in common with the nonpathogen *E. coli* K-12. Fig. 1 presents a Venn diagram demonstrating this comparison method and the results obtained. The net common gene set between the *L. pneumophila* strains and *C. burnetii* was only ~240 genes and may indicate an overlapping set of virulence genes between these pathogens.

Several additional insights that have been derived from genomic analyses include the marked plasticity of the genomes, the evidence of large-scale rearrangements, and the presence of a large number of eukaryotic-like proteins.

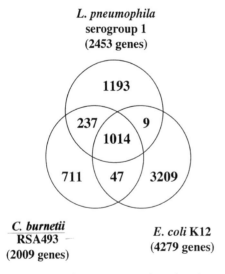

L. pneumophila serogroup 1 (2453 genes)

1193

237 9

1014

711 47 3209

C. burnetii RSA493 (2009 genes)

E. coli K12 (4279 genes)

FIGURE 1 A reductionist approach to identify common genes between the three *L. pneumophila* serogroup 1 strains (Paris, Lens, and Philadelphia-1), *C. burnetii*, and *E. coli* K-12. The Venn diagram represents the three sets of genomes, and the numbers indicate the number of genes in common between different organisms.

L. pneumophila genomic plasticity can be seen at the level of interstrain variability for individual genes. For example, in strain Paris the *rtxA* and *arpB* genes are fused and are followed by a region of about 17 kb of repetitive DNA organized in approximately 30 highly conserved tandem repeats. The same organization is observed in strain Lens, but with two kinds of repeats, which are completely different from those observed in strain Paris. In strain Philadelphia-1, *arpB* is absent, a frameshift disrupts the fusion with the repeat region, and the tandem repeats are highly conserved, but again, the repeat type is different.

In addition to changes seen in individual genes, several large-scale genome rearrangements become apparent when comparing the three sequenced *L. pneumophila* genomes (2, 4, 5). These include positions containing purportedly plasmid-derived elements such as IS elements or the *lvr/lvh* gene cluster that can exist in a plasmid or a chromosomal context. Equally intriguing is a variably placed 100-kb locus rich in genes encoding membrane transporters and surrounded by mobile elements. A surprisingly large number of the genes in this locus are RpoS-dependent heavy metal efflux proteins, as determined by studies using *Legionella* expression microarrays. Thus, these three *L. pneumophila* species seem to possess a complicated evolutionary history, including horizontal transfer of a number of genes or gene cassettes, gene mutation, and gene loss. Several participants of the panel discussion noted the potential for genomic diversity, particularly regarding genetic variations observed among commonly used *L. pneumophila* serogroup 1 strains. Some of these changes may be due to mutations derived by passage in the laboratory, and the panel agreed that original isolates with a minimal passage history, such as *L. pneumophila* strain Lens, should be used whenever possible.

A third major insight obtained by *L. pneumophila* genomic analysis is that this organism encodes an unexpectedly high number and wide variety of eukaryotic-like proteins (3–5). In strain Paris, 31 eukaryotic-like proteins were identified, as well as 28 in strain Lens and 29 in strain Philadelphia-1 (3). Of these, four eukaryotic-like proteins (a sphingosine-1-phosphate lyase, a sphingosine kinase, and two secreted apyrases) represent the first examples of such proteins present in a prokaryote. In addition, a large number of *L. pneumophila* proteins were found to contain motifs predominantly present only in eukaryotes (e.g., ankyrin, Sel-1 [TPR], U-box, and F-box motifs). *L. pneumophila* eukaryotic-like proteins represent a large class of proteins that may be injected into a host cell, perhaps by the Dot/Icm type IV secretion apparatus.

In addition to the three sequenced *L. pneumophila* strains, sequence information will soon be available for *L. pneumophila* strain Corby as well as for the species *L. hackeliae*, *L. micdadei* (Institut for Molecular Biotechnology, IMB Jena, Germany), and an isolate of *L. longbeachae* (Institut Pasteur, Paris, France). Moreover, proposals to sequence a large number of additional *Legionella* strains using a variety of exciting new technologies were discussed. These technologies would have the significant advantage of eliminating the need for shotgun clone libraries, thus saving time and effort (6). For instance, one strategy involves a water-in-oil emulsion where each droplet of water has a high likelihood of containing a single genomic fragment and all the necessary ingredients (primers, nucleotides, etc.) for its amplification. The sequencing reactions occur on beads, which can be collected in arrays on slides and subjected to a number of different micro- or nano-scale technologies for reading the sequences of the attached amplified DNA. These include microscale Sanger reactions, base-by-base extension reactions utilizing fluorescent nucleotides (10) or pyrosequencing (8), and ligation reactions using "coded" degenerate oligonucleotides (12). The new technologies offer the possibility of producing a draft sequence of a bacterial genome on a single slide in one day at very low cost. The availability of a large number of different genome sequences of *Legionella* will clearly benefit the community and pave the way for in-depth comparative genomics. In summary, the field of *Legionella* genomics has only begun and promises a bright future.

REFERENCES

1. **Aurell, H., J. Etienne, F. Forey, M. Reyrolle, P. Girardo, P. Farge, B. Decludt, C. Campese, F. Vandenesch, and S. Jarraud.** 2003. *Legionella pneumophila* serogroup 1 strain Paris: endemic distribution throughout France. *J. Clin. Microbiol.* **41:** 3320–3322.

2. **Brassinga, A. K., M. F. Hiltz, G. R. Sisson, M. G. Morash, N. Hill, E. Garduno, P. H. Edelstein, R. A. Garduno, and P. S. Hoffman.** 2003. A 65-kilobase pathogenicity island is unique to Philadelphia-1 strains of *Legionella pneumophila*. *J. Bacteriol.* **185:**4630–4637.

3. **Bruggemann, H., C. Cazalet, and C. Buchrieser.** 2006. Adaptation of *Legionella pneumophila* to the host environment: role of protein secretion, effectors and eukaryotic-like proteins. *Curr. Opin. Microbiol.*

4. **Cazalet, C., C. Rusniok, H. Bruggemann, N. Zidane, A. Magnier, L. Ma, M. Tichit, S. Jarraud, C. Bouchier, F. Vandenesch, F. Kunst, J. Etienne, P. Glaser, and C. Buchrieser.** 2004. Evidence in the *Legionella pneumophila* genome for exploitation of host cell functions and high genome plasticity. *Nat. Genet.* **36:**1165–1173.

5. **Chien, M., I. Morozova, S. Shi, H. Sheng, J. Chen, S. M. Gomez, G. Asamani, K. Hill, J. Nuara, M. Feder, J. Rineer, J. J. Greenberg, V. Steshenko, S. H. Park, B. Zhao, E. Teplitskaya, J. R. Edwards, S. Pampou, A. Georghiou, I. C. Chou, W. Iannuccilli, M. E. Ulz, D. H. Kim, A. Geringer-Sameth, C. Goldsberry, P. Morozov, S. G. Fischer, G. Segal, X. Qu, A. Rzhetsky, P. Zhang, E. Cayanis, P. J. De Jong, J. Ju, S. Kalachikov, H. A. Shuman, and J. J. Russo.** 2004. The genomic sequence of the accidental pathogen *Legionella pneumophila*. *Science* **305:**1966–1968.

6. **Dressman, D., H. Yan, G. Traverso, K. W. Kinzler, and B. Vogelstein.** 2003. Transforming single DNA molecules into fluorescent magnetic particles for detection and enumeration of genetic variations. *Proc. Natl. Acad. Sci. USA* **100:**8817–8822.

7. **Fraser, D. W., T. R. Tsai, W. Orenstein, W. E. Parkin, H. J. Beecham, R. G. Sharrar, J. Harris, G. F. Mallison, S. M. Martin, J. E. McDade, C. C. Shepard, and P. S. Brachman.** 1977. Legionnaires' disease: description of an epidemic of pneumonia. *N. Engl. J. Med.* **297:**1189–1197.

8. Margulies, M., M. Egholm, W. E. Altman, S. Attiya, J. S. Bader, L. A. Bemben, J. Berka, M. S. Braverman, Y. J. Chen, Z. Chen, S. B. Dewell, L. Du, J. M. Fierro, X. V. Gomes, B. C. Godwin, W. He, S. Helgesen, C. H. Ho, G. P. Irzyk, S. C. Jando, M. L. Alenquer, T. P. Jarvie, K. B. Jirage, J. B. Kim, J. R. Knight, J. R. Lanza, J. H. Leamon, S. M. Lefkowitz, M. Lei, J. Li, K. L. Lohman, H. Lu, V. B. Makhijani, K. E. McDade, M. P. McKenna, E. W. Myers, E. Nickerson, J. R. Nobile, R. Plant, B. P. Puc, M. T. Ronan, G. T. Roth, G. J. Sarkis, J. F. Simons, J. W. Simpson, M. Srinivasan, K. R. Tartaro, A. Tomasz, K. A. Vogt, G. A. Volkmer, S. H. Wang, Y. Wang, M. P. Weiner, P. Yu, R. F. Begley, and J. M. Rothberg. 2005. Genome sequencing in microfabricated high-density picolitre reactors. *Nature* **437:** 376–380.

9. Nguyen, T. M., D. Ilef, S. Jarraud, L. Rouil, C. Campese, D. Che, S. Haeghebaert, F. Ganiayre, F. Marcel, J. Etienne, and J. C. Desenclos. 2006. A community-wide outbreak of Legionnaires disease linked to industrial cooling towers: how far can contaminated aerosols spread? *J. Infect. Dis.* **193:**102–111.

10. Seo, T. S., X. Bai, D. H. Kim, Q. Meng, S. Shi, H. Ruparel, Z. Li, N. J. Turro, and J. Ju. 2005. Four-color DNA sequencing by synthesis on a chip using photocleavable fluorescent nucleotides. *Proc. Natl. Acad. Sci. USA* **102:**5926–5931.

11. Seshadri, R., I. T. Paulsen, J. A. Eisen, T. D. Read, K. E. Nelson, W. C. Nelson, N. L. Ward, H. Tettelin, T. M. Davidsen, M. J. Beanan, R. T. Deboy, S. C. Daugherty, L. M. Brinkac, R. Madupu, R. J. Dodson, H. M. Khouri, K. H. Lee, H. A. Carty, D. Scanlan, R. A. Heinzen, H. A. Thompson, J. E. Samuel, C. M. Fraser, and J. F. Heidelberg. 2003. Complete genome sequence of the Q-fever pathogen *Coxiella burnetii*. *Proc. Natl. Acad. Sci. USA* **100:**5455–5460.

12. Shendure, J., G. J. Porreca, N. B. Reppas, X. Lin, J. P. McCutcheon, A. M. Rosenbaum, M. D. Wang, K. Zhang, R. D. Mitra, and G. M. Church. 2005. Accurate multiplex polony sequencing of an evolved bacterial genome. *Science* **309:**1728–1732.

ENVIRONMENTAL BIOLOGY, DETECTION, PREVENTION, AND CONTROL

CHARACTERIZATION OF SESSILE AND PLANKTONIC *LEGIONELLA PNEUMOPHILA* IN MODEL BIOFILMS

Barry S. Fields and Claressa E. Lucas

90

Legionellae persist in building water systems by colonizing the biofilms present in these environments. These bacteria are more easily detected from swab samples of biofilm in plumbing systems than from flowing water, suggesting that the majority of the legionellae are biofilm associated (9). The bacteria must be released from the biofilm and come in contact with a susceptible host and cause disease. Some outbreaks are caused by massive descalement of biofilm from disruption of plumbing systems during construction or other disturbances (6). The natural mechanism of legionella colonization and release from biofilms is not understood.

The majority of *Legionella*/biofilm studies that have been conducted employ naturally occurring microbial communities. These have evaluated the effect of temperature and surface materials on the growth of *Legionella pneumophila* as well as the effect of biocides on planktonic and sessile bacteria (2, 11, 14). Such studies have the advantage of representing a true and natural microbial community, but all the organisms present have not been identified, and their contribution to the survival and multiplication of legionellae remains unknown. A

biofilm reactor containing defined organisms and *L. pneumophila* in tap water has been constructed to more precisely characterize the interaction of these organisms (7). This biofilm was composed of *Pseudomonas aeruginosa*, *Klebsiella pneumoniae*, and *Flavobacterium* spp. isolated from a water sample containing legionellae and the amoeba, *Hartmannella vermiformis*. This study indicated that *L. pneumophila* may persist in biofilms in the absence of amoebae, but in the model, the amoebae were required for multiplication of the bacteria. The following studies have used this defined biofilm model and a more simplified model to characterize the natural mechanisms of *L. pneumophila* colonization and release from biofilms.

THE ROLE OF *L. PNEUMOPHILA* PILI IN BIOFILM COLONIZATION

A number of bacterial cell structures and factors have been shown to be critical for biofilm formation. These include flagella, quorum-sensing factors, polysaccharides, and pili. For example, the expression of type IV pili is essential for *P. aeruginosa* biofilm colonization and maturation (8). These adhesins are expressed by numerous gram-negative bacteria and are highly conserved in structural subunit, assembly regulation sequences (12). These pili also participate

Barry S. Fields and Claressa E. Lucas Respiratory Diseases Branch, Centers for Disease Control and Prevention, Atlanta, GA 30333.

Legionella: State of the Art 30 Years after Its Recognition
Edited by Nicholas P. Cianciotto et al.
©2006 ASM Press, Washington, D.C.

in cell-cell communication with related species and unrelated microorganisms in these multicellular matrices. In addition, type II secretion systems used to export pilus machinery for assembly may contribute to biofilm development (1, 4). The discovery and characterization of type IV pili (PilE) and prepilin peptidase (PilD) in *L. pneumophila* allowed studies to characterize the role of these components in biofilm formation (4, 10). Strains deficient in the expression of PilE (strain BS100) and PilD (strain NU243) were generated by mini-TN10 insertion (4, 10). These strains, PilE- (BS100) and PilD- (NU243), were compared with wild-type *L. pneumophila* (130b) for their ability to colonize and persist in biofilms.

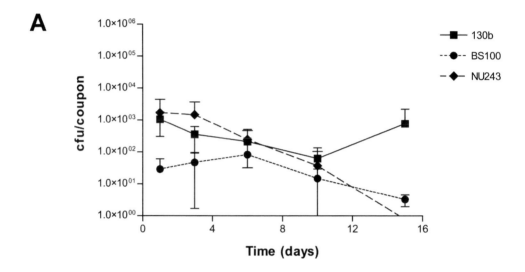

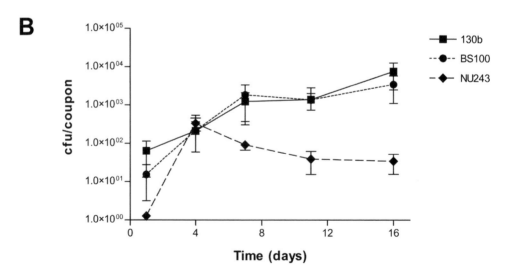

FIGURE 1   (A) Biofilm-associated *L. pneumophila* without amoebae. (B) Biofilm-associated *L. pneumophila* with amoebae.

The biofilm model described by Murga et al. was used to compare these strains in the presence of heterotrophic bacteria (*P. aeruginosa*, *K. pneumoniae*, and *Flavobacterium* spp.) with and without the amoeba *H. vermiformis*. Both strain BS100 and NU243 show a significant decrease in retention in the biofilm in the absence of amoebae (Fig. 1A). In the presence of amoebae, BS100 is retained as well as the wild type (Fig. 1B). Strain NU243 was not retained as well as BS100 or the wild type, presumably because this strain demonstrates limited ability to infect and multiply within amoebae (4). The studies demonstrate that *L. pneumophila* type IV pili and prepilin peptidase play a limited role in the bacteria's colonization of this biofilm model and that the ability to multiply in amoebae is more critical for biofilm colonization and persistence. Based upon these findings, the type IV pilus of *L. pneumophila* does not play as critical a role in biofilm formation as this structure does for *P. aeruginosa* biofilm maturation.

A SIMPLIFIED BIOFILM MODEL

Complex, defined biofilm models incorporating heterotrophic bacteria, *L. pneumophila*, and *H. vermiformis* represent natural environments and are useful for studies such as those described above or for evaluating the efficacy of disinfectants (2). However, the presence of the nucleic acids and proteins of the other genera of bacteria complicate molecular analysis of sessile and planktonic *L. pneumophila*. Characterization of *L. pneumophila* genotypic and phenotypic changes could be more easily studied with a simplified model without other prokaryotes. In order to develop a suitable biofilm model for these studies we attempted to grow *L. pneumophila* alone in the biofilm reactor. The bacteria were harvested from buffered charcoal yeast extract agar plates and inoculated in various broth media. Tap water was pumped into the reactor, creating the flow and eventually diluting out the broth (5). Despite attempts using several media formulations designed to grow *L. pneumophila*, hetertrophic bacteria, or protozoa, *L. pneumophila* could not establish a biofilm in the reactor and no organisms were detected within 1 to 2 days after the flow was established in the reactor (data not shown). This implies that *L. pneumophila* cannot form biofilms in this model in the absence of other organisms.

The next level of complexity is a biofilm model comprised only of *L. pneumophila* and a eukaryotic host. The addition of the amoeba *H. vermiformis* allowed formation of a rudimentary biofilm that could be maintained for at least 41 days (Fig. 2). This simple biofilm

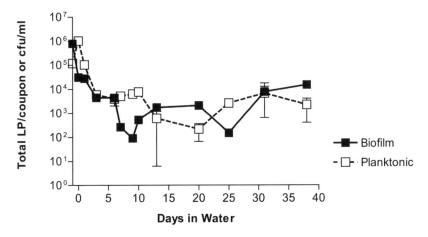

FIGURE 2 *L. pneumophila* concentration in the simplified model comprised of *L. pneumophila* and *H. vermiformis*.

showed some cyclic growth patterns which suggest some cyclic shift between sessile and planktonic *L. pneumophila*. This could be the result of a natural cycle within *L. pneumophila* itself, interaction with *H. vermiformis*, or the result of some external factor such as water chemistry.

CHARACTERIZING THE GROWTH PHASE OF SESSILE AND PLANKTONIC *L. PNEUMOPHILA*

A number of studies have demonstrated that *L. pneumophila* undergoes two growth phases during its life cycle (3, 13). These include a replicative phase characterized by reduced virulence, sodium resistance, lack of cytotoxicity for host cells, osmotic sensitivity, and lack of motility, an infective phase characterized by increased virulence, sodium sensitivity, cytotoxicity, osmotic resistance, and motility. We designed a series of experiments to determine if replicative or infective *L. pneumophila* predominate in the planktonic or biofilm fraction the biofilm reactor. For these experiments we utilize a strain of *L. pneumophila* (130b) that contained a plasmid in which the green fluorescent protein (gfp) was fused to the major flagellin subunit (FlaA) promoter. Thus, we were able to determine not only viable counts of the bacteria but also whether it was in the replicative (nonmotile) of infective (motile) state. The simplified biofilm model was utilized for these studies, and after the *L. pneumophila*/*H.vermiformis* biofilm had been established, the model was challenged by introducing live or heat-killed *P. aeruginosa* and *E. coli*. Viable *L. pneumophila* (total) were enumerated by culture on buffered charcoal yeast extract agar as previously described (5). The percent of motile bacteria (FlaA+) were determined using quantitative microscopy by determining the mean number of fluorescent *L. pneumophila* counted in 10 random fields. The total number of *L. pneumophila* were determined by staining duplicate slides with rabbit anti–*L. pneumophila* fluorescein isothyocyanate labeled sera and counting the same number of microscopic fields. The number of total *L. pneumophila* was

compared to the number of FlaA+ *L. pneumophila* to determine the percent FlaA+.

The total concentration of *L. pneumophila* (viable count) shows an indistinguishable response to challenge with live or killed *P. aeruginosa* or *E. coli* (Fig. 3). In each of these experiments the total number of *L. pneumophila* increased approximately 2 logs in response to the challenge with the other bacteria. In contrast, a significant increase in the percent of FlaA+ *L. pneumophila* occurred in response to challenge with live *P. aeruginosa* (Fig. 4). Challenge with spent culture medium from a *P. aeruginosa* broth culture did not result in an increase in the number of Fla+ *L. pneumophila* (data not shown). This suggests the response to live *P. aeruginosa* may be due to an effector with a short half-life or limited range or may require cell-to-cell contact.

This effect was most dramatic in the effluent fraction of the reactor where the number of Fla+ cells increased by 90%. This demonstrates that a *P. aeruginosa*-specific response results in an increase in the number of infective phase *L. pneumophila*, particularly in the planktonic and effluent fractions. These findings suggest that signals from other genera of bacteria could result in increased numbers of infective phase *L. pneumophila* in water flowing through building water systems. These findings may have implications for preventing release of *L. pneumophila* and disease transmission.

SUMMARY

The mechanism for biofilm colonization by *L. pneumophila* is unique among gram-negative bacteria studied to date. This is demonstrated by the fact that these bacteria are unable to form biofilms or colonize surfaces in the absence of other microorganisms. The addition of host amoeba cells allows *L. pneumophila* to persist in the sessile and planktonic fractions of the biofilm reactor for extended periods of time. Challenging this model with *P. aeruginosa* suggests that *L. pneumophila* may respond to signals from other microorganisms comprising biofilms. These microbial interactions may hold implications in the transmission of *L. pneumophila* and human disease.

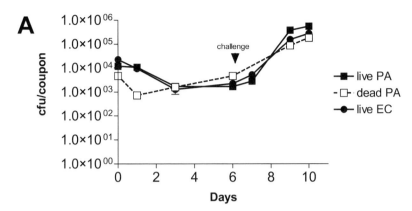

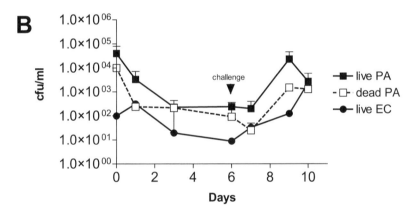

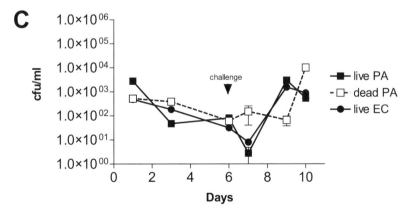

FIGURE 3   Viable counts of *L. pneumophila* (A) in biofilm, (B) planktonic, and (C) in effluent.

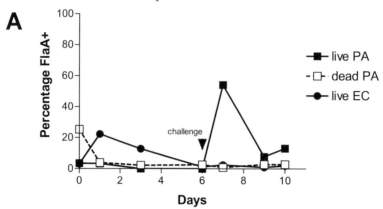

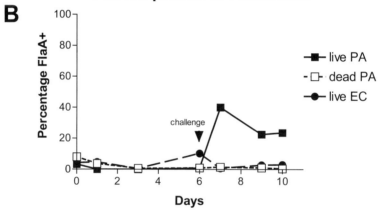

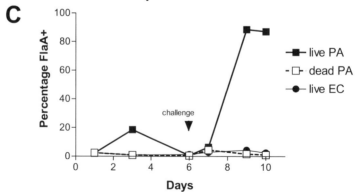

FIGURE 4 FlaA expression (A) in biofilm, (B) planktonic, and (C) in effluent.

ACKNOWLEDGMENT

The authors wish to thank Nicholas P. Cianciotto and Yousef Abu Kwaik for strains NU243 and BS100, respectively. We thank Michele S. Swanson for the pFlaA-gfp plasmid.

REFERENCES

1. **China, B., and F. Goffaux.** 1999. Secretion of virulence factors by *Escherichia coli. Vet. Res.* **30:**181–202.
2. **Donlan. R. M., T. Forster, R. Murga, E. Brown, C. Lucas, J. Carpenter, and B. Fields.** 2005. *Legionella pneumophila* associated with the protozoan *Hartmannella vermiformis* in a model multi-species biofilm has reduced susceptibility to disinfectants *Biofouling* **21:**1–7.
3. **Garduno, R. A., E. Garduno, M. Hiltz, and P. S. Hoffman.** 2002. Intracellular growth of *Legionella pneumophila* gives rise to a differentiated form dissimilar to stationary-phase forms. *Infect. Immun.* **70:**6273–6283.
4. **Liles, M. R., P. H. Edelstein, and N. P. Cianciotto.** 1999. The prepilin peptidase is required for protein secretion by and the virulence of the intracellular pathogen *Legionella pneumophila. Mol. Microbiol.* **31:**959–970.
5. **Lucas, C. E., and B. S. Fields.** Characterization of bacterial growth phase in a *L. pnemophila/Hartmannella vermiformis* biofilm model. Submitted.
6. **Mermel, L. A., S. L. Joesephson, C. H. Giorgio, J. Dempsey, and S. Parenteau.** 1995. Association of Legionnaires' disease with construction: contamination of potable water? *Infect. Contr. Hosp. Epidemiol.* **16:**76–80.
7. **Murga, R., T. S. Forster, E. Brown, J. M. Pruckler, B. S. Fields, and R. M. Donlan.** 2001. The role of biofilms in the survival of *Legionella pneumophila* in a model potable water system. *Microbiology* **147:**3121–3126.
8. **O'Toole, G., H. B. Kaplan, and R. Kolter.** 1999. Biofilm formation as microbial development. *Annu. Rev. Microbiol.* **54:**49–79.
9. **Rogers, J. A., B. Dowsett, P. J. Dennis, J. V. Lee, and C. W. Keevil.** 1994. Influence of temperature and plumbing material selection on biofilm formation and growth of *Legionella pneumophila* in a model potable water system containing complex microbial flora. *Appl. Environ. Microbiol.* **60:**1585–1592.
10. **Stone, B. J., and Y. Abu Kwaik.** 1998. Expression of multiple pili by *Legionella pneumophila*: identification and characterization of a type IV pilin gene and its role in adherence to mammalian and protozoan cells. *Infect. Immun.* **66:**1768–1769.
11. **Storey, M. V., J. Langmark, N. J. Ashbolt, and T. A. Stenstrom.** 2004. The fate of legionellae within distribution pipe biofilms: measurement of their persistence, inactivation, and detachment. *Water Sci. Tech.* **49:**269–275.
12. **Strom, M. S., and S. Lory.** 1993. Structure-function and biogenesis of the type IV pili. *Annu. Rev. Microbiol.* **47:**565–596.
13. **Swanson, M. S., and B. K. Hammer.** 2000. *Legionella pneumophila* pathogenesis: a fateful journey from amoebae to macrophages. *Annu. Rev. Microbiol.* **54:**576–613.
14. **Van der Kooij, D., H. R. Veenendaal, and W. J. H. Scheffer.** 2005. Biofilm formation and multiplication of *Legionella* in a model warm water system with pipes of copper, stainless steel and cross-linked polyethylene. *Water Res.* **39:**2789–2798.

THE AMOEBA *DICTYOSTELIUM DISCOIDEUM* CONTRIBUTES TO *LEGIONELLA* INFECTION

Michael Steinert, Carina Wagner, Marcela Fajardo,
Olga Shevchuk, Can Ünal, Frank Galka, Klaus Heuner,
Ludwig Eichinger, and Salvatore Bozzaro

91

Legionella pneumophila is a microbe that goes astray. Ubiquitously present in aquatic habitats, *L. pneumophila* replicates intracellularly within different species of free-living protozoa. As protozoa are the natural hosts of *Legionella*, the infection of humans is obviously accidental. The different protozoa species in which *Legionella* parasitizes intracellularly provide nutrients, protect the bacteria from adverse conditions, and serve as a vehicle for the colonization of new habitats. Over evolutionary time the protozoa-*Legionella* interaction may have generated a pool of virulence traits which pre-adapted this pathogen for human infection (18).

In recent years the infection cycle of *Legionella* has mainly been investigated on the cellular and subcellular level. It was shown that the prevention of phagocytic killing is mediated by the bacterial type IV Dot/Icm secretion system, which delivers bacterial effector molecules that modulate the endocytic maturation of the host cell. As a result of this reprogramming of the host, the bacteria reside in a phagosome that initially avoids fusion with lysosomes. Later the bacteria replicate in a specialized vacuole derived from the host endoplasmic reticulum (ER) (1). In order to better understand the host side of the underlying complex host-pathogen cross talk we have used the genetically tractable protozoan *Dictyostelium discoideum* as an infection model.

The haploid social amoeba *D. discoideum* is a well-defined model organism which is amenable to diverse biochemical, cell biological, and molecular genetic approaches (20). The entire branch to which this organism belongs is thought to constitute a kingdom in the eukaryote tree of life in its own right besides the plant and animal/fungi kingdoms. The haploid genome is 34 Mbp in size (http://dictybase .org). The six chromosomes carry approximately 13,000 genes, many of which are homologous to those in higher eukaryotes. All of the cytoskeletal elements of *Dictyostelium* are also found in mammalian cells. In addition, *Dictyostelium* shares its chemotactic capacity with leukocytes, and the process of particle uptake in *Dictyostelium* looks remarkably similar to macrophage phagocytosis. The similarities between the *Dictyostelium* and mammalian cells also extend to membrane trafficking, endocytic transit and sorting events (3, 11). Therefore, *Dictyostelium* opens the exciting possibility of investigating and manipulating both sides of the *Legionella*-host interaction.

Michael Steinert, Carina Wagner, Marcela Fajardo, Olga Shevchuk, Can Ünal, Frank Galka, Klaus Heuner, Ludwig Eichinger, and Salvatore Bozzaro Institut für Molekulare Infektionsbiologie, Universität Würzburg, D-97070 Würzburg, Germany.

Legionella: State of the Art 30 Years after Its Recognition
Edited by Nicholas P. Cianciotto et al.
©2006 ASM Press, Washington, D.C.

INVESTIGATIONS OF THE *DICTYOSTELIUM-LEGIONELLA* INTERACTION ON THE CELLULAR LEVEL

The intrinsic features of *Dictyostelium* and a set of well-established molecular tools enabled us to prove that *Dictyostelium* is a representative cellular model for *Legionella* infection (7). Various studies showed that the growth kinetics of pathogenic and apathogenic *Legionella* species as well as *Legionella* mutants in *Dictyostelium* correlate with infections of human macrophage-like cell lines or protozoa (8, 15, 16). Moreover, the intracellular location of the pathogen is similar to what we know from the infection of other host cells. The bacteria reside within a membrane-bound vacuole; they recruit organelles including the ER, and the bacterial replication finally results in host cell lysis. Even the early inhibition of the phagolysosome fusion could be observed in the *Dictyostelium* model (7, 17, 19), and it was further demonstrated that *dotA* mutants are defective in intracellular replication (13). Therefore, this model organism appears to be a valuable model to analyze aspects of *Legionella* infection beyond the cellular level. The research areas presented here include the analysis of the transcriptional host cell response to *Legionella* infection and the use of custom tailored *Dictyostelium* mutant cells to identify determinants of susceptibility and resistance.

TRANSCRIPTIONAL HOST CELL RESPONSE TO *LEGIONELLA* INFECTION

In order to describe the transcriptional host cell response to *Legionella* infection, we employed DNA-microarrays which carried approximately half of the *Dictyostelium* genes (6). For different time points, reverse-transcribed RNA of noninfected *Dictyostelium* cells and infected *Dictyostelium* cells were hybridized to the array, and differentially expressed genes were identified. The number of differentially expressed *Dictyostelium* genes during the time course of infection varied widely, and only a minor fraction of these genes encode for proteins of known function. During early infection more genes were up- than downregulated. At later time points

more balanced numbers of up- and downregulated genes were observed (Table 1). Interestingly, most changes in gene expression occured 24 h postinfection. This approach led to the identification of 731 differentially regulated genes in a time course experiment with *L. pneumophila* and to 131 common genes 24 h after infection in a comparative approach with *L. pneumophila*, *L. hackeliae*, or *L. pneumophila* ΔdotA.

In order to translate a systemic host response into the underlying biological processes, we used enriched gene ontology terms (http://www.geneontology.org/). This approach is based on the hypothesis that regulated genes might belong to identical or similar pathways. The functional annotation of the differentially regulated genes of the time course experiment revealed that host ribosome constituents, protein biosynthesis, hydrolases, lysozyme activity, lipid modifying enzymes, and a number of small calcium-binding proteins were downregulated. Genes encoding enzymes involved in tRNA metabolism and modification, glucose homeostasis, glutamine, pyruvate family amino acid metabolism, nucleotide biosynthesis, as well as cytoskeletal proteins were enriched among the upregulated genes. Apart from a stress response of the host, we also differentially identified regulated genes which had already been described in the context of *Legionella* infection or which have a plausible link to pathogenesis. This is the case for the genes encoding RtoA, ARF1, and β′-COP. RtoA⁻ cells (ratioA mutant) show abnormalities in vesicle fusion in endo- and exocytosis, while the ADP-ribosylation factor ARF1 and the coatomer protein β′-COP regulate and direct vesicular trafficking from the early secretory pathway. Further factors which could influence the fate of the phagosome in the host are certain fatty-acid-modifying enzymes and PIP-6 kinases (6, 9).

Taken together, *Legionella* obviously induces the downregulation of the host-specific metabolism and degradative enzymes and on the other hand induces the upregulation of activities that produce nutrients suitable for the pathogen. The obtained results, which are summarized in Fig. 1, are a good basis for further studies

TABLE 1 Number of differentially expressed *Dictyostelium* genes during the time course of *L. pneumophila* infection

Time point (h)	No. of differentially expressed genes		
	Upregulated	Downregulated	Sum
1	18	1	19
3	83	46	129
6	39	13	52
24	273	325	598
48	42	37	79

and may provide new insights into the function of host activities which promote or restrict the intracellular growth of *L. pneumophila*.

FUNCTIONAL STUDIES WITH CUSTOM TAILORED *DICTYOSTELIUM* CELLS AND INHIBITORS

We have analyzed the effects of host cell mutants and inhibitors on different stages of infection (5). These studies showed that the entry of *L. pneumophila* is actin mediated and that cytoplasmic calcium levels; the cytoskeleton proteins coronin, villidin, and α-actinin/filamin; as well as the calcium-binding proteins of the ER calreticulin and calnexin significantly influence this process. Moreover, we were able to demonstrate that *L. pneumophila* uptake into *D. discoideum* cells involves the β-subunit of heterotrimeric G-proteins and the phopholipase pathway (Fig. 1).

In the following we will focus on the two calcium-binding proteins, calnexin and calreticulin, as well as on the cation transporter

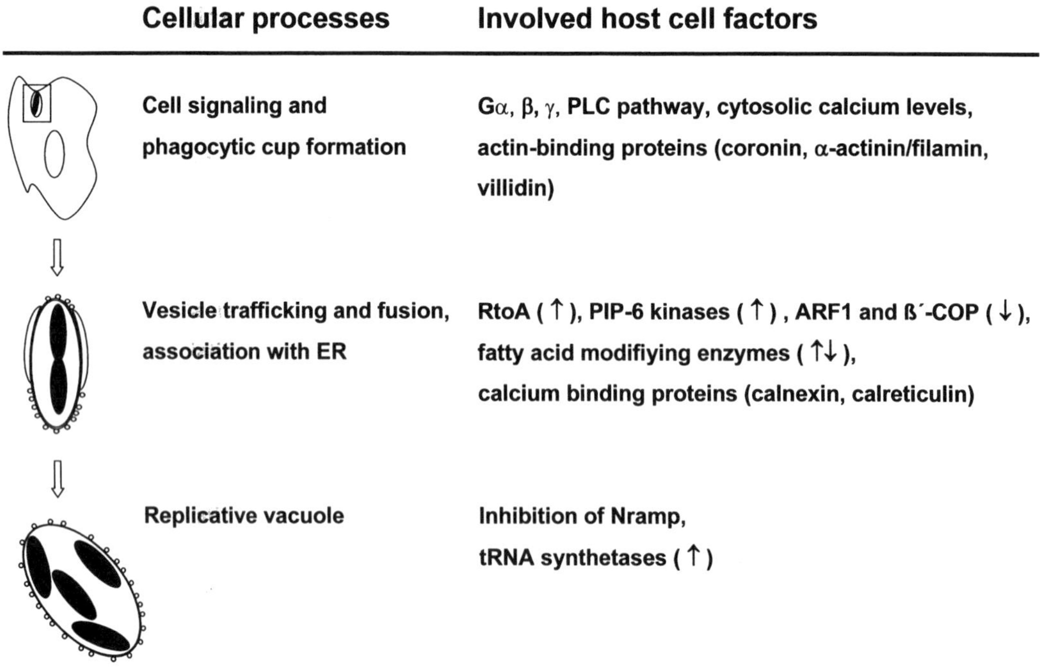

Cellular processes	Involved host cell factors

Cell signaling and phagocytic cup formation — Gα, β, γ, PLC pathway, cytosolic calcium levels, actin-binding proteins (coronin, α-actinin/filamin, villidin)

Vesicle trafficking and fusion, association with ER — RtoA (↑), PIP-6 kinases (↑), ARF1 and ß´-COP (↓), fatty acid modifiying enzymes (↑↓), calcium binding proteins (calnexin, calreticulin)

Replicative vacuole — Inhibition of Nramp, tRNA synthetases (↑)

FIGURE 1 Road map of host cell factors involved in *Legionella* infection. The arrows in brackets indicate the up- and downregulation of the corresponding genes.

Nramp1. Calreticulin is an ER luminal protein and calnexin is an ER-specific type I transmembrane protein (12). Both proteins have been characterized as calcium storage proteins, and several studies have indicated that changes in the concentration of calcium affect ER functions. The phagocytosis rates of *L. pneumophila* into calnexin-minus cells and calreticulin-minus cells were significantly reduced compared to wild-type host cells. By using green fluorescent protein-transformed host cells we analyzed the distribution of calreticulin and calnexin during *Legionella* uptake and intracellular growth (5). A recorded confocal time series showed that both proteins extend into the arms of outgrowing phagocytic cups and permanently decorated the replicative *Legionella* vacuole, whereas the colocalization with *E. coli*-containing vacuoles was only transient. This observation indicates that both proteins could modulate uptake and the formation of an ER-derived *Legionella*-specific vacuole (Fig. 1). A recent study by Lu and Clarke describes the accumulation of calnexin as a Dot/Icm-dependent process. Moreover, the authors suggest that calreticulin and calnexin are not directly incorporated into the membrane of the phagocytic cup (10).

Another functional analysis was performed with the *Dictyostelium* Nramp1 protein (natural resistance associated membrane protein) during *Legionella* infection (14). Nramp1 in mice (*Bcg/Ity/Lsh* locus) controls resistance to infections by *Mycobacterium*, *Salmonella* and *Leishmania*. Additionally, polymorphic variants of the human homologue, which is encoded on the long arm of chromosome 2, are associated with susceptibility to tuberculosis and leprosy. Mammalian Nramp1 decorates endosomes, phagosomes, and postlysosomal vesicles of macrophages, and in *D. discoideum* it has been shown that Nramp1 depletes iron from the phagolysosome in an ATP-dependent process. We could show that Nramp1 gene disruption results in an eightfold higher intracellular growth rate of *L. pneumophila* and a sevenfold higher growth rate of *Mycobacterium avium* compared to infections of *D. discoideum* wild-type host cells. In contrast, overexpression of Nramp1 under the control of a constitutive promoter efficiently protected *Dictyostelium* cells from *L. pneumophila* infection, but not from *M. avium* infection. This intriguing finding was further analyzed by Nramp1 gene expression experiments during *L. pneumophila* and *M. avium* infection. We observed that Nramp1 mRNA slowly disappeared in the half-starving conditions of the medium used in the infection experiments. *L. pneumophila* was able to accelerate this process (Fig. 1), whereas *M. avium* displayed a rather stimulatory effect. Therefore, we conclude that *L. pneumophila* and *M. avium* use different mechanisms to inactivate the antimicrobial activity of Nramp1.

CONCLUSION

Three approaches were used to analyze the complex cross-talk between *Legionella* and *Dictyostelium*. First, *Dictyostelium* microarrays were used to study the systemic host response to *Legionella* infection. Second, fluorescently tagged host cell factors were used to get new insights into the temporal and dynamic interactions between the pathogen and the host. Third, custom tailored mutants and specific inhibitors were used to functionally analyze infection relevant determinants of host susceptibility and resistance. These different approaches contributed to the following road map of infection-relevant host-cell factors: (i) Phagocytic cup formation and the actin-mediated *Legionella* uptake include heterotrimeric G proteins, the phospholipase C pathway, and changes in the cytoplasmic calcium levels and the actin-binding proteins coronin, α-actinin/filamin, and villidin. (ii) The fate of *Legionella*-containing vesicles and association with the ER are likely to be influenced by RtoA, PIP-6 kinases, ARF1, β'-COP, fatty acid-modifying enzymes, and the calcium-binding proteins calnexin and calreticulin. (iii) The inhibition of Nramp1 and the upregulation of tRNA synthetases seem to be important steps in the replicative phase of infection. Of course, further functional studies are needed to refine this road map of host-pathogen interaction. With the determination of the genomes of both the pathogen and the host (2, 4), the availability of a large collection of *Dictyostelium* mutants and of an impressive molecular biological toolbox for *Dictyostelium*, a comprehensive understanding of the respective interactions seems to be at hand.

REFERENCES

1. **Bitar, D. M., M. Molmert, and Y. Abu Kwaik.** 2004. Molecular and cell biology of *Legionella pneumophila*. *Int. J. Med. Microbiol.* **293**:519–527.
2. **Cazalet, C., C. Rusniok, H. Brüggemann, N. Zidane, A. Magnier, L. Ma, M. Tichit, S. Jarraud, C. Bouchier, F. Vandenesch, F. Kunst, J. Etienne, P. Glaser, and C. Buchrieser.** 2004. Evidence in the *Legionella pneumophila* genome for exploitation of host cell functions and high genome plasticity. *Nat. Genet.* **36**:1165–1173.
3. **Duhon, D., and J. Cardelli.** 2002. The regulation of phagosome maturation in *Dictyostelium*. *J. Muscle Cell Motil.* **23**: 803–808.
4. **Eichinger, L., J. A. Pachebat, G. Glöckner, M. A. Rajandream, R. Sucgang, and M. Berriman.** 2005. The genome of the social amoeba *Dictyostelium discoideum*. *Nature* **435**:43–57.
5. **Fajardo, M., M. Schleicher, A. Noegel, S. Bozzaro, S. Killinger, K. Heuner, J. Hacker, and M. Steinert.** 2004. Calnexin, calreticulin and cytoskeleton associated proteins modulate uptake and growth of *Legionella pneumophila* in *Dictyostelium discoideum*. *Microbiology* **150**:2825–2835.
6. **Farbrother, P., C. Wagner, J. Na, B. Tunggal, T. Morio, H. Urshihara, Y. Tanaka, M. Schleicher, M. Steinert, and L. Eichinger.** 2006. *Dictyostelium* transcriptional host cell response upon infection with *Legionella*. *Cell. Microbiol.*, in press.
7. **Hägele, S., R. Köhler, H. Merkert, M. Schleicher, J. Hacker, and M. Steinert.** 2000. *Dictyostelium discoideum*: a new host model system for intracellular pathogens of the genus *Legionella*. *Cell. Microbiol.* **2**:165–171.
8. **Heuner, K., C. Dietrich, C. Skriwan, M. Steinert, and J. Hacker.** 2002. Influence of the alternative σ^{28} factor on virulence and flagellum expression of *Legionella pneumophila*. *Infect. Immun.* **70**:1604–1608.
9. **Li, Z., J. M. Solomon, and R. Isberg.** 2005. *Dictyostelium discoideum* strains lacking the RtoA

protein are defective for maturation of the *Legionella pneumophila* replication vacuole. *Cell. Microbiol.* **7**:431–442.
10. **Lu, H., and M. Clarke.** 2005. Dynamic properties of the *Legionella*-containing phagosomes in *Dictyostelium* amoebae. *Cell. Microbiol.* **7**:995–1007.
11. **Maniak, M.** 2003. Fusion and fission events in the endocytic pathway of *Dictyostelium*. *Traffic* **4**:1–5.
12. **Mäller-Taubenberger, A., A. N. Lupas, H. Li, M. Ecke, E. Simmeth, and G. Gerisch.** 2001. Calreticulin and calnexin in the endoplasmic reticulum are important for phagocytosis. *EMBO J.* **23**:6772–6782.
13. **Otto G. P., M. Y. Wu, M. Clarke, H. Lu, O. R. Anderson, H. Hilbi, H. A. Shuman, and R. H. Kessin.** 2004. Macroautophagy is dispensable for intracellular replication of *Legionella pneumophila* in *Dictyostelium discoideum*. *Mol. Microbiol.* **51**:63–72.
14. **Peracino, B., C. Skriwan, C. Balest, A. Balbo, B. Pergolizzi, A. A. Noegel, M. Steinert, and S. Bozzaro.** 2006. Function and mechanism of action of *Dictyostelium* Nramp1 (Slc11a1) in bacterial infection. *Traffic* **7**:22–38.
15. **Schreiner, T., M. R. Mohrs, R. Blau-Wasser, A. von Krempelhuber, M. Steinert, M. Schleicher, and A. A. Noegel.** 2002. Loss of the F-actin binding and vesicle-associated protein comitin leads to a phagocytosis defect. *Euk. Cell.* **1**:906–914.
16. **Skriwan, C., M. Fajardo, S. Hägele, M. Horn, M. Wagner, R. Michel, G. Krohne, M. Schleicher, J. Hacker, and M. Steinert.** 2002. Various bacterial pathogens and symbionts infect the amoeba *Dictyostelium discoideum*. *Int. J. Med. Micobiol.* **291**:615–624.
17. **Solomon, J. M., and R. R. Isberg.** 2000. Growth of *Legionella pneumophila* in *Dictyostelium discoideum*: a novel system for genetic analysis of host-pathogen interactions. *Trends Microbiol.* **10**: 478–480.
18. **Steinert, M., U. Hentschel, and J. Hacker.** 2002. *Legionella pneumophila*: an aquatic microbe goes astray. *FEMS Microbiol. Rev.* **743**:1–14.
19. **Steinert, M., M. Leippe, and T. Roeder.** 2003. Surrogate hosts: invertebrates as models for studying pathogen-host interactions. *Int. J. Med. Microbiol.* **293**:1–12.
20. **Steinert, M., and K. Heuner.** 2005. *Dictyostelium* as host model for pathogenesis. *Cell. Microbiol.* **7**:307–314.

ACANTHAMOEBA CASTELLANII STRONGLY INCREASES THE NUMBER OF LEGIONELLA PNEUMOPHILA IN MODEL TAP WATER BIOFILMS

P. Declerck, J. Behets, E. Lammertyn, and F. Ollevier

92

Recently we demonstrated through several uptake experiments that some bacterial species such as *Pseudomonas aeruginosa, Escherichia coli, Flavobacterium breve,* and *Aeromonas hydrophila* do not act as competitors on *Legionella pneumophila* uptake by *Acanthamoeba castellanii*, but are able to influence intracellular *Legionella* replication (1). As *Legionella* bacteria survive in aquatic environments by residing in biofilm communities which are grazed by amoebae such as *Acanthamoeba* spp. (5, 6), the current study aimed to provide more insight in the role of *A. castellanii* in the replication and distribution of *L. pneumophila* in substrate-associated biofilms, similar to those in man–made aquatic systems and consisting of the above-mentioned four non-*Legionella* bacterial species.

Biofilm experiments were performed at 35°C in a rotating annular reactor (RAR, Bio-Surface Technologies Corporation, Bozeman, Mont.), mounted with 20 polycarbonate slides (18.75 cm^2), functioning in flow through and fed with dechlorinated tap water (volume renewal: every 2 h). Experiments were repeated at least three times and always consisted of two parts: during the first part, without present *A. castellanii* (days 0 to 36), we investigated if *L. pneumophila* were able to attach and eventually grow in the biofilm. During the second part (days 36 to 42), *A. castellanii* trophozoites were added to the system and the evolution of biofilm-associated *L. pneumophila* was investigated. Biofilm samples were analyzed at specific time intervals by means of classical culture, real-time PCR (3, 7), and fluorescent in situ hybridization (FISH) (2). During the first 9 days of each experiment biofilms consisting of *P. aeruginosa, E. coli, F. breve*, and *A. hydrophila* were grown. When microbial and microscopic analyses showed the presence of a "mature" biofilm, *L. pneumophila* was added to the reactor in a concentration of 5×10^5 cells/ml on day 9, followed 27 days later (day 36) by the addition of *A. castellanii* (10^4 cells/ml). After each inoculation, the reactor operated for 2 days in batch mode, allowing the microorganisms to adhere to the slides.

When *L. pneumophila* was inoculated on day 9, the biofilm contained $\pm$ 10^8 CFU/18.75 cm^2, and analyses of FISH-stained biofilm samples revealed the presence of several microcolonies. Real-time PCR and classical culture results demonstrated the persistence of *Legionella* bacteria in the biofilm during a period

P. Declerck, J. Behets, and F. Ollevier Laboratory of Aquatic Ecology, Zoological Institute, Katholieke Universiteit Leuven, Charles De Bériotstraat 32, B-3000 Leuven, Belgium. *E. Lammertyn* Laboratory of Bacteriology, Rega Institute for Medical Research, Katholieke Universiteit Leuven, Minderbroedersstraat 10, B-3000 Leuven, Belgium.

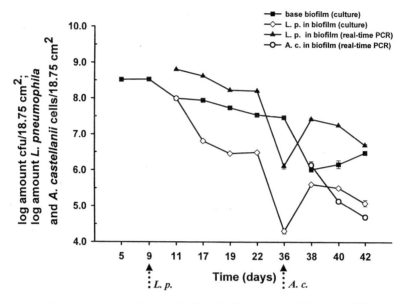

FIGURE 1 Evolution of the base biofilm, biofilm-associated *L. pneumophila,* and *A. castellanii; n* = 2; log x̄ ± 95% confidence intervals. The arrows indicate the inoculation of *L. pneumophila* and *A. castellanii* in the system.

of 11 days (day 11 to day 22). Between day 22 and day 36 the *Legionella* concentration decreased with ± 2 log units. Addition of *A. castellanii* on day 36 resulted within 48 h in a significant decrease of biofilm-associated non-*Legionella* bacteria due to predation, while biofilm-associated *L. pneumophila* increased with 1.5 log units due to intracellular replication in the present *A. castellanii* trophozoites. Additional confocal analyses of FISH-stained biofilm samples confirmed the presence of heavily infected *A. castellanii* trophozoites. Due to lysis of used amoeba hosts by replicated *L. pneumophila,* the *A. castellanii* population started decreasing just 2 days (day 38) after its inoculation. The decrease of amoebae resulted in a renewed decrease of biofilm-associated *Legionella* bacteria and an increase of the non-*Legionella* biofilm species.

Our results clearly show that the presence of *A. castellanii* increases the number of biofilm-associated *L. pneumophila* due to intracellular replication of the bacteria in the amoeba hosts. In that way, *A. castellanii* can play a crucial role in the distribution of *L. pneumophila* in biofilms that are similar to those present in water

distribution systems. This may have serious consequences concerning the human health risk for Legionnaires' disease. The association between Legionnaires' disease and the presence of high numbers of human pathogenic *L. pneumophila* bacteria in man-made water supplies has already been confirmed frequently (4). Problems can arise when concentrated numbers of biofilm-associated *L. pneumophila* become detached from the substratum and mobilized into the bulk water phase, where they have the potential to infect humans. Therefore, our results stress the importance of further obtaining data concerning interactions between *L. pneumophila* and potential amoeba host populations associated with biofilms.

ACKNOWLEDGMENTS

This study was funded by the IWT (GBOU no. 20153).

REFERENCES

1. **Declerck, P., J. Behets, Y. Delaedt, A. Margineanu, E. Lammertyn, and F. Ollevier.** 2005. Impact of non-*Legionella* bacteria on the uptake

and intracellular replication of *Legionella pneumophila* in *Acanthamoeba castellanii* and *Naegleria lovaniensis*. *Microbial Ecol.* **50:**536–549.

2. **Declerck, P., J. Behets, E. Lammertyn, and F. Ollevier.** 2006. Whole cell fluorescent *in situ* hybridization (FISH) of *Legionella* in various kinds of samples. *In* L. O'Connor (ed.). *Diagnostic Bacteriology Protocols*, 2nd ed. Humana Press, Totowa, NJ **345:**175–184.

3. **Herpers, B. L., B. M. de Jongh, K. van der Zwaluw, and E. J. van Hannen.** 2003. Real-time PCR assay targets the 23S-5S spacer for direct detection and differentiation of *Legionella* spp. and *Legionella pneumophila. J. Clin. Microbiol.* **41:**4815–4816.

4. **Meigh, R. E., T. Makin, M. H. Scott, and C. A. Hart.** 1989. *Legionella pneumophila* serogroup 12 pneumonia in a renal transplant recipient: case report and environmental observations. *J. Hosp. Infect.* **13:**315–319.

5. **Noble, J. A., D. G. Ahearn, S. V. Avery, and S. A. Crow.** 2002. Phagocytosis affects biguanide sensitivity of *Acanthamoeba* spp. *Antimicrob. Agents Chemother.* **46:**2069–2076.

6. **Percival, S. L., J. T. Walker, and P. R. Hunter (ed.).** 2000. *In Microbiological Aspects of Biofilms and Drinking Water.* CRC Press, Boca Raton, FL, p. 240.

7. **Rivière, D., F. M. Szczebara, J. M. Berjeaud, J. Frère, and Y. Héchard.** 2006. Development of a real-time PCR assay for quantification of *Acanthamoeba* trophozoites and cysts. *J. Microbiol. Methods* **64:**78–83.

BIOFILM FORMATION OF *LEGIONELLA PNEUMOPHILA* IN COMPLEX MEDIUM UNTER STATIC AND DYNAMIC FLOW CONDITIONS

Jörg Mampel, Thomas Spirig, Stefan S. Weber, Janus A. J. Haagensen, Søren Molin, and Hubert Hilbi

93

Legionella pneumophila, the etiologic agent of the severe pneumonia Legionnaires' disease, replicates in protozoa and colonizes biofilms in natural freshwater or engineered water distribution systems (4, 14, 15). While *L. pneumophila* serves as a paradigm for intracellular bacterial growth within phagocytes, it is less clear whether the bacteria form biofilms or replicate therein in absence of protozoa. To date, the biofilm lifestyle of *L. pneumophila* has been addressed in only a few studies. *L. pneumophila* was shown to persist in biofilms formed by preselected mixtures of defined (9) or undefined heterotrophic bacteria (12) using continuous flow reactors operated with tap water as the sole source of nutrients. Replication of *L. pneumophila* in biofilms formed in poor medium coincided with the presence of amoebae (3, 6, 9). In this study, we analyzed biofilm formation of *L. pneumophila* in rich medium which supports extracellular proliferation of the bacteria (7).

Jörg Mampel, Thomas Spirig, Stefan S. Weber, and Hubert Hilbi Institute of Microbiology, Swiss Federal Institute of Technology (ETH), 8093 Zürich, Switzerland. *Janus A. J. Haagensen and Søren Molin* Molecular Microbial Ecology Group, Bio-Centrum-DTU, Technical University of Denmark, DK-2800 Lyngby, Denmark.

L. PNEUMOPHILA FORMS BIOFILMS UNDER STATIC BUT NOT QUASI-STATIC CONDITIONS IN RICH MEDIUM

To investigate biofilm formation of *Legionella* under static conditions, a 1:1 mixture of *L. pneumophila* labeled with the fluorescent proteins EGFP or DsRed-Express was inoculated in a glass-bottom 35-mm petri dish. The fluorescent proteins were encoded by nonmobilizable derivatives of plasmid pMMB207 and constitutively produced (7). Biofilm formation was analyzed with an inverted confocal microscope (Axiovert 200M, 100 × oil objective Plan Neofluar; Zeiss). After 3 days of growth in *N*-(2-acetamido)-2-aminoethanesulfonic acid-buffered yeast extract (AYE) medium, *L. pneumophila* formed irregular, rather fluffy aggregates (Fig. 1). These aggregates were present in some areas of the glass surface, while in other areas single cells attached to the surface. The green and red fluorescent bacteria interspersed homogenously on the substratum and in the aggregates and did not form extended single-color cell clusters indicative of clonal growth.

Biofilm formation of *L. pneumophila* in rich medium was also quantified by crystal violet staining in upright polystyrene microtiter plates or on polystyrene pins of "inverse" lids under static (no medium exchange) or quasi-static

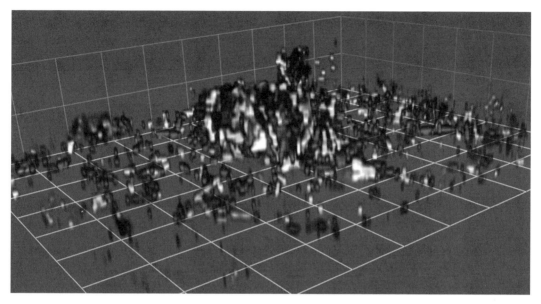

FIGURE 1 Biofilm formation and distribution of EGFP- and DsRedExpress-labeled *L. pneumophila* under static conditions in rich medium. Confocal laser scanning micrograph of a representative biofilm section formed after 3 days in glass-bottom dishes by wild-type *L. pneumophila* JR32 labeled with enhanced green fluorescent protein (EGFP, white) or with the red fluorescent protein DsRedExpress (black). Three-dimensional reconstruction of the image data was performed with the "Volocity" 2.6.1 software (Improvision). The strains were inoculated in a 1:1 mixture and remained homogeneously dispersed throughout the experiment without forming extended patches of clonally grown aggregates. Unit cell: 8.1 μm.

conditions (medium replaced twice a day). Under static conditions, *L. pneumophila* biomass accumulated in the upright as well as the inverse system and coincided with bacterial growth in the broth. However, under quasi-static conditions, where the polystyrene lids were transferred to fresh AYE medium twice a day, neither biomass accumulation nor bacterial growth was observed. In presence of *Acanthamoeba castellanii*, however, *L. pneumophila* replicated, and sessile biomass accumulated under otherwise identical conditions. These results are in agreement with the notion that *L. pneumophila* biomass accumulation is due to recruitment of planktonic (free-swimming) or intracellularly grown cells rather than growth of sessile (surface-attached) bacteria.

BIOFILM FORMATION OF DEFINED *L. PNEUMOPHILA* MUTANTS

Under batch conditions, the laboratory wild-type strain JR32 and 13 environmental isolates formed biofilms to a similar extent, but among 14 clinical isolates, 3 strains produced about 50% less biomass. Interestingly, among defined *L. pneumophila* mutants lacking factors potentially involved in surface attachment and biofilm formation (pili, flagella, Icm/Dot type IV secretion system, Lvh type IV secretion system, stationary phase [RpoS] or alternative [FliA] sigma factors, two-component response regulator LetA, or global repressor CsrA), only a *fliA* mutant strain was significantly affected in biofilm formation under the conditions tested (7). The *L. pneumophila fliA* mutant strain reproducibly accumulated 30% less biomass within 5 days, demonstrating that bacterial factors contribute to biofilm formation. It is noteworthy that mutants lacking *rpoS* or *letA*, both of which are required for the expression of transmissive (virulence) traits of *L. pneumophila*, were not affected in biofilm formation.

ADHERENCE OF *L. PNEUMOPHILA* IN A CONTINUOUS-FLOW CHAMBER SYSTEM

A continuous-flow chamber system was set up (8) to further test the hypothesis that planktonic cells are crucial for biofilm formation on surfaces and to noninvasively monitor in real-time the adherence and biofilm formation of *L. pneumophila* under dynamic flow conditions (7). Each channel (total volume, 160 μl) was equipped with bubble-traps and flow-breaks downstream of a peristaltic pump (IPC-N-16; Ismatec SA). The channels were inoculated by injection of 1:5 to 1:10 dilutions of the virulent *L. pneumophila* strain JR32 (13) constitutively expressing GFP, grown in AYE (30°C, o/n) to mid-log/early stationary growth phase (optical density at 600 nm, 1.5 to 3.0). The bacteria were then allowed to attach for 45 to 60 min without flow. Routine operating conditions were a flow rate of 50 μl/min at 30°C, using a 1:10 diluted AYE medium. An inverted epi-fluorescence microscope (Axioplan 2; Zeiss) equipped with an apotome device was used to monitor the flow chambers.

After inoculation, a monolayer of single dispersed *L. pneumophila* attached to the borosilicate glass (and polystyrene) substratum of the flow chambers in absence but also in presence of flow (50 μl/min). *L. pneumophila* attached to the surfaces but did not further develop a biofilm for up to 14 days under a wide variety of conditions, despite a constant influx of nutrients (AYE, 1:10 diluted). Instead, the bacteria progressively detached from the surface of the channels in a flow rate-dependent manner. At a flow rate of 50 μl/min complete ablation of *L. pneumophila* was observed within 5 to 8 days. This finding correlates with studies suggesting that in stagnant water systems the prevalence of *L. pneumophila* is higher than in systems under continuous flow (1). The inability of *L. pneumophila* to grow and form biofilms under dynamic flow conditions in rich medium is in agreement with a possible requirement for planktonically growing bacteria for biofilm formation but might also be due to mechanical detachment of single bacteria or dilution of a secreted bacterial factor required for growth on surfaces.

After 18 to 48 h we occasionally detected compact (Fig. 2A) or filamentous microcolonies ("streamers," Fig. 2B and C) of green

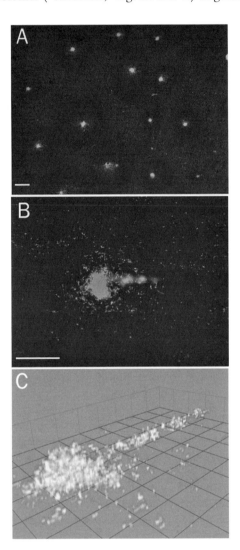

FIGURE 2 Microcolonies formed by *L. pneumophila* in a continuous flow chamber system in rich medium. The fluorescence micrographs show (A) compact microcolonies of GFP-labeled *L. pneumophila* wild-type strain JR32 occasionally formed after 18 h and (B) filamentous microcolonies ("streamers") present after 5 days. (C) Three-dimensional reconstruction of the filamentous microcolony shown in panel B was done as described in the legend to Fig. 1. Bar (A, B), 100 μm. Unit cell (C), 23 μm.

fluorescent protein-labeled *L. pneumophila*. The microcolonies were about 50 μm in diameter, reached heights of 20 μm, and were surrounded by zones of low cell density, creating a halo-like feature around the bacterial aggregates. Most of these microcolonies were stable during the course of the experiment (8 to 14 days). The observation of microcolonies demonstrates the inherent potential of *L. pneumophila* to form three-dimensional cell aggregates in a continuous-flow system. However, the mechanism leading to these microcolonies remains obscure, as they were not reproducibly observed. Notably, even under conditions supporting axenic growth, the morphology of *L. pneumophila* biofilms did not resemble the "mushroom" structures described for *Pseudomonas* spp. (2, 5). Rather, *L. pneumophila* monolayers or microcolonies resemble *Escherichia coli* K-12, which in absence of conjugative pili attaches to surfaces and forms microcolonies without developing towering biofilms (11). Our observations fit the emerging view that surface colonization is a process unique for a specific bacterium (10).

In addition to the formation of *L. pneumophila* monospecies biofilms, we also analyzed colonization by *L. pneumophila* of biofilms formed for 2 days by heterotrophic environmental bacteria. In these capture assays, proliferation of *L. pneumophila* was not observed. *L. pneumophila* persisted in biofilms formed by *Empedobacter breve* or *Microbacterium* sp., but not by other environmental bacteria, while the presence of *Klebsiella pneumoniae* even reduced adherence and accelerated detachment. These results suggest that specific interactions between the bacteria modulate adherence and persistence under continuous-flow conditions.

CONCLUSIONS

L. pneumophila growing in rich medium adheres to polystyrene surfaces and forms monospecies biofilms in upright and on inverse plates under static but not under quasi-static conditions. In a continuous-flow system, *L. pneumophila* is progressively washed out in a flow rate-dependent manner, but interactions with specific bacterial cocultures increase persistence. Our observations indicate that in static systems *L. pneumophila* on its own might accumulate sessile biomass under suitable conditions. However, in dynamic-flow systems *L. pneumophila* apparently relies on an accompanying bacterial biofilm flora and protozoan hosts to persist, replicate, and spread.

ACKNOWLEDGMENTS

This work was supported by the Swiss Commission for Technology and Innovation (CTI/KTI, grant 6629.2 BTS-LS), the National Science Foundation (grant 631-065952), and the Federal Office for Energy (BFE). The Swiss National Reference Center for *Legionella* in Bellinzona provided clinical and environmental *Legionella* isolates.

REFERENCES

1. **Ciesielski, C. A., M. J. Blaser, and W. L. Wang.** 1984. Role of stagnation and obstruction of water flow in isolation of *Legionella pneumophila* from hospital plumbing. *Appl. Environ. Microbiol.* **48:** 984–987.
2. **Costerton, J. W., P. S. Stewart, and E. P. Greenberg.** 1999. Bacterial biofilms: a common cause of persistent infections. *Science* **284:**1318–1322.
3. **Fields, B. S.** 2002. The social life of Legionellae. p. 135–142. *In* R. Marre, Y. Abu Kwaik, C. Bartlett, N. P. Cianciotto, B. S. Fields, M. Frosch, J. Hacker, and P. C. Luck (ed.), *Legionella.* ASM Press, Washington, D.C.
4. **Fields, B. S., R. F. Benson, and R. E. Besser.** 2002. *Legionella* and Legionnaires' disease: 25 years of investigation. *Clin. Microbiol. Rev.* **15:**506–526.
5. **Klausen, M., A. Aaes-Jorgensen, S. Molin, and T. Tolker-Nielsen.** 2003. Involvement of bacterial migration in the development of complex multicellular structures in *Pseudomonas aeruginosa* biofilms. *Mol. Microbiol.* **50:**61–68.
6. **Kuiper, M. W., B. A. Wullings, A. D. Akkermans, R. R. Beumer, and D. van der Kooij.** 2004. Intracellular proliferation of *Legionella pneumophila* in *Hartmannella vermiformis* in aquatic biofilms grown on plasticized polyvinyl chloride. *Appl. Environ. Microbiol.* **70:**6826–6833.
7. **Mampel, J., T. Spirig, S. S. Weber, J. A. J. Haagensen, S. Molin, and H. Hilbi.** 2006. Planktonic replication is essential for biofilm formation of *Legionella pneumophila* in a complex

medium under static and dynamic flow conditions. *Appl. Environ. Microbiol.* **72:**2885–2895.

8. **Moller, S., C. Sternberg, J. B. Andersen, B. B. Christensen, J. L. Ramos, M. Givskov, and S. Molin.** 1998. *In situ* gene expression in mixed–culture biofilms: evidence of metabolic interactions between community members. *Appl. Environ. Microbiol.* **64:**721–732.

9. **Murga, R., T. S. Forster, E. Brown, J. M. Pruckler, B. S. Fields, and R. M. Donlan.** 2001. Role of biofilms in the survival of *Legionella pneumophila* in a model potable-water system. *Microbiology* **147:**3121–3126.

10. **Parsek, M. R., and C. Fuqua.** 2004. Biofilms 2003: emerging themes and challenges in studies of surface–associated microbial life. *J. Bacteriol.* **186:**4427–4440.

11. **Reisner, A., J. A. Haagensen, M. A. Schembri, E. L. Zechner, and S. Molin.** 2003. Development and maturation of *Escherichia coli* K-12 biofilms. *Mol. Microbiol.* **48:**933–946.

12. **Rogers, J., A. B. Dowsett, P. J. Dennis, J. V. Lee, and C. W. Keevil.** 1994. Influence of temperature and plumbing material selection on biofilm formation and growth of *Legionella pneumophila* in a model potable water system containing complex microbial flora. *Appl. Environ. Microbiol.* **60:**1585–1592.

13. **Sadosky, A. B., L. A. Wiater, and H. A. Shuman.** 1993. Identification of *Legionella pneumophila* genes required for growth within and killing of human macrophages. *Infect. Immun.* **61:**5361–5373.

14. **Steinert, M., U. Hentschel, and J. Hacker.** 2002. *Legionella pneumophila*: an aquatic microbe goes astray. *FEMS Microbiol. Rev.* **26:**149–162.

15. **Szewzyk, U., R. Szewzyk, W. Manz, and K. H. Schleifer.** 2000. Microbiological safety of drinking water. *Annu. Rev. Microbiol.* **54:**81–127.

EVALUATION OF SIGNALING BETWEEN *LEGIONELLA PNEUMOPHILA* AND MULTIPLE PROKARYOTES

Stephanie D. Zeigler-Ballerstein and James M. Barbaree

94

The association of legionellae with microorganisms in the aquatic environment has been documented. This relationship includes its cohabitation with protozoa, algae, and prokaryotes within biofilms. While the secretion of amino acids by neighboring microorganisms is a likely contributor to the maintenance of *Legionella*'s viability during the planktonic phase of its life cycle, other factors may play a substantial role (5). Potential candidates are the intercellular signals produced by bacterial species commonly isolated from biofilm communities. Acyl-homoserine lactones (AHSL) and furanosyl borate diesters (AI-2) are two prokaryotic signaling compounds that have been isolated from biofilms (1, 4). In species that produce them, these compounds control multiple metabolic functions. However, the role of AHSL and AI-2 in interspecies (heterospecific) communication has been demonstrated in multiple bacteria but is not well documented. *Pseudomonas aeruginosa*, *Klebsiella pneumoniae*, and *Burkholderia cepacia* have all been shown to respond to signaling molecules that they themselves are incapable of synthesizing (2). For example, *P. aeruginosa* produces two AHSL compounds that control various functions, including biofilm formation and the production of elastase; however, AI-2, which is not synthesized by this microorganism, has been shown to enhance pathogenicity through the upregulation of virulence genes (2). While our data shows that *Legionella* species cannot produce AHSL or AI-2, the likelihood exists that these microorganisms possess a unique intercellular communication system and have the ability to perceive signals produced by other genera of microorganisms as well as potential hosts. Thus, consistent exposure of legionellae to signals circulating throughout biofilms has the potential to impact both their behavior and the maintenance of viability during the planktonic phase of their life cycle. The objective of the following investigation is to evaluate the capacity of *L. pneumophila* to respond to interspecific signals. From evaluation of protein expression after exposure to various bacterial supernatants, multiple proteins were identified as products of *L. pneumophila*'s response to heterospecific signals.

The investigation was initiated with the preparation of 18- to 24-h cell-free supernatants from prokaryotes known to secrete AHSL or AI-2. The microorganisms used in this study included three AI-2 producers (*Esch-*

Stephanie D. Zeigler-Ballerstein and James M. Barbaree Department of Biological Sciences, Auburn University, Auburn, Alabama 36849.

erichia coli O157:H7, *Bacillus subtilis,* and *Streptococcus pyogenes*) and two AHSL producers (*P. aeruginosa* and *B. cepacia*). *L. pneumophila* was grown to stationary phase in *N*-(2-acetamido)-2-aminoethanesulfonic acid-buffered YEB and then resuspended in heterospecific supernatants for 3 h. Following extraction, proteins were resolved via two-dimensional electrophoresis. The first dimension was completed using pH 4 to 7, 18-cm immobilized pH gradient strips; the second dimension was accomplished by applying the strips to 12%, 20 × 20-cm polyacrylmide gels. Gels were stained with SYPRO Ruby, and two-dimensional gels (proteomes) were evaluated manually and by using PDQuest (BioRad). Differentially expressed proteins were selected for further study.

Spots of interest were manually excised from two-dimensional electrophoresis gels, washed, and subjected to in situ digestion with trypsin, and the resultant peptides were desalted and concentrated with C_{18} ZipTips (Millipore) (6). Peptides were mixed with matrix (α-cyano-4-hydroxycinnamic acid) and deposited onto a Micro SCOUT plate (Bruker Daltonics). The peptide masses were analyzed using Microflex MALDI (Bruker Daltonics). The peptide mass fingerprint data retrieved from MALDI analysis was searched against multiple databases using the MASCOT search engine at http:www.matrixscience.com.

In response to 5 heterospecific supernatants, *L. pneumophila* produced an average of 62 more spots than unexposed control cultures. Out of the 50 spots that were selected for MALDI analysis, the BLAST search of the peptide spectra failed to conclusively identify (i.e., a MASCOT score of ≥73) the majority of the proteins. Three of the proteins identified yielded significant val-

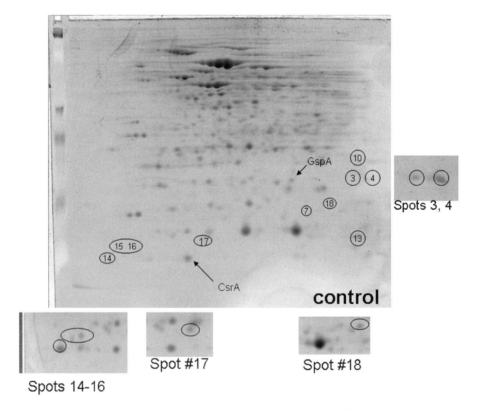

FIGURE 1 Proteome of the control culture with a comparison of differentially expressed proteins.

ues. These proteins (annotated in Fig. 1) were identified as hypothetical proteins lpp1207 (spot 13) and *atpH* (spot 10), and a 24-kDa macrophage-induced protein (spot 4). CspA (a cold shock protein) was also provided as identification for spot 13. Additional *Legionella* proteins, all with peptide scores between 45 and 63 were also noted, with the closest identifications being hypothetical protein lpl1213 (spot 13), Hsp10 (spot 16, a heat shock protein), GroES (spot 17, chaperonin and heat shock protein), and UspA (spots 7 and 18; universal stress protein) (Table 1).

By virtue of enhanced protein expression, the data presented demonstrate the capacity of *L. pneumophila* to detect heterospecific, intercellular signals. The raw data suggest that AI-2 elicits a greater response from *L. pneumophila* than AHSL. AI-2 has been detected in numerous prokaryotes, both gram positive and gram negative, and is commonly referred to as a universal communication signal among microorganisms (4). While the absence of AI-2 in *Legionella*'s supernatant has been noted, the presence of this compound within habitats known to harbor legionellae suggests a predilection for AI-2 signal interception whether they produce this compound or not. This is also substantiated by induction of gene activity (as inferred by the presence of multiple spots in the proteome) upon exposure to AI-2 (and AHSL) supernatants, which strongly infers participation in cross-species communication. Despite the fact that the

majority of the induced spots could not be positively identified or attributed to any known biological function, those that were identified (with some level of confidence) collectively represented a stress response mounted by the bacterium. This conclusion contradicts the original hypothesis where AHSL and AI-2 were thought to the enhance viability of the microorganism. However, this response, which includes the macrophage-induced protein, heat shock protein, universal stress protein, and cold shock proteins, constitutes a set of adverse reactions or stresses rather then the benefit of upregulation of survival or virulence gene expression that is commonly seen in other bacteria. The expression of these genes with regard to heterospecific signals suggests alternative roles for these proteins in a stress response induced by the presence of competitors rather than environmental conditions. In that supernatants were used rather than purified compounds, the response cannot be solely limited to AHSL or AI-2 and could represent a reaction by *L. pneumophila* to a metabolic compound or waste product in the supernatant; however, a consistent response was noted with the use of purified, synthetic AHSL rather than AHSL-laden supernatants (data not shown). Thus, does the detection of AI-2 or AHSL present a significant advantage to *L. pneumophila* in the environment and/or during the genesis of Legionnaire's disease?

The identification of species-specific communication among legionellae remains elu-

TABLE 1 Partial list of proteins identified through MALDI and MASCOT analysis

Spot no.	Protein identification based on peptide mass	MACSOT score[a]
3	24-kDa macrophage-induced protein	43
4	24-kDa macrophage-induced protein	91
7	Hypothetical protein 1p10966; lpp0997; UspA	35
10	Hypothetical protein *atpH*; HspC2	73, 44
13	Hypothetical protein 1pp1207; 1p11213; CspA	110, 76, 92
14	Putative DNA invertase (*P. aeruginosa*)	45
16	Hsp10	56
17	10k-Da chaperonin GroES	61
18	Hypothetical protein 1p10966; 1pp0997; UspA	63

[a]A score of ≥73 is considered a significant match.

sive. However, additional proteomic data has extended the concept of interspecies communication in that *L. pneumophila* can also perceive its host (protozoa). Thus, much work remains to identify potential signal receptors and the genes comprising a proposed AHSL and AI-2 regulon. If such machinery exists, then to what extent can this information be used to control the presence of legionellae in the environment? In summary, the data show that *L. pneumophila* can sense AHSL, AI-2, and/or other intercellular communication signals.

REFERENCES

1. **Davies, D. G., M. R. Parsek, J. P. Pearson, B. H. Iglewski, J. W. Costerton, and E. P. Greenberg.** 1998. The involvement of cell-to-cell signals in the development of a bacterial biofilm. *Science* **280:**295–298.

2. **Duan, K., C. Dammel, J. Stein, H. Rabin, and M. G. Surette.** 2003. Modulation of *Pseudomonas aeruginosa* gene expression by host microflora through interspecies communication. *Mol. Microbiol.* **50:**1477–1491.

3. **Lebeau, I., E. Lammertyn, E. De Buck, L. Maes, N. Geukens, L. Van Mellaert, L. Arckens, J. Anne, and S. Clerens.** 2004. First proteomic analysis of *Legionella pneumophila* based on its developing genome sequence. *Res. Microbiol.* **256:**119–129.

4. **Miller, M. B., and B. L. Bassler.** 2001. Quorum sensing in bacteria. *Annu. Rev. Microbiol.* **55:** 165–199.

5. **Wadowsky, R. M., and R. B. Yee.** 1985. Effects on non-*Legionellaceae* bacteria on the multiplication of *Legionella pneumophila* in potable water. *Appl. Environ. Micobiol.* **49:**1206–1210.

6. **Westermeier, R., and T. Naven.** 2002. *Proteomics in Practice: A Laboratory Manual.* Wiley-VCH.

ANTIMICROBIAL ACTIVITY OF SOME LICHEN EXTRACTS AGAINST *LEGIONELLA PNEUMOPHILA*

Zuhal Zeybek, Nihal Doğruöz, Ayşın Çotuk,
Ali Karagöz, and Ali Aslan

95

Humans have frequently used plants to treat common infectious diseases, and some of these traditional medicines are still part of the habitual treatment of various maladies. It has been reported that 115 articles were published on the antimicrobial activity of medicinal plants in Pub Med during the period between 1966 and 1994, but in the following decade, between 1995 and 2004, 307 were published (12). Lichens are symbiotic organisms of fungi (mycobionts) and algae (phycobionts), comprising about 17,000 species (5). They have long been used in folk medicine in many countries. It has been determined that they show the inhibition effect against many bacteria such as *Bacillus, Pseudomonas, Escherichia coli, Streptococcus, Staphylococcus, Enterococcus,* and *Mycobacterium* (2, 3, 6, 10). However, nothing has been reported about whether lichens have antimicrobial activity against *Legionella pneumophila*. It is known that *L. pneumophila* has been found in natural waters and man-made water systems, and they cause Legionnaires' disease (13). This study in-

vestigated antibacterial activity of some lichens against *L. pneumophila* serogroup (sg) 1 and *L. pneumophila* sg 2 to 14 strains in this study.

The lichen specimens (L1 to L11) used (Table 1) were collected from Artvin, Giresun, and Trabzon provinces, Turkey, in 2003. Samples were dried at room temperature for 48 h and identified according to the method developed by Poelt and Vězda (11). Two separate samples of air-dried and powdered plant material (10 g of each) were extracted with distilled water (80 ml) and 70% ethanol (75 ml) at room temperature using a Waring blender. Plant residues were removed by centrifugation (15,000 rpm, 30 min, 4°C), and the supernatant was filtered and evaporated to dryness under reduced pressure and/or lyophilized. In this way, two different crude extracts were obtained: aqueous extract (AE) and 70% ethanolic extract (EE). AE was dissolved in the medium (Eagle's minimum essential medium, Gibco, EMEM) and EE was dissolved in dimethyl sulfoxide. They were then prepared at 250 µg/ml final concentrations in EMEM. All environmental *L. pneumophila* strains used (B, C, D, and E) that were isolated from our country were grown on buffered charcoal yeast extract agar supplemented with glycine, vancomycin, polymixin, and cycloheximide (7, 13). The inhibition effects of lichen AE and EE

Zuhal Zeybek, Nihal Doğruöz, and Ayşin Çotuk Department of Biology, Istanbul University, Science Faculty, Vezneciler, Istanbul, Turkey 34120. *Ali Karagöz* Department of Molecular Biology and Genetics, Istanbul University, Science Faculty, Vezneciler, Istanbul, Turkey 34120. *Ali Aslan* Department of Biology, Atatürk University, Education Faculty, Erzurum, Turkey, 25240.

TABLE 1 Lichen specimens used for testing antimicrobial activity against *L. pneumophila* strains

Code	Scientific names
L1	*Anaptycia ciliaris*
L2	*Peltigara praetextata*
L3	*Xanthoparmelia tinctina*
L4	*Umbilicaria vellea*
L5	*Peltigara polydactyla*
L6	*Ramalina farinacea*
L7	*Xanthoria elegans*
L8	*Rhizoplaca melanopthalma*
L9	*Cetrelia olivetorum*
L10	*Xanthoria elegans*
L11	*Lecanora muralis*

against *L. pneumophila* sg 1 ATCC 33152 (A), sg 1 (C, E), and sg 2 to 14 (B, D) were performed according to the agar well diffusion method by using buffered charcoal yeast extract agar (1). Wells 6 mm in diameter were punched into the agar and filled with AE, EE, and solvent blanks (distilled water). The plates were then incubated at 37°C for 3 days. The antimicrobial activity was evaluated by measuring the inhibition-zone diameter observed. Also for confirmation, AE of two lichen specimens (L2, L10) were tested against A and B strains for determination of minimum inhibitor concentration and minimum bactericidal concentration (8, 9).

It was found that the antimicrobial activity of the AE was greater than 70% of the activity of EE (Fig. 1 and Fig. 2). The average inhibition zone of AE was determined to be 13 mm (min: 0, max: 27); that of EE was found to be 2.6 mm (min: 0, max: 20). To evaluate the minimum inhibitor and minimum bactericidal concentrations, the upper 15 mm of AE were tested, and it was found that these values were equal and were between 31.25 and 500 μg/ml.

Lichens now constitute important sources for prospecting for new bioactive molecules,

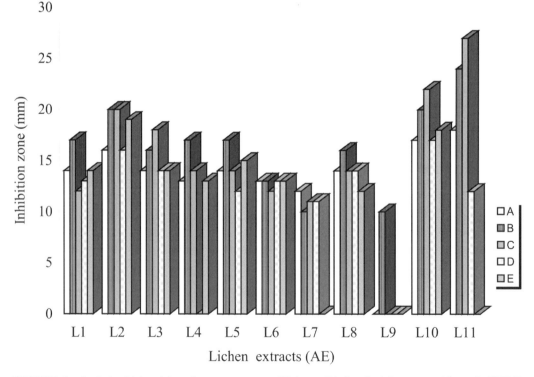

FIGURE 1 Antimicrobial activity of aqueus extract of lichens. (A) Standard *L. pneumophila* sg 1 (ATCC 33152); (B and D) environmental *L. pneumophila* sg 2 to 14; (C and E) environmental *L. pneumophila* sg 1.

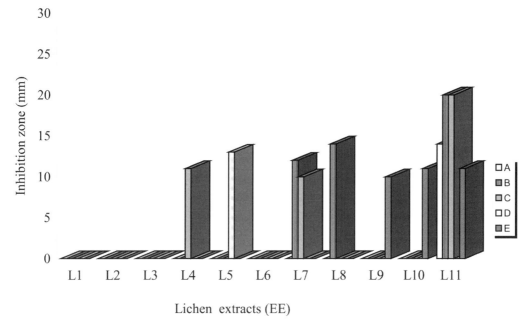

FIGURE 2 Antimicrobial activity of 70% ethanolic extract of lichens. (A) Standard *L. pneumophila* sg 1 (ATCC 33152); (B and D) environmental *L. pneumophila* sg 2 to 14; (C and E) environmental *L. pneumophila* sg 1.

either by the direct use of their secondary metabolites or by employing their biyosynthetic or semisynthetically derived compounds, which are produced with the aim of attaining higher effectiveness, improved absorption, or even decreased toxicity (4).

It has been determined that AEs of lichen have inhibition effects against tested *L. pneumophila* strains. This effect was greater than their EE effect. In contrast to our findings, some investigators reported that the AE of some lichens has inhibition effects against other bacteria such as *E. coli, Salmonella, Bacillus, Staphyloccoccus, Pseudomonas,* and *Listeria* (2, 3). Moreover, Ahmad et al. determined that while the EEs of some lichens have antimicrobial activity, their AEs do not have this effect.

Also, it has been considered that these results can change according to *L. pneumophila* strains and lichen species. Further studies are required on purifying secondary metabolites of lichens and typing of *L. pneumophila* strains on the molecular level to understand the reasons for these different effects.

The establishment disinfection procedure may not prevent the occurrence of *Legionella* in the water system. It is known that Legionnaires' disease frequently appears because of contaminated water systems and it can be treated by antibiotics. However, there are some disadvantages to the use of antibiotics (cost, endotoxic shock, side effects, multiple resistance). Certainly, prevention is more important than the treatment of illness. For this reason we consider that lichens and/or their metabolits may be used to improve building water systems in the future.

REFERENCES
1. **Ahmad, I., Z. Mehmood, and F. Mohammad.** 1998. Screening of some Indian medicinal plants for their antimicrobial properties. *J. Ethnopharm.* **62:**183–193.
2. **Aslan, A., M. Güllüce, and E. Atalan.** 2001. A study of antimicrobial activity of some lichens. *Bull. Pure Appl. Sci.* **20:**23–26.
3. **Esimone, C. O., and M. U. Adikwu.** 1999. Antimicrobial activity and cytotoxicity of *Ramalina farinaceae. Fitoterapia* 428–431.

4. **Gomes, A. T., A. S. Júnior, C. Siedel, E. F. A. Smania, N. K. Honda, F. M. Roese, and R. M. Muzzi.** 2003. Antibacterial activitiy of orsellinates. *Br. J. Microbiol.* **34:**194–196.

5. **Huneck, S.** 1999. The significance of lichens and their metabolites. *Naturwissenschaften* **86:**559–570.

6. **Ingólfsdóttir, K.** 2002. Usnic acid. *Phytochemistry* **61:**729–736.

7. **International Organization for Standardization (ISO).** 1998. Water quality detection and enumeration of Legionella. ISO 11731: 1998 (E).

8. **National Committee for Clinical Laboratory Standards.** 1997. *Methods for Determining Bactericidal Activity of Antimicrobial Agents.* Proposed Guideline. NCCLS document M26-P Villanova.

9. **National Committee for Clincal Laboratory Standards.** 2000. *Methods for Dilution Antimicrobisceptibility Tests for Bacteria that Grow Aerobically,* 5th ed. Approved Standard NCCLS, Document M7-A5, Wayne, Pennsylvania.

10. **Perry, N. G., M. H. Benn, N. J. Brennan, E. J. Burgess, G. Ellis, D. J. Galloway, S. T. Lorimer, and R. S. Tangney.** 1999. Antimicrobial, antiviral, and cyctotoxic activity of New Zealand lichens. *Lichenologist* **31:**627–636.

11. **Poelt, J., and A. Vĕzda.** 1981. Bestimmungsschlüssel Europäisher Flechten. Ergänzungsheft II. *Bibliotheca Lichenologica* **16:**1–390.

12. **Rios, J. L., and M. C. Recio.** 2005. Medicinal plants and antimicrobial activity. *J. Ethnopharmacol.* **100:**80–84

13. **Zeybek, Z., A. Kimiran, and A. Çotuk.** 2003. Distribution of bacteria causing Legionnaires disease at potable water in İstanbul. *Biol. Bratislava* (58) 6.

FIRST REPORT OF AN ANTI-*LEGIONELLA* PEPTIDE PRODUCED BY *STAPHYLOCOCCUS WARNERI*

Yann Héchard, Sébastien Ferraz, Emilie Bruneteau, Michael Steinert, and Jean-Marc Berjeaud

96

Legionella pneumophila is found in freshwater environments associated with biofilms and free-living amoebae (5, 8). This gram-negative bacterium is responsible for a severe bacterial pneumonia called Legionnaires' disease (3). People become infected with *L. pneumophila* after inhaling aerosols of contaminated water droplets (6). The aerosols may be generated by different contaminated devices (e.g. cooling towers, showers). In case of high contamination, treatment such as thermal disinfection, UV irradiation, use of chlorine and derivatives, or metal ionization might be necessary (4). However, these treatments are not fully efficient, and after a while *L. pneumophila* might be found in the treated systems. This ability to resist various treatments has been explained by the intracellular growth of *L. pneumophila* in amoebae (1). The latter are considered to protect the bacteria from treatment, and amoeba-grown bacteria have been shown to be more resistant to chemical disinfectants and treatments with biocides (2). Therefore, it is important to find new

molecules that might be useful for disinfection procedures against *Legionella*. Bacteriocins are antibacterial proteins produced by bacteria (7). Many bacteriocins have been described to be produced by lactic acid bacteria and be active against foodborne pathogens, such as *Listeria monocytogenes* or *Clostridium botulinum*. To our knowledge, no bacteriocin against *Legionella* has been described so far.

Our study described the partial characterization of a new bacteriocin, produced by *Staphylococcus warneri* RK, active against *L. pneumophila*.

CHARACTERIZATION OF THE PRODUCING STRAIN

A bacterial strain, isolated from the environment, displayed an anti-*Legionella* activity by a spot on lawn assay (Fig. 1). The ID 32 Staph and the Vitek galleries were used to identify this strain; they both resulted in a reliable identification of the *S. warneri* species. The strain was named *S. warneri* RK.

CHARACTERIZATION OF THE ANTI-*LEGIONELLA* MOLECULE

In order to characterize the anti-*Legionella* molecule, raw extracts (see above) have been submitted to various proteases (proteinase K 1 mg ml^{-1} or trypsin 1 mg ml^{-1} for 1 h at 37°C) and to heat for 15 min at 70°C, 15 min at

Yann Héchard, Sébastien Ferraz, Emilie Bruneteau, and Jean-Marc Berjeaud   Equipe de Microbiologie, Laboratoire de Chimie de l'Eau et de l'Environnement, UMR CNRS 6008, Université de Poitiers, 40 avenue du recteur Pineau, 86022 Poitiers, France.   *Michael Steinert*   Institut für Molekulare Infektionbiologie, Universität Wurzburg, 97070 Würzburg, Germany.

FIGURE 1 Spot on lawn assay. *L. pneumophila* Lens was spread on BCYE plate before spotting 5 μl of *S. warneri RK* culture. After 3 days of incubation at 37°C, a zone of inhibition of *Legionella* growth was noticed around the *S. warneri RK* colony.

TABLE 1 Antibacterial spectrum of *S. warneri* raw extracts, assessed by well diffusion assays

Bacterial strain	Sensitivity
Bacillus megaterium F04[a]	−
Enterobacter cloacae D03[a]	−
Enterococcus faecalis V583	−
Escherichia coli E01[a]	−
Hafnia alvei 1[a]	−
Legionella bozemanae ATCC 33217	+
Legionella dumoffii ATCC33279	+
Legionella longbeachae ATCC 33484	+
Legionella micdadei ATCC 33218	+
Legionella pneumophila Corby	+
Legionella pneumophila Lens	+
Legionella pneumophila Paris	+
Leuconostoc mesenteroides Y105[a]	−
Listeria monocytogenes EGDe	−
Pseudomonas aeruginosa B06[a]	−
Serratia liquefaciens 6[a]	−
Staphylococcus aureus A15[a]	−
Staphylococcus epidermidis A26[a]	−

[a]Laboratory collection

90°C, or 15 min at 121°C by autoclaving. The protease treatments led to a total loss of inhibitory activity, showing that the anti-*Legionella* molecule is proteinaceous. The heat treatments at 70°C and 90°C did not affect the inhibitory activity, whereas the treatment to 121°C led to a 50% decrease. These results suggest that the molecule might be a peptide rather than a protein, since peptides are less sensitive to high-temperature treatments, mainly because they are able to spontaneously refold.

ANTI-BACTERIAL SPECTRUM

Various bacterial strains, gram positive and gram negative, were used as target organisms to test their sensitivity to *S. warneri* RK by a spot on lawn assay. The results are described in Table 1. All the *Legionella* strains tested, from different species, were sensitive. To the contrary, none of the other strains were sensitive. These results underline that the anti-*Legionella* molecule is highly specific to the *Legionella* genus.

PURIFICATION

S. warneri RK has been cultivated in brain heart infusion at 37°C for 24 h. The culture supernatant has been heated at 70°C for 15 min, leading to the raw extract. The anti-*Legionella* molecule was then purified from the raw ex-

tract during a three-step purification procedure. First, the raw extract was diluted (50:50) with sodium acetate buffer (20 mM, pH 5) and applied to a weak cation exchange chromatography column (Hiprep 16/10 Carboxy-Methyl FF, Amersham Biosciences). The column was washed successively with 100 ml of sodium acetate buffer (20 mM, pH 5) containing 0, 0.1, and 1 M NaCl. The anti-*Legionella* peptide was eluted with 1 M NaCl. Second, the latter fraction was applied onto a solid phase extraction C18 cartridge (Sep-pak plus, Waters), washed successively with 5 ml of 0, 10, 20, 30, 40, and 80% acetonitrile containing 0.1% trifluoroacetic acid. Each fraction was concentrated under vacuum, lyophilized, and resuspended in 1 ml of water. Third, the active fraction (80% acetonitrile) was lyophilized, solubilized with 1 ml of 40% acetonitrile, and injected on a Kromasil C8 reverse-phase HPLC analytical column (5 μm, 100 Å, 4.6 × 250 mm, A.I.T.). Separation was carried out using a water/acetonitrile/trifluoroacetic acid 0.1% (v/v) solvent system. After an initial 5-min wash with

20% acetonitrile, elution was achieved in 40 min at a flow rate of 0.8 ml min^{-1} with a 30-min linear gradient from 20 to 100% acetonitrile, followed by a 10-min wash with 100% acetonitrile. After evaporation under vacuum and lyophilization, each fraction was resuspended in 1 ml of water and tested for anti-*Legionella* activity.

The anti-*Legionella* activity has been eluted, during the last step, at 100% acetonitrile, indicating that the corresponding peptide is highly hydrophobic. This anti-*Legionella*-active fraction was subjected to mass spectrometry analysis, giving two molecular masses of 2613.8 Da and 2449.4 Da. This result shows that two peptides were present in the fraction, one of which likely corresponded to the anti-*Legionella* bacteriocin. Finally, database screening revealed that the molecular mass of none of the previously described bacteriocins corresponds to the molecular masses of these peptides. The anti-*Legionella* molecule is therefore an original peptide.

CONCLUSION

We have described the first anti-*Legionella* peptide, whose activity seems restricted to the *Legionella* genus. This spectrum of activity is intriguing since bacteriocins are usually directed toward species related to the producer. Not only was it not active against the closest strains we have tested, but it was active against *Legionella*, which is a gram-negative genus. This bacteriocin needs to be fully characterized and sequenced to initiate genetic studies. The mode of action is also worth exploring since this bacteriocin might have an original mode of action and be used as an alternative treatment against *Legionella*.

ACKNOWLEDGMENTS

We acknowledge Claire Souchaud and Silke Hammer for technical help.

The work was supported by the Deutsche Forschungsgemeinschaft DFG (SFB 630).

REFERENCES

1. **Abu Kwaik, Y., L. Y. Gao, B. J. Stone, C. Venkataraman, and O. S. Harb.** 1998. Invasion of protozoa by *Legionella pneumophila* and its role in bacterial ecology and pathogenesis. *Appl. Environ. Microbiol.* **64:**3127–3133.
2. **Barker, J., M. R. Brown, P. J. Collier, I. Farrell, and P. Gilbert.** 1992. Relationship between *Legionella pneumophila* and *Acanthamoeba polyphaga*: physiological status and susceptibility to chemical inactivation. *Appl. Environ. Microbiol.* **58:**2420–2425.
3. **Fields, B. S., R. F. Benson, and R. E. Besser.** 2002. Legionella and Legionnaires' disease: 25 years of investigation. *Clin. Microbiol. Rev.* **15:**506–526.
4. **Kim, B. R., J. E. Anderson, S. A. Mueller, W. A. Gaines, and A. M. Kendall.** 2002. Literature review: efficacy of various disinfectants against *Legionella* in water systems. *Water Res.* **36:**4433–4444.
5. **Molofsky, A. B., and M. S. Swanson.** 2004. Differentiate to thrive: lessons from the *Legionella pneumophila* life cycle. *Mol. Microbiol.* **53:**29–40.
6. **Muder, R. R., V. L. Yu, and A. H. Woo.** 1986. Mode of transmission of *Legionella pneumophila*. A critical review. *Arch. Intern. Med.* **146:**1607–1612.
7. **Riley, M. A., and J. E. Wertz.** 2002. Bacteriocins: evolution, ecology, and application. *Annu. Rev. Microbiol.* **56:**117–137.
8. **Steinert, M., U. Hentschel, and J. Hacker.** 2002. *Legionella pneumophila*: an aquatic microbe goes astray. *FEMS Microbiol. Rev.* **26:**149–162.

OCCURRENCE AND DIVERSITY OF *LEGIONELLA PNEUMOPHILA* IN WATER SAMPLES FROM THE BRAZILIAN ENVIRONMENT

Fábio R. S. Carvalho, Annette S. Foronda, and Vivian H. Pellizari

97

Legionella species, well known for their role as pathogenic bacteria causing respiratory disease, are commonly found in natural and man-made water systems. Despite the importance of the surveillance of these bacteria in aquatic systems, few studies have looked at Brazilian environmental samples. *Legionella pneumophila* was first isolated in Brazil by Pereira Gomes and collaborators (5). Pellizari and Martins demonstrated the presence of *Legionella pneumophila* in the boiler and cooling tower of a hospital and public buildings from Sao Paulo (4). Recently, Chedid and collaborators (1) described the occurrence of a community-acquired pneumonia in three patients who were hospitalized in a southern Brazilian general university hospital. As a result of the few data about the presence and diversity of *Legionella* in Brazil, there has been a lack of awareness regarding the health risks of *Legionella* contamination of water distribution systems. In this study, we attempted to detect legionellae from the environment, as man-made systems and natural reservoirs, using different methods to isolate and identify this bacterium.

The diversity of sources and degree of bacterial contamination limit the use of a single method for the recovery of legionellae from environmental samples. Also, the rapid and credible typing of *Legionella* species is very important for assessing community outbreaks and locating sources of infection. In general, *Legionella* species are isolated by culture onto a selective medium buffered charcoal-yeast extract (BCYE) supplemented with L-cysteine and ferric pyrophosphate. In addition, direct fluorescence antibody (DFA) stain and coculture with axenic amoeba are important methods to detect legionellae at low concentrations. In this case, these procedures can be applied in environments with nutritional restriction and submitted to biocide treatment or extreme ranges of pH and temperature. In this study, these techniques were used to screen the occurrence and diversity of *Legionella* strains from the aquatic environment in Brazil. Water samples were collected from the main reservoirs of Sao Paulo, dental units, and cooling towers and boilers from hospitals and industries. A volume of 1-liter from each sample was concentrated by filtration through a 0.45-μm-pore-size polycarbonate membrane, which was submitted to sonication and pretreatment at 50°C for 30 min. After these processes, 0.1 mL of each concentrated sample was plated on αBCYE-GVP (selective medium with glycine, vancomycin, and polymyxin B). Plates were incubated at 37°C for 10 days in a humid

Fábio R. S. Carvalho, Annette S. Foronda, and Vivian H. Pellizari
ICB University of Sao Paulo, São Paulo, SP, Brazil, 05508-900.

chamber. A coculture procedure with axenic culture of *Acanthamoeba castellanii* ATCC 30011 (2) and Direct fluorescent antibody (DFA) stain using two polyvalent fluorescein isothiocyanate-labeled rabbit anti-*Legionella* conjugate pools (*L. pneumophila* serogroups 1 to 14 and non-*pneumophila* species, m-Tech, Atlanta, Ga.) were also applied to concentrated water samples. Colonies grown on BCYE-GVP, but not in the same medium without supplements, were considered "suspected colonies." DNA extraction of the isolates was done for amplification and sequencing the macrophage infectivity potentiator (*mip*) gene according to the protocol described by Ratcliff and collaborators (6).

Legionella species were not isolated from the natural reservoirs analyzed. However, *Legionella* species were detected in 18.36% of the samples evaluated by the different methods. Comparison of the three methods as regards their ability to detect legionellae in the samples studied is shown in Table 1. Culturing is generally accepted as the gold standard for *Legionella* detection from environmental samples. Although the DFA staining does not provide information on cell viability it was useful for the detection and confirmation of *L. pneumophila*. Coculture of sample concentrates with free-living amoebae can improve the sensitivity of culturing and allow detection of *Legionella*-like amoeba pathogens (LLAPs). Nevertheless, in this study we do not detect higher numbers of positive samples by the coculture technique.

The nine isolates obtained in this study were identified as *L. pneumophila* through the sequencing and comparison of the *mip* gene sequences deposited in the GenBank database. Phylogenetic analyses showed that this gene appeared to be relatively stable genetically, with no evidence of homologous recombinants. The phylogenetic tree obtained from the sequences of nine isolates from this study (Fig. 1) showed the presence of three distinct clusters and the diversity of these *L. pneumophila* strains (BR strains). Despite the diversity shown in the phylogenetic tree, the sequence analyses of the *mip* gene were not able to distinguish the differences between *L. pneumophila* serogroups 8 and 12 clustered with strain BR-02 and BR-04.

Detection of legionellae in environmental samples by using PCR and gene probes in conjunction with DFA staining was previously proposed (3). In addition, Sanden and collaborators (7) reported that coculture of water samples with free-living amoebae for several days increased the recovery of *Legionella* spp., which was related to proliferation of amoebae in the samples. Despite the different techniques applied in this study, we were able to detect positive results for the presence of *Legionella*. We agree with the importance of the use of different strategies for these screenings.

TABLE 1 Occurrence of *Legionella* spp. in environmental samples analyzed by different detection methods

Sources	No. of samples	No. of samples positive for *Legionella* spp.		
		Culture on selective medium	DFA staining[a]	Coculture with *A. castellanii*
Natural water reservoir	4	0	0	0
Dental units	10	0	0	0
Cooling towers	13	2	2	2
Water tank reservoir to cooling tower	2	2	2	2
Hot water tank	1	1	1	1
Evaporative condensers	4	0	0	0
Cold water tank reservoir	2	0	0	0
Shower heads	31	4	4	4
Total	67	9	9	9

[a]Applied directly to the concentrated water sample.

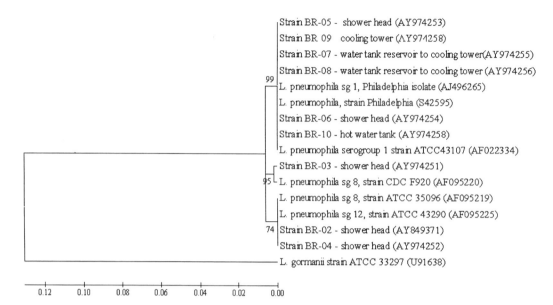

FIGURE 1 Phylogenetic tree showing distances based on the *mip* gene sequences (599 bp) among nine strains obtained in this work and sequences from GenBank (accession numbers in parentheses). The dendrogram was generated by using programs from MEGA software packages version 3.0. Bootstrap analyses were applied to the data, and the values obtained from 10^5 replications are represented at internal nodes.

This study showed that *L. pneumophila* was present in some of the man-made systems analyzed, but this organism was not isolated from the related natural reservoirs that supply the systems. The possible presence of cells in a viable-but-not-culturable state is known. For these reason cultivation-independent methods for screening of *Legionella* are recommended for environmental samples.

REFERENCES

1. **Chedid, M. B. F., D. O. Ilha, M. F. Chedid, P. R. Dalcin, M. Buzzetti, P. J. Saraiva, D. Griza, and S. S. M. Barreto.** 2005. Community-acquired pneumonia by *Legionella pneumophila* serogroups 1-6 in Brazil.

2. **La Scola, B., L. Mezi, P. J. Weiller, and D. Raoult.** 2001. Isolation of *Legionella anisa* using an amoebic coculture procedure. *J. Clin. Microbiol.* **39:**365–366.

3. **Paszko-Kolva, C., H. Yamamoto, M. Shahamat, and R. R. Cowel.** 1993. Polymerase chain reaction, gene probe, and direct fluorescent antibody staining of *Legionella pneumophila* serogroup 1 in drinking water and environmental samples, p. 181–183. *In* J. M. Barbaree, R. F. Breiman, and A. P. Dufour (ed.). *Legionella: Current Status and Emerging Perspectives.* ASM Press, Washington, D.C.

4. **Pellizari, V. H., and M. T. Martins.** 1995. Occurrence of *Legionella* sp in water samples from man-made systems of Sao Paulo—Brazil. *Rev. Microbiol.* **26:**186–191.

5. **Pereira Gomes, J. C., N. A. Mazieri, C. V. Godoy, and S. R. Ados.** 1990. *Legionella pneumophila* associated with acute respiratory insufficiency: first isolation in Brazil. *Rev. Inst. Med. Trop. Sao Paulo* **31:**368–376.

6. **Ratcliff, R. M., J. A. Lanser, P. A. Manning, and M. W. Heuzenroeder.** 1998. Sequence-based classification scheme for the genus *Legionella* targeting the mip gene. *J. Clin. Microbiol.* **36:**1560–1567.

7. **Sanden, G. N., W. E. Morril, B. S. Fields, R. F. Breiman, and J. M. Barbaree.** 1992. Incubation of water samples containing amoebae improves detection of legionellae by the culture method. *Appl. Environ. Microbiol.* **58:**2001–2004.

DIVERSITY OF *LEGIONELLA* SPP. IN ANTARCTIC LAKES OF THE KELLER PENINSULA

Fábio R. S. Carvalho, Fernando R. Nastasi, Rosa C. Gamba, Annette S. Foronda, and Vivian H. Pellizari

98

The ubiquity of *Legionella* species has been reported in different aquatic ecosystems and in a wide range of temperatures (3). The use of culture-independent molecular screening techniques has allowed the description of *Legionella* as a member of the microbial community structure of extreme environments such as polar regions (1, 8). Clone sequences (16S rRNA gene) related to different species of *Legionella* such as *L. rubrilucens* have been identified in Antarctic lakes (1). However, there are no descriptions of isolation of this genera in the Antarctic continent. In this work, we used traditional culture and coculture methods to isolate *Legionella* spp. and independent-culture methods to verify and compare the diversity of the *Legionellaceae* population in the microbial communities of two Antarctic lakes.

Water samples (5 liters) were collected from North and South Lakes, near the Brazilian Scientific Station Comandante Ferraz, located in the Keller Peninsula, King George Island, Antarctica. These samples were transported to the laboratory and concentrated through membrane filtration (0.4-μm-pore-size polycarbonate filters). The filters were placed into 10 ml of each original water sample and shaken vigorously for 1 min. Each sample (0.1 ml) was spread in triplicate directly on buffered charcoal-yeast extract medium (BCYE) containing α-ketoglutarate, L-cysteine, ferric pyrophosphate, glycine, vancomycin, and polymyxin B (αBCYE-GVP) and incubated at room temperature for 10 days. Colonies with ground-glass morphology were subcultured on α-BCYE-GVP and α-BCYE lacking L-cysteine, ferric pyrophosphate, and antimicrobial agents. Total DNA was extracted (QIAamp DNA kit, Qiagen, Hilden) from concentrated lake water samples. Bacterial community structure was analyzed by PCR-denaturing gradient gel electrophoresis (PCR-DGGE) through amplification of the V3 region from the 16S rRNA gene using 338-forward (GC-clamp) and 518-reverse primers (6). In addition, *Legionellaceae* diversity from both Antarctic lakes was analyzed by the amplification and generation of a clone library from an internal fragment of 16S rRNA gene using primers LEG-448 and LEG-85 (7) as described by Calvo-Brado et al. (2). Phylogenetic analyses were conducted using MEGA version 3.0 (4).

Colonies were obtained only from the North Lake samples. Selected colonies were used for total genomic DNA extraction. We observed that subculturing these colonies was only possible in coculture with the axenic free-living

Fábio R. S. Carvalho, Fernando R. Nastasi, Rosa C. Gamba, Annette S. Foronda, and Vivian H. Pellizari ICB University of Sao Paulo, São Paulo, SP, Brazil, 05508-900.

Legionella: State of the Art 30 Years after Its Recognition
Edited by Nicholas P. Cianciotto et al.
©2006 ASM Press, Washington, D.C.

amoeba *Acanthamoeba castellanii* (ATCC 30011) as described by Moffat and Tompkins (5).

DGGE analysis from the water samples showed 28 to 30 bands representing abundant and similar phylotypes between the bacterial structures of both lakes (Fig. 1). The comparison of DGGE patterns among environmental samples, culture, and positive controls (ATCC strains) showed that the *Legionellaceae* population is a member of the microbial community of both Antarctic lakes. In addtion, we confirmed that our cultures were not pure. It is interesting to note the presence of few dominant bands from the cultures in the lower portion of the gels compared to the great number of dominant DGGE bands from North and South Lakes observed in the upper portion of the gels.

In addition, sequence analysis of the 16S rRNA gene clone library showed >90% sequence identities between clones and GenBank sequences, i.e., *L. pneumophila*, *L. lytica*, *L. jeonii*, *L. erythra*, and some strains of *Legionella*-like amoebal pathogens (LLAP-10, LLAP-07,

FIGURE 1 Silver stained DGGE-PCR gel (V3 variable region from 16S rRNA gene). (A through E) *Legionella* strains ATCC 33152, 33153, 33154, 33155, and 33156; (F and G) strains isolated from North Lake; (H and I) total bacterial communities from the North and South Lakes, respectively. White arrows indicate the comparison between bands of ATCC strains, isolated strains, and microbial communities analyzed.

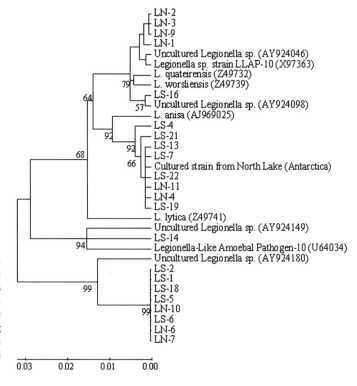

FIGURE 2 Unrooted phylogenetic tree of reference strains, culture obtained in this study, and clones from 16S rRNA gene from North Lake (LN) and South Lake (LS). The dendrogram was generated by using MEGA software package version 3.0. Bootstrap analyses were applied to the data (*n* = 10,000).

LLAP-03, and LLAP-01). Sequences related to an LLPA were previously described in the Antarctic bacterial community composition in a benthic saline pond from the McMurdo Ice Shelf in Ross Sea (9). It is interesting to note that LLAP clones found in South Lake and North Lake Antarctic samples showed similar identities between LLAP sequences (90.3 to 96.9%) reported in a geothermal stream in Yellowstone National Park, suggesting that these strains are present in a wide range of temperatures like other *Legionella* species (8). Phylogenetic inference of these results showed the presence of five main clusters, where three of them showed that North Lake and South Lake clones could be clustered within the same taxon, with the exception of clones LN-11 and LS-4, which were located in single clusters (Fig. 2). These results indicate that the majority of our clones are still unknown.

The DGGE technique demonstrated that *Legionella* is a significant member of the community in both lakes. Such findings could be related to the presence of ferric ions previously determined in the studied areas of Keller Peninsula. Despite the presence of this growth factor, the need of coculturing using free-living amoebae suggests the occurrence of an important symbiotic relationship in Antarctic lakes. The symbiotic association of *Legionella* species and amoebae might indicate the importance of natural hosts in the process of resistance and development of this bacterium in extreme environments, as reported for a pH 2.7 geothermal stream environment (8). The isolation of *Legionella* from these Antarctic lakes will allow future studies in cold-resistence mechanisms of mesophilic bacteria in polar environments.

REFERENCES

1. **Brambilla, E., H. Hippe, A. Hagelstein, B. J. Tindall, and E. Stackenbrandt.** 2001. 16S rDNA diversity of cultured and uncultured prokaryotes of a mat sample from Lake Fryxell, McMurdo Dry Valleys, Antarctica. *Extremophiles* 5:23–33.
2. **Calvo-Brado, L. A., J. A. W. Morgan, M. Sergeant, T. R. Pettitt, and J. M. Whipps.** 2003. Molecular characterization of *Legionella* populations present within slow sand filters used for fungal plant pathogen suppression in horticultural crops. *Appl. Environ. Microbiol.* 69:533–541.
3. **Fliermans, C. B., W. B. Cherry, L. H. Orrison, S. J. Schimdt, D. L. Tison, and D. H. Pope.** 1981. Ecological distribution of Legionella pneumophila. *Appl. Environ. Microbiol.* 41:9–16.
4. **Kumar, S., K. Tamura, and M. Nei.** 2004. MEGA3: integrated software for molecular evolutionary genetics analysis and sequence alignment. *Briefings Bioinformatics* 5:150–163.
5. **Moffat, J. F., and L. S. Tompkins.** 1992. A quantitative model of intracellular growth of *Legionella pneumophila* in *Acanthamoeba castellanii*. *Infect. Immun.* 60:296–301.
6. **Muyzer, G., E. C. De Waal, and A. G. Uitterlinden.** 1993. Profiling of complex microbial populations by denaturing gradient gel electrophoresis analysis of polymerase chain reaction-amplified genes coding for 16S rRNA. *Appl. Environ. Microbiol.* 59:695–700.
7. **Myiamoto, H., H. Yamamoto, K. Arima, J. Fuji, K. Maruta, K. Izu, T. Shiomori, and S. Yoshida.** 1997. Development of a new semi-nested PCR method for detection of *Legionella* species and its application to surveillance of legionellae in hospital cooling tower water. *Appl. Environ. Microbiol.* 63:2489–2494.
8. **Sheehan, K. B., J. M. Henson, and M. J. Ferris.** 2005. *Legionella* species diversity in an acidic biofilm community in Yellowstone National Park. *Appl. Environ. Microbiol.* 71:507–511.
9. **Sjoling, S., and D. A. Cowan.** 2003. High 16SrDNA bacterial diversity in glacial meltwater lake sediment, Bratina Island, Antarctic. *Extremophiles* 7:275–282.

EFFECTS OF SEAWATER CONCENTRATION AND TEMPERATURE ON THE SURVIVAL OF *LEGIONELLA PNEUMOPHILA* SEROGROUP 1

Susan Bennett and Richard Bentham

99

Legionellae are a predominantly aquatic family of bacteria commonly associated with water systems in the built environment. Their presence in engineered water systems has been associated with cases of Legionnaires' disease on numerous occasions. There are substantially fewer reports of the presence and survival of *Legionella* in the marine environment. The presence of *Legionella* in coastal marine environments has been demonstrated (4). The salinity of estuarine and coastal waters may be variable due to terrestrial runoff, and the microbial content of these environments may also be highly influenced by freshwater impacts. The presence of *Legionella* in these environments may be a transient result of such freshwater impacts rather than multiplication within the marine environment.

Heller et al. (3) investigated the survival of *Legionella pneumophila* serogroup (sg) 3 in saline solutions and seawater in laboratory microcosms. The results of these investigations showed that both increasing salinity and temperature reduced survival of the organisms. The limitations of the investigations were that heat sterilization of the seawater would have two major effects:

(i) chemical alteration of the nature of the seawater and (ii) removal of other microflora. *Legionella* grow in conjunction with other microorganisms, and this has been clearly demonstrated in nonsterile tap water cultures (6).

Legionella colonization of water used as cooling for "once through" cooling towers serving industrial power stations has been previously demonstrated (2). *L. pneumophila* sg 1 is the species and serogroup most commonly implicated in cooling tower-associated disease outbreaks. There have been no reports of Legionnaires' disease attributable to seawater-cooled cooling systems, though *Legionella* have been detected in water in proximity to such installations (5).

In these investigations the ability of *L. pneumophila* sg 1 to survive in nonsterile seawater at various temperatures and salinities was investigated. These investigations were carried out in conjunction with limited environmental sampling from seawater-cooled cooling tower effluents.

An *L. pneumophila* sg 1 suspension was added to 1 liter of oceanic sea water samples to obtain final concentrations of approximately 1×10^3 cells per ml. For each of three incubation temperatures (25, 35, and 45°C), three dilutions of seawater and a phosphate buffer control were prepared in triplicate. Dilutions and chemical parameters are tabulated in Table 1.

Susan Bennett City of Charles Sturt, Adelaide, South Australia, Australia, 5001. *Richard Bentham* Department of Environmental Health, Flinders University, GPO Box 2100, Adelaide, South Australia, Australia, 5001.

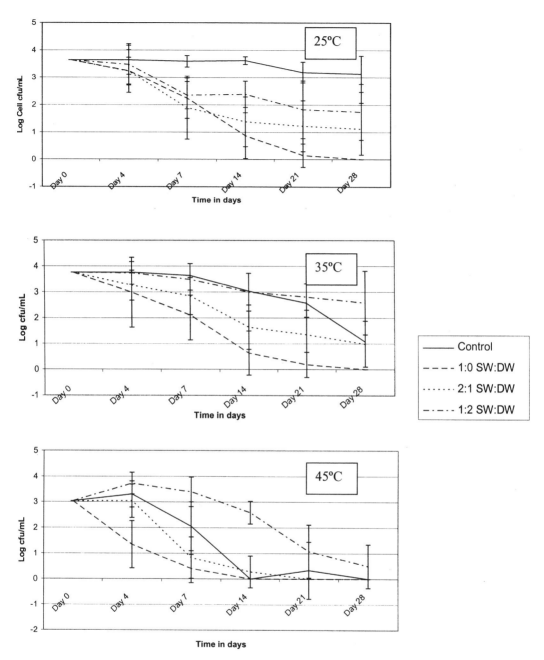

FIGURE 1   Survival of *L. pneumophila* sg 1 in three dilutions of seawater at three temperatures. SW, seawater; DW, deionized water.

TABLE 1 Salinity and chemical parameters in
seawater microcosms

Microcosm[a]	Salinity (20°C) (ppt)	pH	Conductivity (Ms/cm)
1:0 SW:DW	42	7.5	60
2:1 SW:DW	29	7.6	42
1:2 SW:DW	14.5	7.6	22
0.1 M buffer	10	7.2	15

[a]SW, seawater; DW, deionized water.

Legionella were cultured by direct inoculation
of 0.1 ml of sample onto GVPC agar plates followed by incubation at 35°C for 7 days. Representative colony morphologies were enumerated and presumptively confirmed as *Legionella*
species by subculture onto blood agar and
buffered charcoal yeast extract. Samples were
taken at day 0 and then at days 4, 7, 14, 21, and
28. Results are presented in Fig. 1 and were statistically analyzed to determine significance between observed survival rates.

Statistical analysis of the survival rates in the
microcosms was conducted using the Levenes
test of equality of variances and the Post Hoc
Tukey HSD. Statistical analysis of the data
showed significant differences in survival rates
relating to both temperature and salinity. Survival was significantly reduced in undiluted
seawater when compared to the control samples at all temperatures. After the 28-day period survival of *Legionella* in all the seawater dilutions at 25°C was significantly less than the
control. Survival of *Legionella* was significantly
reduced with increasing temperature for all
seawater dilutions and the control. Survival
was most dramatically reduced at 45°C. *Legionella* was not detected by culture in any of
the environmental samples.

The survival of *L. pneumophila* in nonsterile
seawater decreased with increasing salinity.
These results suggested that the organism was
not capable of multiplication in the seawater.
These results confirmed previous investigations using sterile seawater and support the
suggestions that seawater increases stress on
legionella (3,1).

Increasing temperature tended to decrease
the survival rate of the organism. Survival was
best at lower temperatures, as previously observed (5, 3). Survival was most dramatically
reduced at temperatures within the range that
might be associated with once-through cooling operations i.e., 37 to 45°C.

The results of this study suggest that health
risks imposed by *Legionella* in seawater-cooled
cooling systems may be minimal, and colonization of these systems seems unlikely, especially at
warmer operating temperatures. More comprehensive environmental sampling and investigation is necessary to confirm this. The protective
role of amoebae and biofilm in *Legionella* survival and multiplication cannot be disregarded in
this context. Terrestrial runoff and transient survival of *Legionella* is a plausible explanation for
previous reports of relatively high concentrations of *Legionella* in coastal waters (4, 5)

ACKNOWLEDGMENTS

The authors gratefully acknowledge the support of
the South Australian Department of Human Services.

REFERENCES

1. **Dutka, B. J.** 1984. Sensitivity of *Legionella pneumophila* to sunlight in fresh and marine waters.
Appl. Environ. Microbiol. **48:**970–974.
2. **Fliermans, C. B.** 1996. Ecology of Legionella:
from data to knowledge with a little wisdom.
Microb. Ecol. **32:**203–228.
3. **Heller, R., C. Holler, R. Sussmuth, and
K.-O. Gundermann.** 1998. Effect of salt concentration and temperature on survival of *Legionella
pneumophila. Lett. Appl. Microbiol.* **26:**64–68.
4. **Ortiz-Roque C. M., and T. C. Hazen.** 1987.
Abundance and distribution of *Legionellaceae* in
Puerto Rican waters. *Appl. Environ. Microbiol.* **53:**
2231–2236.
5. **Palmer, C. J., Y.-L. Tsai, C. Paszko-Kolva, C.
Mayer, and L. R. Sangermano.** 1993. Detection of *Legionella* species in sewage and ocean water by polymerase chain reaction, direct fluorescent antibody and plate culture methods. *Appl.
Environ. Microbiol.* **59:**3618–3624.
6. **Wadowsky, R. M., T. M. Wilson, N. J. Kapp,
A. J. West, J. M., Kuchta, S. J. States, J. N.
Dowling, and R. B. Yee.** 1991. Multiplication
of *Legionella spp.* in tap water containing *Hartmanella vermiformis. Appl. Environ. Microbiol.* **57:**
1950–1954.

ISOLATION OF *LEGIONELLA* AND AMOEBAE FROM WATER SAMPLES

Laura Franzin, Daniela Cabodi, and Nicoletta Bonfrate

100

Legionellae are ubiquitous in aquatic environments, where they live in symbiosis with amoebae, which serve as a natural host (3, 9). Many studies describe *Legionella's* ability to survive and to multiply in about 30 species of amoebae, especially *Acanthamoeba*, *Naegleria*, and *Hartmanella*, and in two species of ciliates. Interaction with protozoa protects pathogenic bacteria against harsh environmental conditions. In the intracellular environment the virulence of *Legionella* is enhanced and propagation and survival in acquatic environments are increased.

Protozoa may have an important role in the transmission of bacteria from the environment to humans (6, 9). Protozoa's ability to produce membrane–surrounded vesicles of respiratory size containing *Legionella* has been described (2, 9). These bacteria-filled vesicles can be eliminated in the environment (2, 9) and can easily spread in aerosols. Inhalation of these *Legionella*-filled vesicles may be responsible for pulmonary pathology. Recently, other amoeba-resistant microorganisms have been described (e.g., *L. monocytogenes*, *Chlamydia pneumoniae*, *Mycobacterium avium*, *Coxiella burnetii*, etc.). *L. pneumophila* is

the first amoeba-resistant microorganism for which the important roles of free-living amoebae as reservoirs and vectors and an "evolutionary crib" have been reported (5).

In the literature, amoeba isolation from cooling tower water is frequently reported. However, few studies describe colonization and interactions of amoeba with *Legionella* in hospital water supplies (1, 7, 8). The aim of the study was the isolation of *Legionella* from hospital water supplies and from air conditioning systems during periodic controls sometimes performed after water disinfection. The culture of amoebae was also performed, and the interactions with legionellae were studied. A total of 496 water samples were examined. Samples (n = 471; 5,100 ml) were collected from the water supplies of 12 hospitals: 411 from hot water (including hot water tanks) and 60 from cold water. Some of them were collected after disinfection. Twenty-five cold water samples (1,100 ml) were collected from ponds of air conditioning systems.

All samples (5 liters for hospital water and 1 liter for air conditioning system water) were concentrated by filtration (0.2 μm pore size), and *Legionella* culture was determined by quantitative method, as previously described (4). Aliquots of direct, heat-treated, and acid-treated concentrates were plated on buffered

Laura Franzin, Daniela Cabodi, and Nicoletta Bonfrate Specialist Microbiology Research Laboratory, Amedeo di Savoia Hospital, 10149 Turin, Italy.

charcoal yeast extract, BMPA and MWY media. After incubation at 37°C up to 15 days, suspected colonies were typed and quantified.

An aerobic bacterial count was performed on plate count agar medium (pour plate method). The plates were incubated at 37°C for 3 days and 22°C for 5 days. For air conditioning systems, sequential dilutions (10^{-1} to 10^{-6}) of water samples were performed on plate count agar, and the plates were incubated at 30°C for 5 days.

Amoebae were cultured from 100 ml of water, after concentration by centrifugation at $500 \times g$ for 10 min; the pellet (2 to 3 drops) was placed on a nonnutrient agar plate precoated with *Escherichia coli* with 24 h of growth and incubated at 25°C. The plates were observed for the presence of amoeba trophozoites and cysts using the low-power, 10x objective of an inverted microscope daily for 7 days (10). Amoebae were harvested from positive plates, stored at 4°C in Page amoeba saline, and periodically transferred.

Legionella was isolated from 251 of 471 (53.3%) hospital water samples (239 hot and 12 cold), corresponding to 11 of 12 (91.7%) hospitals. The isolates were typed as *L. pneumophila* serogroups 1, 3, 5, 6, and 2-14; *L. bozemanii*, *Legionella* spp.; and fluorescent *Legionella* spp. Amoebae were positive in 82 of 471 (17.4%) samples (74 hot and 8 cold water), corresponding to 11 of 12 (91.7%) hospitals. *Legionella* and amoebae were both present in 55 of 471 (11.7%) samples, corresponding to 10 of 12 (83.3%) hospitals. Amoebae alone were isolated from 27 of 471 (5.7%) samples, while *Legionella* alone were isolated from 196

of 471 (41.6%) samples. One hospital was only contaminated with *Legionella* and another hospital only with amoeba.

Amoeba occurrence was significantly higher ($P < 0.001$) in hospital water samples with *Legionella* counts $>10^3$ CFU/liter than in samples with 20 to 100 CFU/liter (comparison between proportions by z test), as shown in Table 1. *Legionella* cultures from air conditioning system water were negative, but eterotrophic bacteria counts were elevated (up to 2.0×10^5 CFU/ml). Amoebae were found in 16 of 25 (64%) cold water samples: 14 were isolated from 22 air conditioning units and 2 were from 3 cooling towers. The distribution of *Legionella* and amoebae in all water samples in relation to temperature is shown in Fig. 1.

In summary, *Legionella* was isolated from 53.3% of samples collected from 12 hospital water supplies, while amoebae were isolated from 17.4% of the samples. *Legionella* and amoebae were both found in 83.3% of the hospitals. One hospital was only contaminated with *Legionella* and another hospital only with amoebae. The results of this study confirmed those observed by other authors reporting similar data for *Legionella* isolation (55%) (7). However, higher percents ranging from 52 (8) to 68% (7) for amoeba isolation have been reported.

Many studies provide evidence of the important role of protozoa in *Legionella* epidemiology (2, 3, 6, 8). Amoebae favor multiplication of bacteria, increased virulence, enhanced invasion, elevated colonization of pulmonary tissue by inhalation of vesicles filled with *Legionella*, and recolonization of hospital water supplies after treatment.

TABLE 1 Simultaneous occurrence of *Legionella* and amoebae on hospital water samples

Samples	No. (%) of occurrences of *Legionella* at the following CFU/liter		
	20–100	>100–1,000	>1,000
Amoeba positive	6 (8.3)	19 (21.3)	30 (33)
Amoeba negative	66	70	60
Total	72	89	90

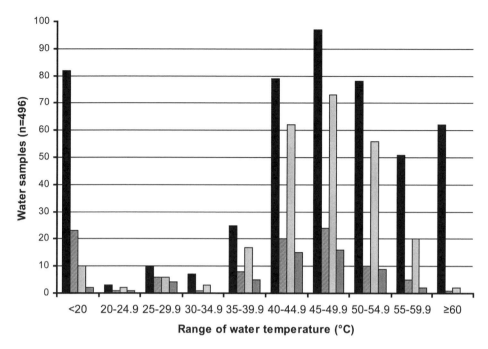

FIGURE 1 Distribution of *Legionella* and amoebae in water samples.

Legionella eradication from the appropriate environmental reservoir is a primary intervention technique employed to control outbreaks of legionellosis. However, eradication is complicated by interactions with protozoa. Legionellae can survive within amoebic cysts and thereby further resist the activities of biocides and physical stresses. For these reasons disinfection of water supplies is not always successful.

In conclusion, in our study the *Legionella* isolation rate was higher than that of amoebae from hospital water supplies, but amoebae were found in higher percents from air conditioning systems (64%, the despite limited number of samples examined) where *Legionella* was absent. In another study of cooling towers the occurrence of amoebae and *Legionella* was reported, respectively, in 87.3 and 36.6% of the samples (L. Cavalie, P. Pernin, L. Garelly, and A. Rambaud, Abstr. 20th Annual Meeting of the European Working Group for *Legionella* Infection, abstr. 19, 2005).

As *Legionellae* multiply intracellulary in amoebae while protozoa contribute to the spread of these bacteria, the presence of amoebae may be considered an important factor promoting the increase of *Legionella* colonization in water systems. Moreover, as free-living amoebae may increase the potential transmission of *Legionella* spp. by producing vesicles filled with bacteria, a possible underestimation of the risk of legionellosis in hospitals by colony plate count methods should be considered (2, 5).

REFERENCES
1. **Barbaree, J. M., B. S. Fields, J. C. Feeley, G. W. Gorman, and W. T. Martin.** 1986. Isolation of protozoa from water associated with a legionellosis outbreak and demonstration of intracellular multiplication of *Legionella pneumophila*. *Appl. Environ. Microbiol.* **51:**422–424.
2. **Berk, S. G., R. S. Ting, G. W. Turner, and R. J. Ashburn.** 1998. Production of respirable vesicles containing live *Legionella pneumophila* cells by two *Acanthamoeba* spp. *Appl. Environ. Microbiol.* **64:**279–286.
3. **Fields, B. S.** 1993. *Legionella* and protozoa: interaction of a pathogen and its natural host, p. 129–136. *In* J. M. Barbaree, R. F. Breiman, and A. P. Dufour (ed.). *Legionella: Current Status and Emerging Perspectives.* ASM Press, Washington, D.C.

4. **Franzin, L., M. Castellani Pastoris, P. Gioannini, and G. Villani.** 1989. Endemicity of *Legionella pneumophila* serogroup 3 in a hospital water supply. *J. Hosp. Infect.* **13:**281–288.

5. **Greub, G., and D. Raoult**. 2004. Microorganisms resistant to free–living amoebae. *Clin. Microbiol. Rev.* **17:**413–433.

6. **Harb, O. S., and Y. A. Kwaik.** 2000. Interaction of *Legionella pneumophila* with protozoa provides lessons. *ASM News* **66:**609–616.

7. **Patterson, W. J., J. Hay, D. V. Seal and J. C. McLuckie.** 1997. Colonization of transplant unit water supplies with *Legionella* and protozoa: precautions required to reduce the risk of legionellosis. *J. Hosp. Infect.* **37:**7–17.

8. **Rohr, U., S. Weber, R. Michel, F. Selenka, and M. Wilhelm.** 1998. Comparison of free-living amoebae in hot water systems of hospitals with isolates from moist sanitary areas by identifying genera and determining temperature tolerance. *Appl. Environ. Microbiol.* **64:**1822–1824.

9. **Rowbotham, T. J.** 1980. Preliminary report on the pathogenicity of *Legionella pneumophila* for freshwater and soil amoebae. *J. Clin. Pathol.* **33:**1179–1183.

10. **Visvesvara, G. S.** 1999. Pathogenic and opportunist free-living amebae, p. 1383–1390. *In* P. R. Murray, E. Y. Baron, M. A. Pfaller, F. C. Tenover, and R. H. Yolken (ed.). *Manual of Clinical Microbiology*, 7th ed. ASM Press, Washington, D.C.

DETECTION AND IDENTIFICATION OF FREE-LIVING PROTOZOA PRESENT IN DRINKING WATER

Rinske Valster, Bart Wullings, Stefan Voost, Geo Bakker,
Hauke Smidt, and Dick van der Kooij

101

Free-living protozoa, e.g., *Acanthamoeba, Naegleria, Saccamoeba, Hartmannella,* and *Vexillifera,* serve as hosts for *Legionella* to proliferate in natural and man-made freshwater environments. These protozoa multiply on biofilms, and their grazing behavior is influenced by the composition and density of the biofilm (3). Protozoa not only provide nutrients for the intracellular growth of *L. pneumophila,* but also form a shelter when environmental conditions become unfavorable. Defining effective measures to prevent proliferation of *L. pneumophila* in tap water installations requires a solid understanding of biotic as well as abiotic factors affecting the persistence and multiplication of this organism, including the presence and behavior of host protozoa. However, information about presence and identity of bacterial free-living protozoa in such installations is still scarce. The objective of this study was to optimize and apply cultivation-independent methods to obtain information about the presence and identity of free-living protozoa in

river water (RW), treated water (TW), and tap water (Tap) in The Netherlands. To our knowledge, this is the first description of protozoal diversity in freshwater by using molecular techniques.

WATER TYPES AND DNA ISOLATION

The examined water samples consist of two RW samples (temperature [T] = 13°C), five TW samples (T = 10 to 15°C; TW 1 and TW 3 were prepared from aerobic groundwater, TW 2 and TW 5 were prepared from anaerobic groundwater, and TW 4 was prepared from surface water), and three Tap samples (T = 16 to 21°C; Tap 1 was taken after 15 min flush-ing, and Tap 2 and Tap 3 were taken from stagnant water). One to two liters of TW and Tap and 20 ml of RW were filtered through a 1.2-μm-pore-size, 55-mm diameter RTTP Isopore Membrane (Millipore, The Netherlands) in a vacuum not exceeding 0.3 bar. DNA was extracted from the filters using the FastDNA SPIN Kit for Soil (Q-Bio-gene, France).

PCR AND T-RFLP ANALYSIS

PCR was performed with a GeneAmp PCR System 9700 (Applied Biosystems) in a reaction mixture (50 μl) containing 10 μl template DNA, each primer at a concentration of 0.5 μM, a 0.2 mM concentration of each

Rinske Valster, Bart Wullings, Stefan Voost, and Dick van der Kooij Kiwa N.V. Water Research, 3430 BB Nieuwegein, The Netherlands.   *Hauke Smidt* Laboratory of Microbiology, Wageningen University, Hesselink van Suchtelenweg 4, 6703 CT Wageningen, The Netherlands.   *Geo Bakker* Vitens Watertechnology, Oude Veerweg 1, 8019 BE Zwolle, The Netherlands.

deoxynucleoside triphosphate, 1.5 mM $MgCl_2$, 2.5 U of *Taq* DNA polymerase (Invitrogen, The Netherlands). 18S rRNA gene fragments were amplified with the universal eukaryotic primers (3'FAM-labeled) Euk1a-f (6) and Euk516-r (1). Amplification conditions were as follows: preheating at 94°C for 130 s, 35 cycles of denaturation at 94°C for 30 s, annealing at 56°C for 45 s, and extension at 72°C for 130 s; and a terminal extension at 72°C for 7 min. Fluorescently labeled PCR products (45 μl) were purified by using the DNA Clean & Concentrator-5 Kit (Zymo Research, Calif.) and redissolved in 20 μl of DNA-free water. The digestion reaction mixture (20 μl) contained 5U of *Hha*I, 2 μl of buffer C, 12.5 μl of water and 5 μl of the PCR product and was incubated at 37°C for 6 h. The mixture was cleaned and redissolved in 15 μl of DNA-free water.

The restriction digest product (5 μl) was mixed with 15 μl of loading buffer (15 μl of Hi-Di formamide and 1 μl of GS-500 ROX [Applied Biosystems] as the internal standard). The injection time was 5 s for analysis of and the run time was 35 min. The fluorescently labeled T-RFs were analyzed by electrophoresis on an ABI PRISM 310 genetic analyzer (Applied Biosystems) in Genescan mode. T-RFs were entered into a genomic fingerprint analysis program, Bionumerics v. 3.5 (Applied Maths, Belgium), and fragment sizes were calculated. Banding patterns were compared using a densitometric curve-based method that evaluates the position of the bands to generate pairwise similarity scores (Pearson coefficient) that were subsequently used for cluster analysis.

CONSTRUCTION AND ANALYSIS OF CLONE LIBRARIES

18s rRNA genes were amplified by the above- described PCR conditions, except that nonlabeled Euk1a primer was used. The PCR products were purified and cloned into *Escherichia coli* JM109 by using the Promega pGEM-T easy Vectors system (Promega, United Kingdom). PCR was performed on cell lysates of ampicillin-resistant transformants by using pGEM-T specific primers T7

and Sp6. Sequence analysis was done by Base-Clear Lab services (Leiden, The Netherlands). The primers used for sequencing the 18S rRNA gene fragment were Euk1a and Euk516, and sequences were compared to sequences in the National Center for Biotechnology Information (NCBI)-database by BLAST searches.

PCR and terminal restriction fragment length polymorphism (T-RFLP) analysis of clones were performed as described above, except that the digestion products were 10-fold diluted and the injection time was 1 s at the T-RFLP-assays.

DIVERSITY AND IDENTITY OF EUKARYOTIC ORGANISMS

The T-RFLP analysis showed a high eukaryotic diversity in all water samples (Fig. 1). The highest diversity was observed in TW 4, TW 5, and Tap 1, while the lowest diversity was observed in RW 2, TW 1, and Tap 3. Clones most closely related to identical genera showed T-RFs with identical lengths (data not shown). Occasionally, clones affiliated with different genera also produced T-RFs with identical lengths (data not shown). Free-living protozoa represented a major part of the total eukaryotic community (Fig. 2). The clone libraries show that the eukaryotic community in surface water, treated water, and tap water consisted, respectively, of 44, 62 and 20% of free-living protozoa. Nematodes (metazoa) were observed in treated water and tap water, but not in river water samples.

Blast analysis of the clones confirmed the presence of a highly diverse group of free-living protozoa in all three water types. In TW 3, prepared from aerobic ground water, 13 genera of free-living protozoa could be distinguished. This protozoa community consisted for 84% of seven protozoa genera (each protozoa genus had a minimal contribution of 7 % to the total protozoa community). The dominant protozoa were uncultured Cercozoan clone (25.4%), *Cercomonas* (14.3%), *Acanthamoeba* (12.7%), *Bodo* (7.9%), *Neobodo* (7.9%), *Rhynchomonas* (7.9%), and *Vexillifera* (7.9%). Many clones had

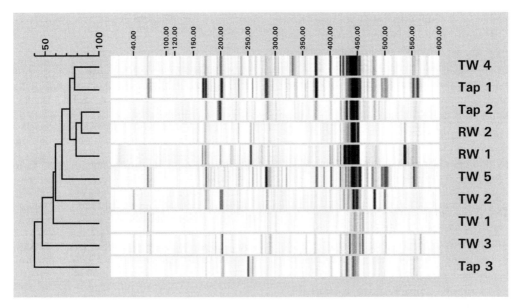

FIGURE 1 Eukaryotic diversity in river water (RW), treated water (TW), and tap water (Tap) as determined by T-RFLP analysis.

similarities lower than 95% to sequences deposited in accessible public databases. Two clone types had the highest similarity to protozoa described as hosts for *Legionella*, viz. *Acanthamoeba* with >95% similarity and *Vex-illifera* with a similarity of >94% to the sequence in the database (7).

CONCLUDING REMARKS

This study shows that T-RFLP assays are a powerful technique to obtain a fast indication of the diversity of the eukaryotic community. Further optimization of T-RFLP assays is needed to achieve better resolution, distributing the T-RFs of dominant free-living protozoa over the T-RFLP pattern from 50 to 500 bp. Clone libraries are a powerful tool to determine the diversity and identity of eukaryotic communities, but this technique is time-consuming. Despite their ecological importance, few studies have dealt with free-living freshwater protozoa, and thus few 18S rRNA gene sequences are

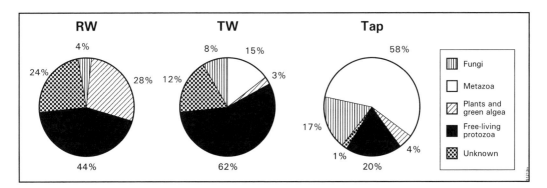

FIGURE 2 Classification of eukaryotic clone libraries made from river water (RW), treated water (TW), and tap water (Tap) collected in The Netherlands.

available. Consequently, many of the clones had less than 95% similarity with sequences in the NCBI-database. We also found that the eukaryotic diversity can be underestimated with T-RFLP assays, as T-RFs of clones most closely related to different genera can have identical lengths. In most eukaryotes 18S rRNA genes are organized in tandem repeat units, but organisms can significantly differ in the genomic copy number of ribosomal RNA genes, and not all eukaryotic organisms are multicellular (5). Hence, clone libraries generated from 18S rRNA genes can give incorrect proportions between different genera in the total eukaryotic community. Still, the clone libraries revealed that free-living protozoa are ubiquitous in river water, treated water, and tap water, and some of these organisms probably serve as hosts for *L. pneumophila*. The application of real-time PCR assays for selected protozoa genera and cultivation methods will further extend our knowledge of free-living protozoa serving as hosts for *L. pneumophila* in tap water installations.

ACKNOWLEDGMENT

This study was conducted within the framework of the Joint Research Program of the water supply companies in The Netherlands and was cofinanced by Delft Cluster.

REFERENCES

1. **Amann, R. I., B. J. Binder, R. J. Olson, S. W. Chisholm, R. Devereux, and D. A. Stahl.** 1990. Combination of 16S rRNA-targeted oligonucleotide probes with flow cytometry for analyzing mixed microbial populations. *Appl. Environ. Microbiol.* **56:**1919–1925.
2. **Fields, B. S., R. F. Benson, and R. E. Besser.** 2002. *Legionella* and Legionnaires' disease: 25 years of investigation. *Clin. Microbiol. Rev.* **15:**506–526.
3. **Hahn, M. W., and M. G. Höfle.** 2001. Grazing of protozoa and its effect on populations of aquatic bacteria. *FEMS Microb. Ecol.* **35:**113–121.
4. **Kuiper, M. W., B. A. Wullings, A. D. L. Akkermans, R. R. Beumer, and D. van der Kooij.** 2004. Intracellular proliferation of *Legionella pneumophila* in *Hartmannella vermiformis* in aquatic biofilms grown on plasticized PVC. *Appl. Environ. Microbiol.* **70:**6826–6833.
5. **Long, E. O., and I. B. Dawid.** 1980. Repeated genes in eukaryotes. *Annu. Rev. Biochem.* **49:**727–764.
6. **Sogin, M. L., and J. H. Gunderson.** 1987. Structural diversity of eukaryotic small subunit ribosomal RNAs. *Ann. NY Acad. Sci.* **503:**125–139.
7. **Steinert, M., U. Hentschel, and J. Hacker.** 2001. *Legionella pneumophila*: an aquatic microbe goes astray. *FEMS Microbiol. Rev.* **26:**149–162.

GROWTH OF *LEGIONELLA* IN NONSTERILIZED, NATURALLY CONTAMINATED BATHING WATER IN A SYSTEM THAT CIRCULATES THE WATER

Katsuhiko Ohata, Kanji Sugiyama, Mitsuaki Suzuki,
Rieko Shimogawara, Shinji Izumiyama,
Kenji Yagita, and Takuro Endo

102

For a long time in Japan, taking a bath in the hot springs has been a popular activity. In recent years, many public bathhouses introduced bathing water circulating systems for extended use, in which a sand filtration unit was installed. However, this resulted in several large-scale outbreaks of legionellosis (1) due to the microbiologically insufficient maintenance of the bathing facilities. Occurrence of *Legionella* in bathing water circulating systems, appears to be common and is a serious public health concern in Japan.

We constructed a life-size model plant of a bathing water circulating system for the simulation experiment. The model plant is composed of a filter (a sand filtration unit filled with three layers of ceramic sand [about 100 kg], with a linear velocity of 25.5m/h), two bathtubs (1 m^3 × 2), an automatic chlorine (in the form of sodium hypochlorite) injector, a hair catcher, a circulating pump, a heater, and a UV light lamp. The water temperature is kept at about 40°C (104°F) and circulated at about 5 m^3/h (Fig. 1).

These experiments are aimed at monitoring changes in the microbial constituents, especially a possible occurrence of *Legionella* in a bathing water circulating system, and developing preventive measures and intervention strategies. Prior to microbial monitoring, a total of either 16 or 40 volunteers were asked to take baths for 10 days (experiment 1) or for 14 days (experiment 2), respectively. Chlorine was added at concentrations ranging from 0.2 to 1.5 mg/liter into the bathing water. After that, in experiment 1, the chlorine injector was turned off and the residual chlorine was degraded completely with the ultraviolet irradiation. Again, five volunteers took a bath, and the ultraviolet irradiation was turned off (day 0). The bathing water was circulated under non-sterilization (1st to 23rd days) and the circulating system was sterilized by 10 ppm chlorine at the end of the experiment (23rd day). In experiment 2, after the residual chlorine was totally degraded with the ultraviolet irradiation, three volunteers took a bath (0 to 1st days). After that, the ultraviolet irradiation was turned off (2nd day) and the bathing water was circulated under nonsterilization (3rd to 31st days). After the bathing water was exchanged (31st day), it was recirculated (31st to 36th days). At the end of the experiment, the system was sterilized by 6% H_2O_2 (36th day).

Katsuhiko Ohata, Kanji Sugiyama, and Mitsuaki Suzuki Shizuoka Institute of Environment and Hygiene, Shizuoka-City, Shizuoka Pref. 420-8637, Japan.   *Rieko Shimogawara, Shinji Izumiyama, Kenji Yagita, and Takuro Endo* Department of Parasitology, National Institute of Infectious Disease, Shinjuku-ku, Tokyo 162-8640, Japan.

Legionella: State of the Art 30 Years after Its Recognition
Edited by Nicholas P. Cianciotto et al.
©2006 ASM Press, Washington, D.C.

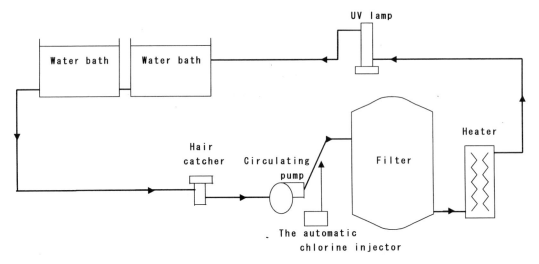

FIGURE 1 A system that circulates bathing water of the model plant. The pointed arrows show the flow of the circulating water.

In these experiments, from the day when the chlorine injector was turned off, microbial monitoring occurred for the number of *Legionella*, total viable bacterial counts, heterotrophic bacteria, and free-living amoebae in the bathing water. This microbial monitoring occurred in the water, in the filter, and in other parts of the model plant, either almost daily or at certain intervals. GVPC Agar (bioMerieux, France) was used for detection of *Legionella*, Standard Method Agar (Nissui Pharmaceutical, Tokyo) for total viable bacterial counts, R2A Agar (Becton Dickinson, USA) for heterotrophic bacteria, and Bacto-Agar (Difico, USA), on which a monolayer of *Escherichia coli* was deposited, for free-living amebae. The $KMnO_4$ consumption test was also applied continuously to the bathing water in experiment 2.

As a result of experiment 1 (Fig. 2), *Legionella* was detected in both the bathing water and the filter water at concentrations of 6.6 ×

10^2 CFU/100 ml on the 3rd day after residual chlorine disappeared. The number of *Legionella* in the bathing water and the filter water increased to 10^5 CFU/100 ml on the 5th day and remained 10^4 to 10^5 CFU/100 ml throughout the experiment. A positive bacterial culture was obtained in the bathing water and the filter water immediately after residual chlorine disappeared, and the number of total bacterial counts reached 10^5 CFU/ml within a day. Then it decreased gradually to 10^3 to 10^4 CFU/ml at the end of the experiment. The number of heterotrophic bacteria was almost the same as the number of total bacterial counts. Free-living amoebae, known to be the hosts of *Legionella* in this environment, were detected 2 days after residual chlorine disappeared. They were calculated to be 12 cells/ml in the bathing water and 11 cells/ml in the filter water. On the 3rd day, the number of amoebae increased to 136 cells/ml and 129

FIGURE 2 The growth of *Legionella,* etc. in the bathing water (A) and the filter water (B) in experiment 1. For 10 days about 16 volunteers took baths under the system with chlorine control, in which the chlorine injector was turned off. After that, five volunteers took baths, and the ultraviolet irradiation was turned off (day 0). The bathing water was circulated under nonsterilization (1st to 23rd days). At the end of the experiment (23rd day) the whole system was sterilized by 10 ppm sodium hypochloride solution. From the day when the chlorine injector was turned off, there was microbial monitoring for the number of *Legionella*, total viable bacterial counts, heterotrophic bacteria, and free-living amoebae.

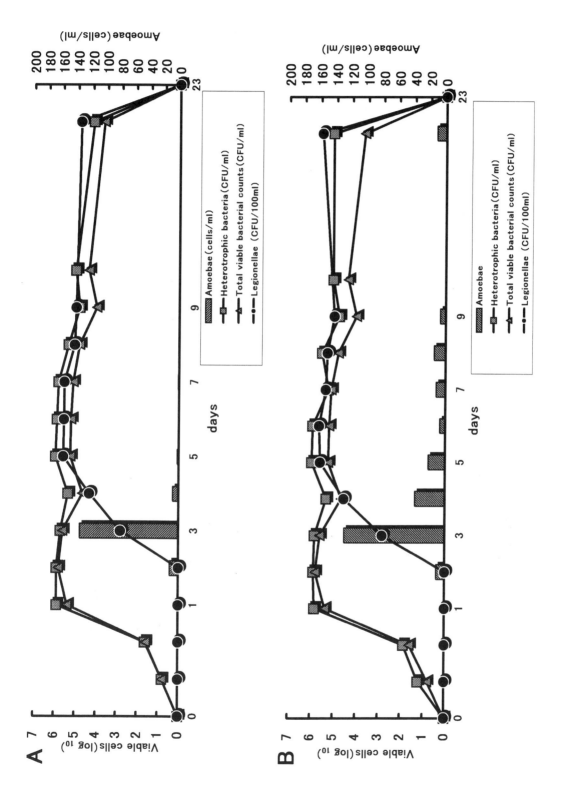

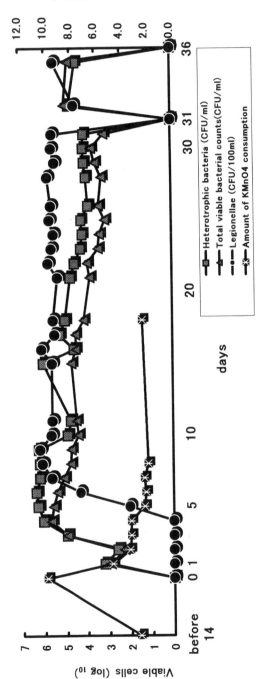

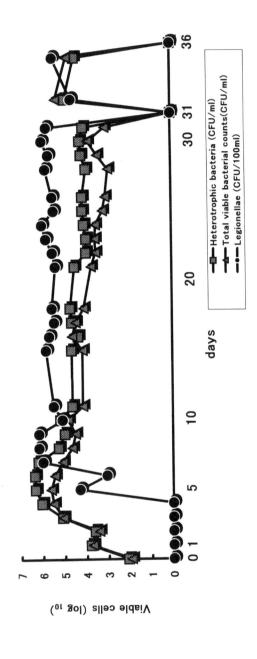

cells/ml, followed by a rapid decrease from the bathing water. They disappeared within 5 days after that, for unknown reasons. The number of amoebae in the filter water fluctuated and amounted to 12 cells/ml at the end of the experiment.

In experiment 2 (Fig. 3), *Legionella* in the bathing water and the filter water became detectable from 3 days after stopping ultraviolet irradiation, which was 5 days after residual chlorine disappeared and the number of *Legionella* detected in the bathing water and the filter water was 1.1×10^2 CFU/100 ml and 2.0×10^4 CFU/100 ml, respectively. The number of *Legionella* had increased to 10^6 CFU/100 ml within 5 days and was 10^5 to 10^6 CFU/100 ml at the end of the experiment. The total viable bacterial counts reached 10^4 to 10^5 CFU/ml in 5 days after the residual chlorine disappeared. The number of heterotrophic bacteria was almost the same as the number of total bacterial counts. A large number of *Legionella*, namely, 1.6×10^7 CFU/g, 3.1×10^5 CFU/g, and 1.8×10^5 CFU/g, was detected from the sand collected from the upper, middle, and lower parts of the filter, respectively. Similarly, 9.1×10^3 cells/g, 1.5×10^3 cells/g, and 1.6×10^3 cells/g of amoebae were detected from the filter. In addition, it is noteworthy that *Legionella* was detected from the hair catcher in the model plant at concentrations around 3.1×10^4 CFU/swab in experiment 1 and, 1.2×10^5 CFU/swab in experiment 2. The amount of $KMnO_4$ consumption in the bathing water was 2.7 mg/liter before use and was calculated to be around 10.1 mg/liter after the use by 40 volunteers under the presence of chlorine residues.

In the present experiments, it was clearly demonstrated that *Legionella* occurred in the bathing water circulating system within a short period in a sequential manner of microbial growth. Namely, concentration of organic matter (dirt) in the bathing water that can be monitored as the $KMnO_4$ consumption value increased in correlation to the number of bathers. The deposited dirt allows bacteria to rapidly undergo multiplication in the bathing water, which consequently supports the occurrence of a large number of host amoebae. The growth of *Legionella* is a manifestation of the extended use of bathing water under inadequate hygienic maintenance. It also turned out that the filter acts as the main hotbed for *Legionella* multiplication (2).

REFERENCES

1. **Sugiyama, K., T. Nishio, Y. Gouda, Z. Fanfei, K. Masuda, and M. Akiyama.** 2000. Relationship between *Legionella* contamination of environmental water and occurrence of legionellosis: an outbreak of legionellosis linked to bath water circulating through a filter and examinations. *Bull. Shizuoka Inst. Environ. Hyg.* **43:**1–4.

2. **Sugiyama, K., T. Okitsu, H. Miyamoto, and N. Nakamura.** 1996. Contamination and disinfection of *Legionella* in bath waters of recirculating systems at homes and large-scale bathing facilities. *Bull. Shizuoka Inst. Environ. Hyg.* **39:**47–52

FIGURE 3 The growth of *Legionella,* etc. in the bathing water (A) and the filter water (B) in experiment 2. For 14 days about 40 volunteers took baths under the system with chlorine control, in which the chlorine injector was turned off. After that, three volunteers took baths (0 to 1st days) and the ultraviolet irradiation was turned off (2nd day). The bathing water was circulated under nonsterilization (3rd to 31st days). After bathing water was exchanged (31st day), it was recirculated (31st to 36th days). At the end of the experiment (36th day) the whole system was sterilized with 6% H_2O_2. From the day when the chlorine injector was turned off, there was microbial monitoring for the number of *Legionella*, total viable bacterial counts, heterotrophic bacteria, and the amount of $KMnO_4$ consumption.

FLUCTUATION IN *LEGIONELLA PNEUMOPHILA* COUNTS IN COOLING TOWERS OVER A 1-YEAR PERIOD

S. Ragull, R. Montenegro, M. García-Núñez, I. Sanchez, A. Soler, N. Sopena, M.L. Pedro-Botet, M. Esteve, and *M. Sabriá*

103

Colonization of cooling towers (CTs) by *Legionella* is frequent and has been implicated in many community outbreaks of Legionnaires´ disease (LD) (1–3). Spanish regulations (5) (RD865/2003) have provided guidelines for action to be taken according to risk categories (Table 1) based on *Legionella* counts in four annual samplings in order to prevent LD outbreaks. However, since fluctuations have been described in *Legionella* counts in CTs, the limit of the categories are difficult to establish. On the other hand, seasonal influences may complicate the extrapolation of results from one area to another.

In Catalonia (northeastern Spain) the environmental temperature oscillates between 4 below zero and 14°C in the winter and between 16 to 35°C in the summer, with four well-defined seasons. In this area, *Legionella* counts in CTs may fluctuate throughout the whole year, with the highest levels being observed during the warmer seasons.

Sonia Ragull, Marian García-Núñez, Inma Sanchez, Anna Soler, Nieves Sopena, Maria Luisa Pedro-Botet, Miguel Sabriá Legionelosis Study Group (GELeg), Infectious Diseases Unit, Hospital Universitario Germans Trias i Pujol, (Autonomous University of Barcelona), 08196 Badalona (Barcelona), Spain. *M. Esteve* Preventive Medicine Unit. Hospital Universitario Germans Trias i Pujol (Autonomous University of Barcelona), 08196 Badalona (Barcelona), Spain. *R. Montenegro* Product Manager-Water Service & Technology Division-Proquimia, 08500 Vic, Barcelona, Spain.

The aim of this study was to describe the fluctuation in *Legionella* counts in CTs over a 1-year period. Fifteen CTs (A through O) located within a 3-km radius in Catalonia were selected. All CTs were subject to the same maintenance regimen to comply with the Spanish Real Decreto 865/2003. The towers were sampled biweekly over 1 year. All CTs were tested between 22 and 26 times. Water samples (volume: 1 liter) were concentrated by filtration and seeded on agar *Legionella* GVPC using established methods (4). The plates were incubated at 36°C for 10 days in a candle jar. Ten colonies of each positive culture were selected and characterized by nutritional requirements and by the immunoglutination latex assay test (OXOID).

Legionella pneumophila was isolated in 13 of 15 CTs (86.6%) (Fig.1). All positive CTs were colonized by *L. pneumophila* serogroup 1. Two CTs were co-inhabited by *L. pneumophila* serogroup 1 and serogroup 2-14. Four CTs were positive only once (range: 66 to 306,666 CFU/liter) (range: 1.8 log-5.5 log CFU/liter). One CT was positive twice (500 CFU/liter, 3,333 CFU/liter) (2.7 log CFU/liter, 3.5 log CFU/liter). In the remaining eight CTs, concentrations of *L. pneumophila* fluctuated during the study period, ranging from negative cultures up to 2×10^6 CFU/liter in some sam-

Legionella: State of the Art 30 Years after Its Recognition
Edited by Nicholas P. Cianciotto et al.
©2006 ASM Press, Washington, D.C.

TABLE 1 Spanish regulation RD 865/2003

Legionella bacteria (CFU/liter)	Action required
100 to <1,000	Review program operation and resample after 15 days.
1,000 to <10,000	Implement corrective action. Review the program operation to identify remedial actions to decrease the *Legionella* counts. Cleaning and disinfection. Resample at 15 days.
>10,000	Implement corrective action. Shut down the system. Before resuming the service, cleaning and chlorine shock should be performed. Resample after 15 days.

ples. Eleven out of 15 CTs (73.3 %) showed counts greater than 1,000 CFU/liter and 8 out of 15 CTs (53.3%) had counts above 10,000 at least once during the study period.

These results show that concentrations of *L. pneumophila* in CTs fluctuate considerably over time, and counts included in the Spanish Regulation (RD 865/2003) within the high-risk range (>10,000 CFU/liter) are common.

This temporal variation of *L. pneumophila* counts in CTs limits the usefulness of infrequent sampling to know the *Legionella* status. More frequent sampling may facilitate better control of inocula and allow appropriate measures to be taken according to the concentrations found.

No seasonal influence was observed with respect to variations in *L. pneumophila* counts. All of the systems would not have been considered a risk at some sample time during the same season.

Frequency of sampling (using rapid techniques: biosensors or reverse transcriptase PCR) and risk-limit categories together with other maintenance and demographic factors should be reconsidered to validate a risk-assessment guidance.

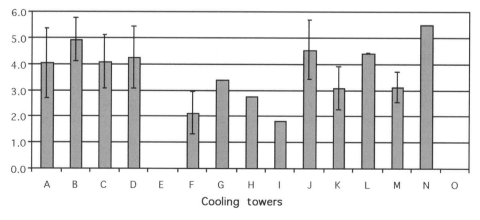

Mean of positive samples

FIGURE 1 Mean of *Legionella*-positive counts (positive samples only)

ACKNOWLEDGMENTS

This work was supported by grant 03/1501 from Fondo de Investigación Sanitaria, Red Respira (RTIC C3/11) from Instituto de Salud Carlos III and by AS-SOCIACIO IBEMI.

REFERENCES:

1. **Fernandez, J. A., P. Lopez, D. Orozco and J. Merino.** 2002. Clinical study of an outbreak of Legionnaires' disease in Alcoy, Southeastern Spain. *Eur. J. Clin. Microbiol. Infect. Dis.* **21:**729–735.

2. **Formica, N., G. Tallis, B. Zwolak, J. Camie, M. Beers, G. Hogg, N. Ryan, and M. Yates.** 2000. Legionnaires' disease outbreak: Victoria´s largest identified outbreak. *Commun. Dis. Intell.* **24:**199–202.

3. **Jansa, J. M., J. A. Cayla, D. Ferrer, J. Gracia, C. Pelaz, M. Salvador, A. Benavides, T. Pellicer, P. Rodríguez, J. M. Garces, A. Segura, J. Guix, and A. Plasencia.** 2002. Barcelona Legionellosis study group. An outbreak of Legionnaires' disease in an inner city district: importance of the first 24 hours in the investigation. *J. Tuberc. Lung Dis.* **6:**831–838.

4. **Sabria, M., M. Garcia-Nuñez, M. L. Pedro-Botet, N. Sopena, J. M.Gimeno, E. Reynaga, J. Morera, and C. Rey-Joly.** 2001. Presence and chromosomal subtyping of Legionella species in potable water systems in 20 hospitals of Catalonia, Spain. *Infect. Control Hosp. Epidemiol.* **22:**673–676.

5. **Real decreto 865**/2003, del 4 de Julio, por el que se establecen los criterios higiénico-sanitarios para la prevención y control de la legionelosis.

GENOTYPIC VARIABILITY AND PERSISTENCE OF *LEGIONELLA PNEUMOPHILA* DNA SUBTYPES IN 23 COOLING TOWERS FROM TWO DIFFERENT AREAS

I. Sanchez, S. Ragull, M. Garcia-Nuñez, N. Sopena,
M. L. Pedro-Botet, R. Montenegro, and M. Sabriá

104

Legionella pneumophila has frequently been found in cooling towers (CT) with different DNA subtypes being implicated in community-acquired *Legionella* infections (2, 4–6).

Both genotypic variability and clonal persistence are concepts of great importance in molecular epidemiology, since they facilitate the search for the foci responsible for sporadic or outbreaks of Legionellosis. Nonetheless, the genetic variability of *L. pneumophila* in cooling towers and the persistence of these subtypes are not completely understood.

The aims of this study were to investigate the genotypic variability of *L. pneumophila* in cooling towers and determine the persistence of the DNA subtypes over time in two different areas of Catalonia, Spain:

- Area A: Sixteen cooling towers within a radius of approximately 40 km were studied.
- Area B: Seven cooling towers within a radius of 1 km were studied (four of which shared the same water distribution system CT-17, CT-18, CT-19 and CT-20).

I. Sanchez, S. Ragull, M. Garcia-Nuñez, N. Sopena, M. L. Pedro-Botet, and M. Sabriá Legionelosis Study Group (GELeg), Infectious Diseases Unit, Hospital Universitario Germans Trias i Pujol (Autonomous University of Barcelona), 08916 Badalona (Barcelona), Spain. *Ramon Montenegro* Water Service & Technology Division, Proquimia, 08500 Vic, Barcelona, Spain.

The cooling tower selection criterion was recovery of *L. pneumophila* on at least two different sampling times (period: >6 months).

Pulsed-field gel electrophoresis was performed according to a previous description (7). Band pattern analysis was performed with Finger Printing II (BioRad) software. Pulsed-field gel electrophoresis isolates were considered to belong to the same genotype if the chromosomal DNA pattern was indistinguishable according to the Tenover criteria (8). The Dice coefficient was used to calculate the similarity between pairs of strains.

In area A, 34 DNA subtypes were obtained from the 16 cooling towers. In six cooling towers we observed a single DNA subtype, in five, two DNA subtypes, in three cooling towers, three DNA subtypes and in two cooling towers we observed fewer than four DNA subtypes (Table 1). Each cooling tower had its own *Legionella* DNA subtypes, which were not shared with any other cooling tower.

In area B, 10 DNA subtypes was obtained among the 7 cooling towers. In three of the seven cooling towers we observed a single DNA subtype, in two we found two DNA subtypes, and in another two, three DNA subtypes were reported (Table 1). Three of the 10 DNA subtypes obtained (BI, BL, and BP) were shared by more than one cooling tower (Table 1).

TABLE 1   Genotypic variability and persistence of each cooling tower

Area	Cooling tower no.	No. of subtypes	DNA subtypes	Persistence (persistent DNA subtypes)
A	1	1	AA	Yes
	2	1	AB	Yes
	3	1	AC	Yes
	4	2	AD / AE	Yes (AD)
	5	2	AF / AG	Yes (AF)
	6	1	AH	Yes
	7	1	AI	Yes
	8	1	AJ	Yes
	9	5	AK / AL / AM / AN / AO	No
	10	4	AP / AQ / AR / AS	Yes (AP)
	11	2	AT / AU	No
	12	2	AV / AW	No
	13	3	AX / AY / AZ	Yes (AX/AY)
	14	2	BA / BB	Yes (BA/BB)
	15	3	BC / BD / BE	Yes (BC/BE)
	16	3	BF / BG / BH	Yes (BF/BG)
B	17	1	BI[a]	Yes
	18	3	BI[a] / BJ / BK	Yes (BI/BK)
	19	2	BL[a] / BM	Yes (BL)
	20	1	BN	Yes
	21	1	BL[a]	Yes
	22	2	BO / BP[a]	Yes (BO)
	23	3	BP[a] / BQ / BR	No

[a]DNA subtypes shared by more than one cooling tower.

When we studied the persistence of the DNA subtypes we found that in 19 of 23 cooling towers (82%) the same DNA subtype was recovered after at least 6 months of follow-up.

These results demonstrate the great genotypic variability of *Legionella* in the cooling towers. A total of 44 molecular patterns were observed among the 23 cooling towers studied (16 towers in area A and 7 in area B). None of the molecular patterns were shared in area A, with each tower having its own molecular pattern. To the contrary, in area B (smaller than area A) some patterns were shared by more than one cooling tower. Of the towers sharing the same water distribution system (CT-17, CT-18, CT-19, and CT-20) only CT- 7 and CT-18 shared molecular patterns. The other cooling towers which had the

same molecular patterns did not have the same water system (CT-19 with CT-21, CT-22 with CT-23) but were close to each other, thereby demonstrating that towers which are in the same area may share molecular patterns regardless of whether they have the same water distribution system.

The patterns were considered to be the same in our study when they were indistinguishable according to Tenover criteria; that is, the patterns had the same number of bands and were the same size. Other authors (1, 3) allow 2 of 3 bands of difference to consider two patterns as identical. This concept is very important in the study of an outbreak since it is essential that the patterns be indistinguishable in order to establish a relationship between clinical and environmental isolates. We found that in 19 of the 23 towers studied at least one of the patterns observed could be recovered after at least 6 months. This is very important since it indicates that the same DNA subtype persists in the same cooling tower over long periods of time. This demonstrates that despite maintenance maneuvers minimizing *Legionella* colonization from the tower at a given time (for example by disinfection), the profile may be recovered after a period of time. This allows us to recognize and to implicate later sources not correctly identified during the evolution of an outbreak.

ACKNOWLEDGMENTS

This work was supported by grant 03/1501 from Fondo de Investigación Sanitaria, Red Respira (RTIC C3/11) from Instituto de Salud Carlos Ill, and by ASSOCIACIO IBEMI.

REFERENCES

1. Aurell, H., J. Etienne, F. Forey, M. Reyrolle, P. Girado, P. Farge, B. Decludt, C. Campese, F. Vandenesch, and S. Jarraud. 2003. Legionella pneumophila serogroup 1 strain Paris: endemic distribution throughout France. *J. Clin. Microbiol.* **41:**3320–3322.

2. Barrabeig, I., M. Garcia-Nuñez, A. Dominguez, A. Rovira, S. Ragull, A. Pedrol, J. M. Elorza, M. L. Pedro-Botet, N. Sopena, and M. Sabria. 2004. Community outbreak of Legionella in Hospitalet (Spain). Usefulness of the epidemiological and molecular data to identify the source. *Clin. Microbiol. Infec.* **(Suppl. 3):**88.

3. Drenning, S. D., J. E. Stout, J. R. Joly, and V. L. Yu. 2001. Unexpected similarity of pulsed-field gel electrophoresis patterns of unrelated clinical isolates of Legionella pneumophila, serogroup 1. *J. Infect. Dis.* **183:**628–632.

4. Fernandez, J. A., P. Lopez, D. Orozco, and J. Merino. 2002. Clinical study of an outbreak of Legionnaires' disease in Alcoy, southeastern Spain. *Eur. J. Clin. Microbiol. Infect. Dis.* **21:**729–735.

5. Formica, N., G. Tallis, B. Zwolak, J. Camie, M. Beers, G. Hogg, N. Ryan, and M. Yates. 2000. Legionnaires disease outbreak: Victoria's largest identified outbreak. *Commun. Dis. Intell.* **24:**199–202.

6. Jansa, J. M., J. A. Cayla, D. Ferrer, J. Gracia, C. Pelaz, M. Salvador, A. Benavides, T. Pellicer, P. Rodriguez, J. M. Garces, A. Segura, J. Guix, and A. Plasencia. 2002. Barcelona Legionellosis study group. An outbreak of Legionnaires' disease in an inner city district: importance of the first 24 hours in the investigation. *J. Tuberc. Lung. Dis.* **6:**831–838.

7. Sabrià, M., M. García-Núñez, M. L. Pedro-Botet, N. Sopena, J. M. Gimeno, E. Reynaga, J. Morera, and C. Rey-Joly. 2001. Presence and chromosomal subtyping of *Legionella* species in potable water systems in 20 hospitals of Catalonia, Spain. *Infect. Control. Hosp. Epidemiol.* **22:**673–676.

8. Tenover, F. C., R. D. Arbeit, R. V. Goering, P. A. Mickelsen, B. E. Murray, D. H. Persing, and B. Swaminathan. 1995. Interpreting chromosomal DNA restriction patterns produced by pulsed-field gel electrophoresis: criteria for bacterial strain typing. *J. Clin. Microbiol.* **33:**2233–2239.

SUITABILITY OF PEPTIDE NUCLEIC ACID PROBES FOR DETECTION OF *LEGIONELLA* IN MAINS DRINKING WATER SUPPLIES

Sandra A. Wilks and C. William Keevil

105

Legionellae are ubiquitous in the environment. They appear to grow best in mixed communities, are thought to be common in biofilms, and have been found to proliferate within certain protozoa, suggesting that they are biofilm species exhibiting facultative intracellular parasitism of grazing eukaryotes and human lung macrophages (10, 12). However, although their widespread distribution is known and accepted, few studies have attempted to quantify environmental populations. The standard detection method for use with environmental samples relies on culturing onto specific agar media following heat/acid selection procedures (ISO 11731). Such an approach has many limitations; environmental populations may be stressed and hence the cells may be viable but noncultivable; if biofilm samples are being analyzed, the bacteria have to be removed, and it can take up to 14 days before colonies appear. Molecular techniques have been applied, but they too have inherent limitations and rarely provide quantitative data (2). Consequently, there is a real need for a rapid and direct detection method.

Fluorescence in situ hybridization (FISH) assays have been well described but have not been widely used for environmental samples. Generally DNA probes are used which target specific sites on the 16S rRNA molecule. The 16S rRNA molecule is known to be highly conserved in evolutionary terms, with variable regions tending to be in inaccessible sites (6). The successful design of a probe relies on target specificity and the ability to cross the cell membrane and reach the target site. To achieve these requirements, specific and stringent hybridization procedures need to be employed. Each DNA probe has its own specific requirements, and the hybridization procedure is highly sensitive to changes in temperature, pH, and ionic conditions (11). This causes problems when working with environmental samples. An alternative is to use peptide nucleic acid (PNA) probes. PNAs are synthetic molecules in which the sugar phosphate backbone has been replaced by 2-aminoethyl-glycine (4). They exhibit sequence-specific recognition of both DNA and RNA, obeying Watson-Crick hydrogen bonding rules (5). However, their unique chemistry gives them several added advantages. First, they tend to be shorter in length, which allows them to cross the cellular membrane with greater ease (11). This means they are not as dependent on solu-

Sandra A. Wilks and C. William Keevil Environmental Healthcare Unit, School of Biological Sciences, University of Southampton, Southampton, SO16 7PX, UK.

bility factors in the hybridization buffers. They also exhibit a greater thermal stability, permitting successful hybridization over a wider temperature range (11). In addition they are resistant to hydrolytic cleavage and ionic changes (11). These characteristics are very important as they result in less stringent requirements for hybridization, which is of particular importance for environmental sampling.

A study (13) has been undertaken to directly compare the binding efficiency of two published DNA probes against two newly designed PNA probes. The DNA probes tested were LEG226 (9) and LegPNE1 (7), which target all species of *Legionella* and only *L. pneumophila,* respectively. They were synthesized with two fluorophores attached: the green fluorophore, carboxyfluorescein, for LEG226 and the red fluorophore, tetramethylrhodamine, for LegPNE1. Two PNA probes were designed based on these published sequences but shortened in length to 15-mer. Ribosomal databases were used to check the specificity of these probes (3). The two PNA probes (PLEG200 targeting all *Legionella* species and PLPNE620 targeting *L. pneumophila* only) were also synthesized with the same fluorophores. Hybridization studies were carried out using both the methods described for the DNA probes (7, 9) and a modified PNA protocol (1). A total of 47 strains of *Legionella* were tested, including 22 strains of *L. pneumophila.*

The general species PNA probe (PLEG200) labeled all strains of *Legionella*. In contrast, the DNA probe (LEG226) did not label all cells. In some cases, labeling was very faint or only a proportion of the population was labeled. In addition, there was clear clumping of the cells and apparent diffuse fluorescence, making results difficult to interpret. Likewise, the specific *L. pneumophila* PNA probe (PLPNE620) showed more consistent and specific labeling when compared to the DNA probe (LegPNE1). PLPNE620 labeled all strains of *L. pneumophila* strongly but did not label non-*pneumophila* strains, with two exceptions: *L. quinlivanii* and *L. longbeachae* serogroup 1. The labeling of *L. quinlivanii* can be explained by strong autoflu-

orescence of the cells in the trimethylrhodamine isocyanate fluorescence channel (i.e., masking the signal from the tetramethylrhodamine fluorophore). *L. longbeachae* did not exhibit any autofluorescence, but ribosomal database analysis did position it as the next closest relative based on this probe sequence. In contrast, LegPNE1 gave very unpredictable results. Strains of *L. pneumophila* were labeled with this probe, but often only faintly, and cells were observed to be clumping with noticeable diffuse fluorescence. Of more concern, this probe labeled eight non-*pneumophila* species, and this could not be explained by autofluorescence.

This study clearly demonstrates the advantages of using PNA probes on laboratory type cultures, but there is a requirement for a method which can be used on environmental samples and, if possible, can be used directly on samples without the need for cells to be removed. To investigate this, sections of mains supply drinking water pipe were obtained from a leading European water company and analyzed. Each cast iron pipe was cut into small (4 cm × 2 cm) coupons (Fig. 1). These are challenging samples to analyze directly, as corrosion deposits can be extensive (Fig. 1). The hybridization protocol (1, 13) was carried out directly onto the coupon surface. Following hybridization, the coupons were examined using an episcopic differential interference contrast/epifluorescence (EDIC/EF) microscope (Best Scientific, UK) (8). This microscope utilizes long working objectives and has a specialized episcopic illumination system, which permits the examination of the surface of solid or opaque materials. Using a combination of EDIC and EF illumination with this hybridization procedure, it was possible to visualize the surface of the pipe, with associated biofilm, and identify labeled cells on the sample (Fig. 2A and B). It has previously not been possible to locate individual cells within a biofilm on such a complex sample without having to employ specialized techniques such as laser confocal scanning microscopy, which itself has some inherent limitations.

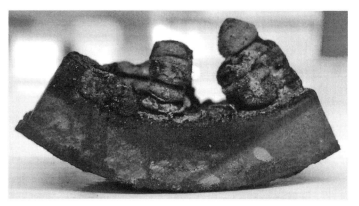

FIGURE 1 A coupon cut from a cast iron pipe from a mains supply drinking water system. The coupon is approximately 4 cm × 2 cm and has considerable corrosion deposits (up to 2 cm in height), illustrating the problems associated with such samples. The pipe was laid in 1940 in an urban area and had been scraped and bitumen lined in 1979. Pipe supplied by a leading European water company.

This work will provide important information on the survival and distribution of *legionellae* in biofilms associated with water systems. The use of PNA probes overcomes many of the problems encountered when using DNA probes and leading to the restricted use of FISH-based assays. The thermal stability and chemical resistance of PNA probes make them a valuable tool for use with complex environmental samples and will permit the simultaneous labeling of several target species. In addition they are able to target inaccessible regions of the 16S rRNA molecule. Both DNA probes tested (LEG226 and LegPNE1) target sequences located in these inaccessible but highly variable regions (6), and this contributed to their inefficiency. The DNA probe, LEG226, (9) was found to be inadequate for use in direct FISH assays, but Wilks & Keevil (13) have clearly illustrated that a PNA probe based on a shortened, but identical, sequence was efficient at labeling the target strains. Additionally, combining the use of PNA probes with the sophisticated EDIC/EF microscope has produced a technique for the direct examination of biofilm samples with minimal disruption on even highly corroded surfaces. It has been shown previously (2) that *legionellae* are widespread in potable water systems, but there is no information on population sizes and the proportion associated with biofilm compared to the bulk water. The PNA-FISH method provides a method to undertake

FIGURE 2 Coupon cut from a mains supply drinking water system, followed by hybridization with the *L. pneumophila*-specific PNA probe (PLPNE620). (A) EDIC image showing the surface of the pipe. (B) Tetramethyl-rhodamine isocyanate channel showing labeled bacteria on pipe surface. Magnification × 1000.

such a study and provide quantification of populations. As outbreaks of Legionnaires' disease continue, it is imperative that we increase our understanding of the distribution and prevalence of this serious human pathogen.

ACKNOWLEDGMENTS

This work has been undertaken as part of a research project which is supported by the European Union within the Fifth Framework Programme, "Energy, Environment and Sustainable Development Programme," no. EVK1-2002-00108. The authors are solely responsible for the work; it does not represent the opinion of the Community and the Community is not responsible for any use that might be made of data appearing herein.

REFERENCES

1. **Azevedo, N. F., M. J. Vieira, and C. W. Keevil.** 2003. Establishment of a continuous model system to study *Helicobacter pylori* survival in potable water biofilms. *Water Sci. Technol.* **47:**155–160.
2. **Brooks, T., R. A. Osicki, V. S. Springthorpe, S. A. Sattar, L. Filion, D. Abrial, and S. Rifford.** 2004. Detection and identification of *Legionella* species from groundwaters. *J. Toxicol. Environ. Health* **67:**1845–1859.
3. **Cole, J. R., B. Chai, R. J. Farris, Q. Wang, S. A. Kulam, D. M. McGarrell, G. M. Garrity, and J. M. Tiedje.** 2005. The Ribosomal Database Project (RDP-II): sequences and tools for high-throughput rRNA analysis. *Nucleic Acids Res.* **3:** D294–D296.
4. **Egholm, M., O. Buchardt, P. E. Nielsen, and R. H. Berg.** 1992. Peptide nucleic acids (PNA). Oligonucleotide analogues with an achiral peptide backbone. *J. Am. Chem. Soc.* **114:**1895–1897.
5. **Egholm, M., O. Buchardt, L. Christensen, C. Behrens, S. M. Freier, D. A. Driver, R. H. Berg, S. K. Kim, B. Nordén, and P. E. Nielsen.** 1993. PNA hybridizes to complementary oligonucleotides obeying the Watson-Crick hydrogen bonding rules. *Nature* **365:**556–568.
6. **Fuchs, B. M., G. Wallner, W. Beisker, I. Schwippl, W. Ludwig, and R. Amann.** 1998. Flow cytometric analysis of the *in situ* accessibility of *Escherichia coli* 16S rRNA for fluorescently labelled oligonucleotide probes. *Appl. Environ. Microbiol.* **64:**4973–4982.
7. **Grimm, D., H. Merkert, W. Ludwig, K.-H. Schleifer, J. Hacker, and B. C. Brand.** 1998. Specific detection of Legionella pneumophila: construction of a new 16S rRNA-targeted oligonucleotide probe. *Appl. Environ. Microbiol.* **64:**2686–2690.
8. **Keevil, C. W.** 2003. Rapid detection of biofilms and adherent pathogens using scanning confocal laser microscopy and episcopic differential interference contrast microscopy. *Water Sci. Technol.* **47:**105–116.
9. **Manz, W., R. Amann, R. Szewzyk, U. Szewzyk, T.-A. Stenström, P. Hutzler, and K.-H. Schleifer.** 1995. *In situ* identification of *Legionellaceae* using 16S rRNA-targeted oligonucleotide probes and confocal laser scanning microscopy. *Microbiology* **141:**29–39.
10. **Rogers, J., and C. W. Keevil.** 1992. Immunogold and fluorescein labeling of *Legionella pneumophila* within an aquatic biofilm visualised by using episcopic differential interference contrast microscopy. *Appl. Environ. Microbiol.* **58:**2326–2330.
11. **Stender, H., M. Fiandaca, J. J. Hyldig-Nielsen, and J. Coull.** 2002. PNA for rapid microbiology. *J. Microbiol. Methods* **48:**1–17.
12. **Surman, S., G. Morton, C. W. Keevil, and R. Fitzgeorge.** 2002. *Legionella pneumophila* proliferation is not dependent on intracellular replication, p. 86–89. *In* R. Marre, (ed.). *Legionella.* ASM Press, Washington, D.C.
13. **Wilks, S. A., and C. W. Keevil.** 2006. Targeting species-specific low-affinity 16S rRNA binding sites by using peptide nucleic acids for detection of legionellae in biofilms. *Appl. Environ. Microbiol.* **72:**5453–5462.

LEGIONELLA DETECTION FROM WATER SAMPLES BY REAL-TIME PCR

Laura Franzin, Daniela Cabodi, and Nicoletta Bonfrate

106

Contamination of hospital water systems with *Legionella* is a well-known cause of nosocomial legionellosis, and rapid identification of the source of infection is essential to prevent further cases. Culture is the "gold standard" for the diagnosis of *Legionella* infection (2, 3, 7) and for the detection from environmental samples. However, several days are required to obtain a positive result, with most *Legionella* colonies being detected within 3 to 5 days. Real-time PCR is a promising technique, confirmed by recent publications reporting high sensitivity and specificity in clinical (5, 8, 10) and in water samples (1, 6, 9).

In this study, the direct detection of *Legionella* from hospital water samples was performed by commercial real-time (RT)-PCR system, that offers rapid results and reduced risk of cross-contamination. The preliminary results of 61 water samples from 6 hospitals and 15 water samples from air conditioning systems are presented here and compared with those of the isolation method.

Culture was performed following a standard protocol (4). Water samples (5 liters) were concentrated by filtration through cellulose acetate

membrane filter (0.2-μm pore size) and resuspended in 10 ml of the same water. Aliquots (0.1 ml) of the untreated, heat-treated (50°C for 30 min) and acid-washed suspensions were plated on buffered charcoal yeast extract, BMPA, and MWY. The plates were incubated at 37°C for 15 days. The strains of *Legionella* isolated were serologically typed by slide agglutination and by immunofluorescence assay.

For direct molecular detection, 1 liter water was filtered using a polyvinylidene difluoride membrane filter (0.45-μm pore size). DNA extraction was performed with QIamp DNA Mini Kit (QUIAGEN). The principle steps were cell lysis with proteinase K, purification on spin column (adsorption onto silica-gel membrane) and elution.

Qualitative and quantitative detection of *L. pneumophila* and *Legionella* spp. were then performed on each DNA extracted by RT-PCR with iCycler and the BIO-RAD reagents (iQ-Check Screen for *Legionella* spp. and for *L. pneumophila*; iQ-Check Quanti for *Legionella* spp. and for *L. pneumophila*) following producer instructions. The genus *Legionella* is identified by amplifying the DNA sequence with specific primers complementary to conserved regions in the 5S ribosomal RNA gene. The amplicon are detected in real time using a 5S rRNA-specific fluorescent probe (Molecular

Laura Franzin, Daniela Cabodi, and Nicoletta Bonfrate Specialist Microbiology Research Laboratory, Amedeo di Savoia Hospital, 10149 Turin, Italy.

Legionella: State of the Art 30 Years after Its Recognition
Edited by Nicholas P. Cianciotto et al.
©2006 ASM Press, Washington, D.C.

Beacon). The identification of *L. pneumophila* is performed using the *mip* gene as a target. The quantitative determination (genomic units [GU]/liter) of *L. pneumophila* and *Legionella* spp. is calculated by comparison to a standard curve based on four DNA standard controls (Qs1 to Qs4), which also provide the quantification range.

Culture was positive (*Legionella* ≥20 CFU/liter) in 28 (45.9%) hospital water samples, and the strains were identified as *L. pneumophila* serogroup 1, 3, 6, and 2-14 and *Legionella* spp. The results of qualitative RT-PCR were as follows: *Legionella* spp. was positive (detection limit: 133 GU/liter) in 60 (98.4%) samples and *L. pneumophila* was positive in 44 (72.1%). Culture and *Legionella* spp PCR were both positive in 27 (44.3%) samples, culture and *L. pneumophila* PCR in 25 (41%). Concordances between culture (detection limit: 20 CFU/liter) and RT-PCR (detection limit: 133 GU/liter) were 44.3% for *Legionella* spp. and 63.9% for *L. pneumophila*. The results of culture and qualitative analysis for *L. pneumophila* by RT-PCR are shown in Table 1. The qualitative detection of *Legionella* spp. and *L. pneumophila* showed that RT-PCR was more sensitive than culture. Isolation and quantitative

TABLE 1 Results of culture and qualitative analysis for *L.pneumophila* by RT-PCR

RT-PCR result	No. (%) of samples with following result		Total
	Positive	Negative	
Positive	25 (41)	19 (31.1)	44 (72.1)
Negative	3 (4.9)	14 (23)	17 (27.9)
Total	28 (45.9)	33 (54.1)	61

PCR methods for enumeration of *Legionella* spp. and *L. pneumophila* showed an association between the methods, but statistical correlation was weak. The results comparison of culture and quantitative PCR methods for enumeration of *L. pneumophila* are shown in Figure 1. All hospitals examined showed some positive samples for culture and PCR, except for one that was culture negative and PCR positive. The culture-positive samples in colonized hospitals ranged from 33.3% to 75%, the samples positive for *L. pneumophila* by RT-PCR ranged from 57.1% to 100%, and those for *Legionella* spp. ranged from 66.7% to 100%. All water samples from air conditioning systems were culture negative (<100 CFU/liter), but RT-PCR was positive in 15 water samples for *Legionella* spp. and in 2 for *L. pneumophila*.

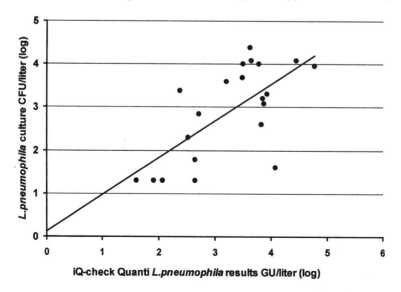

FIGURE 1 Comparison of culture and quantitative PCR methods for enumeration of *L. pneumophila*.

Culture is the gold standard for *Legionella* detection from water. The rate of recovery is usually noticeable less than 100% of the time due to fastidious requirements, overgrowth by other bacteria and damage to legionellae by concentration steps. CFU may also underestimate the true number of bacteria (presence of *Legionella* in clump or in amoebae) (6, 9). In particular, intraamoebal legionellae do not form colonies but are still viable and can reenter a growth phase at any moment, depending on environmental conditions.

RT-PCR offers the advantages of high sensitivity (100%) and rapid results, as the total time required was about 5 h. Nevertheless, the large amount of *Legionella* detected by PCR may also represent nonviable cells or only bacterial DNA which is not infectious to humans or possibly unknown environmental bacteria. Therefore, the high PCR signals should be critically interpreted and do not necessarily represent a health risk for exposed persons (6, 9). In our study, culture was positive in 45.9% of the samples, while RT-PCR was positive in 98.4% for *Legionella* spp. and in 72.1% for *L. pneumophila*. High sensitivity (98.7%) for RT-PCR compared to culture have been previously described for environmental samples (9). In conclusion, *Legionella* detection by RT-PCR from environmental methods seems to be promising. However, at the moment, correlation between the methods results interpretations and public health significance need further evaluations.

ACKNOWLEDGMENTS

We gratefully acknowledge the "Compagnia di San Paolo" Foundation, Turin, Italy for financial support (grant and instrumentations).

REFERENCES

1. **Ballard, A. L., N. K. Fry, L. Chan, S. B. Surman, J. V. Lee, T. G. Harrison, and K. J. Towner.** 2000. Detection of *Legionella pneumophila* using a real-time PCR hybridization assay. *J. Clin. Microbiol.* **38:**4215–4218.
2. **Edelstein, P. H.** 1993. Legionnaires' disease. *Clin. Infect. Dis.* **16:**741–749.
3. **Fields, B. S., R. F. Benson, and R. E. Besser.** 2002. *Legionella* and Legionnaires' disease: 25 years of investigation. *Clin. Microbiol. Rev.* **15:**506–526.
4. **Franzin, L., M. Castellani Pastoris, P. Gioannini, and G. Villani.** 1989. Endemicity of *Legionella pneumophila* serogroup 3 in a hospital water supply. *J. Hosp. Infect.* **13:**281–288.
5. **Hayden, R. T., J. R. Uhl, X. Qian, M. K. Hopkins, M. C. Aubry, A. H. Limper, R. V. Lloyd, and F. R. Cockerill.** 2001. Direct detection of *Legionella* species from bronchoalveolar lavage and open biopsy specimens: comparison of LightCycler PCR, in situ hybridization, direct fluorescence antigen detection, and culture. *J. Clin. Microbiol.* **39:**2618–2626.
6. **Levi, K., J. Smedley, and K. J. Towner.** 2003. Evaluation of a real-time PCR hybridization assay for rapid detection of *Legionella pneumophila* in hospital and environmental water samples. *Clin. Microbiol. Infect.* **9:**754–758.
7. **Murdoch, D. R.** 2003. Diagnosis of *Legionella* infection. *Clin. Infect. Dis.* **36:**64–69.
8. **Templeton, K. E., S. A. Scheltinga, P. Sillekens, J. W. Crielaard, A. P. van Dam, H. Goossens, and E. C. Claas.** 2003. Development and clinical evaluation of an internally controlled, single-tube multiplex real time PCR assay for detection of *Legionella pneumophila* and other *Legionella* species. *J. Clin. Microbiol.* **41:**4016–4021.
9. **Wellinghausen, N., C. Frost, and R. Marre.** 2001. Detection of Legionellae in hospital water samples by quantitative real-time LightCycler PCR. *Appl. Environ. Microbiol.* **67:**3985–3993.
10. **Wilson, D. A., B. Yen-Lieberman, U. Reischl, S. M. Gordon, and G. W. Procop.** 2003. Detection of *Legionella pneumophila* by real-time PCR for the *mip* gene. *J. Clin. Microbiol.* **41:**3327–3330.

EVALUATION OF THE DYNAL BIOTECH *LEGIONELLA* IMMUNOMAGNETIC SEPARATION METHOD VERSUS CONVENTIONAL CULTURE FOR THE ISOLATION OF *LEGIONELLA PNEUMOPHILA* SEROGROUP 1 FROM WATER SAMPLES

Sue M. Mietzner, Janet E. Stout, Jaclynn L. Shannon, Victor L. Yu, and David R. Wareing

107

Environmental monitoring of water systems for *Legionella* is performed for routine microbiological surveillance and outbreak investigations. The recovery of *Legionella* from water samples using conventional culture on selective media can be improved by sample concentration and/or sample decontamination techniques, but there is a reduction in recovery associated with these processes. A new immunoseparation method has been developed to improve the recovery of *Legionella* by concentrating and selectively separating *Legionella* spp. directly from water samples using immunomagnetic separation (IMS). Immunomagnetic separation has been used successfully for detection of other pathogens including *Salmonella* spp. (3). Our study was designed to determine whether IMS could enhance, or replace, steps in the culture method for *Legionella* recovery such that the use of the IMS method would result in greater sensitivity and specificity compared to conventional culture isolation.

We conducted the study in two phases and evaluated the IMS process alone and in combination with concentration and/or acid pre-

Sue M. Mietzner, Janet E. Stout, Jaclynn L. Shannon, and Victor L. Yu Department of Infectious Disease, VA Pittsburgh Healthcare System, Pittsburgh, PA 15240. *David R. Wareing* Microbiology R&D, Invitrogen Corporation, Preston, UK.

treatment in detecting *Legionella* spp. and *L. pneumophila* serogroup 1 (sg 1). Our objectives were to (i) determine concordance between the IMS method and conventional culture for the isolation of *L. pneumophila* using known positive and negative potable water samples; and (ii) compare the sensitivity and specificity of the IMS method versus culture isolation for the recovery of *Legionella* in prospectively tested potable water samples.

WATER SAMPLES AND METHODS

The IMS procedure was performed using the BeadRetriever apparatus (Dynal Biotech Ltd., Preston, United Kingdom) for automated IMS using Dynabeads anti-Legionella (Dynal Biotech Ltd.). Legionella IMS was performed per the manufacturer's instructions. Briefly, reagents and duplicate water samples were added to sterile, disposable tube strips. Up to 15 water samples could be tested in one 45-min run of the BeadRetriever. Sterile disposable tip combs were aseptically inserted in the instrument to protect the sample(s) from the magnetic probes. A 1-ml sample of water, a 100-μl sample buffer, and 50-μl Dynabeads anti-*Legionella* were incubated in tube 1 of the strip. *Legionella*, if present, will bind to the antibody-coated beads. The bead-bacteria complexes were subsequently separated by the application of a magnetic field through each probe. The bead complexes were transferred through tubes 2 and 3 of the strips for washing. The bead complexes were released into 100 μl of wash buffer, which was transferred and spread onto culture media (buffered charcoal yeast extract [BCYE] and BCYE with dyes, glycine vancomycin, and polymixin B [DGVP]).

In phase I, 11 potable water samples were known by previous culture to be positive and negative for *L. pneumophila* sg 1, with concentrations between 0 and >100 CFU/ml. The samples and their filtered concentrates (100 to 10 ml) (*n* = 22) were tested by both methods, untreated and after heat or acid pretreatment. Direct and pretreated samples (0.1 ml) were plated in duplicate to three media: BCYE, DGVP, and CCVC (BCYE with cephalothin, colistin, vancomycin, and cycloheximide) (*n* = 369 plates). Direct and pretreated samples (1.0 ml) were processed by IMS, and 100 μl of bead-bacteria complexes were divided equally (50 μl) between two media: BCYE and GVPC (BCYE with glycine, vancomycin, polymyxin B, and cycloheximide) (*n* = 132 plates). All culture plates were incubated at 37°C and were examined at 4 to 7 days. Latex agglutination (Oxoid Ltd, Basingstoke, England) was used to confirm isolates as *L. pneumophila* sg 1. Isolate identification was confirmed by direct fluorescent antibody testing (M-Technologies, Inc., Alpharetta, Ga.).

In phase II, 84 potable water samples from 14 institutions were submitted to the Special Pathogens Laboratory for analysis and were prospectively tested. Samples were plated to BCYE and DGVP (0.1 ml) directly and after filter concentration as previously described (6). A 100-ml volume was filter concentrated, and the filter was resuspended in 10 ml of the original sample (concentrate). Aliquots of the direct sample (1.0 ml) were processed in duplicate by IMS. The bacteria-bead complexes were plated (100 μl) to BCYE and DGVP. Cultures were examined as in phase I and reported as positive, negative, or uninterpretable (overgrowth of non-*Legionella* bacteria on

both media.) Isolate identification was as described for phase I. We defined a "true positive" as positive by conventional culture. Analysis was performed for samples positive for any *Legionella* species and for samples positive for *L. pneumophila* sg 1 only.

COMPARISON OF CULTURE WITH OR WITHOUT IMS

With the 11 known potable water samples, there was concordance (100% sensitivity) of culture results between the two methods, regardless of concentration or pretreatment of samples. The unconcentrated, untreated samples showed the greatest agreement in CFU/ml (Table 1). The lower limit of detection of the IMS method was ≥20 CFU/ml of *L. pneumophila* sg 1.

The number of colonies per plate was greater with IMS. The mean CFU/plate of *L. pneumophila* sg 1 by culture versus IMS for unconcentrated untreated, heat-treated, and acid treated samples were 22 versus 70; 18 versus 36; and 4 versus 12, respectively. However, when calculated to reflect CFU/milliliter of the original sample, the mean CFU/milliliter by conventional culture of the filtered concentrate was higher than by IMS. Specifically, mean CFU/milliliter of *L. pneumophila* sg 1 by culture versus IMS for untreated, heat-treated, and acid-treated concentrates were 1,503 versus 101; 1,392 versus 118; and 818 versus 177, respectively.

The overall sensitivity/specificity of the *Legionella* IMS method for detecting *Legionella* spp. in the 84 samples tested was 78% and 95%, respectively (Table 2). Isolates recovered from these samples by both methods included *L. pneumophila* serogroups 1, 3, 6, and 12. Some isolates were identified as *L. pneumophila*, not serogroups 1 to 6. *L anisa* was also isolated. Among the 84 samples tested, 17 of 84 were positive for *L. pneumophila* sg 1 by conventional culture. Twelve samples were positive for *L. pneumophila* sg 1 by both methods, and five were positive for *L. pneumophila* sg 1 only by conventional culture (false-negative by IMS) (71% sensitivity for *L. pneumophila* sg 1). For

TABLE 1. Results from the phase I study. The *Legionella* IMS method recovered *Legionella pneumophila* serogroup 1 from potable water samples at concentrations similar to conventional culture

Sample no.	*Legionella* (CFU/ml)					
	Untreated		Heat-treated[a]		Acid treated[b]	
	Culture	IMS	Culture	IMS	Culture	IMS
1	0	0	0	0	0	0
2	0	0	0	0	0	0
3	0	0	0	0	0	0
4	20	126	50	6	10	12
5	690	>200[c]	490	>200	100	24
6	70	>200	30	4	10	0
7	30	20	20	8	20	4
8	0	0	0	0	0	0
9	0	0	0	0	0	0
10	180	98	130	34	30	12
11	300	>200	340	176	70	88

[a]Heat-treated: 30 min. at 50°C.
[b]Acid-treated: Culture—centrifuge 1 ml for 10 min. at 6,000 rpm, remove 0.5 ml and resuspend to 1 ml with acid for 3 min; for IMS = 1:1 acid mix for 5 min.
[c]>200 CFU/ml = maximum value by IMS due to sample processing limits during the phase I study.

one sample, culture was positive for *L. pneumophila* sg 1, but the IMS only detected *L anisa*. The concentration of *L. pneumophila* sg 1 and *L. anisa* in this sample was 1 to 10 CFU/ml. Compared to conventional culture for *L. pneumophila* sg 1 only, the IMS method demonstrated a specificity of 98.5% (65 were negative and 2 were false-positive).

CONCLUSIONS
Phase I of our study showed the results of the IMS method to be concordant with culture in

terms of positive/negative. Phase II showed conventional culture to be more sensitive than IMS for the detection of *L. pneumophila* sg 1.

The sensitivity of IMS for potable water samples was approximately 78%; however, it should be noted that the low number of positive samples (23 of 84) may have caused an underestimation of the sensitivity. In addition, of the five samples that were culture positive/ IMS negative, the concentration of *Legionella* was 1 to 10 CFU/ml in three of the five samples. This concentration is below the stated

TABLE 2 Results from phase II prospective study of potable water samples

Culture result	No. of *Legionella* samples with the following IMS result[a]		Total
	Positive	Negative[c]	
Positive[b]	18	5	23
Negative	3	58	61
Total	21	63	84

[a]IMS had a sensitivity of 78% and a specificity of 95%.
[b]Culture positive, any positive result (direct or filter concentrated).
[c]For 3 of 5 IMS negative samples, the concentration of *Legionella* was 1 to 10 CFU/ml. This is below the stated detection limit for the IMS method.

detection limit for the IMS method. Based on this study, the lower limit of detection of the IMS method for *Legionella* was approximately 20 CFU/ml.

Environmental culture using *Legionella* IMS could be used to determine the risk of legionellosis in the hospital setting. Routine environmental testing for *Legionella* has been recommended by health authorities (1, 5). The Allegheny County Health Department recommends annual environmental testing and defines increased risk of disease with high percent positivity (> 30%) of water outlets. Concentration has not correlated with increased risk, but risk has correlated with percent positivity (2, 4). Therefore, the IMS method could be used to determine percent positivity of water outlets.

The *Legionella* IMS may be an adjunct to conventional culture by concentrating *Legionella* species in a sample without the need for time-consuming filtration.

REFERENCES

1. **Allegheny County Health Department.** 1997. *Approaches to Prevention and Control of Legionella Infection in Allegheny County Health Care Facilities,* 2nd ed., p. 1–15. Allegheny County Health Department, Pittsburgh, Pa. (http://www.legionella.org.)

2. **Best, M., V. L. Yu, J. Stout, A. Goetz, R. R. Muder, and F. Taylor.** 1983. *Legionellaceae* in the hospital water supply: epidemiological link with disease and evaluation of a method of control of nosocomial Legionnaires' disease and Pittsburgh pneumonia. *Lancet* **2:**307-310.

3. **Duncanson, P., D. R. A. Wareing, and O. Jones.** 2003. Application of an automated immunomagnetic separation-enzyme immunoassay for the detection of *Salmonella* spp. during an outbreak associated with a retail premises. *Lett. Appl. Microbiol.* **37:**1–5.

4. **Kool, J. L., D. Bergmire-Sweat, J. C. Butler, E. W. Brown, D. J. Peabody, D. S. Massi, J. C. Carpenter, J. M. Pruckler, R. F. Benson, and B. S. Fields.** 1999. Hospital characteristics associated with colonization of water systems by *Legionella* and risk of nosocomial Legionnaire's disease: a cohort study of 15 hospitals. *Infect. Control Hosp. Epidemiol.* **20:**798–805.

5. **State of Maryland Department of Health and Mental Hygiene.** 2000. *Report of the Maryland Scientific Working Group to Study Legionella in Water Systems in Healthcare Institutions.*

6. **Ta, A. C., J. E. Stout, V. L. Yu, and M. M. Wagener.** 1995. Comparison of culture methods for monitoring *Legionella* species in hospital potable water systems and recommendations for standardization of such methods. *J. Clin. Microbiol.* **33:**2118–2123.

A NOVEL AND RAPID *LEGIONELLA* DETECTION SYSTEM FOR WATER ANALYSIS

Steven Giglio, Paul T. Monis, and Christopher P. Saint

108

The usefulness of sampling waters for the detection of *Legionella* remains a point of discussion among researchers and industry for a variety of reasons. There is still no consensus regarding the infectious dose and environmental levels of *Legionella* necessary for either the spread of disease or the ability of currently reported counts to aid with risk management principles. Data on *Legionella* counts from cooling towers implicated in outbreaks are not readily available, but counts between 1,000 CFU/ml (1) to 100,000 CFU/ml (3) have been found in suspected sources, whereas counts found in potable water supplies in nosocomial settings have been very low (6).

Nonetheless, numerous authorities have prescribed mandatory testing of cooling tower and warm water systems as part of the maintenance of such systems. Moreover, perceptions of adequate maintenance are often rooted in a failure to detect *Legionella* by culture, even when there is no legal requirement for testing, and hence testing of waters for *Legionella* is almost routine.

Currently, most reporting laboratories adhere to Australian/New Zealand standards (AS/NZS 3896:1998) or international standards (ISO 17113:1998) for the quantitative examination of *Legionella* in water. These methods are similar in nature and are traditionally slow, with detection and confirmation of *Legionella* taking from 5 to 14 days, depending on laboratory protocol and the *Legionella* species present. Numerous other technologies exist for the detection of *Legionella* (e.g., direct fluorescence antibodies), but interpretation is difficult because some systems only detect *Legionella pneumophila* serogroup (sg) 1, and others cannot discriminate between viable and nonviable organisms. Here we describe a novel detection system utilizing two liquid media (medium 1 and medium 2, formulations are currently under provisional patent applications) and real-time PCR that is capable of detecting viable *Legionella* in a variety of waters within 2 to 3 days. Over 30 strains of *Legionella* have been tested by this method, and the real-time PCR detection system has been validated by our laboratory (4). We have tentatively named this the GMS method, and it is currently under U.S. and Australian provisional patent applications.

Water samples (1 ml) were inoculated into 49-ml aliquots of liquid medium 1 and incubated at 35°C with shaking at 180 rpm for 4 h. For each sample, the entire contents of liquid

Steven Giglio, Paul T. Monis, and Christopher P. Saint Australian Water Quality Centre, South Australian Water Corporation, PMB 3, Salisbury, SA, 5108, Australia.

Legionella: State of the Art 30 Years after Its Recognition
Edited by Nicholas P. Cianciotto et al.
©2006 ASM Press, Washington, D.C.

medium 1 were aseptically transferred to 50 ml of liquid medium 2 to give a total volume of 100 ml, and a 1-ml aliquot was taken from the mix and stored at 4°C until required (time zero PCR). The resultant mixtures were incubated at 35°C with shaking at 180 rpm for 44 h. For each sample, a 1-ml aliquot was drawn at the end of incubation and diluted 1:100 in sterile phosphate-buffered saline (Fig. 1). Real-time PCR utilizing *Legionella* 16S rRNA primers was performed on a Corbett Rotor-Gene 3000 utilizing SYTO9 as a DNA intercalating dye for PCR as described by Monis et al. (5). The GMS method was challenged with 82 field samples (cooling tower, evaporative tower, and potable waters) and run in parallel to AS/NZS 3896:1998. In addition, the GMS method was subjected to validation with a total of 150 samples as stipulated in the AS/NZS 4659.1:1999 "Guide to Determining the

Equivalency of Food Microbiology Test Methods: Qualitative Test" using low-level spikes of *L. pneumophila* sg 1 and 6, *L. anisa, L. bozemanii,* and *L. micdadei.* The shelf life of the liquid media was determined according to the National Australian Testing Authority and the Australian Society for Microbiology guidelines for medium preparation (2).

Evaluation against AS/NZS 4659 highlighted five samples in which *Legionella* were detected by the GMS method but not the standard method (Table 1). The standard deviation between replicate analyses of a single water sample containing *Legionella* is typically large, and data from our laboratory have shown that 10 replicate analyses of a water sample seeded with 10 CFU of *Legionella* per ml yielded only the one detection when tested by AS/NZS 3896:1998. This may be due to the distribution of the seeded *Legionella* in the wa-

Water sample

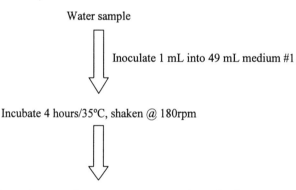

Inoculate 1 mL into 49 mL medium #1

Incubate 4 hours/35°C, shaken @ 180rpm

Transfer all medium #1 to 50 mL medium #2, remove 1 mL aliquot for time zero PCR

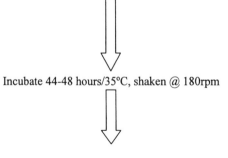

Incubate 44-48 hours/35°C, shaken @ 180rpm

Remove 1 mL and dilute 1/100
Perform *Legionella* 16S rRNA real-time PCR assay

FIGURE 1   Flow chart of the new GMS method.

TABLE 1 Validation of the GMS method against AS/NZS 4659 and AS/NZ 3896:1998

Test result	No. of samples	
	GMS Positive	GMS Negative
AS/NZS 4659 positive	70	0
AS/NZS 4659 negative	5	75
AS/NZ 3896:1998 positive	11	1
AS/NZ 3896:1998 negative	0	70

ter sample or the inability of selective media to recover low-level *Legionella*. Our results, however, suggest that there may be an increased sensitivity when using the GMS method over the current standard method.

Comparison of methods using field samples showed that the GMS method is 100% specific (no false positives) and 92% sensitive, with a single discrepant sample (Table 1). This discrepancy was due to the presence of *Delftia* sp. that overgrew and/or inhibited the growth of *Legionella* in liquid medium 2, with subsequent failure to detect by real-time PCR. The distribution of *Delftia* sp. in water samples has not been extensively studied, and the significance of this discrepant result is uncertain.

The performance of medium 1 and medium 2 remained within quality control limits for a minimum period of 9 weeks when stored at 4°C and in the dark. Parameters that were monitored included pH, growth of control organisms (*Legionella*), inhibition of organisms (e.g, *Escherichia coli*), and physical integrity. These data suggest that the media used in the GMS method would lend itself to storage and quality control procedures currently employed by most microbiology laboratories.

The GMS method described is rapid, simple, cost effective, less laborious than conventional culture, and easy to implement in any modern day microbiology laboratory. It offers multiplatform real-time PCR compatibility and is capable of detecting a wide range of *Legionella* in industrial samples within 2 to 3 days, compared to 5 to 14 days for current culture methods.

The GMS method will allow simultaneous reporting of heterotrophic colony counts (HCC) with *Legionella* detection, as both require 2 to 3 days to complete. This will be particularly advantageous to industry, as administration of remedial actions with both the HCC and *Legionella* results in hand will result in swift administration of the correct remedial actions if necessary. In a potentially hazardous situation, the elimination of the period between receiving a HCC report and a *Legionella* report (typically 5 to 10 days) will lessen the exposure time to susceptible individuals and potentially reduce the number of legionellosis notifications.

REFERENCES

1. **Anonymous.** 1998. *Waters: Examination for Legionellae AS/NZS 3896:1998.* Standards Australia, North Sydney, NSW, Australia.
2. **Arghros, M., G. Douglass, M. Locher, P. Mugg, D. C. Myatt, T. Olma, A. Scholtes, and I. Wilkinson.** 2004. *Guidelines for Assuring Quality of Food and Water Microbiological Culture Media.* Australian Society for Microbiology, Media Quality Control Special Interest Group.
3. **Broadbent, C.** 1996. *Guidance for the Control of Legionella.* National Environmental Health Forum Monographs, Water Series no. 1.
4. **Giglio, S., P. T. Monis, and C. P. Saint.** 2005. *Legionella* confirmation using real-time PCR and SYTO9 is an alternative to current methodology. *Appl. Environ. Microbiol.* **71:**8944–8948.
5. **Monis, P. T., S. Giglio, and C. P. Saint.** 2005. Comparison of SYTO9 and SYBR Green I for real-time polymerase chain reaction and investigation of the effect of dye concentration on amplification and DNA melting curve analysis. *Anal. Biochem.* **340:**24–34.
6. **Torii, K., Y. Iinuma, M. Ichikawa, K. Kato, M. Koide, H. Baba, R. Suzuki, and M. Ohta.** 2003. A case of nosocomial *Legionella pneumophila* pneumonia. *Jpn. J. Infect. Dis.* **56:**101–102.

USE OF REAL-TIME PCR FOR DETECTION AND QUANTIFICATION OF *LEGIONELLA* BACTERIA IN WATER ON THE SCALE OF A WATERSHED: THE VIDOURLE VALLEY

Laurent Garrelly, Celine Minervini, L. Rolland, Séverin Pistre, and Jean-Christian Personné

109

Legionella bacterium is present naturally in ground water (1, 3) and has been described as a new pathogen growing in water distribution systems (10). This pathogen survives in association with biofilms (2, 5, 7) and demonstrates intracellular development in amoebae (6). The existence of a viable-but-nonculturable state (VBNC) of *Legionella* according to the environmental conditions has been described in literature (1, 10). *Legionella* culturing according to the ISO 11730 standard (1998) from cold ground water very often shows negative results. We have observed that culture results are often positive after domestic or industrial use of the distributed water. This increase of *Legionella* contamination is due either to the proliferation of the bacteria in the water distribution system or to the transformation of VBNC or intraamoebaean *Legionella* into culturable *Legionella*.

Real-time PCR is a technology that permits us to quantify, in a rapid, sensitive, and re-

liable way, the genomic units of bacteria in water (8). This method detects and quantifies the total *Legionella* flora present in the sample, independent of the culturable state of the bacteria. Therefore, we have decided to use real-time PCR to study the presence of *Legionella* bacteria in the environment and in water distribution networks. We have quantified *Legionella* spp. and *L. pneumophila* in natural water, water distributed in the urban network following disinfectant treatment, and waste water from domestic and industrial use. Samples were collected along the length of a hydrographical basin crossed by the Vidourle River, a coastal river 100 km long with a karstic and alluvial aquifer (9). The impact of the towns of St. Hyppolite, Sommmières, and Lunel, which are crossed by the river, was the subject of a more in-depth investigation. In total, 31 sampling points gave rise to 48 analyzed samples. The water microbiology was studied by ATP measurements and enumeration of *Legionella* by culture and by real-time PCR. The enumeration by culture was carried out in accordance with the ISO 11 731 (1998) standard. DNA from water samples was extracted using the Aquadien kit (Bio-Rad). Real-time PCR analyses were carried out using iQ-Check Quanti *Legionella* spp. and iQ-Check Quanti *L. pneumophila* kits (Bio-Rad) on the

Laurent Garrelly, Celine Minervini, and L. Rolland R&D Department, Bouisson Bertrand Laboratoires, Parc Euromédecine, 778 rue de la Croix Verte, 34 196 Montpellier, France. *Séverin Pistre and Jean-Christian Personné* Laboratoire Hydrosciences UMR 5569, Maison des Sciences de l'Eau, Université Montpellier II, Place Eugène Bataillon, 34095 Montpellier Cedex, France.

iCycler iQ thermal cycler (Bio-Rad). PCR results are expressed in logarithm of genome units per liter of water sample (log GU/liter). Watergiene (Charm Sciences, Inc.) swabs were used for the ATP measurements. ATP-metry results are expressed in logarithm of relative luminescent units (log RLU).

ANALYSIS OF THE DATA OBTAINED

Thirty natural water samples were collected along the Vidourle river. Culture of *Legionella* did not show any positive results. No *L. pneumophila* was detected in the samples by real-time PCR. *Legionella* spp. PCR results were all positive. The mean value of PCR result was 5, with a standard deviation of 0.8 ($n = 30$; $\bar{x} = 5$; standard deviation [SD] $= 0.8$). The mean ATP-metry result for these samples was 5.9, with a standard deviation of 0.6 ($n = 30$; $\bar{x} = 5.9$; SD $= 0.6$).

Twelve cold water samples were collected from the water distribution system in the towns of St. Hippolyte, Sommières, and Lunel. As in natural water, all samples gave positive *Legionella* spp. PCR results with a mean value of 3.5 and a standard deviation of 0.85 ($n = 12$; $\bar{x} = 3.5$; SD $= 0.85$). Culture results and *L. pneumophila* PCR results were all negative. The mean ATP-metry value for this group of samples was 5.7 ($n = 12$; $\bar{x} = 5.7$; SD $= 0.8$).

Six waste water samples were collected in the three towns. Again, culture and *L. pneumophila* PCR results were all negative, whereas *Legionella* spp. PCR results were all positive, with a mean value of 6.3 and a standard deviation of 0.5 ($n = 6$; $\bar{x} = 6.3$; SD $= 0.5$). The mean ATP-metry result value was 8.1, with a standard deviation of 0.5 ($n = 12$; $\bar{x} = 8.1$; SD $= 0.5$).

Statistical analysis (using Student's law with 5% risk) shows that a significant difference exists between iQ-Check *Legionella* spp. results in natural, distributed, and waste water. The presence of *L. pneumophila* in the natural, distributed, or waste water tested in this study is not significant. Concerning ATP-metry results, a significant statistical difference (using Student's law) exists between natural and waste water and between distributed and waste water, but

no significant difference is observed between natural and distributed water.

CONTAMINATION OF NATURAL, DISTRIBUTED AND WASTE WATER STUDIED BY REAL-TIME PCR AND ATP-METRY

In general, ATP measurements and *Legionella* spp. PCR results were higher in natural water than in distributed water and lower than in waste water (Fig.1). These results demonstrate the obvious efficiency of disinfectant water treatment in reducing *Legionella* contamination. Moreover, used urban water is the basis of a very significant increase in *Legionella* spp. contamination.

The calculated correlation coefficient between ATP-metry and real-time PCR results ($r = 0.56$; 46 degrees of freedom) indicates the existence of a clear positive correlation between both bacterial indicators. This result illustrates the viability of *Legionella* bacteria detected with iQ-Check *Legionella* PCR kits, even though few are observed by culture.

EVOLUTION OF *LEGIONELLA* SPP. CONTAMINATION IN NATURAL WATER, DISTRIBUTED WATER, AND WASTE WATER IN THREE DIFFERENT TOWNS DOWNSTREAM ON THE VIDOURLE RIVER

A significant statistical increase in *Legionella* contamination was observed in the upstream to downstream direction of aquifers along the Vidourle river (Fig. 2). The *Legionella* concentration at the level of the first upstream town (St. Hippolite) is 3.7 log GU/liter, whereas in the second town (Sommières) the contamination is 4.75 log GU/liter, and in the third town (Lunel) the contamination is 5.9 log GU/liter.

The contamination of *Legionella* in the water distribution network in the three towns studied is significantly lower than in ground water. The decrease of *Legionella* contamination in treatment plants demonstrates the major improvement in the quality of water in these stations. Disinfection treatment efficiency varies from one town to another and

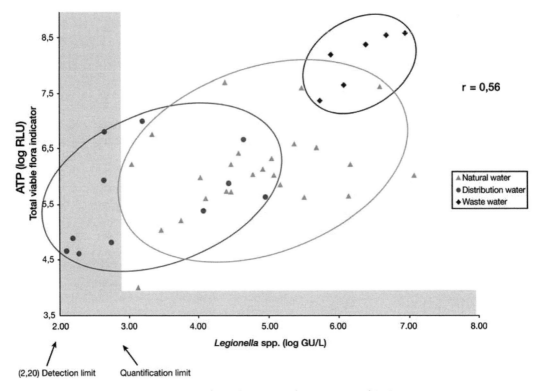

FIGURE 1 Correlation between real-time PCR and ATP-metry.

between the different points of each distribution network.

At the same time, domestic and industrial use contributes significantly to the increase of *Legionella* contamination. Waste water contamination varies between 6 and 7 log GU/liter. By using the real-time PCR technique, the efficiency of water treatment and the role of urban networks in the distribution and cleansing water resources have been successfully demonstrated.

CONTRIBUTION OF THE REAL-TIME PCR AND ATP-METRY IN ENVIRONMENTAL MONITORING

In this study, real-time PCR shed light on the presence of *Legionella* species flora in natural water along Vidourle watershed. Where present,

the *L. pneumophila* concentration was very low. *L. pneumophila* is not the predominant *Legionella* species present in the environment, even though clinically this is the most frequently encountered species (4). The use of the culture method alone for the monitoring of *Legionella* would have led to the conclusion of the absence of the cultivable bacterium in the natural and distributed water of this basin, while PCR analyses give positive results without consideration to the viability of the bacterium. The proliferation of *Legionella* seems to be due mostly to domestic or industrial use. No change from the nonculturable to the culturable state of *Legionella* flora could be demonstrated in this study.

Real-time PCR is an effective method for monitoring *Legionella* contamination in the

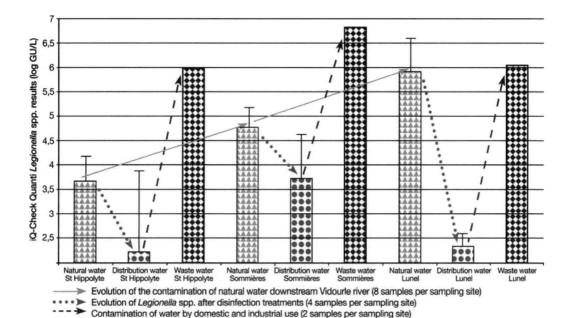

FIGURE 2 Evolution of *Legionella* spp. contamination in three towns.

environment and in urban water networks. This method can be used to evaluate the efficiency of treatment performed in the tap water treatment plants.

Proliferating in biofilms and in amoebae, the *Legionella* species may be a useful indicator for the estimation of the total microbiological quality of water. ATP-metry can also be an interesting tool for the characterization of the microbiological quality of water.

REFERENCES

1. **Brooks, T., R. Osicki, V. Springthorpe, S. Sattar, L. Filion, D. Abrial, and S. Riffard.** 2004. Detection and identification of Legionella species from ground waters. *J. Toxicol. Environ. Health.* **67:**1845–1859.
2. **Calvo-Bado, L. A., W. Morgan, M. Sergeant, T. Pettitt, and J. Whipps.** 2003. Molecular characterization of Legionella populations present within slow sand filters used for fungal plant pathogen suppression in horticultural crops. *Appl. Environ. Microbiol.* **69:**533–541.
3. **Costa, J., I. Tiago, M. S. da Costa, and A. Verissimo.** 2005. Presence and persistence of Legionella spp. in groundwater. *Appl. Environ. Microbiol.* **71:**663–671.
4. **Doleans, A., H. Aurell, M. Reyrolle, G. Lina, J. Freney, F. Vandenesch, J. Etienne, and S. Jarraud.** 2004. Clinical and environmental distributions of Legionella strains in France are different. *J. Clin. Microbiol.* **42:**458–460.
5. **Emtiazi, F., T. Schwartz, S. M. Marten, P. Krolla-Sidenstein, and U. Obst.** 2004. Investigation of natural biofilms formed during the production of drinking water from surface water embankment filtration. *Water Res.* **38:**1197–1206.
6. **Greub, G., and D. Raoult.** 2004. Microorganisms resistant to free-living amoebae. *Clin. Microbiol. Rev.* **17:**413–433.
7. **Murga, R., T. S. Forster, E. Brown, J. M. Pruckler, B. S. Fields, and R. M. Donlan.** 2001. Role of biofilms in the survival of Legionella pneumophila in a model potable-water system. *Microbiology* **147:**3121–3126.
8. **Palmer, C. J., Y. L. Tsai, C. Paszko-Kolva, C. Mayer, and L. R. Sangermano.** 1993. Detection of Legionella species in sewage and ocean water by polymerase chain reaction, direct fluorescent-antibody, and plate culture methods. *Appl. Environ. Microbiol.* **59:**3618–3624.
9. **Personné, J. C., F. Poty, L. Vaute, and C. Drogue.** 1998. Survival, transport and dissemination of Eschericha coli and enterococcci in a fissured environment study of a flood in a karstic aquifer. *J. Appl. Microbiol.* **84:**431–438.
10. **Szewzyk, U., R. Szewzyk, W. Manz, and K. H. Schleifer.** 2000. Microbiological safety of drinking water. *Annu. Rev. Microbiol.* **54:**81–127.

FIELD EVALUATION OF THE BINAX EQUATE TEST KIT FOR ENUMERATION OF *LEGIONELLA PNEUMOPHILA* SEROGROUP 1 IN COOLING WATER SAMPLES

Anita Benovic and Richard Bentham

110

This project investigated the efficacy of detection of *Legionella pneumophila* serogroup 1 (sg 1) in cooling waters using the Binax Equate *Legionella* water test. The test kit offers a rapid detection method (45 min) for *L. pneumophila* sg 1 in environmental water samples. The enzyme-linked immunosorbent assay-based method has a 200-CFU/ml detection limit from a 500 ml filtered sample. In this investigation, 141 cooling water samples from 106 cooling towers were collected and simultaneously subjected to the Binax Equate test method and the Australian Standard culture method for *Legionella* isolation (1).

Samples were also cultured for the presence of *Pseudomonas* spp. including *P. fluorescens*. Additionally, a heat-killed culture of *L. pneumophila* sg 1 at two dilutions (10^6 and 10^5 CFU/ml) was subjected to the Binax Equate test for a 2-week period.

A total of 36.4% of the samples produced a Binax absorbance reading greater than 0.1 optical density at 450 nm (OD_{450}), indicating that they were *L. pneumophila* sg 1 positive. Absorbance readings from the Binax tests ranged between 0 and 1.89 OD_{450}. The distribution of

samples within this range is displayed in Fig. 1. From this figure, it can be seen that almost 50% ($n = 26$) of Binax positive samples had an absorbance reading of less than 0.39 OD_{450}. Of these, 20.8% ($n = 11$) of the samples were slightly above the limit of detection (0.1 to 0.19 OD_{450}). A high proportion of Binax positive results were also between 1.0 and 1.9 OD_{450}.

Legionella species were cultured from 22% ($n = 31$) of all cooling water samples. The results show that out of all of the positive cooling water samples ($n = 31$), 3.2% ($n = 1$) contained *L. pneumophila* sg 1, 12.9% ($n = 4$) contained *L. pneumophila* sg 2-14, 74.2% ($n = 23$) contained autofluorescent *Legionella* spp., and 35.5% ($n = 11$) contained other *Legionella* species. Only one cooling water sample was confirmed to be *L. pneumophila* sg 1 positive by culture methods and subsequent latex agglutination. This sample recorded a concentration of 3.6 × 10^3 CFU/ml by culture and also produced a positive Binax absorbance of 0.62 OD_{450}. For the remaining 52 samples that were Binax positive, *L. pneumophila* sg 1 was not detected by culture. This indicates that 37.1% of all cooling tower samples produced false-positive Binax results. No significant correlation between total *Legionella* isolated by culture methods and Binax absorbance readings was observed. The results

Anita Benovic City of Adelaide, Adelaide, South Australia, Australia, 5000. *Richard Bentham* Department of Environmental Health, Flinders University, GPO Box 2100, Adelaide, South Australia, Australia, 5001.

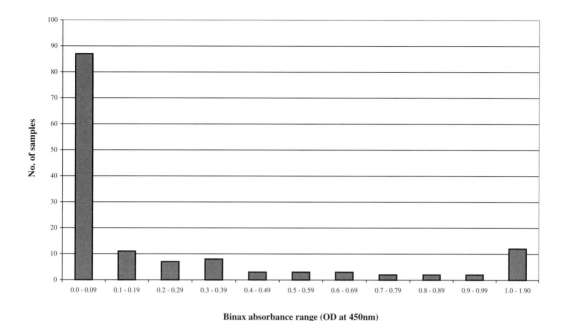

Binax absorbance range (OD at 450nm)

FIGURE 1 Distribution of the Binax absorbance readings of the cooling tower water samples.

show that 13.7% of all Binax negative samples contained culturable *Legionella* spp. and 17.3% of all Binax positive samples contained culturable *Legionella* spp. (not including the *L. pneumophila* sg 1 isolation).

Presumptive *Pseudomonas* spp. were isolated from 42.9% of all cooling water samples. The concentrations ranged between 0 and 2×10^3 CFU/ml. Two samples grew to confluence, indicating very high concentrations of *Pseudomonas* spp. in those samples. Presumptive *P. fluorescens* was isolated from 21.9% of all cooling water samples (Table 1). Statistical analysis of presumptive *Pseudomonas* spp. concentrations and corresponding Binax absorbance readings determined that there was a significant difference between Binax positive and negative samples ($P = 0.1$, student's t test). Further analysis

shows that *Pseudomonas* spp. are more prevalent in Binax positive samples (52.3%; $n = 23$) as opposed to Binax negative samples (36.5%; $n = 27$). The percentage of positive *P. fluorescens* species in Binax positive and negative samples indicates that the distribution was almost identical (i.e., 20.5 and 21.6%, respectively). Statistical analysis of presumptive *P. fluorescens* concentrations and corresponding Binax absorbance readings did not show significant differences between Binax positive and negative samples. This was confirmed when pure cultures of the isolated presumptive *P. fluorescens* produced negative Binax results when subjected to the test.

The Binax absorbance readings decreased after a kill treatment was applied. The rate at which this occurred is displayed in Fig. 2. The results show that the 10^{-1} dilution reduced by

TABLE 1 Summary of microbiological analyses and positive cooling water samples and systems

Microbiological parameters	% of positive cooling water samples	n	% of positive cooling water systems	n
Binax	37.9	140	40.0	105
Legionella culture	15.6	141	14.2	106
Presumptive *Pseudomonas* spp.	42.9	119	45.4	97
Presumptive *Pseudomonas fluorescens*	21.9	119	24.7	97

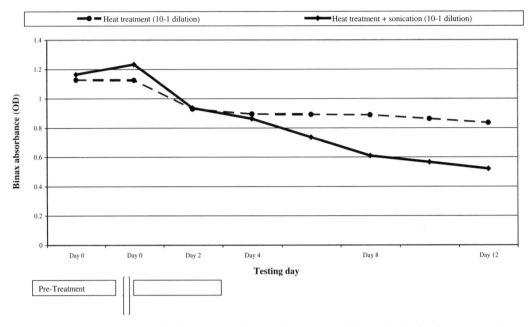

FIGURE 2 The detection of killed *L. pneumophila* sg 1. The *L. pneumophila* sg 1 load of each treatment prior to undergoing treatment was: Heat treatment 10^{-1} dilution = 7.6 × 10^6 CFU; Heat treatment 10^{-2} dilution = 3.85 × 10^5 CFU; Heat treatment + sonication 10^{-1} dilution = 7.65 × 10^6 CFU.

a total of 0.29 OD units in 12 days, while the 10^{-2} dilution reduced by a total of 0.45 OD units in 8 days. Pretreatment (day 0) results show that the heat with sonication treatment of samples produced a slightly higher absorbance than the heat treatment sample. This may be due to the increased *L. pneumophila* sg 1 fragmentation and better homogenization. This result was expected, as it was anticipated that sonication would rupture the cell walls and allow more interaction and binding of the cell surface antigen with the antibodies coated onto the Binax tubes. The posttreatment results for the heat-treated sample showed that the average absorbance decreased following the treatment. Binax positive results were recorded for 5 to 12 days depending upon the initial concentration of the inoculum.

Positive *Legionella* culture results were not correlated with positive Binax Equate results. A large number of false-positive Binax results were recorded whether or not *Legionella* spp. were cultured from the sample. The results suggest that the Binax Equate test may cross-react with environmental *Pseudomonas* spp., giving

false-positive results. Laboratory work discounted *P. fluorescens* as causing cross reaction. Alternatively, the test also recognized heat-killed cells and false-positive results may have been due to the presence of dead *L. pneumophila* in the cooling systems. The results suggest that this system lacks the required specificity for routine use in cooling towers or other water systems. It seems likely that the test method would produce false-positive results in systems where disinfection was applied. False-positive results from cooling tower samples in this study may have been due to the presence of killed *L. pneumophila* sg 1. The results point to the necessity of indication of viability in future environmental tests for *Legionella*.

ACKNOWLEDGMENTS

The authors gratefully acknowledge the support for this study from MedLab Diagnostics, Sydney, Australia, and Binax Inc, Scarborough, Maine.

REFERENCES

1. **Standards Australia.** *AS/NZS 3896:1998:Waters: Examination for Legionellae Including Legionella Pneumophila.* Standards Australia, Sydney Australia.

METHOD DEVELOPMENT FOR *LEGIONELLA* DETECTION IN METALWORKING FLUIDS

Katalin Rossmoore, Leonard Rossmoore, and Christine Cuthbert

Metalworking fluids (MWFs) are highly susceptible to microbiological contamination from the environment due to high water content, available nutrients, and optimum growth temperature for most environmental microorganisms (4,5). Generation of respirable aerosols is also related to routine machining operations with MWFs. According to NIOSH (National Institute for Occupational Safety and Health) estimates (NIOSH 94-120, 1994), approximately 1 million U.S. workers are potentially exposed to MWFs. Outbreaks of Pontiac fever and Legionnaires' disease were associated with MWFs at two unrelated automotive facilities. *Legionella feeleii* and *Legionella pneumophila* serogroup 1 (sg 1) were identified as the probable causative organisms (3).

The objective of this work was to develop a selective and sensitive culture method for detection of *Legionella* in MWFs in order to prevent and control possible occupational health-related problems.

Recovery efficiency was evaluated by spiking MWF field samples, highly contaminated (10^7

to 10^8/ml) with nonlegionella organisms with high levels (10^6 to 10^7/ml) of *L. pneumophila* sg 1 and *L. feeleii*. Several sample treatment conditions to reduce the interference of the high background population, including acidification, heat treatment, and use of antibiotics in the recovery medium, were evaluated (1, 2).

Acid treatment with 1 N HCl to pH 2.1 for 10 min reduced the background population by various degrees for the different fluids. The average reduction for 25 samples was only 1.8 log. Heat treatment conditions at 45, 50, 55, and 60°C were tested for 10, 20, and 30 min. Heat treatment at 55°C for 30 min resulted in adequate reduction of background population; however, the *L. feeleii* population also was sensitive to this treatment.

Heat treatment at 55°C for 30 min combined with plating on antibiotic-containing BCYE-GVPC medium (buffered charcoal yeast extract with glycin, vancomycin HCl, polymixin-B SO_4, and cycloheximide) yielded *Legionella* recovery with a sensitivity limit of 100 *L. pneumophia* sg 1/ml and 1,000 *L. feeleii*/ml. The above culture method is valuable for detecting high concentrations of *L. pneumophila* sg 1 and *L. feeleii* in the presence of high background, nonlegionella population.

Katalin Rossmoore, Leonard Rossmoore, and Christine Cuthbert Biosan Laboratories, Inc., 1950 Tobsal Court, Warren, MI 48091.

As a complement to the above test, additional challenge tests with *L. pneumophila* sg 1 and *L. feeleii* are suggested to predict a MWF ability for supporting legionella growth. MWF field samples were autoclaved and challenged with 10^6 to 10^7/ml of *L. pneumophila* sg 1 and *L. feeleii* in separate experiments. Ten randomly selected MWF field samples were assessed for legionellae survival. The seeded fluids were incubated at 35°C, and samples were plated for survival intermittently for 2 weeks. The results indicated that only 2 of the 10 fluids showed short survival of either *L. pneumophila* sg 1 or *L. feeleii*.

The culture method developed can be used to detect high, potentially dangerous levels of *L. pneumophila* sg 1 and *L. feeleii* in MWFs. The recommended challenge test would indicate the susceptibility of the fluid for *L. pneumophila* sg 1 and *L. feeleii* growth support or survival. This latter phenomenon was found to be dependent on the formulation of the particular MWF in combination with the presence of the heterotrophic background bacterial population.

REFERENCES

1. **Bopp, C. A., J. W. Summer, G. K. Morris, and J. G. Wells.** 1981. Isolation of *Legionella* spp. from environmental water samples by low-pH treatment and use of a selective medium. *J. Clin. Microbiol.* **13:**174–719.

2. **Edelstein, P. H., J. B. Snitzer, and J. A. Bridge.** 1982. Enhancement of recovery of Legionella pneumophila from contaminated respiratory tract specimens by heat. *J. Clin. Microbiol.* **16:**1061–1065.

3. **Herwaldt, L. A., G. W. Gorman, T. McGrath, S. Toma, B. Brake, A. W. Hightower, J. Jones, A. L. Reingold, P. A. Boxer, P. W. Tang, C. W. Moss, H. Wilkinson, D. J. Brenner, A. G. Steigerwalt, and C. V. Broome.** 1984. A new Legionella species, *Legionella feeleii* species nova, causes Pontiac fever in an automobile plant. *Ann. Intern. Med.* **100:**333–338.

4. **Passman, F. J., and H. W. Rossmoore.** 2002. Reassessing the health risks associated with employee exposure to metalworking fluid microbes. *Lubrication Eng.* **58:**30–38.

5. **Rossmoore, H. W., L. A. Rossmoore, and C. E. Young.** 1987. Microbial ecology of an automotive engine plant. *Int. Biodeterioration Biodegradation* **1:**255–268.

RISK ASSESSMENT FOR *LEGIONELLA* IN BUILDING WATER SYSTEMS: MANAGING THE MYTHS

Richard Bentham

112

Outbreaks of disease have always preceded and promoted the development of guidelines and legislation aimed at reducing or removing risk. These legislative actions are necessarily based upon the best available knowledge at the time of promulgation and are often driven by public outrage and political necessity. An unfortunate consequence of these processes is the inclusion of rule of thumb elements that lack any empirical validation. Once incorporated into legislation these elements then acquire credibility through the legislative process but may be inaccurate or baseless. As a result of this unavoidable process, a series of myths can arise that are widely accepted, but unsubstantiated. In this paper, some of these myths and the erroneous interpretation of available empirical data surrounding the control of Legionnaires' disease will be addressed. This paper uses only conference papers presented at the 2000 Ulm *Legionella* meeting (published in 2002) as a means of demonstrating the gap that exists between the current evidence base and the risk management approaches relating to *Legionella* in the built environment. The myths addressed in this paper

Richard Bentham Department of Environmental Health, Flinders University, GPO Box 2100, Adelaide, South Australia, Australia, 5001.

relate to the microbial ecology of building water systems, aspects of *Legionella* virulence and dose response, and the notion of generic risk factors.

TOTAL HETEROTROPHIC PLATE COUNTS (TPCs)

It has been demonstrated that there is no correlation between *Legionella* and TPCs in cooling water systems. This has been interpreted to mean that the TPCs are of little value in assessing risks for *Legionella*. This interpretation is incorrect, as there is an absolute correlation between the presence of other bacteria and the presence of *Legionella*. As organisms that colonize biofilms and predate amoebae, the presence of other bacteria is a prerequisite for *Legionella* colonization (7). TPCs are useful indicators of the microbial status of a system.

Miller and Koebel (8) showed a direct correlation between high TPCs and the presence of *Legionella* in spa pools. This demonstrates that it is not safe to assume that the lack of correlation between *Legionella* and TPCs in cooling towers can be extrapolated into other water systems. TPC evaluations may be a rapid and reliable means to augment *Legionella* culture for risk management in spa pools. *Legionella* have been reported in water systems with high TPCs (e.g., warm water systems), but correlations between these two populations have not been demonstrated (10).

LEGIONELLA SPECIES AND SEROGROUPS

Legionella virulence is extremely variable within and between species and serogroups (10). Virulence may also be a product of environmental influences (7). Cooling tower-associated outbreaks of disease are almost exclusively associated with *Legionella pneumophila*, and for the most part with *L. pneumophila* serogroup 1. Outbreaks of disease from spa pools may be attributed to a wider range of *L. pneumophila* serogroups and other *Legionella* species. Infections from the built environment, and especially nosocomial disease, encompass a much wider range of *Legionella* species (4, 5, 10). The variation in causative organisms between systems is most probably a function of system design and operation. The focus of guidelines on *L. pneumophila* serogroup 1 as the major organism of concern may be warranted for cooling towers but is probably overemphasized in other systems where other species play a significant role in disease (10). The variable virulence of *Legionella* species must also be considered in the light of the immune status of the exposed population. The demography of the exposed population may be a more reliable indicator of risk than the identification of the *Legionella* species present (2).

DOSE RESPONSE

It has been suggested that *Legionella* concentrations in water relate to risk, as they can be used as predictors of dose. The relationship between *Legionella* in water samples and that disseminated in aerosol or aspirates is multifactorial. Dissemination relies on degree of aerosol production, relative humidity, air velocity, UV light incidence, and temperature, to name a few (2). The susceptible population is usually of poor immune statue. As a result, it is difficult to predict a reliable risk relationship from *Legionella* concentrations in water samples with disease causation (1). This is compounded by the variable viru-

lence (see above) and inaccuracy of culture determinations (see below) (2). Dose will vary substantially between exposures to systems. For instance, exposure to a dose of *Legionella* in aerosol from a cooling tower is a far less likely event than exposure to a dose of *Legionella* while using a spa pool. Dose response is likely to be extremely variable between systems and exposed demography with climatic variation (2).

LEGIONELLA TESTING

Legionella testing usually relies on culture from water samples taken from the systems. Action directives based on the colony-forming units have been promoted as reliable risk management tools (3). Several problems are associated with culture result interpretation. First, *Legionella* are transient members of the planktonic population. The primary location of *Legionella* multiplication is at surfaces in biofilm, either as predators of grazing protozoans or as members of the biofilm community (1). This means that the population in water is not a normal distribution, as it is dependent on mechanical and physical disruption and sloughing of biofilm into the aqueous phase (1). Second, *Legionella* culture is notoriously inaccurate and not readily reproducible (2, 6). Culture results must always be interpreted as being an underestimate of the actual numbers of organisms in the sample, and false-negative results are commonplace (1, 6). Third, microbial populations cannot be regarded as significantly different within a single log range. The doubling time of organisms in biofilm or within amoebae, may be such that no significance can be attached to culture results that differ by less than two orders of magnitude (1). These three considerations provide an evidence base which strongly undermines the use of action directives based on colony-forming units. However, it should be noted that "high" counts suggest high risk, though *high* is a relative term dependent upon the system being assessed and the exposed demography (1, 5).

Other methods of *Legionella* detection, for example PCR techniques, are not able to reliably discriminate between viable and nonviable organisms. This is a significant deficiency, as building water systems often employ thermal or chemical disinfection processes which would result in the presence of nonviable *Legionella* in water samples (3).

EVIDENCE-BASED RISK ASSESSMENT

The currently available evidence base demonstrates that health risk is variable between water systems. Often the risk is unique both between and within individual systems (e.g., cooling towers, spas, showers). Factors such as total microbial load, *Legionella* species, disease transmission, and population exposure change dramatically with the nature of the disseminating system (2, 5). Although there are common ecological determinants of *Legionella* growth and infection, the risks cannot rationally be generically applied.

Legionella forms part of a microbial community upon which it is entirely dependent. Sampling of that community can be used to assess the propensity for *Legionella* colonization. Current sampling techniques for *Legionella* are grossly inaccurate and may be misleading due to variations in culturability, virulence, and population distributions (6, 10). The uncertainty surrounding accuracy, reproducibility and implied dose response from *Legionella* testing of water samples suggests that numbers of positive results are better risk indicators than individual culture results (2, 5).

The evidence base also demonstrates that outbreaks of disease follow distinct trends for distinct sources. Wide geographical spread of disease from cooling towers is well documented, while disease from spa pools and building water systems tends to be much more geographically confined (1, 5, 9).

Available reports strongly suggest that good design, knowledge, and maintenance of water systems in the built environment are the most critical factors in preventing disease. An accurate and cost-effective risk assessment can only be made by considering these factors and the exposed demography. Once this is in place, culture for the organism and its microbial

community can be used as secondary control verifications.

REFERENCES

1. **Bentham, R.** 2002. Routine sampling and temporal variation of Legionella concentrations in cooling tower water systems, p. 321–324. *In* R. Marre, Y. Abu Kwaik, C. Bartlett, N. P. Cianciotto, B. S. Fields, M. Frosch, J. Hacker, and P. C. Lück (ed.). *Legionella,* ASM Press, Washington, D.C.
2. **Bentham, R., M. Pradhan, P. Hakendorf, and P. Wilmot.** 2002. Using geographic information systems for risk assessment and control of Legionnaires' disease associated with cooling towers, p. 318–320. *In* R. Marre, Y. Abu Kwaik, C. Bartlett, N. P. Cianciotto, B. S. Fields, M. Frosch, J. Hacker, and P. C. Lück (ed.). *Legionella,* ASM Press, Washington, D.C.
3. **Exner, M., M. Kramer, and S. Pleischl.** 2002. Strategies for prevention and control of Legionnaires' disease in Germany, p. 385–389. *In* R. Marre, Y. Abu Kwaik, C. Bartlett, N. P. Cianciotto, B. S. Fields, M. Frosch, J. Hacker, and P. C. Lück (ed.). *Legionella,* ASM Press, Washington, D.C.
4. **Joseph, C.** 2002. Surveillance of Legionnaires' disease in Europe, p. 311–317. *In* R. Marre, Y. Abu Kwaik, C. Bartlett, N. P. Cianciotto, B. S. Fields, M. Frosch, J. Hacker, and P. C. Lück (ed.). *Legionella,* ASM Press, Washington, D.C.
5. **Kusnetsov, J., M. Tiittanen, S. Mentula, and H. Jousimies-Somer.** 2002. Hot water systems with low concentrations of legionellae may be a risk on cruise ships, p. 349–352. *In* R. Marre, Y. Abu Kwaik, C. Bartlett, N. P. Cianciotto, B. S. Fields, M. Frosch, J. Hacker, and P. C. Lück (ed.). *Legionella,* ASM Press, Washington, D.C.
6. **Lee, J. V., S. Surman, M. Hall, and L. Cuthbert.** 2002. Development of an international external quality assurance scheme for isolation of *Legionella* species from environmental specimens, p. 271–274. *In* R. Marre, Y. Abu Kwaik, C. Bartlett, N. P. Cianciotto, B. S. Fields, M. Frosch, J. Hacker, and P. C. Lück (ed.). *Legionella,* ASM Press, Washington, D.C.
7. **McNealy, T., A. Newsome, R. Johnson, and S. Berk.** 2002. Impact of amoebae, bacteria and Tetrahymena on *Legionella pnemophila* multiplication and distribution in an aquatic environment, p. 170–175. *In* R. Marre, Y. Abu Kwaik, C. Bartlett, N. P. Cianciotto, B. S. Fields, M. Frosch, J. Hacker, and P. C. Lück (ed.). *Legionella,* ASM Press, Washington, D.C.
8. **Miller, R., and D. Koebel.** 2002. Prevalence of Legionella in whirlpool spas: correlation with total bacterial numbers, p. 275–279. *In* R. Marre, Y.

Abu Kwaik, C. Bartlett, N. P. Cianciotto, B. S. Fields, M. Frosch, J. Hacker, and P. C. Lück (ed.). *Legionella,* ASM Press, Washington, D.C.

9. **Pascual, M., O. Ronveaux, K. De Schrijver, and F. Van Loock.** 2002. Legionellosis outbreak at a commercial fair in Kapellen, Belgium, 1999: a case-control study, p. 342–345. *In* R. Marre, Y. Abu Kwaik, C. Bartlett, N. P. Cianciotto, B. S. Fields, M. Frosch, J. Hacker, and P. C. Lück (ed.). *Legionella,* ASM Press, Washington, D.C.

10. **Pringler, N., P. Brydov, and S. Uldum.** 2002. Occurrence of *Legionella* in Danish hot water systems, p. 298–301. *In* R. Marre, Y. Abu Kwaik, C. Bartlett, N. P. Cianciotto, B. S. Fields, M. Frosch, J. Hacker, and P. C. Lück (ed.). *Legionella,* ASM Press, Washington, D.C.

CONTROLLING *LEGIONELLA* IN HOSPITAL WATER SYSTEMS: FACTS VERSUS FOLKLORE

Janet E. Stout

113

Acute-care and long term-care facilities continue to experience cases of hospital-acquired Legionnaires' disease (3, 20). The mortality for this health care-associated disease remains high despite our ability to make a rapid diagnosis and to treat it empirically (3, 24). A number of advisory documents from various health authorities provide guidelines for approaches to prevention (Table 1). Unfortunately, a consensus opinion for prevention of this disease still does not exist and the issue remains unresolved.

One of the major unresolved issues is whether the recommendations found in these guidelines will, if followed, result in the control and prevention of hospital-acquired Legionnaires' disease. An evidence-based approach has been suggested as a way to resolve many of these issues (30). If applied to a guideline, the criteria should be that (i) the recommendations should be prospectively validated under controlled studies using a step-wise approach, (ii) the evaluation should be a prolonged observational period (>1 year) to evaluate the efficacy of the recommendations, and (iii) the recommended approach/actions should achieve the expected result—prevention of the disease through environmental control.

Legionnaires' disease is an environmentally acquired illness. Transmission occurs via exposure to water that contains virulent *Legionella* bacteria. *Legionella* spp. are typically in greatest concentrations in the warm water distribution systems of healthcare facilities. The transmission of *Legionella* typically occurs via aerosolization or aspiration. If the major route of exposure for hospital-acquired Legionnaires' disease is via aerosolization, then it would be prudent to restrict showering for patients. What are the facts? Numerous studies have explored the hypothesis that showering was a mode of transmission for hospital-acquired Legionnaires' disease. Interestingly, all of them failed to link showering to *Legionella* infection (5, 11, 12, 14, 21). In fact, the case-control study that followed the original study that first reported a possible link failed to show that showering was a risk factor (11, 31). Finally, one observational study reported that a patient did not bathe or shower but did ingest tap water during a period of highly impaired cell-mediated immunity (17).

Thus, guidelines that recommend sterile/ bottled water for immune compromised patients are well founded, but those that restrict showering for all patients are not.

Not only should the recommended practice be routinely effective at eliminating or reducing

Janet E. Stout   Department of Infectious Disease, VA Pittsburgh Healthcare System, Pittsburgh, PA 15240.

TABLE 1 Guidelines for prevention of Legionnaires' disease for U.S. health care facilities[a]

State/organization (reference)	Diagnostic testing	Clinical surveillance	Routine environmental testing	Approach to prevention
Allegheny County Health Department 1993/1997 (1)	Active: in-house urinary antigen (UA) testing	If environmental, positive-active clinical surveillance	Yes: Annually. Transplant hospital more often	Consider disinfection if >30% sites positive; empiric antimicrobial therapy macrolide or quinolone
Maryland Health Department (25)	Acute care: UA In-house/transplant hospital: culture on site	Test pneumonia cases for *Legionella*	Yes: routine culture	If cases identified, disinfection recommended
Texas Department of Health (29)	Acute and long term: UA In-house/ transplant hospitals: culture on site	Active case detection after case identified	Routine: no If high risk of cases: yes	Enhanced clinical surveillance and remediation if cases identified
Centers for Disease Control (7)	Routinely test without knowledge of environmental status	Educate rediagnosis/ 400+ beds = UA/ culture in-house	No: unless cases identified or transplant unit	Disinfect only if source identified

[a]Reprinted with permission from Lippincott Williams & Wilkins, Inc. Baltimore, MD. J. E. Stout and V. L. Yu. Hospital-acquired Legionnaires' disease. *Current Opinion in Infectious Diseases.* 2003;16:337–341.

Legionella in the environment, but it should also be practical and achievable.

One recommendation often found in guidance documents is "remove showerheads and aerators monthly for cleaning with chlorine bleach" (9). In a large acute-care hospital, the number of showerheads and faucet aerators can be in the thousands. I suggest that the efficacy of such actions is "folklore." Do the data suggest that this will have any lasting effect on *Legionella* colonization? In a report by Kusnetsov et al., showerheads were opened weekly and taps monthly for mechanical cleaning with a brush and disinfected in 1,000 mg of chlorine per liter (15). The conclusion from this study was that mechanical cleaning and disinfection did not reduce the concentration of *Legionella* in tap and shower waters.

In this same study, the effect of another often-repeated recommendation was tested. Many guidelines will recommend that the circulating hot water temperature be >50°C (2). Will this eliminate *Legionella*? The study by Kusnetsov showed that despite elevated recirculation temperatures (≥60°C), peripheral sites remained heavily colonized (15).

The role of environmental monitoring in *Legionella* prevention has been a source of debate for many years (6, 26, 30). However, there have been several studies that provide evidence for the utility of monitoring in the prevention of hospital-acquired Legionnaires' disease. Two studies from Spain show that *Legionella* colonization was extensive among Barcelona hospitals and that environmental monitoring followed by intensive clinical surveillance identified previously unrecognized cases of hospital-acquired Legionnaires' disease (18, 19). The Allegheny County Health Department recommends periodic environmental monitoring of acute-care facilities as part of their recommended prevention plan. The effect of this approach was recently evaluated, and the results showed a significant decrease in the number of health care-associated cases of Legionnaires' disease after the preventive guideline was in place (24).

Remediation in response to the identification of cases is also included in many guidelines. However, adequate validation of some of these disinfection methods has not been performed. We recommend that each disinfection method undergo a four-step evaluation of effi-cacy (27). This includes (i) demonstrated effi-cacy in vitro, (ii) anecdotal experience in indi-vidual hospitals, (iii) controlled studies of sufficient duration (years) in single hospitals, and (iv) confirmatory reports from multiple hospitals (validation step).

A number of disinfection methods have been used for control of *Legionella* in hospital water systems. These include thermal eradication (heat and flush), hyperchlorination, copper-silver ion-ization, point-of-use filters, and chlorine dioxide (13, 16, 22, 23). Each of these methods has com-pleted some of the evaluation criteria. All four steps of the evaluation criteria have been fulfilled for copper-silver ionization (27).

The original recommendations for perform-ing a thermal eradication recommended multi-ple 30-min flushes of distal outlets with 70°C water (4). The Centers for Disease Control and Prevention recommended that the duration of the heat and flush be ≥5 min (8). A recent eval-uation of the short (5 min) duration thermal eradication was performed in Taiwan. The in-vestigators found that the abbreviated duration of 5 min was ineffective in reducing *Legionella* positivity (10). They also evaluated the replace-ment of faucets and showerheads to prevent *Le-gionella* colonization. This action had no effect on minimizing *Legionella* colonization. This data contradicts the guidance found in the American Society for Heating, Refrigeration, and Air-Conditioning Engineers (ASHRAE) guideline for minimizing *Legionella* in building water systems (2).

Also included in the ASHRAE guideline, as well as other guidance documents, is the re-moval of dead leg sections of pipe. It should be noted that this is an untested and unconfirmed recommendation. That places this recommen-dation in the "folklore" category. Two studies in the literature noted that removal of dead legs had no effect on reducing *Legionella* posi-tivity in hospital water systems (23, 28).

Why have nonscientific and unconfirmed experiences been adopted by so many? There is a general aversion to the concept that the water supply is colonized with a pathogen ("see no evil"). Prospective evaluations have

not been performed because there is an aver-sion to knowing that your actions may have been ineffective. *Legionella* also has the poten-tial to bring with it negative publicity and liti-gation. Finally, many do not culture the water supply after the intervention, and as a result there is no end point for evaluation of efficacy.

The path to effective control of hospital-acquired Legionnaires' disease must be evidence-based. Patients and healthcare facilities suffer when unconfirmed and untested recommenda-tions become part of prevention guidelines. The door is open for validation of recommendations that are as yet unproven. For now, stick with the facts and leave the "folklore" behind.

REFERENCES

1. **Allegheny County Health Department.** 1997. *Approaches to Prevention and Control of Legionella Infection in Allegheny County Health Care Facilities*, 2nd ed., p. 1–15 . Allegheny County Health Department, Pittsburgh, Pa. (http://www.legionella.org).
2. **American Society of Heating, Refrigeration, and Air-Conditioning Engineers.** 2000. *ASHRAE Guideline 12-2000: Minimizing the Risk of Legionellosis Associated with Building Water Systems.* American Society of Heating, Refrigerating, and Air-Conditioning Engineers, Inc.
2. **Benin, A. L., R. F. Benson, and R. E. Besser.** 2002. Trends in Legionnaires' disease, 1980-1998: declining mortality and new patterns of diagnosis. *Clin. Infect. Dis.* **35:**1039–1046.
4. **Best, M. G., A. M. Goetz, and V. L. Yu.** 1984. Heat eradication measures for control of hospital-acquired Legionnaires' disease: implementation, education, and cost analysis. *Am. J. Infect. Control.* **12:**26–30.
5. **Blatt, S. P., M. D. Parkinson, E. Pace, P. Hoffman, D. Dolan, P. Lauderdale, R. A. Zajac, and S. P. Melcher.** 1993. Nosocomial Legionnaires' disease: aspiration as a primary mode of transmission. *Am. J. Med.* **95:**16–22.
6. **Butler, J. C., B. S. Fields, and R. F. Breiman.** 1997. Prevention and control of legionellosis. *Infect. Dis. Clin. Pract.* **6:**458–464.
7. **Centers for Disease Control and Prevention.** 1997. Guidelines for prevention of nosocomial pneumonia. *Morb. Mortal. Wkly. Rep.* **46:**1–79.
8. **Centers for Disease Control and Prevention.** 2004. Guidelines for preventing health-care-associated pneumonia, 2003. *Morb. Mortal. Wkly. Rep.* **53:**1–36.
9. **Centers for Disease Control and the Healthcare Infection Control Practices Advisory**

Committee (HICPAC). 2003. Guidelines for environmental infection control in health-care facilities. *Morb. Mortal. Wkly. Rep.* **52:**1.

10. Chen, Y., Y. Liu, S. S. Lee, H. Tsai, S. Wann, C. Kao, C. Chang, W. Huang, T. Huang, H. L. Chao, C. Li, C. Ke, and Y. E. Lin. 2005. Abbreviated duration of superheat-and-flush and disinfection of taps for *Legionella* disinfection: lessons learned from failure. *Am. J. Infect. Control* 33:606–610.

11. Cordes, L. G., A. M. Wiesenthal, G. W. Gorman, et. al. 1981. Isolation of *Legionella pneumophila* from hospital showerheads. *Ann. Intern. Med.* **94:**195–197.

12. Ezzedine, H., C. VanOssel, M. Delmee, and G. Wauters. 1989. *Legionella* spp. in a hospital hot water system: effect of control measures. *J. Hosp. Infect.* **13:**121–131.

13. Kim, B. R., J. E. Anderson, S. A. Mueller, W. A. Gaines, and A. M. Kendall. 2002. Literature review: efficacy of various disinfectants against *Legionella* in water systems. *Water Res.* **36:**4433–4444.

14. Kool, J. L., A. E. Fiore, C. M. Kioski, E. W. Brown, R. F. Benson, J. M. Pruckler, C. Glasby, J. C. Butler, G. D. Cage, J. C. Carpenter, R. M. Mandel, B. England, and R. F. Breiman. 1998. More than ten years of unrecognized nosocomial transmission of Legionnaires' disease among transplant patients. *Infect. Contr. Hosp. Epidemiol.* **19:**898–904.

15. Kusnetsov, J., E. Torvinen, O. Perola, T. Nousianien, and M. L. Katila. 2003. Colonization of hospital water systems by legionellae, mycobacteria and other heterotrophic bacteria potentially hazardous to risk group patients. *APMIS* **111:** 546–556.

16. Lin, Y. E., R. D. Vidic, J. E. Stout, and V. L. Yu. 1998. *Legionella* in water distribution systems. *J. Am. Water Works Assoc.* **90:**112–121.

17. Mathys, W., M. C. Deng, J. Meyer, and E. Junge-Mathys. 1999. Fatal nosocomial Legionnaires' disease after heart transplantation: clinical course, epidemiology, and prevention strategies for the highly immunocompromised host. *J. Hosp. Infect.* **43:**242–246.

18. Sabria, M., M. Garcia-Nunez, M. L. Pedro-Botet, N. Sopena, J. M. Gimeno, E. Reynaga, J. Morera, and C. Rey-Joly. 2002. Presence and chromosomal subtyping of *Legionella* spp in potable water systems in 20 hospitals of Catalonia, Spain. *Infect. Control Hosp. Epidemiol.* **22:**673–676.

19. Sabria, M., J. M. Modol, M. Garcia-Nunez, E. Reynaga, M. L. Pedro-Botet, J. M. Gimeno, N. Sopena, and C. Rey-Joly. 2004. Environmental cultures and hospital-acquired legionnaires' disease. A 5-year prospective study in 20 hospitals in Catalonia, Spain. *Infect. Cont. Hosp. Epidemiol.* **25:**1072–1076.

20. Seenivasan, M., V. L. Yu, and R. R. Muder. 2005. Legionnaires' disease in long-term care facilities: overview and proposed solutions. *J. Am. Geriatr. Soc.* **53:**875–880.

21. Shands, K., J. Ho, R. Meyer, G. Gorman, et al. 1985. Potable water as a source of Legionnaires' disease. *JAMA* **253:**1412–1416.

22. Sheffer, P. J., J. E. Stout, M. M. Wagener, and R. R. Muder. 2005. Efficacy of new point-of-use water filters for preventing exposure to Legionella and waterborne bacteria. *Am. J. Infect. Control* **33:**S20–S25.

23. Sidari, F. P., and J. M. VanBriesen. 2002. Evaluation of a chlorine dioxide secondary disinfection system. *Water Eng. Manage.* **149:**29–33.

24. Squier, C. L., J. E. Stout, S. Krystofiak, J. McMahon, M. M. Wagener, B. Dixon, and V. L. Yu. 2005. A proactive approach to prevention of healthcare-acquired Legionnaires' disease: the Allegheny County (Pittsburgh) experience. *Am. J. Infect. Control* **33:**360–367.

25. State of Maryland Department of Health and Mental Hygiene. 2000. *Report of the Maryland Scientific Working Group to Study Legionella in Water Systems in Healthcare Institutions.*

26. Stout, J. E., and V. L. Yu. 2001. *Legionella* in the hospital water supply: a plea for decision making based on evidence-based medicine. *Infect. Control Hosp. Epidemiol.* **22:**670–672.

27. Stout, J. E., and V. L. Yu. 2003. Experiences of the first 16 hospitals using copper-silver ionization for *Legionella* control: implications for the evaluation of other disinfection modalities. *Infect. Control Hosp. Epidemiol.* **24:**563–568.

28. Struelens, M. J., N. Maes, F. Rost, A. Deplano, F. Jacobs, C. Liesnard, N. Bornstein, F. Grimont, S. Lauwers, and M. P. McIntyre. 1992. Genotypic and phenotypic methods for the investigation of a hospital-acquired *Legionella pneumophila* outbreak and efficacy of control measures. *J. Infect. Dis.* **166:**22–30.

29. Texas Department of Health. 2002. *Report of the Legionnaiares' Disease Task force.* Austin, TX.

30. Yu, V. L. 1998. Resolving the controversy on environmental cultures for *Legionella*. *Infect. Control Hosp. Epidemiol.* **19:**893–897.

31. Yu, V. L., F. J. Kroboth, J. Shonnard, A. Brown, S. McDearman, and M. H. Magnussen. 1982. Legionnaires' disease: new clinical perspective from a prospective pneumonia study. *Am. J. Med.* **73:**357–361.

STRATEGIES FOR INFECTION CONTROL OF NOSOCOMIAL LEGIONNAIRES' DISEASE: FOUR-YEAR SURVEILLANCE EXPERIENCE IN A TEACHING HOSPITAL IN ITALY

S. Boccia, P. Borella, V. Romano-Spica, P. Laurenti, A. Cambieri, G. Branca, M. Tumbarello, R. Cauda, G. Fadda, and G. Ricciardi

114

The degree of *Legionella pneumophila* contamination in hospital water supplies has been shown to correlate with the incidence of nosocomial Legionnaires' disease (8). Italian guidelines indicate that concentrations of *Legionella* of $\geq 10^4$ CFU/liter of water represent a real hazard, and different European guidelines suggest that measures need to be taken when levels reach 10^3 CFU/liter (7). Despite the knowledge about Legionnaires' disease (LD), the lack of pneumonia surveillance systems in hospitals and the difficulty of making a correct diagnosis render LD underestimated in Italy (5). In accordance with this, and within the context of the Italian Multicentric Study of Legionellosis, we carried out a 4-year active prospective LD surveillance program in a large university hospital in Rome. We also assessed the usefulness of the hospital water monitoring program to predict the risk of nosocomial LD. In addition, the genetic correlation among the environmental *L. pneumophila* isolates collected was assessed by means of pulsed-field gel electrophoresis (PFGE) for epidemiological investigations.

The A. Gemelli is a university hospital with 1,500 patient beds that opened in 1964. The water reaches the hospital by means of one pipeline to a hydrostatic tank and is sent via electric motor pumps through self-cleaning filters to three intermediate tanks. The water doesn't undergo any chlorination after it is collected from the city pipelines. A monthly decontamination procedure of the intermediate tanks is carried out, which consists of mechanically cleaning out the tanks to remove the formed organics, followed by washing out the tanks with sodium hypochlorite.

NOSOCOMIAL PNEUMONIA AND ENVIRONMENTAL SURVEILLANCE

A nosocomial pneumonia surveillance program was launched in June 2001. The protocol suggests that a dedicated medical doctor for each hospital ward should report all patients with a nosocomial pneumonia to the senior nurse serving on the ward. A case of pneumonia was considered to be nosocomial based on

S. Boccia, P. Laurenti, and G. Ricciardi   Institute of Hygiene, Catholic University of Sacred Heart, 00168, Rome, Italy. *P. Borella*   Department of Hygiene, University of Modena and Reggio Emilia, 41100, Modena, Italy.   *V. Romano-Spica*   Hygiene Laboratory, Department of Human Movement and Sport Sciences, University Institute for Movement Sciences, 00194, Rome, Italy.   *A. Cambieri*   Hospital Head Unit, Catholic University of Sacred Heart, 00168, Rome, Italy.   *G. Branca and G. Fadda*   Institute of Microbiology, Catholic University of Sacred Heart, 00168, Rome, Italy.   *M. Tumbarello and R. Cauda*   Institute of Infectious Diseases, Catholic University of Sacred Heart, 00168, Rome, Italy.

Legionella: State of the Art 30 Years after Its Recognition
Edited by Nicholas P. Cianciotto et al.
©2006 ASM Press, Washington, D.C.

a definition reported elsewhere (2). For each of the suspected cases of nosocomial pneumonia a urine and, when possible, a lower respiratory tract secretion sample was sent for legionella analysis to the hospital laboratory of clinical microbiology.

Concerning the environmental investigations, if there is not an epidemic, the Italian guidelines do not suggest either the frequency or the number of sites to be sampled, so we decided to use our own sampling strategy. A water sample for bacteriological analysis was collected every 6 months from selected taps and showers in high- and medium-risk hospital wards, amounting to 85 samples in 4 years. Legionellae were isolated following Italian procedures for isolation of *Legionella* spp. in water (7), and serotyping was performed by agglutination with commercial specific monoclonal rabbit antibodies (Pro-lab Diagnostics, Canada).

L. pneumophila environmental isolates were submitted for PFGE with a contour-clamped homogeneous electric field system (CHEF MAPPER, Biorad) using a known protocol (9). According to Tenover criteria, a pattern designation was assigned if the electrophoretic profile differed by more than three bands (10).

SURVEILLANCE RESULTS

From June 2001 through May 2005, the pneumonia surveillance identified on February

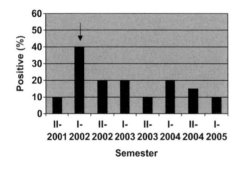

FIGURE 1 Percentages of water cultures positive for *L. pneumophila* per semester. Cultures were performed from high- and medium-risk hospital wards each semester. The arrow indicates the detected case of nosocomial pneumonia due to *L. pneumophila* sg 1 on February 2002.

2002 one nosocomial LD case among 94 nosocomial pneumonia cases (coming mainly from surgical wards and neurotraumatology). Environmental investigations detected *L. pneumophila* in 17 of the 85 (20%) water samples, of which 41% belonged to serogroup (sg) 1, 23.5% to sg 6, 17.3% to sg 2, and the remaining to sg 3, 4, and 9. Fig. 1 shows the percentages of water cultures positive for *L. pneumophila* per semester of surveillance. From June 2001 to May 2005, the *L. pneumophila* count and the percentage of positive locations never exceeded 10^2 CFU/liter and 20%, respectively, except when the LD nosocomial case occurred (positive water samples 40% with a count $<10^2$ CFU/liter). From June 2004 to May 2005, the *L. pneumophila* count reached 10^4 CFU/liter on two occasions, but no cases of LD were detected.

Once the nosocomial LD case had occurred, epidemiological and environmental investigations were performed in the hospital ward involved. The epidemiologic study did not discover any further nosocomial LD cases. Two samples were collected for environmental analyses, one from the tap faucet of the sink and one from the shower of the room where the patient was hospitalized. The latter sample provided a positive result for *L. pneumophila* sg 1 contamination, with a *L. pneumophila* count of $<10^2$ CFU/l. As suggested by the Italian guidelines, a decontamination protocol by "superheat and flush" procedure was carried out in the column 3, which provided water to the interested ward: for three consecutive days, the water temperature was maintained at 70°C and all taps were allowed to run for at least 30 min/day. Subsequent surveillance cultures showed the eradication of *L. pneumophila* from that ward. Unfortunately, it was not possible to genetically compare clinical and environmental isolates since the bacteria was not isolated from the patient.

Among the 17 *L. pneumophila* isolates collected in 4 years surveillance, three main pulsotypes were identified: pulsotype A, which is represented by *L. pneumophila* isolates of sg 1, pulsotype B, which contains *L. pneumophila* isolates of different serogroups (2, 3, 4, 6), and pulsotype C, which is represented by one *L. pneumophila* sg 9 isolate (data not shown). The main PFGE clone, which is represented by pulsotype A, was recovered from November 2001 to May 2005 in different hospital wards.

REMARKS AND CONCLUSION

The low incidence of nosocomial LD cases in our hospital during the study period seems to be correlated to the low contamination level of *L. pneumophila* (20% of positive samples in 4 years) in hospital wards and to a low percentage of positive water samples per semester of surveillance. However, we cannot exclude that an underestimation of the true nosocomial pneumonia cases has occurred due to some clinicians' less than full compliance to the active surveillance protocol.

The debate about the possible link between the contamination level of legionellae in environmental samples and incidence of LD is well known. Many U.S. and Spanish researchers deny the existence of a threshold for hot water systems in health-care facilities, whereas in France, Italy, Switzerland, and the United Kingdom, this level is fixed at 10^3 CFU/l (1, 3, 7). In the absence of nosocomial LD cases, the utility of environmental monitoring is also debated: according to CDC guidelines, primary interventions are not suggested, with the exception of organ transplant units, but several European guidelines, in accordance with the Allegheny County Health Department, suggest carrying out periodic environmental surveillance in high-risk hospital wards (especially in large hospitals) (6). Although unknown predisposing individual factors of the nosocomial LD case could probably explain the reason for its occurrence despite a low water contamination ($<10^2$ CFU/l), it is notable that it happened when 40% of the hospital wards were contaminated. Furthermore, even though the contamination level of *L. pneumophila* in the investigated wards was always low, it was fluctuating in a positive range of contaminated wards of 10 to 40% per semester. This suggests that it is important to repeat the samplings in

order to evaluate the risk. With regard to the control procedures adopted in the hospital to contain water contamination by legionellae, the monthly mechanical cleaning of the intermediate tanks proved to be effective.

Molecular typing also revealed the persistence of one *L. pneumophila* clone across 4 years of surveillance, and that one pulsotype was represented by different serogroups (4).

We can conclude that based on our experience, nosocomial LD is a potentially preventable infection requiring coordinated efforts for its control. Furthermore, *L. pneumophila* counts of less than 10^3 CFU/l are not without risk, and there is not a safe level of colonization, even though a good indicator appears to be the percentage of positive locations, as described by Best (3). An infection control system for nosocomial LD should therefore be based on both environmental and clinical surveillance, together with the appropriate maintenance of the hospital water distribution system.

ACKNOWLEDGMENTS

This work was supported by grant no. MM06172998_006 from the Italian Ministry of University and Research (MIUR).

We are grateful to Rosarita Amore, Gennaro Capalbo, and Katia Del Gigante for their skillful assistance in epidemiological surveillance.

REFERENCES

1. **Allegheny County Health Department.** 1997. *Approaches to Prevention and Control of Legionella Infection in Allegheny County Health Care Facilities*, 2nd ed. Pittsburgh, Pa: Allegheny County Health Department, p. 1–15.
2. **American Thoracic Society.** 1996. Hospital-acquired pneumonia in adults: diagnosis, assessment of severity, initial antimicrobial therapy, and preventive strategies. *Am. J. Crit. Care Med.* **153:** 1711–1725.
3. **Best, M., V. L. Yu, J. Stout, A. Goetz, R. R. Muder, and F. Taylor.** 1983. Legionellaceae in the hospital water-supply. Epidemiological link with disease and evaluation of a method for control of nosocomial Legionnaires' disease and Pittsburgh pneumonia. *Lancet* **2:**307–310.
4. **Boccia, S., R. Amore, A. Stenico, L. Moroder, M. Orsini, V. Romano- Spica, and G. Ricciardi.** 2005. Molecular epidemiology by

automated ribotyping and pulsed field gel electrophoresis of *Legionella pneumophila* environmental isolates representing nine different serogroups.
Epidemiol. Infect. **133**:1097–1105.

5. **Borella, P., M. T. Montagna, V. Romano-
Spica, S. Stampi, G. Stancanelli, M. Triassi,
A. Bargellini, P. Giacobazzi, F. Vercilli,
S. Scaltriti, I. Marchesi, C. Napoli, D. Tato,
G. Spilotros, N. Paglionico, G. Quaranta,
M. Branca, M. Tumbarello, P. Laurenti, U.
Moscato, E. Capoluongo, G. De Luca, P. P.
Legnani, E. Leoni, R. Sacchetti, F. Zanetti,
M. Moro, C. Ossi, L. Lopalco, R. Santarpia,
V. Conturso, G. Ribera d'Alcala, and S.
Montegrosso.** 2003. Environmental diffusion of
Legionella spp and Legionellosis frequency
among patients with pneumonia: preliminary re-
sults of a multicentric Italian survey. *Ann. Ig.* **15**:
493–503.

6. **Centers for Disease Control and Prevention.**
2004. Guidelines for preventing health-care-associated pneumonia, 2003: recommendations of CDC
and the Healthcare Infection Control Practices Advisory Committee. *Morbid. Mortal. Wkly. Rep.* **53**:
10–13.

7. **G.U. 5/5/2000, n.103.** 2000. Superior Institute
of Health and Department of Health. *Guidelines
for Prevention and Control of Legionellosis.*

8. **Kool, J. L., D. Bergmire-Sweat, and J. C.
Butler.** 1999. Hospital characteristics associated
with colonization of water systems by Legionella
and risk of nosocomial legionnaires' disease: a cohort study of 15 hospitals. *Infect. Control Hosp. Epidemiol.* **20**:798–805.

9. **Schoonmaker, D., T. Heimberger, and G.
Birkhead.** 1992. Comparison of ribotyping and
restriction enzyme analysis using pulsed-field gel
electrophoresis for distinguishing *Legionella pneumophila* isolates obtained during a nosocomial
outbreak. *J. Clin. Microbiol.* **30**:1491–1498.

10. **Tenover, F. C., R. B. Arbeit, R. V. Goering,
P. A. Mickelsen, B. E. Murray, D. H. Persing,
and B. Swaminathan.** 1995. Interpreting chromosomal DNA restriction patterns produced by
pulsed-field gel electrophoresis: criteria for bacterial strain typing. *J. Clin. Microbiol.* **33**:2233–2239.

LEGIONELLA CONTAMINATION OF DOMESTIC HOT WATER IN A TERTIARY LEVEL HOSPITAL AND RESULTING INTRODUCTION OF CONTROL MEASURE

Mona Schousboe, Alan Bavis, and Ros Podmore

115

The Canterbury District Health Board serves a population of >400,000 people. It is responsible for several hospitals, one of which is a tertiary level 750-bed hospital, Christchurch Hospital (ChChH). ChChH underwent a total rebuilding program between 1973 and 1993. The construction was carried out in many stages, leading to logistical problems with plant and systems feeding domestic hot water to the site. Our objective for this research is to analyze variables in the rebuilding program resulting in *Legionella* in the domestic hot water system (DHWS), legionella monitoring of water and air-conditioning systems, and systems implemented to control contamination.

The hospital Maintenance and Engineering Department's records of the domestic water circulation system from 1990 to 2005 were searched. The District Health Board Infection Control Committee minutes and annual reports were researched for reports on *Legionella* infection in patients and environmental colonization. Archived laboratory records from 1995 to 2005 for environmental *Legionella* culture of the DHWS and cooling towers were

reviewed for culture results. Standard methods for *Legionella* culture of environmental samples and identification of isolates were followed for both water samples and swabs from showerheads (3).

The maintenance records show that in early 1990, when stage I of the new hospital development was commissioned, the DHW plant was oversized due to plans for future load requirements. A pre-heat coil was installed to utilize the heat from steam trap condensate. Depending on flow rates, the temperature range at the output from this plant was around 60 to 80°C. In 1994, increased flows were required from the DHW calorifiers. An additional heat exchanger was added to the system in the form of a waste heat recovery system using excess heat from the medical air compressors. These steps increased flow rates and lowered the outflow temperature to approximately 45°C. However, water going to the DHWS exited the calorifiers at 60°C and returned at 55°C. *Legionella pneumophila* serogroup 1 (sg 1) was found in the bottom of the hot water calorifiers, the heat exchangers supplying the calorifiers and several other sites of the supply line to the hot water calorifiers. Sites distal to the calorifiers were negative for *L. pneumophila* sg 1 (Fig. 1). Partial redesign of the DHWS replaced the condensate cooler with a plate heat exchanger. The waste heat

Mona Schousboe and Ros Podmore   Canterbury Health Laboratories, P O Box 151, Christchurch, New Zealand.   *Alan Bavis*   Maintenance and Engineering Department, Christchurch Hospital, P O Box 151, Christchurch, New Zealand.

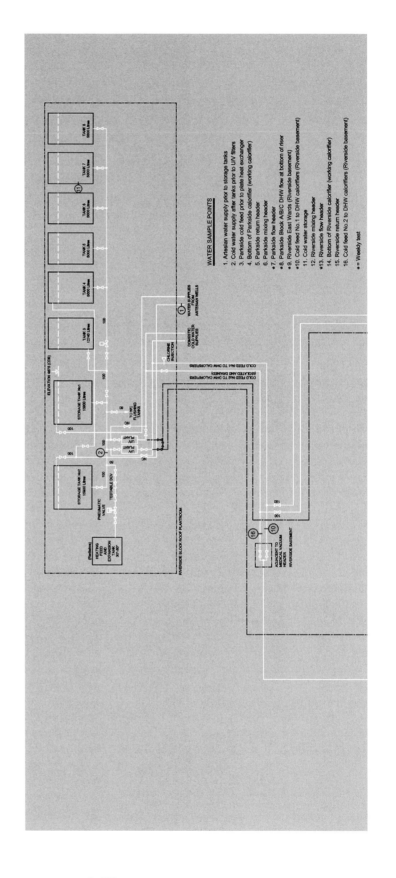

WATER SAMPLE POINTS

1. Artesian water supply prior to storage tanks
2. Cold water supply after tanks prior to U/V filters
3. Parkside cold feed prior to plate heat exchanger
4. Bottom of Parkside calorifier (working calorifier)
5. Parkside return header
6. Parkside mixing header
*7. Parkside flow header
*8. Parkside Block A/B/C DHW flow at bottom of riser
*9. Riverside East Wards (Riverside basement)
*10. Cold feed No.1 to DHW calorifiers (Riverside basement)
11. Cold water storage
12. Riverside mixing header
*13. Riverside flow header
14. Bottom of Riverside calorifier (working calorifier)
15. Riverside return header
16. Cold feed No.2 to DHW calorifiers (Riverside basement)

* = Weekly test

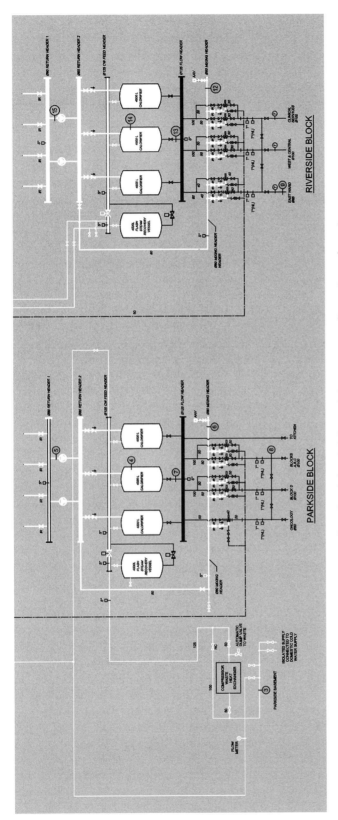

FIGURE 1 Christchurch hospital domestic hot water sampling points schematic.

exchanger was bypassed and removed from the circuit. A regular testing regime was put into place and samples were taken for analysis at several points within the DHW reticulation system (Table 1).

Until December 1998 the reticulation temperature of the DHW was set to 60°C. This was in contravention of the New Zealand Building Code (2), which requires no more than 55°C at each outlet. Consequently, tempering valves were fitted to each outlet. In December 1998, it was decided that the temperature had to be reduced to 55°C. This was achieved by keeping the calorifiers at 70°C and tempering the water to 55°C shortly after it left the calorifiers (Fig. 1). By January 1999, test sites on the DHWS were positive for *L. pneumophila* sg 1. Several ward showers tested positive for *L. pneumophila* sg 1. A chance culture of sandy deposit from a cold water tank storing water from the 82-m deep artesian well serving the hospital was also found to be positive for *L. pneumophila* sg 1 after prolonged incubation.

In an effort to control the contamination of the water supply, a UV irradiation plant was installed on the downstream supply from the cold water storage tanks treating all cold water going to the reticulation system in July 1999 (Fig. 1). A chlorination plant for shock treatment of the DHWS was planned for later installation. The Infection Control Committee advised the clinical staff to test patients with possible nosocomial pneumonia for *L. pneumophila* sg 1 infection by urine antigen tests. In August 1999, a patient was diagnosed with *Legionella* pneumonia and a lung abscess 4 days after admission for respiratory symptoms. The patient had been discharged 5 days earlier after a 12-day admission during which he had received immunosuppressive treatment. Environmental culture found many sites positive for *L. pneumophila* sg 1 including feed water to the calorifiers and a sample taken just after tempering of the water exiting the calorifier. A direct supply line of cold water for tempering was established from the outlet of the UV irradiation plant, and a chlorination system was installed to the calorifiers' cold water supply,

making shock treatment of the DHWS possible. Regular testing regimes were established for ward showers, which were to be tested twice a year, and several sites of the DHWS (Fig. 1). In 2000 multiple sites from cold water sites feeding the calorifiers, showers and thermostatic mixing valve removed from a ward shower were found to be positive for *L. pneumophila* sg 1 (Table 1). It was decided to replace shower thermostatic mixing valves when a showerhead was culture positive. By August that year, all shower thermostatic valves were replaced. The valves were sterilized by autoclaving, but the heat destroyed the seals, which had to be replaced. In September, the temperature of the calorifiers was raised to 80°C and the water tempered to 60°C for circulation. Attention was paid to the total eradication of dead legs in the DHWS. Thermostatic mixing valves protected the user from scalding in certain areas.

From November 2000 to June 2005, the following control regimes were in place. Calorifiers were kept at 80°C, and the hot water circulated at 60°C. The DHWS was shock treated by chlorination to 2 ppm twice a year. Ward showers were cultured twice a year, and if results were positive, the thermostatic mixing valve was changed and the whole DHWS was shock chlorinated for a week. Six sites of the cold water feeding to the calorifiers were cultured monthly, as were seven sites of the hot water system. By 2002, few sites of the DHWS were positive for *L. pneumophila* sg 1. Only one shower was positive in 2004. One site (no. 10 in Fig. 1) on the cold water site was continuously positive until 2003. All sites have been clear of *L. pneumophila* sg 1 since July 2004. One ventilated intensive care unit patient receiving nutrition via a PEG stoma had *L. pneumophila* sg 1 isolated from a tracheal aspirate in March 2002. Ten cooling towers were cultured monthly and were only found positive on three occasions: 1999, 2001, and 2004. On each occasion a different cooling tower was positive (Table 1).

The monitoring of the DHWS showed that when a preheater, installed to utilize waste steam for water heating, had its water temperature reduced from 80°C to 40°C, it became

TABLE 1 Number of cultures positive for *L. pneumophila* sg 1

Site sampled	No. of cultures positive										
	1995	1996	1997	1998	1999	2000	2001	2002	2003	2004	2005
Showers	0	0	0	0	1	13	7	2	6	1	0
Hot water	0	0	0	3	25	2	3	0	0	0	0
Cold water	0	0	0	0	12	36	16	4	4	0	0
Cooling towers	0	0	0	0	1	0	1	0	0	1	0
Total samples	23	21	41	58	329	712	786	551	702	828	663

an ideal incubator for *Legionella* species. The water from pipes supplying the calorifiers was also positive on testing. The supply water remained positive even when the preheater was removed, indicating colonization of the pipes. A chance culture of the cold water storage indicated that the *Legionella* most likely originated from its artesian well source. When it became a requirement to lower the circulating DHWS temperature to 55°C, a water supply for tempering the hot water was obtained from the closest cold water supply, which happened to be the colonized pipes. This resulted in peripheral sites of the DHWS being contaminated. The peripheral contamination with *L. pneumophila* sg 1 was discovered within months, but when a patient was recognized as possibly being infected, extensive changes to the DHWS were carried out. A UV irradiation plant was installed near the exit of the cold water tanks to reduce further colonization. A chlorinator for shock chlorination of the DHWS was installed. The calorifier temperature was increased to 80°C and the water tempered to 60°C for supply and 59°C on return. A new cold water pipe for tempering water was installed directly from the cold water storage tank distal to the UV plant.

An extensive *Legionella* monitoring regime was initiated, involving both the cold water feeds to the calorifiers and the hot water system. It was discovered that if the thermostatic mixing valves in showers had become colonized, showerheads could not return negative cultures until the mixing valve had been changed. If a culture was positive, the DHWS was shock chlorinated to 2 ppm for 1 week. With this intensive treatment of the DHWS, it took 4 years to control the contamination to a level where positive cultures were no longer received.

Several control measures have been researched with the aim of controlling *Legionella* in domestic hot water, but none have been totally successful (1, 6-10). Darelid (4) showed from a 10-year surveillance that keeping the circulating hot water above 55°C could prevent nosocomial Legionnaires' disease. Goetz (6) advocated the use of environmental cultures to alert clinical staff to the risk of nosocomial *Legionella* infection. Only one possible and one definite infection were found in this review. This might be because the circulating water temperature was kept at 60°C except for the year after tempering the water to 55°C. A nosocomial pneumonia study carried out in 1995 did not find any infections with *L. pneumophila* sg 1 (5).

We concluded that tempering the domestic hot water to 55°C from a colonized water supply resulted in widespread contamination of a major hospital DHWS. Introducing several control measures such as engineering changes to the hot water supply, raising the circulating hot water temperature to 60°C, shock chlorinating, and routine surveillance succeeded in controlling the colonization of the domestic hot water in this tertiary level hospital.

REFERENCES

1. **Atlas, R. M.** 1999. *Legionella*: from environmental habitats to disease pathology, detection and control. *Environ. Microbiol.* **1:**283–293.
2. **Building Industry Authority.** 1992. *The New Zealand Building Code Water Supplies Clause G12,* 2nd ed.

3. 1992. Culture of hospital water for members of the family: *Legionellaceae*. Clinical Microbiology Procedures Handbook, Vol. 2: 11.3.

4. **Darelid, J., S. Lofgren, and B. E. Malmvall.** 2002. Control of nosocomial Legionnaires' disease by keeping the circulating hot water above temperature degrees C: experience from a 10-year surveillance programme in a district general hospital. *J. Hosp. Infect.* **30:**213–219.

5. **Everts, R. J, D. R. Murdoch, S. T. Chambers, G. I. Town, S. G. Withington, I. R. Martin, M. J. Epton, C. Frampton, A. Y. Chereshsky, and M. I. Schousboe.** 2000. Nosocomial pneumonia in adult medical and surgical patients at Christ-church Hospital. *N. Z. Med. J.* **113:**221–224.

6. **Goetz, A. M., J. E. Stout, S. L. Jacobs, M. A. Fisher, R. E. Ponzer, S. Drenning, and V. L. Yu.** 1998. Nosocomial Legionnaires' disease discovered in community hospitals following cultures of the water system: seek and ye shall find. *Am. J. Infect. Control.* **26:**8–11.

7. **Isenberg, H. D.** 1992. *Clinical Microbiology Procedures Handbook.* American Society for Microbiology, Washington D.C.

8. **O'Neill, E., and H. Humphreys.** 2005. Surveillance of hospital water and primary prevention of nosocomial legionellosis: what is the evidence? *J. Hosp. Infect.* **59:**273–279.

9. **Sabria, M., and V. L. Yu.** 2002. Hospital-acquired Legionellosis: solution for a preventable infection. *Lancet Infect. Dis.* **2:**368–373.

10. **Thomas, V., T. Bouchez, V. Nicolas, S. Robert, J. F. Loret, and Y. Lévi.** 2004. Amoebae in domestic water systems: resistance to disinfections and implication in *Legionella* persistence. *J. Appl. Microbiol.* **97:**950–963.

REVIEW OF NOSOCOMIAL *LEGIONELLA* OUTBREAKS

Tim Eckmanns, Christiane Reichhardt, Maria Martin,
Frauke Nietschke-Tiemann, and Henning Rüden

116

Legionella are the cause of serious hospital-acquired infections, particularly in immunocompromised patients. About 70% of cases of Legionnaires' diseases (LD) occur during outbreaks (3). During an outbreak, the proportion of health care-associated pneumonia due to LD may be as high as 50% (2). The mortality of LD is about 14% for nosocomial cases (3). The aim of this review is to analyze the evidence of different disinfection methods to stop outbreaks of LD.

LITERATURE RESEARCH AND EXTRACTION

Outbreak database (http://www.outbreak-databse.com), MEDLINE, EMBASE, Meditec, Scisearch, and Biosis were researched for the years 1977 to 2005. The following search terms were used: "legionell* AND outbreak." Articles describing original outbreaks of *Legionella* were selected for the review. All references identified were initially selected on the basis of their titles and/or abstracts. Full reports of potentially relevant publications were obtained and checked for eligibility. Decisions on which trials to include were based on full text articles. The following parameters were extracted: country, year of publication, year of outbreak, duration of outbreak, source of outbreak, facility and department, number of patients involved, number of health care workers involved, mortality, disinfection methods, and success of methods. Two reviewers extracted data independently.

ANALYSIS OF THE LITERATURE

In total, 54 studies from 13 countries were identified. The average time from outbreak year to publication was 4 years. Numbers of outbreaks in 5-year periods are presented in Fig. 1. In comparison, the number of published articles about *Legionella* are shown as well.

The average duration of outbreaks was 12 months. The main sources of the outbreaks were 34 hospital water systems (64.2%), 7 cooling towers (13.2%), and 4 showers (7.5%). Sources mentioned only once were air conditioning, distilled water, transesophageal echocardiography, shock absorbers, and stagnant cylinder. In six (11.3%) papers, no outbreak source is mentioned. Twenty-four percent of the outbreaks happened in ICUs only, 33% only in non-ICUs, and 31% in both (12% not specified). The department most often involved was internal medicine (54%). Other departments were transplantation (19%) and

Tim Eckmanns, Christiane Reichhardt, Maria Martin, Frauke Nietschke-Tiemann, and Henning Rüden Institute of Hygiene and Environmental Medicine, Charité - University Medicine Berlin, 12203 Berlin, Germany.

Legionella: State of the Art 30 Years after Its Recognition
Edited by Nicholas P. Cianciotto et al.
©2006 ASM Press, Washington, D.C.

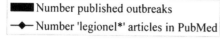

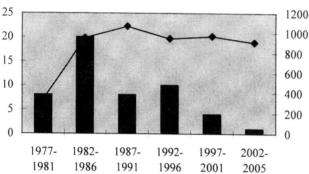

FIGURE 1 Numbers of published outbreaks and published articles about *Legionella* in PubMed in 5-year periods for the 5pyear periods fort he years 1977 to 2005. (Outbreaks lasting over 1 year were counted for the first year of the outrbreak. Three studies gave no outbreak year.)

surgery (13%). Twenty-two percent did not mention a department (the sum is above 100 because some of the outbreaks happened in multiple departments).

In total, 579 (mean, 12.7; median, 8) patients and 220 (mean, 4.1; median, 0) health care workers were involved. Of the infected patients, 140 (17.5%) died.

In the 54 outbreaks, 65 measure to decontaminate the hospital water systems were conducted. Results are given in Table 1.

HOW TO DEAL WITH OUTBREAKS

Legionella outbreaks have a significant important impact on patient morbidity and mortality in hospitals. They also might cause bad publicity. Moreover, they are a long-lasting burden for the hospital. Therefore, it is necessary to terminate outbreaks as fast as possible. In 68% of outbreaks, the source was the hospital water system or showers. In these cases, disinfection methods focussing on the water system should be chosen. The success of these disinfection methods

TABLE 1 Methods used to terminate the outbreaks and percent of success

Method	No. (%) of outbreaks	Percent of outbreaks in which *Legionella* spp. were detected after intervention	
		No	Yes
Chlorination	17 (26.2)	70.6	29.4
Hyperchlorination	4 (6.2)	75	25
Chlorine dioxide	1 (1.5)	100	0
Permanent temperature change	17 (26.2)	58.8	41.2
Temperature change once a month	7 (10.8)	71.4	28.6
Heat and flush	2 (3.1)	50	50
Ultra violet	1 (1.5)	0	100
Sterile water	2 (3.1)	100	0
Filters	1 (1.5)	100	0
Other methods[a]	13 (20)	76.9	23.1
Total	65 (100)	69.2	30.8

[a]Other methods are mostly technical measurements such as plumbing renovation, hot water tank drain, use of electric showers.

depends on the status of the water system and the disinfection method used. Since not all methods succeed in the same proportion, choosing the right method is crucial.

When comparing methods (listed in Table 1) that were mentioned in five or more studies, those listed as "other methods" proved to be most successful in eliminating *Legionella* or LD (76.9%). "Other methods" means technical measures such as plumbing renovation, hot water tank drain, and use of electric showers. This implies that technical measurements are very important factors for termination of outbreaks, despite other effective disinfection methods such as chlorination (70.6%) and elevating temperature once per month (71.4%). The safest interventions are sterile water for patient care or point-of-use filters.

Interestingly, the number of published outbreaks has decreased since the 1980s, whereas the literature about *Legionella* did not. Whether or not this is due to better plumbing design of hospital water systems and more successful disinfection methods remains to be proven.

REFERENCES

1. **Joseph, C. A., J. M. Watson, T. G. Harrison, and C. L. Bartlett.** 1994. Nosocomial Legionnaires' disease in England and Wales, 1980-92. *Epidemiol. Infect.* **112:**329–345.
2. **Centers for Disease Control.** 2004. Guidelines for prevention of healthcare-associated pneumonia, 2003. *Morbid. Mortal. Wkly. Rep.* **53:**1–36.
3. **Benin, A. L., R. F. Benson, and R. E. Besser.** 2002. Trends in Legionnaires disease, 1980-1998: declining mortality and new patterns of diagnosis. *Clin. Infect. Dis.* **35:**1039–1046.

QUANTITATIVE MICROBIAL RISK ASSESSMENT MODEL FOR *LEGIONELLA:* SUMMARY OF METHODS AND RESULTS

Thomas W. Armstrong and Charles N. Haas

117

Beginning with the first reports of *Legionella pneumophila* as the cause of the July 1976 Legionnaires' disease epidemic, there has been conjecture about the infectious dose-response in humans. Since then, quantitative microbial risk assessment (QMRA) techniques have evolved and have been applied mainly to water- and food-borne pathogens. QMRA for *Legionella* may provide a means of setting risk based limits on air concentration or risk-based limits on the *Legionella* concentrations in water.

The work to extend QMRA to *Legionella* involved (i) dose-response modeling data sets for guinea pigs exposed to aerosols, (ii) interspecies extrapolation of the dose-response models to humans using risk assessment techniques, (iii) estimation of human exposures for several well-documented outbreaks, (iv) estimation of human infection rates for those outbreaks using the QMRA models, and (v) comparison of the estimated risks to the reported infection rates for the outbreaks.

ANIMAL MODEL SELECTION

Guinea pigs provide a reasonable animal model and dose-response data on which to base human risk projections due to similarities in the course of the disease in guinea pigs and humans and in vitro uptake and replication rates of *Legionella* in human and guinea pig alveolar macrophages.

Most mouse and rat strains appear to be relatively resistant to *Legionella* infection due to less compliant alveolar macrophages. Limited data suggest that nonhuman primate macrophages are also resistant to *Legionella*, but the data are too limited to support firm conclusions on this point.

ANIMAL DOSE-RESPONSE MODELING

Dose-response modeling was completed using data from multiple published investigations. Models were used to extrapolate the dose-response relationship below the experimental range used to set the model parameters. Model parameters were fit using maximum likelihood techniques.

Exponential and approximate beta-Poisson models were selected for subsequent risk projection work due to mechanistic considerations and low dose extrapolation limited by exposure probability. Model goodness of fit

Thomas W. Armstrong Public and Occupational Health Division, ExxonMobil Biomedical Sciences, Inc., Annandale, NJ 08801. *Charles N. Haas* Drexel University, Civil, Architectural and Environmental Engineering, Philadelphia, PA 19104.

was tested and passed for the data sets employed in the QMRA.

Infectivity (ID) and mortality (LD) dose-response curves in guinea pigs were developed for the selected dose-response models. For the exponential model (equation 1) the best fit for the infectivity data was with $r = 0.06$ and for mortality $r = 1.07 \times 10^{-4}$ These may be used with equation 1 to predict response at other doses.

$$P_{(d)} = 1 - e^{-rd} \tag{1}$$

Where $P_{(d)}$ is the predicted response at dose d and r is the best fit parameter for the model fit to the data set.

The estimated responses are, for guinea pigs, as retained dose in animal lungs:

LD 50%: 6,200 CFU, LD 1%: 92 CFU

ID 50%: 12 CFU, ID 1%: 0.17 CFU

With the assumption of species equivalence and no dose scaling, these results then also represent quantitative estimates of human risk. Dose scaling was not applied largely because of mechanistic considerations, and no interspecies factors (since none were a priori seen as needed) were applied for the guinea pig to human extrapolation. The results suggest these a priori decisions were appropriate.

OUTBREAK EXPOSURE ASSESSMENTS

For a whirlpool spa outbreak, exposures were estimated from aerosol generation information, assumed *L. pneumophila* content in water (based on data from published reports), estimated bacterial content of the aerosol, amount of time workers are in the building, and two distance zones of workers from the whirlpool. A two-zone box model was used to estimate the air concentrations in CFU/m^3, with stochastic input distributions using Monte Carlo simulation to yield results as probability distributions.

- Predicted zone 1 exposures as retained dose: mean, 10 CFU; 95% range 1.3 to 34.
- Predicted zone 2 exposures as retained dose: mean, 7 CFU; 95% range 1.3 to 19.

Exposures for two hot spring spa outbreaks were estimated from the reported water concentrations of Legionella, a water to air bacterial partitioning coefficient, and the estimated time typically spent in a hot spring environment. The distributions of estimated exposures were generated via Monte Carlo simulation.

- Predicted exposures for outbreak 2: mean, 47 CFU; 95% range 24 to 84.
- Predicted exposures for outbreak 3: mean, 2.3 CFU; 95% range 1.1 to 4.1.

RISK ESTIMATION AND EVALUATION RESULTS

The respective exposure distributions were fed into the dose-response model using Monte Carlo simulation. The resulting estimated risk distributions were then compared to the reported rates for the outbreaks. The confidence intervals of the predicted risks generally overlap the confidence intervals on the reported rates of Legionnaires' disease or miss by less than 10 times. This suggests that the model is generally valid. See Table 1 for a summary of the results.

Guinea pig mortality-based modeling predicts rates slightly less than the reported human clinical infection rates and generally better matches the reported human mortality rates. The guinea pig infection data adequately predict the reported subclinical (seroprevalence) rates.

The outbreaks used for the validation involved generally healthy, active adults. Risks may be higher than predicted for more infirm individuals and are likely higher for immune-compromised individuals.

For the outbreaks used for model evaluation, the reported rates of disease span an order of magnitude. The estimated exposures and the QMRA model predictions of disease rates generally held over that range.

The guinea pig data were almost all from exposure to *L. pneumophila* serogroup 1. The outbreaks used for model evaluation also involved *L. pneumophila* serogroup 1. Thus, the QMRA may not extrapolate to other *Legionella* species and strains.

TABLE 1 Summary of results of the evaluation of the calculated versus reported risks for the three outbreaks used for model validation[a]

Outbreak	Subclinical infection	Clinical severity of infection	Mortality
Whirlpool spa <15 M group	x	Reported <10 X higher	x[b]
Whirlpool spa >15 M group	x	Reported <10 X higher	x[b]
Hot spring outbreak 1	x[b]	x	Reported <10 X lower
Hot spring outbreak 2	Reported <10 X lower[b]	x	x

[a]X, the confidence intervals of the predicted risks overlap the corresponding reported rates.
[b]Based on estimated rates since these parameters were not reported for the corresponding outbreak.

The extended abstract presented here cannot cover all of the aspects of the work. Further details with the supporting references noted in the text are provided in the full dissertation. A PDF version of the document (2) is available at http://dspace.library.drexel.edu/handle/1860/615.

REFERENCES

1. **Anonymous.** 2003. Legionellosis, April 1999-December 2002, Japan. *IASR* **24:**27–28.
2. **Armstrong, T. W.** 2005. Microbial Risk Assessment Model for Human Inhalation Exposure to Legionella. Ph. D. thesis. Drexel University, Philadelphia. Available at http://dspace.library.drexel.edu/handle/1860/615.
3. **Baskerville, A., R. B. Fitzgeorge, M. Broster, P. Hambleton, and P. J. Dennis.** 1981. Experimental transmission of Legionnaires' disease by exposure to aerosols of Legionella pneumophila. *Lancet* **2:**1389–1390.
4. **den Boer, J. W., E. P. Yzerman, J. Schellekens, K. D. Lettinga, H. C. Boshuizen, J. E. Van Steenbergen, A. Bosman, H. S. Van den, H. A. Van Vliet, M. F. Peeters, R. J. van Ketel, P. Speelman, J. L. Kool, and M. A. Conyn-van Spaendonck.** 2002. A large outbreak of Legionnaires' disease at a flower show, The Netherlands, 1999. *Emerg. Infect. Dis.* **8:**37–43.
5. **Fitzgeorge, R. B., A. Baskerville, M. Broster, P. Hambleton, and P. J. Dennis.** 1983. Aerosol infection of animals with strains of Legionella pneumophila of different virulence: comparison with intraperitoneal and intranasal routes of infection. *J. Hyg.* **90:**81–89.
6. **Haas, C. N., J. B. Rose, and C. P. Gerba.** 1999. *Quantitative Microbial Risk Assessment.* John Wiley, New York.
7. **Jacobs, R. F., R. M. Locksley, C. B. Wilson, J. E. Haas, and S. J. Klebanoff.** 1984. Interaction of primate alveolar macrophages and Legionella pneumophila. *J. Clin. Invest.* **73:**1515–1523.
8. **Muller, D., M. L. Edwards, and D. W. Smith.** 1983. Changes in iron and transferrin levels and body temperature in experimental airborne legionellosis. *J. Infect. Dis.* **147:**302–307.
9. **Nagelkerke, N. J., H. C. Boshuizen, H. E. de Melker, J. F. Schellekens, M. F. Peeters, and M. Conyn-van Spaendonck.** 2003. Estimating the incidence of subclinical infections with Legionella pneumonia using data augmentation: analysis of an outbreak in The Netherlands. *Stat. Med.* **22:**3713–3724.
10. **Nicas, M., and T. W. Armstrong.** 2003. Computer implementation of mathematical exposure modeling. *Appl. Occup. Environ. Hyg.* **18:**566–571.

RISK OF *LEGIONELLA* IN THE SPA INDUSTRY: INADEQUACY OF CURRENT LEGISLATION COVERING THERMAL WATERS USED FOR MEDICINAL PURPOSES

Vladimir Drasar, Radomir Polcar, and Paul Christian Lück

118

The current spa industry provides a complex range of services comprising both therapy and accommodation. The general interest of spa managers is to attract both patients and vacationers who are looking for recondition stays. The emphasis is therefore laid on the quality of services provided. While the spa quality management is mostly ISO certified, a *Legionella* risk assessment has not yet been included.

Prevention should be based on a close exercise of national health (spa laws and decrees, drinking water) and technical legislation (guidelines for designers, plumbers, manufacturers and service providers). The technical guidelines represent an important foundation for public health authorities to understand how to eliminate or at least reduce conditions supporting *Legionella* colonization in spas.

Thermal water used for medicinal purposes is an ideal environment for *Legionella* (8, 10). The favorable temperatures (ranging from 36 to 41°C); the absence of disinfection that is forbidden by law (2); and particular technolog-

ical operations comprising drawing, collection, storage, and long distance distribution to hydrotherapy hotels all contribute to *Legionella* growth. Finally, particular spa procedures may cause a health risk for clients. Numerous *Legionella* infections linked to thermal springs from France, Spain, Portugal, and Japan have been reported (1, 4–7, 9).

A cluster of travel–associated *L. pneumophila* serogroup (sg) 1 subgroup Philadelphia infections associated with the biggest Bohemian spa was reported to the European Working Group on *Legionella* Infection. The spa medical care there is focused on locomotive disorders and postoperative conditions. Typical clients are senior citizens. Every month, 4 to 8 of these people were found to be immunocompromised. Annually, 18,500 clients attend the spa. There is no off season. Thermal, slightly mineralized, and radioactive water is used in bubble baths, hand and underwater massages, and pressure sprays. The entire spa network consists of seven hotels and surrounding grounds with two decorative fountains (Fig. 1).

The source investigation was facilitated by the clinical isolate of the Philadelphia strain obtained from the last fatal case in Dresden.

The objective of the study was to investigate the overall *Legionella* colonization of thermal water and cold and hot water of particular

Vladimir Drasar   National Legionella Reference Laboratory, Public Health Institute, Vyskov, CZ 682 01, Czech Republic. *Radomir Polcar*   FACTOR.E, s.r.o., Brno, CZ 602 00, Czech Republic.   *Paul Christian Lück*   Legionella Reference Laboratory, Institute of Medical Microbiology and Hygiene, TU Dreseden, Dresden, D-01307, Germany.

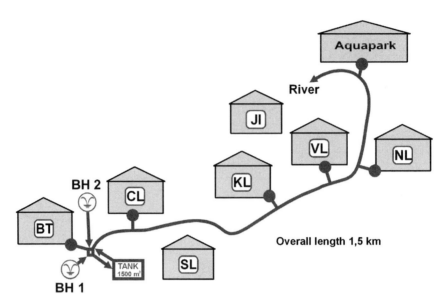

FIGURE 1 Schematic diagram of the thermal water system. Water is drawn from two boreholes (BH1, BH2), pumped into the tank, and distributed to five hotels (BT, CL, KL, VL, NL) and an aquapark.

hotels, including the detailed identification and typing of species and serogroups. The aim was also to evaluate the risk of particular hydrotherapy procedures, to identify the source of the Philadelphia strain in the spa complex, and to propose corrective actions to minimize the risk of future *Legionella* infections.

All hotels were followed up according to European Union guidelines (3). *Legionella* isolates were identified and serotyped by the quantitative micoragglutination test with all known rabbit sera (63 species and serogroups) and by an indirect immunofluorescence test. Detailed typing was carried out by amplification and sequencing of six *L. pneumophila* genes—*dotA, gspA, neuA, ompS,* and *pilE.* Resulting sequences were analyzed with the software BioNumerics (Applied Maths, Kortjik, Belgium).

A detailed follow-up revealed that five hydrotherapy hotels and an aquapark were supplied with the same thermal water. Its distribution system, 90 years old and 1.5 km long, is fed from two boreholes. The water is mixed and then pumped into a big concrete tank. Its three chambers store 1,500 m^3 of water. The

system as a whole has not been treated with any biocide, just an occasional disinfection of the tank was admitted by the staff. The source colonization search found no *Legionella* in water from the hot springs, but the first sampling sites on the delivery pipes were revealed to be positive. The summarizing results are shown in Tables 1 and 2. They confirmed that the thermal water and its pipe work were abundantly colonized by *L. pneumophila* sg1 (four subgroups including Philadelphia); additional serogroups 2, 5, and 6; and a further seven *Legionella* species. *Pseudomonas aeruginosa* and nontuberculous mycobacteria were also isolated. *Legionella* finds from cold and hot water

TABLE 1 Tracing the source of the Philadelphia strain: *Legionella* concetrations in the thremal water system

Source of sample	*Legionella* colonization (CFU/100 ml)
Borehole water	0
Delivery pipes	4–100
Storage tanks	0
Distribution	2–62

TABLE 2 Tracing the source of the Philadelphia strain: *Legionella* species and serogroups isolated the thermal water system

L. pneumophila sg 1
MAb type Philadelphia
MAb France
MAb OLDA
MAb Bellingham
L. pneumophila sg.2, 5, and 6
L. brunensis
L. gratiana
L. maceachernii
L. oakridgensis
L. rubrilucens
L. spiritensis
L. worsleiensis
P. aeruginosa
Nontuberculous mycobacteria
New bacterial species BT-17

systems of all seven spa hotels are given in Table 3. The Philadelphia strains were confirmed in the two hotels where the cases were reported. Again, a broader spectrum of *L. pneumophila* serogroups (1, 2, 3, 5, 6) and five more *Legionella* species were isolated.

An immediate detailed technical inspection was carried out. It commenced with a proper briefing of the spa management and staff concerned about *Legionella*, and the relevant legislation they must comply with. The identification of critical points followed. Short- and long-term corrective actions were proposed, including the introduction of risk assessment documents. Permanent chlorine dioxide disinfection was recommended for the hot water systems of the two hotels. A dilemma caused by the legislation has arisen as to how to treat the thermal water and its distribution system. Oxidizing biocides and ionizers are not permitted by law (2). Regular thermal disinfection is recommended, but its application is often unfeasible for the rapid settling of mineral deposits and scale, which usually clog pipes and valves. Moreover, higher temperatures could adversely affect the mineral make-up of the water. Shot dose disinfection in the off season that lasts only one week did not meet expectations. Although counts of planctonic legionella were reduced, they soon reverted to the original ones, which indicated that *Legionella* living in biofilms survived.

Bubble jet baths and hand rehabilitations represented the highest health risk of the hydrotherapy procedures investigated, because of the proximity of the patient's breathing organs to the bubbling water during the procedure. The bath equipment (mostly very sophisticated) contained many different kinds of tubing and hoses and air and water jets that were almost all colonized with biofilms harboring *Legionella*. Despite regular disinfection after each use, *Legionellae* were often detected from air and water jet swabs.

TABLE 3 Tracing the source of the Philadelphia strain: hot water plumbing systems of particular spa hotels

Spa hotel[a]	*L. pneumophila* strain or serogroup	Other *Legionella* species detected	Total *Legionella* count (CFU/100 ml) in patients' room[b]
BT	Philadelphia-1 MAb type sg 1, 2, 3, 5, 6	*L. rubrilucens*	8,900
CL	Philadelphia-1 MAb type	*L. spiritensis*	32
KL	sg 6	*L. rubrilucens, L. brunensis*	
NL	sg 1 and 5	*L. brunensis, L. geestiana*	
VL	sg 5	*L. brunensis*	
JI★	sg 1	*L. micdadei*	
SL★	sg 1 and 6		

[a] Asterisk indicates thermal water was not delivered.
[b] The rooms where the patients had stayed.

CONCLUSION

The general view of the risk of *Legionella* in spas is based on a basic assumption that there is a source of infection, mode of transmission and susceptible host. The spas meet all these conditions. Thermal and hot water distribution systems and hydrotherapy procedures present the sources, while aerosol inhalation and drinking water appear to be the transmission. Clients are usually elderly people with underlying diseases and impaired immunity. They are more susceptible to infections. The facts entail that they have no escape from the vicious circle, as they are exposed to *Legionella* both from thermal baths and showers using hot water. The current legislation, and not only the Czech one, hampers prevention in practice. Its controversy rests in the necessity to retain the unique curative powers of thermal waters while protecting patients from infection. Section 3 of the Czech legislation (2) says that water for medicinal purposes must be microbiologically safe. It must not contain any bacteria that could present a health risk to human beings. However, section 5 of the same document declares that such waters shall neither be disinfected nor be supplemented with bacteriostatic agents.

To comply with the legislation, both technical and organizational measures had to be adopted. Chlorine dioxide generators were installed on the hot water systems of the two hotels where the clients contracted Legionnaires' disease. A further managerial decision has been made to no longer admit immunosupressed patients. They should be sorted out by admitting physicians. The regulation generated a repercussion to tour operators. They had to be made aware of the ban, with the explanation that all the steps taken have been fully in line with the European Union guidelines to protect client's health.

Czech spas, like elsewhere, are colonized with *Legionella*. Our corrective actions proposed and implemented have proved their worth. No more *Legionella* cases have been reported from the spa since that time. Nevertheless, prevention of sporadic cases in relation to thermal waters still remains an open legislative and health problem to solve.

REFERENCES

1. **Bornstein N., D. Maremet, M. Surgot, M. Novicki, A. Arslan, J. Esteve, and J. Fleurette.** 1989. Exposure to Legionellacae at a hot spring spa: a prospective clinical and serological study. *Epidem. Inf.* **102:**31–36.
2. **Czech Ministry of Health.** 2001. *Decree of the Czech Ministry of Health No. 423/2001: On Spas and Sources.*
3. *European Guidelines for Control and Prevention of Travel Associated Legionnaires' Disease*, 2003 (www .ewgli.org).
4. **Koide, M.** 2002. Hot spring bath as the reservoir of Legionella bacterium. *Intern. Med.* **41:**759.
5. **Miyamoto H., S. Jitsurong, R. Shiota, K. Maruta, S. Yoshida, and E. Yabuuchi.** 1997. Molecular determination of infection source of a sporadic Legionella pneumonia case associated with a hot spring bath. *Microbiol. Immunol.* **41:** 197–202.
6. **Pelaz C., R. Cano, and B. Baldrón.** 2004. Legionella and spas in Spain (1993-2003). Poster No. 4., in 19th EWGLI Meeting Chamonix, Abstract book.
7. **Rocha G., A. Veríssimo, R. Bowker, N. Bornstein, and M. S. da Costa.** 1995. Relationship between *Legionella spp.* and antibody titres at a therapeutic thermal spa in Portugal. *Epidemiol. Infect.* **115:**79–88.
8. **Schaffler-Dullnig K., F. F. Reinthale, and E. Marth.** 1992. Nachweis von Legionellen im thermalwasser. *Zbl. Hyg.* **192:**473–478.
9. **Tominaga M., Y. Aoki, S. Haraguchi, M. Fukuoka, S. Hayashi, M. Tamesada, E. Yabuuchi, and K. Nagasawa.** 2001. Legionnaires' disease associated with habitual drinking of hot spring water. *Intern. Med.* **40:**1064–1067.
10. **Veríssimo A., G. Marrao, F. Gomes da Silva, and M. S. da Costa.** 1991. Distribution of *Legionella spp.* in hydrothermal areas in continental Portugal and the Island of Sao Miguel, Azores. *Appl. Environ. Microbiol.* **57:**2921–2927.

BIOLOGICAL TREATMENT OF INDUSTRIAL WASTEWATER: A POSSIBLE SOURCE OF *LEGIONELLA* INFECTION

Görel Allestam, Birgitta de Jong, and Jonas Långmark

119

This paper summarizes the results of an extensive study of biological treatment plants (BTPs) from 43 paper mills in Sweden. During September and October 2005, each paper mill with a BTP was sampled systematically. High concentrations of *Legionella*, up to 10^9 CFU /liter, were found in investigated treatment plants. In total, 66% of all investigated paper mills were positive for *Legionella*.

In December 2004, a BTP at one paper mill in Sweden was sampled since one of the workers (57 years old, male, smoker) at the plant had developed Legionnaires' disease. This plant was one of several possible sources of the infection. Epidemiological investigation found no legionella in showers at the worker's home. Showers at work had low counts of *Legionella pneumophila* serogroup (sg) 4, which differed from the patient's isolate, which was *L. pneumophila* sg 1 subtype Benidorm. This isolate matched, however, the *Legionella* isolated from the aeration pond of the mill's BTP. Genotyping was performed with amplified fragment length polymorphism according to the protocol of the European Working Group for Legionella Infections. This finding showed that a more systematic investigation was necessary, and this was preformed in cooperation with the Swedish Forest Industries (1). The aim of this investigation was to answer several questions. Does legionella grow in other BTPs from other paper mills or other kinds of industries? What is the impact of the cooling towers? Where in the process does the growth occur? What preventive measures can be taken?

BIOLOGICAL TREATMENT PLANTS

According to the legislation of the Swedish Environmental Protection Agency, industries are obliged to build up biological treatment plants. These are introduced to break down organic substances with high phosphorus and nitrogen content. An efficiency of 50 to 70% chemical oxygen demand reduction is achievable and is accepted. Large amounts of water pass through these systems each day (up to 75,000 m^3). BTPs are open-air systems with ponds as large as a football pitch area (Fig. 1). Visible evaporation clouds can be seen from long distances. The inlet water is chilled down to approximately 37°C either by a heat exchanger or through cooling towers. Urea and phosphate are often added as nutrients for the growth of microflora that is responsible for the

Görel Allestam and Jonas Långmark Department of Parasitology, Mycology and Water and Environmental Microbiology, Swedish Institute for Infectious Disease Control, SE-171 82 Solna, Sweden. *Birgitta de Jong* Department of Epidemiology, Swedish Institute for Infectious Disease Control, SE-171 82 Solna, Sweden.

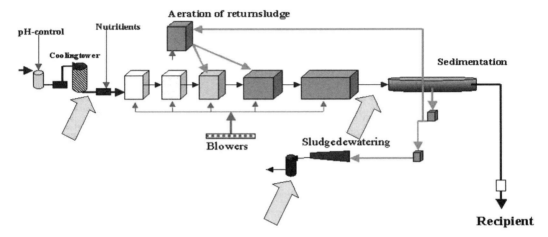

FIGURE 1 Flow chart of a BTP. Big arrows indicate sampling points.

breakdown of organic loads. Huge amounts of air are blown into the ponds, often by a surface pump, which results in massive production of water droplets rich with fibrous matter, foam, and aerosols which can then be spread by winds. The biomass that undergoes sedimentation in secondary ponds is reinoculated into the system by return-loops to stimulate the growth of the microflora. The remaining biomass is dewatered and used as an energy source at the mill and for soil-making outside the plant. As much as 30 tons of biosludge is produced daily and is often mixed with fibrous waste.

Conditions for legionella growth in these systems are almost optimal with a temperature of approximately 37°C, available nutrients and iron sources, and massive aeration. A continuous legionella-culturing system is created.

Every paper mill in Sweden with a BTP was sampled for legionella. In Sweden, 43 out of 66 paper mills have installed BTPs. During September and October 2005, every paper mill with a BTP was sampled systematically. Sample points were cooling tower (if present), outlet of aeration ponds, and biosludge. Analyses were performed on pooled water samples and dewatered sludge. A combined pretreatment of the samples in order to minimize the growth of atypical/heterotrophic microflora was added to the standard method, ISO 11731.

Pretreatment consisted of heat treatment for 30 min at 50°C, followed by twofold dilution in acid buffer pH 2.2 for 5 min for the best effect in reducing background growth. Appropriate dilution series, up to 10^{-6}, were further plated on MWY agar (Oxoid).

FINDINGS
In 65% of the paper mills legionella bacteria were isolated from one or more sampling points, showing that legionella frequently occurs in this kind of system. However, legionella were only found in 3 of 12 cooling towers and in rather low counts. Nevertheless, more than 50% of the aeration ponds were positive in extremely high concentrations (Table 1). Levels above 10^{8}/liter were detected in 14% of the aeration ponds (Fig. 2). Similar numbers were recovered from biosludge. The dominating type in all plants was *L. pneumophila* sg 2-14 while *L. pneumophila* sg 1 was detected in five plants. Mixed legionella populations were found in several biological treatments plants.

Samples of biosludge that had undergone 6 months of a high-temperature composting process showed high counts of *Legionella*, which indicates that the bacteria has a long persistence in this type of material. Numbers as high as $10^{5}/g_{ww}$ in these samples could imply that Legionnaires' disease cases that are re-

TABLE 1 Legionella in biological treatment plants at Swedish paper mills

Sample source	n	No. of *Legionella*-positive samples	Percentage of *Legionella*-positive samples
Cooling tower	12	3	25
Aeration pond	43	24	56
Biosludge	39	21	54
All mills	43	28	65

lated to soil could be associated with residues from biosludge. In 43% of the aeration ponds no legionella was detected. Parallel testing with other techniques such as fluorescence in situ hybridization and Q-PCR are to be introduced and could improve the recovery rate.

The common factor for all BTPs with heavy legionella growth was a temperature of $38 \pm 4°C$, given as a daily mean of the inlet water. When inlet water was below 30°C, no legionella was found (five BTPs). No linear correlation could be found with other factors such as type of treatment plant, retention time in the aeration pond, amount and age of sludge, conductivity, use of iron sulphate for flocculation of the return sludge.

Prior to the described investigation, our laboratory has on few occasions analyzed samples from BTPs of other industries. All but one of these were negative for legionella. The positive one represented a chemical industry that used cellulose as a raw material in its process. Legionella growth was also discovered in a "wastewater basin" at a petrochemical industry during an investigating of an outbreak in Pas-de-Calais, France. The authors discuss that both a cooling tower and the wastewater basin were suspected to be the cause of the outbreak and that aerosols were spread across long distances (3). In our investigation, the cooling towers seem to have little impact on the introduction of legionella downstream into BTPs. Other international studies of biological treatment of sewage have detected legionella with PCR but not by culture (2).

More research is necessary to answer questions such as, What is the risk level for workers and visitors and those who live nearby? How far can legionella be spread through the air? What is the environmental origin of legionella bacteria found in the BTPs of paper mill? Despite the knowledge of the role of protozoa and other bacteria in legionella biology, shouldn't we pay more attention to the temperature of the environment and try to make BTPs efficient at much lower temperatures?

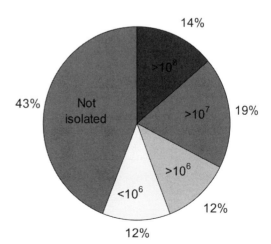

FIGURE 2 Apportioning of legionella concentrations in aeration ponds in BTPs at Swedish paper mills (CFU/liter; $n = 43$).

RISK MANAGEMENT

At all mills, even where legionella have not been isolated, it was recommended that workers wear respiratory protective masks when working in the vicinity of a biological treatment plant. A local risk assessment should be done together with the public health authorities to identify risks for spreading legionella by aerosols and to minimize the spread of disease. Maintenance of cooling towers including regular cleaning is also recommended (http://www.skogsindustrierna.se. 2005. Legionellaförekomst i flertalet bioreningsanläggningar resultat från kartläggning inom massa-och pappersindustrin. PUA meddelande Nr 6. Okt.).

REFERENCES

1. **Allestam, G.** 2005. *Legionella in Biological Treatment Plants*. Tillsyns Nytt, Swedish Environmental Preotection Agency. Nr 2. (In Swedish)

2. **Medema, G., B. Wuilings, P. Roeleveld, and D. Van der Kooij.** 2004. Risk assessment of Legionella and enteric pathogens in sewage treatment works. *Water Sci. Tech. Water Supply* **4:**125–132.

3. **Nhu Nguyen, T. M., D. Ilef,. S. Jarraud, L. Rouil, C. Campese, D. Che, S. Haeghebaert, F. Ganiayre, F. Marcel, J. Etienne, and J. C. Desenclos.** 2006. A community-wide outbreak of Legionnaires' disease linked to industrial cooling towers: how far can contaminated aerosols spread? *J. Infect. Dis.* **193:**102–111.

INHIBITION OF *LEGIONELLA* GROWTH IN CIRCULATING BATHING WATER BY A FILTER REFRESHMENT METHOD USING A HIGH CONCENTRATION OF CHLORINE

Kanji Sugiyama, Katsuhiko Ohata, Mitsuaki Suzuki, Rieko Shimogawara, Shinji Izumiyama, Kenji Yagita, and Takuro Endo

120

In most of hot spring baths in Japan, bathing water is circulated for extended use to conserve hot spring water. In recent years, massive outbreaks of Legionnaires' disease among hot spring bath users have been reported in many districts in Japan (1–3).

Through experiments using a model bathing facility (consisting of a bathtub (2 m^3), a hair catcher, a circulating pump, a filter (ceramic filtering medium, 100 kg, LV: 25.5 m/h), a water heater (40°C), and pipe), we could reproduce the growth of *Legionella* (10^5 to 10^6 CFU/100 ml) naturally under a circulating condition when chlorine was not added into the water (see chapter 102). Furthermore, we found that the filtering medium was the most highly contaminated by *Legionella* among the parts of the circulating system. Thus, the filtering medium itself became a new source resulting in continuous contamination of bathing water by *Legionella* when disinfection of the filtering medium was inadequate, even though the bathing water had been replaced (see chapter 102).

In the present study, we used a bath model to investigate the effectiveness of backwashing the filtering medium using a high concentration of chlorine for disinfection and growth inhibition of *Legionella*. We then assessed the usefulness of this method from the perspective of hygiene control of circulating bathing water.

The procedure for assessing short-term and long-term effectiveness of backwashing using a model bath is as follows.

Short-term effectiveness: After proliferation of *Legionella* in the model bath, backwashing with chlorinated water at concentrations ranging from 5 to 10 mg/liter was performed for 5 min. We called this "the filter refreshment method." The water samples and ceramic sand were collected from the filter unit and analyzed both for *Legionella* and for host amoebae. Bacto-Agar (DIFCO), a monolayer that heat-inactivated *Escherichia coli* is deposited on, was used for isolating amoebae. The effectiveness of backwashing with tap water (0.2 mg/liter of chlorine concentration) was also assessed.

Long-term effectiveness: After depositions of organic substances by bathing in the presence of chlorine in the model bath, the addition of chlorine was stopped while the bath was in use until the chlorine was disappeared.

Kanji Sugiyama, Katsuhiko Ohata, and Mitsuaki Suzuki Department of Microbiology, Shizuoka Institute of Environment and Hygiene, Shizuoka-City, Shizuoka Pref. 420-8637, Japan. *Rieko Shimogawara, Shinji Izumiyama, Kenji Yagita, and Takuro Endo* Department of Parasitology, National Institute of Infectious Disease, Shinjuku-ku, Tokyo 162-8640, Japan.

Chlorine backwashing by the filter refreshment method (5 to 10 mg/liter) was performed once a day for 9 days, and the bathing water and the water from the filter unit were collected every day prior to backwashing to determine the presence of *Legionella* and amoebae.

SHORT-TERM EFFECTIVENESS OF THE FILTER REFRESHMENT METHOD

The number of *Legionella* in the filter water, measured to be 10^3 CFU/100 ml at the beginning of the experiment, decreased with the increase of residual chlorine concentrations

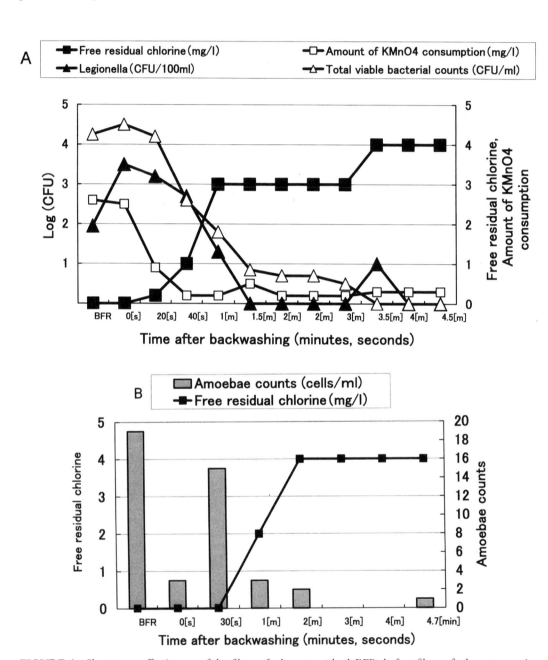

FIGURE 1   Short-term effectiveness of the filter refreshment method. BFR, before filter refreshment; m, minutes; s, seconds.

(4 mg/liter at the maximum) and became unde-tectable 4 min after the backwashing (Fig. 1A). *Legionella* in the filtering medium was also unde-tectable (6.0×10^2/g to 0/g) after being back-washed for 5 min. The amoebae count could also be reduced by the backwashing (Fig. 1B). The amount of dissolved organic materials, repre-sented by the amount of potassium perman-ganate consumed, was also reduced over time af-ter backwashing (Fig. 1A). However, when tap water (0.2 mg/liter residual chlorine concentra-tion) was used for backwashing, *Legionella* was not disinfected effectively and there were no marked differences in *Legionella* counts in the fil-tering medium before and after backwashing. These findings demonstrated that backwashing with chlorinated water at concentrations ranging from 5 to 10 mg/liter was essential for the re-moval of *Legionella* from the filtering medium.

LONG-TERM EFFECTIVENESS OF THE FILTER REFRESHMENT METHOD

The number of *Legionella* in both bathing wa-ter and filter water was maintained at a level lower than 10 to 70 CFU/100 ml by repeated backwashing with chlorinated water alone once a day; the *Legionella* growth was greatly inhibited (Fig.2) compared to that under non-disinfection conditions (see chapter 102). Amoebic growth could be inhibited to a lim-ited level (Fig. 2). On the basis of these results, daily backwashing by the filter refreshment method is considered to be effective for growth inhibition of both *Legionella* and host amoebae in circulating bathing water.

Together with daily use of the filter refresh-ment method, addition of chlorine into the bathing water to a minimum concentration of 0.2 to 0.4 mg/liter may ensure the supply of circulating bathing water with increased mi-crobial safety.

It was also demonstrated that the filter re-freshment method could prevent the deposi-tion of organic substances in the filter medium from the bathers and, thus, reduce the chlorine smell markedly.

The effectiveness of the filter refreshment method has been confirmed in practice by bathhouses serving the public, and the cost has been proven to be inexpensive.

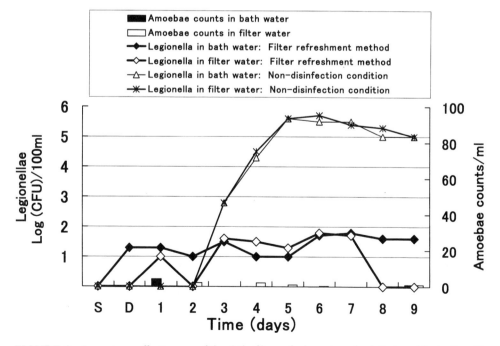

FIGURE 2 Long-term effectiveness of the daily filter refreshment method. S, stop chlorination; D, disappearance of residual chlorine.

The filter refreshment method has been incorporated into the Shizuoka Prefectural Enforcement Ordinances for the Public Bathhouse Law and Hotel Business Law, both of which have been enforced since April 2004.

REFERENCES

1. **Nakamura, H., H. Yagyu, K. Kishi, F. Tsuchida, S. OhIshi, and K. Yamaguchi.** 2003. A large outbreak of Legionnaires' disease due to an inadequate circulating and filtration system for bath water: epidemiologic manifestations. *Intern. Med.* **42:**806–811.

2. **Okada, M., K. Kwano, F. Kura, J. Amemura-Maekawa, H. Watanabe, K. Yagita, T. Endo, and S. Suzuki.** 2005. The largest outbreak of Legionellosis in Japan associated with spa baths: epidemic curve and environmental investigation. *J. Jpn. Assoc. Infect. Dis.* **79:**365–374.

3. **Sugiyama, K., T. Nishio, Y. Goda, K. Masuda, F. Zhang, M. Akiyama, and H. Miyamoto.** 2000. An outbreak of legionellosis linked to bath water circulating through a filter at a spa resort, March–April 2000: Shizuoka. *Infect. Agents Surveillance Rep.* **21:**188.

DISINFECTION OF HOSPITAL WATER SYSTEMS AND THE PREVENTION OF LEGIONELLOSIS: WHAT IS THE EVIDENCE?

Christiane Reichardt, Maria Martin, Henning Rüden, and Tim Eckmanns

121

WHAT ARE WE TALKING ABOUT?

The contaminated hospital water system (HWS) is often identified as the source of *Legionella* spp. (2, 7). In this case, disinfection of the HWS is required, aiming at the prevention of nosocomial Legionnaires' disease (LD). *Legionella* spp. are the cause of serious hospital acquired infections, particulary in immunocompromised patients. Numerous disinfection methods are available, some of them quite costly and time and staff intensive. The decision of what method to use is sometimes difficult, depending on factors such as plumbing status of the HWS, type of patients cared for, and financial and personnel resources, just to mention a few.

The aim of this study is to analyze the literature for evidence of different disinfection methods and their success in preventing nosocomial LD. We systematically searched databases, abstract books, standards books, international guidelines, company information, and review bibliographies focussing on disinfection studies in contaminated HWSs with the outcome LD. Only published articles from peer-reviewed journals, such as case reports, case-control studies, and cohort studies, were included. We developed a rating system for objective analysis of study quality (Table 1).

WHAT DID WE GET?

The literature research search resulted in 970 titles (PubMed, 424; EMBASE, 286; MEDITECT, 103; SCISEARCH, 29; BIOSIS, 70; guidelines and reference books, 58). After screening of titles and abstracts, 113 publications were studied in full text. We found 45 studies describing disinfection methods in HWS, 16 in which the outcome LD was included in this review. Of these 16 studies, 12 are outbreak related. The reviewed publications reported nine different methods: hyperchlorination (reported eight times), technical measures (eight times), shock heat/heat-flush (seven times), continuous chlorination (five times), raising warm-water maintenance temperature to $>55°C$ (five times), copper/silver ions (four times), UV-light (two times), chlorine dioxide (once), and filtering (once).

The quality of the studies reviewed varied greatly. Nine authors reported three or more interventions to achieve disinfection of the HWS (Table 2). Based on our study-quality rating system, we calculated rates for each disinfection method. We summarized the points based on the rate of study quality and divided

Christiane Reichardt, Maria Martin, Henning Rüden, and Tim Eckmanns Institute of Hygiene and Environmental Medicine, Charité-University Hospital Berlin, 12203 Berlin, Germany.

TABLE 1 Rating system for study quality

Outcome rated	Rates
Additional outcome *Legionella* spp. in water pipe system	1 = yes 0 = no
Control group in the study	2 = case-control 1 = historical 0 = non
Routine surveillance for LD according to CDC guidelines	1 = yes 0 = no
Diagnosis of LD according to CDC guidelines	1 = yes 0 = no
Routine surveillance for LD according to CDC guidelines after a successfull intervention	0 = 0–6 months 1 = 6–24 months 2 = >24 months
Clear causality of effect?	2 = yes 0 = no

the sum by the number of times the method was used: UV-light, 6.5; copper/silver ions, 6.3; raising warm water maintenance temperature to >55°C, 6.0; filtering 5.0; shock heat/heat-flush, 5.1; continuous chlorination, 4.8; hyperchlorination, 4.0; technical measures, 4.0; and chlorine dioxide, 2.0.

There is a vast amount of literature out there dealing with *Legionella* spp. We found 970 articles published since 1983 when using the search terms "legionell* AND prevention" in different databases.

The analysis of the studies revealed a great variety of quality, with rates ranging from eight to two (see table 2). Nine studies implemented three or more control measures, which made it very difficult to determine the efficacy of a single method. Moreover, there were only a few prospective studies and no randomized controlled trials. Most papers reported outbreak situations (12 out of 16 articles), focussing mainly on containing the outbreak, rather than on long term effects.

All authors reported a decrease in the number of cases of nosocomial LD, irrespective of the disinfection method chosen. Based on our study quality rating, we analyzed the evidence for each method reported (rates ranging from 9 = excellent to 1 = unacceptable). When methods are sorted by calculated rates, UV

light seems to be the best method, followed closely by copper/silver ions and elevated warm-water maintenance temperature. We found that our rate does not fully reflect the requirements of the different situations and conditions under which a disinfection method might be necessary.

UV light (rate 6.5), for example, was used successfully to prevent contamination of a new HWS (8). In another study, UV-light was implemented for long-term control of low *Legionella* spp. counts after an outbreak was already contained by other disinfection methods (10), as were Cu/Ag ions (1, 3, 11). These measures led to no or significantly fewer cases of LD. Cu/Ag ions (rate 6.3) were also prospectively trialed in a nonoutbreak situation over 1 year (13). Heat/flush (rate 5.1) was used as a short-term measurement in outbreak situations (11, 7), as was hyperchlorination (rate 4.0) (1, 5, 9, 12). Because of quick recolonization, further measures for maintenance of low *Legionella* counts were necessary in most studies. Continuous chlorination (rate 4.8) was used four times after hyperchlorination of the HWS. In all cases, further interventions were necessary (2, 9, 10, 12). Raising the warm-water maintenance temperature above 55°C (rate 6.0) was successfully implemented for maintaining low *Legionella* spp. counts over time (6) and was chosen four times

TABLE 2 Intervention studies with outcome Legionnaires' disease[a]

Author (year of publication)	No. of interventions	Disinfection methods	No. of patients with nosocomial LD before/after intervention	Study quality[b]
Best (1983)★	1	HF	88/12 in 28/17 mo	8
Darelid (2002)★	1	warm-water maint. temp. >55°C	31/4 in 0.4/9 yrs	8
Stout (1998)★	1	Cu/Ag	6/2 per yr	8
Hall (2003)★	1	UV	1 in 13 yrs	7
Matulonis (1993)	5	HCH, CCH, RD, UV, warm-water maint. temp. >60°C	33/3 of 150/201 pts	6
Mietzner (1997)	2	HF, Cu/Ag	2/0	6
Borau (2000)	1	SH/warm-water maint. temp. >55°C	2/0	6
Colbourne (1984)	2	HCH, warm-water maint. temp. >60°C, exchange of rubber components	2/0	5
Ezzedine (1989)	6	CCH, HF, RD/removal mixing tank, flow increase	15/1 in 6/1 yr	5
Helms (1988)	4	HCH, CCH, warm-water maint. temp. >53°C	35/1 per 1000 adm	5
Snyder (1990)	4	HCH, CCH, flow increase, HF	7/0 in 5/17 mo	5
Vonberg (2005)	1	Filter	nm/0	5
MMWR (1997) (hospital Y)	5	HCH, sterile water for nebulizers, HF, Cu/Ag	38/1 in 7/1 yrs	4
Biurrun (1999)	5	HCH, CCH, Cu/Ag, D and exchange of shower heads/aerators, RD, sterile water for nebulizers	1/0	3
MMWR (1997) (hospital X)	6	HCH, HF, RD	8/0 in 9/nm mo	2
Walker (1995)	4	HCH, CHD, D of shower heads, technical measures to assure CHD conc.	3/nm	2

[a]Asterisks indicate study not outbreak related; CCH, continuous chlorination; HCH, hyperchlorination; HF, heat/flush; CHD, chlorine dioxide; SH, shock heat; RD, removal of dead legs; D, disinfection; mo, month; yr, year; pt, patient; adm, admissions; nm, not mentioned.

[b]9, excellent; ≤4, poor.

after the HWS was already disinfected (4, 5, 9, 10). All authors reported a significant decline in the number of cases of LD. No cases of LD were found when filters were installed at distal outlets (rate 5.0) in high-risk patient areas (14). Chlorine dioxide (rate 2.0) was used in one study with the outcome being LD only, where several disinfection methods for LD control were implemented (15). In 43.75% of cases, additional technical measures were necessary to achieve

low *Legionella* counts, indicating the importance of a well-designed and maintained plumbing system (1, 3, 5, 7, 10, 12, 15).

CONCLUSION

Choosing the right disinfection method remains difficult. Our tool allows the reviewer to sort through the vast amount of literature by quality and thereby limit the sample size of the literature to be considered. This work is still in progress, and at this point we would like to emphasize that in outbreak situations, a combined approach of several methods along with technical measures seems to be necessary. Once a system is technically well maintained, and a low *Legionella* spp. count is achieved, a single method, e.g., copper/silver ions or high warm-water maintenance temperature (>55°C at distal outlets) for the prevention of LD might be sufficient.

REFERENCES

1. **Anonymous.** 1997. Sustained transmission of nosocomial Legionnaires' disease: Arizona and Ohio. *Morbid. Mortal. Wkly. Rep.* **46:**416–421.
2. **Best, M., V. L. Yu, J. Stout, A. Goetz, R. R. Muder, and F. Taylor.** 1983. Legionellaceae in the hospital water-supply. Epidemiological link with disease and evaluation of a method for control of nosocomial Legionnaires' disease and Pittsburgh pneumonia. *Lancet* **2:**307–310.
3. **Biurrun, A., L. Caballero, C. Pelaz, E. Leon, and A. Gago.** 1999. Treatment of a Legionella pneumophila-colonized water distribution system using copper-silver ionization and continuous chlorination. *Infect. Control. Hosp. Epidemiol.* **20:** 426–428.
4. **Borau, J., R. T. Czap, K. A. Strellrecht, and R. A. Venezia.** 2000. Long-term control of Legionella species in potable water after a nosocomial legionellosis outbreak in an intensive care unit. *Infect. Control Hosp. Epidemiol.* **21:**602–603.
5. **Colbourne, J. S., D. J. Pratt, M. G. Smith, S. P. Fisher-Hoch, and D. Harper.** 1984. Water fittings as sources of Legionella pneumophila in a hospital plumbing system. *Lancet* **1:**21021–21023.
6. **Darelid, J., S. Lofgren, and B. E. Malmvall.** 2002. Control of nosocomial Legionnaires' disease by keeping the circulating hot water temperature above 55 degrees C: experience from a 10-year surveillance programme in a district general hospital. *J. Hosp. Infect.* **50:**213–219.
7. **Ezzeddine, H., C. Van Ossel, M. Delmee, and G. Wauters.** 1989. Legionella spp. in a hospital hot water system: effect of control measures. *J. Hosp. Infect.* **13:**121–131.
8. **Hall, K. K., E. T. Giannetta, S. I. Getchell-White, L. J. Durbin, and B. M. Farr.** 2003. Ultraviolet light disinfection of hospital water for preventing nosocomial Legionella infection: a 13-year follow-up. *Infect. Control Hosp. Epidemiol.* **24:** 580–583.
9. **Helms, C. M., R. M. Massanari, R. P. Wenzel, M. A. Pfaller, N. P. Moyer, and N. Hall.** 1988. Legionnaires' disease associated with a hospital water system. A five-year progress report on continuous hyperchlorination. *JAMA* **259:**2423–2427.
10. **Matulonis, U., C. S. Rosenfeld, and R. K. Shadduk.** 1993. Prevention of Legionella infections in a bone marrow transplant unit: multifaceted approach to decontamination of a water system. *Infect. Control. Hosp. Epidemiol.* **14:**571–675.
11. **Mietzner, S., R. C. Schwille, A. Farley, E. R. Wald, J. H. Ge, S. J. States, T. Libert, and R. M. Wadowsky.** 1997. Efficacy of thermal treatment and copper-silver ionization for controlling Legionella pneumophila in high-volume hot water plumbing systems in hospitals. *Am. J. Infect. Control* **25:**452–457.
12. **Snyder, M. B., M. Siwicki, J. Wireman, D. Pohlod, M. Grimes, S. Bowman-Riney, and L. D. Saravolatz.** 1990. Reduction in Legionella pneumophila through heat flushing followed by continuous supplemental chlorination of hospital hot water. *J. Infect. Dis.* **162:**127–132.
13. **Stout, J. E., Y. S. Lin, A. M. Goetz, and R. R. Muder.** 1998. Controlling Legionella in hospital water systems: experience with the superheat-and-flush method and copper-silver ionization. *Infect. Control Hosp. Epidemiol.* **19:**911–914.
14. **Vonberg, R. P., D. Rotermund-Rauchenberger, and P. Gastmeier.** 2005. Reusable terminal tap water filters for nosocomial legionellosis prevention. *Ann. Hematol.* **84:**403–405.
15. **Walker, J. T., C. W. Mackernes, D. Mallon, T. Makin, T. Williets, and C. W. Keevil.** 1995. Control of Legionella pneumophila in a hospital water system by chlorine dioxide. *J. Ind. Microbiol.* **15:**384–390.

SIX-MONTH EXPERIENCE OF SILVER-HYDROGEN PEROXIDE TREATMENT FOR *LEGIONELLA* CONTROL IN TWO NURSING HOME WATER SYSTEMS

M. L. Ricci, I. Dell'Eva, M. Scaturro, P. Baruchelli, G. De Ponte, M. Losardo, M. Ottaviani, and F. Guizzardi

122

Elderly people, an ever increasing number of individuals, represent the population most susceptible to Legionnaires' disease (LD). Hospitals and long-term care facilities usually admit this risk group of patients and, for this reason, it is important to periodically control water systems for *Legionella* contamination to prevent the occurrence of LD cases. However, in the absence of LD cases, the detection of environmental *Legionella* and installation of a disinfection system in the water system as primary prevention remains a controversial issue (6). Several outbreaks have been described (3, 4, 5) in long-term care facilities, including the recent and serious one that occurred in Canada in October 2005, where 93 cases and 16 deaths were reported (http://www.promedmail.org).

Different methods have been applied to control *Legionella* in water systems such as copper-silver ionization (9), chlorine dioxide (8), thermal treatments, hyperchlorination, UV radiation (2), etc.

In this study, we evaluated the efficacy of a continuous system based on a silver-hydrogen peroxide mixture in two nursing home water systems.

Two nursing homes (A and B) located in two buildings consisting of four floors and 75 beds and of five floors and 70 beds, respectively, were evaluated in this study. A newly constructed plumbing system was present in nursing home A and an old one in nursing home B, both consisting of galvanized steel iron material.

In December 2004, a continuous silver-hydrogen peroxide treatment system was installed in both buildings. A solution of 10 mg of hydrogen peroxide and 10 μg of silver ions per liter was applied (CILLIT-ALLSIL SUPER25, Cillichemie Italiana, Milano, Italy). Due to the heavy *Legionella* contamination of nursing home A, observed at the beginning of the study, an initial shock treatment was carried out in the hot water system, using a solution of 1.5g/liter hydrogen peroxide and 1.5 mg/liter Ag$^+$ for 6 h. Water sampling conducted before the treatment in nursing home B showed the presence of *Legionella* at a lower concentration, and no shock treatment was performed in this setting.

In both nursing homes, four sample sites were selected, and water samples were collected

M. L. Ricci, M. Scaturro, G. De Ponte, and M. Losardo Department of Infectious, Parasitic end Immune-mediate Diseases, Istituto Superiore di Sanità, Viale Regina Elena 299, 00161Roma, Italy. *I. Dell'Eva, P. Baruchelli, and F. Guizzardi* Azienda Provinciale per i Servizi Sanitari, Direzione Igiene e Sanità Pubblica, Via Piave9, 38100 Trento, Italy. *M. Ottaviani* Istituto Superiore di Sanità, Department of Environment and Primary Prevention, Viale Regina Elena 299, 00161 Roma, Italy.

monthly and analyzed by the regional reference laboratory in Trento. One sample was represented by the circulation system (sample 1) and two (sites 1 and 2) were from showers; sample 3 (site 3) was from a wash basin. Samples were collected from the hot water system and processed in accordance with the standard ISO 11731.

Besides *Legionella* counts, temperature, total aerobic colony count, *Pseudomonas aeruginosa* concentration, and silver and residual hydrogen peroxide concentrations (by iodometric titration) were analyzed. According to the Italian regulations on drinking water, silver concentration, (µg/liter) detected by ICP-OES spectrometry, was maintained at less than or equal to 0.01 mg/liter.

Overall, 44 samples were analyzed in nursing homes A and B. Microbiological investigation performed before water treatment revealed an initial concentration of $2.9 \pm 1.5 \times 10^5$ CFU of *L. pneumophila* serogroup 6 per liter in A and $1.5 \pm 1.3 \times 10^4$ CFU of *L. pneumophila* serogroup 1 per liter in B. The colony counts for *Legionella* dropped below 600 CFU/liter in both buildings as soon as the disinfection system was installed. For the following 6 months a significant reduction of *Legionella* CFUs was achieved in both nursing homes (2.5 log, $P < 0.01$ in A and 1.8 log, $P < 0.01$ in B) as shown in Fig. 1 and Fig. 2.

Some local increase in *Legionella* counts was due to unperformed flushing procedures (e.g., weekly flushing of discontinuously used outlets for several minutes) not revealed in the circulation system where the biocide was always present. Total colony counts performed at 22 and 36°C during all the observation periods was very low in every sampled site (<90 CFU/ml). *P. aeruginosa* concentration was always zero. Water temperature ranged between 44.1 ± 4.8 and 44.8 ± 4.7 in A and B, respectively. The Ag detection was lower or equal to 0.01 mg/liter, in accordance with Italian drinking water regulation.

There are few data on the real incidence of legionellosis in long-term care facilities and further studies are necessary to evaluate the relationship between water system colonization and occurrence of legionellosis in nursing homes (6). However, as already demonstrated for hospitals (10), the presence of *Legionella* in the water systems of nursing homes could be an important risk factor for the occurrence of LD. Recent studies showed a decline of LD in health care facility-acquired cases where environmental monitoring of the water system and installation of a disinfection method were performed (7). Therefore, when the plumbing system is heavily contaminated by *Legionella*, it is important to apply control measures to reduce *Legionella* colonization and infection risk.

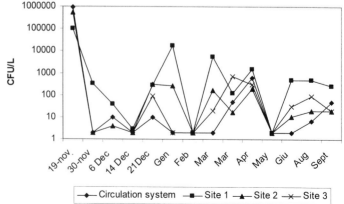

FIGURE 1 Effect of the continuous silver-hydrogen peroxide treatment in nursing home A's hot water system. Shock treatment was performed on November 30; on November 14 the continuous treatment was installed.

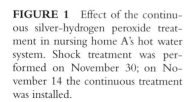

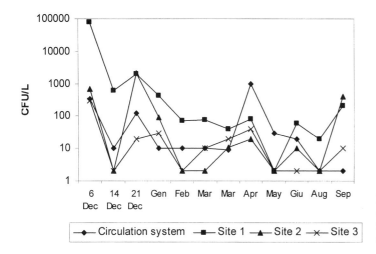

FIGURE 2 Effect of the continuous silver-hydrogen peroxide treatment in nursing home B's hot water system, switched on November 14.

Studies on the efficacy of the silver-hydrogen peroxide disinfection system are currently lacking, and to our knowledge, the present study is the first reported for *Legionella* control in hot water systems. In vitro studies demonstrated that hydrogen peroxide is a weak disinfectant in the absence of a catalyst. However, silver ions at 0.08 mg/liter, are able to reduce a millionfold the number of *Legionella* bacteria in 24 h (1). The combined action of hydrogen peroxide with silver as the catalyst characterize the disinfection power of the mixture used in this study.

Results of this study, although preliminary, show that such disinfection systems can effectively control *Legionella* water system colonization even if some increase of *Legionella* levels at distal sites is observed. This problem is often present because it is not always possible to ensure that the flushing procedure has been carried out. It is important to note that as for the other disinfection systems, numerous factors affect the efficacy of a biocide (temperature, water quality, water system composition and complexity, presence of dead legs, biofilm, etc.), and a definitive eradication of *Legionella* is difficult, if not impossible, to achieve.

In conclusion, the silver-hydrogen peroxide method is easy to apply and maintain, is quite

inexpensive, and, on the basis of these preliminary data, appears to be promising. However, a long-term follow-up is essential to confirm these findings.

REFERENCES

1. **Domingue E. L., R. L. Tyndall, W. R. Mayberry, and O. C. Pancorbo.** 1988. Effects of three oxidizing biocides on *Legionella pneumophila* serogroup 1. *Appl. Environ. Microbiol.* **54:**741–747.
2. **Lin Y. S., J. E. Stout., V. L Yu, and R. D. Vidic.** 1998. Disinfection of water distribution system for *Legionella. Semin. Respir. Infect.*
3. **Loeb M., A. E. Simor, L. Mandell, P. Krueger, M. McArthur, M. James, S. Walter, E. Richardson, M. Lingley, J. Stout, D. Stronach, and A. McGeer.** 1999. Two nursing home outbreaks of respiratory infection with *Legionella sainthelensi. J. Am. Geriatr. Soc.* **47:**547–552.
4. **Maesaki S., S. Kohno, H. Koga, M. Kaku, Y. Yoshitomi, H. Yamada, H. Matsuda, Y. Higashiyama, K. Hara, and M. Seto, et al.** 1992. An outbreak of Legionnaires' pneumonia in a nursing home. *Intern. Med.* **31:**508–512.
5. **Nechwatal R., W. Ehret, O. J. Klatte, H. J. Zeissler, A. Prull, and H. Lutz.** 1993. Nosocomial outbreak of legionellosis in a rehabilitation center. Demonstration of potable water as a source. *Infection* **21:**235–240.
6. **Seenivasan M. H., V. L. Yu, and R. R. Muder.** 2005. Legionnaires' disease in long-term care facilities: overview and proposed solutions. *J. Am. Geriatr. Soc.* **53:**875–880.

7. **Squier C. L., J. E. Stout, S. Krsytofiak, J. McMahon M. M. Wagener, B. Dixon, and V. L. Yu.** 2005. A proactive approach to prevention of health care-acquired Legionnaires' disease: the Allegheny County (Pittsburgh) experience. *Am. J. Infect. Control.* **33:**360–367.

8. **Srinivasan A., G. Bova, T. Ross, K. Mackie, N. Paquette, W. Merz, and T. M. Perl.** 2003. A 17-month evaluation of a chlorine dioxide water treatment system to control *Legionella* species in a hospital water supply. *Inf. Control Hosp. Epidemiol.* **24:**575–579.

9. **Stout J. E., and V. L. Yu.** 2003. Experience of the first 16 hospitals using copper-silver ionization for Legionella control: implications for the evaluation of other disinfection modalities. *Inf. Control Hosp. Epidemiol.* **24:**563–568.

10. **Yu V. L.** 1998. Resolving the controversy on environmental cultures of Legionella: a modest proposal. *Inf. Control Hosp. Epidemiol.* **19:**893–897.

TEMPERATURE REGIMENS VERSUS IONIZATION AND TMVs

John Hayes

123

Thermal eradication of legionella bacteria (superheat and flush) is only a temporary palliative when legionella levels pose an immediate problem. Many institutions do not have the thermal capacity to maintain a high hot water temperature (60°C/140°F) when flushing outlets. The process is both disruptive and expensive. Superchlorination and flush requires a high level of chlorine (20 to 50 ppm free residual chlorine) (2) to be circulated throughout the water system. Old pipework may suffer severe corrosion, and there will be severe disruption to ensure that outlets are not used other than when the flushing regimen takes place. This is an ominous task.

HOT WATER SYSTEMS

Legionella bacteria are inhibited at 60°C (140°F) (5), but it may be very difficult to maintain this water temperature throughout the system due to heat loss from the pipes, as water frequently takes a long and tortuous path from leaving the heating system to its return via the hot water circulation loop. In some cases this may place a strain on the capacity of the water heating system, particularly during peak water usage periods. There is an additional problem

that when maintaining a high water temperature the possibility of scalding has to be considered, particularly when the very young or the elderly and infirm are concerned.

PROBLEMS WITH ANTISCALD THERMOSTATIC MIXING VALVES (TMVs)

It has been shown that prior to fitting TMV devices legionella bacteria had been absent or in very low numbers, but following the installation of these devices, high levels of legionella were detected at outlets (4). This is because whatever preventive measures are applied to the hot water system, there is heavy dilution at the TMV by the introduction of cold water to bring the temperature down to around 45°C (113°F). This produces an ideal environment for the low level of legionella bacteria in the cold water supply to multiply in the outlet pipework, faucets, and showers. The additional problems associated with TMVs are the initial cost of fitting and maintenance, which is time-consuming, particularly when the nonreturn sections of the valve become faulty and cause a cross flow of the hot and cold water into the opposing system, producing a temperature that allows bacteria to multiply. When this happens it is frequently difficult to locate the faulty TMV, again a time-consuming exercise.

John Hayes T P Technology plc, Tarn House, 2-4 Copyground Lane, High Wycombe, Bucks, HP12 3HE, UK.

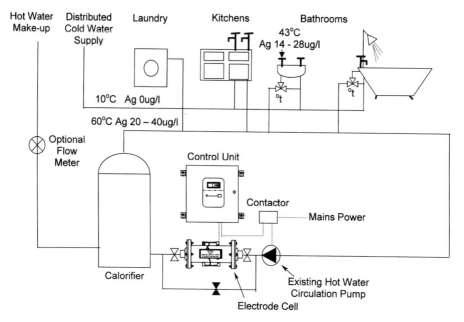

FIGURE 1   Hot water circulation loop operating at 60°C with TMVs.

ALTERNATIVES TO TMVs

Treatment of the hot water may consist of raising the water temperature and/or injecting a chemical such as chlorine, chlorine dioxide, or nonchemical ionization. The downside of chemicals is that they are corrosive to a greater or lesser degree and are volatile, so high levels have to be injected to ensure that effective levels reach all outlets. However, if TMVs are fitted, the secondary side is not treated effectively

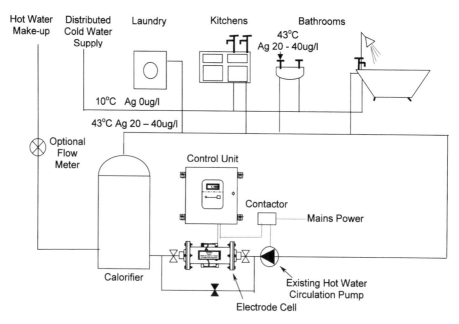

FIGURE 2   Hot water circulation loop operating at 43°C with no TMV's.

due to dilution by the cold water. UV and ozone are not recommended, as these processes are nondispersive, that is, they do not produce a residual treatment (3) that will kill organisms downstream from the equipment and therefore do not attack established biofilm which hosts legionella bacteria. It has been shown that if water is circulated at a nonscalding temperature, there is no requirement for TMVs, provided there is an effective measure of biocidal treatment such as ionization. The ions are nonvolatile, unaffected by temperature, and easily travel to all outlets, and this regimen is far less expensive in both initial cost and maintenance. If TMVs are already fitted, then a treatment regimen can be installed on the incoming cold water town supply. This method ensures that both hot and cold water is treated, and therefore "seeding" from the cold water is eliminated. This is a practical method, particularly in smaller buildings with no hot water circulation. Modern ionization systems can be controlled by water meters so that the low-level dosing is in direct proportion to water usage.

EMERGENCY MEASURES

Micro-pore filters fitted to outlets and showers are very effective at preventing the dissemination of bacteria when present at outlets, particularly in sensitive areas such as intensive care units, operating theaters, and areas where patients have reduced immunity. The filters are very effective but have to be changed regularly—twice a month or more often. This form of protection should not be employed as a long-term solution or replace treatments designed to kill bacteria back-stream rather than let them accumulate in the filters or pipework. This could allow the release of bacteria from any outlet not fitted with a filter. Any relaxation of the maintenance regimen may present serious problems.

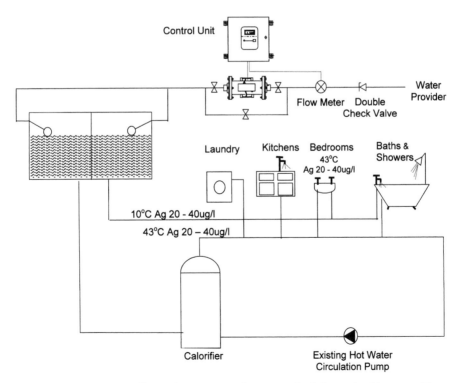

FIGURE 3 Treatment of incoming water supply to cover both hot and cold water with proportional dosing (no TMVs).

CONCLUSION AND DISCUSSION

TMVs were introduced to prevent the scalding of the aged and those unable to react quickly when presented with very hot water. If water temperatures can be reduced and protected by an effective biocidal regimen then the problems and cost associated with TMVs are avoided. Obviously, if the incoming cold water feed to the building is treated effectively, then both hot and cold water supplies are protected, particularly where cold water may rise in temperature due to climatic and ambient temperatures within buildings and encourage legionella bacteria to multiply. The research was pioneered as long ago as 1980 when Gerber (2) and his team at Arizona State University produced the first definitive report from research undertaken to identify the mechanisms by which positively charged ions attack and kill waterborne bacteria. More recently Stout and Yu (5) and their associates produced further evaluations and reports confirming the efficacy of silver/copper ionization. The current systems are the result of continuous research and development, and the latest range of systems include the most powerful available, with power of up to 100 volts at 20 amps. The range of ion emission cells available includes an additional nonemissive platinized titanium electrode in addition to the emissive silver/copper electrodes. In hard water areas this development avoids the problems of the coating of electrodes with calcium and carbonate scale, which limits ion release and previously required frequent electrode cell removal for chemical descaling.

Ionization options include:

- Hot water circulation loop operating at 60°C with TMVs (Fig. 1),
- Hot water circulation loop operating at 43°C with no TMVs (Fig. 2),
- Treatment of incoming water supply to cover both hot and cold with proportional dosing (no TMVs) (Fig. 3).

REFERENCES

1. **Health & Safety Commission (UK).** *Approved Code of Practice and Guidance L8 2000,* p. 48 (ISBN 0-7176-1772-6).
2. **Gerba P. C., and R. B. Thurman.** 1989. The molecular mechanisms of copper and silver on disinfection of bacteria and viruses. *CRC Crit. Rev. Environ. Control* **18**:295–315.
3. **Kramer, S., and S. Leung.** 1997. *Disinfection with Ozone. Water Technology Primer.* Environmental Engineering Management, Civil Engineering Dept., Virginia Tech. http://ewr.cee.vt.edu/environmental/teach/wtprimer/ozone/ozone.html.
4. **Lee, J. V.** 2004. www.show.scot.nhs.uk/scieh/Training/legionella17-03-04/Dr%20john%20Lee.pdf.
5. **Lin Y.-S. E., J. E. Stout, V. L. Yu, and R. D. Vidic.** 1998. Disinfection of water distribution systems for legionella. *Semin. Respir. Infect.* Vol **13**:147–159.
6. **Sheffer P. J., J. E. Stout, M. M. Wagener, and R. R. Muder.** 2005. Efficacy of a new point of use water filter for preventing exposure to Legionella and water-borne bacteria. *Am. J. Infect. Control* **33**:S20–S25.
7. **Stout J. E, and V. L. Yu .** 1994. Controlled evaluation of copper–silver ionization in eradicating Legionella pneumophila from a hospital water distribution system. *J. Infect. Dis.* **169**:919–922.
8. **Vonberg R. P., J. Bruderek, and P. Gastmeier.** 2004. *The Society for Healthcare Epidemiology of America (SHEA), Philadelphia P.A.* Abstract 191.

DESIGN AND REALIZATION OF ZERO-AEROSOL COOLING TOWERS

*Denis Clodic, Assaad Zoughaib, Chantal Maatouk,
Benoît Senejean, and Michèle Merchat*

124

INTRODUCTION

The cooling tower is an essential element for the dissipation of heat by pulverizing water and takes advantage of the wet bulb air temperature (which is usually 5 to 7 K under the dry air temperature). It offers remarkable energy characteristics, its use has been largely developed with the development of air-conditioning systems. It is a reliable, simple, and compact system, but it presents disadvantages which could compromise its use in the future.

The cooling towers gather together many factors for bacteria proliferation and possible transmission to human beings. The biofilm can develop both in the fill and all along the water circuit. Moreover, the usual spray distribution of water as well as the water dispersion on the fill leads to a heavy production of aerosols of various sizes. One of the main issues is knowing the sizes and population of aerosols getting out of the drift eliminators. As it is well known, significant contaminations of *Legionella* have occurred from aerosols released by cooling towers.

Chemical and physical treatments are used to limit the survival and proliferation of *Legionella* bacteria in water circuits, but that is insufficient when water contamination is fast. To eliminate dissemination in the atmosphere, the objective of the presented new system is to eliminate the vectorization of bacteria by eliminating the aerosol production and dispersion from cooling towers.

To eliminate aerosol drift, it is first necessary to find a technique to measure the critical range of size for drops of diameters between 0.5 and 5 μm and their concentration in the atmosphere. A review of the published techniques of aerosol size measurement has been carried out (3).

The selected measurement equipment is based on white light diffraction at 90°. A series of measurements has been carried out on operating cooling towers to characterize the sources that influence the emission of aerosols.

The first laboratory-scale prototype of a zero-aerosol cooling tower has been developed. The key issues that have been analyzed are the management of water distribution in order to control the water film thickness flowing on a reference surface, as well as the velocity of both water and air flows in order to avoid water drift.

Denis Clodic, Assaad Zoughaib, Chantal Maatouk, and Benoît Senejean   Ecole des Mines de Paris, Center for Energy and Processes, 60, boulevard Saint Michel, F -75272 Paris Cedex 06.   *Michèle Merchat*   Climespace, 185, rue de Bercy, F-75012 Paris.

514 ■ CLODIC ET AL.

MEASUREMENT OF DRIFT ON AN OPERATING COOLING TOWER

A review of the published techniques of aerosol size measurement has been carried out. The most suitable technique is an optical technique relevant to flow and able to accurately determine particle size distributions, number density, and total mass (4, 7, 8). New measurement equipment based on white light diffraction has been used to measure both the size of the water aerosols and the population of the different sizes of those particles. This equipment is manufactured by the Welas company (9). This system determines particle size and particle concentration simultaneously. This device simplifies measurements in areas that have been unreachable so far, such as high chimneys. The measuring data are sent to the evaluation unit via an optical fiber cable. The evaluation unit may be placed up to 300 m away from the sensor. This aerosol spectrometer principle is based on the diffraction of white light. The single particles move one by one through a volume illuminated by a homogeneous white light. While moving, the particles provoke a light impulse of a certain intensity. These impulses are based on the "real" Mietheory (6), because their measurement relies on spherical single particles and not on particle collectives. This intensity sent out by the particles is detected by reflecting it by an angle of 90°. The impulse amplitude permits measurement of the particle size. The number of impulses measured per time unit establishes

the particle quantity and therefore the concentration. The Welas can been used for in situ measurement in high concentrations up to 10^6 particles/cm^3. This measurement is achieved with high reliability. This equipment can measure particle ranges between 0.25 and 40 μm, corresponding to the critical particle size able to carry the bacterium.

The first measurements have been realized at the air exit and at 10 cm and 20 cm above the air exit of the cooling tower. These measurements have allowed identification of the influence of the sampling position. In Fig. 1, the left part indicates the distribution of the measured aerosols (the diameter in the x axis and the concentration in the y axis), while the right part is a cumulative curve giving the total concentration. The distributions presented in this figure indicate that the representative aerosol size for the measurements carried out directly at the exit of the cooling tower is 1 μm. It becomes 2 μm at 10 and 20 cm above the exit. In addition, the cumulative distributions show that the number of aerosols detected increases as the sampling position gets higher. The number of particles detected directly at the exit of the cooling tower is about 2,162 particles/cm^3. At 10 cm, the number of particles increases up to 9,788 particles/cm^3, and at 20 cm, it becomes 15,532 particles/cm^3. These phenomena are explained by the plume apparition and the association of its condensed droplets with the drift aerosols. The measurements indicate, therefore, that even with

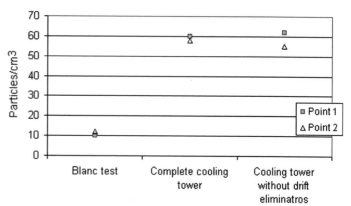

FIGURE 1   Drift eliminator impact.

the use of a drift eliminator, a large number (54 billion particles/h) of micro aerosols are emitted.

Other measurements have been performed to identify the influence of the ventilation velocity in a cooling tower. These measurements have been realized with cold circulating water in the cooling tower, thus avoiding the phenomenon of condensation. The cumulative distribution presented in Fig. 2 indicates that there is no significant difference between the number of particles measured with or without ventilation and that the spraying of water on the labyrinth-like fill is the main source of aerosol entrainments.

Additional measurements have been realized with and without the drift eliminators on the same cooling tower and for the same climatic conditions. The particles measured on the exhaust air does not vary significantly. However, aerosols with large diameter, not measurable by the Welas, are noticed coming out of the cooling tower exit along with the exhaust air. This means the drift eliminators capture only water aerosols of **diameters larger than 1 mm**.

Fig. 1 shows the main results of these measurements. First, measures were realized with no water circulation or air ventilation in the cooling tower: 10 particles/cm^3 were detected. Then the cooling tower was started and pro-

vided with its drift eliminators. The number of particles detected was near 60/cm^3. After that, the drift eliminators were removed from the cooling tower, and other measures show that the number of particles measured was close to the number measured before.

These measurements show that spraying of water on the fill and its bursting are the principal causes of aerosol drift, and that the drift eliminators installed to prevent the emission of aerosols in the exhaust air are not adapted for capturing water aerosols 1 μm in diameter.

A model of aerosol drift in annular two-phase flow has been elaborated by Holowach (5), who considers that liquid flow has a three-dimensional sine wave geometry. The model is developed based on a force balance and stability analysis that has been implemented into a transient three-field (continuous liquid, aerosol, and vapor), two-phase heat transfer and fluid flow system analysis computer code developed by the CEP. This model, applied on a cooling tower fill, shows that aerosol entrainment in two-phase flow due to the impact of the air flow into the liquid cannot be of critical particle size (3). Thus, the two-phase flow is not the fundamental cause of aerosol entrainment, but the nozzles that spray water on the fill are. In addition, the complex geometry of the fill is the origin of water fragmentation

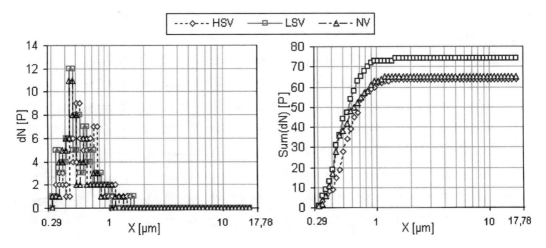

FIGURE 2 Measurement of drift as a function of fan speed.

and contributes to their entrainment in the exhaust air flow.

DESIGN OF A ZERO-AEROSOL COOLING TOWER

The laboratory scale prototype developed in this study is based on a new concept (2) of cooling tower based on laminar liquid flow on a smooth surface, in counter-current with the air flow. The nozzles and the fill of classical cooling towers are eliminated and replaced by a system of water flooding on an inclined plane surface. The basic concepts analyzed were the management of water distribution in order to control the water film thickness flowing on a reference surface, as well as the velocity of both the water and the air flow in order to avoid water drift.

CONCEPT BASIS

A cooling tower is mainly made of a system of water distribution consisting of a fill, a ventilation system, and a water collecting basin. The usual devices for the water distribution on the fill are nozzles discharging water on the fill. This system presents the major deficiency to be aerosol generators.

The risk of aerosol drift in a cooling tower is located in three parts:

- The pulverization and bursting of water,
- The counter-current flow of air and water in the fill,
- The water recovery, which is usually a cross flow between air entering the cooling tower and water leaving the fill.

The measurements done on existing cooling towers and the modeling of counter-current flow show that efforts must be concentrated on water distribution and on water recovery. This requires a water film stuck on the heat-transfer surface to control its thickness and the good distribution on this surface, which constitutes the first stage of prevention of aerosol entrainment. Then it is necessary to control the water velocity on the heat-transfer surface so that the amplitude of the free-flow

waves are sufficiently low so that they cannot be chopped by the air flow. Finally, it is necessary that the water film can be recovered without being crossed by the air flows.

The method elaborated for the water distribution is a flooding water method with controlled film thickness that sticks on the surface. Once the water film, at constant thickness, is homogenously distributed all along the surface, the surface slope, and its properties, will determine the water velocity in conjunction with the air flow. Indeed, water circulates by gravity and thus its movement is uniformly accelerated.

SYSTEM DESCRIPTION

A water distribution system is used for each elementary heat-transfer surface. The feeder tank receives a fraction of the total water flow; its dimensions and its structure make it possible to have a calm water film correctly distributed all over the elementary surface. Water leaves this feeder tank through a lip of distribution, whose opening and length allow precise control of the water film thickness. The association of the feeder tank and the lip of distribution allow distribution of the desired fraction of the water flow all over the heat-transfer surface with a specific water film thickness.

Heat-transfer plates are superposed, and the feeder tank is integrated between these plates. The association of the feeder tank and the lips of distribution on all the plates supply water by a film stuck to the plate.

In an x, y, z reference axis, where x is the axis in the direction of the water flow on the plate, y is the vertical axis in the direction of the plate length, and z is the axis which forms with x a succession of horizontal plates, the plate slope is 4°. For a typical 1.5 m length of the plate, the water speed increases only twice at the edge of the plate. Controlling the effect of gravity acceleration of the water film is essential to maintain a slightly wavy flow with a Reynolds number lower than 270 that defines a wavy laminar flow. Then the wave amplitudes are sufficiently low to not be chopped by the air

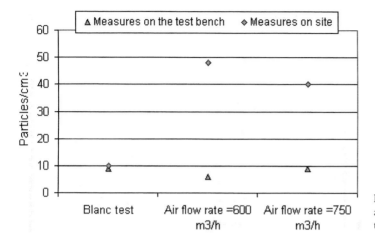

FIGURE 3 Comparison of aerosol drift on a classical cooling tower and on prototype.

flow, thus avoiding the formation of aerosols. Elements of air distribution are formed by two thin plates, delivering the air flow parallel to the water. At the end of the heat transfer plate, a gutter collects the water flow and changes its direction along the y axis. This gutter has a slope of 1° with the y axis. This slope makes it possible to recover each elementary water flow running out on each plate without crossing the air flow, thus avoiding any formation of droplets, by blowing of the air through the water flow.

All the test results show that the new design allows zero or near-zero aerosol emissions. Fig. 3 presents these measurements at the air exit of the test bench.

The test bench is formed with one heat-exchange plate, one gutter, and one feeder tank and distribution system. Many tests have been realized with several water flow rates, several water inlet temperatures, and several air inlet flow rates.

CONCLUSIONS

In this paper, a measurement technique allowing analysis of drift aerosol sizes and populations has been presented. Measurements performed on a cooling tower during operation lead to the conclusion that the aerosol production is associated mainly with the pulverization of water on the fill. The dominant size of the water aerosols

is around 1 μm, and the emissions of aerosols, as measured, are in the range of several million to several billion particles per hour.

A new concept has been developed by the CEP and Climespace, patented to eliminate aerosol production and any direct crossing of air flow and water flow. The key concept is the control distribution of a thin water film on a stack of plates.

The laboratory scale prototype has shown that there are zero aerosol emission and that the heat transfer density is compatible with the actual cooling tower. A scale-1 prototype of 1.5 MW has been built and installed in the center of Paris, and first tests have shown no aerosol production, with energy efficiency similar to the reference cooling towers.

REFERENCES

1. 2004. AHSRAE Handbook. HVAC Systems and Equipment.
2. **Clodic D., A. Zoughaib, M. Merchat, and L. Fassi.** 2004. Procédé et système d'alimentation en eau de tours aéroréfrigérantes. French patent 04/51128 filed on June 8, 2004, and its PCT extension FR/PCT2005/050398.
3. **Dib, J., C. Maatouk, A. Zoughaib, and D. Clodic.** 2004. *Etude et Conception Globale de Tours de Refroidissement sans Panache et à Entraînement de Gouttelettes Limité.* Report for ADEME and Climespace.
4. **Durst, F., and H. Umhauer.** 1975. Local measurements of particle velocity, size distribution

and concentration with a combined laser-Doppler particle sizing system. *In Proceedings of the LDA Symposium*, Copenhagen, p. 430–456.

5. **Holowach, M. J., L. E. Hochreiter, and F. B. Cheung.** 2002. A model for droplet entrainment in heated annular flow. *Int. J.Heat Fluid Flow* **23**:807–822.

6. **Mizutani, Y., H. Kodama, and K. Miyasaka.** 1982. Doppler-Mie combination technique for determination of size velocity correlation of spray droplets. *Combustion Flame* **44**:85–95.

7. **Tayali, N. E., and C. J. Bates.** 1990. Particle sizing techniques in multiphase flows: a review. *Flow Meas. Instrum.* **1**:77–105.

8. **Ungut, A., A. J. Yule, D. S. Taylor, and N. A. Chigier.** 1978. Particle size measurement by laser anemometry. *Energy 2* **6**:330–336.

9. www.palas.de/engl/welas/start.html.

LEGIONELLA POPULATION CONTROL IN COOLING WATER SYSTEMS

Michèle Merchat, Taher. Mamodaly, and Gilles Chaperon

125

The use of a specific chemical product to minimize the population density of bacteria is the required and usual approach to prevent cooling towers from becoming amplifiers of *Legionella*. Sampling the cooling water to control bacterial populations is the most common way to evaluate the water treatment program's efficiency. *Legionella* concentrations are obtained 8 to 10 days after sampling by the culture method measurement. This delay represents an obstacle for risk management. The aim of this study was to compare *Legionella* population counts obtained from different trials on the system to determine the most representative sampling point of the cooling water. Then we checked the results obtained after 3, 5, and 10 days of incubation in order to test the relevancy of these results.

Experiments were performed on Climespace cooling tower plants in Paris, France. On site, the basin towers were connected. Basin water was discharged into a common line to supply water to condensers. Pumps drew warm water from the condenser to a common line which was divided to feed each cooling tower. Make-up water entered directly into the basin of each cooling tower. The feeder was in a corner of the basin.

Each condenser contained an air flush-out on the top and a wash-out on the bottom. In order to define the most representative point of water in cooling tower systems, 10 samples collected from 5 locations were compared. Analyses were performed in compliance with AFNOR T 90 431(French standard method for measuring *Legionella* concentrations). After 3 days the colonies which seemed to have the morphological features of *Legionella* were counted and plated for confirmation, using the same procedure for day 5 and day 10.

The results obtained from the different points showed that the tower basin provides irregular *Legionella* concentrations (Table 1). The poor water quality and the lack of homogeneity from each basin revealed significant fluctuations. *Legionella* were not detected or were slightly concentrated when the tower basin contains make-up water, as illustrated by the conductivity drop off. Then, it is essential to compare this parameter in the basin and in the circuit. The conductivity at 25°C for the make-up is approximately 400 μs/cm, whereas that of the circuit varies between 3,500 and 4,000 μs/cm.

The results obtained from the air flush-out sample show scattered distribution. The most

Michèle Merchat CLIMESPACE, 185 Rue de Bercy, 75012 Paris France. *Gilles Chaperon and Taher Mamodaly* Société CAPSIS, 1, Rue de Terre Neuve, ZA de Courtaboeuf-Bât B, 91940 LES ULIS.

TABLE 1 Representativeness of the place of sampling for *Legionella* analysis

Parameter	*Legionella* concentration (CFU/liter)[a]				
	Basin tower	Warmest water entrance tower	Coldest water exit tower	Air flush-out condenser	Wash-out condenser
Sample no.	10	10	10	10	10
Minimum	ND	5.E3	1.E2	ND	3.E4
Maximum	7.E4	6.E3	2.E3	5.E4	4.E4
Mean	9.E3	5.E3	1.E3	6.E3	3.E4

[a]ND, not detected.

representative sample points were taken from circulating pipe or wash-out at the condenser bottom. There was no influence of the water temperature. *Legionella* were detected in 234 of the 608 samples (38%). Among these 234 samples, *Legionella* are detected after 3 days of culture only for 169 samples (72%), after 5 days of culture for 63 more samples (27%), and after 8 days for 2 more samples (1%).

In 80% of the 608 samples, the provisional result obtained after 3 days of culture did not change, and the final result obtained after 10 days of culture is identical. Therefore, the provisional results obtained after 3 or 5 culture days are very significant on several levels. The implementation of the corrective actions is faster and, at the time of an epidemic, the source of contamination can more quickly be identified.

Therefore, the reduction of the time is significant for the risk management. This is why all the techniques of new analyses aiming at reducing this time must be studied. However,

in the current state of knowledge, only the method by culture provides an interpretable result. It is on the basis of these results that the corrective actions are implemented. Although the concentration is underestimated (since only the cultivable viable part is taken into account), the evaluation of the risk is associated with the results provided by this method of analysis. To date, the use of other techniques which make it possible to evaluate all the *Legionella* (like the technique by PCR) could lead to an abusive use of the chemical treatments generating a significant risk to the environment. Too many people unfortunately use the results obtained by the techniques known as rapid in the same manner as a result obtained by culture.

To be more precise, among all the sample results, *Legionella* was detected in 72% of day 3 samples and in 27% of more day 5 samples. The correlation coefficient between day 3 samples and day 10 samples was about 0.72 (Fig. 1). The correlation coefficient between

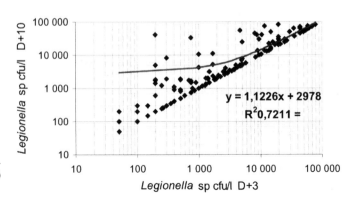

FIGURE 1 Correlation between *Legionella* concentration after 3 and 10 days of culture.

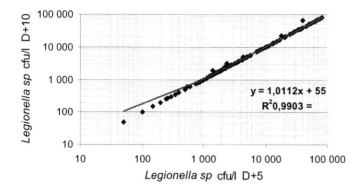

FIGURE 2 Correlation between *Legionella* concentration after 5 and 10 days of culture.

day 5 samples and day 10 samples was about 0.99, and between day 3 samples and day 5 samples was about 0.72 (Fig. 2).

Moreover, it was interesting to notice that *Legionella pneumophila*, which is most frequently the species at the origin of legionellosis cases, grow faster than the other *Legionella* species (data not show).

CONCLUSIONS

This study obviously indicates that sampling has to be carried out accurately in order to represent the circulating water. The critical delay can be reduced to 3 days to improve risk control.

Analysis results obtained from culture methodology after 72 h seem to be reliable and make it possible to shorten the reaction time to decrease *Legionella* concentration in the system. The results obtained in this study show that among the samples for which *Legionella* is confirmed on day 10, 72% revealed colonies on day 3 and show that intermediary data are relevant.

CONTROL OF *LEGIONELLA* PROLIFERATION RISK IN COOLING WATER SYSTEMS

Michèle Merchat and Anabel Deumier

126

Legionnaires' disease is a potentially fatal form of pneumonia caused by the bacterium *Legionella* (2). Although isolated occurrences of Legionnaires' disease constitute the majority of cases, the most serious epidemics of Legionnaires' disease were caused by infected cooling towers.

Present naturally in freshwater in the environment, *Legionella* prefers the cooling tower ecosystem, where there are the best growth conditions. *Legionella* control in cooling water systems depends on many parameters (conception, hydraulic management, maintenance) favorable to biological deposits. Good engineering practices (minimizing system dead legs, reducing cooling tower drift), good water treatment practices (minimizing scale and corrosion of system surfaces), good maintenance (regular inspections and systems cleanouts), and good operating procedures (circulation of all water volume, above all during shock biocide feeding) are necessary to prevent the bacteria proliferation. The use of a specific biocide activity to minimize the population density of *Legionella* is the required and common approach to prevent cooling towers from becoming amplifiers of *Legionella* populations.

The effectiveness of the measures implemented can be assessed by measurement of bacteria concentrations in the water, with a period of time corresponding to the time required for analysis. The use of nonoxidizing and oxidizing biocides mentioned in the literature to be efficient in *Legionella* control is successful, but in a very unsteady way. The risk is not always correctly controlled, and the quantity of biocides fed is steadily increased. The efficiency of the biocides used has often been assessed under laboratory conditions, which are not representative of reality.

The objective of the work was to identify the conditions implemented for cleaning and effective disinfections compatible with the systems and the environment.

A strict follow-up has been performed on three cooling systems after a sample protocol for analysis was implemented. Any variation in the parameters measured (microbiological, physicochemical and environmental parameters) made it possible to identify and destroy *Legionella* growth sources. Experiments were performed on three Climespace cooling tower plants in Paris.

On each site, the basin towers were connected. Basin water was discharged into a common line to supply water to condensers. Pumps drew warm water from the condenser

Michèle Merchat and Anabel Deumier CLIMESPACE, 185 Rue de Bercy, 75012 Paris France.

Legionella: State of the Art 30 Years after Its Recognition
Edited by Nicholas P. Cianciotto et al.
©2006 ASM Press, Washington, D.C.

to a common line which was divided to feed each cooling tower. The make-up water that directly entered the basin of each cooling tower was soft drinking water. Water quality (make-up and circuit) is reported in Table 1.

The study was carried out at three sites, where the power installed is about 8, 32, and 42 MW. The volumes of each site are 50, 300, and 200 m³. The water flow is 1,400 to 4,600 m³/h, and the temperature vary between 27°C for the coldest and 38°C for the hottest. *Legionella* analysis was performed once a week for six years (more than 800 analyses) at the three sites. Samples at room temperature were directly transported to the laboratory. No more than 3 h elapsed between the sampling and the beginning of the analysis.

A test series was performed using different nonoxidizing and oxidizing biocides in situ such as BCDMH, isothiazoline, gluteraldehyde, and DBNPA with or without a biodispersant. The oxidizing biocide feeding was continuous and proportional to the residual contamination level. During the experiments, nonoxidizing biocides and biodispersants were injected in slug-dose twice a week. The time required to inject 100 mg/liter of each biocide molecule was less than 30 min. The biodispersant concentration was about 10 mg/l.

The cooling tower system is an ecosystem in which *Legionella* is always present and can proliferate any time of year. However, *Legionella* concentrations, especially from May to September, were difficult to control. The percentage of samples in which *Legionella* is detected increased after April. In August and September more than 60% of the samples contained viable cultivable *Legionella*. This percentage decreased slowly to reach a minimum of 20%. That is probably related to the mode of exploitation and management of the circuits. The water circulation was not permanent (certain parts of the installation were brought into service according to production needs).

A meticulous analysis of the risks made it possible to identify and manage all the factors of risk of proliferation of *Legionella* (the permanent water circulation, elimination of the dead legs, etc.) (3). The biocide oxidizing was injected without interruption. Twice per week, a nonoxidizing biocide treatment was carried out. The biodispersant was injected 3 hours before the shock in order to permeabilize biological depots. Despite all the means implemented, the *Legionella* concentration in water was not controlled. Biocide activity is influenced and definitely compromised by water quality. Thus, biocide effectiveness is highly affected or even eliminated due to the conditions implemented, but also due to the incompatibility between products. Except for isothiazoline, oxidizing and nonoxidizing biocides are not compatible (5).

As of the injection of biocide not oxidizing shock in the presence of residual oxidants, tests show that residual oxidizing biocide decreases dramatically to zero and does not vary from 3 h to 4 days, depending on the biocide. In same time, the total oxidant concentration increases. This strategy led to an overconsumption of chemicals (in particular, if the biocide injection oxidizing is based on the measurement of residual in water) without great efficiency toward *Legionella* but a higher toxicity for environment. Furthermore, isothiazoline is efficient toward *Legionella* and amoebae only if it is injected without any oxidizing products. The biodispersant allows action against the biofilm. Its effectiveness strongly depends however on the conditions of implementation.

In a high-volume system, the injection of biodispersant in shock is not effective to clean

TABLE 1 Water quality for make-up and in the circuit

Parameter	Make-up	Circuit
pH	7.6	8.5 ± 0.5
TH °F	25.6	2 ± 1
TA °F	0	12 ± 1
TAC °F	20.0	80 ± 5
Oxidizing biocide (mg/liter)	0.08 ± 0.01	0.3–0.6
Polyphosphate (cm³/m³)		84
Conductivity at 25°C (μs/cm)	400 ± 50	2,120
Concentration factor		4 ± 0.5

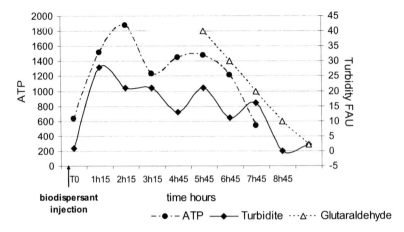

FIGURE 1   Biodispersant effect on water quality.

surfaces. At the time of the injection, the suspended matter is circulating in the water, and settles again after some hours. These materials are not eliminated and are again in suspension in the water at the time of the next shock. The biofilm is most probably deteriorated only on its surface. The risk of contamination of water in circulation is thus significant. In addition, the feeding by shock doses for biodispersant induces a lack of efficiency for biocide because the water turbidity increases due to the biodispersant activity (Fig. 1).

Based on these results, the biodispersant implementation conditions and the feeding methods of the nonoxidizing biocide have been defined precisely to ensure treatment effectiveness. Dosing is continuous for biodispersant and for the oxidizing biocide and is discontinuous for the nonoxidizing biocide (used only when a risk was identified). The injection of biodispersant was split into small quantities in order to limit the foam formation until obtaining residual (5 mg/liter). The turbidity of the water increased quickly and then fell in 2 days. However, the concentration of bacteria, including *Legionella*, remained high for 2 months. The cleaning of the entire surfaces in contact with water is thus not immediate. In a small-volume installation, the injection of biodispersant can be carried out in shock, but the frequency of the injection remains essential to limit the formation of biofilm.

In addition to this this strategy of treatment which makes it possible to avoid the use of nonoxidizing biocides, the circulation of all the volume of water is guaranteed in each installation. Thus, each pump of the installation is in circulation for at least 2 h per 24-h period.

The *Legionella* risk management cannot be carried out without taking account of the protozoa which are resistant to biocides (1, 5). These means implementation makes it possible to fight against the biofilm and thus at the same time against the *Legionella* and the protozoa. In 2004, the number of samples containing *Legionella* decreased from 65 to 5%, and 5,000 liters of nonoxidizing biocide, less than the others years, were used (Fig. 2).

CONCLUSIONS

The effectiveness of the treatments depends on the conditions implemented. A risk of zero does not exist. However, thanks to an approach based on a methodology and accurate and suitable techniques, this very ambitious objective can be attempted through a continuous but asymptotic process so that systems are safer and better controlled. Therefore, it is essential to make use of improved knowledge, new techniques, and feed-back. That means that all

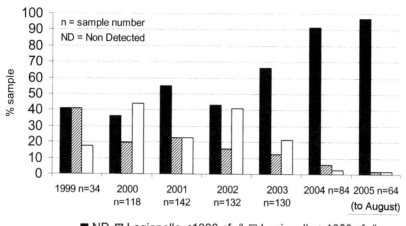

FIGURE 2 *Legionella* sp. concentration by year on three sites.

the actors involved in cooling systems, small or big ones, at any level of the organization, whatever the position, take into account the *Legionella* risk in making decisions and taking action.

To fight continuously against the formation of biofilm seems to be the method most adapted to fight effectively against the proliferation of *Legionella* and limit the use of biocides.

REFERENCES

1. **Berk, S. G., R. S. Ting, G. W. Turner, and R. J. Ashburn.** 1998. Production of respirable vesicles containing live *Legionella pneumophila* cells by two *Acanthamoeba* spp. **64:**1:279–286.

2. **Fraser, D. W., T. R. Tasi, W. Orenstein, W. E. Parkin, H. J. Beecham, and R. G. Sharrar.** 1997. Legionnaires' disease: description of an epidemic of pneumonia *N. Engl. J. Med.* **297:**1189–1197.

3. **Merchat, M.** Guide de formation à la gestion du risque de prolifération des légionelles dans les installations de refroidissement par dispersion d'eau dans un flux d'air (http://www.ecologie.gouv.fr/article.php3?id_article=3734).

4. **Merchat, M.** 2005. Etude de la toxicité des biocides dans les circuits de refroidissement et dans les rejets à l'égout, Rapport d'étude AESN.

5. **Srikanth, S., and S. G. Berk.** 1993. Stimulatory effect of cooling tower biocides on amoebae. *Appl. Environ. Microbiol.* **59:**10:3245–3249.

CONTROL OF *LEGIONELLA* IN LARGE BUILDINGS THROUGH COMMUNITY-WIDE INTRODUCTION OF MONOCHLORAMINE

Matthew R. Moore, Brendan Flannery, Lisa B. Gelling, Michael Conroy, Duc Vugia, James Salerno, June Weintraub, Valerie Stevens, Barry S. Fields, and Richard Besser

127

Monochloramine may be more effective than chlorine at reducing *Legionella* colonization of potable water systems in large buildings, which are key sources of community- and hospital-acquired Legionnaires' disease. Monochloramine has previously been shown to penetrate biofilms better than chlorine (1) and to be associated with a lower risk of Legionnaires' disease in healthcare settings (4, 5). Regulations issued by the Environmental Protection Agency require that U.S. water utilities reduce concentrations of trihalomethanes, carcinogens created through the combination of chlorine and organic compounds. Because monochloramine use is associated with lower concentrations of trihalomethanes, many utilities are converting their residual disinfectant from chlorine to monochloramine. In a small study, we observed that monochloramine conversion was associated with decreased *Legionella* colonization of buildings served by the municipal

Matthew R. Moore, Brendan Flannery, Valerie Stevens, Barry S. Fields, and Richard Besser Centers for Disease Control and Prevention, 1600 Clifton Rd., Atlanta, GA 30333. *Lisa B. Gelling* Epidemiology-Communicable Disease Control, Kaua'i District Health Office, 3040 Umi St., Lihue, HI 96766. *Michael Conroy and James Salerno* San Francisco Public Utilities Commission, 1212 Market St., San Francisco, CA 94102. *Duc Vugia* California Department of Health Services, 850 Marina Bay Parkway, Bldg. P, Richmond, CA 94804. *June Weintraub* City and County of San Francisco, Department of Public Health, 1390 Market St., Suite 910, San Francisco, CA 94102.

water system (6). To determine whether this decreased risk was sustainable over a longer period of time and at a larger number of distal sites, we performed a larger colonization survey in San Francisco before and after conversion to monochloramine (3).

We sampled water from 53 large (≥3 stories) buildings (24 government, 29 private [hotels, offices]) in San Francisco, Calif., every 4 months during February 2003 to September 2004; monochloramine conversion occurred in December 2003. During each of the six rounds of sampling, bulk water was collected from central hot water heaters and four distal sites (e.g., showers, taps) in each building; biofilm swabs were also collected from each distal site. Temperature, pH, and free and total residual chlorine concentrations were measured at the time of collection. *Legionella* were isolated from water samples and biofilm swabs, speciated, and serogrouped using standard methods (2). Laboratory staff were blinded to the identity of buildings in each round. Predictors of colonization were identified using a multivariable model that adjusted for residual disinfectant, building size, temperature, and disruptions in service. We conducted enhanced surveillance for laboratory-confirmed Legionnaires' disease in San Francisco before and after the conversion to monochloramine.

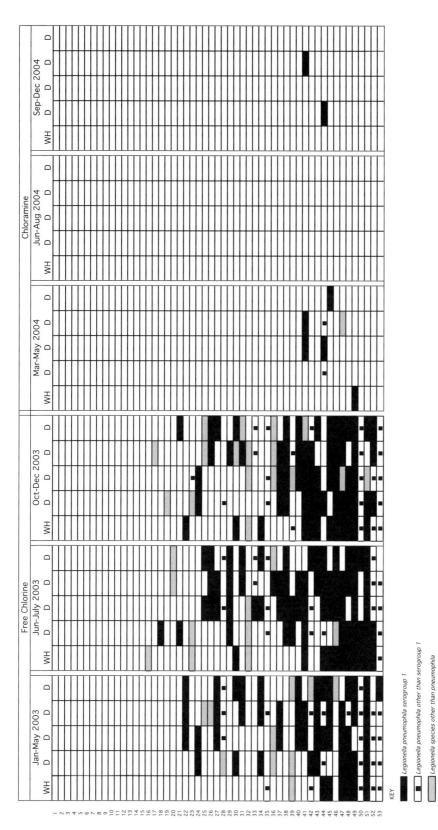

FIGURE 1 *Legionella* colonization of water heaters and distal sites sampled in San Francisco buildings (in rows), by residual disinfectant and sampling interval. *Legionella* species are represented by different patterns. Each row represents a single building and each cell, the results of *Legionella* culture for a site within the building. WH, water heater; D, distal site.

KEY

■ *Legionella pneumophila* serogroup 1

▪ *Legionella pneumophila* other than serogroup 1

▨ *Legionella* species other than *pneumophila*

In the chlorine phase, 37 (70%) buildings were colonized at least once in ≥1 sampling site. In the monochloramine phase, 5 (9%) buildings were colonized ($P < 0.001$), and all had been extensively colonized during the chlorine phase. The overall prevalence of *Legionella* in samples collected was reduced from 25% (352 of 1,405 samples) in the chlorine phase to 0.9% (12 to 1,417) in the monochloramine phase ($P < 0.001$) (Fig. 1). *L. pneumophila* serogroup 1 (sg 1) accounted for 63 and 67% of isolates in the chlorine and monochloramine phases, respectively.

In the chlorine phase, 29% (45 of 157) of samples collected from water heaters yielded *Legionella*; fewer than 1% (1 of 159) of water heaters sampled during the monochloramine phase were positive ($P < 0.001$). After controlling for water heater temperature, building height, and disruptions in service, monochloramine use was associated with a 96% reduction in the prevalence of water heater colonization ($P < 0.001$). We observed decreased risk of colonization of hot water heaters with increasing temperature, from 21 to 24% when the temperature was below 39°C, to 18% at 40 to 49°C, to 1% when the temperature was ≥50°C ($P = 0.001$ for trend). Controlling for other factors, *Legionella* colonization of water heaters was 3 times more likely in buildings >10 stories tall than in buildings 3 to 10 stories tall ($P = 0.002$) and 2.3 times more likely when disruptions in water supply were reported in the 3 months before sampling ($P = 0.003$). *Legionella* colonization of water heaters increased the likelihood of colonization at distal sites 1.7-fold, ($P < 0.001$) and independently, taller buildings were 1.8 times more likely to have *Legionella* recovered from distal sites ($P < 0.001$). Similar to our findings with hot water heaters, the use of monochloramine was associated with a 96% reduction in the risk of colonization at distal sites ($P < 0.001$).

Conversion of the residual disinfectant to monochloramine in San Francisco's municipal water system virtually eliminated *Legionella* colonization in large buildings served by the system. This occurred despite over half of the buildings being colonized in at least one site and despite many buildings being persistently colonized at multiple sample points before the introduction of monochloramine.

Few cases of Legionnaires' disease were diagnosed during our study: 1 case in 2002 (the year before our study), 0 cases during the chlorine phase, and 1 case during the monochloramine phase. Review of requests for *Legionella* diagnostic testing at local clinical laboratories revealed few urine antigen tests and 40 to 60 *Legionella* culture requests each month at 2 laboratories. We were unable to show an impact on Legionnaires' disease in San Francisco due to the low number of diagnosed cases.

Monochloramine holds promise as a measure to prevent community-acquired Legionnaires' disease. Because of the association between *Legionella* colonization of potable water systems in hospitals and increased risk of healthcare-associated Legionnaires' disease, direct introduction of monochloramine into hospital water systems might offer another intervention for prevention of healthcare-associated Legionnaires' disease.

REFERENCES

1. **Cunliffe, D. A.** 1990. Inactivation of Legionella pneumophila by monochloramine. *J. Appl. Bacteriol.* **68**:453–459.
2. **Fields, B. S., R. F. Benson, and R. E. Besser.** 2002. Legionella and Legionnaires' disease: 25 years of investigation. *Clin. Microbiol. Rev.* **15**:506–526.
3. **Flannery, B., L. B. Gelling, D. J. Vugia, J. M. Weintraub, J. J. Salerno, M. J. Conroy, V. A. Stevens, C. E. Rose, M. R. Moore, B. S. Fields, and R. E. Besser.** 2006. Reduction of Legionella colonization after conversion to monochloramine for residual disinfection of municipal drinking water. *Emerg. Infect. Dis.* In press.
4. **Heffelfinger, J. D., J. L. Kool, S. Fridkin, V. J. Fraser, J. Hageman, J. Carpenter, and C. G. Whitney.** 2003. Risk of hospital-acquired Legionnaires' disease in cities using monochloramine versus other water disinfectants. *Infect. Control Hosp. Epidemiol.* **24**:569–574.
5. **Kool, J. L., J. C. Carpenter, and B. S. Fields.** 1999. Effect of monochloramine disinfection of municipal drinking water on risk of nosocomial Legionnaires' disease. *Lancet* **353**:272–277.
6. **Moore, M. R., M. Pryor, B. Fields, C. Lucas, M. Phelan, and R. E. Besser.** 2006. Introduction of monochloramine into a municipal water system: impact on colonization of buildings by Legionella spp. *Appl. Environ. Microbiol.* **72**:378–383.

EFFICACY OF MONOCHLORAMINE AGAINST SURFACE-ASSOCIATED *LEGIONELLA PNEUMOPHILA* IN A COOLING TOWER MODEL SYSTEM

Irfan Türetgen and Ayşin Çotuk

128

Cooling towers can act as microbe reservoirs and have the potential to develop infectious concentrations of *Legionella pneumophila*, the bacteria responsible for causing Legionnaires' disease (2, 7, 9). Legionella counts increases where biofilm and warm water are present. Previous studies have shown a number of Legionnaires' disease outbreaks that originate from cooling towers. Therefore, biocides are employed to reduce the potential for the development of biofilm on equipment surfaces (4). Chlorine has been used as a popular disinfectant for controlling bacterial growth in drinking water and cooling towers (5), but monochloramine was found more stable than free chlorine, even at high temperatures and elevated pH values (6). Moreover, it exhibits greater biofilm penetration and lower levels disinfection by-products and is less corrosive than free chlorine (3). A further disadvantage of free chlorine is its reaction with natural organic matter to form trihalomethanes, which are considered to be carcinogens. Chloramines do not form these products to the same degree. A previous study (6) showed that disinfection of biofilms with 1.5 ppm monochlo-

ramine for 180 min resulted in 3 log reductions of heterotrophic bacteria in cooling tower water. In this study, the efficacy of monochloramine against mature biofilms formed in a model recirculating water system was compared on different surfaces under identical conditions.

A laboratory-scale cooling tower was designed to simulate a wet evaporative cooling system, which was experimentally seeded with *L. pneumophila*. At the beginning of the experiment, the recirculating system was experimentally infected with standard strain (ATCC 33152) suspension (1 ml *L. pneumophila* inoculum of 10^5 cell/ml) and operated continuously until all experiments had been completed. Rather low inoculum was added to the model system to mimic the natural entry of *L. pneumophila* from the water supply. Throughout the experiment, the water temperature was kept constant at 30°C. Rather low final inoculum was provided to the model system to mimic the natural entry of *L. pneumophila* from supply water. Blowdown optimization was conducted by measuring the total dissolved solids of recirculating water. Coupons represented the fill material, on which cooling water is generally distributed. All the materials are in rigid form, certified, and commercially available. Biofilm had been allowed to grow on coupons, and the model system was disinfected after a 180-day

Irfan Türetgen and Ayşin Çotuk Department of Biology, Faculty of Science, Istanbul University, 34118 Vezneciler, Istanbul, Turkey.

period with monochloramine. A residual of 1.5 ppm was maintained for 180 min and then neutralized with sodium thiosulfate. Three coupons of each material were removed from the basin and dip-rinsed in sterile phosphate buffer to remove unattached cells. Biofilms on surfaces were scraped by sterile scalpel, suspended in phosphate buffer, and vortexed for 60 s. Copper, stainless steel, galvanized steel, polyvinyl chloride (PVC), polyethylene (PE), polypropylene (PP), and glass (as control) were tested for heterotrophic bacteria and *L. pneumophila* counts on their surfaces.

For heterotrophic plate count (HPC), 10-fold diluted biofilm homogenates and bulk water were spread-plated (0.1 ml) onto R2A agar plates and incubated at 28°C for 10 days. HPC determinations were performed by triplicate analyses. For the enumeration of *L. pneumophila*, biofilm homogenate was treated with acid solution for 15 min (KCl-HCl solution, pH 2.2) to reduce overgrowth of commensal flora. Pretreated and untreated samples were inoculated (0.1 ml) onto alpha ketoglutarate-supplemented buffered charcoal-yeast extract. Analyses were performed in triplicate. To de-

tect the entry of wild *L. pneumophila* bacterium from the water supply, several isolates were taken at different sampling times and compared using randomly amplified polymorphic DNA PCR fingerprinting analysis (1). Plate counts were $\log_{10}$ transformed and standard errors of the means were calculated. Differences between materials were tested by one-way analysis of variance followed by Student Newman Keul's post hoc test. Significant differences are reported at the $P < .05$ level.

Before disinfection, significantly high heterotrophic bacteria were observed on galvanized steel (Fig. 1). Low numbers of *L. pneumophila* were present on PVC and PP. The bacterial densities in biofilms from different materials after disinfection (180 min, 1.5 ppm dose) suggested a material-dependent activity of monochloramine. Four log reductions of heterotrophic bacteria were recorded on glass, copper, and stainless steel surfaces (Fig. 1). Three log reductions were recorded on PVC, PE, copper, and stainless steel surfaces, whereas a slight reduction of *L. pneumophila* was observed on PP, glass, and galvanized steel surfaces (Fig. 2). Monochloramine was found ineffective against

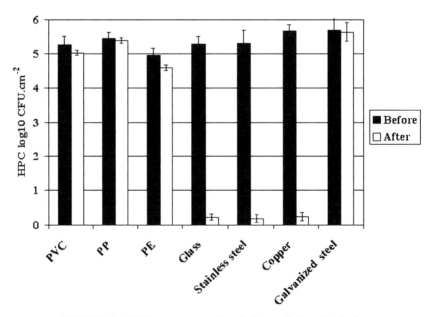

FIGURE 1 HPC counts on surfaces before and after disinfection.

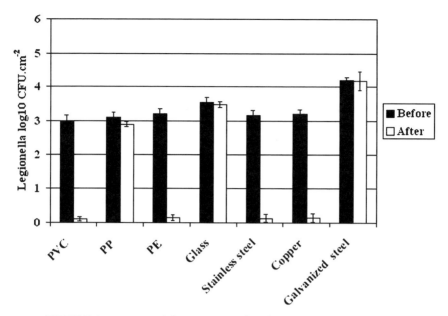

FIGURE 2   *L. pneumophila* counts on surfaces before and after disinfection.

microorganisms on PP and galvanized steel surfaces. Four log reductions were recorded against planktonic heterotrophs and *L. pneumophila* in bulk water. The 180-day period was run in duplicate, and no significant differences were found between two periods in regard to bacterial accumulation on surfaces and disinfection efficacies. No entry of wild *L. pneumophila* was detected from make-up water. There were clear differences in the biofilm formation on different surfaces, and biofilms on all surfaces reached a steady state after 120 days (data not shown).

Results indicated that monochloramine shows material-dependent activity in the cooling tower model system and it has long residual activity at high temperatures and pH levels, leading to improved performance in recirculating water. The results indicated no significant differences on PP and galvanized steel surfaces. This observation supports the material-dependent activity of monochloramine and it could be explained by the formation of dissimilar biofilms on different surfaces, which affects the architecture of the biofilm matrix. Surfaces in contact with water can contribute significantly to biofilm formation when these surfaces release biodegradable components. Therefore, choosing the appropriate material will reduce biofilms, thereby reducing the risk of associated illnesses (8). Since cooling towers are ideal incubators, *L. pneumophila* could multiply in a very short time, and an infectious dose level for *L. pneumophila* has not been established, so low numbers of *L. pneumophila* pose a risk for public health. Chemical disinfection is strongly recommended from the beginning of the tower operation to prevent or reduce the occurrence of Legionnaires' disease (5). Monochloramine disinfection is inexpensive and could be used in automatic injection devices instead of chlorine.

ACKNOWLEDGMENTS

The present work was supported by the Research Fund of Istanbul University. Project No. 641/02092005. The model system was donated by Dizayn Teknik Plastic Pipes & Fittings Co.

REFERENCES

1. **Bansal, N. S., and F. McDonell.** 1997. Identification and DNA fingerprinting of Legionella strains by randomly amplified polymorphic DNA analysis. *J. Clin. Microbiol.* **35:**2310–2314.

2. **Bentham, R. H.** 2000. Routine sampling and the control of *Legionella* spp. in cooling tower water systems. *Curr. Microbiol.* **41:**271–275.

3. **Campos, C., J. F. Loret, A. J. Cooper, and R. F. Kelly.** 2003. Disinfection of domestic water systems for *Legionella pneumophila. J. Water Suppl.* **52:**341–354.

4. **Donlan, R., R. Murga, J. Carpenter, E. Brown, R. Besser, and B. Fields.** 2002. Monochloramine disinfection of biofilm-associated *Legionella pneumophila* in a potable water model system, p. 406–410. *In* R. Marre, Y. A. Kwaik, and B. Fields (ed.). *Legionella,* ASM Press, Washington, D.C.

5. **Fliermans, C. B., G. E. Bettinger, and A. W. Fynsk.** 1982. Treatment of cooling tower systems containing high levels of *Legionella pneumophila, Water Res.* **16:**903–909.

6. **Türetgen, I.** 2004. Comparison of free residual chlorine and monochloramine for efficacy against biofilms in model and full scale cooling towers. *Biofouling* **20:**81–85.

7. **Türetgen, I., E. I. Sungur, and A. Cotuk.** 2005. Enumeration of *Legionella pneumophila* in cooling tower water systems. *Environ. Monitor. Assess.* **100:**53–58.

8. **Türetgen, I., and A. Cotuk.** Formation of bacterial biofilms in model recirculating water system. *J. Environ. Micropaleo. Microbiol. Meiobenth,* in press.

9. **Yamamoto, H., M. Sugiura, E. Kusunoki, T. Ezaki, M. Ikedo, and E. Yabuuchi.** 1992. Factors stimulating propagation of legionellae in cooling tower water. *Appl. Environ. Microbiol.* **58:**1394–1397.

MONOCHLORAMINE TREATMENT INDUCES A VIABLE-BUT-NONCULTURABLE STATE INTO BIOFILM AND PLANKTONIC *LEGIONELLA PNEUMOPHILA* POPULATIONS

Laëtitia Alleron, Jacques Frère, Nicole Merlet, and Bernard Legube

129

Legionella pneumophila, the causative agent of an atypical severe pneumonia in humans called Legionnaires' disease, is ubiquitous in aquatic environments. This gram-negative bacterium is normally found in water as an intracellular pathogen of ameobae. As a free-living organism, *L. pneumophila* can persist for long periods in biofilms commonly found in man-made water systems such as plumbing systems, air-conditioning equipment, or whirlpool spas (4). Grazing amoebae should consume these sessile *L. pneumophila* in biofilm communities and allow them to multiply (3). It is widely admitted that biofilms play a critical role in the persistence of these bacteria within water systems. Biofilm provides shelter and nutrients for the embedded community and prevents disinfectants from gaining access to the bacteria through the exopolysaccharide matrix (4).

The strategies of *L. pneumophila* to adapt and resist stressful environmental conditions include not only interaction with amoebae and biofilm localization but also the ability to enter in a viable-but-nonculturable (VBNC) state. Cells in a VBNC state fail to grow on the routine bacteriological media on which they would normally grow. They typically exhibit very low levels of metabolic activity but can recover their culturability in favorable conditions. On resuscitation, these cells again became culturable (12). Different stressful conditions, such as poor nutrient conditions, heat and concentrated salts of hot spring waters, or chlorine treatment, can lead to a VBNC state in *L. pneumophila* populations (2, 11, 13). Addition of protozoa, e.g., the amoeba *Acanthamoeba castellanii*, to *L. pneumophila* in a VBNC state, resulted in the resuscitation of these bacteria to a culturable state (13). Such observations demonstrate that VBNC *L. pneumophila* regain pathogenic potential with a reactivation in amoebae and are therefore a public health concern. Many strategies have been used to eradicate *Legionella* in water and plumbing systems, such as chlorination, chloramination, overheating, and UV irradiation of the water. Treatments of contaminated systems by these strategies have been successful, but only for short periods, after which the bacteria have again been found in these sources (1, 14). Eradication of *L. pneumophila* may require continuous treatment. Some authors suggested that continuous treatments with monochloramine have a better impact than free chlorine alone on *Legionella* eradication (8, 9) and on biofilm formation

Laëtitia Alleron and Jacques Frère UMR CNRS 6008, Pôle Biologie Santé, 40 av. du Recteur Pineau, 86022 Poitiers Cedex, France. *Nicole Merlet and Bernard Legube* UMR CNRS 6008, ESIP, 40 av. du Recteur Pineau, 86022 Poitiers Cedex, France.

(10). Monochloramine's biocidal action is slower than that of free chlorine, but it is more stable, and a disinfecting residual activity can be maintained over long distances in a water distribution system (7, 9). Such an impact was also observed in biofilm-associated *L. pneumophila* in a potable water model and cooling towers systems (6, 15).

The aim of our work was to study the impact of monochloramine on sessile and planktonic *L. pneumophila* populations. For this purpose, we used the two recently sequenced Paris and Lens strains (5) obtained from the CNRL (Centre National de Référence des Légionelles, Lyon, France).

SCREENING VBNC AFTER MONOCHLORAMINE TREATMENTS

We have produced *Legionella* biofilms on glass beads (0.5 mm diameter) in buffered-yeast extract (BYE) broth. Bacteria were grown for one week at 37°C, under static conditions. Planktonic cells were obtained in same growth conditions without glass beads. Planktonic and sessile cells were washed twice with sterile water and then treated with 0.25 to 10 mg liter^{-1} monochloramine solutions. Solution concen-

trations were measured using the DPD method. After a 1-h treatment at room temperature, sterile sodium sulfite was added. Glass beads were transferred into a resting medium consisting of BYE broth diluted 10^3-fold and incubated at 20°C for various durations. We used 10^3 diluted BYE broth as the resting medium in order to obtain total organic carbon concentration close to that of surface water (2 mg liter^{-1} total organic carbon). *L. pneumophila* was unable to grow in this diluted medium but was able to survive for several weeks without detectable culturability decline (data not shown).

Sessile bacteria were collected from glass beads by sonication and enumerated on buffered charcoal yeast extract (BCYE) agar. We looked for VBNC cells in samples containing no culturable bacteria. Enumerations of viable and dead bacteria were determined by epifluorescence microscopy using the BacLight LIVE/DEAD Bacterial Viability Kits (Molecular Probe). Viable bacteria appear green and membrane damaged bacteria appear red. Enumeration results are presented Fig. 1 for samples kept 6 days in the resting medium after treatments of biofilms by monochloramine. We observed a loss of cultivability for monochloramine doses

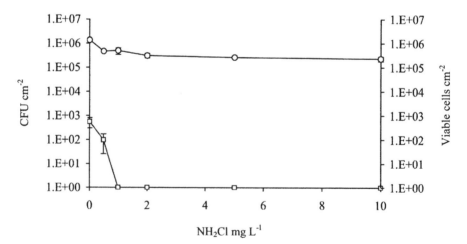

FIGURE 1 Culturability and viability of *L. pneumophila* Paris after monochloramine treatment. After monochloramine treatments, biofilms were transferred into resting medium (see text for details) and incubated at 20°C for 6 days. Squares indicate culturable cells enumerated on BCYE agar, and circles, viable cells enumerated by epifluorescence using the BacLight Kit, as cells cm^{-2} (Molecular Probes).

greater or equal to 1 mg liter^{-1} but a very low decrease in viable cells. Similar results were obtained for samples kept 12 or 20 days in the resting medium (data not shown). These results suggest that these samples contained VBNC cells. Nevertheless, green cells detected using BacLight staining indicate only that those cells possess integer membranes. In order to detect metabolic activity in VBNC cells, planktonic *L. pneumophila* Lens cells were treated with 0.75 mg liter^{-1} of monochloramine in order to obtain no culturable cells. After 8 days in the resting medium, we were able to detect esterase activity and membrane integrity by fluorescence microscopy using the ChemChrom V6 substratum (Chemunex) in samples without culturable cells (Fig. 2). This demonstrates that VBNC cells produced by monochloramine retained metabolic activity.

Steinert et al. (13) have observed that addition of amoebae to VBNC *L. pneumophila* led to resuscitation of these bacteria to a culturable state. Then, in order to check VBNC resuscitation, sessile bacteria collected after sonication were cocultivated with *A. castellanii* ATCC 50739. After 3 and 7 days of coincubation, the amoebae were lysed using centrifugation and vortexing. *L. pneumophila* culturability was thus determined on BCYE agar. Infection attempts were realized with biofilms kept for 30 days in resting medium (Table 1). Amoeba addition to untreated biofilms led to an important increase of CFU, but not for biofilms treated with doses higher than 1 mg liter^{-1} of monochloramine, even after 7 days of coincubation. Recovery of culturability was observed for sessile bacteria treated with 1 mg liter^{-1} of monochloramine and coincubated with *A. castellanii*. However, when the same treated bacteria were not coincubated with the amoebae, no culturable cells were detected. We were not able to resuscitate sessile bacteria treated with higher monochloramine doses, despite detection of viable cells in these samples. It may be possible that these cells are progressively degenerating since they have more oxidative damages than those treated with only 1 mg liter^{-1} of monochloramine or that they required more than 30 days to resume these repairs to be able to infect amoeba.

CONCLUSION

Legionella samples treated with monochloramine contained viable and dead bacteria. We have observed that moderate monochloramine treatments of *L. pneumophila* biofilm could

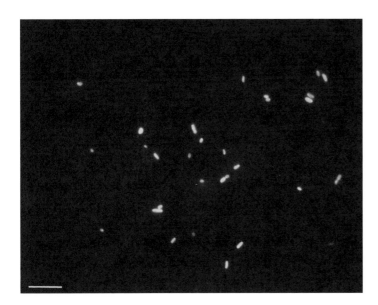

FIGURE 2 Esterase activity of *L. pneumophila* Lens 8 days after a 0.75mg liter^{-1} monochloramine treatment. The white line represents 10 μm.

TABLE 1 Influence of *Acanthamoeba castellanii* on culturability of sessile *L. pneumophila* Paris treated with monochloramine

Monochloramine treatment (mg liter^{-1})[a]	CFU obtained on BCYE-α agar[b]	
	Without amoeba	With amoeba[c]
Untreated	Yes	Yes
1	No	Yes
2	No	No
5	No	No
10	No	No

[a]Biofilms were processed with the indicated monochloramine doses and kept 30 days in the resting medium.
[b]Yes, colonies were observed on BCYE-α agar; no, no culturable cells were obtained on BCYE-α medium.
[c]Bacteria from the biofilms were cocultivated with *Acanthamoeba castellannii* for 3 days at 37°C.

induce VBNC cell formation. These nonculturable *L. pneumophila* are able to recover their culturability after amoeba infection. These organisms are commonly found in water systems, suggesting that our observations should be effective in the natural environment. We have treated *L. pneumophila* biofilms only for short periods. It will be necessary to investigate continuous treatments with low monochloramine doses as is recommended for municipal drinking water (9) to know if these conditions also produce VBNC *L. pneumophila*.

ACKNOWLEDGMENT

L. Alleron was supported by a grant of the Région Poitou-Charentes (France).

We are sincerely grateful to Michael Steinert for providing the *A. castellanii* strain. Many thanks to Yann Héchard for his help in amoeba experiments.

REFERENCES

1. **Abu Kwaik, Y., C. Venkataraman, O. S. Harb, and L. Y. Gao.** 1998. Signal transduction in the protozoan host *Hartmannella vermiformis* upon attachment and invasion by *Legionella micdadei*. *Appl. Environ. Microbiol.* **64:**3134–3139.
2. **Bej, A. K., M. H. Mahbubani, and R. M. Atlas.** 1991. Detection of viable *Legionella pneumophila* in water by polymerase chain reaction and gene probe methods. *Appl. Environ. Microbiol.* **57:**597–600.
3. **Bitar, D. M., M. Molmeret, and Y. A. Kwaik.** 2005. Structure-function analysis of the C-terminus

4. **Borella, P., E. Guerrieri, I. Marchesi, M. Bondi, and P. Messi.** 2005. Water ecology of *Legionella* and protozoan: environmental and public health perspectives. *Biotechnol. Annu. Rev.* **11:**355–380.
5. **Cazalet, C., C. Rusniok, H. Bruggemann, N. Zidane, A. Magnier, L. Ma, M. Tichit, S. Jarraud, C. Bouchier, F. Vandenesch, F. Kunst, J. Etienne, P. Glaser, and C. Buchrieser.** 2004. Evidence in the *Legionella pneumophila* genome for exploitation of host cell functions and high genome plasticity. *Nat. Genet.* **36:**1165–1173.
6. **Donlan, R., R. Murga, J. Carpenter, E. Brown, R. Besser, and B. Fields.** 2002. Monochloramine disinfection of biofilm-associated *Legionella pneumophila* in a potable water model system, p. 406–410. *In* R. Marre, Y. Abu-Kwaik, C. Bartlett, N. P. Cianciotto, B. S. Fields, M. Frosch, J. Hacker, and P. C. Lück (ed.), *Legionella*. ASM Press, Washington, D.C.
7. **Hoebe, C. J., and J. L. Kool.** 2000. Control of legionella in drinking-water systems. *Lancet* **355:**2093–2094.
8. **Kool, J. L.** 2002. Control of *Legionella* in drinking water systems: impact of monochloramine, p. 411–418. *In* R. Marre, Y. Abu-Kwaik, C. Bartlett, N. P. Cianciotto, B. S. Fields, M. Frosch, J. Hacker, and P. C. Lück (ed.), *Legionella*. ASM Press, Washington, D.C.
9. **Kool, J. L., J. C. Carpenter, and B. S. Fields.** 1999. Effect of monochloramine disinfection of municipal drinking water on risk of nosocomial Legionnaires' disease. *Lancet* **353:**272–277.
10. **Momba, M. N., and M. A. Binda.** 2002. Combining chlorination and chloramination processes

for the inhibition of biofilm formation in drinking surface water system models. *J. Appl. Microbiol.* **92:**641–648.

11. **Ohno, A., N. Kato, K. Yamada, and K. Yamaguchi.** 2003. Factors influencing survival of Legionella pneumophila serotype 1 in hot spring water and tap water. *Appl. Environ. Microbiol.* **69:**2540–2547.

12. **Oliver, J. D.** 2005. The viable but nonculturable state in bacteria. *J. Microbiol.* **43:**93–100.

13. **Steinert, M., L. Emody, R. Amann, and J. Hacker.** 1997. Resuscitation of viable but non-culturable *Legionella pneumophila* Philadelphia JR32 by *Acanthamoeba castellanii. Appl. Environ. Microbiol.* **63:**2047–2053.

14. **Thomas, V., T. Bouchez, V. Nicolas, S. Robert, J. F. Loret, and Y. Levi.** 2004. Amoebae in domestic water systems: resistance to disinfection treatments and implication in Legionella persistence. *J. Appl. Microbiol.* **97:**950–963.

15. **Türetgen, I.** 2004. Comparison of the efficacy of free residual chlorine and monochloramine against biofilms in model and full scale cooling towers. *Biofouling* **20:**81–85.

PREVENTING LEGIONELLOSIS WITH HAZARD ANALYSIS AND CONTROL SYSTEMS

William F. McCoy

130

Most cases of legionellosis result from exposure to building water systems containing *Legionella*. There are two hazard analysis and control systems recommended for use today in legionellosis prevention: water safety plans (WSP) and hazard analysis critical control point (HACCP) plans. They are conceptually and functionally identical. Either system can be equally used to prevent legionellosis associated with building water systems. Hazard analysis and control plans should be simple, cost-effective, and easy to implement. Otherwise, people will not use them.

PUTTING IT INTO PERSPECTIVE: WATER MANAGEMENT PLANS AND BUDGETS

The energy cost of inefficiently operating water utilities is very high. For many facilities, the water system is assumed to be hazard-free and therefore very little money, if any, is spent on water safety. However, every facility should have a water management plan because the cost of water and energy has significantly increased in recent years and because there is a duty to provide hazard-free water to occupants of the building. Buying water and then paying

to dispose of it is a significant part of every facility budget. In many regions, the cost to dispose water to the sewer exceeds the cost to purchase the water. Fig. 1 conceptually illustrates components of the water management plan and budget.

Inefficiency costs in water-related operations are often highest when safety is poorest. This is especially true if water and energy waste is due to microbial fouling in the system. The best first step to improve water system safety is therefore to reduce inefficiencies in the system. The best next step is to allocate some of the savings to safety improvements. In Fig. 1, the size of the dollar sign above "Water Safety Plan" can be increased at the expense of a decrease in the size of the dollar sign above "Cost of Water and Water Utility Operations." Allocation of the budget in this way is a practical means to economically justify reducing waste and increasing safety of facility water systems.

Reducing inefficiencies in the water system and in the water treatment program can be achieved through better use of water preconditioning, more economical and more effective water treatment practice, and closer attention to energy conservation. Savings from better water management can then be used to improve the safety of the water system. Prevention of waterborne disease is most effectively

William F. McCoy   Phigenics, LLC, 1701 Quincy Ave., Suite 32, Naperville, IL 60540.

Water Management Plan and Budget

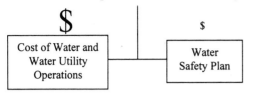

FIGURE 1 Every facility should have a water management plan and budget. All facilities spend money to cover the cost of water, its disposal, and the costs of running utilities. Most facilities spend little or nothing for water safety.

achieved by use of hazard analysis and control systems.

THE WATER SAFETY PLAN (WSP)

The World Health Organization has established a hazard analysis and control scheme for engineered water systems called the Water Safety Plan (10). The conceptual components of the system are establishment of a team to develop the plan, hazard analysis, risk characterization for each hazard, establishment of control limits, validation that control limits are effective, management systems for support, and independent surveillance to verify that the plan is implemented.

Using WSPs to Prevent Legionellosis

In its publication in preparation, Legionella *and the Prevention of Legionellosis*, the use of the WSP hazard analysis and control system is highly promoted by the World Health Organization (10). It is a scientifically based approach

FIGURE 2 Schematic representation of the World Health Organization hazard control scheme (10).

to hazard analysis and control. Fig. 2 is a schematic representation of the WSP.

Developing a WSP to Prevent Legionellosis

For hazard analysis and control schemes to be useful, they must be practical, easy to develop and cost-effectively implemented. A systematic approach to developing the plan is given in Fig. 3.

Prominently discussed in the WHO document on WSPs (10) is the concept of validation, which is the investigative work necessary to show that control limits are quantitatively effective under operating conditions in the system. This is usually achieved with well-directed use of pathogen analytical testing. Independent surveillance that the plan is being implemented (verification) is also prominently required in WSPs.

THE HACCP SYSTEM

HACCP system is a scientifically based method to prevent hazards from harming people. It has been proven effective for controlling biological hazards in the United States and globally. The system is conceptually simple, cost-effective, practical and quantitative.

In France, the HACCP system was specified in 2005 as the prefered hazard analysis and control system for preventing legionellosis associated with cooling towers (7).

For prevention of waterborne disease, the HACCP system is being applied or considered for water systems all over the world including in Australia and New Zealand (8), Belgium (2), Finland (3), Germany (6), Italy (1), South Africa (5), Sweden (9) and the United Kingdom (4).

At the time of this writing, the HACCP system is not yet widely used to prevent legionellosis in America. However, since so many kitchens in American facilities of interest (such as hospitals, hotels, and restaurants) are already preventing foodborne illnesses by implementing HACCP plans, it will be practical to extend those plans to prevent waterborne disease, including legionellosis, in their water systems.

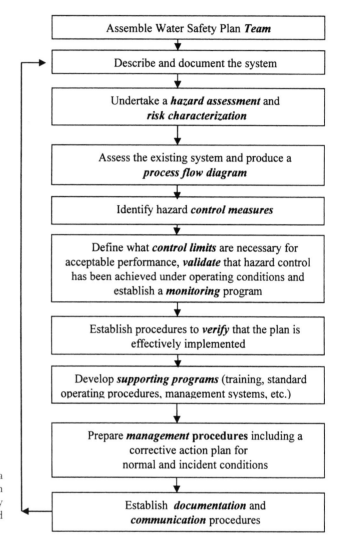

FIGURE 3 Steps in the development of a WSP as recommended by the World Health Organization. Text in bold indicates key concepts in the WSP hazard analysis and control system.

Most HACCP plans at the food service level in America are used voluntarily. Many restaurants and food service kitchens (such as in hospitals) routinely implement HACCP plans to prevent foodborne pathogens from causing disease in their facilities. Development of HACCP plans in the food processing/manufacturing industry is driven either by a legal mandate or by buyers of food who require suppliers to have such systems in place. HACCP systems are a means of ensuring safe foods and ingredients in the processing sector.

Many HACCP plans to control foodborne pathogens in America are required by regulation published in the Code of Federal Regulations (9 CFR 417). The United States Department of Agriculture requires food production plants to use HACCPs. The Food and Drug Administration requires seafood processors to use HACCPs. The FDA also requires HACCP plans in certain canned-goods production processes to prevent botulism.

In response to quantitatively documented success in the United States, the World Health

Organization has adopted HACCP as the preferred recommended system for preventing foodborne disease worldwide.

HACCP System Principles

The system is conceptually simple. It is comprised of seven principles:

1. Conduct a hazard analysis.

2. Identify critical control points (the last control point to prevent, eliminate, or reduce the hazard).

3. Establish critical limits for each critical control point.

4. Establish a monitoring plan for critical limits at critical control points.

5. Establish corrective actions for each critical limit.

6. Establish procedures to document all activities and results.

7. Establish procedures to confirm that the plan actually works under operating conditions (validation), is being implemented properly (verification), and is periodically reassessed.

In practice, a few preliminary steps are necessary before these seven principles can be applied. They are:

- Assemble an HACCP team including one HACCP-trained person (e.g., a certified HACCP manager/consultant, a trained employee, or other resource, for example, from a local college).
- Identify the use and users of the product (in this case, water), especially to determine at-risk consumers.
- Identify all uses for the product (in this case, water) at the facility.
- Develop process flow diagrams to describe how the product is processed in the facility.
- Verify that process flow diagrams are accurate by on-site audit.

It is advised that there be at least one HACCP-trained person on the HACCP team who is independent of the facility owner or management.

An excellent authoritative source for details is the *Guidebook for the Preparation of HACCP Plans* available online at www.fsis.usda.gov/Science/HACCP_Models/index.asp.

CONCLUSIONS

Most legionellosis is the result of exposure to *Legionella*-containing building water. It is the most serious cause of pneumonia people get from water in the buildings they occupy. Legionellosis prevention is a matter of effective water treatment and building water system management. Building water system management is the process of cost-efficiently optimizing operations and improving safety. It is proposed that hazard analysis and control is the most effective means to improve safety of building water systems. There are two functionally and conceptually equivalent systems: WSP and HACCP. Potentially, thousands of cases of legionellosis could be prevented by implementing quantitative hazard analysis and control plans for building water system management.

REFERENCES

1. **Angelillo, I. F., N. M. Viggiani, R. M. Greco, and D. Rito.** 2001. HACCP and food hygiene in hospitals: knowledge, attitudes, and practices of food-services staff in Calabria, Italy. Collaborative Group. *Infect. Control Hosp. Epidemiol.* **22**:363–369.

2. **Dewettinck, T., E. Van Houtte, D. Geenens, K. Van Hege, and W. Verstraete.** 2001. HACCP (hazard analysis and critical control points) to guarantee safe water reuse and drinking water production: a case study. *Water Sci. Technol.* **43**:31–38.

3. **Horman A., R. Rimhanen-Finne, L. Maunula, C. H. von Bonsdorff, J. Rapala, K. Lahti, and M. L. Hanninen.** 2004. Evaluation of the purification capacity of nine portable, small-scale water purification devices. *Water Sci. Technol.* **50**:179–183.

4. **Howard, G.** 2003. Water safety plans for small systems: a model for applying HACCP concepts for cost-effective monitoring in developing countries. *Water Sci. Technol.* **47**:215-220.

5. **Jagals, C., and P. Jagals.** 2004 Application of HACCP principles as a management tool for monitoring and controlling microbiological hazards in water treatment facilities. *Water Sci. Technol.* **50**:69–76.

6. **Kistemann, T., S. Herbst, F. Dangendorf, and M. Exner.** 2001. GIS-based analysis of drinking-water supply structures: a module for microbial risk assessment. *Int. J. Hyg. Environ. Health* **203:** 301–310.

7. **Ministère de l'Écologie et du Développement Durable.** 2005. *Guide méthodologique pour la Réalisation d'une Analyse de Risque de Prolifération de Légionelles dans les Installations de Refroidissement par Dispersion d'Eau dans un Flux d'Air.* Ministère de l'Écologie et du Développement Durable, République Française. www1.environnement .gouv.fr/IMG/pdf/GUIDEWEB_analyse_de_risq ue_de_proliferation_des_legionelles.pdf.

8. **Nadebaum, P., M. Chapman, S. Ortisi, and A. Baker.** 2003 Application of quality management systems for drinking water quality *Water Supply* **3:**359–364.

9. **Westrell, T., C. Schonning, T. A. Stenstrom, and N. J. Ashbolt.** 2004. QMRA (quantitative microbial risk assessment) and HACCP (hazard analysis and critical control points) for management of pathogens in wastewater and sewage sludge treatment and reuse. *Water Sci. Technol.* **50:** 23–30.

10. **WHO (World Health Organization).** 2004. *Guidelines for Drinking Water Quality*, 3rd ed., chapter 4. World Health Organization.

ENVIRONMENTAL SAMPLING DATA TO DETERMINE RISK: A UNITED KINGDOM PERSPECTIVE

Susanne Surman-Lee and Richard Bentham

131

This paper presents the panel responses to questions posed to the round table discussion "Environmental Sampling Data to Determine Risk." Questions are presented in italics.

Question: *Could risk assessment determine whether an institution needs to monitor its water systems for Legionella? Is sampling and testing necessary for all institutions?*

A risk assessment (RA) should be carried out in all premises that have public access and should include all healthcare premises, factories, office blocks, etc. It is now over 5 years since the United Kingdom Health and Safety Commission published the Approved Code of Practice and Guidance L8, which requires RA in premises where the Health and Safety at Work Act Applies (in practice, in all public buildings). It is cause for concern that many premises have still not had RAs carried out, including premises with high-risk patients, or if they have, the RA is inadequate or urgent remedial actions have been identified but not acted upon. There are still many premises, es-

pecially healthcare buildings, which are chronically colonized with legionellae. A thorough RA should identify if any remedial measures are required and if additional water treatment is necessary. The World Health Organization has established a hazard analysis and control scheme for engineered water systems called the Water Safety Plan (2). The conceptual components of the system are:

- Establishment of a team to develop the plan
- Hazard analysis and cause and risk characterization for each hazard
- Determination of critical limits and target and corrective actions
- Establishment of control limits and parameters for monitoring to verify that controls are effective
- Management systems for support and independent surveillance and audit to verify that the plan is implemented and effective

An RA should include a thorough site survey for all water systems in use, cooling towers, hot and cold water systems, spa pools, irrigation systems, etc. and should take into account the population using the premises. Following the water safety plan approach, understanding the system design complete with the effect that any modification may have on the system is a first approach to assessment of the system.

Susanne Surman-Lee Health Protection Agency, London Regional Food Water & Environmental Microbiology Laboratory, Centre for Infections, 61 Colindale Avenue, London NW9 5EQ, United Kingdom. *Richard Bentham* Department of Environmental Health, Flinders University, GPO Box 2100, Adelaide, South Australia, Australia, 5001.

TABLE 1 Water safety plan approach for a hot water system (after reference 2)

Hazard event	Cause	Risk	Critical limits (CFU/liter)	Target action	Control limits (CFU/liter)	Monitoring measure	Corrective action
Legionella detected in hot water system	Inadequate temperature control Presence of biofilm due to high nutrients Inappropriate materials used. No or too little biocide used Stagnation or low flow due to insufficient usage and/or design/engineering failures	Moderate	$>10^4$	Use good quality source water Use materials which do not encourage microbial growth Maintain a biocide residual Minimize aerosol production Raise temperature in hot water system and maintain circulating hot water temperatures above 50°C	$>10^3$	Measure hot and cold water temperatures on flow and return to heat source and in storage tanks Monitor temperature at outlets. Monitor biocide levels, check usage In high-risk areas sample for *Legionella*	Replacement of water system components with materials more resistant to colonization Improve engineering to prevent stagnation and/or remove unused outlets Install appropriate microbial control systems Improve insulation

Ideally, there should be an up-to-date schematic diagram. In practice this is rarely available, so a physical inspection of all systems is necessary. This should identify system usage and detect problems which may act as a reservoir of biofilm containing legionellae or supporting microorganisms. Monitoring temperatures at each outlet is simple and is the most useful tool when carrying out an RA of hot and cold water systems to identify areas of low flow. A well-engineered system which is used consistently and maintains hot water to each outlet at 50°C and a cold water system maintained below 20°C will successfully control legionellae. It is not sufficient just to monitor flow and return temperatures of the heat source, especially in large buildings as localized areas where temperatures are not maintained will not be identified. Even small parts of a system which are colonized can compromise legionellae control of the whole.

Lack of financing is often cited as a reason why remedial work has not been carried out. Experience has shown that, particularly in healthcare premises, much money is wasted and patients put at risk by inappropriate "knee jerk reactions" to test results, with chemical control systems being installed without first ensuring that the engineering is optimized. Inefficiency costs in water-related operations are often highest when safety is poorest. In many cases, areas of poor flow (dead legs, e.g., bathrooms not used, blind ends), are found which act as foci for continued system infection. Getting the initial design right, using appropriate materials, and ensuring the system is used efficiently and maintained with legionellae control in mind should mean that sampling and testing are not necessary. However, on many occasions when investigating single cases or outbreaks, deficiencies in engineering have only been brought to light by poor *Legionella* test results. Sampling for legionellae should never be carried out in isolation, but is a useful tool to verify that controls systems are working especially if used in conjunction with temperature monitoring. The latter can be automated and linked to a building management system.

Rather than determining whether *Legionella* testing is appropriate, RA can help determine the required extent and frequency of testing and promote a proactive rather than reactive approach to system management.

Question: *Why can culture not be replaced by more rapid and sensitive detection methods such as real time PCR?*

PCR can be used with the following provisos: there are still some issues around the sensitivity and specificity of PCR for legionellae, especially around the target sequences for the genus. Genus sequences are not specific enough, and where these are used, it is possible that they will detect both nonlegionellae bacteria and legionellae species which are not yet cultivable by conventional media. This makes both interpretation and comparison with culture results difficult. The target sequences for *Legionella pneumophila* are more specific, but non-*pneumophila* species pose a significant health risk in many settings. There are potential problems with inhibitors such as rust and humic acids. These are surmountable if the sample is cleaned up using one of the many available kits prior to testing and a positive control sequence is included in each test to detect inhibitors. Care must be taken when choosing columns to include both positive and negative controls, as problems with contamination with Legionella DNA have been highlighted in the past. (1). In outbreak situations PCR is very useful to rapidly eliminate negative sources within the suspect area, allowing resources to be more effectively targeted. Regardless, culture is still essential to obtain isolates for typing.

The presence of low concentrations of nonviable Legionella can be expected even in well-managed and disinfected systems (especially cooling water systems). Nonculture based methods must be able to reliably differentiate between viable and nonviable cells. The lack of sensitivity and the possibility of viable nonculturable Legionella in some circumstances mean that, outside outbreak investigations, culture is not sufficiently reliable or reproducible enough to be used as anything other than verification of a system management program.

Question: *Given the successes and failures of various disinfection methods, is there a way to predict which technology might be most appropriate for a given institution?*

So many factors affect the ability of a particular system to control microbial growth that is very difficult to predict what will and will not work. It is perhaps easier to predict what may not work in a particular situation or what may not work without system modifications; e.g., copper-silver ionization works best in softened water when the electrodes do not get scaled up. Residence time, organic load, and water chemistry are all factors that influence disinfectant choice that should be identified in the RA. In the absence of a "generic" system, it is unwise to adopt a generic approach to disinfection. Some systems will, by their nature of operation, preclude some disinfectant approaches, e.g., high temperature for cooling towers, nonoxidizing biocides for potable water. No disinfection method will work unless any design and/or engineering problems have been removed. For example, oxidizing biocides will be compromised by high levels of organic matter unless these are removed by filtration. A common problem when installing a new disinfection system in a chronically colonized building is that disinfectant concentrations are dosed without any verification that effective levels are being achieved at the outlets. Low disinfectant levels will not cope with high organic loads. In these situations the whole system—to the outlets—needs pretreating with the equivalent of a hyper-chlorination for a sustained period of time (hours or sometimes days).

Question: *Does installation of thermostatic mixing valves necessarily compromise microbial control?*

The problems associated with thermostatic mixing valves (TMVs) are mainly associated with the fact that the temperature downstream of these is within the ideal range for microbial growth. This is especially a problem in older institutions where not only does the TMV itself support growth, but there is also a considerable length of pipework after the device, with no controls in place. The situation can be improved by having a program of cleaning and disinfecting the TMV, showers, hoses, heads, etc., but this is costly in staff time. In high-risk areas, point-of-use filtration will protect vulnerable patients, but again these filters are expensive and need to be replaced frequently depending on the water quality. In high-risk areas such as intensive care facilities, it is arguable that the risk to the patient is too high to ignore these options. There are new TMVs where the outlet is at the point of use and some where it is possible to pasteurize the valve on a regular basis. The use of TMVs does not necessarily compromise microbial control, but adds an extra challenge to system management and adds complexity to disinfection. If necessary, microbial control of TMVs can be achieved by routine maintenance and flushing of the outlets.

Question: *What extent of testing would you see as necessary for validation of a new technology or field application?*

This is an interesting question and quite pertinent. If the laboratory-based validation tests are satisfactory (these must always be carried out first), then the next step is field testing. Primary field testing of new technology should never be attempted in an area where there are high-risk patients. Problem systems should be avoided, ideally with a cooling tower. For example, you would choose a site where there are well-controlled towers with their own water system but with the same water supply, same atmospheric conditions, design, and treatment regime to date, etc. A period of intensive monitoring should be undertaken including aerobic colony counts (not dipslides) combined with *Legionella* testing, ideally using both culture and PCR. This should occur before the trial begins to build up a thorough picture of the microbial growth patterns during changes in atmospheric conditions. Once this data has been analyzed, it can be used to set the acceptable deviations in counts for that system. When the new technology is installed, then an intensive monitoring (daily at first) program should begin. PCR results are valuable at this stage to show any deviations from the accepted range before the system gets out of control if the new technology does not work. Once confidence in the new system is established, the monitoring regimen can be reduced to normal monitoring practices.

Question: *Several groups suggest taking preventative action based on detecting certain concentrations of Legionella or detecting a certain percentage of positive sites. These approaches treat all species and/or serogroups of Legionella equally. However, we know that various strains are vastly different in the amount of disease they cause. How could this information be incorporated into a prevention strategy? What kind of studies would be required to obtain this information?*

Even now, the infective dose for someone to contract Legionnaires' disease is not known. This is because of the range of predisposing factors which are known to make one person more susceptible than another. We know, for example, that men are more likely to contract Legionnaires' disease than women. However, it would be very difficult to justify treating systems used only by women differently from those used by men. Where sources are sampled that are actively causing cases, such as in large point source outbreaks, then the concentrations of legionellae in those systems are usually in excess of 10,000 CFU/liter. However, when the system is disinfected or receives a low mains water residual, cases have occurred following a short period of stagnation (days). Exposure varies between sources, e.g., a shower compared to a spa pool. It follows that the dose received will vary and cannot easily be extrapolated from concentrations in the water. Legionellosis prevention is a matter of effective water treatment and building water system management. Testing for the organism is therefore verification of prevention and not a control strategy.

We do know that mAb 2 positive (Pontiac) strains are more likely to cause community-acquired Legionnaires' disease than other le-

gionellae, but we also know that there have been Legionnaires' disease cases in the immunocompromised caused by at least 20 other *Legionella* including non-*pneumophila* species. There could be many people in the community who would be at risk.

The approach in the United Kingdom is that if a system supports the growth of *Legionella,* irrespective of serogroup, then there is the potential for a species to grow that could cause Legionnaires' disease. We certainly have some indication that a building can have a stable population of several species of legionellae. It is not known if these will preclude any incomers of a more virulent nature taking hold and replacing these resident strains. This could be an interesting area of research. As it is possible to achieve control in water systems with appropriate engineering, temperature control, and, if necessary, disinfection systems, it would be a brave move to attempt control in this way and convince the courts that this was a sensible strategy.

Question: *Can risk assessment for different institutions be standardized to any degree without negatively impacting the efficacy of this approach? How could one measure the impact/cost effectiveness of a risk assessment program?*

This is many questions in one. One must be careful with standardizing approaches, although it is common sense that if a checklist approach is used, this must not exclude the risk assessor from having an open mind and lateral thinking approach to look for the unusual. There is always someone out there with a money-saving idea which could impact on *Legionella* control, e.g., diverting warm overflow water into the cold water storage tank or installing a central heating radiator in a cold water storage tank to prewarm the water. It is not acceptable to carry out a generic risk assessment approach to cover buildings, even those constructed at the same time and of the same design; inevitably someone will alter something: install or remove outlets, etc. which would compromise the generic risk assessment. Measuring the cost/impact of the RA programme is more difficult in the absence of a case of disease. One death is very costly, not only emotionally, but has in instances caused businesses to go under. A thorough risk assessment can save money by identifying areas that are potential causes of *Legionella* growth in a system (see earlier answer) so that appropriate action can be taken instead of trying to treat a poorly designed/engineered system which will never be satisfactorily controlled by expensive chemical treatments.

The ecological determinants of *Legionella* colonization are common to all systems. Temperature, nutrients, presence of other microorganisms, and stagnation are all major predictors of *Legionella* growth. The extent to which each of these predictors influences *Legionella* growth varies with system operation and design. For instance, cooling towers are unable to operate outside the optimal temperature range for *Legionella*, and hot water systems never have organic loads comparable to cooling towers or spa pools. These variations hold for different designs of similar systems. No two hot water systems are the same. As a result, ecological determinants can be readily standardized across a wide range of systems, but risk assessment cannot.

Measuring the cost of a preventative strategy is always difficult. In the absence of an established disability adjusted life years value for legionellosis, it is not possible to establish burden of disease and thus make estimates of global cost. It is clear that in most cases, current reports underestimate the actual prevalence of disease. From a single institution viewpoint it is hard to assess the financial impact of a single case, let alone an outbreak. Certainly, community-based outbreaks have resulted in multimillion dollar legal claims, which may be sufficient justification for many for implementation of a risk-based management plan.

REFERENCES

1. **van der Zee, A., M. Peeters, C. de Jong, H. Verbakel, J. W. Crielaard, E. C. J. Claas, and K. E. Templeton.** 2002. Novel PCR-probe assay for detection of and discrimination between Legionella pneumophila and other Legionella species in clinical samples *J. Clin. Microbiol.* **40:** 1124–1125.
2. **WHO (World Health Organization).** 2004. *Guidelines for Drinking Water Quality*, 3rd ed., chapter 4. World Health Organization.

AUTHOR INDEX

SUBJECT INDEX

A

A/J mouse model, 211
Abscess, 11
Acanthamoeba castellani, 201–206, 347–349, 395–396
Acanthamoeba polyphaga, 186
Acidithiobacillus ferrooxidans, 205
Actin cytoskeleton, 258
Adult respiratory distress syndrome (ARDS), 9, 10
Aerosolized water, 149, 150, 184
AEX. *See* Anion exchange chromatography
AFLP technique, 129
Age, 23, 29, 109
Agrobacterium tumefaciens, 187, 189, 192, 194–197
AIDS, 31
Alcoholism, 23, 109
Alleles, 95, 112, 159, 160, 164, 166
Allelic vector exchange, 339–342
Allentown subgroup, 154
Amoebae
  in biofilms, 383–386
  Dictyostelium discoideum, 390–394
  isolation of, 423–425
Amoebae plate test (APT), 203–204, 206
Amoxicillin-clavulanate, 40
Amphotericin B, 147
Ancillary treatments, 86
Andalusia, Spain, 119, 120, 122, 123
Animal dose-response modeling, 486–487
Animal model selection, 486
Anion exchange chromatography (AEX), 238, 239
Antarctica, 417–419
Anti-*Legionella* peptides, 411–413
Antibiotic therapy
  for immunocompromised patients, 40–41
  indications for, 85
  initiation of, 29
  Legionnaires' disease with, 8–10
Antibiotics, in vitro activities of, 43–46
Antigenic diversity, 76–78

Antigenuria, 18–19
Antiretroviral therapy, 31
Antisense strategy, 336–338
APACHE score, 13
Apoptosis induction, 283–286
APT. *See* Amoebae plate test
Apyrexia, 29
Aquarium outbreak, 101, 102, 104
ARDS. *See* Adult respiratory distress syndrome
Aspergillus, 11
Aspergillus fumigatus, 146, 147
Aspiration, 29, 146–148
Australia, 97, 100–104, 251
  outbreaks, 102
  potting-mix risk study in, 149–150
  rising notifications, 100–101
  risk profile recognition, 102, 103
  urinary antigen testing, 102, 103
Austria, 97, 153, 154
Axenic growth, 343–345
Azithromycin, 9–11, 40–41, 43–46

B

Bacteremia, 5
Bacterial transmission, 43
Balearic Islands, 119, 120, 122, 123
Barcelona, Spain, 25, 28
Barrow, UK, 98
Barrow-in-Furness, England, 105, 106, 152
Bartonella tribocorum, 192
Bathing water
  filter refreshment using high chlorine concentration in, 497–500
  growth of *Legionella* in, 431–435
BBC outbreak, 106
BCYE agar. *See* Buffered charcoal yeast extract agar
Belgium, 92
Bellingham 1 subgroup, 130
Bellingham subgroup, 63–66, 70, 154
Benidorm subgroup, 63–66, 68, 70

553